The Nidoviruses: Toward Control of SARS and other Nidovirus Diseases

eeting Participants
by M. Buchmeier and K. Holmes

Stanley Perlman
Kathryn V. Holmes

Editors

The Nidoviruses:

Toward Control of SARS and other Nidovirus Diseases

With 96 Illustrations

Editors
Stanley Perlman, M.D., Ph.D.
Department of Pediatrics
University of Iowa
Iowa City, IA 52242
USA
Stanley-Perlman@uiowa.edu

Kathryn V. Holmes, Ph.D.
Department of Microbiology
University of Colorado Health Sciences Center at Fitzsimons
Aurora, CO 80045-8333
USA
Kathryn.Holmes@ucHSC.edu

Proceedings of the Xth International Nidovirus Symposium "Toward Control of SARS and other Nidovirus Diseases," held in Colorado Springs, Colorado, June 25-30, 2005.

Library of Congress Control Number: 2005939039

ISBN-10: 0-387-26202-4 e-ISBN: 0-387-33012-7
ISBN-13: 978-0387-26202-4

Printed on acid-free paper.

9 8 7 6 5 4 3 2 1

springer.com

PREFACE

This book summarizes the keynote and plenary speeches and posters of the "Xth International *Nidovirus* Symposium: Toward Control of SARS and Other *Nidovirus* Diseases" that was held in Colorado Springs, Colorado, June 25–30, 2005. The nine previous meetings of scientists investigating the molecular biology and pathogenesis of *coronaviruses*, *toroviruses*, *arteriviruses*, and *okaviruses* were generally held every 3 years since the first meeting was convened in Wurzburg, Germany, in October, 1980. The Xth International Symposium was held just 2 years after the IXth International Symposium (*Nido2003*) in The Netherlands, because of the tremendously increased research on *nidoviruses* that resulted from the discovery that the global epidemic of severe acute respiratory syndrome (SARS) in 2002–2003 was caused by a newly discovered *coronavirus* called SARS-CoV. A record 225 scientists from 14 countries attended the Xth International *Nidovirus* Symposium, and important advances in every aspect of *nidovirus* molecular biology and pathogenesis were reported and discussed. The meeting was divided into 12 sessions, with keynote speakers providing a general review of research pertinent to each one. This volume is a collection of scientific papers presented at the symposium.

Once a *coronavirus* was recognized as the etiological agent of SARS, intensive work by many investigators resulted in determination of the sequence of the virus, engineering of reverse genetics systems, and identification of the host cell receptor used by the virus. With the increased interest in *coronaviruses*, new members of the family associated with human disease were identified. Most notably, HCoV-NL63 and HCoV-HKU1 were recently recognized as important agents of human upper and lower respiratory tract disease. With the identification of new members of the *nidovirus* family, it became important to determine the relationship between these newly recognized viruses and previously classified *nidoviruses*. The *nidovirus* group of the International Committee for Taxonomy of Viruses proposed a taxonomic tree of the *nidoviruses* that is reproduced here (Figure 1). The structure of the viral nucleocapsid, number of subgenomic RNAs and length of the plus strand RNA genomes are strikingly different for each of the *nidoviruses*, although their replication strategies are very similar. This information, coupled with sequencing data, is used to place the newly identified viruses into the pre existing data set. As examples, HCoV-NL63 has been classified as a group 1b *coronavirus*, while SARS-CoV is tentatively classified as a distant member of the group 2 family (group 2b). Other newly identified *nidoviruses*, including those infecting bats, have been similarly analyzed and classified.

The first sessions of the meeting covered "Viral RNA Synthesis." Given the large size of the *nidovirus* replicase gene (gene 1ab, more than 20,000 *nucleotides* for *coronaviruses*)

NEW TAXONOMY

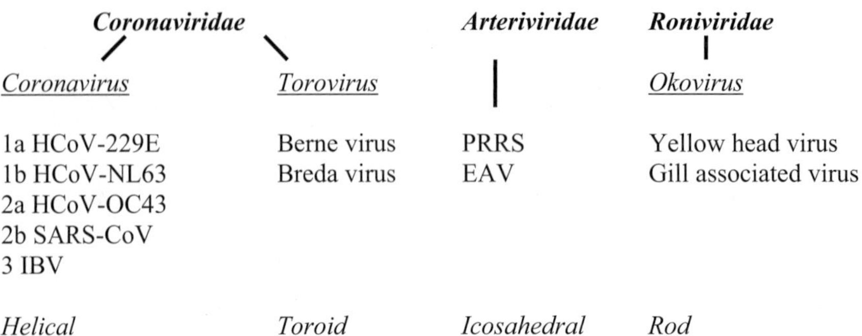

Figure 1. Proposed taxonomy. HCoV-human coronavirus; SARS-CoV, severe acute respiratory syndrome–coronavirus; IBV, infectious bronchitis virus; PRRS, porcine respiratory and reproductive virus; EAV, equine arteritis virus.

and the observation that the gene product is co-translationally cleaved into many proteins, it has been a challenge to determine the functions of individual cleavage products in virus replication. *In silico* analyses suggested roles for these proteins, and advances in genetic manipulations of the viruses coupled with confocal analyses and X-ray crystallography have provided insight into their structures and functions. Certain functional domains of the replicase polyprotein are expressed only by some *coronaviruses*, whereas other domains, such as a uridylate-specific endoribonuclease (NendoU), are encoded in both *coronaviruses* and *arteriviruses*. While the exact functions of these proteins in nidovirus replication need to be determined, much progress has been made in delineating structural domains in some of them and solving their structure.

Much work in the recent past has focused on the structure and function of *nidovirus* structural and nonstructural proteins, encoded downstream of the replicase gene. Each *nidovirus* has an apparently unique set of genes encoding nonstructural proteins that are interspersed with structural genes at the 3' end of the genome. Nothing is known about where the genes encoding these proteins came from and how they were inserted into *coronavirus* genomes. These non structural proteins are apparently not required for virus production *in vitro*, but several contribute to diseases in the infected host. The structures and functions of structural and nonstructural proteins were the topic of "Protein Synthesis, Structure and Processing."

All *nidoviruses* bud intracellularly ("Viral Assembly and Release"). An active area of investigation is to determine the viral and host factors important for virus egress from the cell. Curiously, although the E protein, which has ion channel activity, is believed to be the nidus for virus particle formation, E is not essential for assembly of all *coronaviruses*, because virus-like particles form in its absence. The structures of several *nidoviruses* are being elucidated, which will facilitate understanding of the functions of individual proteins in virion formation.

Nidoviruses use a variety of host cell receptors to enter infected cells ("Viral Entry"), via binding to the virus surface (S) glycoprotein and/or the hemagglutinin esterase glycoproteins found on *coronaviruses* and *toroviruses*. Most group 1 *coronaviruses* enter via interactions with aminopeptidase. An exception is HCoV-NL63, which uses angiotensin converting enzyme 2 (ACE2) as its host cell receptor. SARS-CoV also uses ACE2 to enter cells. Regions on ACE2 important for SARS-CoV entry have been delineated, and the role of host lectins, such as CD209L, in facilitating *coronavirus* entry has also been established. The crystal structure of the receptor binding domain of the SARS S protein bound to ACE2 has also been solved; this structure will be useful not only for understanding virus entry but also for design of antiviral therapies. Elegant studies have also delineated amino acid substitutions in the SARS-CoV S protein that were selected during the 2002–2003 epidemic and facilitated binding to human ACE2, permitting human-to-human spread of the virus and increasing virulence in humans. It is also clear that cleavage of the S protein is critical for fusion of viral and host cell membranes. In many *nidoviruses*, the S protein is cleaved during exit, often by furin or a related serine protease, whereas in others, including SARS-CoV, cleavage occurs during entry and is mediated by cathepsins in endosomes.

Understanding the mechanisms by which nidoviruses cause disease in the infected animal was a major focus of the symposium. Several non-human coronaviruses, including mouse hepatitis virus (MHV) and feline peritonitis virus (FIPV), have been intensively studied for years and are known to cause disease that is partly due to immunopathology. The pathogenesis of these infections was discussed in "Pathogenesis of Non-Human *Coronaviruses*." All strains of MHV uses CEACAM1 to enter cells, but different strains exhibit differences in tissue tropism. For example, MHV-1, a strain that has not been intensively studied in the past, preferentially infects the lower respiratory tract and may serve as a useful model for SARS. *Nidoviruses* also modulate the expression of host cell RNA and protein, presumably to enhance their ability to replicate in infected cells. Induction of immunomodulatory molecules, such as induction of a novel prothrombinase by MHV-3 infection, may also result in severe disease in the infected host.

Arteriviruses cause important diseases, such as equine arteritis and porcine reproductive and respiratory syndrome, in animals. Studies of their replication and pathogenesis have been facilitated by the development of infectious cDNA clones, as described in "Pathogenesis of *Arteriviruses* and *Toroviruses*." These studies will lead to development of vaccines and therapeutics for these important veterinary pathogens.

Prior to the isolation of SARS-CoV as the etiological agent of SARS, human coronaviruses (HCoV-OC43 and HCoV-229E) were known to cause respiratory tract infections, and occasionally to be associated with outbreaks of diarrhea. The identification of SARS-CoV, HCoV-NL63, and HCoV-HKU1 increased the interest in pathogenesis of human *coronavirus* infections ("Pathogenesis of Human *Coronaviruses*"). Several animals can be infected with SARS-CoV, but none of them reproducibly develops the pulmonary disease observed in infected humans. SARS-CoV–infected ferrets are considered the most promising of the available animal models for SARS. Other approaches include infection of human airway cells with SARS-CoV or with retroviruses pseudotyped with the SARS-CoV S protein, because these cells are primary targets for the virus in infected humans. Another approach for delineating the functions of some SARS-CoV nonstructural proteins is to develop chimeric *coronaviruses* of lab animals that express individual SARS-CoV proteins. The interest in

SARS-CoV and HCoV-NL63 has spilled over into research into the pathogenesis of HCoV-229E and HCoV-OC43. These two viruses show striking differences in their ability to cross species. HCoV-229E infects only humans, and infection of mice transgenic for the virus receptor (human *aminopeptidase*) is not robust. In contrast, HCoV-OC43 readily adapts to infect other species and in mice causes a profound infection of neurons after only a few *in vivo* passages.

The papers comprising the final section, "Vaccines, Antiviral Drugs, and Diagnostics," reflect the importance of SARS-CoV as a human pathogen. Efforts to develop inactivated, subunit vaccines and live attenuated vaccines are underway. Testing of these vaccines will benefit from the development of an animal model for SARS. Passive immunization with anti-SARS-CoV antibody may also be used during an epidemic, and human monoclonal antibodies that neutralize the virus have been developed. Crystal structures of proteins, such as the SARS-CoV main protease, will also lead to development of drugs that inhibit SARS-CoV replication with minimal effect on host cell functions. Finally, antisense RNA and siRNA methodologies are being developed as novel approaches to SARS therapy.

The organizers of the meeting wish to thank all of those who helped to make the meeting a success. Vince Santoscoy, Jan Harkin, and Heather Williams of Resort Management Associates, Inc., were a huge help in organizing the meeting and with the registration of attendees. Kathi L. Basso and Stu Woods of Cheyenne Mountain Resort also helped with the on-site arrangements. David Leake and Laverle Crist designed the meeting Web site. We thank Katherine O'Malley and Jason Netland for their help during the meeting and Neal Perlman for help with design of the meeting logo. This book could not have been completed without the help of Julie Nealson. We also thank our sponsors, Pfizer Animal Health and Fort Dodge Animal Health, for their generous contributions. The planning committee and convenors helped to organize the sessions, select topics and speakers, and lead wide-ranging discussions. Finally, we thank all of the attendees who presented their research in plenary speeches and posters and contributed to the discussions that were a vital part of the successful meeting.

University of Iowa
University of Colorado

Stanley Perlman,
Kathryn V. Holmes,
Editors
January 2006

CONTENTS

I. VIRAL RNA SYNTHESIS

V. PATHOGENESIS OF NON-HUMAN CORONAVIRUSES

University of Bologna, Bologna, Italy

quin, Université Catholique de Louvain, Brussels, Belgium

ont, Washington University School of Medicine, St. Louis, Missouri

ieman, University of North Carolina, Chapel Hill, North Carolina

aga, National Hospital Organization Kinki-Chuo Chest Medical Center, Sakai, pan

shi, National Institute of Infectious Diseases, Tokyo, Japan

awa, National Hospital Organization Kinki-Chuo Chest Medical Center, Sakai, apan

JCSMR, ANU, ACT, Australia

gnon, INRS-Institut Armand Frappier, Laval, Québec, Canada

alán, Centro Nacional de Biotecnología, CSIC, Madrid, Spain

. Gallagher, Loyola University Chicago Stritch School of Medicine, Maywood, Illinois

Garcia, Northwestern University Feinberg School of Medicine, Chicago, Illinois

. Garry, Tulane University Health Sciences Center, New Orleans, Louisiana

zón, Campus Universidad Autónoma, Cantoblanco, Madrid, Spain

Geier, University of Erlangen-Nürnberg, Nürnberg, Germany

. Giedroc, Texas A&M University, College Station, Texas

Gillim-Ross, New York State Department of Health, Albany, New York

M. Goletz, Loyola University Medical Center, Maywood, Illinois

Gonzales, University of Southern California, Los Angeles, California

nder E. Gorbalenya, Leiden University Medical Center, Leiden, The Netherlands

Gorczynski, University of Toronto, Toronto, Canada

hel L. Graham, Vanderbilt University, Nashville, Tennessee

mas Gramberg, University of Erlangen-Nürnberg, Nürnberg, Germany

vid Grant, University of Toronto, Toronto, Canada

unshan Gui, Drug Discovery Center, Shanghai, China

ephan Günther, Bernhard-Nocht-Institute, Hamburg, Germany

eyin Guo, Wuhan University, Wuhan, People's Republic of China

CONTRIBUTORS

Brian D. Adair, The Scripps Research Institute, La Jolla, California

Fernando Almazán, Centro Nacional de Biotecnología, CSIC, Madrid, Spain

Sara Alonso, Centro Nacional de Biotecnología, CSIC, Madrid, Spain

Enrique Álvarez, Centro Nacional de Biotecnología, CSIC, Madrid, Spain

Yasushi Ami, National Institute of Infectious Diseases, Tokyo, Japan

Kanchan Anand, University of Lübeck, Lübeck, Germany

William D. Arndt, Arizona State University, Tempe, Arizona

Yasuko Asahi-Ozaki, National Institute of Infectious Diseases, Tokyo, Japan

Ehtesham Baig, University of Toronto, Toronto, Canada

Susan C. Baker, Loyola University Chicago Stritch School of Medicine, Maywood, Illinois

Udeni B. Balasuriya, University of Kentucky, Lexington, Kentucky

Ralph S. Baric, University of North Carolina, Chapel Hill, North Carolina

Naina Barretto, Loyola University Chicago Stritch School of Medicine, Maywood, Illinois

Paul Bates, University of Pennsylvania School of Medicine, Philadelphia, Pennsylvania

Mara Battilani, University of Bologna, Italy

Beverley E. Bauman, Cornell University, Ithaca, New York

John Bechill, Loyola University Chicago, Maywood, Illinois

Valerie Bednar, Arizona State University, Tempe, Arizona

Cornelia C. Bergmann, Lerner Research Institute, Cleveland Clinic Foundation, Cleveland, Ohio

Ben Berkhout, University of Amsterdam, Amsterdam, The Netherlands

Richard K. Bestwick, AVI Biopharma Inc., Corvallis, Oregon

Katrina Bicknell, University of Reading, Reading, United Kingdom

Andrew Blount, Arizona State University, Tempe, Arizona

Joseph A. Boscarino, Loyola University Medical Center, Maywood, Illinois

Berend Jan Bosch, Utrecht University, Utrecht, The Netherlands

Joseph W. Brewer, Loyola University Chicago, Maywood, Illinois

Paul Britton, Institute for Animal Health, Compton Laboratory, Newbury, Berkshire, United Kingdom

Sarah M. Brockway, Vanderbilt University Medical Center, Nashville, Tennessee

Maarten Brom, University of Pennsylvania School of Medicine, Philadelphia, Pennsylvania

Gavin Brooks, University of Reading, Reading, United Kingdom

Michael J. Buchmeier, The Scripps Research Institute, La Jolla, California

Susan E. Burkett, University of North Carolina, Chapel Hill, North Carolina

Kelly M. Burkhart, Boehringer Ingelheim Vetmedica, Incorporated, Ames, Iowa

Jane C. Burns, University of California San Diego School of Medicine, La Jolla, California

Renaud Burrer, The Scripps Research Institute, La Jolla, California

Jagdish Butany, University of Toronto, Toronto, Canada

Noah Butler, University of Iowa, Iowa City, Iowa

Yingyun Cai, University of Arkansas for Medical Sciences, Little Rock, Arkansas

Jay G. Calvert, Pfizer Animal Health, Kalamazoo, Michigan

Carmen Capiscol, Centro Nacional de Biotecnología, CSIC, Madrid, Spain

Rosa Casais, Institute for Animal Health, Compton Laboratory, Newbury, Berkshire, United Kingdom

Dave Cavanagh, Institute for Animal Health, Compton Laboratory, Newbury, Berkshire, United Kingdom

Luisa Cervantes, Kantonal Hospital St. Gallen, St. Gallen, Switzerland

Lili Chen, Drug Discovery Center, Shanghai, China

Pei-Jer Chen, National Taiwan University College of Medicine, Taipei, Taiwan

Yu Chen, Wuhan University, Wuhan, People's Republic of China

Zhongbin Chen, Loyola University Chicago Stritch School of Medicine, Maywood, Illinois

Hyeryun Choe, Harvard Medical School, Boston, Massachusetts

J. Christopher-Hennings, South Dakota State University, Brookings, South Dakota

Victor C. Chu, Cornell University, Ithaca, New York

Sara Ciulli, University of Bologna, Italy

Ellen W. Collisson, Texas A&M University, College Station, Texas

Jean-Paul Coutelier, Université Catholique de Louvain, Brussels, Belgium

Oswald R. Crasta, Virginia Polytechnic Institute and State University, Blacksburg, Virgionia

Jennifer Crookshank, University of Toronto, Toronto, Ca

Myron Cybulsky, University of Toronto, Toronto, Canada

Michael J. Czar, Virginia Polytechnic Institute and State Un

Marc Davis, Institute for Animal Health, Compton Laboratory

Marta L. DeDiego, Centro Nacional de Biotecnología, CSIC, M

Nadine DeAlbuquerque, University of Toronto, Toronto, Canad

Peter L. Delputte, Ghent University, Merelbeke, Belgium

Daphne E. deMello, St. Louis University Health Science Center, S

Damon J. Deming, University of North Carolina, Chapel Hill, Nort

Mark R. Denison, Vanderbilt University Medical Center, Nashville,

Marc Desforges, INRS-Institut Armand Frappier, Laval, Québec, Can

Michael S. Diamond, Washington University School of Medicine, St. L

Eric F. Donaldson, University of North Carolina, Chapel Hill, North Car

Brian K. Dove, University of Leeds, Leeds, United Kingdom

Christa Drexler, Intervet International B.V., The Netherlands

Christian Drosten, Bernhard-Nocht-Institute, Hamburg, Germany

Lance D. Eckerle, Vanderbilt University Medical Center, Nashville, Tenness

Luis Enjuanes, Centro Nacional de Biotecnología, CSIC, Madrid, Spain

Klara K. Eriksson, Kantonal Hospital St. Gallen, St. Gallen, Switzerland

Gary Ewart, Biotron Ltd., ACT, Australia

Kay S. Faaberg, University of Minnesota, Saint Paul, Minnesota

Ying Fang, South Dakota State University, Brookings, South Dakota

Michael Farzan, Harvard Medical School and New England Primate Research Cent Massachusetts

Min Feng, Texas A&M University, College Station, Texas

A Damon Ferguson, Cornell University, Ithaca, New York

Witold Filipowicz, Friedrich Miescher Institute for Biomedical Research, Basel, Switzer

Eleanor Fish, University of Toronto, Toronto, Canada

Johannes Forster, St. Josefs Hospital, Freiburg, Germany

Ambra Foschi,

Stéphanie Fra

Daved H. Fre

Matthew B. F

Yukari Fuku Osaka, Ja

Shuetsu Fuk

Izumi Furu Osaka,

Peter Gage

Mylène Ga

Carmen G

Thomas

Francesc

Robert

Ana Ga

Martin

David

Laura

Jeffre

J.-M.

Alex

Reg

Rac

Th

Da

C

S

Marlena Habal, University of Toronto, Toronto, Canada

Jun Han, University of Minnesota, Saint Paul, Minnesota

Ayako Harashima, National Institute of Infectious Diseases, Tokyo, Japan

Laura Harmon, University of Arkansas for Medical Sciences, Little Rock, Arkansas

Sally Harrison, University of Leeds, Leeds, United Kingdom

Stephen C. Harrison, Harvard Medical School and Children's Hospital, Boston, Massachusetts, and Howard Hughes Medical Institute

Hideki Hasegawa, National Institute of Infectious Diseases, Tokyo, Japan

Tsutomu Hashikawa, RIKEN, Wako, Japan

Satomi Hashimoto, National Hospital Organization Kinki-Chuo Chest Medical Center, Sakai, Osaka, Japan

Lia Haynes, Centers for Disease Control and Prevention, Atlanta, Georgia

Yuxian He, New York Blood Center, New York, New York

Mark Heise, University of North Carolina, Chapel Hill, North Carolina

Lindsay K. Heller, New York State Department of Health, Albany, New York

Erin M. Hemmila, University of Colorado Health Sciences Center at Fitzsimons, Aurora, Colorado

Georg Herrler, Institut für Virologie, Tierärztliche Hochschule Hannover, Hannover, Germany

Tobias Hertzig, University of Würzburg, Würzburg, Germany

Melissa Hickey, University of Iowa, Iowa City, Iowa

Rolf Hilgenfeld, University of Lübeck, Lübeck, Germany

Norio Hirano, Iwate University, Morioka, Japan

Julian A. Hiscox, University of Leeds, Leeds, United Kingdom

Robert S. Hodges, University of Colorado Health Sciences Center, Aurora, Colorado

Douglas C. Hodgins, University of Guelph, Guelph, Ontario, Canada

Teri Hodgson, Institute for Animal Health, Compton Laboratory, Newbury, Berkshire, United Kingdom

Heike Hofmann, University of Erlangen-Nürnberg, Germany

Brenda G. Hogue, Arizona State University, Tempe, Arizona

Kathryn V. Holmes, University of Colorado Health Sciences Center at Fitzsimons, Aurora, Colorado

Randall K. Holmes, University of Colorado Health Sciences Center, Aurora, Colorado

Megan W. Howard, University of Colorado Health Sciences Center, Aurora, Colorado

Bilan Hsue, Stratagene, La Jolla, California

I-Chueh Huang, Harvard Medical School, Boston, Massachusetts

Kelley R. Hurst, New York State Department of Health, Albany, New York

Snawar Hussain, Wuhan University, Wuhan, People's Republic of China

Antonio Iglesias, Max Planck Institute for Neurobiology, München, Germany

Gabriele Ihorst, University Hospital, Freiburg, Germany

Isao Ishida, Kirin Brewery Co., Shibuya, Tokyo, Japan

Koji Ishii, National Institute of Infectious Diseases, Tokyo, Japan

Shigeyuki Itamura, National Institute of Infectious Diseases, Tokyo, Japan

Patrick L. Iversen, AVI Biopharma Inc., Corvallis, Oregon

Naoko Iwata, National Institute of Infectious Diseases, Tokyo, Japan

Miwa Izumiya, National Hospital Organization Kinki-Chuo Chest Medical Center, Sakai, Osaka, Japan

Hélène Jacomy, INRS-Institut Armand Frappier, Laval, Québec, Canada

Maarten F. Jebbink, University of Amsterdam, Amsterdam, The Netherlands

Scott A. Jeffers, University of Colorado Health Sciences Center at Fitzsimons, Aurora, Colorado

Hong Peng Jia, University of Iowa, Iowa City, Iowa

Hualiang Jiang, Chinese Academy of Sciences, Shanghai, China

Michael G. Jobling, University of Colorado Health Sciences Center, Aurora, Colorado

Rika Jolie, Pfizer Animal Health, Kalamazoo, Michigan

Ariel Jones, Arizona State University, Tempe, Arizona

Dalia Jukneliene, Loyola University Chicago Stritch School of Medicine, Maywood, Illinois, and Beijing Institute of Radiation Medicine, Beijing, China

Noriko Kanamaru, National Hospital Organization Kinki-Chuo Chest Medical Center, Sakai, Osaka, Japan

Hyojeung Kang, Texas A&M University, College Station, Texas

Nadja Karl, University of Würzburg, Würzburg, Germany

Tetsuo Kase, Osaka Prefectural Institute of Public Health, Higashinari-ku, Osaka, Japan

Yoko Kita, National Hospital Organization Kinki-Chuo Chest Medical Center, Sakai, Osaka, Japan

Steven B. Kleiboeker, University of Missouri, Columbia, Missouri

Cheri A. Koetzner, New York State Department of Health, Albany, New York

Andrew D. Kroeker, AVI Biopharma Inc., Corvallis, Oregon

T. W. Kuijper, Academic Medical Center, Amsterdam, The Netherlands

Lili Kuo, New York State Department of Health, Albany, New York

Ichiro Kurane, National Institute of Infectious Diseases, Tokyo, Japan

Sachiko Kuwayama, National Hospital Organization Kinki-Chuo Chest Medical Center, Sakai, Osaka, Japan

Thomas E. Lane, University of California, Irvine, California

Lewis L. Lanier, University of California, San Francisco, California

Thao Le Thi Phuong, Université Catholique de Louvain, Brussels, Belgium

Changhee Lee, University of Guelph, Guelph, Ontario, Canada

Julian L. Leibowitz, Texas A&M University System-HCS, College Station, Texas

Gary Levy, University of Toronto, Toronto, Canada

Fang Li, Harvard Medical School and Children's Hospital, Boston, Massachusetts

Lichun Li, Texas A&M University, College Station, Texas

Qisheng Li, Nanyang Technological University, Singapore, and Institute of Molecular and Cell Biology, Proteos, Singapore

Wenhui Li, Harvard Medical School and New England Primate Research Center, Southborough, Massachusetts

Zhaoyang Li, Wuhan University, Wuhan, People's Republic of China

Ying Liao, Nanyang Technological University, Singapore, and Institute of Molecular and Cell Biology, Proteos, Singapore

Ding X. Liu, National University of Singapore, Nanyang Technological University, Singapore, and Institute of Molecular and Cell Biology, Singapore

Hao Liu, University of Toronto, Toronto, Canada

Pinghua Liu, Texas A&M University System-HCS, College Station, Texas

Yin Liu, University of Arkansas for Medical Sciences, Little Rock, Arkansas

Melissa B. Lodoen, University of California, San Francisco, California

Dwight C. Look, University of Iowa, Iowa City, Iowa

Lisa A. Lopez, Arizona State University, Tempe, Arizona

Xiaotao Lu, Vanderbilt University Medical Center, Nashville, Tennessee

Burkhard Ludewig, Kantonal Hospital St. Gallen, St. Gallen, Switzerland

Xiaomin Luo, Drug Discovery Center, Shanghai, China

Carolyn E. Machamer, Johns Hopkins University School of Medicine, Baltimore, Maryland

N. James MacLachlan, University of California, Davis, California

Katherine C. MacNamara, University of Pennsylvania, Philadelphia, Pennsylvania

Ana M. Maestre, Campus Universidad Autónoma, Cantoblanco, Madrid, Spain

Reinhard Maier, Kantonal Hospital St. Gallen, St. Gallen, Switzerland

Divine Makia, Kantonal Hospital St. Gallen, St. Gallen, Switzerland

Karen E. Malone, University of Southern California, Los Angeles, California

Justin Manuel, University of Toronto, Toronto, Canada

Dominique Markine-Goriaynoff, Université Catholique de Louvain, Brussels, Belgium

Andrea Marzi, University of Erlangen-Nürnberg, Nürnberg, Germany

Robert J. Mason, National Jewish Medical and Research Center, Denver, Colorado

Paul S. Masters, New York State Department of Health, Albany, New York

Makoto Matsumoto, Otsuka Pharmaceutical Co., Ltd., Tokushima, Japan

Shutoku Matsuyama, National Institute of Infectious Diseases, Tokyo, Japan

Paul B. McCray, Jr., University of Iowa, Iowa City, Iowa

Lisa J. McElroy, Cornell University, Ithaca, New York

Ian McGilvray, University of Toronto, Toronto, Canada

Willie C. McRoy, University of North Carolina, Chapel Hill, North Carolina

Peter Meerts, Ghent University, Merelbeke, Belgium

Andrew D. Mesecar, Beijing Institute of Radiation Medicine, Beijing, China

Jeroen R. Mesters, University of Lübeck, Lübeck, Germany

Tina Miletti, INRS-Institut Armand Frappier, Laval, Québec, Canada

Jason J. Millership, Texas A&M University System-HCS, College Station, Texas

Ronald A. Milligan, The Scripps Research Institute, La Jolla, California

Gerald Misinzo, Ghent University, Merelbeke, Belgium

Tanya A. Miura, University of Colorado Health Sciences Center, Aurora, Colorado

Tatsuo Miyamura, National Institute of Infectious Diseases, Tokyo, Japan

Tetsuya Mizutani, National Institute of Infectious Diseases, Tokyo, Japan

Ralf Moll, University of Lübeck, Lübeck, Germany

Jose L. Moreno, Centro Nacional de Biotecnología, CSIC, Madrid, Spain

Luigi Morganti, University of Bologna, Italy

Shigeru Morikawa, National Institute of Infectious Diseases, Tokyo, Japan

Thomas E. Morrison, University of North Carolina, Chapel Hill, North Carolina

Eric C. Mossel, Colorado State University, Fort Collins, Colorado

Hong M. Moulton, AVI Biopharma Inc., Corvallis, Oregon

Yumiko Muraki, National Hospital Organization Kinki-Chuo Chest Medical Center, Sakai, Osaka, Japan

Andrei Musaji, Université Catholique de Louvain, Brussels, Belgium

Noriyo Nagata, National Institute of Infectious Diseases, Tokyo, Japan

Kazuhide Nakagaki, Nippon Veterinary and Animal Science University, Tokyo, Japan

Keiko Nakagaki, National Institute of Infectious Diseases, Tokyo, Japan

Takenori Nakayama, Kobe Steel Ltd., Kobe, Japan

Hans J. Nauwynck, Ghent University, Merelbeke, Belgium

Sonia Navas-Martín, Drexel University College of Medicine, Philadelphia, Pennsylvania

Eric A. Nelson, South Dakota State University, Brookings, South Dakota

Jason Netland, University of Iowa, Iowa City, Iowa

Benjamin W. Neuman, The Scripps Research Institute, La Jolla, California

Tatsuji Nomura, Central Institute for Experimental Animals, Kawasaki, Kanagawa, Japan

Takato Odagiri, National Institute of Infectious Diseases, Tokyo, Japan

Chika Okada, National Hospital Organization Kinki-Chuo Chest Medical Center, Sakai, Osaka, Japan

Masaji Okada, National Hospital Organization Kinki-Chuo Chest Medical Center, Sakai, Osaka, Japan

Yoshinobu Okuno, Osaka Prefectural Institute of Public Health, Higashinari-ku, Osaka, Japan

Heidi Olivares, Loyola University Chicago Stritch School of Medicine, Maywood, Illnois

Emily R. Olivieri, New York State Department of Health and State University of New York, Albany, New York

Fernando Osorio, University of Nebraska, Lincoln, Nebraska

Ji'an Pan, Wuhan University, Wuhan, China

Asit K. Pattnaik, University of Nebraska, Lincoln, Nebraska

Joseph S. Malik Peiris, The University of Hong Kong, Hong Kong, China

Andrew Pekosz, Washington University School of Medicine, St. Louis, Missouri

Yu Peng, Wuhan University, Wuhan, People's Republic of China

Stanley Perlman, University of Iowa, Iowa City, Iowa

C. J. Peters, University of Texas Medical Branch, Galveston, Texas

Gudula Petersen, Wyeth Pharma, Münster, Germany

Lecia Pewe, University of Iowa, Iowa City, Iowa

M. James Phillips, University of Toronto, Toronto, Canada

Raymond J. Pickles, University of North Carolina, Chapel Hill, North Carolina

Jaime Pignatelli, Campus Universidad Autónoma, Cantoblanco, Madrid, Spain

Yvonne Piotrowski, University of Lübeck, Lübeck, Germany

Stefan Pöhlmann, University of Erlangen-Nürnberg, Nürnberg, Germany

Rajesh Ponnusamy, University of Lübeck, Lübeck, Germany

Leo L. M. Poon, The University of Hong Kong, Pokfulam, Hong Kong SAR, China

Santino Prosperi, University of Bologna, Italy

Yinghui Pu, University of Arkansas for Medical Sciences, Little Rock, Arkansas

Anjan Purkayastha, Virginia Polytechnic Institute and State University, Blacksburg, Virginia,

Ákos Putics, University of Würzburg, Würzburg, Germany

Krzysztof Pyrc, University of Amsterdam, Amsterdam, The Netherlands

Joel D. Quispe, The Scripps Research Institute, La Jolla, California

Chandran Ramakrishna, University of Southern California, Los Angeles, California

Zihe Rao, Tsinghua University, Beijing, China

Kiira Ratia, Beijing Institute of Radiation Medicine, Beijing, China

Mark L. Reed, University of Leeds, Leeds, United Kingdom

S. L. Reed, University of California San Diego School of Medicine, La Jolla, California

Juan Reguera, Centro Nacional de Biotecnología, CSIC, Madrid, Spain

M. Teresa Rejas, Campus Universidad Autónoma, Cantoblanco, Madrid, Spain

Julia D. Rempel, University of Manitoba, Winnipeg, Manitoba, Canada

Xiaofeng Ren, Institut für Virologie, Tierärztliche Hochschule Hannover, Hannover, Germany

Andrew J. Rennekamp, University of Pennsylvania, Philadelphia, Pennsylvania

Anjeanette Roberts, NIAID, National Institutes of Health, Bethesda, Maryland

Rhonda S. Roberts, University of North Carolina, Chapel Hill, North Carolina

Barry Rockx, University of North Carolina, Chapel Hill, North Carolina

Dolores Rodríguez, Campus Universidad Autónoma, Cantoblanco, Madrid, Spain

Michael B. Roof, Boehringer Ingelheim Vetmedica, Incorporated, Ames, Iowa

Peter M. Rottier, Utrecht University, Utrecht, The Netherlands

Raymond R. R. Rowland, Kansas State University, Manhattan, Kansas

Anne H. Rowley, Northwestern University Feinberg School of Medicine, Chicago, Illinois

Masayuki Saijo, National Institute of Infectious Diseases, Tokyo, Japan

Bruno Sainz, Jr., The Scripps Research Institute, La Jolla, California

Takehiko Saito, National Institute of Infectious Diseases, Tokyo, Japan

Yayoi Sakaguchi, National Hospital Organization Kinki-Chuo Chest Medical Center, Sakai, Osaka, Japan

Mitsunori Sakatani, National Hospital Organization Kinki-Chuo Chest Medical Center, Sakai, Osaka, Japan

Tetsutaro Sata, National Institute of Infectious Diseases, Tokyo, Japan

Shigehiro Sato, Iwate Medical University, Morioka, Japan

Yuko Sato, National Institute of Infectious Diseases, Tokyo, Japan

Alessandra Scagliarini, University of Bologna, Bologna, Italy

Elke Scandella, Kantonal Hospital St. Gallen, St. Gallen, Switzerland

Scott R. Schaecher, Washington University School of Medicine, St. Louis, Missouri

Barbara Schelle, University of Würzburg, Würzburg, Germany

Susan K. Schommer, University of Missouri, Columbia, Missouri

Meagan E. Schroeder, Texas A&M University, College Station, Texas

Christel Schwegmann-Weβels, Institut für Virologie, Tierärztliche Hochschule Hannover, Hannover, Germany

Joao C. Setubal, Virginia Polytechnic Institute and State University, Blacksburg, Virginia

Itay Shalev, University of Toronto, Toronto, Canada

Timothy Sheahan, University of North Carolina, Chapel Hill, North Carolina

Jianhua Shen, Drug Discovery Center, Shanghai, China

Xu Shen, Chinese Academy of Sciences, Shanghai, China

Lei Shi, University of Iowa, Iowa City, Iowa

H. Shike, University of California San Diego School of Medicine, La Jolla, California

C. Shimizu, University of California San Diego School of Medicine, La Jolla, California

Stanford T. Shulman, Northwestern University Feinberg School of Medicine, Chicago, Illinois

Stuart G. Siddell, University of Bristol, Bristol, United Kingdom

Graham Simmons, University of Pennsylvania School of Medicine, Philadelphia, Pennsylvania

Amy C. Sims, University of North Carolina, Chapel Hill, North Carolina

Jutta Slaby, University of Würzburg, Würzburg, Germany

M. K. Smith, University of Colorado Health Sciences Center, Aurora, Colorado

Eric E. Snyder, Virginia Polytechnic Institute and State University, Blacksburg, Virginia

Bruno W. Sobral, Virginia Polytechnic Institute and State University, Blacksburg, Virginia

Isabel Sola, Centro Nacional de Biotecnología, CSIC, Madrid, Spain

Jennifer Sparks, Vanderbilt University, Nashville, Tennessee

Kelly-Anne Spencer, University of Leeds, Leeds, United Kingdom

Steven M. Sperry, Vanderbilt University Medical Center, Nashville, Tennessee

Alexander Stang, Ruhr-University Bochum, Bochum, Germany

David A. Stein, AVI Biopharma Inc., Corvallis, Oregon

Julien R. St-Jean, INRS-Institut Armand Frappier, Laval, Québec, Canada

Stephan A. Stohlman, University of Southern California, Los Angeles, California

Kanta Subbarao, NIAID, National Institutes of Health, Bethesda, Maryland

Klaus Sure, Ruhr-University Bochum, Bochum, Germany

Kazumitsu Suzuki, National Institute of Infectious Diseases, Tokyo, Japan

Tetsuro Suzuki, National Institute of Infectious Diseases, Tokyo, Japan

Fumihiro Taguchi, National Institute of Infectious Diseases, Tokyo, Japan

Hideharu Taira, Iwate University, Morioka, Japan

Hiroko Takai, National Hospital Organization Kinki-Chuo Chest Medical Center, Sakai, Osaka, Japan

Toshitada Takemori, National Institute of Infectious Diseases, Tokyo, Japan

CONTRIBUTORS

Brian D. Adair, The Scripps Research Institute, La Jolla, California

Fernando Almazán, Centro Nacional de Biotecnología, CSIC, Madrid, Spain

Sara Alonso, Centro Nacional de Biotecnología, CSIC, Madrid, Spain

Enrique Álvarez, Centro Nacional de Biotecnología, CSIC, Madrid, Spain

Yasushi Ami, National Institute of Infectious Diseases, Tokyo, Japan

Kanchan Anand, University of Lübeck, Lübeck, Germany

William D. Arndt, Arizona State University, Tempe, Arizona

Yasuko Asahi-Ozaki, National Institute of Infectious Diseases, Tokyo, Japan

Ehtesham Baig, University of Toronto, Toronto, Canada

Susan C. Baker, Loyola University Chicago Stritch School of Medicine, Maywood, Illinois

Udeni B. Balasuriya, University of Kentucky, Lexington, Kentucky

Ralph S. Baric, University of North Carolina, Chapel Hill, North Carolina

Naina Barretto, Loyola University Chicago Stritch School of Medicine, Maywood, Illinois

Paul Bates, University of Pennsylvania School of Medicine, Philadelphia, Pennsylvania

Mara Battilani, University of Bologna, Italy

Beverley E. Bauman, Cornell University, Ithaca, New York

John Bechill, Loyola University Chicago, Maywood, Illinois

Valerie Bednar, Arizona State University, Tempe, Arizona

Cornelia C. Bergmann, Lerner Research Institute, Cleveland Clinic Foundation, Cleveland, Ohio

Ben Berkhout, University of Amsterdam, Amsterdam, The Netherlands

Richard K. Bestwick, AVI Biopharma Inc., Corvallis, Oregon

Katrina Bicknell, University of Reading, Reading, United Kingdom

Andrew Blount, Arizona State University, Tempe, Arizona

Joseph A. Boscarino, Loyola University Medical Center, Maywood, Illinois

Berend Jan Bosch, Utrecht University, Utrecht, The Netherlands

Joseph W. Brewer, Loyola University Chicago, Maywood, Illinois

Paul Britton, Institute for Animal Health, Compton Laboratory, Newbury, Berkshire, United Kingdom

Sarah M. Brockway, Vanderbilt University Medical Center, Nashville, Tennessee

Maarten Brom, University of Pennsylvania School of Medicine, Philadelphia, Pennsylvania

Gavin Brooks, University of Reading, Reading, United Kingdom

Michael J. Buchmeier, The Scripps Research Institute, La Jolla, California

Susan E. Burkett, University of North Carolina, Chapel Hill, North Carolina

Kelly M. Burkhart, Boehringer Ingelheim Vetmedica, Incorporated, Ames, Iowa

Jane C. Burns, University of California San Diego School of Medicine, La Jolla, California

Renaud Burrer, The Scripps Research Institute, La Jolla, California

Jagdish Butany, University of Toronto, Toronto, Canada

Noah Butler, University of Iowa, Iowa City, Iowa

Yingyun Cai, University of Arkansas for Medical Sciences, Little Rock, Arkansas

Jay G. Calvert, Pfizer Animal Health, Kalamazoo, Michigan

Carmen Capiscol, Centro Nacional de Biotecnología, CSIC, Madrid, Spain

Rosa Casais, Institute for Animal Health, Compton Laboratory, Newbury, Berkshire, United Kingdom

Dave Cavanagh, Institute for Animal Health, Compton Laboratory, Newbury, Berkshire, United Kingdom

Luisa Cervantes, Kantonal Hospital St. Gallen, St. Gallen, Switzerland

Lili Chen, Drug Discovery Center, Shanghai, China

Pei-Jer Chen, National Taiwan University College of Medicine, Taipei, Taiwan

Yu Chen, Wuhan University, Wuhan, People's Republic of China

Zhongbin Chen, Loyola University Chicago Stritch School of Medicine, Maywood, Illinois

Hyeryun Choe, Harvard Medical School, Boston, Massachusetts

J. Christopher-Hennings, South Dakota State University, Brookings, South Dakota

Victor C. Chu, Cornell University, Ithaca, New York

Sara Ciulli, University of Bologna, Italy

Ellen W. Collisson, Texas A&M University, College Station, Texas

Jean-Paul Coutelier, Université Catholique de Louvain, Brussels, Belgium

Oswald R. Crasta, Virginia Polytechnic Institute and State University, Blacksburg, Virgionia

Jennifer Crookshank, University of Toronto, Toronto, Canada

Myron Cybulsky, University of Toronto, Toronto, Canada

Michael J. Czar, Virginia Polytechnic Institute and State University, Blacksburg, Virginia

Marc Davis, Institute for Animal Health, Compton Laboratory, Newbury, Berkshire, United Kingdom

Marta L. DeDiego, Centro Nacional de Biotecnología, CSIC, Madrid, Spain

Nadine DeAlbuquerque, University of Toronto, Toronto, Canada

Peter L. Delputte, Ghent University, Merelbeke, Belgium

Daphne E. deMello, St. Louis University Health Science Center, St. Louis, Missouri

Damon J. Deming, University of North Carolina, Chapel Hill, North Carolina

Mark R. Denison, Vanderbilt University Medical Center, Nashville, Tennessee

Marc Desforges, INRS-Institut Armand Frappier, Laval, Québec, Canada

Michael S. Diamond, Washington University School of Medicine, St. Louis, Missouri

Eric F. Donaldson, University of North Carolina, Chapel Hill, North Carolina

Brian K. Dove, University of Leeds, Leeds, United Kingdom

Christa Drexler, Intervet International B.V., The Netherlands

Christian Drosten, Bernhard-Nocht-Institute, Hamburg, Germany

Lance D. Eckerle, Vanderbilt University Medical Center, Nashville, Tennessee

Luis Enjuanes, Centro Nacional de Biotecnología, CSIC, Madrid, Spain

Klara K. Eriksson, Kantonal Hospital St. Gallen, St. Gallen, Switzerland

Gary Ewart, Biotron Ltd., ACT, Australia

Kay S. Faaberg, University of Minnesota, Saint Paul, Minnesota

Ying Fang, South Dakota State University, Brookings, South Dakota

Michael Farzan, Harvard Medical School and New England Primate Research Center, Southborough, Massachusetts

Min Feng, Texas A&M University, College Station, Texas

A Damon Ferguson, Cornell University, Ithaca, New York

Witold Filipowicz, Friedrich Miescher Institute for Biomedical Research, Basel, Switzerland

Eleanor Fish, University of Toronto, Toronto, Canada

Johannes Forster, St. Josefs Hospital, Freiburg, Germany

Ambra Foschi, University of Bologna, Bologna, Italy

Stéphanie Franquin, Université Catholique de Louvain, Brussels, Belgium

Daved H. Fremont, Washington University School of Medicine, St. Louis, Missouri

Matthew B. Frieman, University of North Carolina, Chapel Hill, North Carolina

Yukari Fukunaga, National Hospital Organization Kinki-Chuo Chest Medical Center, Sakai, Osaka, Japan

Shuetsu Fukushi, National Institute of Infectious Diseases, Tokyo, Japan

Izumi Furukawa, National Hospital Organization Kinki-Chuo Chest Medical Center, Sakai, Osaka, Japan

Peter Gage, JCSMR, ANU, ACT, Australia

Mylène Gagnon, INRS-Institut Armand Frappier, Laval, Québec, Canada

Carmen Galán, Centro Nacional de Biotecnología, CSIC, Madrid, Spain

Thomas M. Gallagher, Loyola University Chicago Stritch School of Medicine, Maywood, Illinois

Francesca Garcia, Northwestern University Feinberg School of Medicine, Chicago, Illinois

Robert F. Garry, Tulane University Health Sciences Center, New Orleans, Louisiana

Ana Garzón, Campus Universidad Autónoma, Cantoblanco, Madrid, Spain

Martina Geier, University of Erlangen-Nürnberg, Nürnberg, Germany

David P. Giedroc, Texas A&M University, College Station, Texas

Laura Gillim-Ross, New York State Department of Health, Albany, New York

Jeffrey M. Goletz, Loyola University Medical Center, Maywood, Illinois

J.-M. Gonzales, University of Southern California, Los Angeles, California

Alexander E. Gorbalenya, Leiden University Medical Center, Leiden, The Netherlands

Reg Gorczynski, University of Toronto, Toronto, Canada

Rachel L. Graham, Vanderbilt University, Nashville, Tennessee

Thomas Gramberg, University of Erlangen-Nürnberg, Nürnberg, Germany

David Grant, University of Toronto, Toronto, Canada

Chunshan Gui, Drug Discovery Center, Shanghai, China

Stephan Günther, Bernhard-Nocht-Institute, Hamburg, Germany

Deyin Guo, Wuhan University, Wuhan, People's Republic of China

Yuji Takemoto, National Hospital Organization Kinki-Chuo Chest Medical Center, Sakai, Osaka, Japan

Pierre J. Talbot, INRS- Institut Armand Frappier, Laval, Québec, Canada

James P. Tam, Nanyang Technological University, Singapore, and Institute of Molecular and Cell Biology, Proteos, Singapore

Jinzhi Tan, Chinese Academy of Sciences, Shanghai, China

Takao Tanaka, National Hospital Organization Kinki-Chuo Chest Medical Center, Sakai, Osaka, Japan

Chandra Tangadu, Loyola University Chicago Stritch School of Medicine, Maywood, Illinois

Masato Tashiro, National Institute of Infectious Diseases, Tokyo, Japan

Volker Thiel, Kantonal Hospital St. Gallen, St. Gallen, Switzerland

Gaëtan Thirion, Université Catholique de Louvain, Brussels, Belgium

Po Tien, Wuhan University, Wuhan, People's Republic of China

Koujiro Tohyama, Iwate Medical University, Morioka, Japan

Kathryn Trandem, University of Iowa, Iowa City, Iowa

Emily A. Travanty, University of Colorado Health Sciences Center, Aurora, Colorado

Brian Tripet, University of Colorado Health Sciences Center, Aurora, Colorado

S. I. Tschen, University of Southern California, Los Angeles, California

Sonia Tusell, University of Colorado Health Sciences Center, Aurora, Colorado

Yasuko Tsunetsugu-Yokota, National Institute of Infectious Diseases, Tokyo, Japan

Klaus Überla, Ruhr-University Bochum, Bochum, Germany

Makoto Ujike, National Institute of Infectious Diseases, Tokyo, Japan

Lia van der Hoek, University of Amsterdam, Amsterdam, The Netherlands

Eric M. Vaughn, Boehringer Ingelheim Vetmedica, Incorporated, Ames, Iowa

Sandhya Verma, Arizona State University, Tempe, Arizona

Koen H. G. Verschueren, University of Lübeck, Lübeck, Germany

Matthias G. von Herrath, La Jolla Institute of Allergy and Immunology, La Jolla, California

Kevin B. Walsh, University of California, Irvine, California

Siao-Kun Wan Welch, Pfizer Animal Health, Kalamazoo, Michigan

Jieru Wang, National Jewish Medical and Research Center, Denver, Colorado

Xiao X. Wang, National University of Singapore, Singapore

Yue Wang, University of Minnesota, Saint Paul, Minnesota

Rie Watanabe, National Institute of Infectious Diseases, Tokyo, Japan

Anja Wegele, University of Erlangen-Nürnberg, Germany

Susan R. Weiss, University of Pennsylvania, Philadelphia, Pennsylvania

David E. Wentworth, New York State Department of Health and State University of New York, Albany, New York

Tiana C. White, Arizona State University, Tempe, Arizona

Gary R. Whittaker, Cornell University, Ithaca, New York

John V. Williams, Vanderbilt University Medical Center, Nashville, Tennessee

Lauren Wilson, ANU Medical School, ACT, Australia

Christine Wohlford-Lenane, University of Iowa, Iowa City, Iowa

Tom Wolfe, La Jolla Institute of Allergy and Immunology, La Jolla, California

Sek M. Wong, National University of Singapore, Singapore

Ying Wu, Wuhan University, Wuhan, China

Han Xiao, Nanyang Technological University, Singapore, and Institute of Molecular and Cell Biology, Proteos, Singapore

Jing Xu, Wuhan University, Wuhan, People's Republic of China

Max Xuezhong, University of Toronto, Toronto, Canada

Kyoko Yamada, National Hospital Organization Kinki-Chuo Chest Medical Center, Sakai, Osaka, Japan

Yasuko K. Yamada, National Institute of Infectious Diseases, Tokyo, Japan

Naoki Yamamoto, Tokyo Medical and Dental University, Bunkyo-ku, Tokyo, Japan

Yalin Yang, Wuhan University, Wuhan, People's Republic of China

Rong Ye, New York State Department of Health, Albany, New York

Ye Ye, Arizona State University, Tempe, Arizona

Mark Yeager, The Scripps Research Institute, La Jolla, California

Dongwan Yoo, University of Guelph, Guelph, Ontario, Canada

Shigeto Yoshida, Jichi Medical School, Minamikawachi-machi, Tochigi, Japan

Yoshiyuki Yoshinaka, Tokyo Medical and Dental University, Bunkyo-ku, Tokyo, Japan

Craig Yoshioka, The Scripps Research Institute, La Jolla, California

Jae-Hwan You, University of Leeds, Leeds, United Kingdom

Soonjeon Youn, Johns Hopkins University School of Medicine, Baltimore, Maryland

Boyd Yount, University of North Carolina, Chapel Hill, North Carolina

Dongdong Yu, University of Arkansas for Medical Sciences, Little Rock, Arkansas

Aya Zamoto, National Institute of Infectious Diseases, Tokyo, Japan

Li Zhang, University of Toronto, Toronto, Canada

Xuming Zhang, University of Arkansas for Medical Sciences, Little Rock, Arkansas

Haixia Zhou, University of Iowa, Iowa City, Iowa

Hongqing Zhu, University of Arkansas for Medical Sciences, Little Rock, Arkansas

Ying Zhu, Wuhan University, Wuhan, China

John Ziebuhr, University of Würzburg, Würzburg, Germany

Sonia Zúñiga, Centro Nacional de Biotecnología, CSIC, Madrid, Spain

I. VIRAL RNA SYNTHESIS

THE CORONAVIRUS REPLICASE: INSIGHTS INTO A SOPHISTICATED ENZYME MACHINERY

John Ziebuhr*

1. INTRODUCTION

The order *Nidovirales* currently comprises three families of plus-strand RNA viruses, the *Coronaviridae*, *Arteriviridae*, and *Roniviridae*, which are thought to have a common phylogenetic origin.[1–3] The proposed close phylogenetic relationship between the three virus families is mainly based upon common features that discrimate nidoviruses from other RNA plus-strand RNA viruses, including (i) the production of a nested set of 3'-coterminal subgenome-length RNAs,[4] (ii) the use of ribosomal frameshifting into the –1 reading frame to express the key replicative functions,[5] and (iii) the conservation of a set of functional domains that are arranged in the nonstructural polyproteins in the following order (from N- to C-terminus): two-β-barrel-fold main protease, RNA-dependent RNA polymerase of superfamily 1 (RdRp), zinc finger-containing helicase of superfamily 1, and endoribonuclease.[6–9] Despite their common ancestry, the nidovirus families differ considerably with regard to (i) their genome sizes (ranging from 12.7 to 31.3 kilobases), (ii) their structural proteins, and (iii) the conservation of several RNA-processing enzymes.[3]

2. ORGANIZATION AND EXPRESSION OF THE CORONAVIRUS REPLICASE GENE

With genome sizes of between 27.3 and 31.3 kilobases, coronaviruses represent the largest nonsegmented RNA viruses currently known. About two-thirds of their genomes (>20,000 nucleotides) are devoted to encoding the viral replicase, which mediates viral RNA synthesis but probably has many more functions. The replicase is encoded by the replicase gene. This gene is composed of two large open reading frames (ORFs), designated ORF1a and ORF1b, that are located at the 5'-end of the genome. The upstream ORF1a encodes a polyprotein of 450–500 kDa, termed polyprotein (pp) 1a, whereas ORF1a and ORF1b together encode pp1ab (750–800 kDa) (Fig. 1). Expression of the C-terminal ORF1b-encoded portion of pp1ab depends on a (–1) ribosomal frameshift. As a consequence, the key replicative enzymes (RdRp, helicase, and others) are expressed at a significantly lower level than are ORF1a-encoded protein functions.[5,10] The coronavirus replication complex is bound to intracellular membranes through several ORF1a-encoded hydrophobic domains (reviewed in Refs. 3 and 9) and, besides replicase

*University of Würzburg, D-97078 Würzburg, Germany and Queen's University, Belfast, United Kingdom, BT9 7BL

gene-encoded proteins, includes several cellular proteins and the viral nucleocapsid protein.[11–13]

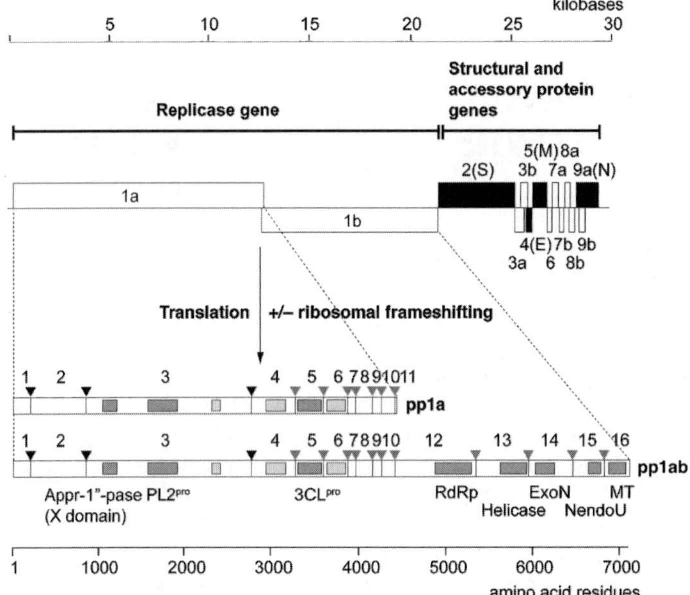

Figure 1. SARS-CoV genome organization. The putative functional open reading frames (ORFs) in the 29.7-kb genome of SARS-CoV are indicated (reviewed in Ref. 14). Translation of the genomic RNA gives rise to two replicative polyproteins, pp1a and pp1ab, that mediate the key functions required for SARS-CoV genome replication and synthesis of 8 major subgenomic RNAs encoding the structural proteins (shown in black) and several SARS-CoV-specific accessory proteins. Expression of ORF1b sequences requires a programmed ribosomal frameshift into the –1 reading frame during translation of the genome RNA, which occurs just upstream of the ORF1a translation stop codon. The replicative polyproteins are extensively processed by viral proteases. The processing end-products of pp1a are designated nonstructural proteins (nsp) 1 to nsp11 and those of pp1ab are designated nsp1 to nsp10 and nsp12 to nsp16. Cleavage sites that are processed by the 3C-like protease ($3CL^{pro}$), which is also called main protease (M^{pro}), are indicated by gray arrowheads, and sites that are processed by the papain-like protease, $PL2^{pro}$, are indicated by black arrowheads. The major replicative domains are shown in dark gray and transmembrane domains predicted to anchor the viral replicase to intracellular membranes are indicated in light gray. Appr-1"-pase, ADP-ribose 1"-phosphatase; $PL2^{pro}$, papain-like cysteine protease 2; $3CL^{pro}$, 3C-like main protease; RdRp, RNA-dependent RNA polymerase; ExoN, 3'-to-5' exoribonuclease; NendoU, nidoviral uridylate-specific endoribonuclease; MT, putative ribose-2'-*O* methyltransferase.

3. PROTEASES AND PROTEOLYTIC PROCESSING

Activation of the coronavirus replication complex involves extensive proteolytic processing of the replicase polyproteins to produce 16 (in IBV: 15) mature products called nonstructural proteins (nsp) 1 to 16 (reviewed in Refs. 3 and 15). The processing involves two different types of proteases. The key enzyme is the so-called 3C-like protease ($3CL^{pro}$), also called main protease (M^{pro}), which cleaves the central and

C-terminal regions of pp1a and pp1ab at 11 conserved sites.[15] 3CL^{pro} is a cysteine protease featuring a Cys-His catalytic dyad and a three-domain structure.[16,17] The N-terminal domains I and II adopt a two-β-barrel-fold structure that resembles the structures of serine proteases of the chymotrypsin superfamily. The C-terminal entirely α-helical domain III is critically involved in proteolytic activity and dimerization. It is now generally accepted that the 3CL^{pro} dimer represents the active form of the protease.[18,19] Intermolecular interactions within the dimer have been suggested to keep the active center (in particular, the S1 subsite) in a proteolytically competent conformation.[17,20] Coronavirus main proteases have conserved substrate specificities [(L,I)–Q↓(S,A,G)] and, accordingly, similar substrate-binding pocket structures, as has been shown by crystal structure analyses of main proteases from different coronavirus genetic groups.[17,21–23] Based on this information, protease inhibitors have been developed that effectively inhibit a broad range of coronavirus main proteases.[23] Potentially, these compounds may be useful for treatment of known or newly emerging coronavirus infections.

The N-terminal regions of the coronavirus polyproteins, which are poorly conserved among the coronavirus groups I, II, and III, are cleaved at two (in IBV) or three sites (in all other coronaviruses) by one (IBV and SARS-CoV) or two zinc-finger-containing papain-like cysteine proteases called PL1^{pro} and PL2^{pro}.[24–32] PL^{pro} cleavage sites are usually flanked by small residues (mainly Gly and Ala).[15] PL1^{pro} and PL2^{pro} probably evolved by gene duplication in one of the ancestors of the present-day coronaviruses.[29] For MHV, it was recently demonstrated that cleavage of the nsp1|nsp2 site and, even more suprisingly, the C-terminal nsp1 and the entire nsp2 sequence are not required for MHV replication *in vitro*.[33–35]

4. POLYMERASE AND HELICASE ACTIVITIES

The coronavirus RdRp has been classified as an outgroup of the RdRp superfamily 1.[36] It resides in the ORF1b-encoded nonstructural proteins 12, as was predicted about 15 years ago[6] and now demonstrated for a bacterially expressed form of the SARS-CoV nsp12.[37] The RdRp catalytic domain occupies the C-terminal two-thirds of nsp12,[6] but also the N-terminal part of nsp12 seems to contribute to activity.[37] Based on MHV co-immunoprecipitation data, the N-terminal part of nsp12 has been suggested to interact with other replicase subunits, specifically with nsp5 (3CL^{pro}), nsp8, and nsp9,[38] which is in line with the recently published crystal structure of an nsp7–nsp8 complex from SARS-CoV.[39] The X-ray crystallography study by Zhai et al.[39] provides first insights into the sophisticated architecture of the coronavirus replicase. It shows that eight molecules of each nsp7 and nsp8 assemble into a hexadecameric supercomplex that forms a hollow, cylinder-like structure, which, because of its internal dimensions and electrostatic properties, seems to be capable of encircling RNA. This nsp7–nsp8 supercomplex, in turn, interacts with nsp9, a small single-stranded RNA-binding protein,[40,41] and interactions of nsp7–nsp8(–nsp9) with yet other replicase subunits (including nsp12) are likely. On the basis of structural and functional data, it has been speculated that the nsp7–nsp8 complex might act as an important RdRp cofactor, for example, to increase the RdRp processivity.[39] Evidence for a critical involvement of the C-terminal pp1a processing products (nsp7–10) in viral RNA synthesis also comes from work on MHV temperature-sensitive mutants, which has shown that nsp10, a Cys/His-rich protein, has a distinct role in minus-strand RNA synthesis.[42]

After the RdRp, helicases are the second best conserved enzymes of plus-strand RNA viruses. The coronavirus helicase resides in nsp13 and has been classified as belonging to the helicase superfamily 1[6,43] (Fig. 1). Coronavirus helicases and all their nidovirus homologs are linked to an N-terminal zinc-binding domain (ZBD) involving 12–13 conserved Cys/His residues.[6] The conservation of a ZBD-associated superfamily 1 helicase is considered one of the major genetic markers of nidoviruses. Over the past years, the helicase activities of HCoV-229E and, to a lesser extent, of SARS-CoV have been characterized.[44–48] Mutagenesis and biochemical data revealed that the ZBD is essential for the enzymatic activities of both the coronavirus and arterivirus helicases.[49] Coronavirus helicases proved to have multiple enzymatic activities, including nucleic acid-stimulated NTPase, dNTPase, and RNA (plus DNA) duplex-unwinding activities.[44,47,48] Coronavirus (and arterivirus) helicases were demonstrated to unwind double-stranded RNA substrates with 5'-to-3' polarity, that is, they move in a 5'-to-3' direction along the strand to which they initially bind.[44,50] The polarity of coronavirus helicase activities contrasts with that of the helicases of flavi-, pesti-, and hepaciviruses, which all operate in the opposite direction,[51] indicating that the biological functions of helicases from the *Nidovirales* and *Flaviviridae* might differ fundamentally from each other. Coronavirus (nidovirus) helicases belong to the few helicases that act on both RNA or DNA substrates with nearly equal efficacy. Given that the helicase is part part of the (cytoplasmic) coronavirus replicase complex, it seems unlikely that the observed DNA duplex-unwinding activity is of biological significance. Coronavirus helicases are able to unwind up to several hundred base pairs of double-stranded nucleic acid in a processive manner,[47,48] supporting a role as a "replicative" helicase that separates regions of double-stranded RNA that the polymerase might encounter during RNA synthesis.

Coronavirus helicases are able to hydrolyze essentially all natural nucleotides to fuel their translocation along nucleic acids and concomitant unwinding of duplex RNA (and DNA) structures.[44–48] Besides their NTPase and dNTPase activities, coronavirus helicases were shown to possess yet another phosphohydrolyase activity, namely an RNA 5'-triphosphatase activity that employs the active site of the (d)NTPase activity. It is reasonable to believe that the nsp13-associated RNA 5'-triphosphatase catalyzes the first reaction in the synthesis of coronavirus 5' RNA cap structures.[47,48] Given the multiple enzymatic activities of coronavirus helicases and the previously observed diverse roles of the arterivirus helicase in viral replication, transcription and virion biogenesis,[52] it is reasonable to believe that coronavirus helicases are involved in more than one metabolic pathway in the viral life cycle.

5. RNA-PROCESSING ENZYMES

A detailed sequence analysis by Gorbalenya, Snijder and colleagues of the SARS-CoV genome revealed that the enzymology of coronavirus RNA synthesis may be significantly more complex than previously thought.[8] In this seminal study published in 2003, as many as five novel enzymatic activities were identified in the genomes of coronaviruses and, to a varying extent, other nidoviruses (Fig. 1). The putative activities included (i) a 3'-to-5' exonuclease (ExoN) associated with nsp14, (ii) an endoribonuclease (NendoU) associated with nsp15, (iii) an S-adenosyl methionine-dependent ribose 2'-*O*-methyltransferase (MT) associated with nsp16, (iv) an ADP-ribose 1"-phosphatase (Appr-1"-pase) associated with the so-called "X domain" of nsp3, and (v) a cyclic phosphodiesterase (CPDase), which is only conserved in group II coronaviruses (excluding SARS-CoV). Four of the activities were found to be conserved in all coronaviruses, suggesting that they have an essential role in the coronaviral life cycle. Recently, the predicted exo- and endoribonuclease activities were established and characterized using bacterially expressed forms of the ExoN from SARS-CoV (Minskaia, Hertzig, Gorbalenya, Campanacci, Cambillau, Canard, and Ziebuhr, unpublished data), and the NendoU from HCoV-229E, SARS-CoV, and IBV.[53,54] We have recently demonstrated by reverse genetics using the HCoV-229E infectious clone that the ExoN, NendoU, and MT domains are all essential for the production of virus progeny. Substitutions of predicted active-site residues resulted in diverse defects in viral RNA synthesis and virus production[54] (Hertzig, Ulferts, Schelle, and Ziebuhr, unpublished data). The underlying mechanisms for the observed defects are currently being studied in detail.

The observed pattern of conservation in different nidovirus families suggests a functional hierarchy for the newly identified RNA-processing activities, with the manganese ion-dependent uridylate-specific endoribonuclease, NendoU, playing a central role. This enzyme is universally conserved in nidoviruses and, besides the ZBD-associated helicase, represents a genetic marker of nidoviruses that is also reflected by its previous designation as "nidovirus-specific conserved domain" [55,56]. As mentioned above, NendoU has been demonstrated by site-directed mutagenesis of the full-length HCoV-229E clone to be essential for coronaviral RNA synthesis. However, we recently obtained mutagenesis data that argue for a more complex scenario regarding the activities of NendoU. Thus, for example, we found that several single-residue substitutions that abolished the NendoU activity in an *in vitro* RNase assay gave rise to viable (albeit low-titer) virus when transferred to the HCoV-229E infectious clone (Hertzig, Ulferts and Ziebuhr, unpublished data). These data indicate that NendoU might have more than one activity and/or cleave more than one substrate that likely differs from that used in our *in vitro* assays. Clearly, more work will be required to elucidate the function(s) of NendoU in the nidoviral life cycle.

In view of the very similar gene expression and RNA synthesis strategies employed by the various nidovirus families and genera, it is intriguing that only one (namely NendoU) of the newly identified nidovirus RNA-processing activities is conserved in arteriviruses. The basis for this differential conservation pattern of RNA-processing enzymes is currently unclear but might indicate that the enzymology involved in viral genome-length and subgenome-length RNA synthesis varies to some extent among the nidovirus families. Alternatively, members of the various nidovirus families/genera might interact differentially with yet-to-be-defined RNA-processing pathways of the host cell.

It is also worth remembering that arterivirus genomes are about 2 times smaller than other nidovirus genomes and it is possible that (some of) the extra domains conserved in corona- and toroviruses (and, to a lesser extent, in roniviruses) are required to replicate genomes of this unique size. The coronavirus ExoN activity was recently shown by sequence analysis and site-sirected mutagenesis to be related to cellular enzymes of the DEDD exonuclease superfamily (Minskaia, Hertzig, Gorbalenya, Campanacci,

Cambillau, Canard, and Ziebuhr, unpublished data). This exonuclease family includes cellular 3'-to-5' exonucleases involved in proofreading, repair, and/or recombination, and it is tempting to speculate that nidoviruses use their ExoN domains in similar processes, for example, to keep the error frequency of their low-fidelity RdRp below a critical thereshold.

In contrast to the essential RNA-processing activities (ExoN, NendoU, and MT), the nsp3 X domain-associated Appr-1"-pase activity was found to be dispensable for viral replication *in vitro*.[57] Substitutions of putative active-site residues of the Appr-1"-pase domain, which have been confirmed to abolish enzymatic activity *in vitro*, gave rise to viable virus. The virus had no apparent defects in viral RNA synthesis and grew to the same titers as the wild-type virus.[57] Similarly, MHV reverse genetics data demonstrated that coronaviruses tolerate specific substitutions and even deletions in the replicase gene. For example, cleavage of the nsp1|nsp2 cleavage site was shown to be dispensable for viral replication, and even deletion of the C-terminal part of nsp1 and the entire nsp2, respectively, had only minor effects on viral replication.[33–35] By contrast, several other mutations/deletions were tolerated *in vitro* but caused attenuation in the natural host.[58] Taken together, the data suggest that coronavirus replicases have evolved to include a number of nonessential functions. The fact that some of the "nonessential" protease cleavages and enzymatic activities are conserved in coronaviruses or in specific coronavirus genetic groups suggests that they provide a selective advantage only in the host.

6. CONCLUDING REMARKS

Although much has been learned about coronavirus replicase organization, localization, proteolytic processing, and viral replicative enzymes, there are still major gaps in our knowledge. Inspired by the SARS epidemic in 2003, numerous studies aiming at the development of coronavirus vaccines and antivirals have recently been published, and there is no doubt that coronavirus research has gained momentum over the past few years. Given the rapidly accumulating biochemical, structural, and genetic information on coronaviruses, a more detailed understanding of the molecular mechanisms involved in coronaviral RNA synthesis can be expected to emerge in the near future. It will be of particular interest to identify those proteins that are involved in the specific mechanisms of coronavirus RNA synthesis, such as the production of a nested set of subgenome-length RNAs and the replication of RNA genomes of unprecedented size. Furthermore, studies on coronavirus-encoded RNA-processing activities and their cellular homologs might reveal interesting insights into the relationship (or interplay) of coronaviral and cellular RNA metabolism pathways. In the long run, the unique structural properties of several conserved coronavirus replicative enzymes may lead to the development of selective enzyme inhibitors and possibly even drugs suitable to treat coronavirus infections of humans and animals.

7. ACKNOWLEDGMENT

My work is supported by the Deutsche Forschungsgemeinschaft and the European Commission.

8. REFERENCES

1. Cavanagh, D., 1997, *Nidovirales*: a new order comprising *Coronaviridae* and *Arteriviridae*, *Arch. Virol.* **142**:629.
2. González, J. M., Gomez-Puertas, P., Cavanagh, D., Gorbalenya, A. E., and Enjuanes, L., 2003, A comparative sequence analysis to revise the current taxonomy of the family *Coronaviridae*, *Arch. Virol.* **148**:2207.
3. Siddell, S. G., Ziebuhr, J., and Snijder, E. J., 2005, Coronaviruses, toroviruses, and arteriviruses, in: *Topley and Wilson's Microbiology and Microbial Infections, 10th Edition*, B. W. J. Mahy and V. ter Meulen, eds., Hodder Arnold, London, p. 823.
4. Spaan, W., Delius, H., Skinner, M., Armstrong, J., Rottier, P., Smeekens, S., van der Zeijst, B. A., and Siddell, S. G., 1983, Coronavirus mRNA synthesis involves fusion of non-contiguous sequences, *EMBO J.* **2**:1839.
5. Brierley, I., Boursnell, M. E., Binns, M. M., Bilimoria, B., Blok, V. C., Brown, T. D., and Inglis, S. C., 1987, An efficient ribosomal frame-shifting signal in the polymerase-encoding region of the coronavirus IBV, *EMBO J.* **6**:3779.
6. Gorbalenya, A. E., Koonin, E. V., Donchenko, A. P., and Blinov, V. M., 1989b, Coronavirus genome: prediction of putative functional domains in the non-structural polyprotein by comparative amino acid sequence analysis, *Nucleic Acids Res.* **17**:4847.
7. Gorbalenya, A. E., 2001, Big nidovirus genome. When count and order of domains matter, *Adv. Exp. Med. Biol.* **494**:1.
8. Snijder, E. J., Bredenbeek, P. J., Dobbe, J. C., Thiel, V., Ziebuhr, J., Poon, L. L., Guan, Y., Rozanov, M., Spaan, W. J., and Gorbalenya, A. E., 2003, Unique and conserved features of genome and proteome of SARS-coronavirus, an early split-off from the coronavirus group 2 lineage, *J. Mol. Biol.* **331**:991.
9. Ziebuhr, J., 2005, The coronavirus replicase, *Curr. Top. Microbiol. Immunol.* **287**:57.
10. Herold, J., and Siddell, S. G., 1993, An 'elaborated' pseudoknot is required for high frequency frameshifting during translation of HCV 229E polymerase mRNA. *Nucleic Acids Res* **21**:5838.
11. Almazán, F., Galán, C., and Enjuanes, L., 2004, The nucleoprotein is required for efficient coronavirus genome replication, *J. Virol.* **78**:12683.
12. Schelle, B., Karl, N., Ludewig, B., Siddell, S. G., and Thiel, V., 2005, Selective replication of coronavirus genomes that express nucleocapsid protein, *J. Virol.* **79**:6620.
13. Shi, S. T, and Lai, M. M., 2005, Viral and cellular proteins involved in coronavirus replication, *Curr. Top. Microbiol. Immunol.* **287**:95.
14. Ziebuhr, J., 2004, Molecular biology of severe acute respiratory syndrome coronavirus, *Curr. Opin. Microbiol.* **7**:412.
15. Ziebuhr, J., Snijder, E. J., and Gorbalenya, A. E., 2000, Virus-encoded proteinases and proteolytic processing in the Nidovirales, *J. Gen. Virol.* **81**:853.
16. Anand, K., Palm, G. J., Mesters, J. R., Siddell, S. G., Ziebuhr, J., and Hilgenfeld, R., 2002, Structure of coronavirus main proteinase reveals combination of a chymotrypsin fold with an extra alpha-helical domain, *EMBO J.* **21**:3213.
17. Anand, K., Ziebuhr, J., Wadhwani, P., Mesters, J. R., and Hilgenfeld, R., 2003, Coronavirus main proteinase ($3CL^{pro}$) structure: basis for design of anti-SARS drugs, *Science* **300**:1763.
18. Shi, J., Wei, Z., and Song, J., 2004, Dissection study on the severe acute respiratory syndrome 3C-like protease reveals the critical role of the extra domain in dimerization of the enzyme: defining the extra domain as a new target for design of highly specific protease inhibitors. *J. Biol. Chem.* **279**:24765.
19. Hsu, W. C., Chang, H. C., Chou, C. Y., Tsai, P. J., Lin, P. I., and Chang, G. G., 2005, Critical assessment of important regions in the subunit association and catalytic action of the severe acute respiratory syndrome coronavirus main protease, *J. Biol. Chem.* **280**:22741.
20. Chen, S., Chen, L., Tan, J., Chen, J., Du, L., Sun, T., Shen, J., Chen, K., Jiang, H., and Shen, X., 2005, Severe acute respiratory syndrome coronavirus 3C-like proteinase N terminus is indispensable for proteolytic activity but not for enzyme dimerization. Biochemical and thermodynamic investigation in conjunction with molecular dynamics simulations, *J. Biol. Chem.* **280**:164.
21. Hegyi, A., and Ziebuhr, J., 2002, Conservation of substrate specificities among coronavirus main proteases, *J. Gen. Virol.* **83**:595.
22. Yang, H., Yang, M., Ding, Y., Liu, Y., Lou, Z., Zhou, Z., Sun, L., Mo, L., Ye, S., Pang, H., Gao, G. F., Anand, K., Bartlam, M., Hilgenfeld, R., and Rao, Z., 2003, The crystal structures of severe acute respiratory syndrome virus main protease and its complex with an inhibitor, *Proc. Natl. Acad. Sci. USA* **100**:13190.

23. Yang, H., Xie, W., Xue, X., Yang, K., Ma, J., Liang, W., Zhao, Q., Zhou, Z., Pei, D., Ziebuhr, J., Hilgenfeld, R., Yuen, K. Y., Wong, L., Gao, G., Chen, S., Chen, Z., Ma, D., Bartlam, M., and Rao, Z., 2005, Design of wide-spectrum inhibitors targeting coronavirus main proteases, *PLoS Biol.* **3**:e324.
24. Baker, S. C., Yokomori, K., Dong, S., Carlisle, R., Gorbalenya, A. E., Koonin, E. V., and Lai, M. M., 1993, Identification of the catalytic sites of a papain-like cysteine proteinase of murine coronavirus, *J. Virol.* **67**:6056.
25. Herold, J., Gorbalenya, A. E., Thiel, V., Schelle, B., and Siddell, S. G., 1998, Proteolytic processing at the amino terminus of human coronavirus 229E gene 1-encoded polyproteins: identification of a papain-like proteinase and its substrate, *J. Virol.* **72**:910.
26. Lim, K. P., and Liu, D. X., 1998, Characterization of the two overlapping papain-like proteinase domains encoded in gene 1 of the coronavirus infectious bronchitis virus and determination of the C-terminal cleavage site of an 87-kDa protein, *Virology* **245**:303.
27. Kanjanahaluethai, A., and Baker, S. C., 2000, Identification of mouse hepatitis virus papain-like proteinase 2 activity, *J. Virol.* **74**:7911.
28. Lim, K. P., Ng, L. F., and Liu, D. X., 2000, Identification of a novel cleavage activity of the first papain-like proteinase domain encoded by open reading frame 1a of the coronavirus Avian infectious bronchitis virus and characterization of the cleavage products, *J. Virol.* **74**:1674.
29. Ziebuhr, J., Thiel, V., and Gorbalenya, A. E., 2001, The autocatalytic release of a putative RNA virus transcription factor from its polyprotein precursor involves two paralogous papain-like proteases that cleave the same peptide bond, *J. Biol. Chem.* **276**:33220.
30. Thiel, V., Ivanov, K. A., Putics, Á., Hertzig, T., Schelle, B., Bayer, S., Weissbrich, B., Snijder, E. J., Rabenau, H., Doerr, H. W., Gorbalenya, A. E., and Ziebuhr, J., 2003, Mechanisms and enzymes involved in SARS coronavirus genome expression, *J. Gen. Virol.* **84**:2305.
31. Harcourt, B. H., Jukneliene, D., Kanjanahaluethai, A., Bechill, J., Severson, K. M., Smith, C. M., Rota, P. A., and Baker, S. C., 2004, Identification of severe acute respiratory syndrome coronavirus replicase products and characterization of papain-like protease activity, *J. Virol.* **78**:13600.
32. Han, Y. S., Chang, G. G., Juo, C. G., Lee, H. J., Yeh, S. H., Hsu, J. T., and Chen, X., 2005, Papain-like protease 2 (PLP2) from severe acute respiratory syndrome coronavirus (SARS-CoV): expression, purification, characterization, and inhibition, *Biochemistry* **44**:10349.
33. Denison, M. R., Yount, B., Brockway, S. M., Graham, R. L., Sims, A. C., Lu, X., and Baric, R. S., 2004, Cleavage between replicase proteins p28 and p65 of mouse hepatitis virus is not required for virus replication, *J. Virol.* **78**:5957.
34. Brockway, S. M., and Denison, M. R., 2005, Mutagenesis of the murine hepatitis virus nsp1-coding region identifies residues important for protein processing, viral RNA synthesis, and viral replication, *Virology* **340**:209.
35. Graham, R. L., Sims, A. C., Brockway, S. M., Baric, R. S., and Denison, M. R., 2005, The nsp2 replicase proteins of murine hepatitis virus and severe acute respiratory syndrome coronavirus are dispensable for viral replication, *J. Virol.* **79**:13399.
36. Koonin, E. V., 1991, The phylogeny of RNA-dependent RNA polymerases of positive-strand RNA viruses, *J. Gen. Virol.* **72**:2197.
37. Cheng, A., Zhang, W., Xie, Y., Jiang, W., Arnold, E., Sarafianos, S. G., and Ding, J., 2005, Expression, purification, and characterization of SARS coronavirus RNA polymerase, *Virology* **335**:165.
38. Brockway, S. M., Clay, C. T., Lu, X. T., and Denison, M. R., 2003, Characterization of the expression, intracellular localization, and replication complex association of the putative mouse hepatitis virus RNA-dependent RNA polymerase, *J. Virol.* **77**:10515.
39. Zhai, Y., Sun, F., Li, X., Pang, H., Xu, X., Bartlam, M., and Rao, Z., 2005, Insights into SARS-CoV transcription and replication from the structure of the nsp7-nsp8 hexadecamer, *Nat. Struct. Mol. Biol.* Epub ahead of print.
40. Egloff, M. P., Ferron, F., Campanacci, V., Longhi, S., Rancurel, C., Dutartre, H., Snijder, E. J., Gorbalenya, A. E., Cambillau, C., and Canard, B., 2004, The severe acute respiratory syndrome-coronavirus replicative protein nsp9 is a single-stranded RNA-binding subunit unique in the RNA virus world, *Proc. Natl. Acad. Sci. USA* **101**:3792.
41. Sutton, G., Fry, E., Carter, L., Sainsbury, S., Walter, T., Nettleship, J., Berrow, N., Owens, R., Gilbert, R., Davidson, A., Siddell, S., Poon, L. L., Diprose, J., Alderton, D., Walsh, M., Grimes, J. M., and Stuart, D. I., 2004, The nsp9 replicase protein of SARS-coronavirus, structure and functional insights, *Structure (Camb)* **12**:341.
42. Siddell, S., Sawicki, D., Meyer, Y., Thiel, V., and Sawicki, S., 2001, Identification of the mutations responsible for the phenotype of three MHV RNA-negative ts mutants, *Adv. Exp. Med. Biol.* **494**:453.

43. Gorbalenya, A. E., Koonin, E. V., Donchenko, A. P., and Blinov, V. M., 1989a, Two related superfamilies of putative helicases involved in replication, recombination, repair and expression of DNA and RNA genomes, *Nucleic Acids Res.* **17**:4713.
44. Seybert, A., Hegyi, A., Siddell, S. G., and Ziebuhr, J., 2000a, The human coronavirus 229E superfamily 1 helicase has RNA and DNA duplex-unwinding activities with 5'-to-3' polarity, *RNA* **6**:1056.
45. Seybert, A., and Ziebuhr, J., 2001, Guanosine triphosphatase activity of the human coronavirus helicase, *Adv. Exp. Med. Biol.* **494**:255.
46. Tanner, J. A., Watt, R. M., Chai, Y. B., Lu, L. Y., Lin, M. C., Peiris, J. S., Poon, L. L., Kung, H. F., and Huang, J. D., 2003, The severe acute respiratory syndrome (SARS) coronavirus NTPase/helicase belongs to a distinct class of 5' to 3' viral helicases, *J. Biol. Chem.* **278**:39578.
47. Ivanov, K. A., Thiel, V., Dobbe, J. C., van der Meer, Y., Snijder, E. J., and Ziebuhr, J., 2004b, Multiple enzymatic activities associated with severe acute respiratory syndrome coronavirus helicase, *J. Virol.* **78**:5619.
48. Ivanov, K. A., and Ziebuhr, J., 2004, Human coronavirus 229E nonstructural protein 13: characterization of duplex-unwinding, nucleoside triphosphatase, and RNA 5'-triphosphatase activities, *J. Virol.* **78**:7833.
49. Seybert, A., Posthuma, C. C., van Dinten, L. C., Snijder, E. J., Gorbalenya, A. E., and Ziebuhr, J., 2005, A complex zinc finger controls the enzymatic activities of nidovirus helicases, *J. Virol.* **79**:696.
50. Seybert, A., van Dinten, L. C., Snijder, E. J., and Ziebuhr, J., 2000b, Biochemical characterization of the equine arteritis virus helicase suggests a close functional relationship between arterivirus and coronavirus helicases, *J. Virol.* **74**:9586.
51. Kadaré, G., and Haenni, A. L., 1997, Virus-encoded RNA helicases, *J. Virol.* **71**:2583.
52. van Dinten, L. C., van Tol, H., Gorbalenya, A. E., and Snijder, E. J., 2000, The predicted metal-binding region of the arterivirus helicase protein is involved in subgenomic mRNA synthesis, genome replication, and virion biogenesis, *J. Virol.* **74**:5213.
53. Bhardwaj, K., Guarino, L., and Kao, C. C., 2004, The severe acute respiratory syndrome coronavirus Nsp15 protein is an endoribonuclease that prefers manganese as a cofactor, *J. Virol.* **78**:12218.
54. Ivanov, K. A., Hertzig, T., Rozanov, M., Bayer, S., Thiel, V., Gorbalenya, A. E., and Ziebuhr, J., 2004a, Major genetic marker of nidoviruses encodes a replicative endoribonuclease, *Proc. Natl. Acad. Sci. USA* **101**:12694.
55. Snijder, E. J., den Boon, J. A., Bredenbeek, P. J., Horzinek, M. C., Rijnbrand, R., and Spaan, W. J., 1990, The carboxyl-terminal part of the putative Berne virus polymerase is expressed by ribosomal frameshifting and contains sequence motifs which indicate that toro- and coronaviruses are evolutionarily related, *Nucleic Acids Res.* **18**:4535.
56. den Boon, J. A., Snijder, E. J., Chirnside, E. D., de Vries, A. A., Horzinek, M. C., and Spaan, W. J., 1991, Equine arteritis virus is not a togavirus but belongs to the coronaviruslike superfamily, *J. Virol.* **65**:2910.
57. Putics, A., Filipowicz, W., Hall, J., Gorbalenya, A. E., and Ziebuhr, J., 2005, ADP-ribose-1"-monophosphatase: a conserved coronavirus enzyme that is dispensable for viral replication in tissue culture, *J. Virol.* **79**:12721.
58. Sperry, S. M., Kazi, L., Graham, R. L., Baric, R. S., Weiss, S. R., and Denison, M. R., 2005, Single-amino-acid substitutions in open reading frame (ORF) 1b-nsp14 and ORF 2a proteins of the coronavirus mouse hepatitis virus are attenuating in mice, *J. Virol.* **79**:3391.

BIOCHEMICAL ASPECTS OF CORONAVIRUS REPLICATION

Luis Enjuanes, Fernando Almazán, Isabel Sola, Sonia Zúñiga, Enrique Alvarez, Juan Reguera, and Carmen Capiscol*

1. INTRODUCTION

Extensive morphological and biochemical changes occur in coronavirus (CoV)-infected cells. Nevertheless, there is a limited knowledge of the biochemical events occurring in host cells and in the biochemistry of the infection. Infections by CoVs cause alterations in host cells in transcription and translation patterns, in the cell cycle, in the cytoskeleton, and in apoptosis pathways. In addition, in the host, CoV infection may cause inflammation, alterations of the immune response, of cytokine and chemokine levels, of interferon (INF)-induced gene expression and of stress responses, and modification of coagulation pathways. This chapter will focus on selected biochemical aspects of CoV replication and transcription with special attention to the interaction between cell and viral factors.

2. INFLUENCE OF VIRAL AND CELLULAR PROTEINS IN CoV REPLICATION

2.1. Nuclear Localization of CoV Proteins

There are at least three CoV proteins that have been localized within the nucleus of infected cells: nucleoprotein (N), 3b, and nsp1 (Table 1). The nucleolus has been implicated in many aspects of cell biology that include functions such as ribosomal rRNA synthesis and ribosome biogenesis, gene silencing, senescence, and cell cycle regulation.[1-5] The nucleolus contains different factors including nucleolin, fibrillarin, spectrin, B23, rRNA, and ribosomal proteins S5 and L9. Viruses interact with the nucleolus and its antigens; viral proteins co-localize with factors such as nucleolin, B23, and fibrillarin and cause their redistribution during infection.[2] N proteins from CoV genus α

*Centro Nacional de Biotecnología, CSIC, Darwin, 3, Cantoblanco, 28049 Madrid, Spain.

Table 1. NiV proteins in the nucleus of infected cells.

PROTEIN	VIRUS	REFERENCE
N	IBV	Chen, *et al.*, 2003
	MHV	Wurm, *et al.*, 2001
	SARS-CoV	Timani, *et al.*, 2004
	TGEV	Wurm, *et al.*, 2001
	PRRSV	Rowland, *et al.*,1999
	EAV	Tijms, *et al.*, 2002
3b	SARS-CoV	Yuan, *et al.*, 2005
nsp1	EAV	van de Meer, *et al.*, 1999

(transmissible gastroenteritis virus, TGEV),[6] β (mouse hepatitis virus, MHV, and severe and acute respiratory syndrome coronavirus, SARS-CoV),[6-8] and γ (infectious bronchitis virus, IBV),[9] and also from two arteriviruses [porcine respiratory and reproductive syndrome virus (PRRSV) and equine arteritis virus (EAV)][10,11] localize in the nucleolus, and this may be a common feature among nidovirus N proteins that influences host cell proliferation.[6] The association of N protein with the nucleus may be cell dependent. In fact, until now, N protein has been located in the nucleus of LLC-PK1 and Vero cells transfected with plasmids expressing N protein[6], but not in ST cells transfected with plasmids expressing N protein or infected with TGEV[12]. Overall, these data indicate that the presence of N protein in the nucleus of the infected cells might be of functional significance.

SARS-CoV ORF 3b encodes a protein of 154 amino acids, lacking similarities to any known protein. Protein 3b is predominantly localized in the nucleolus. A functional nuclear localization signal is located in amino acids 134 to 154. Ectopic over-expression of protein 3b in Vero, 293, and COS-7 cells induced cell cycle arrest at Go/G1 phase.[13] EAV nsp1 also has been localized in the nucleus.[10,14] Therefore, in total, at least three nidovirales proteins (N, 3b and nsp1) have been detected in the nucleus of infected cells, suggesting that nidovirales may modify cell behavior through the nucleus.

2.2. CoV Genome Replication

In CoV replication, recognition of RNA genome 5' and 3' ends by viral and cellular proteins is most likely essential. Furthermore, the interaction of these ends probably is a requirement for replication and transcription, as these are processes that must be initiated at the 3' end of the genome, and it has been shown that these processes are influenced by sequences mapping at the 5' end of the genome.[15,16] There may be a direct interaction between the CoV 5' end 3' ends, as predicted for MHV and TGEV RNA genomes in the absence of protein using computer programs.[17,18] Nevertheless, this direct interaction seems unlikely inside cells in the presence of cell and viral proteins. In fact, during genome synthesis, the first end synthesized (3') will be most likely immediately folded and nonspecifically covered by proteins, such as the N protein or nsp9, or by proteins binding specific RNA motifs with characteristic secondary structures. In fact, the postulated cross-talk between 5' and 3' ends and CoV replication has been shown in our laboratory.[19] Precipitation of digoxigeninated 3'-ends by biotinylated 5'-ends using streptavidin sepharose beads and, the reverse (precipitation of digoxigeninated 5'-ends with biotinylated 3'-ends) have been shown. Cross-talk between the 5' and 3' ends of TGEV genome has been observed for CoV genome ends and only require the presence of cell proteins.

The N protein probably has a prominent role in CoV replication as it influences many viral and cellular processes. The role of the N protein is most likely constrained by its propensity to self-assemble to form the capsid and also by its phosphorylation state. N protein activity has to be a consequence of its interaction with other viral and cellular proteins and with virus and host cell nucleic acids. CoV N protein is associated with the replicase complex in double-membrane structures derived from the endoplasmic reticulum[20] and also binds to genome RNA forming the nucleocapsid.[21-23] The nucleocapsid binds to the M protein carboxy-terminus in the endoplasmic reticulum (ER), Golgi and intermediate compartment (ERGIC) membranes.[21,22,24,25] Particles bud as immature virions with annular large nucleocapsids. Immature virions are transported through the Golgi compartment, where a major rearrangement of the nucleocapsid takes place, giving rise to secretory vesicles containing mature virions with electrondense cores.[21,24,26]

The N protein has a variable size in different CoVs (Fig. 1). Self-interactions of N proteins was first described in MHV.[27] The N protein has conserved secondary structures, including highly conserved α helices, and a highly conserved serine-rich domain including the repetitive sequence SSDNSRSRSQSRSRSR[12,28] (Fig. 1). Within this area, several active N protein domains have been mapped, such as the RNA binding domain of the IBV genome,[29] the oligomerization binding domain (amino acids 184-196),[30,31] and the M protein binding domain (amino acids 168-208),[32] which is also part of the N protein oligomerization domain. These protein sequences may be crucial in maintaining the N protein in a correct conformation. In fact, deletion of the 168-208 aa region results in the complete loss of N protein dimerization.

Phosphorylation has been shown to cause conformational changes in MHV N protein structure.[33] TGEV and IBV phosphoserine residues have been mapped within the CoV N protein primary and secondary structures. TGEV N protein serines 9, 156, 254, and 256 are phosphorylated in infected cells,[12,28] while in IBV, N protein phosphorylation sites have been localized to serines 190, 192, 379, and threonine 378.[29] CoVs N proteins present a conserved pattern of secondary structural elements, and a strong correlation has been observed between the MHV N protein three-domain organization and the predicted structure. N protein domains I and III are the most unstructured and divergent between CoV, while domain II is more conserved. Interestingly, TGEV N serines 156, 254, and 256 were localized to domain II, adjacent to conserved secondary elements $\beta 3$, $\beta 6$, and $\alpha 7$, respectively. Therefore, phosphorylation in these serine residues could affect the structure of these secondary elements by the introduction of negative charges in a basic environment[34,35] and affect N protein RNA binding activity.

IBV N protein phosphorylation has been localized in sites distinct from those identified in TGEV, based on sequence comparisons. This apparent discrepancy could be explained by intrinsic differences between CoV species. The relevance of the identified TGEV N protein phosphoserines has been analyzed by site-directed mutagenesis using the TGEV infectious cDNA clone.[36] Mutagenesis of all four TGEV phosphorylated serines to alanine did not prevent TGEV rescue from infectious cDNA nor lead to a significant TGEV titer reduction. This mutation may affect the binding of CoV N protein to RNA mediated by the amino terminus of this protein.[29] Additional work is in progress to study the role of TGEV N protein phosphorylation.

The requirement of N protein for virus replication and transcription is debated. Certain observations suggest that N protein plays a role in replication,[37-43] while others using either CoV[44] or arterivirus systems[45] claim that N protein is not essential. Using three TGEV derived replicons, two containing N gene in *cis*, and another one lacking this gene (Fig. 2), it has been clearly shown that TGEV replicon in the absence of N protein, provided either in *cis* or in *trans*, resulted in 50-fold greater levels of a reporter subgenomic RNA (gene 7) than background levels (Fig. 2).[46] Interestingly, when N protein is provided in *cis*, replication-transcription increases 100-fold over the levels in the absence of N protein. If N protein is exclusively provided in *trans*, replication-transcription levels increased 10-fold more (i.e., 1000-fold over levels in the absence of N protein). If N protein is in addition provided in *cis*, amplification levels do not increase over those reached when the N protein is only provided in *trans*. Two groups have shown that background levels of CoV transcription have been observed in the absence of N protein. [43,46] Nevertheless, a substantial increase

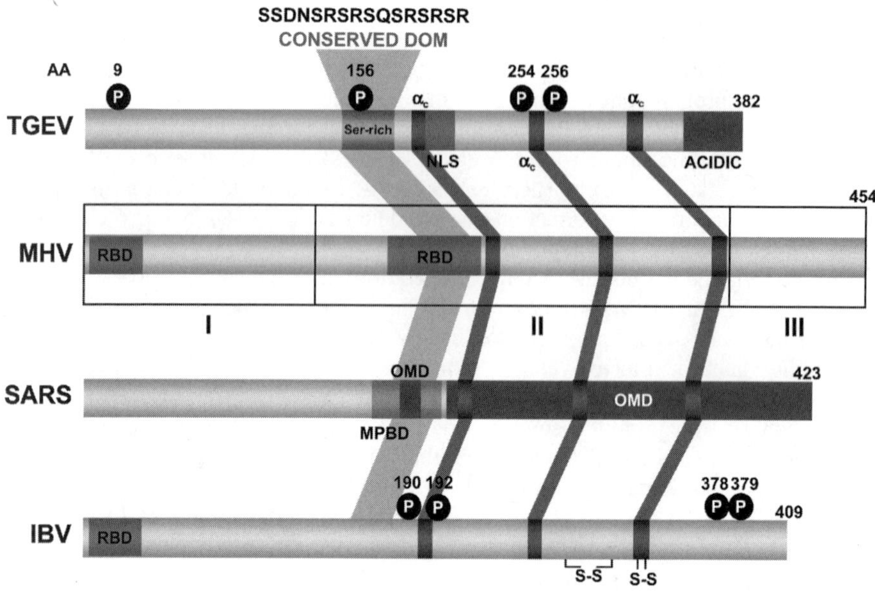

Figure 1. Scheme of N protein from different coronaviruses. The organization of N protein from four representative CoVs of genus α (TGEV), genus β (MHV and SARS-CoV), and genus γ (IBV) is indicated (not to scale). Conserved predicted structural elements are joined by gray shadowing zones. The three-domain organization proposed for MHV N protein by P. Master's group is indicated as open boxes over the MHV N protein (I, II, and III). P, phosphorylation sites. αc, protein domains with highly conserved alpha structure. AA, amino acid. NLS, nuclear localization signal. RBD, RNA binding domain. OMD, oligomerization domain. MPBD, M protein binding domain. S-S, disulfide bridge.

in CoV transcription is observed by providing N protein either in *cis* or in *trans*. The increase in reporter gene expression could be due to an increase in the replication, in the transcription levels, or to both. There is a general agreement that the presence of N protein enhances the rescue of infectious virus from cDNA clones generated from different CoVs, such as IBV,[47] human coronavirus (HCoV)-229E,[43,48] and TGEV using RNA *in vitro* transcripts[49] or replicons.[46]

3. CoV TRANSCRIPTION

CoV transcription, and in general transcription in the *Nidovirales* order, is an RNA-dependent RNA process which includes a discontinuous step during the production of subgenomic mRNAs.[50, 51] This transcription process ultimately generates a nested set of subgenomic mRNAs that are 5'- and 3'-coterminal with the virus genome. The common 5'-terminal leader sequence, 93 nucleotides (nt) in TGEV, is fused to the 5' end of the mRNA coding sequence (body) by a discontinuous transcription mechanism. Sequences preceding each gene represent signals for synthesis of subgenomic mRNAs (sgmRNAs).

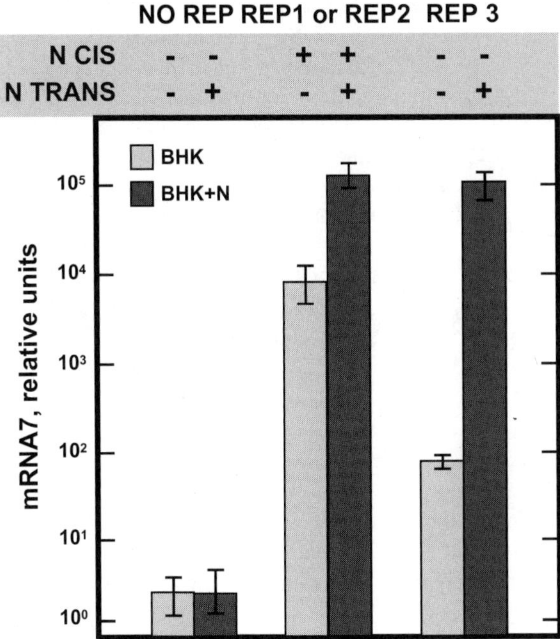

Figure 2. Expression of gene 7 in the presence of N protein. To study the effect of N protein in TGEV-derived replicon activity, the amount of mRNA7, expressed as relative units, was determined by real-time RT-PCR with specific oligonucleotides in RNA samples isolated from standard BHK-pAPN (BHK) or BHK-pAPN expressing N protein (BHK + N). Cells were transfected with either a non replicative cDNA clone (NO REP), two replicons that express the N protein (REP 1 and REP 2), or a replicon that does not encode protein N (REP 3). N protein was provided in *cis* by the replicons, or in *trans* using a Sinbis virus replicon when indicated (+). -, absence of N protein. The results indicate the mean values from three experiments, with standard deviations shown as error bars.

These are the transcription-regulating sequences (TRSs) that include a conserved core sequence (CS; 5'-CUAAAC-3'), identical in all TGEV genes (CS-B), and the 5' and 3' flanking sequences (5' TRS and 3' TRS, respectively) that regulate transcription[52] (Fig. 3). As this CS sequence is also found at the 3' end of the leader sequence (CS-L), it could base pair with the nascent minus strand complementary to each CS-B (cCS-B). In fact, the requirement for base pairing during transcription has been formally demonstrated in arteriviruses[53,54] and CoVs by experiments in which base pairing between CS-L and the complement of CS-B was engineered in infectious genomic cDNAs.[55] The canonical CS was nonessential for the generation of subgenomic mRNAs (sgmRNAs), but its presence led to transcription levels at least 10^3-fold higher than its absence. The data obtained are compatible with a model of transcription that includes three steps (Fig. 3): (i) formation of 5'-3' complexes in the genomic RNA, (ii) scanning of base pairing of the nascent (-) RNA strand by the TRS-L, and (iii) template switch during synthesis of the negative strand to complete the (-) sgRNA. This template switch takes place after copying the CS sequence and was predicted *in silico* based on a high base pairing score between the nascent (-) RNA strand and the TRS-L.

The role in transcription of four nucleotides immediately flanking the CS both at the 5' and 3' ends has been studied using a transcriptionally inactive canonical CS (CS-S2) internal to the S gene.[56] The rationale for selecting 5' and 3' TRS flanking sequences consisting of four nucleotides comes from the results of an *in silico* analysis showing that to predict both viral mRNAs and alternative mRNAs at noncanonical junction sites, an optimal TRS-L should include the CS plus four nucleotides flanking the CS at both ends. These predictions have been supported by experimental data performed by reverse genetic analysis of the sequences immediately flanking CS-S2. A good correlation was observed between the free energy of the TRS-L and cTRS-B duplex formation and the levels of subgenomic mRNA-S2, demonstrating that base pairing between leader and body beyond the CS was a determinant in the regulation of CoV transcription. In TRS mutants with increasing complementarity between TRS-L and cTRS-B, a tendency to reach a plateau in ΔG values was observed, suggesting that a more precise definition of the TRS limits might be proposed, consisting of the central CS and approximately four nucleotides flanking 5' and 3' the CS. Sequences downstream of the CS exert a stronger influence on the template-switching decision in accordance with a model of polymerase strand-transfer and template-switching during minus strand synthesis.

According to the working model of transcription proposed by our laboratory (Fig. 3), the first step is the interaction of the leader TRS with a complex presumably formed by the replicase, the helicase, the nascent RNA of negative polarity, and other viral and cellular proteins involved in transcription. Candidate proteins have been reported by several laboratories. On the viral side, essential proteins in transcription are the RNA-dependent RNA-polymerse (RdRp), and the helicase (Hel). In addition, N protein probably increases basal transcription (see above) and nsp1 has clearly been involved in arterivirus transcription.[57,58] It has also been suggested that NendoU may play a role by specifically cutting double stranded RNA generated (transcriptive intermediates) during the synthesis of the nascent RNA of negative polarity. NendoU nuclease has a strong preference for cleavage at GU(U) sequences in double-stranded RNA substrates.[59,60] It has been suggested that the GU(U) sequence at the 3' terminus of nascent minus-strand RNAs, which corresponds to conserved AAC nucleotides in the core of the CoV gene TRSs elements, might be substrate of this activity; therefore, NendoU activity might be involved in the transcription of subgenomic mRNAs. Data from our laboratory in which we analyzed around 90 different sgRNAs generated during the mutagenesis of a TGEV CS,[55] and also from other laboratories,[43] support the functional relevance of the AAC sequence in transcription, but further studies are required to provide a direct link to the activities of enzymes such as the uridylate-specific endoribonuclease.

In addition, cell proteins most likely play a role in CoV transcription regulation. In fact, there are data directly involving heterogeneous nuclear ribonucleoprotein (hnRNP) A1,[61,62] hnRNP I (PTB),[63] and the elongation factor eEF-1 in CoV transcription.[64,65] Furthermore, proteins such as p100 kDa coactivator and annexin A2 may be involved in CoV transcription as we have shown that these proteins bind to TRS sequences.[56] In the arteriviruses, the p100 kDa coactivator interacts with nsp1 involved in transcription and may regulate this activity.[58]

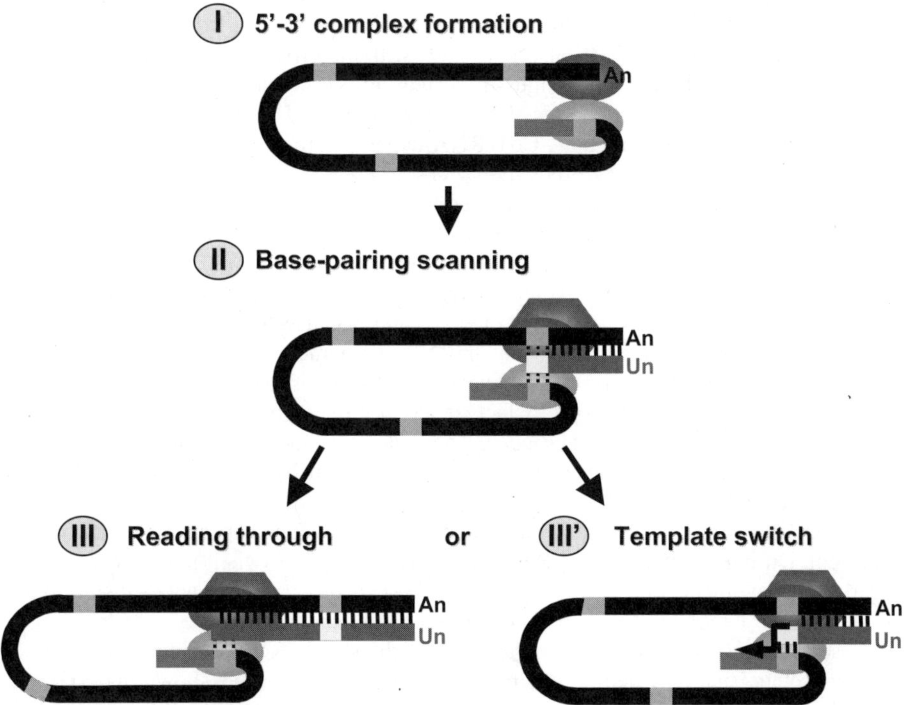

Figure 3. Three-step working model of coronavirus transcription**.** (I) 5'-3' complex formation step. Proteins binding the 5' and 3' end TGEV sequences are represented by ellipsoids. Leader sequence is indicated with dark gray bars, CS sequences are indicated with a clear bar. An, poly A tail. (II) Base pairing scanning step. Minus strand RNA is in a lighter color compared with positive-strand RNA. The transcription complex is represented by the hexagon. Vertical dotted bars represent scanning of base pairing by the TRS-L sequence in the transcription process. Vertical solid bars indicate complementarity between the genomic (gRNA) and the nascent minus strand. Un, poly U tail. (III) The synthesis of the negative strand can continue to make a longer sgRNA (III left), or a template switch step can take place (III right) as indicated in the text. The thick arrow indicates the switch in the template made by the transcription complex to complete the synthesis of (-) sgRNA.

CoV genomic RNA interacts with at least three proteins: hnRNP A1,[61,62,66,67] PTB,[68,69] and N[37,42,70] and may mediate the formation of complexes between the leader TRS and the transcription complex at the body TRS. Formation of cyclic complexes could in principle be mediated by the interaction among these proteins. In fact, binding between hnRNPA1 and PTB,[71,72] hnRNP A1 and N protein,[73,74] and PTB and N protein[63] have been documented. Many of these biochemical interactions have been reported in the past eight years and their role can now be reinterpreted in the context of the transcription model implying discontinuous RNA synthesis during the production of the negative strand. A role in transcription has been assigned to different proteins:

(i) hnRNPA1 protein binds to the complementary strand (negative-polarity) of the MHV leader (cL) and to TRS sequences, particularly the consensus (3'-AGAUUUG-5') sequence of MHV RNA located at the 3' end of the genome.[62] Site-specific mutations of TRSs inhibited the mRNA transcription from MHV DI RNA, in direct proportion to the extent of reduction of hnRNPA1 binding to cL.[62,75] The effect of hnRNP A1 on MHV RNA transcription was further confirmed in cell lines.[67] Direct evidence for a functional role for hnRNP A1 in MHV synthesis has been demonstrated in MHV-infected cells.[76] Binding of hnRNP A1 to a TRS also correlates with the efficiency of transcription from that TRS.[62,75] In addition, a C-terminus-truncated hnRNPA1 mutant exhibited dominant-negative effects on viral genomic RNA replication and subgenomic transcription. Therefore, hnRNP A1 may regulate CoV RNA-dependent transcription.

(ii) The N protein of MHV binds the UCUAAAC sequence of the leader RNA, and it has been suggested that N protein is involved in MHV RNA transcription.[37,42] The role of N protein in MHV RNA replication has also been shown in an *in vivo* replication system.[38] These findings suggest that both cellular hnRNPA1 and viral N protein are components of the MHV replication and transcription complex. As hnRNPA1 interacts with some serine-arginine (SR)-rich proteins,[73] and because N protein also contains an SR motif,[77,78] it has been proposed that hnRNPA1 interacts directly with N protein to bring the leader RNA to the CS sequence of the template RNA for initiation of subgenomic mRNA transcription. In fact, it has been shown[74] that N protein interacts with hnRNPA1 both *in vivo* and *in vitro*. The data was confirmed using the two-hybrid system. In agreement with these results, we have shown by Far-Western blotting that TGEV N protein binds PTB.[79]

Template switch during transcription may be aided by the chaperone activity of CoV N protein. During negative-strand RNA synthesis, a template switch is required to add a negative copy of the leader to the negative strand. This process represents a displacement of the former template RNA by another one, the leader sequence. These types of processes need to overcome an energy barrier threshold. RNA chaperones are RNA binding proteins that may help to overcome this threshold. We have shown that TGEV N protein is an RNA chaperone that is also active in viral RNA annealing (Zuñiga *et al.* in this book[79]).

(iii) PTB binds to the c3'-untranslated region (UTR) of MHV inducing a conformational change in RNA structure.[68] Mutations of the PTB-binding site in either 5'-leader or the sequences complementary to 3'-UTR inhibited replication and transcription of MHV genomic and defective-interfering (DI) RNA, in direct proportion to the extent of reduction of PTB binding, suggesting that PTB plays a role in regulating viral RNA synthesis. Thus, the interaction of N protein with PTB may modulate transcription.[63]

(iv) SYNCRIPT (p70) is a member of the hnRNP family and localizes largely in the cytoplasm. The p70 cross-linked to MHV positive- or negative-strand RNA. The p70-binding site was mapped to the leader sequence of the 5'-UTR, requiring the UCUAA repeat sequence. Overexpression of p70 inhibited syncytium formation induced by MHV. Furthermore, downregulation of the endogenous p70 with a specific short iRNA delayed MHV RNA synthesis. These results suggest that p70 may be directly involved in MHV RNA replication as a positive regulator.[80]

(v) EAV nsp1 has been proposed to couple genome replication and transcription.[81] Nsp-1 has been shown to interact with p100,[58] and nsp1-p100 interactions have been speculated to be important for viral sgRNA synthesis, either directly or by recruiting a p100 binding protein to the viral RdRp complex. Alternatively, nsp1 might modulate transcription in the infected cell, explaining why the protein is targeted to the nucleus.[10]

4. CONCLUSIONS

The precise role of the described protein and other viral and cellular proteins needs to be confirmed in the context of discontinuous transcription during the synthesis of the negative strand, giving special attention to intermediates of the replicase processing and to proteins associated with membrane structures located in the cytoplasm. Functional proteomics could be of great help in this complicated task. In addition, the establishment of *in vitro* replication and transcription systems will help to clarify the mechanisms involved in CoV replication.

We thank D. Dorado and M. González for technical assistance. This work was supported by grants from the Comisión Interministerial de Ciencia y Tecnología (CICYT) of the Department of Education and Science (Spain), Fort Dodge Veterinaria and the European Communities (Frame VI Life Sciences, projects DISSECT ref. SP22-CT-2004-511060, and TIPP ref. QLK2-CT-2002-01050). CC has received a fellowship from the Education Department of the Community of Madrid.

5. REFERENCES

1. M. Carmo-Fonseca, L. Mendes-Soares, and I. Campos, To be or not to be in the nucleolus, *Nat. Cell Biol.* **2**, E107-E112 (2002).
2. J. A. Hiscox, The nucleolus - a gateway to viral infection? *Arch. Virol.* **147**, 1077-1089 (2002).
3. M. O. Olson, M. Dundr, and A. Szebeni, The nucleolus: an old factory with unexpected capabilities, *Trends Cell Biol.* **10**, 189-196 (2000).
4. T. Pederson, The plurifunctional nucleolus, *Nucleic Acids Res.* **26**, 3871-3876 (1998).
5. T. Pederson and J. C. Politz, The nucleolus and the four ribonucleoproteins of translation, *J. Cell Biol.* **148**, 1091-1095 (2000).
6. T. Wurm, H. Chen, T. Hodgson, P. Britton, G. Brooks, and J. A. Hiscox, Localization to the nucleolus is a common feature of coronavirus nucleoproteins, and the protein may disrupt host cell division, *J. Virol.* **75**, 9345-9356 (2001).
7. J. H. You, B. K. Dove, G. Howell, P. Heinen, M. Zambon, and J. A. Hiscox, Sub-cellular localisation of the severe acute respiratory syndrome coronavirus viral RNA binding protein, nucleoprotein, *J. Gen. Virol.* in press (2005).
8. K. A. Timani, L. Ye, Y. Zhu, Z. Wu, and Z. Gong, Cloning, sequencing, expression, and purification of SARS-associated coronavirus nucleocapsid protein for serodiagnosis of SARS, *J. Clin. Virol.* **30**, 309-312 (2004).

9. H. Chen, B. Coote, S. Attree, and J. A. Hiscox, Evaluation of a nucleoprotein-based enzyme-linked immunosorbent assay for the detection of antibodies against infectious bronchitis virus, *Avian Pathol.* **32**, 519-526 (2003).
10. M. A. Tijms, Y. van der Meer, and E. J. Snijder, Nuclear localization of non-structural protein 1 and nucleocapsid protein of equine arteritis virus, *J. Gen. Virol.* **83**, 795-800 (2002).
11. R. R. Rowland, R. Kervin, C. Kuckleburg, A. Sperlich, and D. A. Benfield, The localization of porcine reproductive and respiratory syndrome virus nucleocapsid protein to the nucleolus of infected cells and identification of a potential nucleolar localization signal sequence, *Virus Res.* **64**, 1-12 (1999).
12. E. Calvo, D. Escors, J. A. Lopez, et al., Phosphorylation and subcellular localization of transmissible gastroenteritis virus nucleocapsid protein in infected cells, *J. Gen. Virol.* **86**, 2255-2267 (2005).
13. X. Yuan, Z. Yao, Y. Shan, et al., Nucleolar localization of non-structural protein 3b, a protein specifically encoded by the severe acute respiratory syndrome coronavirus, *Virus Res.* in press (2005).
14. Y. van der Meer, E. J. Snijder, J. C. Dobbe, et al., Localization of mouse hepatitis virus nonstructural proteins and RNA synthesis indicates a role for late endosomes in viral replication, *J. Virol.* **73**, 7641-7657 (1999).
15. R. van der Most, W. Luytjes, S. Rutjes, and S. J. M. Spaan, Translation but not the encoded sequence is essential for the efficient propagation of the defective interfering RNAs of the coronavirus mouse hepatitis virus, *J. Virol.* **69**, 3744-3751 (1995).
16. Y.-J. Lin, C. L. Liao, and M. M. C. Lai, Identification of the cis-acting signal for minus-strand RNA synthesis of a murine coronavirus: implications for the role of minus-strand RNA in RNA replication and transcription, *J. Virol.* **68**, 8131-8140 (1994).
17. J. Holt, J. Y. Sgro, M. Zuker, and A. Palmenberg, Computer folding of full-length viral genomes: a new toolkit for automated analysis of RNAs longer than 10,000 bases, Seventh International Symposium on positive strand RNA viruses, San Francisco, California, USA (2004).
18. J. Y. Sgro, J. Holt, M. Zuker, and A. Palmenberg, RNA folding of the complete SARS and MHV coronavirus genomes, Seventh International Symposium on positive strand RNA viruses, San Francisco, California, USA (2004).
19. C. Galan, F. Almazan, and L. Enjuanes, Cross-talk between the 5'- and 3'-ends of coronavirus genome, submitted (2005).
20. R. Gosert, A. Kanjanahaluethai, D. Egger, K. Bienz, and S .C. Baker, RNA replication of mouse hepatitis virus takes place at double-membrane vesicles, *J. Virol.* **76**, 3697-3708 (2002).
21. D. Escors, J. Ortego, H. Laude, and L. Enjuanes, The membrane M protein carboxy terminus binds to transmissible gastroenteritis coronavirus core and contributes to core stability, *J. Virol.* **75**, 1312-1324 (2001).
22. K. Narayanan, A. Maeda, J. Maeda, and S. Makino, Characterization of the coronavirus M protein and nucleocapsid interaction in infected cells, *J. Virol.* **74**, 8127-8134 (2000).
23. L. S. Sturman, K. V. Holmes, and J. Behnke, Isolation of coronavirus envelope glycoproteins and interaction with the viral nucleocapsid, *J. Virol.* **33**, 449-462 (1980).
24. I. J. Salanueva, J. L. Carrascosa, and C. Risco, Structural maturation of the transmissible gastroenteritis coronavirus, *J. Virol.* **73**, 7952-7964 (1999).
25. C. Risco, M. Muntión, L. Enjuanes, and J. L. Carrascosa, Two types of virus-related particles are found during transmissible gastroenteritis virus morphogenesis, *J. Virol.* **72**, 4022-4031 (1998).
26. J. Ortego, D. Escors, H. Laude, and L. Enjuanes, Generation of a replication-competent, propagation-deficient virus vector based on the transmissible gastroenteritis coronavirus genome, *J. Virol.* **76**, 11518-11529 (2002).
27. S. G. Robbins, M. F. Frana, J. J. McGowan, J. F. Boyle, and K. V. Holmes, RNA-binding proteins of coronavirus MHV: detection of monomeric and multimeric N protein with an RNA overlay-protein blot assay, *Virology* **150**, 402-410 (1986).
28. P. Britton, Coronavirus motif, *Nature* **353**, 394 (1991).
29. H. Chen, A. Gill, B. K. Dove, et al., Mass spectroscopic characterization of the coronavirus infectious bronchitis virus nucleoprotein and elucidation of the role of phosphorylation in RNA binding by using surface plasmon resonance, *J. Virol.* **79**, 1164-1179 (2005).
30. R. He, F. Dobie, M. Ballantine, et al., Analysis of multimerization of the SARS coronavirus nucleocapsid protein, *Biochem. Biophys. Res. Commun.* **316**, 476-483 (2004).
31. M. Surjit, B. Liu, S. Jameel, V. T. Chow, and S. K. Lal, The SARS coronavirus nucleocapsid protein induces actin reorganization and apoptosis in COS-1 cells in the absence of growth factors, *Biochem. J.* **383**, 13-18 (2004).
32. R. He, A. Leeson, M. Ballantine, et al., Characterization of protein-protein interactions between the nucleocapsid protein and membrane protein of the SARS coronavirus, *Virus Res.* **105**, 121-125 (2004).

33. S. A. Stohlman, J. O. Fleming, C. D. Patton, and M. M. C. Lai, Synthesis and subcellular localization of the murine coronavirus nucleocapsid protein, *Virology* **130**, 527-532 (1983).
34. M. M. Parker and P. S. Masters, Sequence comparison of the N genes of five strains of the coronavirus mouse hepatitis virus suggests a three domain structure for the nucleocapsid protein, *Virology* **179**, 463-468 (1990).
35. P. S. Masters, Localization of an RNA-binding domain in the nucleocapsid protein of the coronavirus mouse hepatitis virus, *Arch. Virol.* **125**, 141-160 (1992).
36. E. Alvarez, M. L. DeDiego, D. Escors, and L. Enjuanes, Role of transmissible gastroenteritis Coronavirus N protein phosphorylation in virus replication, submitted (2005).
37. R. S. Baric, G. W. Nelson, J. O. Fleming, et al., Interactions between coronavirus nucleocapsid protein and viral RNAs: implications for viral transcription, *J. Virol.* **62**, 4280-4287 (1988).
38. S. R. Compton, D. B. Rogers, K. V. Holmes, D. Fertsch, J. Remenick, and J. J. McGowan, In vitro replication of mouse hepatitis virus strain A59, *J. Virol.* **69**, 2313-2321 (1987).
39. Y.-N. Kim and S. Makino, Characterization of a murine coronavirus defective interfering RNA internal *cis*-acting replication signal, *J. Virol.* **69**, 4963-4971 (1995).
40. H. Laude and P. S. Masters, in: *The Coronaviridae*, edited by S. G. Siddell (Plenum Press, New York, 1995) pp. 141-158.
41. G. W. Nelson, S. A. Stohlman, and S. M. Tahara, High affinity interaction between nucleocapsid protein and leader/intergenic sequence of mouse hepatitis virus RNA, *J. Gen. Virol.* **81**, 181-188 (2000).
42. S. A. Stohlman, R. S. Baric, G. N. Nelson, L. H. Soe, L. M. Welter, and R. J. Deans, Specific interaction between coronavirus leader RNA and nucleocapsid protein, *J. Virol.* **62**, 4288-4295 (1988).
43. B. Schelle, N. Karl, B. Ludewig, S. G. Siddell, and V. Thiel, Selective replication of coronavirus genomes that express nucleocapsid protein, *J. Virol.* **79**, 6620-6630 (2005).
44. V. Thiel, J. Herold, B. Schelle, and S. G. Siddell, Viral replicase gene products suffice for coronavirus discontinuous transcription, *J. Virol.* **75**, 6676-6681 (2001).
45. R. Molenkamp, H. van Tol, B. C. Rozier, Y. van der Meer, W. J. Spaan, and E. J. Snijder, The arterivirus replicase is the only viral protein required for genome replication and subgenomic mRNA transcription, *J. Gen. Virol.* **81**, 2491-2496 (2000).
46. F. Almazan, C. Galan, and L. Enjuanes, The nucleoprotein is required for efficient coronavirus genome replication, *J. Virol.* **78**, 12683-12688 (2004).
47. R. Casais, V. Thiel, S. G. Siddell, D. Cavanagh, and P. Britton, Reverse genetics system for the avian coronavirus infectious bronchitis virus, *J. Virol.* **75**, 12359-12369 (2001).
48. V. Thiel, N. Karl, B. Schelle, P. Disterer, I. Klagge, and S. G. Siddell, Multigene RNA vector based on coronavirus transcription, *J. Virol.* **77**, 9790-9798 (2003).
49. K. M. Curtis, B. Yount, and R. S. Baric, Heterologous gene expression from transmissible gastroenteritis virus replicon particles, *J. Virol.* **76**, 1422-1434 (2002).
50. S. G. Sawicki and D. L. Sawicki, A new model for coronavirus transcription, *Adv. Exp. Med. Biol.* **440**, 215-220 (1998).
51. M. M. C. Lai and D. Cavanagh, The molecular biology of coronaviruses, *Adv. Virus Res.* **48**, 1-100 (1997).
52. S. Alonso, A. Izeta, I. Sola, and L. Enjuanes, Transcription regulatory sequences and mRNA expression levels in the coronavirus transmissible gastroenteritis virus, *J. Virol.* **76**, 1293-1308 (2002).
53. A. O. Pasternak, E. van den Born, W. J. M. Spaan, and E. J. Snijder, Sequence requirements for RNA strand transfer during nidovirus discontinuous subgenomic RNA synthesis, *EMBO J.* **20**, 7220-7228 (2001).
54. G. van Marle, J. C. Dobbe, A. P. Gultyaev, W. Luytjes, W. J. M. Spaan, and E. J. Snijder, Arterivirus discontinuous mRNA transcription is guided by base pairing between sense and antisense transcription-regulating sequences, *Proc. Natl. Acad. Sci. USA* **96**, 12056-12061 (1999).
55. S. Zúñiga, I. Sola, S. Alonso, and L. Enjuanes, Sequence motifs involved in the regulation of discontinuous coronavirus subgenomic RNA synthesis, *J. Virol.* **78**, 980-994 (2004).
56. I. Sola, J. L. Moreno, S. Zúñiga, S. Alonso, and L. Enjuanes, Role of nucleotides immediately flanking the transcription-regulating sequence core in coronavirus subgenomic mRNA synthesis, *J. Virol.* **79**, 2506-2516 (2005).
57. L. C. van Dinten, J. A. den Boon, A. L. M. Wassenaar, W. J. M. Spaan, and E. J. Snijder, An infectious arterivirus cDNA clone: identification of a replicase point mutation that abolishes discontinuous mRNA transcription, *Proc. Natl. Acad. Sci. USA* **94**, 991-996 (1997).
58. M. A. Tijms and E. J. Snijder, Equine arteritis virus non-structural protein 1, an essential factor for viral subgenomic mRNA synthesis, interacts with the cellular transcription co-factor p100, *J. Gen. Virol.* **84**, 2317-2322 (2003).
59. K. A. Ivanov, T. Hertzig, M. Rozanov, et al., Major genetic marker of nidoviruses encodes a replicative endoribonuclease, *Proc. Natl. Acad. Sci. USA* **101**, 12694-12699 (2004).

60. E. J. Snijder, P. J. Bredenbeek, J. C. Dobbe, et al., Unique and conserved features of genome and proteome of SARS-coronavirus, and early split-off from the coronavirus group 2 lineage, *J. Mol. Biol.* **331**, 991-1004 (2003).
61. G. Zhang, V. Slowinski, and K. A. White, Subgenomic mRNA regulation by a distal RNA element in a (+) - strand RNA virus, *RNA* **5**, 550-561 (1999).
62. H. P. Li, X. Zhang, R. Duncan, L. Comai, and M. M. C. Lai, Heterogeneous nuclear ribonucleoprotein A1 binds to the transcription-regulatory region of mouse hepatitis virus RNA, *Proc. Natl. Acad. Sci. USA* **94**, 9544-9549 (1997).
63. K. S. Choi, P. Huang, and M. M. Lai, Polypyrimidine-tract-binding protein affects transcription but not translation of mouse hepatitis virus RNA, *Virology* **303**, 58-68 (2002).
64. S. Banerjee, K. Narayanan, T. Mizutani, and S. Makino, Murine coronavirus replication-induced p38 mitogen-activated protein kinase activation promotes interleukin-6 production and virus replication in cultured cells, *J. Virol.* **76**, 5937-5948 (2002).
65. T. Mizutani, S. Fukushi, M. Saijo, I. Kurane, and S. Morikawa, Phosphorylation of p38 MAPK and its downstream targets in SARS coronavirus-infected cells, *Biochem. Biophys. Res. Commun.* **319**, 1228-1234 (2004).
66. T. Furuya and M. M. C. Lai, Three different cellular proteins bind to complementary sites on the 5'-end-positive and 3'-end negative strands of mouse hepatitis virus RNA, *J. Virol.* **67**, 7215-7222 (1993).
67. S. T. Shi, P. Huang, H. P. Li, and M. M. C. Lai, Heterogeneous nuclear ribonucleoprotein A1 regulates RNA synthesis of a cytoplasmic virus, *EMBO J.* **19**, 4701-4711 (2000).
68. P. Huang and M. M. C. Lai, Polypyrimidine tract-binding protein binds to the complementary strand of the mouse hepatitis virus 3' untranslated region, thereby altering RNA conformation, *J. Virol.* **73**, 9110-9116 (1999).
69. H.-P. Li, P. Huang, S. Park, and M. M. C. Lai, Polypyrimidine tract-binding protein binds to the leader RNA of mouse hepatitits virus and serves as a regulator of viral transcription, *J. Virol.* **73**, 772-777 (1999).
70. C. Galan, L. Enjuanes, and F. Almazan, A point mutation within the replicase gene differentially affects coronavirus genome versus minigenome replication, submitted (2005).
71. A. L. Bothwell, D. W. Ballard, W. M. Philbrick, et al., Murine polypyrimidine tract binding protein. Purification, cloning, and mapping of the RNA binding domain, *J. Biol. Chem.* **266**, 24657-24663 (1991).
72. M. M. C. Lai, Cellular factors in the transcription and replication of viral RNA genomes: a parallel to DNA-dependent RNA transcription, *Virology* **244**, 1-12 (1998).
73. L. Cartegni, M. Maconi, E. Morandi, F. Cobianchi, S. Riva, and G. Biamonti, hnRNP A1 selectively interacts through its Gly-rich domain with different RNA-binding proteins, *J. Mol. Biol.* **59**, 337-348 (1996).
74. Y. Wang and X. Zhang, The nucleocapsid protein of coronavirus mouse hepatitis virus interacts with the cellular heterogeneous nuclear ribonucleoprotein A1 in vitro and in vivo, *Virology* **265**, 96-109 (1999).
75. X. Zhang and M. M. C. Lai, Interactions between the cytoplasmic proteins and the intergenic (promoter) sequence of mouse hepatitis virus RNA: correlation with the amounts of subgenomic mRNA transcribed, *J. Virol.* **69**, 1637-1644 (1995).
76. X. Zhang, H. P. Li, W. Xue, and M. M. C. Lai, Formation of a ribonucleoprotein complex of mouse hepatitis virus involving heterogeneous nuclear ribonucleoprotein A1 and transcription-regulatory elements of viral RNA, *Virology* **264**, 115-124 (1999).
77. D. Peng, C. A. Koetzner, and P. S. Masters, Analysis of second-site revertants of a murine coronavirus nucleocapsid protein deletion mutant and construction of nucleocapsid protein mutants by targeted RNA recombination, *J. Virol.* **69**, 3449-3457 (1995).
78. D. Peng, C. A. Koetzner, T. McMahon, Y. Zhu, and P. Masters, Construction of murine coronavirus mutants containing interspecies chimeric nucleocapsid proteins, *J. Virol.* **69**, 5475-5484 (1995).
79. S. Zúñiga, I. Sola, J. L. Moreno, and L. Enjuanes, Coronavirus nucleocapsid protein is an RNA chaperone. Implications for transcription regulation, submitted (2005).
80. K. S. Choi, A. Mizutani, and M. M. C. Lai, SYNCRIP, a member of the heterogeneous nuclear ribonucleoprotein family, is involved in mouse hepatitis virus RNA synthesis, *J. Virol.* **78**, 13153-13162 (2004).
81. M. A. Tijms, L. C. van Dinten, A. E. Gorbalenya, and E. J. Snijder, A zinc finger-containing papain-like protease couples subgenomic mRNA synthesis to genome translation in a positive-stranded RNA virus, *Proc. Natl. Acad. Sci. USA* **98**, 1889-1894 (2001).

A PREVIOUSLY UNRECOGNIZED UNR STEM-LOOP STRUCTURE IN THE CORONAVIRUS 5' UNTRANSLATED REGION PLAYS A FUNCTIONAL ROLE IN REPLICATION

Pinghua Liu, Jason J. Millership, Lichun Li, David P. Giedroc, and Julian L. Leibowitz*

1. INTRODUCTION

Cis-acting sequences in the MHV and bovine coronavirus (BCoV) 5' UTR are required for defective interfering RNA replication, subgenomic RNA synthesis, and presumably also for virus genome replication.[1-3] For the BCoV 5' UTR, the minimum free energy structure is predicted to contain three stem-loops.[3] We have analyzed the entire 5' UTR sequences of nine group 1 and group 2 coronaviruses, including the newly discovered SARS and HKU1 coronaviruses, using consensus, covariation secondary structure predictions. Our analysis shows that the predicted secondary structures of all coronavirus 5' UTRs are strikingly similar and contain three or four stem-loops, including a previously unrecognized highly conserved UNR stem-loop.

Computer-assisted modeling predicts an invariant and previously unrecognized UNR stem-loop among nine coronavirus UTR sequences, denoted SL2. This predicted conserved structure encompasses nucleotides 42-56 for both MHV and the SARS-coronavirus. NMR spectroscopy of the 16-nt RNA $SL2^{sars}$ reveals spectral features consistent with a UNR hairpin loop. Reverse genetics studies revealed that SL2 is required for MHV replication; MHV genomes containing a substitution of the required U48 with C in the UNR loop (U48C) were not viable. RT-PCR analysis of the U48C mutant indicated that negative sense genome sized RNAs were present in cells electroporated with this mutant; however, neither positive nor negative sense subgenomic RNAs were detected. Mutations that destabilized the stem of SL2 were viable but had moderately to severely impaired replication phenotypes. Mutants that maintained the stem-loop structure replicated similarly to wild-type MHV. These genetic data strongly support the existence of the predicted UNR stem loop and its functional importance in viral replication.

---

* Pinghua Liu, Jason J. Millership, Julian L. Leibowitz, Texas A&M University System-HCS, College Station, Texas. Lichun Li, David P. Giedroc, Texas A&M University, College Station, Texas.

2. MATERIALS AND METHODS

The RNA secondary structure prediction algorithm Vienna RNA 1.5[4] was used to predict the secondary structures of group 1 and group 2 coronavirus 5' UTRs. A reverse genetic system based on *in vitro* assembly of cloned cDNAs (A-G) was used to recover wild-type MHV-A59 1000 and mutant viruses.[5] To construct mutants in SL2, a series of overlapping oligonucleotides spanning the sequence between MluI and SacII sites (1 to 106 nts) in plasmid A were synthesized and ligated together to form a 135-bp DNA fragment with MluI and Sac II sites at its 5' and 3' ends, respectively. The assembled fragment containing either the wild-type sequence, or with mutations in SL2, was ligated into plasmid A. cDNA corresponding to the entire MHV-A59 genome were assembled by *in vitro* ligation, transcribed *in vitro*, and the transcribed mutant genomic RNA or wild-type MHV-A59-1000 RNA was electroporated into BHK-R cells to recover infectious virus as described.[5] Viruses were plaque purified and expanded once in DBT cells. Mutants were tested three times, including three blind passages of lysates from electroporated cultures without recovering infectious virus, before being declared non-viable. Total cellular RNAs were extracted 4 and 8 hours post-electroporation and assayed by nested RT-PCR to detect negative sense genome sized RNA and positive and negative sense subgenomic RNAs, as described.[6]

3. RESULTS

A series of three stem-loops denoted I, II, and III had been predicted in the BCoV 5' UTR. These predictions were supported in part by nuclease mapping experiments and by DI replication assays.[3,7] However, it was puzzling to us that the predicted stem-loops I and II of BCoV are not conserved amongst group 2 coronaviruses. Here we used Vienna RNA 1.5 [4] to predict the 5' UTR secondary structures of nine coronaviruses (summarized in Table 1). The results of Vienna RNA folding predictions carried out for MHV and SARS-CoV are shown in Figure 1. All coronavirus 5' UTR secondary structural models are strikingly similar, and are characterized by three major helical stems, denoted SL1, SL2, and SL4. Some sequences show a fourth stem-loop, denoted SL3, in which the leader TRS (TRS-L) is folded into a hairpin loop. SL2 is absolutely conserved and previously unrecognized. The (C/U)**UUG(**U/C) pentaloop sequence is the most highly conserved contiguous run of nucleotides in the entire 5' UTR outside of the core TRS-L, and covariation analysis clearly reveals that this loop is always stacked on a 5-bp helix. Analysis of the entire ≈30 kb MHV and SARS-CoV genomes reveals that SL2-like stem loops are extremely rare (appearing just 3 and 5 other times, respectively); this suggests an important role in coronavirus replication.

The (C/U)UUG(U/C) sequence of SL2 has all the features of a classical **UNR** (**U_0•N_{+1}•R_{+2}**) hairpin loop, in which a UNR triloop stacks on a Y:Y, Y:A (Y-pyrimidine) or G:A noncanonical base pair.[8] The basic structural feature of the U-turn architecture is a sharp turn in the phosphate backbone between U_0 and N_{+1} (*i.e.,* a uridine (U)-turn, first

Table 1. Predicted 5' UTR secondary structure for group 1 and 2 coronaviruses.

Viruses	Predicted secondary structure			
	SL1	SL2	SL3	SL4
HCoV-OC43	Predicted	Strongly predicted	Predicted	Predicted
BCoV	Predicted	Strongly predicted	Unfolded at 37°C	Predicted
MHV	Predicted	Strongly predicted	Not predicted	Predicted
HKU1	Predicted	Strongly predicted	Unfolded at 37°C	Predicted
SARS-CoV	Predicted	Strongly predicted	Predicted	Predicted
HCoV-NL63	Predicted	Strongly predicted	Not predicted	Predicted
HCoV-229E	Predicted	Strongly predicted	Not predicted	Predicted
TGEV	Predicted	Strongly predicted	Not predicted	Predicted

identified in the anticodon and T-loops of tRNA). In a UNR loop, U_0 is stacked on the noncanonical base pair and is engaged in two critical hydrogen bonds: the U_0 imino proton donates a hydrogen bond to the nonbridging phosphate oxygen following R_{+2}, and the U_0 2'-OH proton donates a hydrogen bond to the N7 of R_{+2}. Substitution of U_0 with any other nucleotide will abrogate formation of these hydrogen bonds and therefore destabilize the loop.

A 16-nt RNA termed $SL2^{sars}$ *in vitro* transcribed by SP6 RNA polymerase was subjected to NMR spectroscopy. ^{1}H-^{15}N HSQC and ^{2}J HNN-COSY spectra acquired with uniformly ^{13}C,^{15}N-labeled $SL2^{sars}$ (Fig. 2A, B) reveal the expected correlations for all five base pairs in the stem as well as two upfield-shifted ^{1}H-^{15}N correlations (Fig. 2B); their detection in the uridine-only ^{13}C,^{15}N-[U]-labeled sample reveal that they must correspond to U3-H3 cross-peaks (Fig. 2C) indicative of noncanonical hydrogen bonding. These uridines likely correspond to U48 and U51 (Fig. 1B).

Based on the predicted UNR stem-loop, a series of mutations were introduced into the SL2 region of MHV. Introduction of a conservative U48C substitution into the MHV genome resulted in a nonviable genome, a result consistent with our prediction of an invariant U at position 48. At 4 and 8 h post-electroporation of U48C or wild-type genomes, total RNA was extracted and analyzed by nested RT-PCR. As shown in Figure 3, negative sense genome sized RNAs were present in cells electroporated with U48C; however, neither positive nor negative sense subgenomic RNA6 and RNA7 were detected. A similar result was also obtained for RNA3 (not shown).

Consistent with the predicted structure of the UNR loop, the U49A mutant was viable and produced a virus with a near normal plaque size (Figs 4A-B). Three other mutant viruses, C45G, G53C, and C45G/G53C, with substitutions in the SL2 helical stem were recovered, although their plaque sizes are different. Mutants C45G and G53C, predicted to destabilize SL2, both form smaller plaques than wild-type virus. The double mutant C45G/G53C restores the SL2 stem and forms plaques slightly larger than wild-type MHV-A59. One step growth curves confirmed the growth phenotypes of these viruses (Figs. 4C-D). Denaturing gel electrophoresis of metabolically labeled RNA prepared from cells infected with wild-type MHV or various SL2 mutants demonstrated that a similar ratios of genome and subgenomic RNAs were synthesized, but that the amount of RNA synthesized correlated with virus growth phenotype. Thus cells infected with the G53C mutant synthesized much less RNA than wild-type MHV-A59 1000, cells infected with the C45G/G53C and U49A mutants synthesized near wild-type levels of

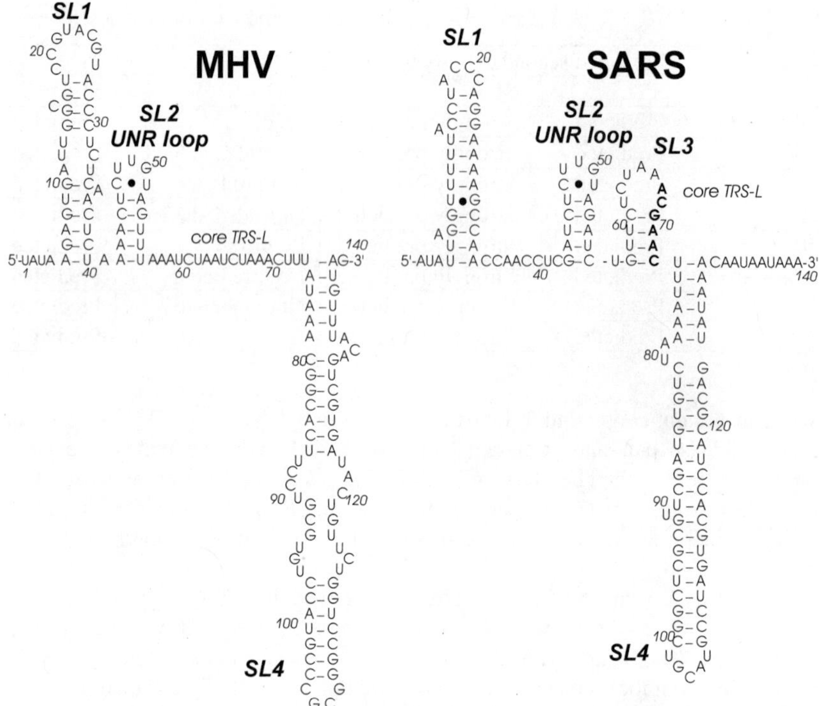

Figure 1. Predicted 5' UTR secondary structures of MHV and SARS-CoV.

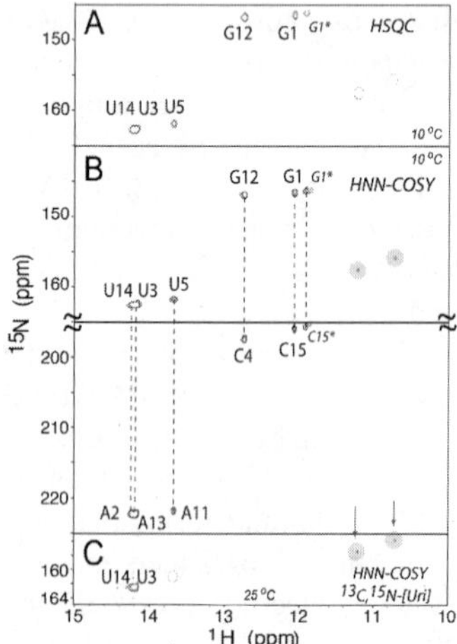

Figure 2. (A) ^{1}H-^{15}N HSQC and (B) ^{2}J HNN-COSY spectra acquired with uniformly ^{13}C,^{15}N-labeled SL2^{sars}. (C) ^{2}J HNN-COSY spectrum acquired for a uridine-only ^{13}C,^{15}N-[U]-labeled sample. Conditions: ^{1}H frequency 600 MHz, pH 6.0, 0.1 M KCl, 5 mM $MgCl_2$, 10–25°C.

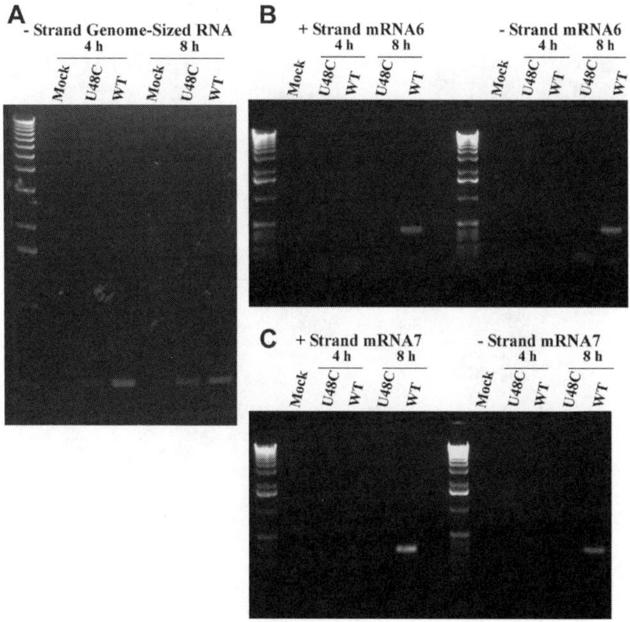

Figure 3. Nested RT-PCR analysis of cells electroporated with wild-type or U48C mutant genomes.

RNA, and cells infected with the C45G mutant synthesized somewhat less RNA than wild-type infected cells (data not shown). Taken together, these results are consistent with a role for SL2 in MHV RNA replication and transcription.

4. DISCUSSION

The genetic and NMR studies presented above support the proposed secondary structure model for the coronavirus 5' UTR, and the functional importance of SL2 in subgenomic RNA synthesis. All coronaviruses contain similar SL1, SL2, and SL4 stem-loops. This covariation-based model differs in several respects from an earlier minimum free energy structure predicted for the BCoV 5' UTR.[3] The highly conserved SL2 structure we have studied here is not present in the prior BCoV model. Our genetic and biophysical studies support the prediction of SL2 being a UNR U-turn stem-loop. One feature shared by both models is a stem-loop designated as stem-loop III in the earlier BCoV model, which corresponds to stem-loop 4B in our model. The previous BCoV minimum free energy model is also supported by nuclease mapping studies, particularly for stem-loop III, and by genetic studies using DI RNAs. This raises the remote possibility, not addressed here, that the two competing confirmations may exist in equilibrium and perhaps have different functional roles in coronavirus replication.

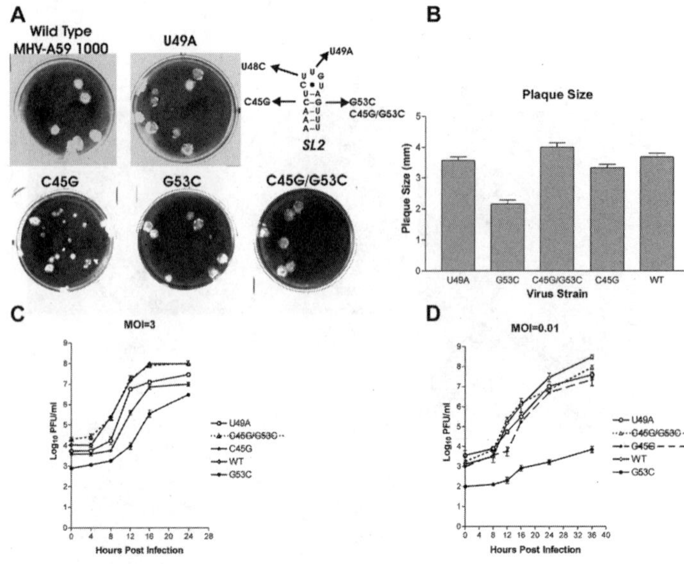

Figure 4. Characterization of growth phenotypes for viable SL2 mutant viruses.

5. ACKNOWLEDGMENTS

This work was supported by NIH grants AI051493 (J.L.L.) and AI040187 (D.P.G.).

6. REFERENCES

1. R.-Y. Chang, M. A. Hofmann, P. B. Sethna, and D. A. Brian, A cis-acting function for the coronavirus leader in defective interfering RNA replication, *J. Virol.* **68**, 8223-8231 (1994).
2. Y. Kim, Y. S. Jeong, and S. Makino, Analysis of *cis*-acting sequences essential for coronavirus defective interfering RNA replication, *Virology* **197**, 53-63 (1993).
3. S. Raman, P. Bouma, G. D. Williams, and D. A. Brian, Stem-loop III in the 5' untranslated region is a cis-acting element in bovine coronavirus defective interfering RNA replication, *J. Virol.* **77**, 6720-6730 (2003).
4. I. L. Hofacker, Vienna RNA secondary structure server, *Nucleic Acids Res.* **31**, 3429-3431 (2003).
5. B. Yount, M. R. Denison, S. R. Weiss, and R. S. Baric, Systematic assembly of a full-length infectious cDNA of mouse hepatitis virus strain A59, *J. Virol.* **76**, 11065-11078 (2002).
6. R. F. Johnson, M. Feng, P. Liu, J. J. Millership, B. Yount, R. S. Baric, and J. L. Leibowitz, The effect of mutations in the mouse hepatitis virus 3'(+)42 protein binding element on RNA replication, *J. Virol.* (in press).
7. R.-Y. Chang, R. Krishnan, and D. A. Brian, The UCUAAAC promoter motif is not required for high-frequency leader recombination in bovine coronavirus defective interfering RNA, *J. Virol.* **70**, 2720-2729 (1996).
8. R. R. Gutell, J. J. Cannone, D. Konings, and D. Gautheret, Predicting U-turns in ribosomal RNA with comparative sequence analysis, *J. Mol. Biol.* **300**, 791-803 (2000).

REGULATION OF CORONAVIRUS TRANSCRIPTION: VIRAL AND CELLULAR PROTEINS INTERACTING WITH TRANSCRIPTION-REGULATING SEQUENCES

Sonia Zúñiga, Isabel Sola, Jose L. Moreno, Sara Alonso, and Luis Enjuanes*

1. INTRODUCTION

The last step in the current model of coronavirus transcription is a template switch during synthesis of the negative strand, to complete the minus sgRNA.[1] It was shown that the free energy of duplex formation between leader transcription-regulating sequence (TRS-L) and the nascent negative-strand plays a crucial role in template switch and is the driving force of coronavirus transcription.[2,3] This step requires overcoming an energy threshold. Coronavirus nucleoprotein (N) plays a structural role in virus assembly and has also been shown to be important in RNA synthesis.[4] In addition, template switching, an obligatory step in CoV transcription, needs to overcome an energy threshold. Therefore, we asked whether RNA chaperones are involved in transcription and, most importantly, if N is an RNA chaperone.

RNA chaperones are proteins that bind RNA with broad specificity and that rescue RNAs trapped in unproductive folding states.[5-8] One of their main characteristics is that, once the RNA has been folded, they are no longer needed and, therefore, they can be removed without altering RNA conformation. There are three RNA chaperone activities easily evaluable *in vitro*: (i) enhancement of RNA ribozyme cleavage, (ii) rapid and accurate RNA-RNA annealing, and (iii) facilitation of RNA strand transfer and exchange.

RNA chaperones decrease the activation energy required for a transition between two states, which is energetically favored. Template switch during coronavirus transcription could be interpreted as a transition between two states: in the first one, a duplex between the nascent minus RNA strand and the genomic positive RNA used as template is formed; in the second one, the nascent RNA strand is paired with the TRS of the leader. Therefore, RNA chaperones could be involved in template switch by decreasing the energy required for the transition from the first to the second duplex (Fig. 1).

---

* Centro Nacional de Biotecnología, CSIC, Darwin, 3, Cantoblanco, 28049, Madrid, Spain.

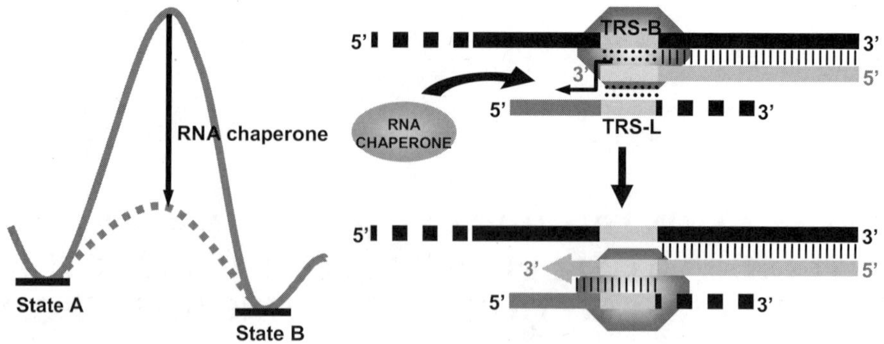

Figure 1. Tentative RNA chaperone involvement in template switch during coronavirus transcription. Left panel, scheme illustrates the action of RNA chaperones. Right panel, elements involved in the template switch step of coronavirus transcription.

Up to now, there are just three RNA chaperones described and all are nucleocapsid proteins from three RNA viruses: (i) retrovirus, the best one analyzed being that of human immunodeficiency virus (HIV-1),[9,10] (ii) hepatitis delta virus (HDV),[11,12] and (iii) hepatitis C virus (HCV).[13] We thought that coronavirus N proteins are good candidates to be RNA chaperones. We used transmissible gastroenteritis virus (TGEV) as a model to investigate this possibility. No RNA chaperone activity can be predicted based on domain conservation. Nevertheless, it was recently reported that RNA chaperones are the protein class with the highest frequency of containing long intrinsically disordered regions.[14] Structural analyses of coronavirus N proteins showed that they also fulfill this criterion.

2. MATERIALS AND METHODS

2.1. Protein Expression and Purification

TGEV N gene, nucleotides 26917 to 28065 from the genome (GeneBank accession number AJ271965), was cloned into the pGEX-4T-2 vector (Amersham Biosciences). Plasmid pET28a-PTB[15] was a generous gift from D. Black (Howard Hughes Medical Institute, UCLA). *Escherichia coli* cells, strain BL21(DE3)pLys (Novagen), were transformed with plasmids pGEX4T2-N or pET28a-PTB. GST-N fusion protein was purified using Glutathione Sepharose 4B (Amersham Biosciences) according to the manufacturer's specifications. His-PTB protein was purified as previously described.[15]

2.2. Electrophoretic Mobility Shift Assay (EMSA)

RNA-protein binding reactions were performed by incubating 10 or 1 pmol of biotinylated RNA with 300 ng of recombinant purified protein in binding buffer (12% glycerol, 20 mM TrisHCl pH 7.4, 50 mM KCl, 1 mM EDTA, 1 mM $MgCl_2$, 1 mM DTT) for 30 min at 25ºC. Reactions were loaded on a 4% non denaturing PAGE. After electrophoresis, the gel was blotted onto positively charged nylon membranes (BrightStar-Plus,

Ambion) following the manufacturer's instructions. Detection of the biotinylated RNA was performed using the BrightStar BioDetect kit (Ambion). When indicated, recombinant protein was preincubated with mAb 30 minutes at 4°C.

2.3. *In Vitro* Self-cleavage of RNA

pBdASBVd[A28][16] was a generous gift from J.A. Daròs and R. Flores (Plant Molecular and Cell Biology Institute, UPV). *In vitro* transcription, cleavage, and electrophoresis of dimeric ASBVd (+) RNA was performed as previously described[16] except that the RNA was labeled with biotin. Densitometric analysis of the bands from three different experiments was performed using Quantity One 4.5.1 Software (BioRad).

3. RESULTS AND DISCUSSION

The functionality of purified TGEV N protein on RNA binding was evaluated by EMSA. Recombinant N protein was incubated with biotinylated RNA oligonucleotides representing viral TRSs or a cellular RNA. A band shift appeared in all cases, indicating that N protein binds RNA nonspecifically, as expected (data not shown). To map the RNA binding domain in the N protein, a set of monoclonal antibodies (mAbs), generated in our laboratory, was used.[17] In similar EMSA experiments, it was found that some of the mAbs recognizing the amino terminus of the protein significantly blocked N-RNA binding, while mAbs recognizing the carboxy terminus did not, and a supershift band appeared in these cases (data not shown). A mAb from each set was used in subsequent experiments.

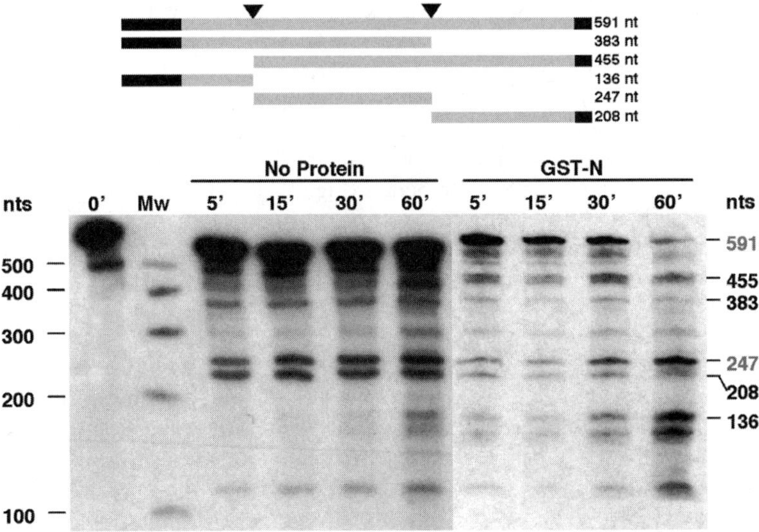

Figure 2. Hammerhead ribozyme self-cleavage. Upper panel, scheme of the substrate (591 nt) and the self-processed products. Low panel, time-course in cleavage conditions.

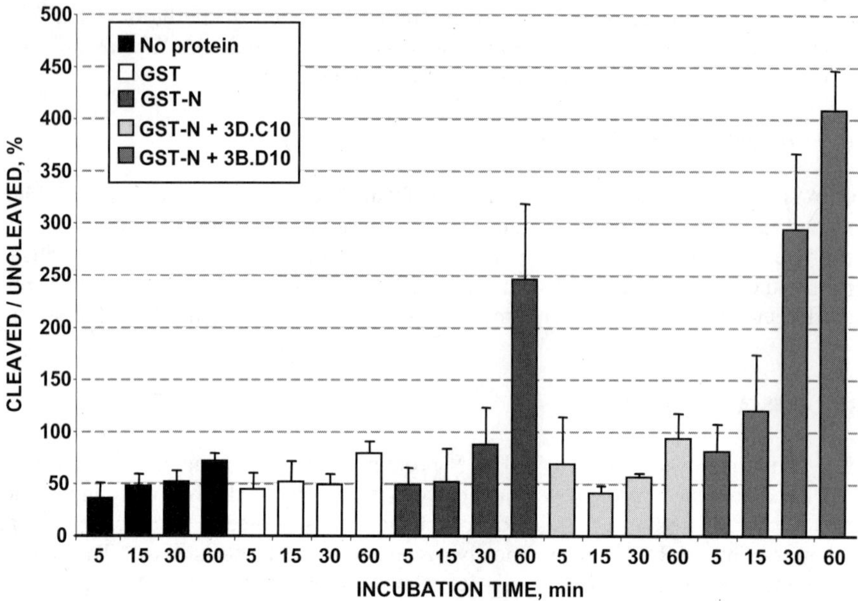

Figure 3. Recombinant TGEV N protein enhances ribozyme self-cleavage. Densitometric quantification of time-course of ribozyme self-cleavage reactions. Error bars indicate the standard deviation from three independent experiments.

Once the functionality of the recombinant N protein was assessed, an advanced RNA chaperone assay was performed. Avocado sunblotch viroid (ASBVd) dimer RNA was used in a hammerhead ribozyme self-cleavage assay.[16] This RNA has two ribozyme cleavage sites and must be properly folded for the cleavage reaction to take place. In the absence of protein, under cleavage conditions, all processed products appeared but there was no progression in the cleavage reaction with incubation time. In contrast, in the presence of recombinant N protein, cleavage products appeared, with a significant decrease in the amount of uncleaved products (Fig. 2). This result strongly suggested that TGEV N protein is an RNA chaperone.

Similar experiments were performed with several controls, and the bands corresponding to the uncleaved substrate and the completely cleaved product were quantified (Fig 3). In the absence of protein, there were no changes in the cleavage reaction with incubation time. The same result was obtained in the presence of the control GST protein. In the presence of recombinant N protein, the ratio of cleaved to uncleaved product increased more than threefold compared with reactions lacking the N protein. This enhancement of the cleavage reaction was due to the N protein, because preincubation of the GST-N with a mAb that blocked RNA-protein binding also blocked activity in the cleavage reaction. The levels of cleavage product obtained were similar to those observed in the absence of N protein. Similarly, N protein preincubated with a mAb that did not block N-RNA binding enhanced the cleavage reaction, and the ratio of cleaved to uncleaved product was more than fivefold compared with the protein-free reactions. The difference between results obtained with GST-N protein alone or with the mAb present was probably explained by an aggregation effect. The enhancement produced by the N protein was not simply due to its RNA binding ability, as another RNA binding protein, polypyrimidine tract binding protein (PTB), did not exert any effect on the cleavage reaction (data not shown). These results clearly indicate that TGEV N protein is a RNA chaperone.

However, this is a heterologous system and therefore, preliminary annealing experiments were performed using a biotinylated TRS-L and a unlabeled cTRS-7. In the presence of N protein, at 25ºC or 37ºC, the amount of dsRNA was higher than that obtained from protein-free reactions, as confirmed by quantifying the gel bands (data not shown). Even in the presence of magnesium, which stabilizes dsRNAs, the effect of N protein was still noted and confirmed by quantification of gel bands. These results strongly suggest that the CoV N protein promotes the annealing of viral TRSs.

The next step will be to study the role of this RNA chaperone activity *in vivo*, in coronavirus transcription.

4. REFERENCES

1. S. G. Sawicki and D. L. Sawicki, A new model for coronavirus transcription, *Adv. Exp. Med. Biol.* **440**, 215-220 (1998).
2. I. Sola, J. L. Moreno, S. Zúñiga, S. Alonso, and L. Enjuanes, Role of nucleotides immediately flanking the transcription-regulating sequence core in coronavirus subgenomic mRNA synthesis, *J. Virol.* **79**, 2506-2516 (2005).
3. S. Zúñiga, I. Sola, S. Alonso, and L. Enjuanes, Sequence motifs involved in the regulation of discontinuous coronavirus subgenomic RNA synthesis, *J. Virol.* **78**, 980-994 (2004).
4. F. Almazán, C. Galán, and L. Enjuanes, The nucleoprotein is required for efficient coronavirus genome replication, *J. Virol.* **78**, 12683-12688 (2004).
5. G. Cristofari and J. L. Darlix, The ubiquitous nature of RNA chaperone proteins, *Prog. Nucleic Acid Res. Mol. Biol.* **72**, 223-268 (2002).
6. D. Herschlag, RNA chaperones and the RNA folding problem, *J. Biol. Chem.* **270**, 20871-2084 (1995).
7. J. R. Lorsch, RNA chaperones exist and DEAD box proteins get a life, *Cell* **109**, 797-800 (2002).
8. R. Schroeder, A. Barta, and K. Semrad, Strategies for RNA folding and assembly, *Nat. Rev. Mol. Cell. Biol.* **5**, 908-919 (2004).
9. E. L. Bertrand and J. J. Rossi, Facilitation of hammerhead ribozyme catalysis by the nucleocapsid protein of HIV-1 and the heterogeneous nuclear ribonucleoprotein A1, *EMBO. J.* **13**, 2904-2912 (1994).
10. Z. Tsuchihashi and P. O. Brown, DNA strand exchange and selective DNA annealing promoted by the Human Immunodeficiency Virus Type I nucleocapsid protein, *J. Virol.* **68**, 5863-5870 (1994).
11. Z. S. Huang and H. N. Wu, Identification and characterization of the RNA chaperone activity of hepatitis delta antigen peptides, *J. Biol. Chem.* **273**, 26455-26461 (1998).
12. Z. S. Huang, W. H. Su, J. L. Wang, and H. N. Wu, Selective strand annealing and selective strand exchange promoted by the N-terminal domain of hepatitis delta antigen, *J. Biol. Chem.* **278**, 5685-5693 (2003).
13. G. Cristofari, R. Ivanyi-Nagy, C. Gabus, et al., The hepatitis C virus Core protein is a potent nucleic acid chaperone that directs dimerization of the viral (+) strand RNA in vitro, *Nucleic Acids Res.* **32**, 2623-2631 (2004).
14. R. Ivanyi-Nagy, L. Davidovic, E. W. Khandjian, and J. L. Darlix, Disordered RNA chaperone proteins: from functions to disease, *Cell. Mol. Life Sci.* **62**, 1409-1417 (2005).
15. J. Xie, J. Lee, T. L. Kress, K. L. Mowry, and D. L. Black, Protein kinase A phosphorylation modulates transport of the polypyrimidine tract-binding protein, *Proc. Natl. Acad. Sci. USA* **100**, 8776-8781 (2003).
16. J. A. Daròs and R. Flores, A chloroplast protein binds a viroid RNA in vivo and facilitates its hammerhead-mediated self-cleavage, *EMBO. J.* **21**, 749-759 (2002).
17. J. M. Martín-Alonso, M. Balbín, D. J. Garwes, L. Enjuanes, S. Gascón, and F. Parra, Antigenic structure of transmissible gastroenteritis virus nucleoprotein, *Virology* **188**, 168-174 (1992).

DEUBIQUITINATING ACTIVITY OF THE SARS-CoV PAPAIN-LIKE PROTEASE

Naina Barretto[1], Dalia Jukneliene[1], Kiira Ratia[2], Zhongbin Chen[1,3], Andrew D. Mesecar[2] and Susan C. Baker[1]

1. INTRODUCTION

The SARS-CoV replicase polyproteins are processed into 16 nonstructural proteins (nsps) by two viral proteases; a 3C-like protease (3CLpro) and a papain-like protease (PLpro). PLpro processes the amino-terminal end of the replicase polyprotein to release nsp1, nsp2, and nsp3. In this study, we identified a 316 amino acid core catalytic domain for SARS-CoV PLpro that is active in trans-cleavage assays. Interestingly, bioinformatics analysis of the SARS-CoV PLpro domain suggested that this protease may also have deubiquitinating activity because it is predicted to have structural similarity to a cellular deubiquitinase, HAUSP (herpesvirus-associated ubiquitin-specific-protease). Using a purified preparation of the catalytic core domain in an *in vitro* assay, we demonstrate that PLpro has the ability to cleave ubiquitinated substrates. We also established a FRET-based assay to study the kinetics of proteolysis and deubiquitination by SARS-CoV PLpro. This characterization of a PLpro catalytic core will facilitate structural studies as well as high-throughput assays to identify antiviral compounds.

IDENTIFYING A PLpro CATALYTIC CORE DOMAIN

Proteolytic processing of the coronavirus replicase polyprotein by 3CLpro and the papain-like proteases is essential for the formation of the mature replicase complex. In light of its role in processing the amino-terminal end of the replicase polyprotein, PLpro is a potential target for the development of antiviral drugs. To identify a catalytically active core domain for PLpro, we performed deletion analysis from the C- and N-terminii of a region of SARS-CoV nsp3, which we had previously shown to be proteolytic active. We generated four N-terminal, and four C-terminal constructs and

[1]Loyola University Chicago Stritch School of Medicine, Maywood, IL, [2]Center for Pharmaceutical Biotechnology and Department of Medicinal Chemistry and Pharmacognosy, University of Illinois at Chicago, [3]Department of Biochemistry and Molecular Biology, Beijing Institute of Radiation Medicine, Beijing, China.

tested them for proteolytic activity in a trans-cleavage assay. The region from amino acids 1541-1855 was the smallest fragment with proteolytic activity.[3]

To identify the important features of this segment, we aligned the amino acid sequences of 15 papain-like protease domains from 8 different coronaviruses with that of SARS-CoV PLpro1541-1855. While there is only an 18–32% identity between the sequences, there are some important conserved features, including the catalytic cysteine and histidine residues (identified as C1651 and H1812 for SARS-CoV PLpro)[2] and a conserved Zn-binding finger, which has been shown to be important for proteolytic activity of HCoV-229E PLP1.[4] In addition, we identified an aspartic acid residue (D1826 in SARS-CoV PLpro), which is conserved amongst all the coronavirus papain-like proteases. We found that PLpro in which D1826 was replaced by an alanine had no detectable proteolytic activity in the cellular trans-cleavage assay,[3] indicating that this residue may play an important role in proteolytic activity. We predict that this residue forms a part of a catalytic triad with the catalytic cysteine and histidine residues, similar to the active sites of other papain-like proteases.[5] Ultimately, crystallographic data will be critical for a complete understanding of the coronavirus PLpro active site.

DEUBIQUITINATING ACTIVITY OF SARS-CoV PLpro.

Interestingly, an independent bioinformatics study predicted that SARS-CoV PLpro may have structural similarity to a cellular deubiquitinating enzyme, HAUSP (herpesvirus-associated ubiquitin-specific protease).[6,7] On the basis of this similarity in molecular architecture and the predicted specificity of PLpro for cleavage after a diglycine residue, Sulea and co-workers predicted that PLpro has deubiquitinating (DUB) activity.

To test this prediction we used our experimentally determined catalytic core domain in an in vitro assay for deubiquitination.[7] Briefly, wild-type PLpro and mutated versions where C1651 or D1826 was replaced by alanine, were expressed in *E. coli* BL21 and purified to homogeneity.[3] In the assay, purified WT PLpro was incubated alone (Fig. 1, lane 2) or with di-ubiquitin substrate (Boston Biochem) (lane 3) for 1 hour at 37°C in a buffer containing BSA. In lanes 4 and 5, purified mutant PLpro D1826A and PLpro C1651A respectively were incubated with the substrate. WT PLpro cleaved the substrate completely, while the C1651A mutant had no activity. As expected, the D1826A mutant showed a small amount of processing activity. This is the first evidence that SARS-CoV PLpro is able to cleave a ubiquitinated substrate.

To study the kinetics of proteolysis and deubiquitination by PLpro, we established a FRET-based assay using a substrate [E-EDANS]RELNGGAPI-[KDABCYL]S for proteolysis. This substrate is based on the naturally occurring nsp1/nsp2 and nsp2/nsp3 cleavage sites for PLpro. A similar assay using the commercially available substrate, ubiquitin-AMC (Boston Biochem) was used to measure the kinetics of deubiquitination. The substrates were incubated with the purified proteins and the reaction was allowed to go to completion. The fluorescence released was measured on a Cary Eclipse fluorescence spectrophotometer.

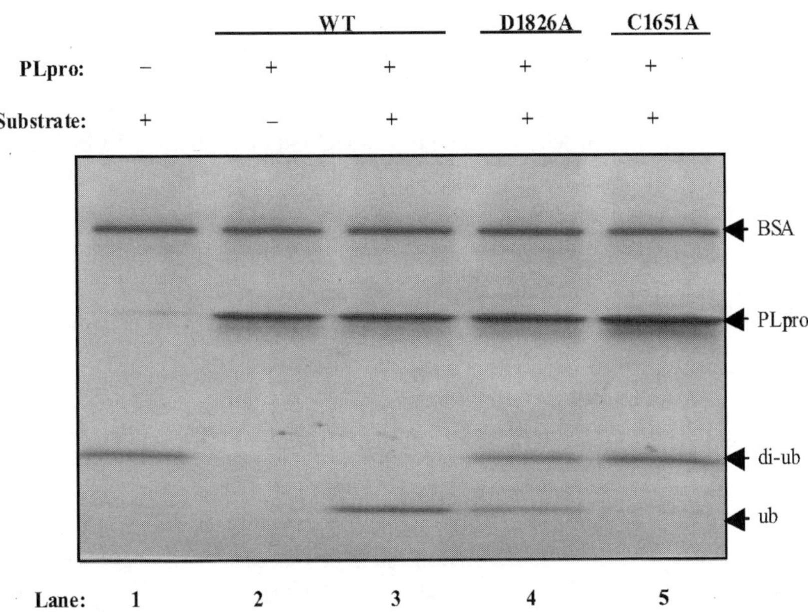

Figure 1. Deubiquitinating activity of SARS-CoV PLpro. Purified WT or mutant PLpro was incubated with the di-ubiquitin substrate for 1 hour at 37°C. Products were resolved by electrophoresis on a 10–20% gradient polyacrylamide gel, which was stained with Coomassie blue.

Table 1. Apparent k_{cat}/K_m (k_{app}) values obtained in FRET-based assays for peptide hydrolysis or deubiquitination for SARS CoV PLpro1541-1855 and mutants. Assay as described in Ref. 3.

Protease **SARS-CoV PLpro1541-1855**	**Peptide hydrolysis** k_{app} min^{-1}µM^{-1}	**deubiquitination** k_{app} min^{-1}µM^{-1}
WT	(2.44+/–0.03) x10^{-2}	4.48+/–0.11
C1651A	No activity	No activity
D1826A	(2.3+/–0.2) x10^{-4}	<1% activity of WT

Since the enzyme could not be saturated with substrate, pseudo-first rate order kinetics were used to calculate the apparent $k_{cat}/\ k_m$ (k_{app}), indicated in Table 1. WT PLpro1541-1855 cleaved the ubiquitin substrate with faster kinetics than the peptide substrate, though this might reflect in vitro conditions and the substrates used. However in a similar assay, HAUSP cleaved ubiquitin-AMC with a k_{cat}/K_m of 13 $min^{-1}mM^{-1}$(Hu et al., 2002), indicating that PLpro has higher deubiquitinating activity than this cellular deubiquitinase.

The mutant D1826A hydrolyzed peptide substrate at approximately 1% of the efficiency of the WT, which is consistent with an assisting role for this residue in a catalytic triad with the catalytic cysteine and histidine residues. The C1651A mutant showed no cleavage activity with either substrate, demonstrating the essential role of the catalytic cysteine residue (Johnston et al., 1997).

A "SIGNATURE SEQUENCE" FOR DEUBIQUITINATION ACTIVITY

Sulea and co-workers noted "structural signatures for strict specificity" (Sulea et al., 2005) present in HAUSP and also in some of the coronavirus papain-like proteases, which form part of the substrate-binding site. In the case of SARS-CoV PLpro the residues Y1804 and Y1813 occlude the substrate-binding site, imposing the requirement for the small diglycine residues at the cleavage site (Sulea et al., 2005). Our alignment of the amino acid sequences of 16 papain-like proteases from nine coronaviruses shows that the "signature sequence" is present in twelve of these sequences. This suggests that these proteases may also have deubiquitinating activity, though this remains to be experimentally verified.

```
              DUB signature seq    catalytic his
                               *
SCoV PLp1804  YT---GNYQCGH-Y-T-HITAKETLYR----<32>IKP-------------%ID
MHVJ P2 1867  FT---GG-SVGH-Y-T-HVKCKPKYQL----<35?>YY-------------32
BCoV P2 1822  FK---GD-KVGH-Y-V-HVKCEQSYQL----<35>YY------------- 31
OC43 P2 1822  FI---GD-KVGH-Y-V-HVKCEQSYQL----<35>YY------------- 30
HKU  P2 1907  FM---G-VGVGH-Y-T-HLKCGSPYQH----<32>LTNY----------- 28
229E P2 1857  FS---GPVDKGH-Y-TVYDTAKKSMY-----<30>VK------------- 22
TGEV P2 1736  YS---GSNRNGH-Y-T-YYDNRNGLV-----<25>KKPQAEERPKNCAFNK 22
IBV  PLp1490  FV---GSTNSGHCY-T-QAAGQ----A----<30>SLPV----------- 22
NL63 P2 1825  YTTFSGSFDNGH-Y-VVYDAANNAVY-----<28>VPTIVSEK------- 21
TGEV P1 1237  YT---GTTQNGH-Y-M--VDDIEHGYC----<28>EKPKQEFKVEKVEQQ- 21
229E P1 1197  FR---GAVSCGH-YQT-NIYSQNLC------<37>IKNTVD--------- 20
NL63 P1 1204  YL---GVKGSGH-Y-------QTNLYSFNKA<33>VKPFAVYKNVK----- 19
|HKU  P1 1306  VD---VNVC--H-S-V-AVIGDE---Q----<36>ITPNVCF--------- 20
|MHVJ P1 1267  VN------DCHS-M-A-VVDGKQ--------<38>ITPNVCF--------- 18
|OC43 P1 1199  KR---IVYKAAC-V-V-DVNDSHSMAV----<43>ITPNVCF--------- 18
|BCoV P1 1199  KR---SVYKAAC-V-V-DVNDSHSMAV----<43>ITPNVCF--------- 18
```

Figure 2. Multiple sequence alignment of *DUB signature sequences* from coronavirus papain-like proteases. The papain-like protease domain amino acid sequences (two domains termed as P1 and P2; one domain termed as PLpro) of nine coronaviruses were aligned using the ALIGN program (SciEd). The residues proposed to be part of the substrate binding site for deubiquitination are boxed. Identical residues are highlighted in light gray. The catalytic histidine residue is boxed in thick black or highlighted in gray. Papain-like proteases which are predicted to lack deubiquitinating activity are indicated by a thick vertical line. Abbreviationas are as follows: SARS CoV- Severe acute respiratory syndrome coronavirus, Urbani strain (AY278741); MHVJ- Mouse hepatitis virus, strain JHM (NC_001846); BCoV- bovine coronavirus (NC_003045); HCoV-OC43-Human coronavirus OC43 (AY585228); HCoV-229E- Human coronavirus 229E (X69721); HCoV-NL63- Human coronavirus NL63 (NC_005831); TGEV- Transmissible gastroenteritis virus of pigs (Z34093); aIBV- avian infectious bronchitis virus (NC_001451); HCoV-HKU1- Hong Kong University coronavirus 1 (NC_006577).

FUTURE DIRECTIONS

Here, we have identified a core catalytic domain for SARS-CoV PLpro that is capable of processing both the amino-terminal end of the replicase polyprotein and ubiquitinated substrates. The role of this deubiquitinating activity in SARS-CoV-infected cells is unknown. Future experiments will focus on determining if this SARS-CoV PLpro DUB is active in virus-infected cells, investigating the significance of DUB activity in the viral life cycle, as well as identifying possible viral and cellular targets.

ACKNOWLEDGMENTS

This work was funded by Public Health Service research grant AI45798 (to S.C.B.) and P01AI060915 (to S.C.B. and A.D.M). N.B. was supported by Training Grant T32 AI007508.

REFERENCES

1. Kim J.C., R.A. Spence, P.F. Currier, X. Lu and M.R. Denison. 1995. Coronavirus protein processing and RNA synthesis is inhibited by the cysteine proteinase inhibitor E64d. Virology **208:**1-8.
2. Harcourt, B.H., D. Jukneliene, A. Kanjanahaluethai, J. Bechill, K.M. Severson, C.M. Smith, P.A. Rota and S.C. Baker. 2004. Identification of severe acute respiratory syndrome coronavirus replicase products and characterization of papain-like protease activity. J. Virol. **78:**13600-13612.
3. Barretto, N., D. Jukneliene, K. Ratia, Z. Chen, A.D. Mesecar and S.C. Baker. The papain-like protease of Severe Acute Respiratory Syndrome (SARS) coronavirus has deubiquitinating activity. J. Virol. **79**: 15189-15198.
4. Herold J., S.G. Siddell and A.E. Gorbalenya. 1999. A human RNA viral cysteine proteinase that depends on a unique Zn^{2+} binding finger connecting the two domains of a papain-like fold. J. Biol. Chem. **274:**14918-14925.
5. Johnston, S.C., C.N. Larsen, W.J. Cook, K.D. Wilkinson, and C.P. Hill. 1997. Crystal structure of a deubiquitinating enzyme (human UCH-L3) at 1.8 A resolution. EMBO J. **16:**3787-3796.
6. Sulea T., H.A. Lindner, E.O. Purisima and R. Menard. 2005. Deubiquitination, a new function of the severe acute respiratory syndrome coronavirus papain-like protease? J. Virol. **79:**4550-4551.
7. Hu, M., P.Li, M. Li, W. Li, T. Yao, J.W. Wu, W. Gu, R.E. Cohen and Y. Shi. 2002. Crystal structure of a UBP-family deubiquitinating enzyme in isolation and complex with ubiquitin aldehyde. Cell. **111:**1041-1054.

NUCLEOCAPSID PROTEIN EXPRESSION FACILITATES CORONAVIRUS REPLICATION

Barbara Schelle, Nadja Karl, Burkhard Ludewig, Stuart G. Siddell, and Volker Thiel*

1. INTRODUCTION

The coronavirus nucleocapsid (N) protein has been implicated in a number of functions. As a structural protein, it forms a ribonucleoprotein complex with genomic RNA. However, it has also been described as an RNA-binding protein[1] that might be involved in coronavirus RNA synthesis.[2] Here, we used a reverse genetic approach to elucidate N protein function(s) in coronavirus replication and transcription. We could show that human coronavirus 229E (HCoV-229E) vector RNAs are greatly impaired in their ability to replicate if they lack the N gene. In contrast, efficient replication was observed if vector RNAs express N protein. Noteworthy, transcription of subgenomic mRNAs was readily detectable, irrespective of the presence or absence of N protein. Finally, by modifying the transcription signal required for the synthesis of N protein mRNA in the HCoV-229E genome, we could demonstrate selective replication of genomes that are able to express the N protein. Therefore, we conclude that the coronavirus N protein is involved in genome replication.

2. RESULTS

2.1. Analysis of Vector RNA Replication and Transcription

To study the role of N protein in coronavirus replication and transcription, we made use of our reverse genetic system for HCoV-229E.[3] We produced two vector RNAs, HCoV-vec-1 and HCoV-vec-GN, that both contain the replicase gene and the gene encoding for the green fluorescent protein (GFP). HCoV-vec-GN encodes in addition the N protein, whereas HCoV-vec-1 is lacking the N gene (Figure 1a). We previously

* Barbara Schelle and Nadja Karl, University of Würzburg, Würzburg, Germany. B. Ludewig and V. Thiel, Kantonal Hospital St. Gallen, 9007 St. Gallen, Switzerland. S.G. Siddell, University of Bristol, Bristol, United Kingdom.

showed that HCoV-vec-1 RNA mediates the transcription of a subgenomic mRNA, encoding GFP.[4] Thus, we concluded, that the replicase gene products suffice for coronavirus transcription. We also observed that only a small number of green fluorescent cells (< 0.1%) can be detected upon electroporation of HCoV-vec-1 RNA into BHK-21 cells. However, if a synthetic mRNA encoding the HCoV-229E N protein was co-electroporated into BHK-21 cells, the number of green fluorescent cells significantly increased (3.4%, Figure 1b). When we electroporated HoV-vec-GN RNA into BHK-21 cells, we again could observe less than 0.1% green fluorescent cells, and elevated numbers of GFP-expressing cells could only be detected if a N protein mRNA is co-electroporated (Figure 1b). Thus, we conclude that co-electroporation of mRNA encoding the HCoV-229E N protein with HCoV-229E-based vector RNAs encoding GFP increased the number of green fluorescent cells, even if the N protein was encoded by the vector RNA (e.g., HCoV-vec-GN). We interpret these results that the presence of N protein early after transfection of vector RNAs is important in our system for the formation of a functional replicase/transcriptase complex.

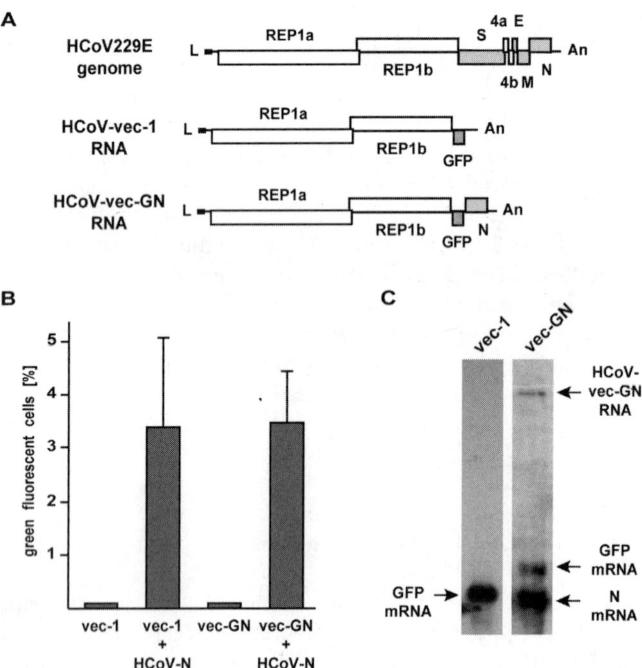

Figure 1. Transfection of vector RNAs. (A) The structural relationship of HCoV-229E and vector RNAs HCoV-vec-1 and HCoV-vec-GN are shown. Open reading frames are indicated as boxes designated by encoded gene products. L, leader RNA; An, poly(A) sequence. (B) Vector RNA HCoV-vec-1 or HCoV-vec-GN was transfected into BHK-21 cells with or without synthetic N protein mRNA as indicated. (C) Northern blot analysis of BHK-21 cells that have been transfected with N protein mRNA and vector RNAs HCoV-vec-1 or HCoV-vec-GN, respectively. Full-length vector and subgenomic mRNAs are indicated (arrows).

Although there was no difference in the number of green fluorescent cells after transfection of HCoV-vec-1 and HCoV-vec-GN RNA into BHK cells, we observed that HCoV-vec-GN-transfected cells displayed relatively intense fluorescence already 16 h post-transfection. In contrast, the intensity of green fluorescence of HCoV-vec-1-transfected cells was lower and only became apparent 48 h post-transfection. To test whether this finding was related to different replication or transcription levels of vector RNAs, we analyzed the vector-specific RNAs in transfected BHK-21 cells by Northern blot. To obtain a sufficient number of green fluorescent cells that allow for the detection of vector-specific RNAs by Northern blot analysis, we co-electroporated synthetic N protein mRNA with vector RNAs HCoV-vec-1 and HCoV-vec-GN, respectively. After co-transfection of HCoV-vec-1 RNA with N protein mRNA, we could detect a faint signal for a subgenomic mRNA encoding GFP (data not shown). To increase the number of green fluorescent cells that have been transfected with HCoV-vec-1 and N protein mRNA, we sorted green fluorescent cells prior to the isolation of poly(A)-containing RNA. With this strategy, an HCoV-vec-1 derived subgenomic mRNA encoding GFP was readily detectable, but we were unable to detect the full-length HCoV-vec-1 RNA (Figure 1c). This contrasted with the co-transfection of HCoV-vec-GN RNA and N protein mRNA, where we could easily detect (even without sorting green fluorescent cells) the full-length vector RNA and two mRNAs encoding GFP and N protein (Figure 1c). These results show that vector RNAs encoding a functional N gene were able to transcribe and replicate RNA, whereas vector RNAs lacking the N gene were able to transcribe RNA but replication of full-length vector RNA was not detectable.

Taken together, it appears that early after transfection of vector RNAs, the N protein (provided by co-transfected N protein mRNA) may be important for the establishment of a functional replicase/transcriptase complex; however, sustained N protein expression (if the N protein is expressed by the vector itself) is required for efficient replication. Accordingly, it is tempting to speculate that in natural infections, N proteins, associated with the viral genome in the ribonucleoprotein complex, may be important early in the infection to establish a functional replication/transcription complex and sustained N protein expression by the transcription of viral N protein mRNA may be required for efficient genome replication.

2.2. Selective Replication of HCoV-229E Genomes That Express N Protein

In order to test the hypothesis that N protein expression is required for efficient genome replication, we constructed recombinant HCoV-229E genomes that had been modified at the TRS of the N gene (TRS-N), a *cis*-acting RNA element that is required for the production of a subgenomic N mRNA encoding the N protein. As illustrated in Figure 2a, we modified the authentic TRS-N core sequence (UCUAAACU) to contain a stretch of three random nucleotides (UCU*NNN*CU). Thus, we constructed, by *in vitro* transcription, a population of 64 different full-length recombinant HCoV-229E genomes. These RNA molecules were transfected into BHK-21 cells (which are not susceptible to HCoV-229E infection), and after 3 days we isolated poly(A)-containing RNA (Figure 2b). By RT-PCR, we compared the sequence of the TRS-N region of the "input" genomes (*in vitro* transcription products) and the "re-isolated" genomes (poly(A)-containing RNA isolated 3 days post-transfection). This analysis revealed that the re-isolated genomes have clearly undergone selection, because the nucleotides at the randomized positions have shifted to a predominance of adenines (data not shown).

However, at one position within the randomized sequence, there was also a prominent uridine peak detectable. To determine the sequence of the TRS-N region on individual genomes, we cloned the RT-PCR products and determined the sequences of 44 individual clones corresponding to the input genomes and 41 individual clones corresponding to the re-isolated genomes. The result of this analysis is shown in Figure 2c. We could detect an increased number of genomes that contained the authentic (wild-type) TRS-N amongst re-isolated genomes (increase from 4.5% to 9.8%). Similarly, we observed an increased number of genomes containing the uridine within the TRS-N (NNU; increase from 20.5% to 41.5%). These two groups obviously had undergone a positive selection during amplification in BHK-21 cells. Genomes that contained one nucleotide difference compared to the authentic TRS-N or leader-TRS remained approximately at the same level (31.8% and 29.3% for input and re-isolated RNAs, respectively). Genomes that contained sequences not matching to the groups mentioned above had presumably undergone a negative selection, because their percentage dropped from 43.2% in the input genomes to 19.5% in the re-isolated genomes.

The observation that specific HCoV-229E genomes had undergone positive selection in our assay indicates that these genomes replicated preferentially. According to our hypothesis, this might be related to their ability to express N protein. Therefore, we specifically amplified, by RT-PCR, the subgenomic mRNA encoding N protein that had been produced in transfected cells. Again, the RT-PCR product was cloned, and the sequences of individual clones, corresponding to the sequences of subgenomic N protein mRNAs at the leader-body fusion sites, were determined. As expected, most N protein mRNAs (> 60%) contained either, the authentic TRS-N sequence (UCUAAACU) or the sequence of the leader TRS (UCUCAACU), confirming that these TRS elements were efficient in directing the synthesis of subgenomic mRNAs. In addition to these wild-type TRS elements, we could determine nine different N protein mRNA sequences. As shown in Figure 2d, eight of these subgenomic N protein mRNA sequences match with genomes that contain either the NNU sequence at the TRS-N (group of positively selected genomes) or only one nucleotide difference compared to the TRS-N or leader TRS (group of genomes that remained on the same level in our assay). Interestingly, only one subgenomic N protein mRNA has been detected that does not correspond to the groups of genomes that had undergone positive selection or remained at the same level during their passage in BHK-21 cells.

In summary, we could demonstrate that a population of genomes containing a stretch of randomised nucleotides within the TRS-N had undergone a selection process during the passage in BHK-21 cells. Furthermore, the vast majority of subgenomic mRNAs produced during the passage in BHK-21 cells correspond at the TRS-N region to the sequences of positively selected genomes or genomes that remained at the same level. Therefore, these data provide genetic evidence that HCoV-229E genomes that are able to express N protein confer a selective advantage for replication in nonpermissive cells.

3. CONCLUSIONS

Our data conclusively demonstrate that at least one structural protein, the N protein, is involved in coronavirus genome replication. This conclusion is based upon a reverse genetic analysis of HCoV-229E vector RNAs that showed that (i) transcription of subgenomic mRNAs can take place in the absence of N protein[4,5] and (ii) efficient

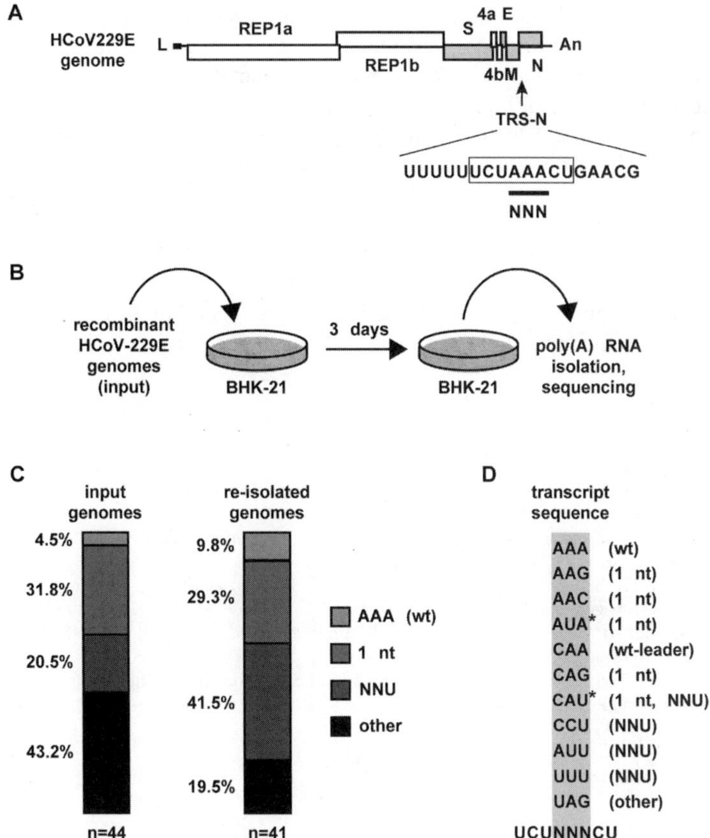

Figure 2. Selective replication of HCoV-229E genomes. (A) HCoV-229E nucleotides at the TRS-N core sequence (boxed) are shown together with the structure of the HCoV-229E genome. The authentic HCoV-229E TRS-N core sequence (UCUAAACU; randomized nucleotides [AAA] are underlined) was changed to contain a stretch of three random nucleotides (NNN). (B) Illustration of the experimental procedure. (C) Representation of genome sequences amongst input genomes and re-isolated genomes derived from 44 and 41 individual plasmid clones, respectively. The recombinant genomes were placed in four groups: group 1, recombinant genomes with the HCoV-229E wild-type sequence (AAA); group 2, recombinant genomes with a 1-nucleotide (1 nt) change compared to the TRS-N or leader TRS sequence; group 3, recombinant genomes containing a U nucleotide at the third randomized nucleotide (NNU); group 4, recombinant genomes that do not match to groups 1 to 3. The percentage of each group in the population of input and re-isolated genomes are indicated. (D) Sequences of subgenomic N protein mRNA detected in poly(A)-containing RNA 3 days post-transfection. The sequences determined at the randomized stretch of three nucleotides are shown (shaded in gray) together with corresponding groups of recombinant genomes in parentheses.

replication of vector RNAs is dependent on N protein expression. Furthermore, our conclusion is corroborated by the genetic evidence that the ability of recombinant HCoV-229E genomes to express N protein confer a selective advantage for genome replication. Thus, our data provide substantial evidence for a functional role of the N protein in coronavirus RNA synthesis. However, the nature of N protein function(s) in coronavirus RNA synthesis remains to be determined. Future studies are needed to elucidate at which process(es) the N protein interferes in coronavirus RNA synthesis (e.g., initiation of negative-strand synthesis, recognition of TRS elements, regulation of replication/ transcription levels) and how these function(s) are mediated (e.g., interaction with viral RNA, replicative proteins, or host cell proteins). It also remains to be determined whether distinct enzyme complexes, involved in replication and transcription, respectively, may exist and whether the N protein has a regulatory role in these complexes.

4. ACKNOWLEDGMENTS

This work was supported by the Swiss National Science Foundation, the Gebert-Rüf oundation, Switzerland, and the Deutsche Forschungsgemeinschaft.

5. REFERENCES

1. P. S. Masters, Localization of an RNA-binding domain in the nucleocapsid protein of the coronavirus mouse hepatitis virus, *Arch. Virol.* **125**, 141-160 (1992).
2. F. Almazan, C. Galan, and L. Enjuanes, The nucleoprotein is required for efficient coronavirus genome replication, *J. Virol.* **78**, 12683-12688 (2004).
3. V. Thiel, J. Herold, B. Schelle, and S. G. Siddell, Infectious RNA transcribed in vitro from a cDNA copy of the human coronavirus genome cloned in vaccinia virus, *J. Gen. Virol.* **82**, 1273-1281 (2001).
4. V. Thiel, J. Herold, B. Schelle, and S. G. Siddell, Viral replicase gene products suffice for coronavirus discontinuous transcription, *J. Virol.* **75**, 6676-6681 (2001).
5. V. Thiel, N. Karl, B. Schelle, P. Disterer, I. Klagge, and S. G. Siddell, Multigene RNA vector based on coronavirus transcription, *J. Virol.* **77**, 9790-9798 (2003).

NON STRUCTURAL PROTEINS 8 AND 9 OF HUMAN CORONAVIRUS 229E

Rajesh Ponnusamy, Jeroen R. Mesters, John Ziebuhr, Ralf Moll, and Rolf Hilgenfeld*

1. INTRODUCTION

The genome of human coronavirus 229E (HCoV-229E) consists of 27, 277 nucleotides. The viral replicase gene comprises two large, overlapping open reading frames (ORFs), ORF1a and ORF1b.[1] ORF1a encodes the polyprotein pp1a with a calculated molecular mass of 454 kDa. The downstream ORF1b is expressed as a fusion protein with pp1a by a mechanism involving a (–1) ribosomal frame shift during translation.[1,2] The ORF1a/1b gene product has a calculated molecular mass of 754 kDa and is referred to as polyprotein 1ab (pp1ab). pp1a and pp1ab are processed by two virus-encoded, papain-like proteases and the main protease M^{pro}, resulting in at least 16 non structural proteins (Nsps).[3] Several or all of these nonstructural proteins build the replicase complex, probably mediating all the functions necessary for polyprotein processing, viral transcription, and replication.[4] The 3′ region of the coronaviral ORF1a encodes a set of relatively small polypeptides (Nsp6 to Nsp11), of which only SARS-CoV Nsp9 has had a function assigned, i.e., as ssDNA/RNA-binding protein.[5,6] In mouse hepatitis CoV, several of these polypeptides colocalize with other components of the viral replication complex in the perinuclear region of the infected cell.[7] Thus, the HCoV-229E polypeptides Nsp 6, 7, 8, and 10 are probably involved, directly or indirectly, in the viral replication complex. In this communication, we will describe the expression of genes coding for HCoV-229E Nsp8 and Nsp9, as well as the purification and biophysical characterization of the proteins. The two proteins are shown to bind tRNA using zone-interference electrophoresis and fluorescence spectroscopy.

* Rajesh Ponnusamy, Jeroen R. Mesters, Ralf Moll, and Rolf Hilgenfeld, University of Lübeck, 23538 Lübeck, Germany. John Ziebuhr, University of Würzburg, 97078 Würzburg, Germany.

2. MATERIALS AND METHODS

2.1. Cloning

Genes coding for Nsp8 and Nsp9 of HCoV–229E were amplified by polymerase chain reaction from virus-derived cDNA fragments. The amino-acid sequences encompass pp1a residues 3630-3824 for Nsp8, with an additional methionine at the N-terminus and six histidines at the C-terminus, and residues 3825-3933 for Nsp9, with an additional methionine, six histidines and a His-tag cleavage site for M^{pro} at the N-terminus. The nsp8 PCR product was cloned into the pET11a expression vector resulting in pETHCoV-229E/nsp8. The nsp9 PCR product was cloned into pET15b resulting in pETHCoV-229E/nsp9.

2.2. Protein Production, Purification, and Characterization

Nsp8- and Nsp9-encoding plasmids were transformed in competent *E. coli* B834 (DE3) and *E. coli* Tuner (DE3) pLacI strains, respectively (Novagen). Cultures were grown in TY medium at 37°C until cells reached an O.D. of 0.4 at 660 nm. Cells were then induced with 1 mM IPTG and grew for a further 5 h for Nsp8 and 4 h for Nsp9 at 37°C. The cells were then harvested by centrifugation at 5500 rpm for 30 min at 4°C. The resulting pellets were frozen at -20°C. For lysis, the cell pellets were resuspended in 50 mM Tris-HCl and 300 mM NaCl pH 7.5 (25°C). Cells were broken by ultrasonification on ice after adding glycerol (1%, v/v, for Nsp8, and 10%, v/v, for Nsp9), 20 mM imidazole, 0.01% (w/v) *n*-octyl-β-glucoside (Bachem) (final concentrations). To optimize the solubilization of overproduced Nsp8 and Nsp9, a sparse matrix screen of buffer composition was applied.[8] The sample was ultracentrifuged at 30,000 rpm for 1 h at 4 °C (Kontron TGA centrifuge, TFT 45.94 rotor). The supernatant was applied to a His Trap HP column (1 ml, Amersham Pharmacia) with a flow rate of 1 ml/min. After washing with 50 mM Tris-HCl, 20 mM imidazole, 300 mM NaCl, pH 7.5 (25°C), the protein was eluted with a linear gradient ranging from 20 mM up to 500 mM imidazole. The engineered N-terminal His-tag contained a SARS-CoV M^{pro} cleavage site. Therefore, by incubating Nsp9 with M^{pro} in a molar ratio of 100:1 for 16 h at 37°C in a reduced state (5mM DTT), the His tag was cleaved off from the Nsp9. Protein was blotted and detected with anti-tetra-histidine antibodies (Dianova) and anti-mouse IgG-alkaline phosphatase conjugate (Sigma).

2.3. Zone-Interference Gel Electrophoresis

The zone-interference gel electrophoresis device was constructed as described.[9] The 1% (w/v) agarose gel was prepared in 20 mM Tris acetate, pH 8.3, 50 mM NaCl, and 3.5 mM $MgCl_2$ (electrophoresis buffer). Hundred microliter samples with increasing tRNA concentration were applied to the extended zone slot (sample buffer: 10% DMSO in electrophoresis buffer). Ten microliter of Nsp8 with a final concentration of 10 μM in sample buffer was loaded into the small slot of the device. Gels were run at 200 mA for 2 h at 4°C in electrophoresis buffer. The temperature of the gel during the run was measured to be around 25°C. The gel was then stained, destained, and stored as described.[9] K_d values were calculated using the equation:

$$[(d_{exp} - d_M) / [L] = - (d_{exp} - d_{ML}) / K_d]$$

(d_{exp} , migration distance of protein with varying tRNA concentrations; d_{ML}, migration distance of the complex macromolecule (M) and ligand (L); d_M , migration distance of macromolecule; [L], ligand concentration in μM). (d_{exp} – d_M) / [L] was plotted against d_{exp} with a slope of -1/ K_d.

The zone-interference gel electrophoresis apparatus was reconstructed for HCoV-229E Nsp9. The protein was positively charged at pH 8.3. So, protein was loaded at the anodic side into the small slot and tRNA was loaded at the cathodic side into the extended slot. During the run, protein and tRNA migrated in opposite directions and crossed each other in the gel.

2.4. Fluorescence Measurement

Fluorescence emission spectra were recorded in 50 mM Tris, 100 mM NaCl, pH 8.5, at room temperature, using a Cary Eclipse fluorescence spectrometer. Proteins were added to final concentrations of 5 μM. Spectra were recorded under identical spectrometer settings using an excitation wavelength of 280 nm. Fluorescence titration experiments were carried out using the excitation wavelength of 280 nm and emission wavelengths of 330 nm (for Nsp8) or 350 nm (for Nsp9). Fluorescence was measured in the presence of increasing concentrations of *E. coli* tRNA.

3. RESULTS

3.1. Expression of HCoV-229E nsp8 and nsp9

HCoV-229E Nsp8 and Nsp9 were successfully produced under heterologous conditions at about 10–20 mg protein per liter expression culture. The proteins exhibited apparent molecular masses of about 23 kDa and 15 kDa, respectively, under denaturing conditions in SDS-PAGE (Fig. 1, A1 and B1). Owing to the attached hexahistidine tag, the Nsp8 and Nsp9 could be readily detected in immunoblots using an anti-histidine antibody (Fig. 1, A2 and B2). Around 90% of HCoV-229E Nsp9 molecules had their N-terminal His tag removed using SARS-CoV M^{pro} (Lane 5, Fig. 1, B1 and B2).

3.2. RNA Binding Using Zone-Interference Gel Electrophoresis

HCoV-229E Nsp8 and Nsp9 bind to deacetylated tRNA. The zone-interference gel pattern showed that the Nsp8 binding to the negatively charged tRNA increased considerably the electrophoretic mobility of the Nsp8 in a complex with tRNA, which is formed in a concentration-dependent manner (Fig. 2A). *E. coli* elongation factor EF-Tu was used as negative control and dissociated rapidly from deacetylated tRNA, resulting in no net increase of electrophoretic mobility.[9] The K_d value for the Nsp8/tRNA complex was determined as 4 μM (Fig. 2B). HCoV-229E Nsp8 also showed binding to different polyribonucleotides (figure not shown).

HCoV-229E Nsp9 also interacted with tRNA. In this case, the protein is positively charged at pH 8.3 in the electrophoresis buffer. Therefore, it was loaded at the anodic side and the tRNA loaded at the cathodic side. The protein and tRNA migrated in the opposite direction. Nsp9 binding to the negatively charged tRNA decreased the migration of the protein to the cathodic side. Hence, the Nsp9 electrophoretic mobility was increasingly retarded with increasing tRNA concentrations (Fig. 3).

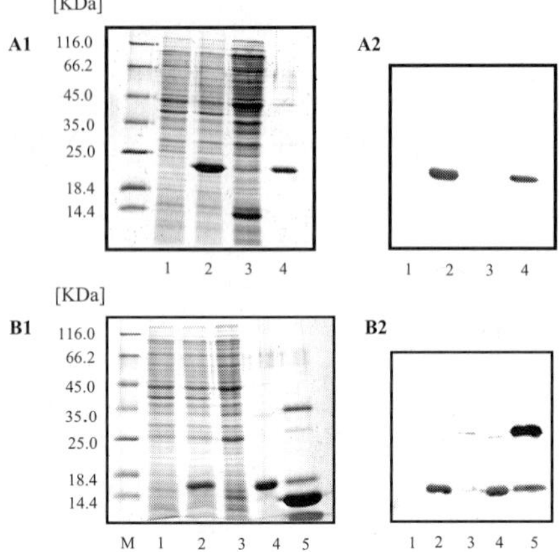

Figure 1. Expression of HCoV-229E nsp8 and nsp9. A1 & B1: SDS-PAGE of Nsp8 and Nsp9, respectively; A2 & B2: Western blot of Nsp8 and Nsp9, respectively; detection using an anti-(His)$_4$-antibody. Lane 1: lysate prior induction, lane 2: lysate after induction, lane 3: flow through of NiNTA-chromatography, lane 4: eluted Nsp8 and Nsp9, respectively, lane 5: His tag-cleaved Nsp9 with SARS-CoV M^{pro} (with His-tag).

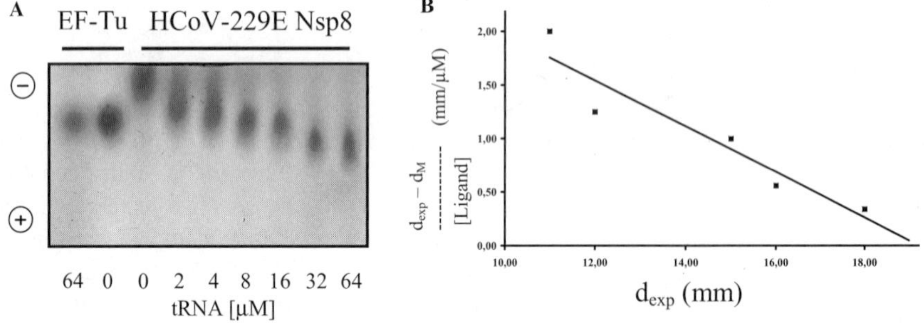

Figure 2. A: Nsp8 zone-interference gel electrophoresis with varying tRNA concentrations. B: K_d determination of the Nsp8/tRNA complex.

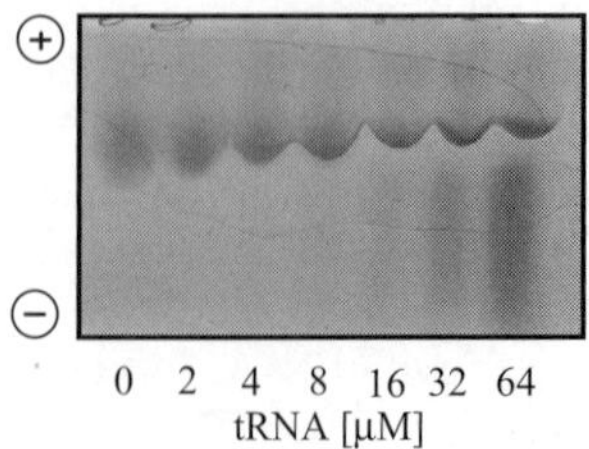

Figure 3. HCoV-229E Nsp9 zone-interference gel electrophoresis with varying tRNA concentrations.

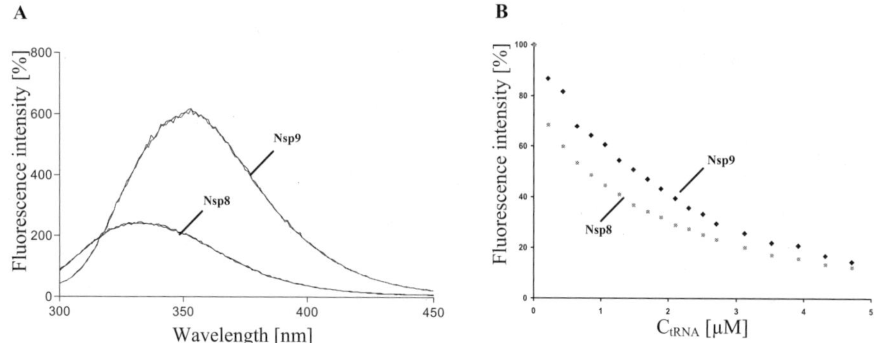

Figure 4. A: Fluorescence emission spectra of HCoV-229E Nsp8 and Nsp9. B: Tryptophan fluorescence decrease of HCoV-229E Nsp8 and Nsp9 in the presence of tRNA.

3.4. Fluorescence Quenching

The HCoV-229E Nsp8 protein displayed an emission maximum around 332 nm due to its tryptophan fluorescence (Fig. 4 A). In comparison, the HCoV-229E Nsp9 emission maximum is red-shifted to 350 nm by excitation at 280 nm. Therefore, the single tryptophan in Nsp9 is more exposed to the hydrophilic environment than the two tryptophans in Nsp8. Remarkably, the tryptophan fluorescence of the two nonstructural proteins was quenched by increasing concentrations of *E. coli* tRNA (Fig. 4 B). At 5 µM tRNA, the intensity of the emitted light was only 20% of the fluorescence in the absence of tRNA. The tryptophan fluorescence quenching clearly demonstrated binding of the two proteins to tRNA in the lower micromolar range. Due to the obvious decrease of the tryptophan fluorescence it might be speculated that these residues in the two proteins are involved in complex formation. This interaction study strongly supports results deduced from the gel shift experiment outlined above.

4. DISCUSSION

Human coronavirus 229E Nsp8 and Nsp9 were overproduced in a highly purified soluble form. The nonstructural proteins exhibited unspecific interaction with tRNA as shown by zone-interference gel electrophoresis and fluorescence spectroscopy. Nsp8 has an affinity in the low micromolar range. It is also tempting to speculate that tRNAs of the host cell might interact with these non structural proteins, which would suggest virus-induced modifications in the host´s translational processes.

5. ACKNOWLEDGMENTS

This project is supported by the Sino-European Project on SARS Diagnostics and Antivirals (SEPSDA, contract no. SP22-CT-2004-003831; www.sepsda.info) and by VIZIER (contract no. LSHG-CT-2004-511960; www.vizier-europe.org), both funded by the European Commission.

6. REFERENCES

1. J. Herold, T. Raabe, B. Schelle-Prinz, and S. G. Siddell, Nucleotide sequence of the human coronavirus 229E RNA polymerase locus, *Virology* **195,** 680-691 (1993).
2. J. Herold and S. G. Siddell, An elaborated pseudoknot is required for high frequency frameshifting during translation of HCV 229E polymerase mRNA, *Nucl. Acids Res.* **21,** 5838-5842 (1993).
3. J. Ziebuhr, V. Thiel, and A. E. Gorbalenya, The autocatalytic release of a putative RNA virus transcription factor from its polyprotein precursor involves two paralogous papain-like proteases that cleave the same peptide bond, *J. Biol. Chem.* **276**, 33220-33232 (2001).
4. J. Ziebuhr, The coronavirus replicase, *Curr. Top. Microbiol. Immunol.* **287**, 57-94 (2005).
5. M. P. Egloff, F. Ferron, V. Campanacci, S. Longhi, C. Rancurel, H. Duartre, E. J. Snijder, A. E. Gorbalenya, C. Cambillau, and B. Canard, The severe acute respiratory syndrome-coronavirus replicative protein nsp9 is a single-stranded RNA-binding subunit unique in the RNA virus world, *Proc. Natl. Acad. Sci. USA* **101**, 3792-3796 (2004).
6. G. Sutton, E. Fry, L. Carter, S. Sainsbury, T. Walter, J. Nettleship, N. Barrow, R. Owens, R. Gilbert, A. Davidson, S. Siddell, L. L. M. Poon, J. Diprose, D. Alserton, M. Walsh, J. M. Grimes, and D. I. Stuart, The nsp9 replicase protein of SARS-coronavirus, structure and functional insights, *Structure* **12**, 341-353 (2004).
7. A. G. Bost, R. H. Carnahan, X. T. Lu, and M. R. Denison, Four proteins processed from the replicase gene polyprotein of mouse hepatitis virus colocalize in the cell periphery and adjacent to sites of virion assembly, *J. Virol.* **74**, 3379-3387 (2000).
8. G. Lindwall, M.-F. Chau, S. R. Gardner, and L. A. Kohlstadt, A sparse matrix approach to the solubilization of overexpressed proteins, *Protein Engineering* **13**, 67-71 (2000).
9. J. P. Abrahams, B. Kraal, and L. Bosch, Zone-interference gel electrophoresis: a new method for studying weak protein-nucleic acid complexes under native equilibrium conditions, *Nuclic. Acids Res.* **16**, 10099-10108 (1988).

EFFECTS OF MUTAGENESIS OF MURINE HEPATITIS VIRUS NSP1 AND NSP14 ON REPLICATION IN CULTURE

Lance D. Eckerle, Sarah M. Brockway, Steven M. Sperry, Xiaotao Lu, and Mark R. Denison*

1. INTRODUCTION

The 32-kb positive-strand RNA genome of murine hepatitis virus (MHV) contains a replicase gene (gene 1) that comprises two-thirds (22kb) of the genome, is the 5' most gene, and is translated from two overlapping open reading frames (ORFs 1a and 1b) and processed to yield intermediate and mature nonstructural proteins (nsps). For the group 2 coronaviruses such as MHV, as well as for SARS-CoV, at least 16 mature nsps are processed co- and post-translationally from the gene 1 polyprotein by two or three proteinases expressed as part of the polyprotein.[1,2] The intermediate and mature nsps are thought to be essential for replication. These include demonstrated or predicted functions such as RNA polymerase and RNA helicase (nsp12 and nsp13), as well as recently predicted functions in RNA synthesis, modification, or processing such as ADP ribosylation, exonuclease, endoribonuclease, and RNA methyltransferase (nsp3, nsp14, nsp15, and nsp16, respectively). Finally, there are several nsps with no demonstrated or predicted functions, such as nsp1 and nsp2. To test the requirements for nsp1 and nsp14 in replication and to probe their functions, deletions or mutations were engineered into the viral genome in nsp1 and nsp14 and mutant viruses were analyzed for virus viability, replication, protein expression, and RNA synthesis. The results demonstrate that deletions and substitutions in nsp1 are tolerated in viable mutants, including deletion of the carboxy-terminal half of nsp1. Nsp14 appears to be essential for replication in culture but can tolerate substitution of tyrosine 414, as well as deletion of flanking cleavage sites. Together, these results show the ability to generate mutations in each of these proteins and recover viable mutants.

2. MATERIALS AND METHODS

2.1. Generation of Recombinant Viruses

Mutations and deletions of nsp1 and nsp14 were introduced into MHV as previously described.[3,4] Briefly, mutations were engineered into cloned MHV genome fragments at

*Vanderbilt University Medical Center, Nashville, Tennessee 37232-2581.

locations shown in Table 1, and the cloned and mutated cDNA fragments were digested, assembled into full-length genome cDNA *in vitro*, and transcribed into full-length genome RNA that was electroporated into replication-permissive BHK cells expressing the MHV receptor (BHK-R cells). Electroporated cells were monitored for cytopathic effect (CPE, syncytia) beginning at 24 h p.i., and both supernatant media and cells were passed onto new monolayers of BHK-R or murine delayed brain tumor (DBT) cells.

2.2. Virus Infection and Growth Experiments

Virus stocks were obtained from infected cell supernatants and titer determined on DBT cells. For growth experiments, DBT cells were infected with wild-type virus and with nsp1 and nsp14 mutants at an MOI of 5 pfu/cell (high MOI-single cycle) or 0.01 pfu/cell (low MOI-multiple cycle). Media supernatant was obtained at intervals from 0

Table 1. Mutations and deletions in MHV nsp1 and nsp14.

Nsp1 Mutations	**Amino Acid**	**Supernatant Virus**
VUSB 1	K4A, K6A	Yes
VUSB 3	E26A, K27A	Yes
VUSB 4	E46A, K48A	Yes
VUSB 5	H57A	Yes
VUSB 6	R64A, E69A	No
VUSB 7	R78A, E69A	No
VUSB 8	K88A, E90A	Yes
VUSB 9	K132A, R133A	Yes
VUSB 13	R207A, R208A	No
VUSB 15	E220A, D221A	No
VUSB 17	K231A, R233A	Yes
Δ124-242	Δ124 - 242	Yes
Nsp14 Mutations	**Amino Acid**	**Supernatant Virus**
VUSS 6	nsp13, Δ Q600	Yes
VUSS 17	nsp14, Δ Q521	Yes
VUSS 8	D90A, E92A	Yes
VUSS 11	C205A, C208A	No
VUSS 9	D243A	No
VUSS 10	D272A	No
VUSS 3	Y414 (wt)	Yes
VUSS 13	Y414S	Yes
VUSS 14	Y414A	Yes
VUSS 16	S412Y, Y414H	Yes
VUSS 20	Y414T	Yes
VUSS 21	Y414K	No
VUSS 22	Δ Y414	No

to 48 h p.i. and used for plaque assay on DBT cells. Extent of CPE was assessed by cell loss and percent of monolayer involved in virus-induced syncytia formation.

3. RESULTS AND DISCUSSION

3.1. Deletions in nsp1 Demonstrate Requirement for Amino-Terminal Two-Thirds of nsp1 for Replication and Carboxy-Terminal Third for Processing of nsp1

Deletions were introduced into the nsp1 coding sequence at the amino terminal, middle and carboxy-terminal third of the protein domain. In addition, GFP coding sequence was substituted for the nsp1 coding sequence in cloned cDNA. When infectious genome RNA containing these changes was electroporated into permissive cells, only the carboxy-terminal nsp1 deletion allowed recovery of an infectious mutant virus (nsp1 Δ 124-242). The recovered virus had slightly impaired peak titers and viral RNA synthesis compared with parental wild-type virus, but otherwise was indistinguishable in growth kinetics and RNA species generated.[5] The deletion was engineered to retain the proximal (P5-P1) residues of the nsp1-nsp2 cleavage site, and the mutant virus had similar timing and extent of cleavage at the cleavage site between nsp1 and nsp2 as wild type. These results suggested that RNA or protein determinants in the amino-terminal two-thirds of nsp1 are essential for replication in culture, and that the amino-terminal protein determinants are important for the timing or extent of cleavage at the nsp1-nsp2 cleavage site.

3.2. Mutations in nsp1 Demonstrate Requirements for Specific Residues in Virus Viability in Culture

Based on the above results, systematic mutagenesis of clustered-charged residues was performed, both within the putative essential amino-terminal two-thirds of nsp1 as

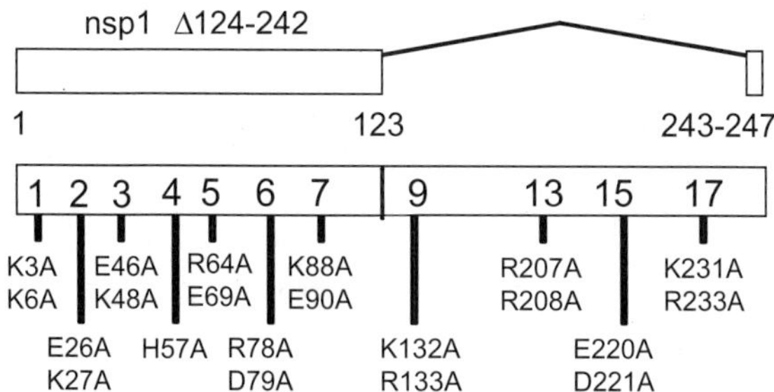

Figure. 1. Nsp1 mutations and deletions. The organization of nsp1 is shown. Top schematic shows the deletion of aa 124 to 242. The bottom schematic shows clustered-charge to alanine mutations. Mutants from Table 1 are shown by VUSB number inside the box.

well as the dispensable carboxy-terminal third of the nsp1 protein domain. Residues were prioritized for substitution based on clustering of charge and across the protein sequence. Most alanine substitutions were tolerated for virus viability and replication in culture (Table 1 and Fig. 1). These included multiple substitutions in the amino-terminal half of nsp1. The exceptions were alanine substitutions VUSB5 (R64A, E69A) and VUSB6 (R78A, D79A), which did not allow recovery of infectious virus from supernatants of infected cells. Interestingly, we identified substitutions in the carboxy-terminal half of nsp1 that also were lethal for recovery of infectious mutant viruses, VUSB13 (R207A, R208A) and VUSB15 (E220A, D221A). This was surprising in light of the ability to delete this portion of the protein domain in viable mutants. Together, these results demonstrate significant flexibility in the amino acid sequence and lack of critical functions for a majority of charged residues. However, the results also suggest that specific residues may be critical for virus survival in culture, possibly due to protein interactions, folding or function. Finally, the results suggest that within apparently dispensable protein domains there may exist protein structure or function determinants that in the context of the intact protein may dramatically impact virus replication.

3.3. Amino Acid Substitutions or Deletions in nsp14 Putative Catalytic Residues or Zn Finger are Highly Deleterious or Lethal for Replication in Culture

The nsp14 protein is conserved in location, size, and significantly in amino acid sequence among all groups of coronaviruses, and it is predicted to be an exoribonuclease of the DE-D-D superfamily of exonucleases.[2] We have previously demonstrated that substitution of nsp14 Tyr414 by His (Y414H) does not affect virus viability or replication in culture but abolishes virulence in mice.[4] Significantly, the Tyr414 residue is 100% conserved in all sequenced coronaviruses. We sought to determine if nsp14 is required for virus viability in culture and if it affects virus growth. We performed systematic mutagenesis of the cloned cDNA genome Fragment F containing the nsp14 coding sequence and used the mutated clones to assemble genome cDNA. We targeted the flanking cleavage sites, predicted catalytic residues and the putative Zn finger motif within nsp14 with substitutions. In addition, multiple substitutions at Tyr414 were introduced to determine if this residue was tolerant of all changes and if different substitutions had distinct effects on replication.

When alanine was substituted for putative catalytic Asp242 or Asp272 residues, CPE was not detected in electroporated cells, nor was infectious virus recovered. Interestingly, when the Asp89 and Glu91 residues were both substituted with Ala, limited CPE was observed in electroporated cells, but this was not sustained on passage of the entire electroporated cell monolayer nor on overlay of fresh DBT cells. Infectious virus

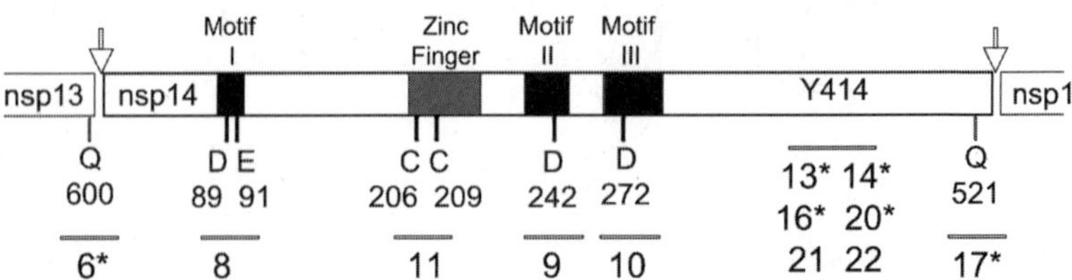

Figure 2. Nsp14 motifs and mutations. The organization of nsp14 is shown, with putative exonuclease motifs I, II, and III as well as a possible Zn finger motif. Amino- and carboxy-terminal cleavage sites are indicated by arrows. Below schematic are residues and positions substituted as described in Table 1 and text: Q – Gln, D – Asp, E – Glu, C – Cys. Mutants from Table 1 are shown by VUSS number below the bars. * indicates viable mutants.

was recovered from initial electroporated cells and had a profoundly reduced titer of 70 pfu per ml. Although the virus could form plaques, it could not be passaged, suggesting a highly impaired replication machinery. Finally, no substitutions of Ala for Cys residues of the predicted Zn finger structure within nsp14 (Cys206, Cys209) were tolerated for productive virus infection in culture. Together, all of these results support the conclusion that nsp14 performs functions essential in the generation of infectious virus either within the cell or in viral RNA synthesis, and that the Asp242, Asp272, Cys206, and Cys209 residues are indispensable for virus replication. The results also suggest that the Asp89 and Glu91 residues may have some structural or functional flexibility, but that changes are highly deleterious and cannot support survival over multiple rounds of replication. Future experiments will involve Ala substitutions at residues 89 and 91 alone or exchange of Asp89 with Glu and of Glu91 with Asp to retain charge.

3.4. Nsp14 Tolerates Multiple Substitutions at Tyr414, but Not Lysine Substitution or Deletion

Because the substitution at Tyr414 by His had no effect on replication in culture but abolished virulence in mice,[4] we engineered different substitutions or deletion at the Tyr414 residue to determine if it had any requirement in replication. Engineered substitutions for Tyr414 by Ser, Ala, or Thr each resulted in productive infection and infectious virus with growth indistinguishable from wild-type recombinant virus or virus with a Tyr 414 His substitution, as did a double substitution of Tyr414His and Ser412Tyr. Surprisingly, a Tyr414Lys substitution did not allow recovery of infectious virus from electroporated cells. Deletion of Tyr414 also abolished production of infectious virus in culture. These results demonstrate that while Tyr414 may have great flexibility, it is intolerant of deletion or specific substitutions, suggesting that it may be important for specific viral functions or possibly protein interactions.

3.5. Amino Acid Deletions at Cleavage Sites Flanking nsp14 Allow Productive, but Impaired Virus Growth in Culture

Having shown that nsp14 has essential and dispensable residues and that the protein may serve roles in both replication and pathogenesis, we next sought to determine if intact processing of the protein was required for virus viability or normal replication. Deletions of P1- Gln residues in the flanking cleavage sites between nsp13-14 (VUSS6) and nsp14-15 (VUSS17) were engineered in the infectious clone cDNA and used to generate full-length genome RNA for electroporation into permissive cells. In doing so, the nsp13-14 cleavage site deletion resulted in loss of the carboxy-terminal Gln600 residue in nsp13, while the nsp14-15 cleavage site deletion resulted in loss of the carboxy-terminal Gln521 residue in nsp14. Both cleavage site deletions resulted in syncytia in electroporated cells and recovery of infectious virus from the supernatant media. While both cleavage site deletion mutants were impaired in extent of CPE and virus growth, they differed in the degree of replication impairment. Specifically, the VUSS17 mutant had a slight delay in growth but attained wild-type peak titers, while the VUSS6 mutant showed delays of more than 4 h in peak titer during high MOI infection (5 pfu/cell) and never attained wild-type peak titers. The result suggests that incomplete processing of nsp14 from the flanking proteins alters the replication efficiency of the viruses. However, it is also possible that the observed effects result from the deletion of the carboxy-terminal glutamine residue of nsp13 (VUSS6) or nsp14 (VUSS17).

4. SUMMARY

For nsp1, the fact that the carboxy-terminal but not the amino-terminal half of the protein can be deleted suggests that there may be specific and distinct domains within the protein or that the entire protein is dispensable but that the RNA encoding the amino-terminal half of nsp1 cannot be deleted. The identification of specific required residues support the conclusion that it is the portion of the protein that is required for replication.

The results of mutagenesis of the nsp14 coding region and flanking cleavage sites also provided important new insights into this protein and its requirements. Our previous study raised the question as to the essential nature of nsp14 in replication. The results of this study show that putative active site residues cannot be substituted without loss of replication in culture. Interestingly, mutagenesis of Tyr414 showed that while this residue can tolerate a number of substitutions, it was intolerant of Lysine or deletion. The results suggest that nsp14 is required for replication. However, whatever functions nsp14 serves appear to be retained by noncleaved or partially processed nsp14, since abolition of either the amino-terminal or carboxy-terminal cleavage site allowed recovery of viable virus.

5. ACKNOWLEDGMENTS

This work was supported by National Institutes of Health grants RO1 AI26603 (M.R.D.) and the Cellular and Molecular Microbiology Training Grant T32 AI049824 (L.D.E). Support was provided by Public Health Service Grant CA68485 for the Vanderbilt DNA Sequencing Shared Resource.

6. REFERENCES

1. Schiller, J. J., A. Kanjanahaluethai, and S. C. Baker, 1998, Processing of the coronavirus MHV-JHM polymerase polyprotein: identification of precursors and proteolytic products spanning 400 kilodaltons of ORF1a. *Virology* **242**:288-302.
2. Snijder, E. J., P. J. Bredenbeek, J. C. Dobbe, V. Thiel, J. Ziebuhr, L. L. Poon, Y. Guan, M. Rozanov, W. J. Spaan, and A. E. Gorbalenya, 2003, Unique and conserved features of genome and proteome of SARS-coronavirus, an early split-off from the coronavirus group 2 lineage. *J Mol Biol* **331**:991-1004.
3. Yount, B., M. R. Denison, S. R. Weiss, and R. S. Baric, 2002, Systematic assembly of a full-length infectious cDNA of mouse hepatitis virus strain A59. *J Virol* **76**:11065-11078.
4. Sperry, S. M., L. Kazi, R. L. Graham, R. S. Baric, S. R. Weiss, and M. R. Denison, 2005, Single-amino-acid substitutions in open reading frame (ORF) 1b-nsp14 and ORF 2a proteins of the coronavirus mouse hepatitis virus are attenuating in mice. *J Virol* **79**:3391-3400.
5. Brockway, S. M., Lu, X. T., Peters, T. R., Dermody, T. S., and Denison, M. R. 2004. Intracellular localization and protein interactions of the gene 1 protein p28 during mouse hepatitis virus replication. *J Virol* **78**:11551-11562.
6. Denison, M.R., Spaan, J.M., van der Meer, Y., Gibson, C.A., Sims, A.C., Prentice, E., and Lu, X.T, 1999, The putative helicase of the coronavirus mouse hepatitis virus is processed from the replicase gene polyprotein and localizes in complexes that are active in viral RNA synthesis. *J Virol* **73**:6862-6871.
7. Ziebuhr, J., E. J. Snijder, and A. E. Gorbalenya, 2000, Virus-encoded proteinases and proteolytic processing in the Nidovirales. *J Gen Virol* **81**:853-879.

MUTATIONAL ANALYSIS OF MHV-A59 REPLICASE PROTEIN-NSP10

Eric F. Donaldson, Amy C. Sims, Damon J. Deming, and Ralph S. Baric*

1. INTRODUCTION

Mouse hepatitis virus (MHV) replication is initiated by the translation of the large 22-kb replicase gene to yield two large polyproteins from two overlapping open reading frames (ORF1a and ORF1b). Approximately 70% of the time ORF1a is translated to produce a 495-kDa polyprotein. About 30% of the time, ribosomal slippage results in an ORF1ab 803-kDa polyprotein via a ribosomal frameshift. These polyproteins are autocleaved both co- and post-translationally by two papain-like proteinases (PLP1 and PLP2) and a main proteinase (Mpro) to generate a series of intermediates and precursors that are completely processed into ~16 nonstructural proteins (nsp). These proteins include the three viral proteinases, an RNA-dependent RNA polymerase (RdRp), an RNA helicase, four putative RNA-processing enzymes, and approximately seven proteins of unknown functions.[1,2]

Confocal immunofluorescence studies have demonstrated that many of these replicase proteins co-localize with the viral RdRp to intracellular double-membrane vesicles, where together they form the viral replication complex.[3] One of these proteins, nsp10, a 15-kDa product (139 amino acids) encoded at the 3' end of ORF1a, has been shown by yeast two-hybrid and coimmunoprecipitation analyses to interact with itself, nsp7 (p10), and nsp1 (p28)[4]; by IFA to co-localize to the same region as the replication complexes[5]; and by gene knockout to be required for replication.

Our initial approach was to delete nsp10, followed by altering the nsp9/10 cleavage signal. To further investigate the role of nsp10 in viral replication, we employed a reverse genetics approach to introduce 17 mutations (alanine substitutions) into key residue positions predicted to knockout specific protein domains or charged amino acid pairs or triplets. Here we report on the preliminary characterization of 19 nsp10 mutants.

---

*University of North Carolina, Chapel Hill, North Carolina 27599.

2. METHODS

A full-length molecular clone developed by our laboratory[6] was used to engineer the appropriate mutations in the nsp10 sequence. Briefly, primers were designed that incorporated type IIS restriction enzyme sites that introduced at least 2 changes within the codon being targeted. The mutants were then cloned into TopoXL vector and sequenced to verify that the correct changes were incorporated. The full-length infectious clone was assembled as previously described,[6] incorporating each mutant fragment. Full-length cDNA constructs were transcribed and transfected into 10^6 baby hamster kidney (BHK-MHVr) cells expressing the MHV receptor. Transfected BHKs were then poured onto delayed brain tumor cells (DBTs) and cultures were incubated at 37°C for 24–72 hours. Flasks were examined at regular intervals for cytopathic effect (CPE), and viable mutants were verified by reverse transcriptase PCR (RT-PCR) of subgenomic RNA using primers targeting the leader sequence and the 5' end of the N-glycoprotein gene. Plaque purified viruses were sequenced to confirm that the correct mutations were present in the recombinant virus.

Viral titers were determined by plaque assay at a multiplicity of infection (MOI) of 0.2 using DBT cells with time points of 2, 4, 8, 12, and 16 hours. The cells from the 8-hour time point were harvested in Trizol reagent and total RNA isolated.

Quantitative real time RT-PCR was conducted using SYBR green to detect subgenomic copy number in RNA harvested from cells infected at MOI 0.2 with primers optimized to detect ~120 nucleotides of mRNA-7 (nucleocapsid gene).

3. RESULTS

To determine if nsp10 is required for viral replication, we engineered a mutant that deleted nsp10 from ORF1a, while preserving the ribosomal frameshift. This deletion resulted in a lethal phenotype (data not shown), suggesting that nsp10 is essential for *in vitro* growth or that nsp10 deletion altered efficient processing of the ORF1a/b polyprotein to produce a lethal phenotype.

To determine if cleavage of nsp10 is required for replication, we ablated the cleavage junction between nsp9 and nsp10 by replacing a tyrosine essential for cleavage at position 1 of the cleavage motif (TVRLQ | AGTAT) with an alanine residue (TVRL<u>A</u> | AGTAT). Growth curve analysis showed that this mutant initially demonstrated a 1 to 1.5 $\log_{10}$ reduction in replication, which increased to wild-type kinetics with passage. However, Western blot analysis using anti-sera directed against nsp9 and nsp10 confirmed that cleavage of nsp9-nsp10 was ablated, resulting in a single ~30-kDa product instead of the 12-kDa and 15-kDa products found in wild type infections. The cleavage was not rescued by passage, and no compensatory mutations were discovered in the 3000 nucleotides that comprise the 3' end of ORF1a. Further, immunofluorescence confirmed that the 30-kDa product co-localized to sites of viral replication (data not shown).

To further investigate the role of nsp10 in viral replication, we employed a reverse genetics approach whereby we introduced alanine substitutions into key residues found in protein domains or charged amino acid pairs and triplets, and characterized the resulting mutants.

Bioinformatic predictions, using Prosite[7] to predict conserved protein domains and predictors of natural disordered regions (PONDR)[8] to determine putative disordered

Table 1. Domains predicted by Prosite and PONDR that fall within a highly active region of nsp10, which was targeted by site-directed mutagenesis.

Position	Amino acids	Domain
63-66	TnqD	CKII phosphorylation[a]
56-58	TiK	PKC phosphorylation[b]
50-55	GTgmAI	Myristoylation[c]
52-57	GMaiTI	Myristoylation[c]
70-75	GAsvCI	Myristoylation[c]
55-66	ITIKPEATTNQD	Disorder[d]

[a] Casein kinase II phosphorylation site. [b] Protein kinase C phosphorylation site. [c] N-Myristoylation site.
[d] Natural disordered region.

domains, identified a highly active region from amino acids 50 to 75 that contains multiple domains and a disordered region (Table 1). Although the disordered region was not conserved at the amino acid level, the same amino acids were predicted by PONDR to be disordered in all coronaviruses, suggesting a functional role for the region. This site was targeted for mutagenesis as a potential protein:protein or protein:nucleic acid interaction site.

Nine mutations predicted to knockout specific protein domains and the disordered region were identified and introduced into the MHV infectious clone. In addition, eight scanning alanine mutations targeting charged amino acid pairs or triplets were introduced into the clone. Of the 17 mutations made by site directed mutagenesis (Figure 1), six were viable, three were debilitated, and eight were lethal (Figure 1). The viable mutants all showed CPE, the debilitated mutants showed no sign of CPE, but low levels of subgenomic RNAs were detectable via real-time PCR, and the lethal mutants had no CPE and no detectable subgenomic RNA.

Next, we predicted the structure of nsp10 using a program called Rosetta,[9] and mapped the mutations onto the putative structure (data not shown). Interestingly, all of the viable mutants mapped to the N-terminal half of the predicted structure, suggesting that the C-terminal portion of nsp10 encodes critical residues for viability (Figure 1).

Further, most of the mutations in the C-terminal half of nsp10 occurred in predicted turns, suggesting that conservation of the structure is essential for viral replication. In contrast, both mutations predicted to interfere with turns in the N-terminal half of nsp10 resulted in debilitated phenotypes.

Next we analyzed growth kinetics and subgenomic RNA synthesis of the viable mutants via plaque assay (Table 2) and quantitative RT-PCR (Figure 2). U1 grew to similar titers as wild type at all time points, while E1, E2, and E3 grew 1–2 logs lower at all time points. U3 and U4 showed an approximate 2-log reduction at all time points (Table 2).

As demonstrated by quantitative RT-PCR, subgenomic RNA synthesis was greatly reduced in E1. E2 and E3 generated more subgenomic RNAs than E1; and U1, U3, and U4 generated wt levels of subgenomic RNAs at the 8-hour time point (Figure 2).

This trend was verified by Northern blot analysis, which showed subgenomic copy numbers to range from lowest to highest: E1, E2, and E3, U1 and U3, U4, and then wild type (data not shown).

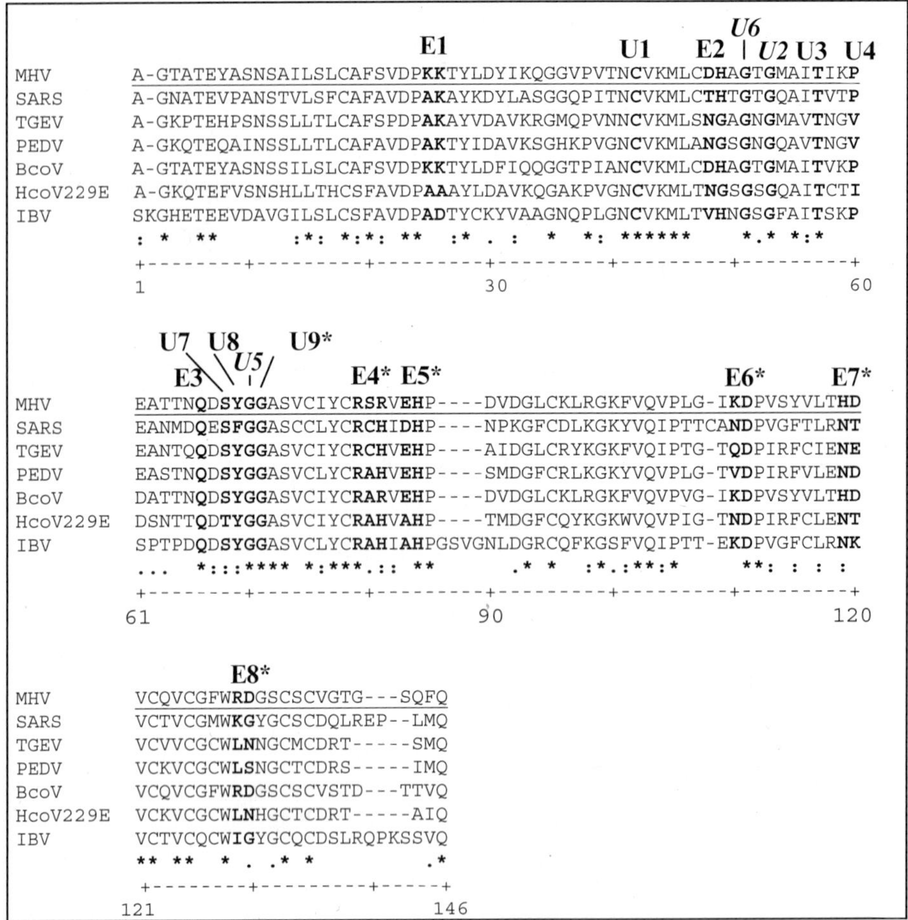

Figure 1. Alignment and position of lethal versus viable mutants of nsp10. A multiple alignment of representative coronavirus nsp10 amino acid sequences was generated using ClustalX1.83 with default parameters. Seventeen mutations are noted in bold and labeled by type and phenotype. U-protein domain mutations, and E-charged amino acid pairs or triplets. *Italics*=debilitated, *=lethal, *****=identical, **:**=conserved, **.**=similar. SARS-CoV, NP_828868.1; TGEV, NP_840008.1; PEDV, NP_839964.1; BCoV, NP_742137.1; HCoV-229E, NP_835351.1; MHV, NP_740615.1; IBV, NP_740628.1.

4. DISCUSSION

The role of the four small c-terminal ORF1a replicase proteins in coronavirus replication and transcription is unknown, and we have taken a genetic approach to study the putative role of nsp10 in these processes. Previous immunofluorescence studies have demonstrated that nsp10 likely plays a role in viral RNA synthesis, as it appears to co-localize to sites of the viral replication complex.[5] A ts mutant LA3, defective in RNA synthesis, contains a mutation in nsp10, suggesting a critical role in RNA synthesis.[10]

Table 2. Titers of the viable mutants at 8, 12, and 16 hours postinfection.

Virus	8 hr	12 hr	16 hr
WT	3.90E+06	5.45E+07	2.85E+08
E1	3.25E+03	2.50E+06	2.80E+07
E2	5.25E+03	5.60E+06	4.00E+07
E3	3.75E+03	6.75E+06	5.10E+07
U1	1.50E+05	2.85E+07	8.00E+07
U3	9.00E+03	2.93E+05	2.10E+06
U4	3.75E+03	5.53E+05	5.00E+06

We have shown that deleting nsp10 results in a lethal phenotype, which suggests that nsp10 is essential for viral replication, although it remains possible that the nsp10 deletion results in conformationally altered precurser polyproteins that are resistant to proteolytic processing. This seems less likely as the C-terminus of nsp10 was highly intolerant of mutagenesis, suggesting a direct and critical role in RNA synthesis.

Ablating the cleavage signal between nsp9/10 resulted in a viable but attenuated replication phenotype. Upon passage, virus revertants emerged with wild-type replication phenotypes, but interestingly compensating mutations did not rescue cleavage and did not occur in the nsp7-10 region of ORF1a. We are currently investigating the adaptive changes that occurred in this mutant.

Bioinformatic analysis has shown that nsp10 is highly conserved among the coronavirus genus (51–60% identical), second only to the RdRp (61–70% identity). This suggests that the two proteins have co-evolved, and provides further support for a critical role for nsp10 in coronavirus replication (Figure 1).

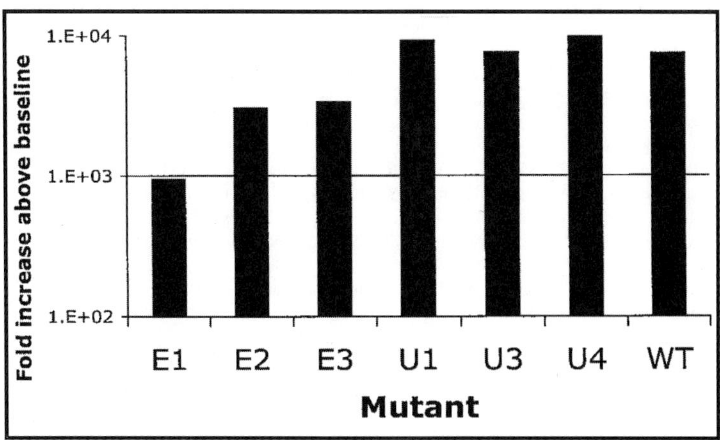

Figure 2. Copy number of subgenomic mRNA-7 as determined by quantitative real-time PCR of the viable nsp10 mutants at 8 hours postinfection and M.O.I of 0.2.

Mapping the 17 mutations onto the putative structure of nsp10 revealed three important predictions. First, the highly active site (amino acids 50-75) appears to extend outward as a loop from the structure, allowing it to potentially act as a contact site to form dimers, as has been demonstrated to occur in nsp10 of infectious bronchitis virus *in vitro*,[11] or interact with other replicase proteins or viral RNA. Second, mutations that occur in predicted turns are more frequently lethal than those that occur in loops and α-helices. And most importantly, mutations that occur in the C-terminal portion of nsp10 are all lethal, suggesting a highly conserved structure required for viral replication.

This observation is consistent with the fact that the cleavage mutant is viable, as hypothetically the C-terminal portion of nsp10 could still fold into its native structure even if cleavage did not occur, and therefore function is not completely ablated.

Analysis of the growth kinetics and subgenomic RNA synthesis of the viable mutants demonstrated that nsp10 has a global effect on RNA synthesis, with 5-of-6 mutants showing reduced growth and some showing reduced subgenomic RNA synthesis at each time point. In fact, preliminary results using quantitative RT-PCR to determine copy number at each time point suggests that the viable nsp10 mutants are particularly defective in genomic RNA synthesis (data not shown). We are currently confirming this observation.

5. REFERENCES

1. M. M. Lai and D. Cavanagh, The molecular biology of coronaviruses, *Adv. Virus Res.* **48**, 1-100 (1997).
2. E. J. Snijder, P. J. Bredenbeek, J. C. Dobbe, V. Thiel, J. Ziebuhr, L. L. Poon, Y. Guan, M. Rozanov, W. J. Spaan, and A. E. Gorbalenya, Unique and conserved features of genome and proteome of SARS-coronavirus, an early split off from the coronavirus group 2 lineage, *J. Mol. Biol.* **331**, 991-1004 (2003).
3. S. T. Shi, J. J. Schiller, A. Kanjanahaluethai, S. C. Baker, J. W. Oh, and M. M. Lai, Colocalization and membrane association of murine hepatitis virus gene 1 products and De novo-synthesized viral RNA in infected cells, *J. Virol.* **73**, 5957-5969 (1999).
4. S. M. Brockway, X. T. Lu, T. R. Peters, T. S. Dermody, and M. R. Denison, Intracellular localization and protein interactions of the gene 1 protein p28 during mouse hepatitis virus replication, *J. Virol.* **78**, 11551-11562 (2004).
5. A. G. Bost, R. H. Carnahan, X. T. Lu, and M. R. Denison, Four proteins processed from the replicase gene polyprotein of mouse hepatitis virus colocalize in the cell periphery and adjacent to sites of virion assembly, *J. Virol.* **74**, 3379-3387 (2000).
6. B. Yount, M. R. Denison, S. R. Weiss, and R. S. Baric, Systematic assembly of a full-length infectious cDNA of mouse hepatitis virus strain A59, *J. Virol.* **76**, 11065-11078 (2002).
7. C. J. A. Sigrist, L. Cerutti, N. Hulo, A. Gattiker, L. Falquet, M. Pagni, A. Bairoch, and P. Bucher, PROSITE: a documented database using patterns and profiles as motif descriptors, *Brief Bioinform.* **3**, 265-274 (2002).
8. P. Romero, Z. Obradovic, C. R. Kissinger, J. E. Villafranca, and A. K. Dunker, Identifying disordered regions in proteins from amino acid sequences, *Proc. I.E.E.E. International Conference on Neural Networks*, pp. 90-95 (1997).
9. C. Bystroff and Y. Shao, Fully automated ab initio protein structure prediction using I-SITES, HMMSTR and ROSETTA, *Bioinformatics*, **18**, S54-S61 (2002).
10. D. R. Younker and S. G. Sawicki, Negative strand RNA synthesis by temperature-sensitive mutants of mouse hepatitis virus, *Adv. Exp. Med. Biol.* **440**, 221-226 (1998).
11. L. F. Ng and D. X. Liu, Membrane association and dimerization of a cysteine-rich, 16-kilodalton polypeptide released from the C-terminal region of the coronavirus infectious bronchitis virus 1a polyprotein, *J. Virol.* **76**, 6257-6267 (2002).

THE NSP2 PROTEINS OF MOUSE HEPATITIS VIRUS AND SARS CORONAVIRUS ARE DISPENSABLE FOR VIRAL REPLICATION

Rachel L. Graham, Amy C. Sims, Ralph S. Baric, and Mark R. Denison*

1. INTRODUCTION

The positive-strand RNA genome of the coronaviruses is translated from ORF1 to yield polyproteins that are proteolytically processed into intermediate and mature nonstructural proteins (nsps). Murine hepatitis virus (MHV) and severe acute respiratory syndrome coronavirus (SARS-CoV) polyproteins incorporate 16 protein domains (nsps), with nsp1 and nsp2 being the most variable among the coronaviruses and having no experimentally confirmed or predicted functions in replication. To determine if nsp2 is essential for viral replication, MHV and SARS-CoV genome RNA was generated with deletions of the nsp2 coding sequence (MHVΔnsp2 and SARSΔnsp2). Infectious MHVΔnsp2 and SARSΔnsp2 viruses were recovered from electroporated cells. The Δnsp2 mutant viruses lacked expression of both nsp2 and an nsp2-nsp3 precursor, but cleaved the engineered chimeric nsp1/3 cleavage site as efficiently as the native nsp1-nsp2 cleavage site. Replication complexes in MHVΔnsp2-infected cells lacked nsp2 but were morphologically indistinguishable from those of wild-type MHV by immunofluorescence. These results demonstrate that while nsp2 of MHV and SARS-CoV is dispensable for viral replication in cell culture, deletion of the nsp2 coding sequence attenuates viral growth and RNA synthesis. These findings also provide a system for the study of determinants of nsp targeting and function.

2. MATERIALS AND METHODS

2.1. Generation of Recombinant Viruses

Deletions of the nsp2 coding sequence were engineered into the infectious cDNAs of MHV and SARS-CoV by deleting nt 951-2705 (aa Val248-832Ala) for MHV and nt

* Vanderbilt University, Nashville, Tennessee 37232.

805-2718 (aa Ala181-818Gly) for SARS-CoV. Mutant viruses were then generated as previously described.[1,2]

2.2. Virus Infection, Immunoprecipitation, and Immunoblot

DBT cells were infected with MHV at an MOI of 5 pfu/cell. At 4.5 h p.i., cells were incubated in medium lacking methionine and cysteine and supplemented with Actinomycin D. [^{35}S]-methionine/cysteine was added at 6 h p.i., and cells were then harvested at 10 h p.i. Lysates were then immunoprecipitated using antibodies against nsp1, nsp2, nsp3, and nsp8. Proteins were then resolved and visualized by SDS-PAGE and fluorography.

Vero-E6 cells were infected with SARS-CoV at an MOI of 1 PFU/cell. At 12 h p.i., cells were harvested and lysates resolved by SDS-PAGE. Separated proteins were transferred to nitrocellulose, and nsp1, nsp2, nsp3, and nsp8 antibodies were used to detect viral proteins as described.[3]

2.3. Immunofluorescence

DBT cells were grown on glass coverslips as previously described.[4] Cells were then infected for 6.5 h, fixed and permeablized in -20°C methanol, and processed for immunofluorescence as described.[5]

2.4. Confocal Microscopy

All cell images were acquired on a Zeiss 510 LSM laser scanning confocal microscope. Images were acquired using lasers at 488 nm (green) and 543 nm (red) and using a 40X, 1.3 NA oil immersion objective. Images were processed using Adobe Photoshop CS.

3. RESULTS AND DISCUSSION

3.1. Deletion of Nsp2 Demonstrates Polyprotein Tolerance to Change

The deletion of the nsp2 coding sequence from both MHV and SARS-CoV yielded infectious virus. Retention of the engineered deletion was confirmed by sequencing of viral RNA from progeny viruses as well as by detection of viral proteins from infected cells by immunoprecipitation (MHV, Fig. 1) and immunoblot (SARS-CoV, data not shown). No additional mutations were noted in the regions sequenced, consisting of bases ~500 nt 5' and 3' of the deletion for MHV and ~250 nt 5' and 3' of the deletion for SARS-CoV.

Deletion of the nsp2 coding region of these two viruses is, to our knowledge, the first demonstration of the deletion of the coding region of an entire protein domain from a RNA virus polyprotein. This deletion illustrates a previously unrealized flexibility within the coronavirus polyprotein.

3.2. Deletion of Nsp2 Does Not Interfere with Polyprotein Processing

To determine if the deletion of the nsp2 coding region altered processing within the replicase polyprotein, MHV-infected cells were radiolabeled with [^{35}S]-methionine/cysteine, lysed, and viral proteins were detected by immunoprecipitation (Fig. 1). While MHVΔnsp2 mutant virus produced no detectable nsp2 protein, all other tested proteins (nsp1, nsp3, and nsp8) were processed comparably to wild type, demonstrating that deletion of nsp2 from the viral polyprotein did not inhibit polyprotein translation or processing. Notably, infection with the MHVΔnsp2 virus did not produce the high molecular weight (~275 kDa) protein that has been tentatively identified in previous studies as the nsp2-3 precursor protein.[6,7] Similarly, lysates generated from SARS-CoV infected cells were probed for nsp1, nsp2, nsp3, and nsp8 proteins by immunoblot (data not shown). As with MHV, the SARS-CoVΔnsp2 virus did not produce detectable nsp2 protein, though nsp1, nsp3, and nsp8 were all expressed and processed comparably to wild type.

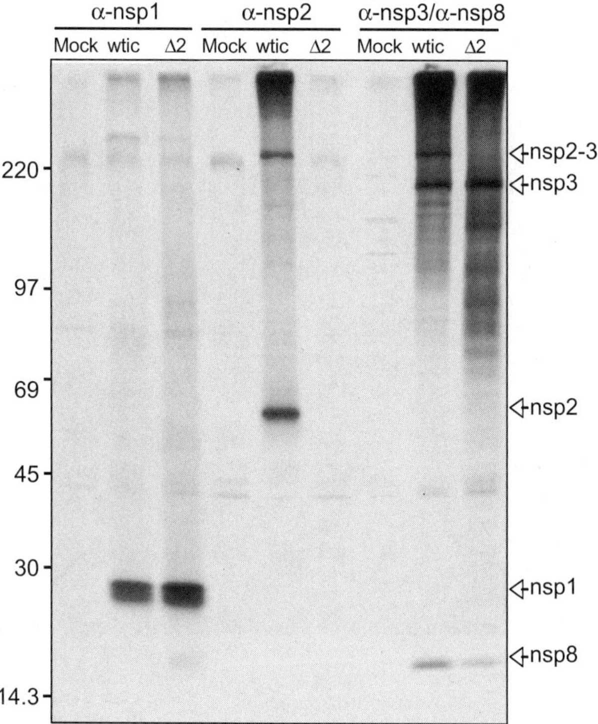

Figure 1. Expression and processing in MHVΔnsp2 mutant. DBT cells were mock-infected, infected with recombinant wild-type MHV (wtic), or MHVΔnsp2 (Δ2), radiolabeled with [^{35}S]-methionine/cysteine, lysed, and immunoprecipitated with antibodies against the indicated proteins. Proteins were then resolved by SDS-PAGE and imaged by fluorography.

3.3. Replication Complexes Form in the Absence of Nsp2

To determine if replication complexes formed in cells infected with the MHVΔnsp2 mutant virus, DBT cells were infected with either wild-type or MHVΔnsp2 for 6.5 h, fixed, and stained by indirect immunofluorescence for the viral proteins nsp2, nsp8, N (nucleocapsid), which localize to punctate cytoplasmic foci (replication complexes), and M, a marker for sites of viral assembly (Fig. 2). Cells infected with MHVΔnsp2 viruses produced no detectable nsp2. However, nsp8 and N proteins both localized to punctate cytoplasmic focal patterns indistinguishable from wild-type viral replication complexes as well as colocalized with each other. This staining pattern was distinct from the staining pattern of the viral M protein. These results suggest that the MHVΔnsp2 virus was capable of forming replication complexes in the absence of the nsp2 protein.

4. SUMMARY

The results presented here demonstrate that the MHV and SARS-CoV nsp2 proteins are not required for the production of infectious virus, for polyprotein expression or processing, or for viral replication complex formation in cell culture. The nsp2 protein

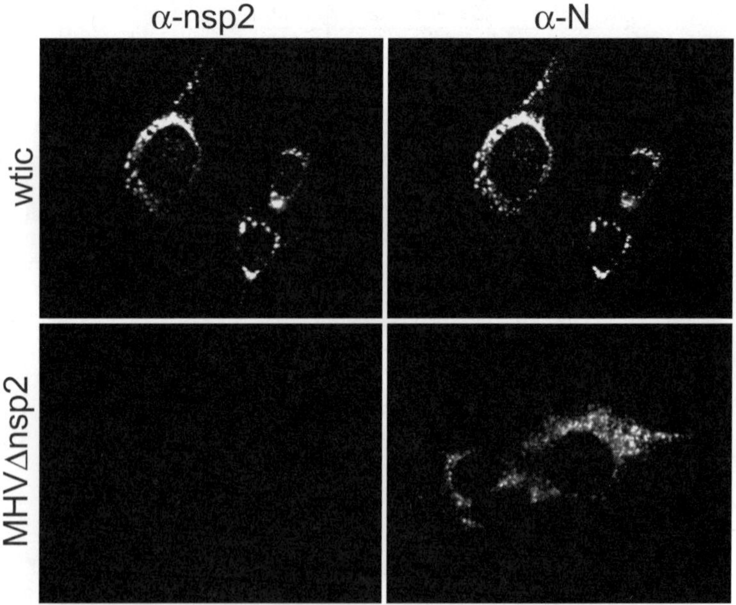

Figure 2. Replication complex formation in MHVΔnsp2 mutant. DBT cells on glass coverslips were infected with either recombinant wild-type MHV (wtic) or MHVΔnsp2 for 6.5 h, then fixed and permeablized in -20°C methanol. Cells were then stained using antibodies against nsp2 and nucleocapsid protein (N), a marker for replication complexes.

domain resides in a region of the coronavirus replicase that is relatively nonconserved across coronaviruses. In fact, the size and amino acid sequence variability of nsp2 across the different coronaviruses has led some investigators to speculate that the nsp2 protein, along with the nsp1 and nsp3 proteins, may play host- and/or cell-specific roles in the virus life cycle.[8–10] While this may be the case, it should be noted that nsp2, in some form, exists in all coronaviruses studied to date and likely plays a pivotal role in the viral life cycle. A previous study from our laboratory identified a coronavirus replicase protein that plays an important role in viral pathogenesis.[11] Such may prove to be the case for nsp2, as well. Alternatively, beacuse nsp2 exists as a detectable precursor protein nsp2-3 prior to processing of nsp2 and nsp3 into mature proteins, nsp2 may play a critical adaptor/regulatory role for nsp3 function. Importantly, the viruses produced in this study provide a system by which the role of the nsp2 protein in viral infection can be characterized.

5. ACKNOWLEDGMENTS

This work was supported by National Institutes of Health grants RO1 AI26603 (M.R.D.), RO1 AI059136-01 (R.S.B.), and PO1 AI059443-01 (R.S.B. and M.R.D.) and the Training Grant for Cellular, Biochemical, and Molecular Sciences, Vanderbilt University School of Medicine 5T32GM00855 (R.L.G.). Additional support was provided by the Public Health Service award CA68485 for the Vanderbilt DNA Sequencing Shared Resource and the Molecular Imaging Shared Resource of the Vanderbilt-Ingram Cancer Center.

6. REFERENCES

1. Yount, B., Denison, M. R., Weiss, S. R., and Baric, R. S., 2002, Systematic assembly of a full-length infectious cDNA of mouse hepatitis virus strain A59, *J. Virol.* **76**, 11065-11078.
2. Yount, B., Curtis, K. M., Fritz, E. A., Hensley, L. E., Jahrling, P. B., Prentice, E., Denison, M. R., Geisbert, T. W., and Baric, R. S., 2003, Reverse genetics with a full-length infectious cDNA of severe acute respiratory syndrome coronavirus, *Proc. Natl. Acad. Sci. USA* **100**, 12995-13000.
3. Prentice, E., McAuliffe, J., Lu, X., Subbarao, K., and Denison, M. R., 2004, Identification and characterization of severe acute respiratory syndrome coronavirus replicase proteins, *J. Virol.* **78**, 9977-9986.
4. Bost, A. G., Carnahan, R. H., Lu, X. T., and Denison, M. R., 2000, Four proteins processed from the replicase gene polyprotein of mouse hepatitis virus colocalize in the cell periphery and adjacent to sites of virion assembly, *J. Virol.* **74**, 3379-3387.
5. Denison, M. R., Spaan, J. M., van der Meer, Y., Gibson, C. A., Sims, A. C., Prentice, E., and Lu, X. T, 1999, The putative helicase of the coronavirus mouse hepatitis virus is processed from the replicase gene polyprotein and localizes in complexes that are active in viral RNA synthesis, *J. Virol.* **73**, 6862-6871.
6. Denison, M. R., Hughes, S. A., and Weiss, S. R., 1995, Identification and characterization of a 65-kDa protein processed from the gene 1 polyprotein of the murine coronavirus MHV-A59, *Virology* **207**, 316-320.
7. Schiller, J. J., Kanjanahaluethai, A., and Baker, S. C., 1998, Processing of the coronavirus mhv-jhm polymerase polyprotein: identification of precursors and proteolytic products spanning 400 kilodaltons of ORF1a, *Virology* **242**, 288-302.
8. de Vries, A. A. F., Horzinek, M. C., Rottier, P. J. M., and deGroot, R. J., 1997, The genome organization of the nidovirales: similarities and differenced between arteri-, toro, and coronaviruses, *Sem. Virology* **8**, 33-47.
9. Ziebuhr, J., Snijder, E. J., and Gorbalenya, A. E., 2000, Virus-encoded proteinases and proteolytic processing in the Nidovirales, *J. Gen. Virol.* **81**, 853-879.

10. Ziebuhr, J., Thiel, V., and Gorbalenya, A. E., 2001, The autocatalytic release of a putative RNA virus transcription factor from its polyprotein precursor involves two paralogous papain-like proteases that cleave the same peptide bond, *J. Biol. Chem.* **276**, 33220-33232.
11. Sperry, S. M., Kazi, L., Graham, R. L., Baric, R. S., Weiss, S. R., and Denison, M. R., 2005, Single-amino-acid substitutions in open reading frame (ORF) 1b-nsp14 and ORF 2a proteins of the coronavirus mouse hepatitis virus are attenuating in mice, *J. Virol.* **79**, 3391-4000.

MOLECULAR DISSECTION OF PORCINE REPRODUCTIVE AND RESPIRATORY VIRUS PUTATIVE NONSTRUCTURAL PROTEIN 2

Kay S. Faaberg, Jun Han, and Yue Wang*

1. INTRODUCTION

Porcine reproductive and respiratory syndrome virus (PRRSV) belongs to the family *Arteriviridae* in the order *Nidovirales*.[1] The virus is now known to consist of two different genotypes based on the finding that the prototypes viruses, Lelystad virus (European-like, Type 1[2]) and VR-2332 (North American-like, Type 2[3]), display only approximately 60% nucleotide identity. PRRSV has since been shown to consist of multiple virus isolates that vary within each genotype as much as 20% in nucleotide composition[4,5] (Faaberg, unpublished data). Within this genetic backdrop, seemingly novel field isolates of PRRSV suddenly appeared in the southwestern region of the State of Minnesota, U.S.A., in 2002. Phylogenic analysis, based on the ORF5 gene (encoding the viral attachment protein) of 916 unique PRRSV isolates, revealed that these isolates were most similar to those found in Canada in the early 1990s[6,7] (Faaberg, unpublished data). The PRRSV isolates were determined to have the restriction fragment length polymorphism pattern designated 1-8-4[8] and thus were named MN184 isolates. In order to examine the MN184 isolates more closely, we determined the full-length nucleotide sequences of two field isolates differing in apparent virulence. MN184 isolate comparison revealed that differences existed throughout the genome, most notably in nonstructural protein (Nsp) 2, for which no function has been assigned, except by comparison with the genome of equine arteritis virus.[9–11] The comparison of the MN184 field isolates to the prototypic strain VR-2332,[3] and to the first Type 1 strain seen in the United States, EuroPRRSV,[5] was then completed.

*University of Minnesota, Saint Paul, Minnesota 55108.

2. METHODS

2.1. PRRSV Strains

MN184 field isolates A (moderate clinical signs) and B (severe clinical signs) were obtained from Kurt Rossow, D.V.M., Ph.D., at the Minnesota Veterinary Diagnostic Laboratory (MVDL) after a single round of PRRSV amplification on freshly isolated porcine alveolar macrophages (PAM). Strains VR-2332 (U87392)[3,12] and EuroPRRSV (AY366525)[5] have been described previously.

2.2. Determination of Complete Genomes of MN184A and MN184B Isolates

Viral RNA was purified from infected PAM supernatant using the QIAamp viral RNA kit (Qiagen). RNA was converted to DNA using random hexamers and sequence-specific forward and reverse primers by One-Step RT-PCR (Qiagen). 5'- and 3'- rapid amplification of cDNA ends (RACE) was performed using 5' and 3'-Full Race Core Set (TaKaRa Bio Inc.) on purified viral RNA. PCR primers were derived from several sources, including those described for amplification of strain VR-2332,[3,12] strain JA142 (AY424271)[5] and newly generated MN184 sequence. The detailed primer set used to delineate the MN184 genome will be described elsewhere. The individual nucleotide sequences were assembled using the SeqMan II program in the Lasergene software suite (Version 6; DNASTAR, Inc.). A minimum of three-fold sequence coverage of each genome was obtained.

2.3. Genome Analysis

The complete nucleotide sequences for all four PRRSV genomes were analyzed using the Genetics Computer Group Wisconsin Package (GCG, Version 10.3-UNIX; Accelrys, Inc.). The genomes of MN184A and MN184B have been deposited in GenBank.

3. RESULTS

3.1. Complete Nucleotide Sequences of MN184 Isolates

The MN184 isolates were amplified only by a single round of growth on freshly isolated PAM, the host cell, in order to identify as much variation in the nucleotide sequences as possible. This variation, as suggested by nucleotide degeneracy at individual and discrete nucleotides during sequence analysis of individual PCR product tracefiles, was considerable. Notably, isolate MN184B, which produced severe clinical signs in the field, exhibited much more variation than isolate MN184A.

The complete genomes were found to be identical in length (15,019 bases excluding the polyA tail) and the shortest genome identified to date, including the Type 1 strain EuroPRRSV which possesses only 15,047 bases. However, the isolates were more closely related to Type 2 strains, the shortest of which had been the Chinese strain HB-2

(15,398 bases)[13]. The two MN184 isolates were 97.8% identical in nucleotide sequence, and genome-wide possessed only 326 nucleotide differences which included the degenerate sites mentioned above (Table 1).

3.2. Comparison with Prototypic PRRSV Strains VR-2332 and EuroPRRSV

The two MN184 isolates displayed considerable genetic distance from both strain EuroPRRSV (~57%) and strain VR-2332 (~85%). These differences were seen throughout the genome but were greatest in the regions encoding putative Nsp2 (94.5% identity) and the viral attachment protein (open reading frame 5; 97.7% identity). The identified Nsp2 region of ORF1a (the replicase polyprotein) contained the majority of the nucleotide degeneracy seen when analyzing the complete genomes and also included a large deletion when compared with strain VR-2332. We chose this PRRSV genomic region for further bioinformatic analysis.

3.3. The Genomic Region Encoding Nsp2 Contains Several Putative Domains

The putative Nsp2 region, originally identified as spanning amino acids (a.a.) 384-1363,[14] was projected to include a.a. 384-1578 of the strain VR-2332 ORF1a protein through genetic analysis of both *Coronaviridae* and *Arteriviridae*.[15] Nsp2 has been previously shown to be the key region of length difference between Type 1 and Type 2 isolates[3] and also revealed by several investigators to vary extensively between North American-like Type 2 isolates[3,13,14,16] as well as between North American Type 1 isolates.[5,17] Comparative sequence analysis with other Nidoviruses has identified a cysteine protease domain near the N-terminal end.[9] The amino acid makeup of this region is over 10% proline, and contains many PxxP motifs, the signature binding motif of Src homology 3 (SH3) domains, which suggests that Nsp2 may be involved in signal transduction mechanisms. For Type 1 strains, there is a leucine zipper motif (Lx6Lx6Lx6L) near C-terminal end.[5] Several B-cell epitopes were identified for Nsp2 using bacteriophage display.[18]

Further bioinformatic analysis of Nsp2 of strain VR-2332 using SignalP 3.0 (www.cbs.dtu.dk/services/SignalP/)[19] and Interproscan (www.ebi.ac-.uk/InterProScan/)[20] revealed the existence of several other domains.

Table 1. Nucleotide similarity and divergence of the four PRRSV genomes.

		Percent Nucleotide Similarity			
		VR-2332	MN184A	MN184B	EuroPRRSV
Divergence	VR-2332	-	84.5	84.7	56.6
	184A	14.9	-	97.8	57.6
	184B	13.9	1.3	-	57.4
	EuroPRRSV	61.0	61.6	61.0	-

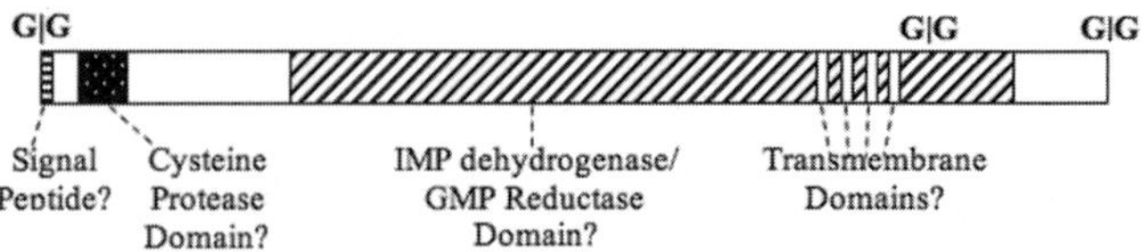

Figure 1. Putative Nsp2 protein is predicted to contain several domains. The first G|G represents the putative site of Nsp1β/Nsp2 cleavage at nt 383|384 and the predicted Nsp2 cleavage site at 1363 (second G|G[14]) or at a.a. 1578 (third G|G[15]). A signal peptide at a.a. 22, an equine arterivirus Nsp2-type cysteine protease domain (a.a. 45-152),[9] an IMP dehydrogenase/GMP reductase domain (a.a. 284-1092), and four transmembrane domains located at a.a. 875-895, 910-930, 959-979, 988-1008 were identified.

4. CONCLUSIONS

The identification of the MN184 isolates in Minnesota 10 years after similar strains were identified in Quebec, Canada, was unexpected. The MVDL has collected over 4400 ORF5 field isolate sequences from the swine producing regions of United States and Canada, generated in the course of diagnostic analysis, suggesting that these novel isolates appeared suddenly and with no known direct origin. The complete genome sequence determination and analysis of two of these isolates, MN184A of reported moderate virulence and MN184B of reported severe virulence, was undertaken to attempt to determine regions on the genome coding for virulence factors. Two genetic regions were revealed by sequence comparison and found to code for Nsp2 (94.5% identity) and ORF5 (97.7% identity), but the number of identified differences noted genome wide (326 nucleotides) precluded immediate identification of such virulence determinants. Nucleotide degeneracy in Nsp2 may be a factor in predicting virulence, but this must be addressed in separate publication.

The complete genomes of MN184A and MN184B consist of 15019 bases and thus identify the smallest PRRSV isolates to date. These Type 2 isolates were found to diverge from strain VR-2332 nucleotide sequence by approximately 15%. To begin to analyze the region found to account for the most sequence divergence between the MN184 isolates (Nsp2), we submitted the Nsp2 region of strain VR-2332 for bioinformatic analysis. The Nsp2 protein had been previously characterized to have over 10% proline, several PxxP motifs, a leucine zipper motif[5] and several B-cell epitopes.[18] However, further analysis identified a potential signal sequence, an unusual IMP dehydrogenase/GMP reductase domain and four transmembrane domains. We are now poised to begin molecular exploration of this extremely variable region of the PRRSV replicase.

5. REFERENCES

1. Cavanagh, D., 1997, Nidovirales: a new order comprising Coronaviridae and Arteriviridae, *Arch. Virol.* **142**:629-633.
2. Meulenberg, J, J., Hulst, M. M., de Meijer, E. J., Moonen, P. L., den Besten, A., de Kluyver, E. P., Wensvoort, G., and Moormann, R. J., 1993, Lelystad virus, the causative agent of porcine epidemic abortion and respiratory syndrome (PEARS), is related to LDV and EAV, *Virology* **192**:62-72.
3. Nelsen, C. J., Murtaugh, M. P., and Faaberg, K. S., 1999, Porcine reproductive and respiratory syndrome virus comparison: divergent evolution on two continents, *J. Virol.* **73**:270-280.

4. Kapur, V., Elam, M. R., Pawlovich, T. M., Murtaugh M. P., 1996, Genetic variation in porcine reproductive and respiratory syndrome virus isolates in the midwestern United States, *J. Gen. Virol.* **77**:1271-1276.
5. Ropp, S. L., Wees, C. E., Fang, Y., Nelson, E. A., Rossow, K. D., Bien, M., Arndt, B., Preszler, S., Steen, P., Christopher-Hennings, J., Collins, J. E., Benfield, D. A., and Faaberg, K. S., 2004, Characterization of emerging European-like porcine reproductive and respiratory syndrome virus isolates in the United States, *J. Virol.* **78**:3684-3703.
6. Mardassi, H., Athanassious, R., Mounir, S., Dea, S., 1994, Porcine reproductive and respiratory syndrome virus: morphological, biochemical and serological characteristics of Quebec isolates associated with acute and chronic outbreaks of porcine reproductive and respiratory syndrome, *Can. J. Vet. Res.* **58**:55-64.
7. Murtaugh, M. P., Yuan, S., Faaberg, K. S., 2001, Appearance of novel PRRSV isolates by recombination in the natural environment, *Adv. Exp. Med. Biol.* **494**:31-36.
8. Wesley, R. D., Mengeling, W. L., Lager, K. M., Clouser, D. F., Landgraf, J. G., Frey, M. L., 1998, Differentiation of a porcine reproductive and respiratory syndrome virus vaccine strain from North American field strains by restriction fragment length polymorphism analysis of ORF 5, *J. Vet. Diagn. Invest.* **10**:140-144.
9. Snijder, E. J., Wassenaar, A. L., Spaan, W. J., and Gorbalenya, A. E., 1995, The arterivirus Nsp2 protease. An unusual cysteine protease with primary structure similarities to both papain-like and chymotrypsin-like proteases, *J. Biol. Chem.* **270**:16671-16676.
10. Pedersen, K. W., van der Meer, Y., Roos, N., and Snijder, E. J., 1999, Open reading frame 1a-encoded subunits of the arterivirus replicase induce endoplasmic reticulum-derived double-membrane vesicles which carry the viral replication complex, *J. Virol.* **73**:2016-2026.
11. Snijder, E. J., van Tol, H., Roos, N., and Pedersen, K. W., 2001, Non-structural proteins 2 and 3 interact to modify host cell membranes during the formation of the arterivirus replication complex, *J. Gen. Virol.* **82**:985-994.
12. Yuan, S., Mickelson, D. M., Murtaugh, M. P., and Faaberg, K. S., 2001, Complete genome comparison of porcine reproductive and respiratory syndrome virus parental and attenuated strains, *Virus Res.* **74**:99-110.
13. Gao, Z. Q., Guo, X. and Yang, H. C., 2004, Genomic characterization of two Chinese isolates of porcine respiratory and reproductive syndrome virus, *Arch. Virol.* **149**:1341-1351.
14. Allende, R., Lewis, T. L., Lu, Z., Rock, D. L., Kutish, G. F., Ali, A., Doster, A. R., and Osorio, F. A., 1999, North American and European porcine reproductive and respiratory syndrome viruses differ in non-structural protein coding regions, *J. Gen. Virol.* **80**:307-315.
15. Ziebuhr, J., Snijder, E. J., and Gorbalenya, A. E., 2000, Virus-encoded proteinases and proteolytic processing in the *Nidovirales, J. Gen. Virol.* **81**:853-879.
16. Shen, S., Kwang, J., Liu, W., and Liu, D. X., 2000, Determination of the complete nucleotide sequence of a vaccine strain of porcine reproductive and respiratory syndrome virus and identification of the Nsp2 gene with a unique insertion, *Arch. Virol.* **145**:871-883.
17. Fang, Y., Kim, D. Y., Ropp, S., Steen, P., Christopher-Hennings, J., Nelson, E. A., and Rowland, R. R., 2004, Heterogeneity in Nsp2 of European-like porcine reproductive and respiratory syndrome viruses isolated in the United States, *Virus Res.* **100**:229-235.
18. Oleksiewicz, M. B., Botner, A., Toft, P., Normann, P., and Storgaard, T., 2001, Epitope mapping porcine reproductive and respiratory syndrome virus by phage display: the nsp2 fragment of the replicase polyprotein contains a cluster of B-cell epitopes, *J. Virol.* **75**:3277-3290.
19. Bendtsen, J. D., Nielsen, H., von Heijne, G., and Brunak, S., 2004, Improved prediction of signal peptides: SignalP 3.0, *J. Mol. Biol.* **340**:783-795.
20. Apweiler, R., Attwood, T. K., Bairoch, A., Bateman, A., Birney, E., Biswas, M., Bucher, P., Cerutti, L., Corpet, F., Croning, M. D., Durbin, R., Falquet, L., Fleischmann, W., Gouzy, J., Hermjakob, H., Hulo, N., Jonassen, I., Kahn, D., Kanapin, A., Karavidopoulou, Y., Lopez, R., Marx, B., Mulder, N. J., Oinn, T. M., Pagni, M., Servant, F., Sigrist, C. J., and Zdobnov, E. M., 2001, The InterPro database, an integrated documentation resource for protein families, domains and functional sites, *Nucleic Acids Res.*, **29**:37-34.

DIFFERENTIAL ROLE OF N-TERMINAL POLYPROTEIN PROCESSING IN CORONAVIRUS GENOME REPLICATION AND MINIGENOME AMPLIFICATION

Carmen Galán, Luis Enjuanes, and Fernando Almazán*

1. INTRODUCTION

Processing of replicase polyproteins is a crucial step in the coronavirus life cycle, providing active components of the replication-transcription complex.[1] Coronavirus ORF1a encodes one or two papain-like proteinases (PLPs) that are responsible for the cleavage at the N-proximal region of the replicase. Although PLP-1 cleavage products have been identified in HCoV-229E virus-infected cells, no specific roles or functional requirements have been studied so far for the HCoV-229E or TGEV proteins p9 (nsp1) and p87 (nsp2).

In the present study, we characterize by reverse genetics the effect of a point mutation at position 637, which mapped in the putative PLP-1 cleavage site at the p9/p87 junction, on TGEV and minigenome replication. A correlation was found between predicted cleaving and noncleaving mutations and the different nucleotides selected for virus or minigenome replication, respectively.

2. MATERIALS AND METHODS

2.1. Plasmid Constructs and Recovery of Viruses from the cDNA Clones

Point mutations at genome position 637 were engineered in pBAC-TGEV^{FL} [2] using an overlapping PCR strategy. BHK cells stably transformed with the porcine aminopeptidase N gene were transfected with the cDNA clones using Lipofectamine 2000 (Invitrogen). After a 6-h incubation period at 37ºC, cells were trypsinized and plated over a confluent monolayer of swine testis (ST) cells. Cell supernatants were harvested for titration at 24, 36, and 48 h post-transfection.

* Centro Nacional de Biotecnología, CSIC, Darwin 3, Cantoblanco, 28049 Madrid, Spain.

The cDNAs encoding TGEV-derived RNA minigenomes DI-C (9.7 kb) and M33 (3,3 kb) were previously described.[3] The M33L minigenome was derived from the M33 minigenome including a 16-bp linker sequence. M33L and DI-C cDNAs with point mutations at position 637 were generated by restriction fragment exchange from plasmids with the corresponding mutations. For the *trans*-cleavage assay, the PLP-1 domain (Glu859 to Ser1315) from pp1a, and the N-terminal 610 amino acids with either Gly (637 G) or Asp (637 A) at position 108, were cloned into pcDNA3 (Invitrogen).

2.2. Minigenome Rescue Quantification

T7-driven *in vitro* transcripts of M33L minigenome mutants were transfected in ST cells previously infected with TGEV PUR46-MAD strain, and five serial passages were performed on fresh ST cells. Total RNA from each passage was used as template for real-time RT-PCR reactions (Q-RT-PCR) with minigenome-specific primers.

2.3. *Trans*-Cleavage Assay

TGEV PLP-1 proteinase and pp1a substrates were expressed using the TNT T7-coupled reticulocyte lysate system (Promega). Different enzyme to [^{35}S]Met-labeled-substrate [E/S] ratios were incubated overnight at 30ºC. Cleavage reactions were resolved by SDS-10% PAGE and processed for fluorography.

3. RESULTS AND DISCUSSION

3.1. Mutations at Genome Position 637 Severely Affect rTGEV Recovery

During the construction of the TGEV full-length cDNA clone, a point mutation that was present in the defective minigenome DI-C (637-T) was maintained as a genetic marker. Interestingly, while other markers remained stable in the recombinant virus rescued (rTGEV), nucleotide 637 reproducibly reverted to the parental virus sequence G or C, indicating a strong selective pressure at this position.

To study the role of nucleotide 637 in virus replication, TGEV cDNA mutants at this position were generated. Nucleotide substitution of the viral sequence G at position 637 by C, T, or A produces an amino acid change at position 108 of the pp1a from Gly to Ala, Val, or Asp, respectively. Virus recovery efficiency (Fig. 1A) and plaque morphology (Fig. 1B) were analyzed after cDNA transfection. No differences between the viruses carrying a G or C at position 637 were detected. However, a reduction of three logarithmic units and small plaque morphology was observed for viruses with T or A at position 637. Mutations that severely affected virus recovery from the cDNA correlated with more drastic amino acid substitutions. No correlation was found between RNA secondary structure predictions of the mutants and virus phenotypes (data not shown), indicating that the effect of mutations at position 637 was most likely at the protein level.

3.2. Effect of Mutations at Position 637 on Minigenome Rescue Efficiency

To study the sequence requirements at position 637 for minigenome replication, M33L minigenome mutants at this position were generated, and their rescue efficiency was determined by Q-RT-PCR (Fig. 1C). Controls of RNA transfection (RNA input) and background levels (Mock; TGEV; NTC) were included. Only M33L minigenomes with A or T at position 637 were efficiently rescued, in contrast with the sequence requirements at the same position for efficient virus recovery from the full-length cDNA clone. A late increase in the RNA accumulation observed for M33L-637G and C was due to a genotypic reversion at position 637 to A. No differences in virus titers that could explain minigenome phenotypes were observed (data not shown). The same mutational analysis with the RNA DI-C (9.5 Kb) showed identical results to that of M33L (3.3 Kb) indicating that the same sequence requirements at nucleotide 637 were necessary for both DI-C and M33L minigenome amplification regardless of their differences in size (data not shown).

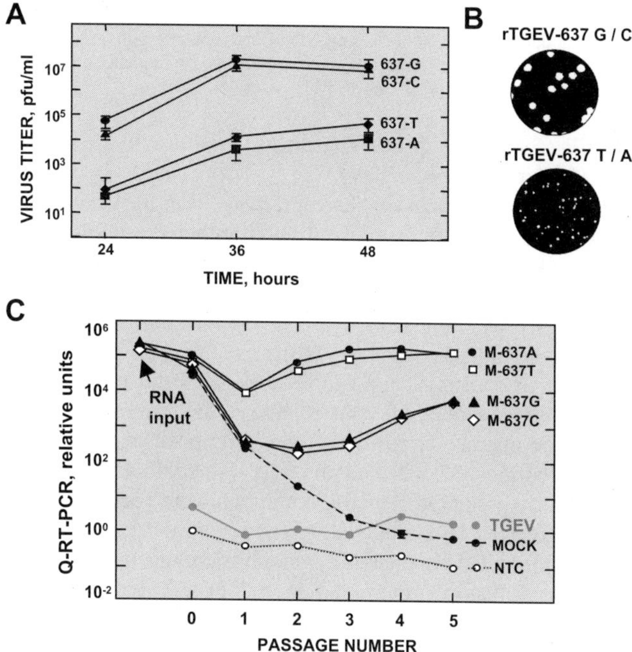

Figure 1. Phenotypes of rTGEV viruses or minigenome mutants. Effect of mutations at position 637 on virus rescue (A), virus plaque morphology (B), and minigenome amplification (C).

3.3. Mutations at Nucleotide 637 Affect N-terminal Replicase Processing by PLP-1

Sequence alignments reported for the HCoV-229E and TGEV replicase polyproteins, proposed two potential N-terminal cleavage sites in the TGEV pp1a,[4] although no experimental data have been provided to date. Mutations at nucleotide 637 affects the nature of amino acid 108 that occupies the P1' or P3 residues relative to the Thr107/Gly108 or Gly110/Ala111 putative cleavage sites, respectively, according with

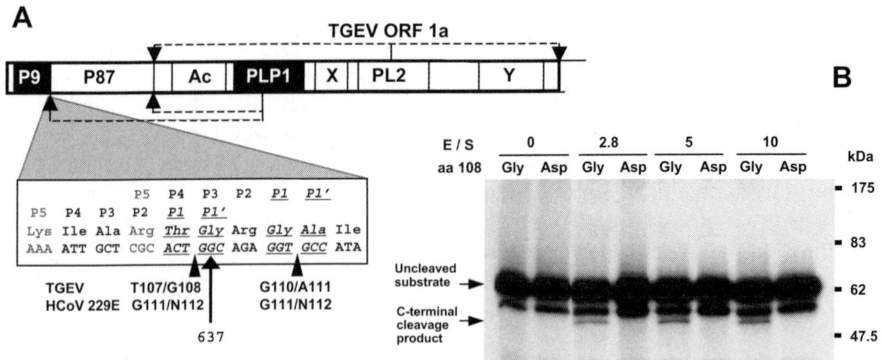

Figure 2. Effect of point mutations at nucleotide 637 on N-terminal replicase processing. (A). Scheme showing the two possible N-terminal cleavage sites (black arrowheads) predicted by sequence alignment with the HCoV-229E and the position of nucleotide 637. (B). *Trans*-cleavage assay.

these predictions (Fig. 2A). In both cases, a drastic amino acid change at position 108 could affect PLP1- mediated processing.

To support this hypothesis a *trans*-cleavage assay was performed with the TGEV PLP-1 domain and a N-terminal pp1a substrate with either the wild-type sequence G at position 637 or an A at the same position, leading to the least conservative amino acid change (Gly108Asp) (Fig. 2B). In the absence of enzyme, only the 67-kDa substrate and a minor protein, probably resulting from a premature termination event, were detected. When the PLP-1 was included, a processed form of the substrate appeared with the substrate containing a Gly residue at position 108, with the expected size of the C-terminal cleavage product generated after the release of the N-terminal p9 protein, but not with the substrate presenting an Asp residue at the same position. These results indicated that mutations at nucleotide 637 affected the PLP-1-mediated N-terminal polyprotein cleavage *in vitro*. Because predicted cleaving mutations were required for virus recovery, we propose a critical role of this processing event for viral replication. In contrast, the minigenomes selected predicted noncleaving mutations, suggesting that the processing of the minigenome-encoded fusion protein led to the generation of pp1a products that interfere with minigenome amplification. Further analysis will be required to assign specific functions to the N-terminal replicase proteins in TGEV replication.

4. REFERENCES

1. J. Ziebuhr, E. J. Snijder, and A. E. Gorbalenya, Virus-encoded proteinases and proteolytic processing in the *Nidovirales*, *J. Gen. Virol.* **81**:853-879 (2000).
2. F. Almazán, J. M. González, Z. Pénzes, et al., Engineering the largest RNA virus genome as an infectious bacterial artificial chromosome, *Proc. Natl. Acad. Sci. USA* **97**, 5516-5521 (2000).
3. A. Izeta, C. Smerdou, S. Alonso, et al., Replication and packaging of transmissible gastroenteritis coronavirus-derived synthetic minigenomes, *J. Virol.* **73**:1535-1545 (1999).

4. J. Herold, A. E. Gorbalenya, V. Thiel, B. Schelle, and S. G. Siddell, Proteolytic processing at the amino terminus of human coronavirus 229E gene 1-encoded polyproteins: identification of a papain-like proteinase and its substrate, *J. Virol*. **72**:910-918 (1998).

IDENTIFICATION AND CHARACTERIZATION OF SEVERE ACUTE RESPIRATORY SYNDROME CORONAVIRUS SUBGENOMIC RNAs

Snawar Hussain, Ji'an Pan, Jing Xu, Yalin Yang,Yu Chen, Yu Peng, Ying Wu, Zhaoyang Li, Ying Zhu, Po Tien, and Deyin Guo*

1. INTRODUCTION

Severe acute respiratory syndrome (SARS) is an atypical form of pneumonia that was first recognized in Guangdong Province, China, in November 2002, and its causative agent was identified as novel coronavirus (SARS-CoV).[1-3] Coronaviruses are the largest RNA viruses, containing a single-stranded, plus-sense RNA ranging from 27 kb to 31.5 kb in size. The two large open reading frames (ORFs) (1a and 1b) at the 5'-end of the genome encode the viral replicase and are translated directly from the genomic RNA, while 1b is expressed by –1 ribosomal frameshifting.[4] The 3'-one third of the genome comprises the genes encoding structural proteins S, E, M, and N and a number of auxiliary proteins of unknown function. These proteins are translated through 6-9 nested and 3'-coterminal subgenomic RNAs (sgRNAs).

The genomes of many SARS-CoV isolates have been sequenced, and consist of approximately 29,700 nucleotides.[5-7] Fourteen ORFs have been identified, of which 12 are located in the 3'-proximal one-third of the genome.[4,5] The exact mechanisms of expression of the 3'-proximal ORFs are unknown but on the analogy with other coronaviruses, these ORFs are predicted to be expressed through a set of sgRNAs.[6] Identification of SARS-CoV sgRNAs in infected cells and characterization of molecular details of the leader-body fusion in the sgRNAs will help elucidate the regulatory mechanism of SARS-CoV transcription and replication. This knowledge will be useful in the development of antiviral therapeutic agents and vaccine for treatment and prevention of this newly emerged disease.

*Wuhan University, Wuhan, People's Republic of China.

2. IDENTIFICATION OF NOVEL SUBGENOMIC RNAS

Northern blot, RT-PCR, and DNA sequencing revealed the existence of ten subgenomic RNAs including two novel subgenomic RNAs named 2-1 and 3-1. The leader-body fusion site (ACGAgC) of subgenomic RNA 2-1 has one nucleotide mismatch (lowercase) with SARS-CoV leader core sequence (CS-L) ACGAAC and is located inside the S gene, 384 nucleotides downstream from the authentic CS (ACGAAC) for mRNA 2/S. The second novel subgenomic RNA (3-1) corresponded to the 3b ORF that was predicted to be expressed from mRNA 3. The leader-body fusion site (AaGAAC) for subgenomic mRNA 3-1 is 10 nucleotides upstream of AUG start codon of ORF 3b and has a mismatch (lowercase) with the SARS-CoV leader core sequence (ACGAAC).

To determine whether these new subgenomic RNAs are functional messages, the 5′-end (containing leader sequence plus downstream 150 nucleotides in case of sgRNA 2-1 and 400 nucleotides in case of sgRNA 3-1) were fused with green fluorescent protein (GFP) gene, and sgRNA codon usage was indirectly determined by the expression of the reporter gene. Strong fluorescence was observed in cells transfected with the construct in which the ORF2b was fused in-frame with GFP (Figure 1A). Western blot with anti-GFP monoclonal antibody confirmed the existence of a 32 kDa band corresponding to the fusion protein. A 27 kDa band corresponding to the wild-type GFP resulting from downstream AUG by leaky scanning was also detected in cells transfected with in-frame and out-frame construct (Figure 1B, upper panel).

Cells transfected with the ORF3b in-frame construct displayed weak fluorescence while no fluorescence was observed in cells transfected with the ORF3b out-frame construct (Figure 1A). Western blot analysis confirmed the existence of a 42, kDa band corresponding to 3b-GFP fusion protein (Figure 1B, lower panel). These results indicated that sgRNA 2-1 and 3-1 could function in the environment of the cell.

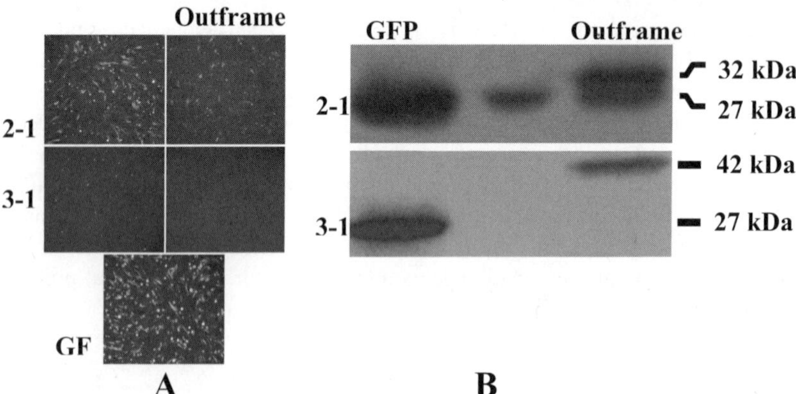

Figure 1. Translatability of novel subgenomic RNAs 2-1 and 3-1. (A) The ORF 2b (sgRNA 2-1) and 3b (sgRNA 3-1) were fused in-frame and out of frame at the 5′-end of GFP gene. The expression of GFP-fusion protein was qualitatively assessed by fluorescent microscopy (magnification: X200). (B) Proteins were extracted from transfected cells and separated by 12% SDS-PAGE. GFP-Fusion proteins were detected by anti-GFP monoclonal antibody.

3. EXPRESSION PROFILE OF SARS-COV SUBGENOMIC RNAS

The 5'-ends (including leader sequence and 200-400 nucleotides of the 5'-end of body sequence) of all 10 (2/S, 2-1, 3, 3-1, 4/E, 5/M, 6, 7, 8, 9/N) subgenomic RNAs were amplified by RT-PCR and fused to the 5'-end of the GFP reporter gene. The AUG initiator codon usage of subgenomic RNAs was indirectly assessed by level of expression of reporter gene.

Strong fluorescence was observed in the cells transfected with fusion constructs containing the 5'-ends of subgenomic RNAs 2-1, 3, 4/E, 5/M, 6 and 9/N, whereas relatively weak fluorescence was observed in cells transfected with fusion constructs containing 5'-ends of subgenomic RNAs 2, 3-1, 7 and 8 (Figure 2A).

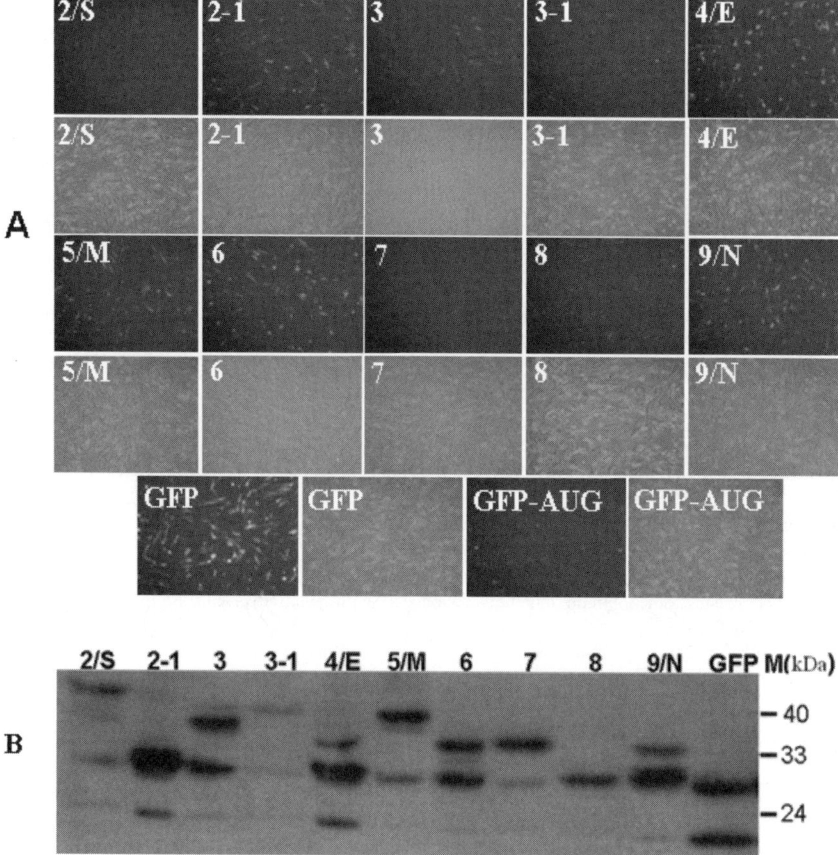

Figure 2. Expression of GFP-fusion protein in BHK cells. (A) The ORFs of all ten subgenomic were fused in-frame with GFP gene. The expression of GFP-fusion proteins was qualitatively assessed by fluorescent microscopy (magnification: X200). (B) Proteins were extracted from transfected cells and separated by 12% SDS-PAGE. GFP-Fusion proteins were detected by anti-GFP monoclonal antibody. Names of the individual subgenomic RNAs are marked on top and molecular weight (kDa) is marked on right side of image.

A single band of fusion protein was detected by Western blot in cells transfected with constructs containing the 5'-end of sgRNA 2, 2-1, 3, 3-1, 4, 6 and 7, whereas, in cells transfected with mRNA 5 and 9, beside the expected fusion protein band, a low molecular weight band was also detected that may result from a downstream AUG codon (Figure 2B).

We also detected some minor low molecular weight bands in some cases, possibly resulting from downstream AUG codons by leaky scanning. Cells transfected with subgenomic RNA8-GFP fusion construct displayed fair amount of fluorescence but in Western blot only a 27 kDa band of wild-type GFP was detected. Taken together-we showed that nine out of ten subgenomic RNAs could be functional messages *in vivo;* however, existence of proteins encoded by these subgenomic in SARS-CoV infected cells is yet to be determined.

In summary, ten subgenomic RNAs including two novel subgenomic RNAs (2-1/3-1) were identified in SARS-CoV infected cells by Northern blot, RT-PCR, and DNA sequencing. Nine out of 10 subgenomic RNAs were potentially functional messages in SARS-CoV infected cells. The initiator AUG codon of subgenomic RNA 8 (ORF8b) was inactive in our experimental set-up.

4. REFERENCES

1. R. A. Fouchier, R. Kuiken, M. Schutten, et al., Aetiology: Koch's postulates fulfilled for SARS virus, *Nature* **423**, 240 (2003).
2. T. G. Ksiazek, D. Erdman, C. S. Goldsmith, et al., A novel coronavirus associated with severe acute respiratory syndrome. *N. Engl. J. Med.* **348**, 1953-1966 (2003).
3. J. S. Peiris, C. M. Chu, V. C. Cheng, et al., Clinical progression and viral load in a community outbreak of coronavirus-associated SARS pneumonia: a prospective study, *Lancet* **361**, 1767-1772 (2003).
4. V. Thiel, K. A. Ivanov, A. Putics, et al., Mechanisms and enzymes involved in SARS coronavirus genome expression, *J. Gen. Virol.* **84**, 2305-2315 (2003).
5. M. A. Marra, S. J. Jones, C. R. Astell, et al., The genome sequence of the SARS-associated coronavirus, *Science* **300**, 1399-1404 (2003).
6. P. A. Rota, M. S. Oberste, S. S. Monroe, et al., Characterization of a novel coronavirus associated with severe acute respiratory syndrome, *Science* **300**, 1394-1399 (2003).
7. Y. J. Ruan, C. L. Wei, A. L. Ee, et al., Comparative full-length genome sequence analysis of 14 SARS coronavirus isolates and common mutations associated with putative origins of infection, *Lancet* **361**, 1779-1785 (2003).

IDENTIFICATION AND CHARACTERIZATION OF A UNIQUE RIBOSOMAL FRAMESHIFTING SIGNAL IN SARS-CoV ORF3A

Xiao X. Wang, Ying Liao, Sek M. Wong, and Ding X. Liu*

1. INTRODUCTION

Severe acute respiratory syndrome coronavirus (SARS-CoV) 3a gene encodes a protein of 274 amino acids. Recent studies showed that 3a is a minor structural protein.[1] A heterogeneous population of sgRNA 3 transcripts, containing one, two, or three nucleotide insertion in a six T stretch located 18 nucleotides downstream of the 3a initiation codon, was identified in SARS-CoV-infected cells as well as in the sera of SARS patients.[2] Here we report that a +1/-1 frameshifting event occurs in the insertion site. A mechanism of simultaneous slippage at both P and A sites may account for the frameshifting event.

2. MATERIALS AND METHODS

Construction of plasmids: Wild-type SARS 3a cDNA was amplified by PCR and was digested with EcoRV and EcoRI. The digested fragment was cloned into the two sites of pFlag vector, generating pF-3a/6T. Constructs with 7T and 8T were made by site-directed mutagenesis.

PCR fragment of EGFP from pEGFP-C1 (Clontech) was digested with BglII and EcoRV and cloned into these two sites on pSARS-3a/7T, generating pEGFP-3a. Deletions in EGFP or 3a regions were made by overlapping PCR.

In vitro transcription, translation, and transient expression of 3a and its mutants in Cos-7 cells: One microgram of plasmid DNA was transcribed and translated in a TnT coupled *in vitro* translation system (Promega) in rabbit reticulocyte lysate (RRL). The polypeptides were labeled with ^{35}S-methionine.

* Xiao X. Wang, Sek M. Wong, Ding X. Liu, National University of Singapore, 117543. Ying Liao, Ding X. Liu, Nanyang Technological University, Singapore, 637551. Ding X. Liu, Institute of Molecular and Cell Biology, Singapore, 138673.

SARS-CoV 3a sequences placed under the control of a T7 promoter were transiently expressed in mammalian cells using the recombinant vaccinia virus (vTF7-3) system. In this study, the transfection reagent used was Effecten Transfection Kit (Qiagen).

SDS-polyacrylamide gel electrophoresis (SDS-PAGE) and Western blot: Electrophoresis of viral polypeptides was performed on SDS-12% polyacrylamide gels. After transferring to PVDF membrane by Semi-dry transfer (Bio-Rad), proteins were detected with 3a antibody raised in rabbits, anti-FLAG antibody (Sigma), and His-tag monoclonal antibody (Santa Cruz).

3. RESULTS

3.1. A Frameshifting Event Occurs in SARS-CoV ORF3a

Full-length cDNA covering the 3a and 3b region with six, seven and eight Ts were translated *in vitro* (Fig. 1a). A band equivalent to the full-length 3a protein was expressed from all three clones (Fig. 1a, lanes 1–3). To further confirm the 3a expression in intact cells, pF-3a with a Flag tag at the N terminal (Fig. 1b) was expressed in Cos-7 cells and detected by Western blot. The 3a expression was detected in all three clones except in control (Fig. 1b, lanes 1–3). These results suggested that a +1 frameshift in ORF3a with seven Ts and a –1 frameshift in ORF3a with eight Ts occurred during translational elongation in order to maintain the full length ORF3a expression.

3.2. Identification of the Slippage Site by Mutagenesis Studies

To characterize the slippage site, point mutations of the seven Ts (from T to C) were made, giving rise to pF-3a/7T(M1-M7). In addition, the nucleotide immediately downstream the seven T (A) was also mutated to C, giving rise to pF-3a/7T(M8). *In vitro* expression (Fig. 2a) showed that the full-length 3a was expressed in pF-3a/7T (lane 1). Mutation of any of the seven Ts significantly decreased the full-length 3a expression (lanes 2–8). However, the A to C mutation did not affect the full-length 3a production (lane 9).

Similar expression patterns were observed in Cos-7 cells transfected with these constructs (Fig. 2b). The full-length 3a was detected in pF-3a/7T (lane 1) and M8 (lane 9) but not in other mutants (lanes 2–8), confirming that mutation of any T to C could greatly reduce the full-length 3a expression.

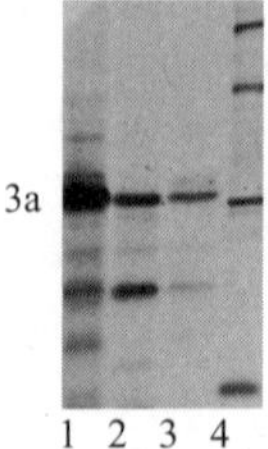

Figure 1a. *In vitro* expression of 3a/6T (1), 7T (2), and 8T (3). 4: markers.

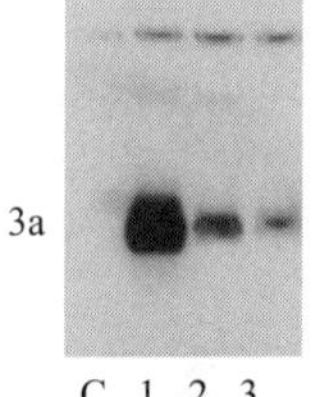

Figure 1b. Expression of 3a/6T (1), 7T (2) and 8T (3) in Cos-7 cells. C: mock transfection.

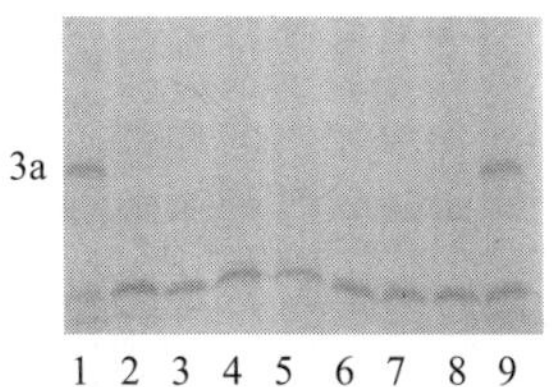

Figure 2a. *In vitro* expression of ORF3a with wild-type (1) and mutant 7T constructs (2–9).

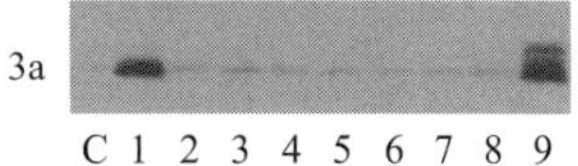

Figure 2b. Expression of ORF3a with wild-type (1) and mutant 7T constructs (2–9) in Cos-7 cells. C: mock transfection.

3.3. Deletion Analysis of Sequences Upstream of the Slippage Site

To analyze the effect of the upstream sequences on the frameshifting efficiency, the 5' end of 3a/7T was fused with the EGFP gene in frame. This extension expected to produce a 28-kDa termination product. The shifted full-length protein (FS) was approximately 55 kDa. *In vitro* translation showed the detection of a major band of 28 kDa and a minor band of 55 kDa (Fig. 3a, lane 1). The expression of the 55-kDa frameshifting product in intact cells was also detected by Western blot using anti-3a antibody (Fig. 3b, lane 1). Deletions of 63 (pEGFPΔ1-3a), 123 (pEGFPΔ2-3a) and 183 (pEGFPΔ1-3a) nucleotides, respectively, in the C terminal region of EGFP were made and expressed both *in vitro* and in intact cells. No substantial changes of the frameshifting efficiency was observed (Fig. 3a & 3b, lanes 2–4), indicating that the upstream sequence renders no obvious effect.

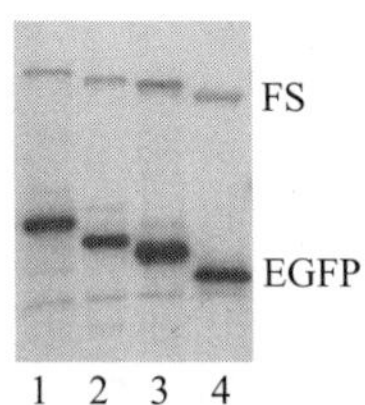

Figure 3a. *In vitro* expression of pEGFP-3a/7T (1) and the 5' deletion constructs pEGFPΔ1-3a (2), pEGFPΔ2-3a (3), and pEGFPΔ3-3a (4).

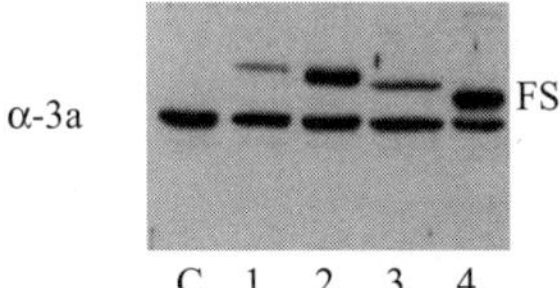

Figure 3b. Expression of pEGFP-3a/7T (1) and the 5' deletion constructs pEGFPΔ1-3a (2), pEGFPΔ2-3a (3), and pEGFPΔ3-3a (4) in Cos-7 cells. C: mock transfection.

3.4. Deletion Analysis of the Downstream Sequence

Analysis of the folding of the sequence downstream of the seven Ts showed the presence of two potential stem loops, forming from nucleotides 39 to 56 (starting from AUG for 3a) and nucleotides 62 to 88. The loop regions could partially pair with the downstream sequences and form potential pseudoknot structures. To analyze the effects of these regions on the frameshifting efficiency, five deletion constructs, pEGFP3aΔ1 to Δ5, were made based on pEGFP-3a/7T (Fig. 4a). *In vitro* translation of these deletion constructs (Fig. 4b) revealed the detection of both the termination products (EGFP) and the full length frameshifted products (FS). Compared to pEGFP-3a/7T, only Δ4 showed a slightly decreased detection of the FS product (Fig. 4b, lane 5), while other deletions

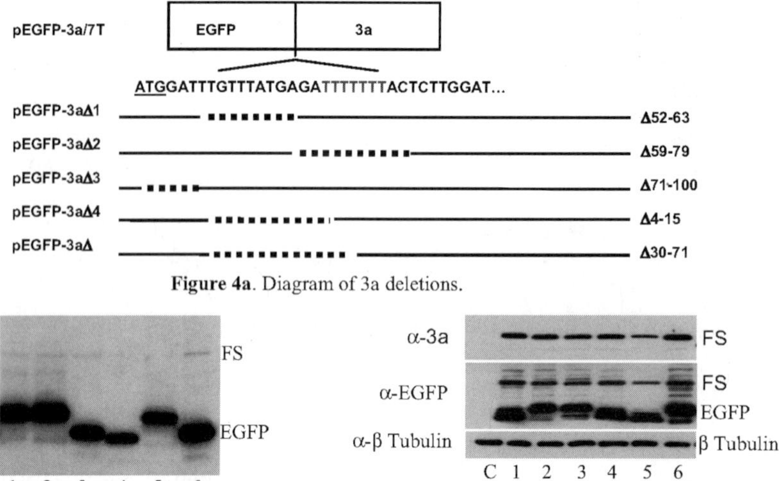

Figure 4a. Diagram of 3a deletions.

Figure 4b. In vitro expression of EGFP-3a/7T (1) and EGFP-3aΔ1 to 5 (2–6).

Figure 4c. Expression of EGFP-3a/7T (1) and EGFP-3aΔ1 to 5 (2–6). C: mock transfection.

didn't have significant effect on the frameshifting efficiency. Similar expression profiles were observed in Cos-7 cells expressing the deletion constructs (Fig. 4c).

4. DISCUSSION

In this study, we present evidence demonstrating that expression of full-length 3a with a single nucleotide insertion required a +1 frameshift at TTT TTT T, characterized as the slippage site. In the event of +1 frameshifting at the slippage site, the $\text{PhetRNA}_{\text{AAA}}$ could form perfect pair in the +1 frame with TTT. As mutation of any single T at the slippage site significantly impaired the frameshifting efficiency, it suggested that a double slippage mechanism is responsible for this frameshifting event and an exact base pairing is required for the efficient frameshifting. No other stimulators were found for this frameshifting event. This lack of additional stimulators was similar to the previous study with the Herpes simplex virus TK gene.[3] This may also explain why no frameshifting event was observed when imperfect base-pairing with the shifted frame was introduced.

5. REFERENCES

1. Ito, N., Mossel, E. C., Narayanan, K., Popov, V. L., Huang, C., Inoue, T., Peters, C. J., and Makino, S., 2005, Severe acute respiratory syndrome coronavirus 3a protein is a viral structural protein, *J. Virol.* **79**:3182-3186.
2. Tan, T. H., Barkham, T., Fielding, B. C., Chou, C. F., Shen, S., Lim, S. G., Hong, W., Tan, Y. J., 2005, Genetic lesions within the 3a gene of SARS-CoV, *Virol. J.* **2**:51.
3. Horsburgh, B., Kollmus, H., Hauser, H., and Coen, D., 1996, Translational recoding induced by G-rich mRNA sequences that form unusual structures, *Cell* **86**:949-959.

ADP-RIBOSE -1"-PHOSPHATASE ACTIVITIES OF THE HUMAN CORONAVIRUS 229E AND SARS CORONAVIRUS X DOMAINS

Ákos Putics, Jutta Slaby, Witold Filipowicz, Alexander E. Gorbalenya, and John Ziebuhr*

1. INTRODUCTION

Coronavirus RNA synthesis is mediated by the viral replicase, a huge multienzyme complex comp. of several cellular proteins and up to 16 viral nonstructural proteins (nsp1–16). For the majority of these proteins, the available functional and structural information is extremely limited.[1] Coronavirus nsp3, the largest viral subunit of the coronavirus replicase, has been predicted to contain several conserved domains, including an N-terminal domain enriched in Glu and Asp residues ("acidic domain"), one or two papain-like proteases (PL1^{pro} and PL2^{pro}), the X domain, and a C-terminal conserved domain ("Y domain") containing putative transmembrane and metal ion-binding domains.[2] The X domain has been predicted to be a phosphatase that converts ADP-ribose-1"-monophosphate (Appr-1"-p) to ADP-ribose (Appr).[3] Appr-1"-p is a downstream metabolite of cellular tRNA splicing. It is generated from ADP-ribose-1", 2"-cyclic-phosphate (Appr>p) by a cyclic phosphodiesterase (CPDase) activity.[4,5] Coronavirus X domain homologs are conserved in very few plus-strand RNA viruses, excluding the closely related arteri- and roniviruses.[3,6] There is a large number of poorly characterized cellular homologs that constitute the so-called macrodomain protein family. Recently, the crystal structure of one of these homologs, the *Archeoglobus fulgidus* AF1521 protein, has been determined[7] and the *Saccharomyces cerevisiae* Poa1p has been demonstrated to have Appr-1"-pase activity.[8] To gain insight into the biochemical properties of coronavirus X domains, we expressed and characterized recombinant forms of the human coronavirus 229E (HCoV-229E) and severe acute respiratory coronavirus (SARS-CoV) X domains.

* Ákos Putics, Jutta Slaby, John Ziebuhr, University of Würzburg, Würzburg, Germany, D-97078. Witold Filipowicz, Friedrich Miescher Institute for Biomedical Research, Basel, Switzerland, CH-4058. Alexander E. Gorbalenya, Leiden University Medical Center, Leiden, The Netherlands, NL-2333 ZA.

2. MATERIALS AND METHODS

Cloning and expression of coronavirus X domains: X domain coding sequences (HCoV-229E nucleotides 4085 to 4600 and SARS-CoV nucleotides 3262 to 3783, respectively) were amplified by reverse transcription-PCR using poly(A) RNA isolated from HCoV-229E-infected MRC-5 cells and SARS-CoV-infected Vero cells, respectively. Each of the PCR reverse primers contained a translation stop codon followed by an EcoRI restriction site. The PCR products were treated with T4 DNA polymerase, polynucleotide kinase, and EcoRI, and ligated with XmnI/EcoRI-digested pMal-c2 plasmid DNA (800-64S; New England Biolabs). The resulting plasmids encoded fusion proteins consisting of the *E. coli* maltose-binding protein (MBP) and the respective coronavirus X domain. Using PCR-based methods, a mutant derivative of the HCoV-229E X domain was generated. This protein contained substitutions (by Ala) of the pp1a/pp1ab residues, Asn1302 and Asn1305. *E. coli* TB1 cells transformed with the appropriate plasmid were grown to an OD_{595} of 0.5, and protein expression was induced with 1 mM IPTG for 3 hours at 18°C. The MBP-X fusion proteins were purified by amylose-affinity chromatography and cleaved with factor Xa. The protein purification and storage buffer contained 20 mM Tris-HCl (pH 7.5), 200 mM NaCl, 1 mM EDTA, and 1 mM dithiothreitol.

Appr-1"-pase activity assay: To produce Appr-1"-p, chemically synthesized Appr>p was treated with *Arabidopsis thaliana* cyclic nucleotide phosphodiesterase as described previously.[4] The purified coronavirus X domains (0.5 μM) were incubated with Appr-1"-p (2 mM) in buffer containing 35 mM Tris-HCl (pH 7.5), 0.005% Triton X-100, 0.5 mM EDTA, 100 mM NaCl, and 0.5 mM dithiothreitol at 30°C for 3 hours. In control reactions, 0.5 U/μl alkaline phosphatase from calf intestine (Roche) was used. Reactions products were separated by cellulose thin-layer chromatography [saturated $(NH_4)_2SO_4$/3M sodium acetate/isopropyl alcohol (80:6:2)] and visualized under UV light.

3. RESULTS AND DISCUSSION

3.1. The HCoV-229E nsp3-Associated X Domain Is a Phosphatase That Converts Appr-1"-p to Appr

The phylogenetic relationship of coronavirus X domains with cellular proteins of the macrodomain family[3] and the recent biochemical characterization of two of the cellular homologs[8,9] suggested that coronavirus X domains might be enzymes that dephosphorylate Appr-1"-p to Appr. To confirm this hypothesis, the HCoV-229E X domain was expressed as an MBP-fusion protein in *E. coli* and its activity was examined in an *in vitro* Appr-1"-pase assay. Based on sequence aligments with other coronavirus X domains[2] and many more cellular homologs and guided by the crystal structure of the *Archeoglobus fulgidus* AF1521 protein,[7] the functional HCoV-229E X domain was predicted to encompass the pp1a/1ab residues Glu 1265 to Val 1436. Based on our prediction that Asn 1302 and Asn 1305 residues are part of the X domain's active site, we expressed a mutant protein in which these two Asn residues were replaced with Ala and used it as a negative control in subsequent experiments. Following IPTG-induced protein expression in *E. coli*, the fusion proteins were purified by amylose-affinity chromatography and cleaved with factor Xa to separate the X domains from MBP. SDS-PAGE analysis of cell lysates obtained from IPTG-induced or noninduced bacteria as well as protein samples from the amylose-affinity purification and factor Xa cleavage (Fig. 1A and 1B) showed that sufficient amounts of soluble HCoV-229E X domain could be obtained for biochemical studies (Fig. 1A and 1B). To test the Appr-1"-pase activity, the factor Xa-cleaved proteins were incubated with Appr-1"-p and the products were analyzed by thin-layer chromatography. As Fig. 1C illustrates, Appr-1"-p was processed by the HCoV-229E X domain (but not the negative control protein containing two putative active-site substitutions) to a product that comigrated with Appr. As expected, the same product was seen when the positive control, alkaline phosphatase, was used (Fig. 1C). Taken together, the data show that (i) the HCoV-229E nsp3-associated X domain has Appr-1"-pase activity and (ii) Asn residues 1302 and 1305 are essential for enzymatic activity.

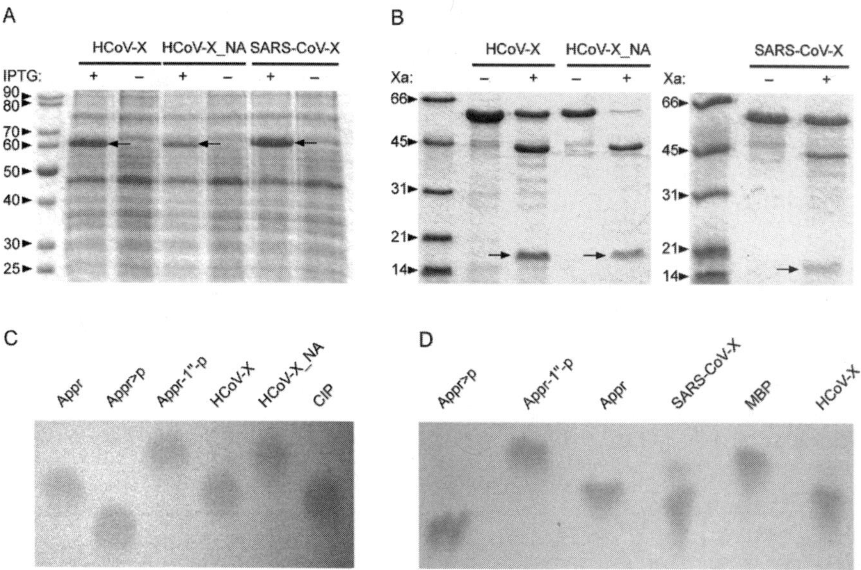

Figure 1. Expression, purification, and ADP-ribose-1"-phosphatase activity of the HCoV-229E and SARS-CoV X domains. A, The HCoV-229E X domain (HCoV-X), a mutant form of the HCoV-229E X domain containing two presumed active-site Asn substitutions (HCoV-X_NA, see text for details), and the SARS-CoV X domain (SARS-CoV-X) were expressed in *E. coli* as MBP fusion proteins. Total lysates obtained from IPTG-induced and noninduced *E. coli* TB1 cells transformed with the appropriate expression plasmids were analyzed in a 12.5% SDS-polyacrylamide gel which was stained with Coomassie brilliant blue. Arrows indicate the overexpressed MBP-X fusion proteins. B, Factor Xa cleavage of amylose affinity-purified MBP-X domain fusion proteins. Arrows indicate the proteolytically released X domains. C, Cellulose thin-layer chromatography analysis of the Appr-1"-pase activities of HCoV-X, HCoV- X_NA, and calf intestine phosphatase (CIP). D, Analysis of the Appr-1"-pase activities of SARS-CoV-X, maltose-binding protein (MBP), and HCoV-X. Markers: Appr>p, Appr-1"-p, and Appr.

3.2. Appr-1"-pase Activity of the SARS-CoV X Domain

The pp1a/pp1ab N-terminal region, which also includes nsp3, is the most divergent part of coronavirus replicase polyproteins.[1,3] Consistent with this observation, SARS-CoV features an nsp3 subdomain organization that significantly differs from that of most other coronaviruses. Thus, for example, SARS-CoV employs only one papain-like protease to process the N-proximal pp1a/pp1ab region.[10] Furthermore, next to the X domain, there is a domain called SUD that is not conserved in other coronaviruses. To investigate whether, in this very divergent sequence context, the X domain of SARS-CoV retained its enzymatic activity, we expressed this protein (SARS-CoV pp1a/pp1ab residues Glu 1000 to Lys 1173) as an MBP-fusion protein using the above described protocols. As shown in Fig. 1A and 1B, the SARS-CoV X domain could be expressed and purified in a soluble form. Following proteolytic release from the MBP fusion protein using factor Xa, the SARS-CoV X domain was incubated with Appr-1"-p and shown to dephosphorylate this substrate effectively (Fig. 1D). The reaction product comigrated with the product generated by the activity of the HCoV-229E X domain which, in previous experiments, had been confirmed to be Appr (data not shown). As expected, the negative control, MBP, had no activity on this particular substrate, confirming that the observed activity was mediated by the viral rather than co-purified bacterial protein(s). Taken together, the data suggest that most (if not all) coronavirus X domains mediate highly specific phosphatase activities whose biological significance remains to be investigated.

4. REFERENCES

1. Ziebuhr, J., 2005, The coronavirus replicase, *Curr. Top. Microbiol. Immunol.* **287**:57-94.
2. Ziebuhr, J., Thiel, V., and Gorbalenya, A. E., 2001, The autocatalytic release of a putative RNA virus transcription factor from its polyprotein precursor involves two paralogous papain-like proteases that cleave the same peptide bond, *J. Biol. Chem.* **276**:33220-33232.
3. Snijder, E. J., Bredenbeek, P. J., et al., 2003, Unique and conserved features of genome and proteome of SARS-coronavirus, an early split-off from the coronavirus group 2 lineage, *J. Mol. Biol.* **331**:991-1004.
4. Genschik, P., Hall, J., and Filipowicz, W., 1997, Cloning and characterization of the Arabidopsis cyclic phosphodiesterase which hydrolyzes ADP-ribose 1″,2″-cyclic phosphate and nucleoside 2′,3′-cyclic phosphates, *J. Biol. Chem.* **272**:13211-13219.
5. Kumaran, D., Eswaramoorthy, S., Studier, F. W., and Swaminathan, S., 2005, Structure and mechanism of ADP-ribose-1″-monophosphatase (Appr-1″-pase), a ubiquitous cellular processing enzyme, *Protein Sci.* **14**:719-726.
6. Gorbalenya, A. E., Koonin, E. V., and Lai, M. M., 1991, Putative papain-related thiol proteases of positive-strand RNA viruses. Identification of rubi- and aphthovirus proteases and delineation of a novel conserved domain associated with proteases of rubi-, alpha- and coronaviruses, *FEBS Lett.* **288**:201-205.
7. Allen, M. D., Buckle, A. M., et al., 2003, The crystal structure of AF1521 a protein from Archaeoglobus fulgidus with homology to the non-histone domain of macroH2A, *J. Mol. Biol.* **330**:503-511.
8. Shull, N. P., Spinelli, S. L., and Phizicky, E. M., 2005, A highly specific phosphatase that acts on ADP-ribose 1″-phosphate, a metabolite of tRNA splicing in Saccharomyces cerevisiae, *Nucleic Acids Res.* **33**: 650-660.
9. Martzen, M.R., McCraith, S. M., et al., 1999, A biochemical genomics approach for identifying genes by the activity of their products, *Science* **286**: 1153-1155.
10. Thiel, V., Ivanov, K. A., et al., 2003, Mechanisms and enzymes involved in SARS coronavirus genome expression, *J. Gen. Virol.* **84**:2305-2315.

NONSTRUCTURAL PROTEINS OF HUMAN CORONAVIRUS NL63

Yvonne Piotrowski, Lia van der Hoek, Krzysztof Pyrc, Ben Berkhout, Ralf Moll, and Rolf Hilgenfeld*

1. INTRODUCTION

In March 2004, a new human coronavirus was identified in The Netherlands. Named HCoV-NL63, it was found to cause acute respiratory disease in both children below the age of 1 and immunocompromised adults.[1] HCoV-NL63 belongs to the first of the three groups the coronaviruses have been subdivided into. Its plus-strand RNA genome consists of 27,553 nucleotides and a poly-A tail. At variance with typical group 1 coronaviruses, HCoV-NL63 has an additional 179-amino acid residue domain in the S-protein and only one open reading frame (ORF) instead of two between the S and the E gene.[1] Very recently, it was found that the HCoV-NL63 spike-protein binds to the SARS-CoV receptor angiotensin-converting enzyme 2 (ACE-2) but not to CD13 as other group I coronaviruses.[2] The genome of all known coronaviruses contains two ORFs that encode nonstructural proteins, and these are followed by the genes encoding the four structural proteins. Nonstructural proteins play an essential role in the replication and transcription of the virus genome as well as in polyprotein processing. Elucidating their structures and functions will pave the way for anticoronaviral drug discovery.[3,4]

2. MATERIALS AND METHODS

Cloning: To create expression clones of the genes encoding the non structural proteins of HCoV-NL63, we decided to apply TOPO cloning in combination with bacteriophage lambda recombination technology (Gateway technology by Invitrogen) instead of the commonly used T4-DNA ligase. In the TOPO reaction, the insert, modified by adding a 4-nucleotide sequence (CACC) at the 5' end, has been cloned using topoisomerase I in the correct orientation into the entry vector pENTR/D-TOPO. In the following LR recombination reaction, the insert was transferred from the entry vector

*Yvonne Piotrowski, Ralf Moll, Rolf Hilgenfeld, University of Lübeck, 23538 Lübeck, Germany. Lia van der Hoek, Krzysztof Pyrc, Ben Berkhout, University of Amsterdam, 1105 AZ Amsterdam, The Netherlands.

into the expression vector pEXP1-DEST using clonase. It contains the bacteriophage lambda recombination proteins integrase (Int) and excisionase (Xis) and the *E. coli*-encoded protein integration host factor (IHF), and promotes *in vitro* recombination between an entry clone (*att*L-flanked gene) and the *att*R-containing destination vector to generate an *att*B-containing expression clone.

Protein production, purification, and characterization: Expression was performed in *E. coli* BL21 Gold (DE3). Cultures were grown in YT medium. Expression was induced at OD_{660nm} = 0.4 by adding 1 mM IPTG. For test expression of the generated constructs, cells were harvested by centrifugation after 6 hours of incubation either at 37°C or 20°C. Samples were analyzed by SDS-PAGE and immunoblotting. For large-scale protein production of HCoV-NL63 Nsp9, cultures were incubated at 37°C for 6 hours, the bacterial pellets were resuspended in 50 mM Tris, 500 mM NaCl, 20 mM imidazole, 0.1 mM β-mercaptoethanol, pH 7.5, 1% Tween 20, 10 mg lysozyme/l cultivation volume, incubated on ice for 30 minutes and disrupted by French Press. After ultracentrifugation, the protein in the supernatant was purified by Ni-NTA-agarose chromatography.

The protein was eluted using a linear gradient from 20 mM to 500 mM imidazole with 50 mM Tris, 500 mM NaCl, 0.1 mM β-mercaptoethanol, pH 7.5. The pooled fractions containing the protein were dialyzed against 50 mM Tris, 100 mM NaCl, pH 7.5, and concentrated in an Amicon ultrafiltration cell.

Spectroscopic investigations. Dynamic light scattering: Dynamic light scattering (DLS) is a method for measuring the size of molecules and particle dispersion. It is useful as a pre-screen to find optimum protein crystallization conditions. A low polydispersity (width of the size distribution) is promising for crystal growth. The spectra were recorded using a Laser-Spectroscatter 201 (RiNA GmbH Netzwerk RNA Technologies). Twenty microlites of the protein solution, as used for the crystallization attempts, were measured in quartz glass cuvettes (light path: 10 mm, broadness of light accessible window: 1.5 mm, Hellma).

Fluorescence spectrometry: RNA binding to Nsp9 was followed by measuring fluorescence quenching of the single tryptophan residue in the protein using a Cary Eclipse fluorescence spectrometer (Varian). Spectra were measured using an excitation wavelength of 280 nm and emission wavelengths between 290 and 540 nm.

The titration experiment was carried out in a 500-μl quartz fluorescence cuvette containing 10 μM protein in 50 mM Tris, 100 NaCl, pH 8.5. Increasing amounts of a 20-mer oligoribonucleotide were added starting with 0.4 μM.

3. RESULTS AND DISCUSSION

3.1. Cloning and Expression of Nonstructural Proteins of HCoV-NL63

Eight out of 13 cDNAs encoding nonstructural proteins have been cloned into the entry vector pENTR/D-TOPO. Using bacteriophage lambda recombination technology, seven out of them could be recombined into the expression vector pEXP1-DEST. Until now, 6 genes encoding nonstructural proteins of HCoV-NL63 have been expressed successfully. Figure 1 displays the expression of Nsp3, Nsp7, Nsp8, Nsp9, and Nsp10, while the expression of Nsp5 failed using the method described.

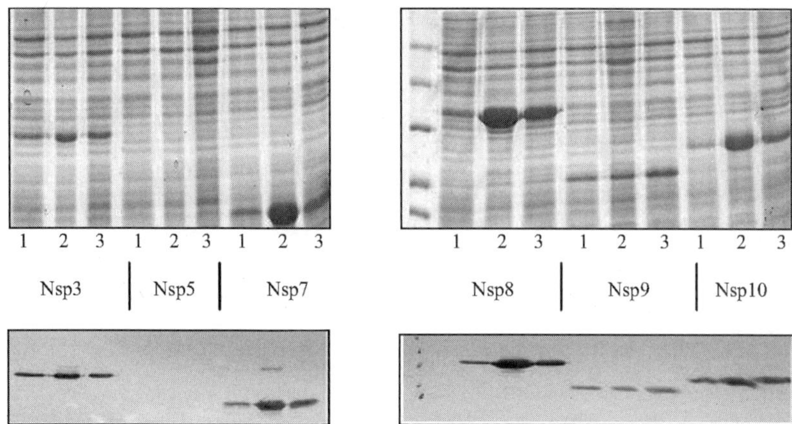

Figure 1. SDS-PAGE (above) and immunoblot with the anti-tetra-His-antibody (below), demonstrating the expression of the non structural proteins 3, 7, 8, 9 and 10. 1: before induction, 2: after 6 hours incubation at 37°C, 3: after 6 hours incubation at 20°C.

3.2. Nonstructural Protein 9 (Nsp9)

Nsp9 of HCoV-NL63 consists of 158 amino-acid residues and has a molecular weight of 17.5 kDa and a theoretical pI of 8.47. The N-terminal His-tag allows single-step purification using Ni-NTA-agarose chromatography. The purified Nsp9 can be concentrated up to 2.3 mg/ml. The size of Nsp9 aggregates can be reduced by addition of detergents such as *n*-octyl-β-glucoside, as shown with the DLS (Fig. 2A/B).

The single tryptophan residue of Nsp9 is partially exposed to solvent, as found in SARS-CoV Nsp9.[5] Due to this tryptophan, the fluorescence spectrum of the NL63 Nsp9 clearly displays an emission maximum at 352 nm (Fig. 3A). It can be almost completely quenched by adding 5.3 μM of a 20-mer oligoribonucleotide (Fig. 3B). Apparently, the tryptophan interacts with the ribonucleotide ligand suggesting that HCoV-NL63 Nsp9 is a single-stranded RNA-binding protein. The emission maximum did not change, indicating that the tryptophan residue remains in a polar environment in the presence of the oligoribonucleotide.

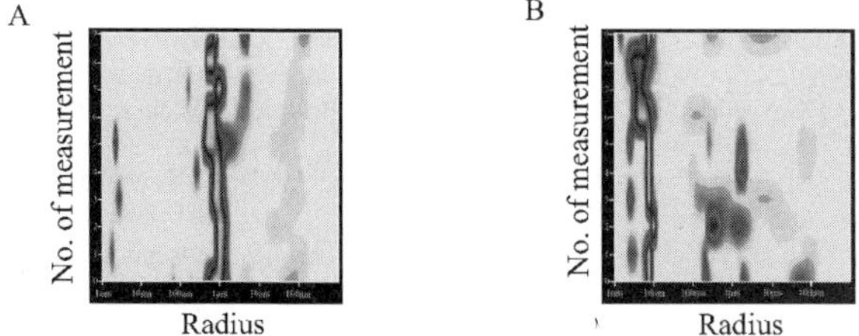

Figure 2. DLS spectrum before (A) and after incubation of the protein solution with 1% *n*-octyl-β-glucoside (B).

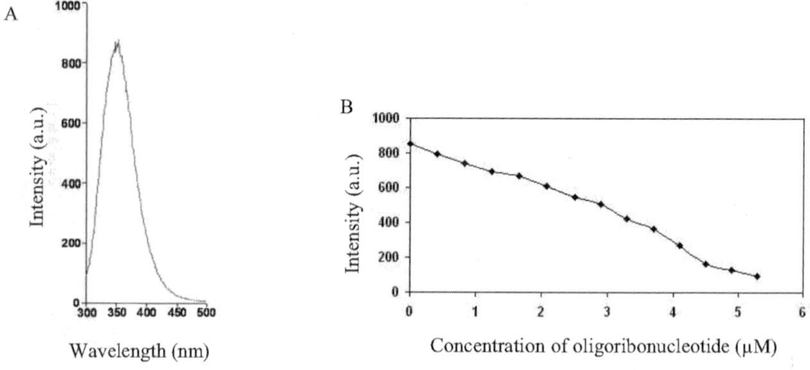

Figure 3. (A) Fluorescence spectrum of NL63 Nsp9. (B) Quenching with increasing amounts of a 20-mer oligoribonucleotide. Excitation wavelength: 280 nm.

4. CONCLUSIONS

Genes coding for 6 out of 13 nonstructural proteins of the human coronavirus NL63 have been successfully expressed as soluble proteins. The purified proteins are being used for biophysical studies and X-ray crystallography to elucidate the structure and function of the HCoV-NL63 replicase complex. Combining the Gateway technology with TOPO cloning provides an easy and fast way of cloning without the requirement of ligase and post-PCR procedures.

5. ACKNOWLEDGMENTS

This project is being supported by VIZIER (Comparative structural genomics of viral enzymes involved in replication; contract no. LSHG-CT-2004-511960; www.vizier-europe.org), an integrated project funded by the European Commission.

6. REFERENCES

1. L. van der Hoek, et al., Identification of a new human coronavirus, *Nat. Med.* **10**, 368-373 (2004).
2. H. Hofmann, et al., Human coronavirus NL63 employs the severe acute respiratory syndrome coronavirus receptor for cellular entry, *Proc. Natl. Acad. Sci. USA* **102**, 7988-7993 (2005).
3. K. Anand, H. Yang, M. Bartlam, Z. Rao, and R. Hilgenfeld, in: *Coronaviruses with special emphasis on first insights concerning SARS*, edited by A. Schmidt, M. H. Wolff, and O. Weber, (Birkhäuser, Basel, 2005), pp. 173-199.
4. D. A. Groneberg, R. Hilgenfeld, and P. Zabel, Molecular mechanisms of severe acute respiratory syndrome (SARS), *Respir. Res.* **6**, 8 (2005).
5. M.-P. Egloff, et al., The severe acute respiratory syndrome-coronavirus replicative nsp9 is a single-stranded RNA-binding subunit unique in the RNA virus world, *Proc. Natl. Acad. Sci. USA* **101**, 3792-3796 (2004).

MHV-A59 ORF1A REPLICASE PROTEIN NSP7-NSP10 PROCESSING IN REPLICATION

Damon J. Deming, Rachel L. Graham, Mark R. Denison, and Ralph S. Baric*

1. INTRODUCTION

The highly conserved region at the carboxy-terminus of the MHV replicase ORF1a polyprotein is processed by the main protease (M^{pro}) into mature products including nsp7-nsp10, which are associated with replication complexes and presumably involved with RNA synthesis.[1-4] The exact function of these proteins, in either their pre- or postcleaved forms, is unknown. However, disruption of M^{pro} cleavage during any stage of the infection cycle blocks replication, suggesting that constitutive proteolytic processing of the nonstructural polyprotein is a requirement for efficient transcription.[5] In this report, we describe preliminary data defining the requirement for the proteolytic processing of the nsp7-nsp10 proteins in MHV-A59 replication. Through use of an efficient MHV-A59 reverse genetics system,[6] we ablated each of the M^{pro} cleavage sites associated with the nsp7-nsp10 cassette, and evaluated whether the mutated genome was capable of supporting a viable virus, and if so, characterized the M^{pro} processing of the mutated protein, transcription function, and *in vitro* growth fitness.

2. RESULTS AND DISCUSSION

The M^{pro} targets amino acid sequences Q-S/N/A, cleaving after the essential Glu at position 1 (P1).[7] To evaluate the importance of cleavage on virus replication, the cleavage sites flanking nsp7, nsp8, nsp9, and nsp10 were individually disrupted by mutating the P1 Glu to an Ala (Table 1). However, there are two potential M^{pro} cleavage sites at the nsp7/8 interface, an LQA (present at positions P5-P3) and LQS (P2-P1'). Although the LQS has been shown to be cleaved during M^{pro} processing,[3] it is possible that the upstream LQA site is also functional in the presence or absence of the LQS site. To address this possibility, both sites were mutated, either individually (nsp7*8.A and

* Damon J. Deming, Ralph S. Baric, University of North Carolina, Chapel Hill, North Carolina. Rachel L. Graham, Mark R. Denison, Vanderbilt University Medical Center, Nashville, Tennessee.

Table 1. Amino-acid sequences for the nsp7-nsp10 Mpro cleavage sites for MHV-A59 (top) and mutant viruses (bottom).

	MHV-A59 M^{pro} Cleavage Sites									
	P5	P4	P3	P2	P1	P1'	P2'	P3'	P4'	P5'
nsp6/7	V	S	Q	I	**Q**	S	R	L	T	D
nsp7/8[a]	L	**Q**	A	L	**Q**	S	E	F	V	N
nsp8/9	T	V	V	L	**Q**	N	N	E	L	M
nsp9/10	T	V	R	L	**Q**	A	G	T	A	T
nsp10/11	G	S	Q	F	**Q**	S	K	D	T	N
	Mutant M^{pro} Cleavage Sites									
nsp6*7	V	S	Q	I	**A**	S	R	L	T	D
nsp7*8.A[b]	L	**A**	A	L	Q	S	E	F	V	N
nsp7*8.B[c]	L	Q	A	L	**A**	S	E	F	V	N
nsp7*8.A+B[d]	L	A	A	L	**A**	S	E	F	V	N
nsp8*9	T	V	V	L	**A**	N	N	E	L	M
nsp9*10	T	V	R	L	**A**	A	G	T	A	T
nsp10*11	G	S	Q	F	**A**	S	K	D	T	N

[a] There are two potential M^{pro} cleavage sites, an LQA and LQS.
[b] The upstream LQA site is mutated.
[c] The downstream LQS site is mutated.
[d] Both sites are mutated.

nsp7*8.B) or in combination (nsp7*8.A+B). The viability of the seven MHV-A59 mutants were tested by transferring the mutations (or combination of mutations) to the MHV-A59 infectious clone, driving full-length transcripts, and electroporating the RNA into cells. Viability was confirmed by the formation of CPE and detection of leader-containing transcripts by RT-PCR.

Three constructs produced replicating viruses. Of the viable mutants, the MHV-nsp7*8.A and MHV-nsp7*8.B cleavage mutants replicated to wild-type titers while MHV-nsp9*10 replication was reduced by about two logs. The genetic stability of the attenuated MHV-nsp9*10 mutant virus was analyzed after 15 serial passages on DBT cells. Plaque purified passage 15 virus, MHV-nsp9*10p15, displayed improved fitness of *in vitro* growth, as demonstrated by the near wild-type titers. Surprisingly, the recovered virus did not revert to wild-type sequence at the nsp9/nsp10 M^{pro} cleavage site, indicating that an as of yet unidentified mutation(s) has compensated for the virus's inability to properly process the nsp9-nsp10 precursor protein.

Immunoprecipitation (i.p.) using either anti-nsp8 (for the nsp7*8 mutants) or anti-nsp10 antibody (for the nsp9*10 mutants)[1] and Western blot analysis was used to verify whether or not mutation of the P1 Glu prevented cleavage of the viable viruses (data not shown). Bands corresponding to nsp10 (15-kDa protein) were absent in lysates of MHV-nsp9*10 and MHV-nsp9*10p15. However, a band of approximately 37 kDa, which is the predicted size for uncleaved nsp9-nsp10, was present in the mutants, but absent in the wild-type control cells. Analysis of the i.p.-Western blot suggested that MHV-A59 was able to use either of the two potential M^{pro} cleavage sites located at the nsp7/nsp8 site, likely accounting for the wild-type growth kinetics. Bands of approximately 22 kDa were precipitated with anti-nsp8 antibody for the wild-type control, MHV-nsp7*8.A, and MHV-nsp7*8.B. Notably, the MHV-nsp7*8.B band was slightly larger than that of the

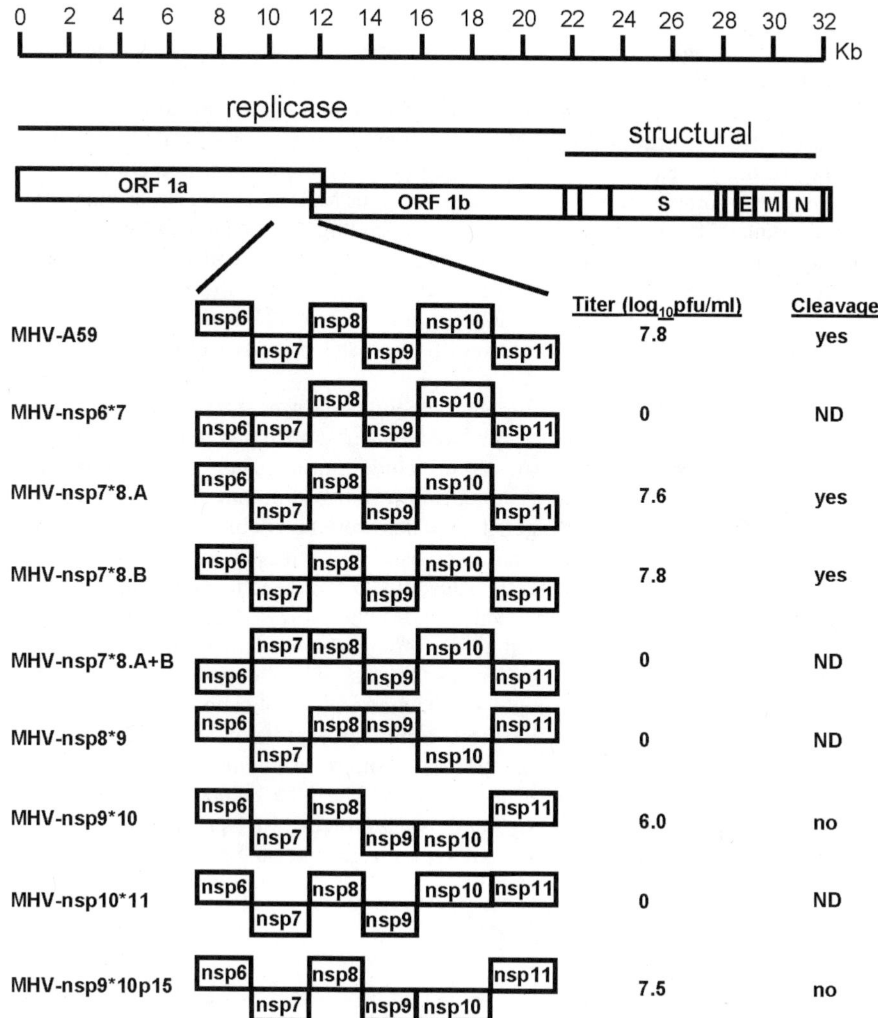

Figure 1. Summary of MHV-A59 and mutant M^pro processing of the nsp7-nsp10 cassette, viral titers, and verification of cleavage by immunoprecipitation and Western blot analysis of viable viruses. The picture of the nsp6-nsp11 proteins illustrates the predicted effects of mutating the P1 Glu M^pro cleavage sites, with uncleaved proteins forming a fusion protein. Peak log titers and the results of i.p.-Western blot analysis for M^pro cleavage at the site of mutation are also presented (yes = identification of proteolytically processed protein; no = identification of a fusion protein unprocessed by M^pro; ND = viral protein not detected due to lack of productive infection).

control or the first nsp7*8 mutant, consistent with the prediction that the LQA site was cleaved and yielded an nsp8 three amino acids larger than that of the LQS cleaved protein.

The cleavage site mutants were found to be similar to MHV-A59 in their transcriptional activity and cellular localization (results summarized in Figure 1). Northern blots hybridized with an RNA probe complementing the 5' end of the N-gene showed no notable alteration in either the pattern or relative amounts of subgenomic to genomic RNA in mutant and control viruses. The distribution of the mutant proteins within cells was compared with their wild-type counterparts. The nsp7-nsp10 proteins are known to colocalize with sites of viral replication while being excluded from regions of virion assembly.[1,2] In order to determine if ablation of the M^{pro} processing interferes with the ability of the protein to traffic into the replication complex, an immunofluorescence (IFA) study was completed. Cells infected with either MHV-A59, mock, or mutant virus were dual-stained for either nsp8 or nsp10 (depending on the mutant, as above) and nucleocapsid (N), which co-localizes with sites of active viral replication, or membrane (M), which is targeted to regions of virus assembly. Regardless of the virus, the nsp8 and nsp10 colocalized with N and was separate from M (data not shown). We were unable to find a difference between the localization of the nsp8 or nsp10 proteins between the mutants or wild-type.

Cleavage site ablation resulted in lethal, debilitated, or near wild-type viability, and mutation of most of the M^{pro} cleavage sites was lethal (summarized in Figure 1). Lethality could be due to disruption of nsp7-10 proteolytic processing causing a failure of precursor, intermediate, or mature protein function within the replication complex. However, not all of the M^{pro} cleavage site mutants were nonviable. Based on the genetic analysis, MHV-A59 has two functional nsp7-nsp8 M^{pro} cleavage sites, LQA and LQS, and disruption of either of these potential sites fails to affect replication competence, M^{pro} cleavage pattern, or cellular localization *in vitro*. Possible *in vivo* effects need to be studied. In contrast, simultaneous mutation of both sites was lethal. The only other M^{pro} cleavage site harboring a viable mutation was that shared by the nsp9 and nsp10 proteins. In this case, the mutant virus was highly attenuated in its replication efficiency and was unable to proteolytically process the fused nsp9-nsp10 protein. Serial passage of this virus restored wild-type replication but did so without reverting the mutated cleavage site or the ability to process the nsp9-nsp10 protein. The data demonstrate that with the exception of cleavage between the nsp9 and nsp10 proteins, M^{pro} processing of the nsp7-nsp10 cassette is essential in coronavirus RNA transcription and replication.

3. REFERENCES

1. A. G. Bost, R. H. Carnahan, X. T. Lu, and M. R. Denison, Four proteins processed from the replicase gene polyprotein of mouse hepatitis virus colocalize in the cell periphery and adjacent to sites of virion assembly, *J. Virol.* **74**, 3379-3387 (2000).
2. A. G. Bost, E. Prentice, and M. R. Denison, Mouse hepatitis virus replicase protein complexes are translocated to sites of M protein accumulation in the ERGIC at late times of infection, *Virology* **285**, 21-29 (2001).
3. X. T. Lu, A. C. Sims, and M. R. Denison, Mouse hepatitis virus 3C-like protease cleaves a 22-kilodalton protein from the open reading frame 1a polyprotein in virus-infected cells and in vitro, *J. Virol.* **72**, 2265-2271 (1998).
4. Y. van der Meer, et al., Localization of mouse hepatitis virus nonstructural proteins and RNA synthesis indicates a role for late endosomes in viral replication, *J. Virol.* **73**, 7641-7657 (1999).
5. J. C. Kim, R. A. Spence, P. F. Currier, X. Lu, and M. R. Denison, Coronavirus protein processing and RNA synthesis is inhibited by the cysteine proteinase inhibitor E64d, *Virology* **208**, 1-8 (1995).
6. B. Yount, M. R. Denison, S. R. Weiss, and R. S. Baric, Systematic assembly of a full-length infectious cDNA of mouse hepatitis virus strain A59, *J. Virol.* **76**, 11065-11078 (2002).
7. H. J. Lee, et al., The complete sequence (22 kilobases) of murine coronavirus gene 1 encoding the putative proteases and RNA polymerase, *Virology* **180**, 567-582 (1991).

STEM-LOOP 1 IN THE 5' UTR OF THE SARS CORONAVIRUS CAN SUBSTITUTE FOR ITS COUNTERPART IN MOUSE HEPATITIS VIRUS

Hyojeung Kang, Min Feng, Meagan E. Schroeder, David P. Giedroc, and Julian L. Leibowitz*

1. INTRODUCTION

A novel coronavirus, SCoV, was the cause of a highly lethal outbreak of respiratory disease in the winter of 2002–2003. SCoV was initially thought to represent a new coronavirus subgroup but more recent phylogenetic analyses have placed SCoV within the group 2 coronaviruses as an early split-off from the group 2 branch, which includes MHV.[3] This study investigated the possibility that the SCoV UTRs, or predicted secondary structural elements within the SCoV 5' UTR, could functionally substitute for their MHV counterparts.

Consensus secondary structural models for the 5'-most 150 nts of the 5' untranslated region (UTR) of mouse hepatitis virus (MHV) and severe acute respiratory syndrome coronavirus (SCoV) were generated with ViennaRNA 1.5 and PKNOTS. The 5' UTR of both viruses is predicted to contain three major helical stem-loop structures, designated SL1, SL2, and SL4. Full-length cDNAs of the MHV genome and MHV/SCoV chimeras were assembled *in vitro* and transcribed to generate chimeric genomes. Replacement of the entire MHV 5' UTR with the SCoV 5' UTR resulted in nonviable genomes. In contrast, the SCoV SL1 and the SCoV 3' UTR can functionally substitute for their MHV counterparts. These chimeric viruses formed smaller plaques and grew more slowly and to lower titer than the parental MHV-A59. A59/SCoV-5' UTR/MHV-TRS and A59/SCoV-SL4-AUG chimeric viruses directed the synthesis of minus-strand genome-sized RNA, but subgenomic RNAs (sgRNAs) were not made.

2. MATERIALS AND METHODS

The A plasmid of the MHV reverse genetic system[5] was utilized as a basis for constructing a fusion of the SCoV 5' UTR to the MHV gene 1 coding sequence. The

* Hyojeung Kang, Min Feng, Julian L. Leibowitz, Texas A&M University-HSC, College Station, Texas, 77843. Meagan E. Schroeder, David Giedroc, Texas A&M University, College Station, Texas, 77843.

strategy employed to construct this fusion exploited “No see’m” technology to eliminate BsmBI restriction sites engineered into the ends of DNA fragments by PCR.[5] An oligonucleotide assembly strategy was used to replace predicted stem-loop structures in MHV with their SCoV counterparts in plasmid A,[5] and to replace the SCoV TRS in the MHV/SCoV-5’ UTR chimera. Genome RNAs were assembled from cloned cDNAs and transcribed *in vitro* from the assembled cDNAs and the transcripts electroporated into BHK-R cells to recover infectious virus as described previously.[5] To determine the virus specific RNAs produced by nonviable chimeras, total RNAs were extracted 8 and 24 hours post electroporation and analyzed by nested RT-PCR.

3. RESULTS

Consensus secondary structural models for the 5’-most 150 nts of the 5’ UTR from MHV and SCoV as well as other coronaviruses were predicted using ViennaRNA. The secondary structural models were strikingly similar, characterized by three major helical stem-loops, denoted as SL1, SL2, and SL4, respectively (Fig. 1A).

MHV/SCoV chimeric genomes (Fig. 1B) were constructed as described in “Materials and Methods”. Only cultures electroporated with A59/SCoV-3’ UTR and A59/SCoV-SL1 chimeric genomes developed cytopathic effect (CPE). Virus was recovered from media, plaque purified, and expanded in DBT cells (Table 1). Average plaque sizes of A59/SCoV-SL1 and A59/SCoV-3’ UTR chimeric viruses were 1.5 mm and 1.4 mm in diameter, respectively. These sizes corresponded to 54% ($P < 0.05$) and 50% ($P < 0.05$) of the average plaque size of MHV-A59-1000 (2.8 mm in diameter) (Table 1). The chimeric viruses achieved maximal titers approximately 50-fold less than those achieved by wild-type MHV-A59-1000 (Fig. 2).

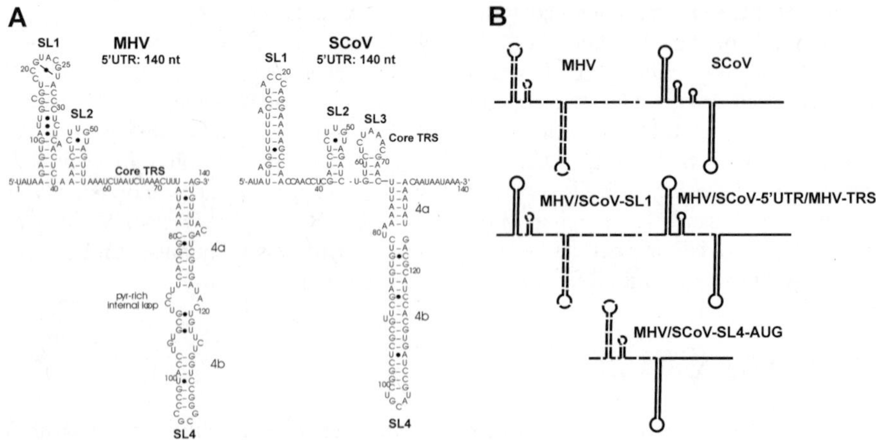

Figure 1. Predicted secondary structures of the first 140 nucleotides of MHV and SCoV 5’ UTRs (A) and schematic diagram of recombinant genomes (B).

Table 1. Characterization of MHV and SARS chimeric mutants.

Mutant virus	Viablilty*, plaque size (mm)	Minus-strand genomic RNA	Minus-strand subgenomic RNA	Plus-strand subgenomic RNA
A59-1000	Yes, 2.8±0.1	Yes	Yes	Yes
MHV/SCOV-5' UTR	No	No	No	No
MHV/SCOV-3' UTR	Yes, 1.4±0.1	ND**	ND	ND
MHV/SCOV-5'&3' UTRs	No	ND	ND	ND
MHV/SCOV-5' UTR/MHV-TRS	No	Yes	No	No
MHV/SCOV-SL1	Yes, 1.5±0.1	ND	ND	ND
MHV/SCOV-SL4-AUG	No	Yes	No	No

*Nonviable genomes failed to produce virus in three independent experiments.
** Not determined.

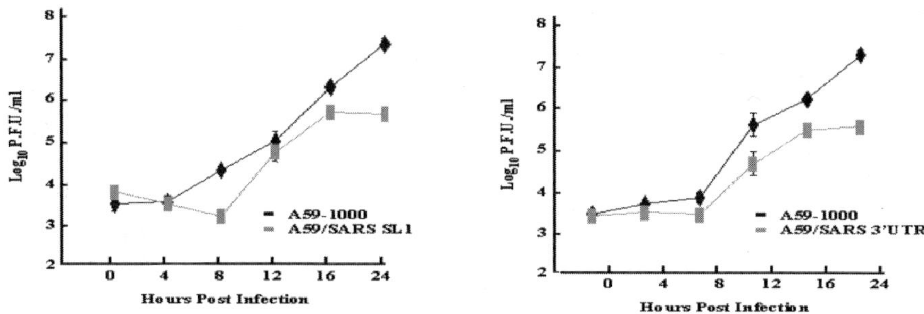

Figure 2. Growth kinetics of MHV/SCoV-SL1 and MHV/SCoV-3' UTR chimeric viruses.

Nested RT-PCR assays reveal that A59/SCoV-5' UTR/MHV-TRS and A59/SCoV-SL4-AUG chimeric genomes directed the synthesis of genome-sized minus-strand RNA. In contrast, cells electroporated with A59/SCoV-5' UTR chimeric genomes failed to produce minus-strand genome-sized RNA (Table 1). We also investigated whether the nonviable chimeric genomes directed the synthesis of sgRNA7. A59/SCoV-5' UTR, A59/SCoV-5' UTR/MHV-TRS, and A59/SCoV-SL4-AUG produced neither minus-strand sgRNA7 nor plus-strand mRNA7 (Table 1).

4. DISCUSSION

This study showed that the entire 3' UTR and SL1 of 5' UTR of SCoV could functionally substitute for their counterparts in MHV. The functional substitution of the SCoV 3' UTR for the MHV 3' UTR is consistent with a previous study that used targeted RNA recombination to isolate a similar 3' UTR chimeric virus.[1] Although the SL1s of SCoV and MHV have just 47.7% sequence identity, SL1 of SCoV is capable of forming a stem-loop structure similar to that of MHV SL1 (Fig. 1A). Substitution of the SCoV SL1 for the MHV SL1, which also increases the spacing between the predicted MHV SL1 and SL2 by 2 nts, did not affect viral viability, but it did decrease viral replication efficiency. These results suggest that SL1 may well play same role in replication of SCoV and MHV and supports our structural model (Fig. 1A).[2] Significant sequence and/or secondary structural differences 3' to SL4 in the MHV and SCoV genomes might account for the failure to recover MHV/SCoV SL4-AUG chimeric viruses.

Failure of A59/SCoV-5' UTR and A59/SCoV-5' UTR/MHV-TRS chimeras to replicate is consistent with a previous study that reported that sequences downstream of the core TRS sequence of *tranmissible gastroenteritis virus* exert a strong influence on template-switching during minus-strand sgRNA synthesis.[4] The three-step working model of coronavirus transcription[6] postulates that the correct 5' leader TRS is required for synthesis of minus-strand sgRNA and subsequent mRNA synthesis. Our data indicate the TRS may function in the synthesis of minus strand genome RNA as well, perhaps by mediating formation of 5'-end 3'-end complex, either by RNA-RNA, protein-RNA or protein-protein interactions. Interestingly, while A59/SCoV-5' UTR/MHV-TRS and A59/SCoV-SL4-AUG chimeric viruses are capable of synthesizing genome-sized minus-strand RNA, they could not produce minus-strand sgRNAs. This suggests a defect in template switching, perhaps due to a mismatch of sequences 3' to the MHV TRS and/or improper structural presentation of the TRS in these chimeras.

5. ACKNOWLEDGMENTS

This work was supported by NIH grants AI051493 (J.L.L.) and AI040187 (D.P.G.).

6. REFERENCES

1. S. J. Goebel, J. Taylor, and P. S. Masters, The 3' cis-acting genomic replication element of the severe acute respiratory syndrome coronavirus can function in the murine coronavirus genome, *J. Virol.* **78**, 7846-7851 (2004).
2. I. L. Hofacker, Vienna RNA secondary structure server, *Nucleic Acids Res.* **31**, 3429-3433 (2003).
3. E. J. Snijder, P. J. Bredenbeek, J. C. Dobbe, V. Thiel, J. Ziebuhr, L. L. Poon, Y. Guan, M. Rozanov, W. J. Spaan, and A. E. Gorbalenya, Unique and conserved features of genome and proteome of SARS-coronavirus, an early split-off from the coronavirus group 2 lineage, *J. Mol. Biol.* **331**, 991-1004 (2003).
4. I. Sola, J. L. Moreno, S. Zuñiga, S. Alonso, and L. Enjuanes, Role of nucleotides immediately flanking the transcription-regulating sequence core in coronavirus subgenomic mRNA synthesis, *J. Virol.* **79**, 2506-2514 (2005).
5. B. Yount, M. R. Denison, S. R. Weiss, and R. S. Baric, Systematic assembly of a full-length infectious cDNA of mouse hepatitis virus strain A59, *J. Virol.* **76**, 11065-11075 (2002).
6. S. Zuñiga, I. Sola, S. Alonso, and L. Enjuanes, Sequence motifs involved in the regulation of discontinuous coronavirus subgenomic RNA synthesis, *J. Virol.* **78**, 980-999 (2004).

TRANSCRIPTIONAL REGULATION OF RNA3 OF INFECTIOUS BRONCHITIS VIRUS

Soonjeon Youn, Ellen W. Collisson, and Carolyn E. Machamer*

1. INTRODUCTION

One of the unique properties of *Nidovirus* replication is production of a nested set of subgenomic RNAs by an unusual discontinuous transcription process.[1] These subgenomic RNAs are subsequently translated into viral proteins; regulation of subgenomic RNA transcription thus controls the timing and levels of viral protein expression during viral replication. Coronaviruses, which belong to the *Nidovirales* order, produce 6 to 9 subgenomic RNAs, depending on the virus. One of the established subgenomic RNA transcription regulation models for coronavirus involves sequences referred to as transcription regulation sequences (TRS). The TRS is well conserved among same strains of viruses, is located upstream of each open reading frame, and regulates subgenomic RNA production.[2] The TRS also has high sequence homology with the 3' end of the leader sequence at the 5' end of the genome, and the sequence homology between the body and leader TRS is an important factor regulating subgenomic RNA transcription.[3] Studies with transmissible gastroenteritis virus showed that not only the homology between leader and body TRS is critical, but also sequences flanking the TRS contribute to the transcription of subgenomic RNAs.[4,5]

Infectious bronchitis virus (IBV) is a Group 3 coronavirus, members of which infect avian species. IBV produces 6 subgenomic RNAs. Even though the TRS is known, importance of TRS itself or adjacent sequences have not been addressed. In this study, using an infectious cDNA clone of IBV and targeted mutagenesis, we show that the sequences surrounding the IBV TRS contribute to subgenomic RNA regulation.

* Soonjeon Youn, Carolyn E. Machamer, Johns Hopkins University School of Medicine, Baltimore, Maryland, 21205. Ellen W. Collisson, Texas A & M University School of Veterinary Medicine, College Station, Texas, 77843.

2. MATERIALS AND METHODS

Site-directed mutations near TRS3 were introduced using PCR-based site-directed mutagenesis, and recombinant IBV containing these mutations were constructed using an infectious cDNA clone of IBV as described previously.[6] The subgenomic RNA profile of IBV and recombinant IBVs were compared by Northern blot analysis. Total cellular RNAs were extracted from infected Vero cells and electrophoresed on 1% denaturing agarose gel containing 2.2M formaldehyde. A ^{32}P-labeled anti-sense IBV 3' UTR probe was used to detect the subgenomic RNAs of IBV.

3. RESULTS

3.1. Four Nucleotide Changes Downstream of the RNA3 TRS Attenuates RNA Transcription

A canonical dilysine endoplasmic reticulum (ER) retrieval signal was previously identified in near the C-terminus of the IBV S protein.[7] To address the significance of this trafficking signal to IBV infection, we used an infectious cDNA clone of IBV.[6] We produced a recombinant virus with a mutation in the ER retrieval signal of the S protein, which is encoded by subgenomic RNA2 (IBV-S2A, Fig. 1). IBV-S2A gave an interesting phenotype: not only did it produce less total virus and form bigger plaques (as expected), but there was also significantly impaired virus release from infected cells (which was unexpected). Further analysis of subgenomic RNA by Northern blot showed that cells infected with IBV-S2A had significantly less RNA3 transcript than those infected with the parental IBV (Fig. 2B). The reduction in RNA3 levels was detected at both early and late times post-infection, and was mirrored by decreased E protein expression, which is one of the ORFs encoded by RNA3. Interestingly, the ER retrieval signal of IBV S is located 10 nucleotides downstream of RNA3 core TRS (Fig. 1).

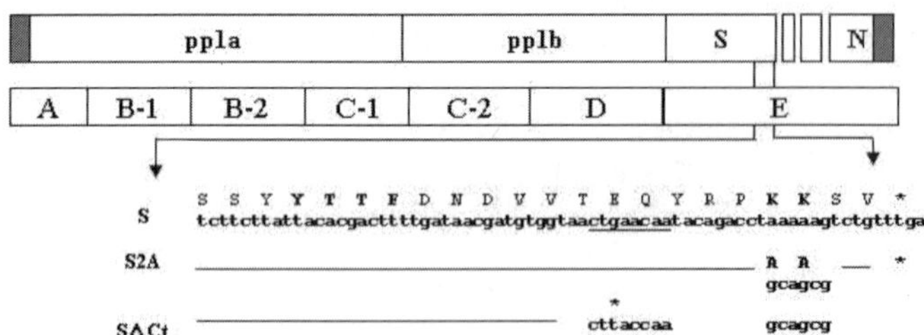

Figure 1. Schematic of recombinant IBV construction and introduced mutations. Canonical dilysine ER retrieval signal in the IBV S C-terminus (bold) was replaced with two alanines (IBV-S2A). A recombinant virus with a premature stop codon upstream of the ER retrieval signal but containing the alanine codons was rescued serendipitously (SΔCt). The TRS of RNA3 is underlined. Asterisks indicate stop codons. Lines represent nucleotide and amino acid sequences that are identical.

3.2. Two Different Contexts of Core TRS and Transcription Regulation

There are not many complete genomic sequences of IBV available. However, based on the available sequence data, the TRS of IBV is well conserved among different strains. Interestingly, unlike other coronaviruses, the IBV genome has two different TRS core sequences (Fig. 2A). The major one is CTTAACAA, which is used as the core TRS for the leader and subgenomic RNA4, 5 and 6. The minor TRS is CTGAACAA, which acts as the core TRS for subgenomic RNA2 and 3. Except for the genomic RNA, RNAs transcribed under the control of the major TRS are produced at higher levels than those under control of the minor TRS (RNA2 and 3, Fig. 2B). One of the mutants rescued while studying the effects of the IBV S ER retrieval signal on IBV infection had a nucleotide change in the TRS. This mutation encodes a premature stop codon in IBV S protein, resulting in a truncated protein lacking the ER retrieval sequence (IBV-SΔCt, Fig. 1). The mutation in this virus also changed the RNA3 TRS from CTGAACAA to CTTAACAA (resulting in a perfect match with the leader TRS). However, Northern blot analysis of RNA from cells infected with this mutant showed that the change to the major TRS did not overcome the RNA3 transcription attenuation caused by the downstream nucleotide changes (data not shown).

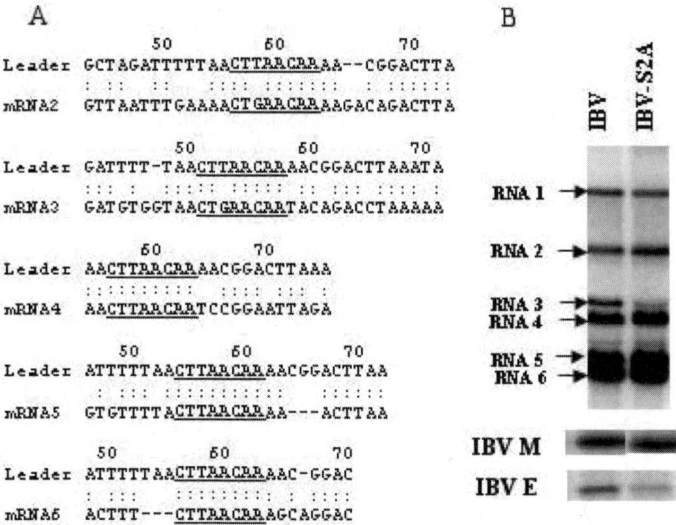

Figure 2. (A) Sequence alignment of each subgenomic RNA TRS with the 3' end of the leader core TRS. The TRS of RNAs 4, 5, and 6 shows perfect sequence homology with the leader core TRS, by contrast with the TRS of RNAs2 and 3. (B) Vero cells were infected with IBV or IBV-S2A and total RNAs were extracted from the infected cells and subjected to Northern blot analysis using a ^{32}P random primed anti-sense IBV 3' UTR RNA as a probe (upper panel). Immunoblot for IBV M and IBV E proteins was also performed from the same samples (lower panel). Four nucleotide changes downstream of the RNA3 TRS result in less RNA3 transcript compared with the RNA3 transcript levels in cells infected with the parental IBV. Reduced RNA3 transcript was confirmed by lower expression level of E protein.

4. DISCUSSION

The genome of IBV is smaller than other coronaviruses (27.6 vs. 29 to 31 kb). However, all of the necessary information is accommodated by overlapping most of the ORFs. Except for RNA1 and RNA5, the TRS for each ORF contains codons that conserve not only the TRS but also the amino acid sequence of the protein encoded on the upstream ORF. For example, IBV S protein (encoded by RNA2) contains targeting signals on its cytoplasmic tail. One of these signals is an ER retrieval signal, which is potentially important for viral assembly. Interestingly, this ER retrieval signal is 10 nucleotides downstream from the RNA3 TRS. Disruption of the ER retrieval signal by 4 nucleotide changes also attenuated transcription of the downstream subgenomic RNA (RNA3) with a corresponding decrease in E protein expression. This is the first study showing that for IBV, both the TRS and neighboring sequences affect subgenomic RNA transcription.

When we isolated the recombinant virus that lacked the ER retrieval signal, we identified a mutant that had a premature stop codon upstream of the ER retrieval signal. This mutation created a perfect match between the leader TRS and the TRS for RNA3. However, the downstream nucleotide changes (in the ER retrieval sequence) remained and perhaps explain why this mutation did not rescue the RNA3 level to normal. Interestingly, the IBV field isolate Mass 41 has the same premature stop codon in the S gene that resides in the RNA3 TRS core sequence of IBV-SΔCt, rendering a perfect match in body and leader TRS for RNA3. However, this virus conserved the nucleotide sequence in the ER retrieval signal. It will be interesting to see if the Mass 41 strain of IBV produces a normal or even higher level of subgenomic RNA3 compared with other strains of IBV.

5. ACKNOWLEDGMENTS

This work was supported by National Institutes of Health grant R01 GM64647.

6. REFERENCES

1. Sawicki, S. G., and Sawicki, D. L., 2005, Coronavirus transcription: a perspective, *Curr. Top. Microbiol. Immunol.* **287**:31.
2. Zuniga, S., Sola, I., Alonso, S., and Enjuanes, L., 2004, Sequence motifs involved in the regulation of discontinuous coronavirus subgenomic RNA synthesis, *J. Virol.* **78**:980.
3. van Marle, G., Dobbe, J. C., Gultyaev, A. P., Luytjes, W., Spaan, W. J., and Snijder, E. J., 1999, Arterivirus discontinuous mRNA transcription is guided by base pairing between sense and antisense transcription-regulating sequences, *Proc. Natl. Acad. Sci. USA* **96**:12056.
4. Alonso, S., Izeta, A., Sola, I., and Enjuanes, L., 2002, Transcription regulatory sequences and mRNA expression levels in the coronavirus transmissible gastroenteritis virus, *J. Virol.* **76**:1293.
5. Sola, I., Moreno, J. L., Zuniga, S., Alonso, S., and Enjuanes, L., 2005, Role of nucleotides immediately flanking the transcription-regulating sequence core in coronavirus subgenomic mRNA synthesis, *J. Virol.* **79**:2506.
6. Youn, S., Leibowitz, J. L., and Collisson, E. W., 2005, In vitro assembled, recombinant infectious bronchitis viruses demonstrate that the 5a open reading frame is not essential for replication, *Virology* **332**:206.
7. Lontok, E., Corse, E., and Machamer, C. E., 2004, Intracellular targeting signals contribute to localization of coronavirus spike proteins near the virus assembly site, *J. Virol.* **78**:5913.

II. PROTEIN SYNTHESIS, STRUCTURE, AND PROCESSING

STRUCTURE, EXPRESSION, AND INTRACELLULAR LOCALIZATION OF THE SARS-CoV ACCESSORY PROTEINS 7a AND 7b

Andrew Pekosz, Scott R. Schaecher, Michael S. Diamond, Daved H. Fremont, Amy C. Sims, and Ralph S. Baric*

1. INTRODUCTION

The virus responsible for severe acute respiratory distress syndrome (SARS) has been identified as a coronavirus (SARS-CoV), and the genomic structure of several human and animal isolates has been determined.[1,2] Consistent with other members of the *Nidoviradae*, the genome consists of several genes that are highly conserved and represent proteins with essential, basic functions in the viral life cycle.[3] In addition to these essential genes, SARS-CoV encodes several accessory genes that are believed to be nonessential but important for viral replication *in vivo*.

Among these accessory genes, gene 7 encodes two major open reading frames, ORF-7a and ORF7b. ORF7a is predicted to be a 122-amino-acid type I integral membrane protein with a cleavable N-terminal signal peptide and a C-terminal membrane spanning domain. ORF7b is predicted to encode a 44-amino-acid, extremely hydrophobic protein. The stop codon for ORF7a overlaps the start codon of ORF7b.

Previously, we and others have confirmed the expression of ORF7a in SARS-CoV infected cells;[4–6] however, there is a discrepancy in the subcellular localization ascribed to the protein. ORF7a has been demonstrated in one study to colocalize with endoplasmic reticulum–Golgi intermediate compartment (ERGIC)-specific markers and possess an endoplasmic reticulum–retrieval motif in the cytoplasmic tail.[7]

However, experiments in our laboratories have localized ORF7a with Golgi-specific markers.[6] In addition, overexpression of ORF7a has been reported to induce apoptosis in a variety of transformed cell lines.[7] As of yet, no particular function has been assigned to the protein but the X-ray crystallographic structure of the extracellular domain has been determined.[6] The expression of the ORF7b protein has yet to be demonstrated in SARS-CoV infected cells.

We have continued our studies on the expression and function of the proteins encoded by gene 7 and identified an endoplasmic réticulum (ER) export signal residing in

*Andrew Pekosz, Scott R. Schaecher, Michael S. Diamond, Daved H. Fremont, Washington University School of Medicine, St. Louis, Missouri 63110. Amy C. Sims, Ralph S. Baric, University of North Carolina, Chapel Hill, North Carolina 27599-7435.

the ORF7a cytoplasmic tail. In addition, we have verified the expression of ORF7b in SARS-CoV–infected cells. Finally, we have begun to characterize the replication and cytopathicity of a recombinant SARS-CoV encoding GFP in place of gene 7.

2. MATERIALS AND METHODS

2.1. Cell Culture, Transfection, and Mutagenesis

Vero cells were cultured at 37°C in a humidified incubator with 5% CO2 as previously described.[8] The cells were plated onto glass coverslips and incubated overnight at 37°C. Transfections were performed using LT-1 (Mirrus) transfection reagent at a ratio of 1 μg plasmid DNA to 4 μl of transfection reagent as previously described.[6,9] Cells were analyzed at 18–20 hours post transfection.

The ORF7a-GFP fusion protein has been described previously.[6] The localization of the ORF7a-GFP fusion protein is indistinguishable from that of the ORF7a protein (data not shown). Site-directed mutagenesis was performed using standard PCR techniques.

2.2. Confocal Microscopy

Indirect immunofluorescence for the Golgi protein Golgin-97 was carried out using a mouse monoclonal antibody (Molecular Probes, diluted 1:50). Coverslips were incubated with phosphate buffered saline (PBS) containing 3% normal goat serum (Sigma), 0.5% bovine serum albumin (BSA; Sigma), and 0.1% saponin (Sigma) for 30 minutes in primary antibody for 1 hour and in secondary antibody (goat anti-mouse IgG conjugated to AlexaFluor 594 (Molecular Probes; diluted 1:500) containing To-Pro-3 (Molecular Probes, diluted 1:500). A Zeiss LSM510 confocal microscope was used to acquire all images.

2.3. Viruses and Infections

Recombinant SARS-CoV containing a complete genome (rSARS WT) or with the gene 7 replaced with the GFP open reading frame (rSARS GFPΔORF7) were generated by reverse genetics.[10] Virus was diluted in DMEM containing 5% fetal bovine serum (FBS) and Vero cells were infected at a multiplicity of infection (MOI) of approximately 0.01 for 1 hour at 37°C. After extensive washing, the cells were incubated with DMEM containing 5% FBS at 37°C. At the indicated times post infection, the virus-infected cell supernatant was removed and stored at -70C. Infectious virus was quantified by plaque assay on Vero cells.[6]

2.4. Western Blotting

Vero cells were infected with an MOI of approximately 5.0 of rSARS WT or rSARS GFPΔORF7. At 24 hours postinfection, the cells were lysed with 1% sodium dodecyl sulfate (SDS) in water. Polypeptides were separated on a 17.5% acrylamide/4M urea SDS-PAGE and transferred to PVDF membranes. The membranes were blocked in PBS containing 0.3% Tween-20 and 5% dry milk powder, followed by antibodies recognizing ORF7b (rabbit polyclonal; diluted 1:1,000), the SARS-CoV nucleocapsid (N) protein (mouse monoclonal 87-A1; diluted 1:2,000) or beta-actin (mouse monoclonal; diluted 1:500; Abcam). After incubation with secondary antibodies conjugated to horseradish peroxidase that recognized either rabbit (diluted 1:7,500; Jackson Laboratories) or mouse (diluted 1:7,500; Jackson Laboratories) IgG, the membranes were developed using Amersham ECL-Plus substrate and quantitated by phosphorimager analysis.

3. RESULTS AND DISCUSSION

3.1. Intracellular Localization of ORF7a

We have demonstrated previously that ORF7a does not reach the plasma membrane after expression from cDNA or in SARS-CoV–infected cells.[6] However, there are contradictory data published with respect to the precise intracellular localization of the protein. The localization of the ORF7a protein was determined by transfecting Vero cells with a cDNA expression plasmid encoding an ORF7a-GFP fusion protein and co-localizing the protein with the Golgi-localized protein Golgin-97 (Figure 1). An identical localization pattern was seen in SARS-CoV virus-infected cells and cells transfected with the ORF7a cDNA alone. The data indicate the ORF7a protein targets primarily to the Golgi apparatus and not to the endoplasmic reticulum or intermediate compartment.

3.2. Identification of an ER Export Motif

The cytoplasmic tail of the ORF7a protein contains a motif R/K-X-R/K (corresponding to amino acids 118–120) that has been demonstrated to mediate COPII-dependent transport out of the ER.[11] To ascertain whether this sequence plays a role in ORF7a export out of the ER, the cDNA encoding the ORF7a-GFP fusion protein was altered from codons encoding Lys to ones encoding Ala at amino acids 118 and 120. The intracellular localization of the mutated proteins with respect to Golgin-97 was then determined. Figure 1 shows that mutation of these amino acids leads to a loss of Golgin-97 co-localization. The protein appears to co-localize with markers for the ER (data not shown), indicating the introduced mutations may be hindering the transport of the protein out of the ER. The OR7aF-GFP K118A K120A still reacts with a conformation-dependent monoclonal antibody,[6] indicating the mutations have not resulted in misfolding of the protein (data not shown). Taken together, the data in Figure 1 indicates the ORF7a protein is targeted to the Golgi apparatus and utilizes the COPII transport machinery to exit the ER.

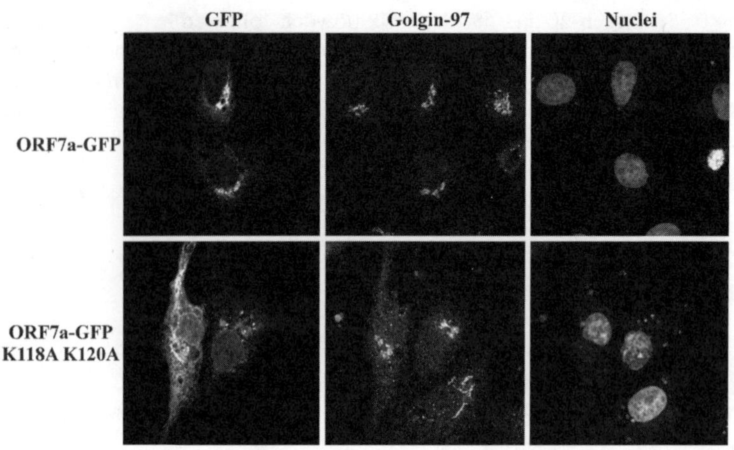

Figure1. Intracellular localization of ORF7aGFP. Vero cells expressing ORF7a-GFP were immunostained for the Golgi marker Golgin-97. The ORF7a-GFP protein co-localizes with Golgin-97. Substituting alanine for lysine at positions 118 and 120 of ORF7a-GFP eliminates a COPII transport sequence and results in a loss of Golgi localization.

3.3. Vero Cell Growth Kinetics of SARS-CoV Lacking Gene 7

The location of gene 7 in the SARS-CoV genome and its lack of conservation among other coronaviruses implies that the gene may be an accessory gene that is not essential for virus replication but important for infection of the natural host for the virus. To test this hypothesis, a recombinant SARS-CoV was generated in which the gene 7 coding region was replaced with that of GFP (rSARS GFPΔORF7ab). The replication of this virus was compared with its recombinant parental virus by infecting Vero cells at an MOI-0.01, harvesting infected cell supernatants at various times postinfection and determining the infectious, virus titer by plaque assay on Vero cells. The parental recombinant and the gene 7 replacement virus displayed identical replication kinetics and reached comparable peak titers (Figure 2), indicating the gene 7 products were not essential for SARS-CoV replication.

3.4. Expression of ORF 7b in SARS-CoV–Infected Cells

Sequence analysis indicates the SARS-CoV gene 7 potentially encodes for two proteins: ORF7a and ORF7b. The start codon of ORF7b is located far downstream of the 5' end of the only mRNA associated with the gene 7 transcription start site, and therefore is not in a very good context to initiate translation.[3,12] Nevertheless, we cloned the

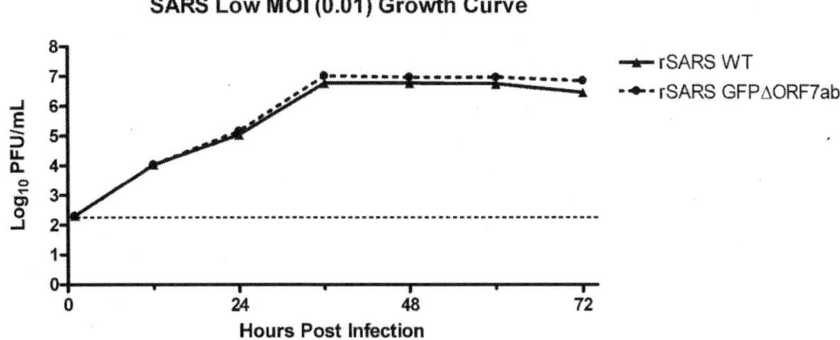

Figure 2. Replication kinetics of wild-type recombinant SARS-CoV compared with a recombinant virus containing GFP in place of gene 7.

predicted ORF7b open reading frame into a bacterial expression vector as a GST-fusion protein and immunized rabbits with the resulting protein. The antiserum was used to probe lysates from mock, SARS-CoV, or SARS-CoV GFPΔORF7ab cells. A band corresponding to the predicted molecular weight of the ORF7b protein was detected in SARS-CoV–infected cells but not in mock- or SARS-CoV GFPΔORF7ab-infected cells (Figure 3), indicating the 7b protein was in fact synthesized in virus-infected cells. The blot was probed with antibodies to the N protein in order to verify the expression of a known viral protein, as well as antibodies to beta-actin, which serves as the protein loading control. The data indicate that the SARS-CoV gene 7 does indeed encode two distinct proteins, ORF7a and ORF7b.

4. CONCLUSIONS

Our data indicate the two proteins predicted to be encoded by gene 7 of SARS-CoV are in fact authentic viral proteins. ORF7a is a type I transmembrane protein that localizes to the Golgi and requires COPII-mediated export out of the ER. As we have not yet ascribed a particular function to this protein, it is not yet clear why the protein possess these two properties. It is intriguing to speculate that ORF7a is perhaps interacting and retaining host cell proteins in the Golgi, in order to prevent their transport to the plasma membrane or other organelles. ORF7b is a highly hydrophobic protein, and preliminary data suggests it associates with cellular membranes (data not shown). It will also be important to determine whether ORF7a or ORF7b are structural components of the virion or simply nonstructural proteins.

Our data on the replication of a gene 7 replacement SARS-CoV indicate the gene 7 products are not essential, nor do they alter virus replication on Vero cells. It is important to emphasize that these results have only been demonstrated in Vero cells, and it is possible that infection and replication of the SARS-CoV gene 7 replacement virus may be altered on other cell lines. It is more likely that these proteins are required for efficient replication of SARS-CoV in the yet to be identified natural host of the virus. The use of animal models such as nonhuman primates, ferrets, mice, or hamsters may also be useful in elucidating a role for the gene 7 protein products in viral pathogenesis.

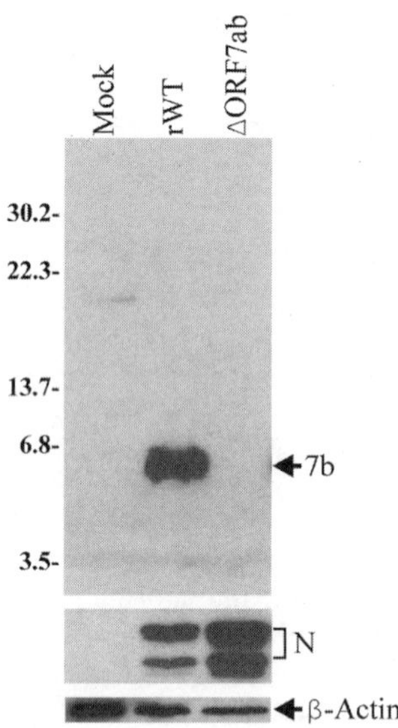

Figure 3. Western blotting for ORF7b expression in SARS-CoV–infected cells.

5. REFERENCES

1. Drosten, C., Gunther, S., et al., 2003, Identification of a novel coronavirus in patients with severe acute respiratory syndrome, *N. Engl. J. Med.* **348**:1967-1976.
2. Rota, P. A., Oberste, M. S., et al., 2003, Characterization of a novel coronavirus associated with severe acute respiratory syndrome, *Science* **300**:1394-1399.
3. Snijder, E. J., Bredenbeek, P. J., et al., 2003, Unique and conserved features of genome and proteome of SARS-coronavirus, an early split-off from the coronavirus group 2 lineage, *J. Mol. Biol.* **331**:991-1004.
4. Fielding, B. C., Tan, Y.-J., et al., 2004, Characterization of a unique group-specific protein (U122) of the severe acute respiratory syndrome coronavirus, *J. Virol.* **78**:7311-7318.
5. Chen, Y. Y., Shuang, B., et al., 2005, The protein X4 of severe acute respiratory syndrome-associated coronavirus is expressed on both virus-infected cells and lung tissue of severe acute respiratory syndrome patients and inhibits growth of Balb/c 3T3 cell line, *Chin. Med. J. (Engl.)* **118**:267-274.
6. Nelson, C. A., Pekosz, A., et al., 2005, Structure and intracellular targeting of the SARS-coronavirus Orf7a accessory protein, *Structure (Camb.)* **13**:75-85.
7. Tan, Y.-J., Fielding, B. C., et al., 2004, Overexpression of 7a, a protein specifically encoded by the severe acute respiratory syndrome coronavirus, induces apoptosis via a caspase-dependent pathway, *J. Virol.* **78**:14043-14047.
8. McCown, M., Diamond, M. S., et al., 2003, The utility of siRNA transcripts produced by RNA polymerase i in down regulating viral gene expression and replication of negative- and positive-strand RNA viruses, *Virology* **313**:514-524.
9. Pekosz, A., and Lamb, R. A., 1999, Cell surface expression of biologically active influenza C virus HEF glycoprotein expressed from cDNA, *J. Virol.* **73**:8808-8812.
10. Yount, B., Curtis, K. M., et al., 2003, Reverse genetics with a full-length infectious cDNA of severe acute respiratory syndrome coronavirus, *Proc. Natl. Acad. Sci. USA* **100**:12995-13000.
11. Giraudo, C. G., and Maccioni, H. J., 2003, Endoplasmic reticulum export of glycosyltransferases depends on interaction of a cytoplasmic dibasic motif with Sar1, *Mol. Biol. Cell* **14**:3753-3766.
12. Hussain, S., Pan, J. A., et al., 2005, Identification of novel subgenomic RNAs and noncanonical transcription initiation signals of severe acute respiratory syndrome coronavirus, *J. Virol.* **79**:5288-5295.

SUMOYLATION OF THE NUCLEOCAPSID PROTEIN OF SEVERE ACUTE RESPIRATORY SYNDROME CORONAVIRUS BY INTERACTION WITH UBC9

Qisheng Li, Han Xiao, James P. Tam, and Ding X. Liu*

1. INTRODUCTION

Severe acute respiratory syndrome coronavirus (SARS-CoV) encodes a highly basic nucleocapsid (N) protein of 422 amino acids. Similar to other coronavirus N proteins, SARS-CoV N protein is predicted to be phosphorylated and may contain nucleolar localization signals (NuLs), RNA binding domain, and regions responsible for self-association and homo-oligomerization.[1–4] In this study, we identified Ubc9, a host protein involved in sumoylation, as a binding partner of the N protein in a yeast two-hybrid screen. This interaction was verified by GST pull-down assay, coimmunoprecipitation and colocalization of the two proteins in cells. Subsequent biochemical characterization studies demonstrate that SARS-CoV N protein is post-translationally modified by covalent attachment to the small ubiquitin-like modifier (SUMO). The major sumoylation site was mapped to the ^{62}lysine residue of the N protein. Further expression and characterization of wild-type N protein and K62A mutant reveal that sumoylation of the N protein drastically promotes its homo-oligomerization. This is the first report showing that a coronavirus N protein undergoes post-translational modification by sumoylation and the functional implication of this modification in the formation of coronavirus ribouncleoprotein complex, virion assembly, and virus-host interactions.

2. MATERIALS AND METHODS

Transient expression of viral protein in HeLa cells: Constructs containing plasmid DNA under the control of a T7 promoter were transiently expressed in mammalian cells using the recombinant vaccinia virus (vTF7-3) system as described before.[5] In this study, the transfection reagent used was Effectene (Qiagen).

*Nanyang Technological University, Singapore, 637551, and Institute of Molecular and Cell Biology, Proteos, Singapore, 1386731.

Immunoprecipitation: Transiently transfected HeLa cells in 100-mm dishes were lysed in 1 ml of lysis buffer with 0.5% protease inhibitor cocktail (Sigma). The lysates were centrifuged at 12,000 rpm for 20 min at 4°C. The supernatants were added with anti-His (Qiagen), anti-SUMO-1 (Zymed), or anti-Flag M2 (Stratagene) antibodies at 4°C for 2 h. Protein-A agarose beads (40 μl) (KPL) were added to the lysates and incubated with rolling for 1 h at 4 °C. The beads were collected by centrifugation and washed three times with RIPA buffer. Proteins binding to the beads were eluted by adding 2x SDS loading buffer and analyzed by Western blotting with anti-Flag antibody.

Expression of GST fusion protein and GST pull-down assay: The SARS-CoV N protein was cloned into pGEX-5X-1 and expressed as GST-N fusion protein in *E. coli* BL21 cells. Both GST-N and GST alone were purified by affinity chromatography using glutathione-Sepharose 4B (Amersham Pharmacia Biotech).

In vitro translation of Ubc9 was carried out using the T7-coupled rabbit reticulocyte lysate system in the presence of [^{35}S] methionine (Promega). For the binding assay, GST alone or GST-Ubc9 fusion proteins were prebound to glutathione-Sepharose beads. Five microliters of ^{35}S-labeled Ubc9 was then added, and incubation was continued for at least 2 h. The beads were washed five times. Labeled proteins bound on the beads were analyzed by SDS-PAGE. Radiolabeled bands were visualized by autoradiography.

Indirect immunofluorescence: SARS-CoV N protein and Ubc9 were transiently expressed in HeLa cells. After rinsing with PBS, cells were fixed with 4% paraformaldehyde for 15 min and permeabilized with 0.2% Triton X-100, followed by incubation with specific antibodies diluted in fluorescence dilution buffer at room temperature for 2 h. Cells were then washed with PBS and incubated with FITC- or TRITC-conjugated anti-rabbit or anti-mouse secondary antibodies (Dako) in fluorescence dilution buffer at 4°C for 1 h before mounting. All images were taken using a Zeiss LSM510 META laser scanning confocal microscope.

Construction of plasmids: Plasmid pcDNA3.1-N, which covers the SARS-CoV N sequence, was constructed by cloning an *Eco*RI/*Not*I digested PCR fragment into *Eco*RI/*Not*I digested pcDNA3.1(+). The PCR fragment was generated using primers (5'-CGGAATTCCGATGTCTGATAATG GACCC-') and (5'-AATAAATAGCGGCCGCTGCCTGAGTTG AATC-3'). pFlag-N was created by cloning a *Pst*I/*Eco*RI digested PCR fragment into *Pst*I/*Eco*RI digested pKT0-Flag. Plasmid pGEX-N was made by cloning a *Bam*HI- and *Eco*RI-digested PCR fragment into *Bam*HI/*Eco*RI digested pGEX-5X-1 (Pharmacia). The K62A mutant was introduced by two rounds of PCR as described before.[6] Human Ubc9 cDNA was inserted into pKT0-Flag at the PstI/EcoRI cloning sites. Primers used were (5'-AACTGCAGC ATGTCGGGGATCGCCCTCAGC-3') and (5'-CGGAATTCCGTTATGAGGGCGCAAAC TTCTT-3'). SUMO-1 was amplified from a human cDNA library derived from HeLa cells by PCR with primers (5'-TATCGGATCCCATGTCTGACCAGGCAAAACC-3') and (5'-CGGATC CTCGAGCTAAACTGTTGAATGACCCCCCGT-3'). The PCR product was digested with *Bam*HI and *Xho*I and cloned into *Bam*HI/*Xho*I digested pcDNA3.1(+) to generate pcDNA3.1-SUMO-1. All constructs were confirmed by automated nucleotide sequencing.

3. RESULTS

3.1. Identification of Ubc9 as an Interacting Protein of SARS-CoV N

To identify host proteins that interact with the N protein, a yeast two-hybrid screen of a HeLa cDNA library was performed. We obtained 20 independent clones corresponding to Ubc9 among a total of 24 positive clones.

This interaction was first tested by GST pull-down assays. As shown in (Fig. 1a, lanes 3 and 4), ^{35}S-labeled Ubc9 associated with GST-N fusion protein but not with GST alone.

In a subsequent coimmunoprecipitation experiment, an 18-kDa band that represents the precipitated Ubc9 could only be detected in cells coexpressing His-tagged N and Flag-tagged Ubc9 (Fig. 1b, lanes 4–6).

Immunofluorescence analyses revealed that the N protein was distributed throughout the cytoplasm and the nucleolus (Fig. 1c). Ubc9 was present more abundantly in the nucleus than in the cytoplasm (Fig. 1c). A merged picture revealed that Ubc9 colocalized with the N protein in the cytoplasm and the nucleolus (Fig. 1c).

3.2. Post-translational Modification of SARS-CoV N Protein by Sumoylation

After confirming the interaction between the N protein and Ubc9, we then examined whether the N protein is modified by sumoylation. As can be seen in Fig. 2a, in addition to three major isoforms of N protein that were detected under all conditions, a protein species of approximately 65-kDa was detected in cell lysates prepared with lysis buffer containing two isopeptidase inhibitors, iodoacetamide (IAA) and *N*-ethylmaleimide (NEM) (Fig. 2a, lanes 2, 3, and 5). Coexpression of N protein with SUMO-1 led to the

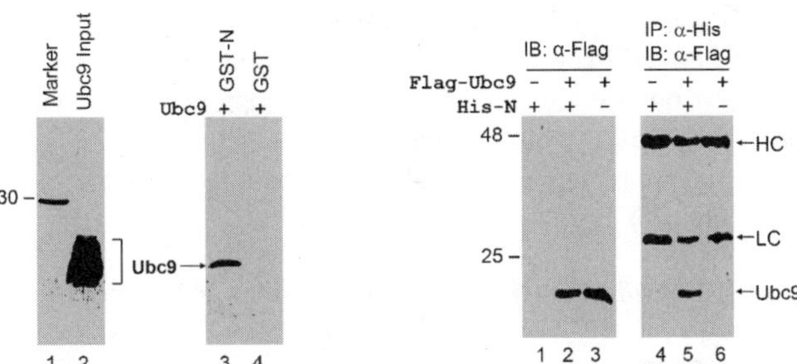

Figure 1a. GST pull-down assay.

Figure. 1b. Co-IP assay.

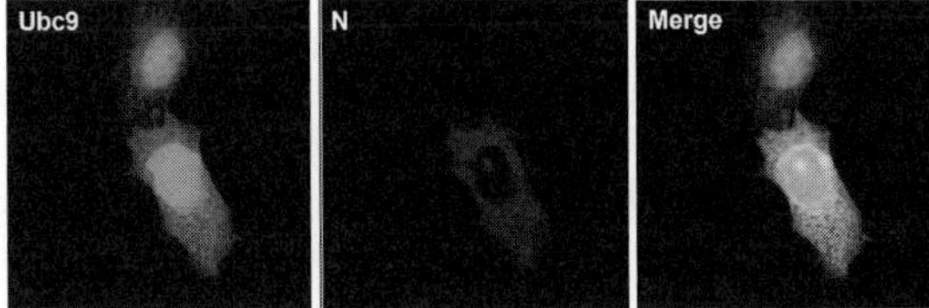

Figure 1c. Subcellular localization of N protein and Ubc9.

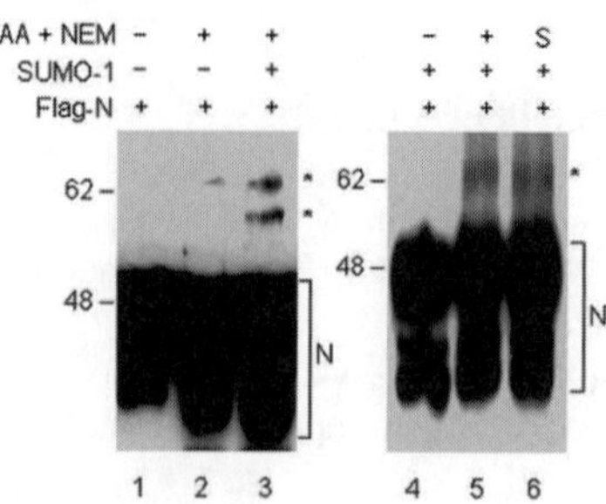

Figure 2a. Analysis of sumoylation of N protein by Western blot.

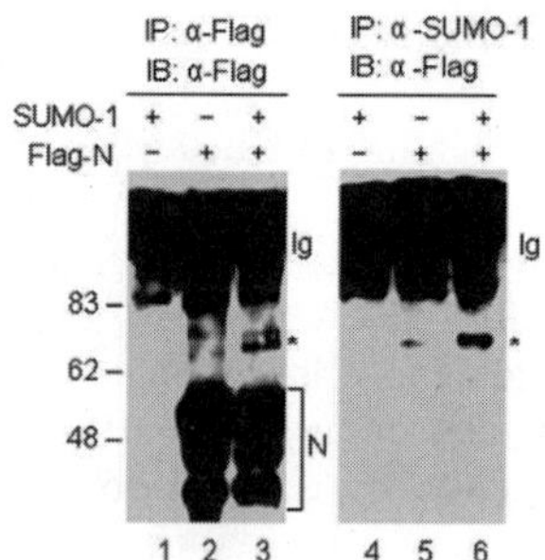

Figure 2b. Analysis of sumoylation of N protein by immunoprecipitation.

detection of significantly more 65-kDa species (Fig. 2a, lane 3). The 65-kDa band was also detected when cells were lysed directly with preheated SDS loading buffer (Fig. 2a, lane 6).

Using immunoprecipitation, we confirmed that the 65-kDa band represented the sumoylated N protein. Anti-Flag antibody precipitated the 65-kDa species from cells transfected with pFlag-N (Fig. 2b, lanes 2 and 3). Analysis of the anti-SUMO-1 precipitates by Western blotting with anti-Flag antibody showed that only the 65-kDa band was detected (Fig. 2b, lanes 5 and 6). Once again, coexpression of N protein with SUMO-1 greatly increased the detection of the 65 kDa species (Fig. 2b, lanes 3 and 6).

3.3. Mapping of the Sumoylation Site on SARS-CoV N Protein

Analysis of the N protein sequence showed that one lysine residue at amino acid position 62, K62, lies roughly within the consensus SUMO-1 modification sequence (GKEE). To determine whether this lysine was responsible for the modification of N protein by sumoylation, it was mutated to an Ala by site-directed mutagenesis. As shown in Fig. 3, similar amounts of the three isoforms of N protein were detected from cells transfected with either wild-type or mutant N constructs (Fig. 3, lanes 1 and 2). The 65-kDa sumoylated band was detected from cells transfected with wild-type N protein only (Fig. 3, lane 1); no 65-kDa sumoylated form was detected from cells expressing the K62A mutant (Fig. 3, lane 2). These results demonstrated that the K62 residue is the major sumoylation site of N protein.

3.4. Promotion of Homo-oligomerization of SARS-Cov N Protein by Sumoylation

To study the effects of sumoylation on the homo-oligomerization of N protein, cells expressing N protein alone or together with SUMO-1 were analyzed. As shown in Fig. 4, Western blot analysis of cells expressing wild-type N protein showed the detection of the 65-kDa sumoylated band (Fig. 4A and B, lanes 1 and 2). In addition, two bands of approximately 85 and 175 kDa were detected (Fig. 4A and B, lanes 1 and 2). Based on their apparent molecular masses, they may represent dimers and tetramers, respectively, of the N protein.

Coexpression of wild-type N protein with SUMO-1 dramatically increased the detection of the 65-kDa sumoylated band and the 85-kDa/175-kDa oligomers (Fig. 4B, lane 3 and), Co-expression of the K62A mutant with SUMO-1, once again, showed no detection of the 65-kDa sumoylated N protein (Fig. 4B, lane 5). Interestingly, only a trace amount of the 85-kDa and 175-kDa species was detected (Figure 4B, lane 5). These results suggest that abolishment of sumoylation of the N protein by mutating the k62 sumoylation site significantly decreases homo-oligomerization of the protein.

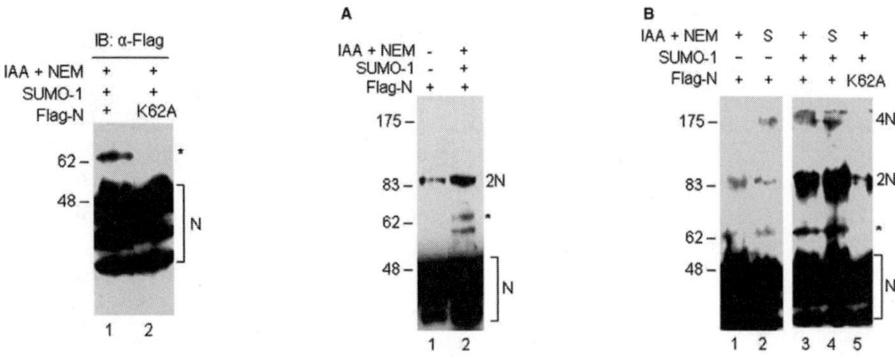

Figure 3. Mapping the major sumoylation site on N protein.

Figure 4. Analysis of the homo-oligomerization of SARS-CoV N protein.

4. DISCUSSION AND CONCLUSION

In this study, we identified Ubc9 as a host protein that interacts specifically with SARS-CoV N protein. This interaction was verified both *in vivo* and *in vitro*. Furthermore, we showed that, in addition to phosphorylation, the N protein was modified by covalent attachment of SUMO to its lysine 62 residue. Evidence provided demonstrated that sumoylation may promote homo-oligomerization of the protein.

SUMO-1 conjugation of a protein (sumoylation) is a highly regulated process in all eukaryotes, involved in diverse regulatory events such as nuclear transport, transcriptional regulation, chromosome segregation, and cell-cycle control.[7–11]

Recent studies have shown that the C-terminal one-third region is essential for self-association and multimerization of the SARS-CoV N protein.[4,12] Data reported in this study demonstrate that sumoylation of the SARS-CoV N protein dramatically enhances the homo-oligomerization of the protein. Promotion of oligomerization of protein by sumoylation has been speculated for a pathogenic protein, Huntingtin.[13] Because self-association and homo-oligomerization of N protein are essential for the assembly of nucleocapsid core, it suggests that sumoylation would play an important role in the SARS-CoV replication cycles. Systematic testing of this possibility would rely on the availability of an infectious cloning system, as developed by Yount et al.[14]

The failure to detect the sumoylated dimer in this study is unexpected, considering that sumoylation was shown to promote dimerization of the N protein. Two possibilities have been considered. First, sumoylation is a highly reversible process. The current data showed that only a small proportion of the N protein was dimerized compared with the monomers, and a certain proportion of the sumoylated dimer may be reversed during sample preparation and detection. The second possibility is that the sumoylated N protein may be not directly involved in the formation of dimmers and other oligomers. Instead, it may target the N protein to different cellular compartments and facilitate the oligomerization of the N protein. Further studies are required to address these possibilities.

Site-directed mutagenesis studies mapped the ^{62}lysine residue as a major site for covalent attachment of SUMO to the protein. We do not know whether other minor sumoylation sites may exist in the SARS-CoV N protein. Potential sumoylation at these minor positions would compensate the effect of K62A mutation. Sumoylation of protein at multiple sites was recently reported for several viral and host proteins. As the SARS-CoV N protein contains a total of 27 lysine residues and no any other lysine residue is located in a consensus sequence context for sumoylation, it would be difficult to further define these sites, if any, by a conventional mutagenesis approach.

5. REFERENCES

1. Lai, M. M., and Cavanagh, D., The molecular biology of coronaviruses, *Adv. Virus Res.* **48**:1-100 (1997).
2. Wurm, T., Chen, H., Britton, P., Brooks, G., and Hiscox, J. A., Localization to the nucleolus is a common feature of coronavirus nucleoproteins and the protein may disrupt host cell division, *J. Virol.* **75**:9345-9356 (2001).
3. Huang, Q., Yu, L., et al., Structure of the N-terminal RNA-binding domain of the SARS CoV nucleocapsid protein, *Biochemistry* **43**:6059-6063 (2004).
4. Yu, I. M., Gustafson, C. L. T., et al., Recombinant SARS coronavirus nucleocapsid protein forms a dimmer through its C-terminal domain, *J. Biol. Chem.* **280**:23280-23286 (2005).
5. Liu, D. X., Cavanagh, P. G., and Inglis, S. C., A polycistronic mRNA specified by the coronavirus infectious bronchitis virus, *Virology* **184**:531-544 (1991).
6. Liu, D. X., Xu, H. Y., and Brown, T. D. K., Proteolytic processing of the coronavirus infectious bronchitis virus 1a Polyprotein: identification of a 10-kilodalton polypeptide and determination of its cleavage sites, *J. Virol.* **71**:1814-1820 (1997).
7. Muller, S., Hoege, C., Pyrowolakis, G., and Jentsch, S., SUMO, ubiquitin's mysterious cousin, *Nat. Rev. Mol. Cell Biol.* **2**:202-210 (2001).
8. Seeler, J. S., and Dejean, A., Nuclear and unclear functions of SUMO, *Nat. Rev. Mol. Cell Biol.* **4**:690-699 (2003).
9. Pichler, A., and Melchior, F., Ubiquitin-related modifier SUMO1 and nucleocytoplasmic transport, *Traffic* **3**:381-387 (2002).
10. Verger, A., Perdomo, J., and Crossley, M., Modification with SUMO. A role in transcriptional regulation, *EMBO Rep.* **4**:137-142 (2003).
11. Wilson, V. G., and Rangasamy, D., Viral interaction with the host cell sumoylation system, *Virus Res.* **81**:17-27 (2001).
12. Surjit, M., Liu, B., Kumar, P., Chow, V. T. K., and La, S. K., The nucleocapsid protein of the SARS coronavirus is capable of self-association through a C-terminal 209 amino acid interaction domain, *Biochem. Biophys. Res. Commun.* **317**:1030-1036 (2004).
13. Steffan, J. S., Agrawal, N., et al., SUMO modification of Huntingtin and Huntington's disease pathology, *Science* **304**:100-104 (2004).
14. Yount, B., Curtis, K. M., et al., Reverse genetics with a full-length infectious cDNA of severe acute respiratory syndrome coronavirus, *Proc. Natl. Acad. Sci. USA* **100**:12995-13000 (2003).

IMPORTANCE OF MHV-CoV A59 NUCLEOCAPSID PROTEIN COOH-TERMINAL NEGATIVE CHARGES

Valerie Bednar, Sandhya Verma, Andrew Blount, and Brenda G. Hogue*

1. INTRODUCTION

Coronaviruses are enveloped and have a single-stranded, positive-sense genome approximately 26–30 kb in length, the largest of all the RNA viruses. All coronoviruses contain at least four structural proteins: three envelope proteins, the membrane (M), spike (S), and envelope (E) proteins, and a 50–60 kDa phosphorylated nucleocapsid (N) protein.[1] The N proteins of all coronaviruses range between 375 and 455 amino acids and are phosphorylated. N protein is a multifunctional viral gene product. In virus-infected cells, N protein binds to the genomic RNA to form a helical ribonucleoprotein (RNP) complex. The N protein also plays a yet undetermined role(s) in transcription and/or replication, and possibly in translational control. N is a highly basic protein that contains a large number of potential phosphorylation sites. The protein has a high concentration of serine residues (7–11%). N consists of three conserved structural domains, two are basic and one, the carboxy terminal domain, is acidic.[2, 3] A number of conserved negatively charged amino acids are located in the carboxy-terminal domain III of the protein. These residues were previously hypothesized to play a role in N-M protein interactions during assembly.[4-6] The residues could alternatively serve as contributors to the general overall functional structure of the protein. Conceivably, the residues could be important for any of the functions that the protein provides during the virus life cycle.

As part of our goal to understand the functional importance of the charged residues in domain III, a series of N mutants were made and studied in the context of the viral genome using a mouse hepatitis coronavirus (MHV-CoV A59) infectious clone. We found that aspartic acids (D) 440 and 441 are functionally important. Viable viruses were recovered when either residue was changed singly to positively charged arginine (R) but not when both residues were changed to alanine (A). Analysis of a large number of plaque purified viruses from the panel of charged single and neutral double mutants revealed that, in addition to the introduced mutations at positions 440 and/or 441, nearly all had new amino acid changes within the N gene. All of these compensating changes were concentrated primarily in one region further toward the amino end of domain III. A

* Arizona State University, Tempe, Arizona.

few viruses were recovered that retained the arginine substitution at position 441 and no other changes. All of these $D_{441}R$ mutants exhibited a strongly crippled phenotype. Overall the results suggest that the negative charges at positions 440 and 441 are important for one or more of the functions of N.

2. MATERIALS AND METHODS

Plasmids containing seven cDNA fragments (A–G) constituting the entire MHV genome were kindly provided by Dr. Ralph Baric, University of North Carolina at Chapel Hill.[7] All original mutants were constructed using GeneEditor site-directed mutagenesis system (Promega) according to the manufacturer's instructions. The four reconstructed N double mutants were mutagenized by whole plasmid PCR using Pfu polymerase (Stratagene). Upon completion of the mutagenesis, all clones were confirmed by sequencing. All full-length infectious clones were assembled using a protocol basically as previously described.[7, 8] After electroporation, all mutant viruses were plaque purified and multiple plaques were picked for each mutation except $DD_{440\text{-}441}RR$, which failed to generate visible fusion foci on mouse L2 cells and was ultimately deemed not viable. Plaques were passaged on L2 cells and the presence of mutations was confirmed by extracting total RNA using Ambion's RNAqueous-4PCR, reverse transcribed using Invitrogen's Superscript RT kit, and amplified using Ambion's SuperTaq Plus polymerase according to manufacturer's directions. The entirety of the E, M, and N genes of all mutants was sequenced. All growth kinetics experiments were carried out in mouse 17Cl1 cells infected at a multiplicity of infection of 5. Cell culture supernatants were collected at 1, 4, 8, 12, and either 16 and 20 or 18 and 24 hpi. Titers were determined by plaque assay on L2 cells. Overlays were removed at 48 hpi, and plaques were fixed and stained with crystal violet in ethanol.

3. RESULTS AND DISCUSSION

As part of our studies to better understand the many functions of the N protein, we chose to focus on the high concentration of negative charges in domain III to gain insight into the role of the carboxy tail. We focused on this region because of the high conservation of the negative charges in this domain throughout the family, yet its inability to be exchanged between different group II coronaviruses.[9] Additionally, the protein participates in a number of different protein-protein interactions such as homo-oligermization and interactions with the M protein.[4, 5, 10-12] N may also directly affect host cell function, as SARS N protein has been shown to activate cellular transcription factors and affect signal transduction pathways.[13] In addition to its various protein-protein interactions, N also interacts with RNA.[14-16] Through N's interactions with the RNA, it is thought to have a role in both genome replication and/or transcription.[17, 18] The helical nucleocapsid's interactions with other viral proteins indicate that N also has an important role in packaging and assembly of the virion. Charged residues often mediate protein-protein interactions and can also affect the tertiary structure of a protein and therefore may affect any number of these functions that N serves.

To study the requirement of these negative charges, nine N mutant viruses were generated: $D_{440}R$, $D_{441}R$, $DD_{440\text{-}441}RR$, $DD_{440\text{-}441}EE$, $DD_{440\text{-}441}AA$, $D_{446}A$, $D_{451}A$, $D_{451}E$, and $EDD_{449\text{-}451}AAA$. Mutant virus full-length cDNA clones were assembled. Viral RNA was transcribed and transfected into cells. All viable viruses were plaque purified, and the retention of the introduced mutations was confirmed by RT-PCR and sequence analysis.

Negatively charged residues 446 and 449–451 appear not to be absolutely required for N functionality. Following transfection, viable viruses were easily recovered, all mutants were plaque purified and multiple plaques of each mutant were followed for five passages. Sequence analysis of passage five viruses confirmed the stability of the introduced mutations and that no additional changes had arisen in the E, M, or N genes. Furthermore, $D_{446}A$, $D_{451}A$, $D_{451}E$, and $EDD_{449\text{-}451}AAA$ displayed growth characteristics and plaque size and morphology like the wild-type virus.[13]

However, negatively charged residues at positions 440–441 appear to be important for N protein function. Following transfection, centers of fusion were observed for all mutant viruses. Of the five mutant viruses, viable virus was easily recovered for four of them. Following electroporation of mutant virus $DD_{440\text{-}441}RR$, limited characteristic cytopathic effects (CPE), including fusion, was observed. The introduction of two positively charges residues at positions 440–441 appears to be lethal to the virus, as multiple attempts to recover the $DD_{440\text{-}441}RR$ virus were unsuccessful. Although the virus did not tolerate double mutations at residues 440–441 to positive charges, double neutral or other negative charge changes were tolerated. $DD_{440\text{-}441}EE$ grew to a titer comparable with wild-type and displayed plaque size, morphology and growth characteristics similar to wild-type virus, all without additional changes in the E, M, or N genes.[13] This strongly suggests that any negative charge in those positions is favorable, but that aspartic acid residues specifically are not required. The removal of both negative charges at positions 440–441, however, does not appear to be tolerated as well. After transfection and plaque purification of $DD_{440\text{-}441}AA$, multiple plaques were followed through five passages. All $DD_{440\text{-}441}AA$ plaques analyzed retained the introduced AA mutation and exhibited an additional change of $SR_{424\text{-}425}GG$ in the N gene (Table 1). Analysis of $DD_{440\text{-}441}AA$ plaques with $SR_{424\text{-}425}GG$ revealed that the mutant viruses exhibited plaque size, morphology, and growth characteristics indistinguishable from the wild-type virus.[13]

When either D_{440} or D_{441} was replaced by positively charged arginine, additional changes within N were observed. No $D_{440}R$ mutant viruses were recovered that did not contain additional changes within N. Some $D_{441}R$ mutant viruses were recovered that had no additional changes. The most prominent compensating change seen was the replacement of R_{425} with glycine. Also of note was the replacement of A_{436} with aspartic acid. Analysis of the growth characteristics and plaque morphology of the plaqued viruses strongly suggested that the new changes were compensating changes that were increasing the viability of the mutant viruses. Further analysis of the additional changes indicated that replacement of the R_{425} with glycine in the $D_{440}R$ and $D_{441}R$ single mutant backgrounds or replacement of A436 with aspartic acid are indeed important compensating changes.[13]

Our mutagenic studies described here have highlighted key negatively charged residues in the carboxy tail of domain III of MHV N protein. The aspartic acid residues at positions 440–441 appear to be critical residues as changes at these positions are tolerated less well than changes of other negatively charged residues in this region. This is further supported by the lethality of $DD_{440\text{-}441}RR$ mutant. Our inability to successfully passage $DD_{440\text{-}441}RR$ strongly suggests that at least one negative charge at position 440 or 441 is

Table 1. $D_{440}R$, $D_{441}R$, and $DD_{440\text{-}441}AA$ recovered plaque purified mutant viruses with summary of additional amino acid changes observed in the N gene.[a]

WT	415					420					425					430					435					440					445					450				
	P	K	S	S	V	Q	R	N	V	S	R	E	L	T	P	E	D	R	S	L	L	A	Q	I	L	D	D	G	V	V	P	D	G	L	E	D	D	S	N	V
$D_{440}R$											G															R														
											G															R														
											G															R														
											G															R														
											G															R														
											G															R														
											G															R														
																					F					R														
															N											R														
															N											R														
															N											R														
											G															R														
$D_{441}R$																						D					R													
																						D					R													
											G																R													
											G																R													
											G																R													
											G																R													
																											R													
																											R													
											G																R													
											G																R													
		Q																									R													
								D																			R													
																											R				Q									
																											R													
																											D													
											G																R													
																											R													
											G																R													
																									V		R													
$DD_{440\text{-}441}AA$										G	G															A	A													
										G	G															A	A													
										G	G															A	A													
										G	G															A	A													
										G	G															A	A													
										G	G															A	A													
										G	G															A	A													

[a] Mutants with charge changes at positions 440 and/or 441 are indicated in the left column. Each line is representative of sequence analysis of a single plaque recovered from the parental mutant.

required for proper N protein function. Furthermore, analysis of the isolated plaques possibly indicates that the positioning of a negative charge at position 440 may be more important than at position 441. All of the plaques that were isolated from the $D_{440}R$ mutant also had at least one new change that provided the virus with a growth advantage. Roughly one-quarter of the plaques analyzed for $D_{441}R$ retained the original mutation with no additional changes. Although these plaques were genetically stable after five passages, they were nonetheless severely crippled and only grew to titers 3–4 orders of magnitude lower than the wild-type counterpart. Without any additional changes, $D_{441}R$ produces plaques significantly smaller than the wild-type virus, and a rudimentary analysis of its kinetics again confirmed the severely crippling effect (data not shown). The larger number of additional changes observed for the $D_{441}R$ mutant than for $D_{440}R$ further suggests that replacement of a negative charge with a positive charge at position 441 may be more easily compensated for than when the charge is place at position 440.

Taken all together, our results indicate that the negatively charged carboxy tail is important for some aspect of the viral life cycle. Maintenance of the overall negative charge of the domain appears to be important because the vast majority of compensating changes seen reduced the net charge of the carboxy tail. The frequency with which R_{425} was replaced with glycine may indicate an effect on the tertiary structure. The removal of a positive charge at position 425 could affect the folding of the N protein. Due to its small size, glycines are implicated in affecting a protein's tertiary structure by allowing its surrounding environment more flexibility.[19]

4. ACKNOWLEDGMENTS

This work was supported by Public Health Service grant (AI54704), from the National Institute of Allergy and Infectious Diseases. We thank Ralph Baric for providing us with the MHV infectious clone.

5. REFERENCES

1. K. V. Holmes and M. M. C. Lai, in: *Coronaviridae: The Viruses and Their Replication*, edited by D. Knipe and P. Howley (Lippincott Williams & Wilkins, Philadelphia, 2001), pp. 1163-1185.
2. M. M. Parker and P. S. Masters, Sequence comparison of the N genes of five strains of the coronavirus mouse hepatitis virus suggests a three domain structure for the nucleocapsid protein, *Virology* **179**, 463-468 (1990).
3. H. Laude and P. S. Masters, in: *The Coronaviridae*, edited by S. G. Siddell (Plenum Press, New York, 1995), pp. 141-163.
4. L. Kuo and P. S. Masters, Genetic evidence for a structural interaction between the carboxy termini of the membrane and nucleocapsid proteins of mouse hepatitis virus, *J. Virol.* **76**, 4987-4999 (2002).
5. K. Narayanan, A. Maeda, J. Maeda, and S. Makino, Characterization of the Coronavirus M protein and nucleocapsid interaction in infected cells, *J. Virol.* **74**, 8127-8134 (2000).
6. D. Escors, J. Ortego, H. Laude, and L. Enjuanes, The membrane m protein carboxy termins binds to transmissible gastroenteritis coronavirus core and contributes to core stability, *J. Virol.* **75**, 1312-1324 (2001).
7. B. Yount, M. R. Denison, S. R. Weiss, and R. S. Baric, Systematic assembly of a full-length infectious cDNA of mouse hepatitis virus strain A50, *J. Virol.* **76**, 11065-11078 (2002).
8. S. Verma, V. Bednar, A. Blount, and B. G. Hogue, Identification of functionally important negatively charged amino acids within domain III of mouse hepatitis coronavirus A59 nucleocapsid protein, *J. Virol.* **80**, 4344-4355 (2006).
9. D. Peng, C. A. Koetzner, T. MCMahon, Y. Zhu, and P. S. Masters, Construction of murine coronavirus mutants containing interspecies chimeric nucleocapsid protein, *J. Virol.* **69**, 5475-5484 (1995).
10. S. G. Robbins, M. F. Frana, J. J. McGowan, J. F. Boyle, and K. V. Holmes, RNA-binding proteins of coronavirus MHV: detection of monomeric and multimeric N protein with an RNA overlay-protein blot assay, *Virology* **150**, 402-410 (1986).
11. K. Narayanan, K. H. Kim, and S. Makino, Characterization of N protein self-association in coronavirus ribonucleoprotein complexes, *Virus Res.* **98**, 131-140 (2003).
12. I. M. Yu, C. L. T. Gustafson, J. Diao, J. W. Burgner, Z. Li, J. Zhang, and J. Chen, Recombinant SARS coronavirus nucleocapsid protein forms a dimer through its C-terminal domain, *J. Biol. Chem.* Accepted Manuscript (2005).
13. R. He, A. Leeson, A. Andonov, Y. Li, N. Bastien, J. Cao, C. Osiowy, F. Dobie, T. Cutts, M. Ballantine, and X. Li, Activation of AP-1 signal transduction pathway by SARS coronavirus nucleocapsid protein, *Biochem. Biophys. Res. Commun.* **311**, 870-876 (2003).
14. G. W. Nelson, S. A. Stohlman, and S. M. Tahara, High affinity interactioin between nucleocapsid protein and leader/intergenic sequence of mouse hepatitis virus RNA, *J. Gen. Virol.* **81**, 181-188 (2000).
15. R. Molenkamp and W. J. M. Spaan, Identification of a specific interaction between the coronavirus mouse hepatitis virus A59 nucleocapsid protein and packaging signal, *Virology* **239**, 78-86 (1997).

16. S. A. Stohlman, R. S. Baric, G. N. Nelson, L. H. Soe, L. M. Welter, and R. J. Deans, Specific interaction between coronavirus leader RNA and nucleocapsid protein, *J. Virol.* **62**, 4288-4295 (1988).
17. F. Alamazán, C. Galán, and L. Enjuanes, The Nucleoprotein is required for efficient coronavirus genome replication, *J. Virol.* **78**, 12683-12688 (2004).
18. B. Schelle, N. Karl, B. Ludewig, S.G. Siddell and V. Thiel, Selective replication of coronavirus genomes that express nucleocapsid protein, *J. Virol.* **79**, 6620-6630 (2005).
19. D. Oh, S. Y. Shin, S. Lee, J. H. Kang, S. D. Kim, P. D. Ryu, K. S. Hahm, and Y. Kim, Role of the hinge region and the tryptophane residue in the synthetic antimicrobial peptides, cecropin A(1-8)-magainin 1(1-12) and its analogues, on their antibiotic activities and structures, *Biochemistry* **39**, 11855-11864 (2000).

EXPRESSION AND STRUCTURAL ANALYSIS OF INFECTIOUS BRONCHITIS VIRUS NUCLEOPROTEIN

Kelly-Anne Spencer and Julian A. Hiscox*

1. INTRODUCTION

The coronavirus nucleoprotein (N protein) is one of the most abundantly expressed viral proteins in an infected cell, with the principal function of binding the viral RNA genome to form the ribonucleocapsid structure (RNP) and forming the viral core. N protein also has roles in viral replication, transcription, and translation as well as modulating cellular processes. Although coronavirus N proteins have the potential to be phosphorylated at multiple serine residues, mass spectroscopic analysis of both the avian infectious bronchitis virus (IBV) and porcine transmissible gastroenteritis virus (TGEV) have shown that N protein is phosphorylated at only three or four residues.[1,2] In the case of IBV N protein, these map to predicted casein kinase II sites.[2] Based on amino acid sequence comparisons, three conserved regions have been identified in the murine coronavirus, mouse hepatitis virus (MHV) N protein.[3] In general, other coronavirus N protein would appear to follow this pattern. Of the three regions, for MHV, region two has been shown to bind both coronavirus and non-coronavirus RNA sequences, whereas regions one and three for IBV N protein have been shown to bind to RNA.[4]

We hypothesize that phosphorylation of N protein controls the RNA binding activity and differential phosphorylation of the protein alters its structure to expose necessary binding motif(s). In support of this hypothesis, antibodies studies suggested that phosphorylation of N protein led to substantial conformation change,[5] and using surface plasmon resonance we have recently shown that phosphorylation of IBV N protein was involved in the recognition of viral RNA from nonviral RNA.[2]

To investigate this hypothesis, we have developed methodologies to determine whether the overall conformation of N protein changes upon binding viral RNA and what regions may be involved in this process. Our technique is to express either wild-type IBV N protein or its three subregions (termed NI, NII, and NIII) and measure changes in structure upon RNA binding using circular dichroism (CD). CD spectroscopy is a form of light adsorption spectroscopy that measures the difference in absorbance of right- and

*University of Leeds, Leeds, LS2 9JT, United Kingdom.

left-circularly polarized light. In proteins, the major chromophores are the amide bonds of the peptide backbone and the aromatic side chains. Polypeptides and proteins have regions where the peptide chromophores are in highly ordered arrays; as a consequence, many common secondary structure motifs, such as the α-helix, ß-pleated sheet, ß-turn and random coil, have very characteristic CD spectra.[6] CD also allows the detection of gross protein conformational changes; it can therefore be utilized to monitor changes in secondary structure upon ligand binding, multimer formation, and to analyze the protein in a variety of environments. CD was used to analyze structural changes that may occur when IBV N protein binds to models of viral RNA and to map any conformational changes to specific regions of the protein. Here we report our initial studies optimizing this system and our preliminary results.

2. MATERIALS AND METHODS

2.1. IBV N Protein Expression and Purification

IBV N protein and the three regions were expressed in Tuner (DE3) pLacI IPTG inducible *E. coli* to produce recombinant protein possessing an N-terminal hexa-histidine tag. Cell lysate containing recombinant protein was purified using nickel chelating chromatography; purified proteins were eluted using increasing concentrations of imidazole. Further purification was performed by separating proteins according to their size using gel filtration chromatography.

2.2. SDS-PAGE and Western Blot Analysis

Protein purity was analyzed by separation on NuPage Bis-Tris 10% pre-cast SDS-PAGE gels (Invitrogen), and proteins were visualized by staining with Coomassie. Proteins were transferred onto PVDF membrane for Western blot analysis and detected using ECL (Amersham/Pharmacia) according to the manufacturer's instructions.

2.3. Circular Dichroism Spectroscopy

CD experiments were performed on a Jasco J715 spectrophotometer. Measurements were taken in the far-UV (190–260 nm) and the CD signal recorded in a 1-mm path-length cell using a protein concentration of 0.2 mg/ml. Protein samples to be analyzed were dialyzed into sodium phosphate buffer, pH7.2. RNA was added at a 1:1 molar ratio of protein to RNA.

3. RESULTS AND DISCUSSION

3.1. Extraction and Purification of IBV N Protein Regions

IBV N regions were purified as described above. However, due to the differing properties of each protein, some subtle changes were made. NI is readily expressed and soluble in large quantities, with recoveries of the order of 10 mg obtained per 500 ml of

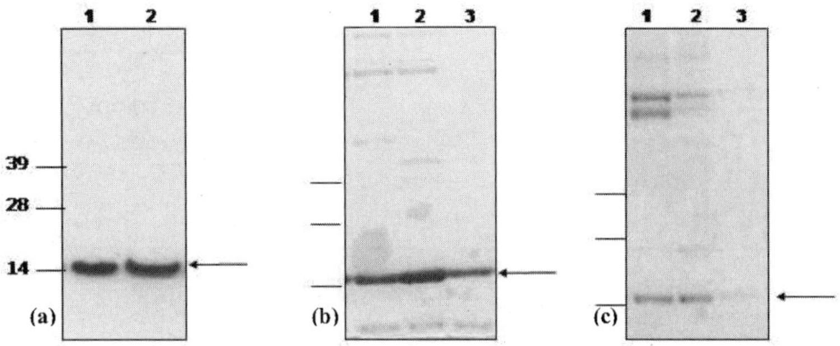

Figure 1. Coomassie-stained SDS-PAGE analysis of IBV N protein regions. (a) Purified region NI obtained by eluting protein from Ni^{2+} column using 150nm imidazole (lane 1) or 200 mM imidazole (lane 2). (b) Purified region NII obtained by eluting protein from Ni^{2+} column using 150nM imidazole (lane 1), 200 mM imidazole (lane 2), and by gel filtration chromatography (lane 3). (c) Purified region NIII obtained by eluting protein from Ni^{2+} column using 150 nm imidazole (lane 1), 200 mM imidazole (lane 2), and by gel filtration chromatography (lane 3). Apparent molecular weight markers (kDa) are shown to the left. Each IBV N protein region is indicated by an arrow.

culture. When purifying NI, high percentage purity was achieved from nickel chelation chromatography alone (Fig. 1a) eliminating the need for gel filtration chromatography. However, both NII and NIII could only be expressed at low levels. Relatively pure (approximately 80%) NII could be produced at lower concentrations (up to 800 μg/ml) after a two-step chromatography strategy, although some low molecular weight contaminants remained after gel filtration (Fig. 1b). NIII was purified using nickel chelating chromatography, after which the protein readily precipitates out of solution making it unsuitable for subsequent purification and dialysis steps (Fig. 1c). Western blotting was performed with antibody specific for IBV to verify the identity of the purified proteins (data not shown).

3.2. Effect of RNA Binding on IBV N Protein Structure

Using CD, we compared the structural changes of IBV N upon binding two models of the IBV genome. The first was RNA synthesized by runoff transcription from pCD-61 (generously provided by Dr. Paul Britton and Dr. Dave Cavanagh) to generate a 6.1-kb analogue of the IBV genome, and the second a synthetic RNAmer that was identical to the 5' end of IBV mRNA 3 up to, and including, the translation initiation codon for gene 3a; both of these targets were used in previous binding studies.[2] Various concentrations of IBV N protein were analyzed in the far UV and averaged. Due to the nature of CD it is difficult to assign definitive structures, however, when delineating the spectrum of IBV N protein the peak between 190 and 200 nm is indicative of ß-sheet and the dip at 220 nm can be due to α-helical secondary structures (Fig. 2). Other features of the spectrum, including the trough at around 210 nm, are consistent with random coil motifs. Upon the addition of CD61 or leader RNA, subtle changes can be seen in the overall shape of the spectra, in particular the negative signal at 210 nm becomes weaker and a stronger negative trough can be seen between 230 and 250 nm (Fig. 2). These changes occur upon binding both RNA's, therefore the 5' leader RNA was used for all further CD analysis.

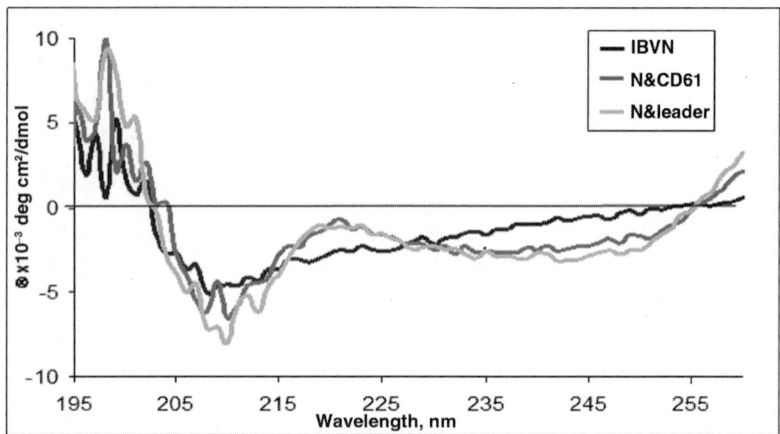

Figure 2. Purified recombinant IBV N protein was analyzed using far-UV circular dichroism spectroscopy. Measurements were taken of IBV N protein without the addition of RNA (black line), in the presence of CD61, an IBV genomic RNA analogue (gray line), and a synthetic RNA model of the 3' leader sequence (light gray line).

3.3. Structural Changes Can Be Mapped to Specific Regions of IBV N

IBV N has a molecular weight of approximately 45 kDa. It is therefore likely that it contains more than one type of secondary structure motif, such as a mixture of α-helices and ß-sheets, or ß-sheets involving ß-turns, together these motifs generate "noisy" spectra. In order to overcome this problem, IBV N was broken down into its three regions, NI, NII, and NIII, and analyzed by CD. The far UV CD-spectrum of NI, the N-terminal region of IBV N protein, can be seen in Fig. 3a. The curve is consistent with a random coil with the major trough at 210 nm. Upon the addition of leader RNA, no change can be seen in the overall shape of the spectrum, indicating that the secondary structure of IBV NI doesn't alter in the presence of viral RNA. NII, the central 15 kDa of IBV N, conversely undergoes a substantial conformational change in the presence of RNA. The far UV CD-spectrum of NII protein in the absence of RNA has a trough at around 215 nm which is indicative of random coil or α-helical secondary structures. Addition of leader RNA results in changes in the spectrum, namely the strong negative trough at 215 nm is lower in intensity and the peak at 235 nm shifts along nearer to 220 nm; these changes suggest that the protein-RNA complex may contain more ß-turns and sheets than that of protein alone (Fig. 3b). NIII, the C-terminal region of IBV N, was analyzed in the far UV. Due to the nature of this protein and problems with purity, strong CD signals proved difficult to obtain, and the resulting spectrum are therefore less definitive and assigning secondary structure problematic. As can be seen on the NIII spectrum (Fig 3c), a strong negative signal is obtained upon the addition of viral RNA,

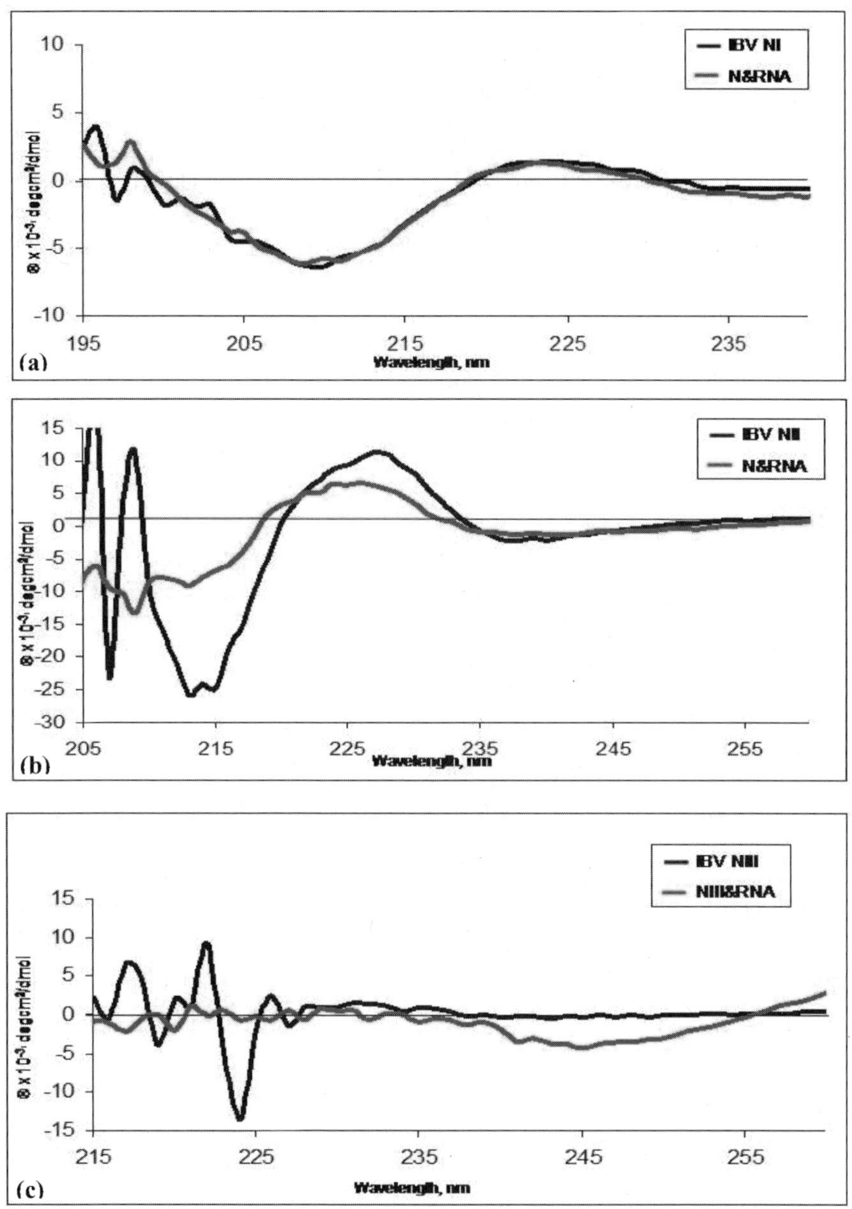

Figure 3. Far-UV spectra of IBV N regions (a) IBV NI, (b) IBV NII, (c) IBV NIII. Black line: protein without the addition of RNA, gray line: protein in a 1:1 ratio with viral RNA.

indicating that this interaction may stabilize the protein and prevent it from precipitating out of solution. The overall spectrum shows some features consistent with ß-sheet motifs, such as the negative signal between 230 and 250 nm, and the spectrum obtained is very weak in intensity, a feature often observed for ß-sheet proteins.[7]

In conclusion, our data would support the hypothesis that N protein undergoes conformational change upon binding viral RNA. These changes would appear to be confined to regions II and III, where RNA binding motifs may be located. Certainly our mass spectroscopic study indicated that regions II and III contained two phosphorylation sites each, whereas region I contained no such sites.

4. ACKNOWLEDGMENTS

This work was supported by the BBSRC and Guildhay Ltd. J.A.H. and K.A.S. would like to thank Prof. Sheena Radford and Dr. Sue Jones for their help and expertise with CD spectroscopy.

5. REFERENCES

1. E. Calvo, D. Escors, J. A. Lopez, J. M. Gonzalez, A. Alvarez, E. Arza, and L. Enjuanes, Phosphorylation and subcellular localization of transmissible gastroenteritis virus nucleocapsid protein in infected cells, *J. Gen. Virol.* **86**, 2255-2267 (2005).
2. H. Chen, A. Gill, B. K. Dove, S. R. Emmett, F. C. Kemp, M. A. Ritchie, M. Dee, and J. A. Hiscox, Mass spectroscopic characterisation of the coronavirus infectious bronchitis virus nucleoprotein and elucidation of the role of phosphorylation in RNA binding using surface plasmon resonance, *J. Virol.* **79**, 1164-1179 (2005).
3. P. S. Masters, Localization of an RNA-binding domain in the nucleocapsid protein of the coronavirus mouse hepatitis virus, *Arch. Virol.* **125**, 141-160 (1992).
4. M. L. Zhou and E. W. Collinson, The amino and carboxyl domains of the infectious bronchitis virus nucleocapsid protein interact with 3' genomic RNA, *Virus Res.* **67**, 31-39 (2000).
5. S. A. Stohlman, J. O. Fleming, C. D. Patton, and M. M. C. Lai, Synthesis and subcellular-localization of the murine coronavirus nucleocapsid protein, *Virology* **130**, 527-532 (1983).
6. N. J. Greenfield, Methods to estimate the conformation of proteins and polypeptides from circular dichroism data, *Anal. Biochem.* **235**, 1-10 (1996).
7. R. W. Woody, Circular dichroism, *Methods Enzymol.* **246**, 34-71 (1995).

MOUSE HEPATITIS VIRUS INFECTION ACTIVATES THE IRE1/XBP1 PATHWAY OF THE UNFOLDED PROTEIN RESPONSE

John Bechill, Zhongbin Chen, Joseph W. Brewer, and Susan C. Baker*

1. INTRODUCTION

The infection and replication of coronaviruses is likely to impose significant stress on the host cell exocytic pathway because of the extensive use of intracellular membranes for assembly of replication complexes and viral particles. MHV utilizes the ER to generate membrane-associated replication complexes, termed double membrane vesicles,[1] and to assemble progeny virus particles. We hypothesized that this extensive use of the ER induces the unfolded protein response. The UPR is a multifaceted signaling pathway that is triggered by perturbations in the normal ER environment (reviewed in Ref.2). The UPR emanates from the ER membrane and has the capacity to increase expression of ER resident chaperones and folding enzymes, to facilitate disposal of misfolded protein, to downregulate protein synthesis, to regulate production of membrane components necessary for expansion of the secretory pathway, and to regulate both cell cycle progression and cell death. Recently, investigators have reported that hepatitis C virus[3,4] and human cytomegalovirus[5] activate and regulate aspects of the UPR. One arm of the UPR, the Ire1-XBP1 pathway, leads to increased ER chaperone levels, upregulation of lipid biosynthesis, and alterations in protein degradation and synthesis, all of which might influence MHV replication. We found that MHV infection activates induces Ire1-mediated splicing of XBP1 mRNA, thereby resulting in synthesis of the active transcription factor, XBP1(S). To investigate the role of the Ire1-XBP1 pathway in MHV infection, we used RNA interference to generate XBP1-silenced cell lines (XBP1si). As expected, XBP1si cells exhibited reduced levels of XBP1 mRNA during MHV infection; however, XBP1(S) protein accumulated in the MHV-infected XBP1si cells. These data indicate that MHV infection is a potent activator of the Ire1/XBP1 pathway and suggest that XBP1(S) protein is stabilized in MHV-infected cells. Future studies will determine if MHV replication regulates the function of XBP1(S) as a transcriptional activator.

*John Bechill, Zhongbin Chen, Joseph W. Brewer, Susan C. Baker, Loyola University Chicago, Maywood, Illinois. Zhongbin Chen, Beijing Institute of Radiation Medicine, Beijing, China.

2. MATERIALS AND METHODS

Virus and cells: MHV strain A59 was propagated as previously described.[6] DBT cells were cultured in MEM supplemented with 5% tryptose phosphate broth, 2% penicillin/streptomycin, 2% L-glutamine, and 5% fetal calf serum.

RNA isolation and Northern blot analysis: RNA was isolated from untreated or 2 ug/ml tunicamycin-treated DBT cells using Qiashredder and RNeasy columns according to the manufacturer's instructions (Qiagen). Total RNA (10 ug) was separated by electrophoresis on a formaldehyde agarose gel, transferred to a nitrocellulose membrane, and probed with radio-labeled DNA specific for XBP1, ERdj4 and ChoB as described in Ref. 7.

XBP1si cell lines: XBP1-specific oligonucleotide primers designed to generate shRNA (short hairpin RNA) that targets XBP1 for RNAi-mediated degradation were synthesized and cloned into the pU6 expression vector (Biomyx). Primer UPR10: 5'-TTTGAGTCAAACTAACGTGGTAGTGACTTCCTGTCATCACTACCACGTTAGTT TGACTCTTTT-3' and primer UPR11: 5'-CTAGAAAAGAGTCAAACTAACGTGGT AGTGATGACAGGAAGTCACTACCACGTTAGTTTGACT were annealed and cloned into the BbsI and XbaI sites of the pU6 vector. The resulting plasmid DNA was designated pU6-XBP1si1. DBT cells were transfected with pU6-XBP1si1 DNA and stable cell lines were selected using 90 ug/ml G418 (Invitrogen). Clonal cell lines were isolated and tested to determine the level of XBP1 mRNA. A cell line with greater than 80% reduction in XBP1 mRNA was designated XBP1si, and a cell line with wild-type levels of XBP1 mRNA was designated as a control cell line (Con).

Real-time RT-PCR: Real-time RT-PCR was performed on RNA isolated from untreated, tunicamycin treated and MHV infected cells using Taqman reagents and probes according to the manufacturer's instructions (Applied Biosystems). For detection of XBP1: forward primer 5'-GCCATTGTCTGAGACCACCTT-3', reverse primer 5'-TCTGTACCAAGTGGAGAAGACATG-3', Taqman probe fam-TGCCTGCTGGACGC TCACAGTGAC-3'. For detection of MHV N: forward primer 5'-ATCCCGTGGGCC AAATAATCG-3', reverse primer 5'-TTAGCCAAAACAAGAGCAGCAATT-3', Taqman probe: fam-AAGCAGTTCCAACCAGCGCCAGCC-3'. The relative concentration of ribosomal RNA was determined using the ribosomal RNA detection system (Applied Biosystems).

Western blotting: Whole cell lysates were prepared by scraping the cells in lysis buffer A (4% sodium dodecyl sulfate, 3% dithiothreitol, 40% glycerol, 0.065 M Tris, pH 6.8) and passing the lysate through a 25-gauge needle to shear the DNA released from the nucleus. Lysates were subjected to electrophoresis on 10% polyacrylamide gels and transferred to nitrocellulose membranes (Optitran BA-S, Schleicher and Schuell). Proteins were detected by Western blotting (Western Lightning Chemiluminescence Reagent Plus, Perkin Elmer Life Sciences) using anti-calnexin (kindly provided by Linda Hendershot, St. Jude Children's Research Hospital, Memphis, TN), anti-D14 (detects p22 also termed nsp8[1]) and anti-XBP1 (Santa Cruz).

3. RESULTS AND DISCUSSION

Using RT-PCR analysis to investigate splicing of XBP1 mRNA and Western blotting to detect XBP1(S), we found that MHV infection activates the IRE1-XBP1 pathway

(Bechill et al., in preparation). MHV replication induces Ire1-mediated splicing of XBP1 mRNA, thereby resulting in synthesis of the transcription factor, XBP1(S). These results indicate that MHV replication is indeed a potent activator of the Ire1/XBP1 pathway of the UPR.

The Ire1-XBP1 pathway has previously been linked to lipid biosynthesis and expansion of the ER,[8,9] both of which might be critical to MHV replication. Therefore, we wanted to determine if knocking down the level of XBP1 mRNA would have an effect on MHV replication. To reduce the level of XBP1 mRNA in cells, we generated a stable cell line, designated XBPsi, that expresses a shRNA targeting XBP1 for RNAi-mediated degradation (as described in "Materials and Methods"). To test whether the XBP1si cells efficiently reduced steady-state levels of XBP1 mRNA, we compared the levels of XBP1 mRNA in untreated cells and cells treated with tunicamycin for 6 or 12 hours (Fig. 1). Northern blot analysis shows that XBP1 mRNA levels are induced by treatment with tunicamycin in DBT cells and control cells. In contrast, the XBP1si cell line has reduced amounts of XBP1 mRNA under all conditions tested. Furthermore, the mRNA level of the XBP1-responsive gene, ERdj4 is also reduced in the XBP1si cell line (Fig. 1, center panel). Overall, we conclude that the XBP1si cell line has reduced levels of XBP1 mRNA and exhibits reduced levels of mRNA of XBP1 responsive genes such as ERdj4 after tunicamycin treatment.

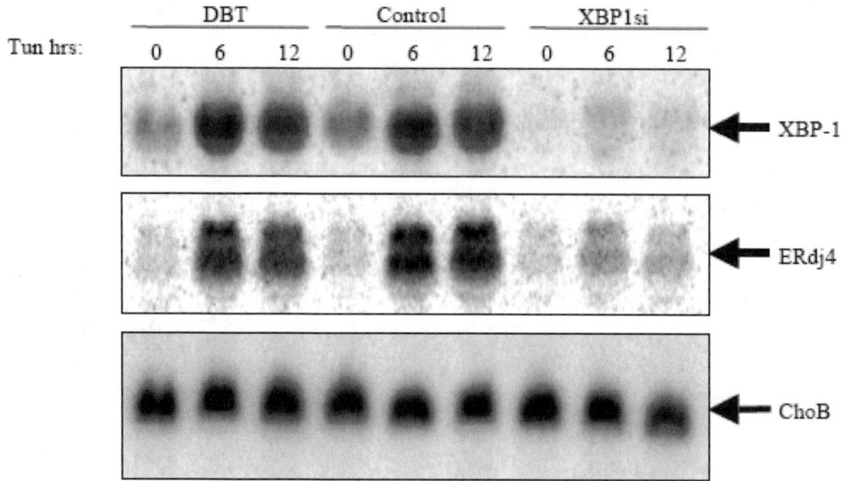

Figure 1. XBP1si cells have reduced levels of XBP1 mRNA and exhibit a reduced response to tunicamycin. Northern blot analysis of total RNA isolated from DBT cells, control and XBP1si cell lines. RNA (10 μg) was subjected to electrophoresis on a formaldehyde-agarose gel, transferred to nitrocellulose and subjected to hybridization with radiolabeled probes to XBP1, ERdj4, and ChoB (a mitochondrial RNA used as a loading control). Hybridization was detected by phosphoimage analysis using a Typhoon Imager.

Next, we wanted to determine if the replication of MHV was altered in the XBP1si cells. We hypothesized that if MHV activation of the Ire1/XBP1 pathway was beneficial to replication, then reduced levels of XBP1 mRNA and protein may inhibit or delay MHV RNA synthesis. XBP1si and control cells were infected with MHV-A59 at MOI 1, RNA was isolated at hourly intervals after infection, and the level of MHV nucleocapsid mRNA was measured by real-time RT-PCR analysis (Fig. 2a). We found that the level of MHV replication (as monitored by accumulation of nucleocapsid RNA) was delayed in the XBP1si cells (6- and 8-hour time points), but ultimately rose to the same level as detected in the control cells (10- and 12-hour time points). Furthermore, the production of virus particles detected in the supernatant was essentially identical from the two cell lines (data not shown). To determine if the Ire1/XBP1 pathway was activated in the XBP1si cells, we generated whole cell lysates and performed Western blotting to examine the level of XBP1(S) protein. Surprisingly, we detected XBP1(S) in both MHV-infected control and MHV-infected XBP1si cells (Fig. 2b). One possible explanation for the presence of the XBP1(S) protein in the XBP1-silenced cells would be that silencing was diminished by MHV infection. To determine if MHV infection affected the level of silencing, we measured the level of XBP1 mRNA in control and XBP1si cell during the time course of MHV infection (Fig. 2c). We found that the silencing of XBP1 was maintained throughout the time course of MHV infection. Therefore, MHV infection does not impede the silencing of the XBP1 mRNA. Overall, we found that MHV infection is a potent inducer of the Ire1/XBP1 pathway of the UPR and that XBP1(S) protein is detected during MHV infection, even in XBP1si cell lines where the XBP1 mRNA concentration is reduced. These data suggest that XBP1(S), typically a short-lived protein, is stabilized in MHV-infected cells.

4. CONCLUSIONS AND FUTURE DIRECTIONS

We found that MHV infection is a potent activator of the Ire1/XBP1 pathway of the unfolded protein response. Indeed, MHV activation of Ire1/XBP1 pathway is detected even in cell lines with reduced expression of XBP1 mRNA (XBP1si cells). Future experiments will assess the turnover of XBP1(S) in MHV-infected cells and determine if XBP1(S) induces UPR responsive genes during MHV infection.

5. ACKNOWLEDGMENTS

This work was funded by Public Health Service research grant AI45798 to S.C.B. J.B. was supported by Training Grant T32 AI007508.

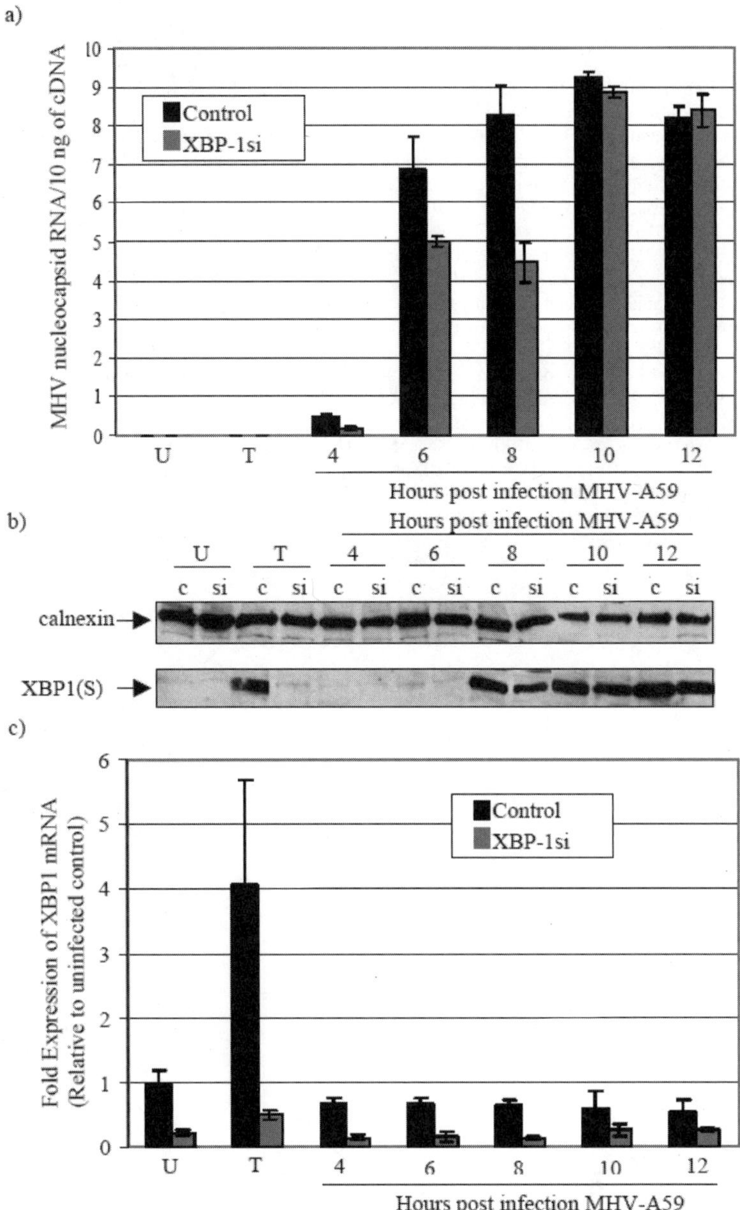

Figure 2. MHV replicates and induces XBP1(S) protein expression in the XBP1si cell line. (a) Real-time RT-PCR analysis of MHV nucleocapsid mRNA levels detected in MHV-infected control and XBP1si cells. (b) MHV infection activates the Ire1/XBP1 pathway in both control and XBPsi cells. Whole cell lystates were prepared at the time indicated and subjected to electrophoresis on a 10% polyacrylamide gel, transferred to nitrocellulose, and calnexin and XBP1(S) proteins detected by Western blotting. (c) RNA silencing is maintained during MHV infection. Real-time RT-PCR analysis of XBP1 mRNA levels during MHV infection of control and XBP1si cells.

6. REFERENCES

1. Gosert, R., Kanjanahaluethai, A., Egger, D., and Baker, S. C., 2002, RNA replication of mouse hepatitis virus takes place at double-membrane vesicles, *J. Virol.* **76**:3697-3708.
2. Rutkowski, D. T., and Kaufman, R. J., 2004, A trip to the ER: coping with stress, *Trends Cell Biol.* **14**:20-28.
3. Tardif, K. D., Mori, K., Kaufman, R. J., and Siddiqui, A., 2004, Hepatitis C virus suppresses the Ire-XBP1 pathway of the unfolded protein response, *J. Biol. Chem.* **279**:17158-17164.
4. Tardif, K. D., Waris, G., and Siddiqui, A., 2005, Hepatitis C virus, ER stress, and oxidative stress, *Trends Microbiol.* **13**:159-163.
5. Isler, J. A., Skalet A., and Alwine, J. C., 2005, Human cytomegalovirus infection activates and regulates the unfolded protein response, *J. Virol.* **79**:6890-6899.
6. Schiller, J. J., Kanjanhaluethai, A., and Baker, S. C., 1998, Processing of the coronavirus MHV-JHM polymerase polyprotein: identification of precursors and proteolytic products spanning 400 kilodaltons of ORF1a, *J. Virol.* **242**:288-302.
7. Gass, J. N., Gifford, N. M., and Brewer, J. W., 2002, Activation of an unfolded protein response during differentiation of antibody-secreting B cells, *J. Biol. Chem.* **277**:49047-49054.
8. Shaffer, A. L., Shapiro-Shelef, M., Iwakoshi, N. N., Lee, A., Qian, S., Zhao, H., Yu, X., Yang, L., Tan, B. K., Rosenwald, A., Hurt, E. M., Petroulakis, E., Sonenberg, N., Yewdell, J. W., Calame, K., Glimcher, L. H., and Staudt, L. M., 2004, XBP1, downstream of BIMP-1, expands the secretory apparatus and other organelles, and increases protein synthesis in plasma cell differentiation, *Immunity* **21**:81-93.
9. Sriburi, R., Jackawski, S., Mori, K., and Brewer, J. W., 2004, XBP1: a link between the unfolded protein response, lipid biosynthesis, and biogenesis of the endplasmic reticulum, *J. Cell Biol.* **167**:35-41.

THE NUCLEAR LOCALIZATION SIGNAL OF THE PRRS VIRUS NUCLEOCAPSID PROTEIN MODULATES VIRAL REPLICATION *IN VITRO* AND ANTIBODY RESPONSE *IN VIVO*

Changhee Lee, Douglas C. Hodgins, Jay G. Calvert, Siao-Kun Wan Welch, Rika Jolie, and Dongwan Yoo*

1. INTRODUCTION

The PRRS virus N protein is a multifunctional protein of 123 amino acids. It is a homodimeric serine phosphoprotein with unknown function.[1] The N protein is self-interactive by noncovalent interactions at amino acids 30–37 through RNA bridging[2] and, as the N protein migrates to the ER and Golgi complex, becomes disulfide-linked via the cysteine residue at position 23.[2] The cysteine-mediated N-N homodimerization is essential for virus infectivity.[3] The N protein is present mainly in the perinuclear region of infected cells but is also specifically localized in the nucleus and nucleolus.[4] A "pat7" nuclear localization signal (NLS) has been identified at positions 41 to 47 (PGKKNKK) and is functional and sufficient for N accumulation to the nucleolus.[5] The N protein nuclear translocation is importin-α and -β dependent.[5] The N protein is an RNA binding protein, and the RNA binding domain has been mapped to the region of amino acids 37–57, which overlaps the NLS sequence.[6] In the nucleus, the N protein colocalizes and interacts with the small nucleolar RNA-associated protein fibrillarin,[6] implicating a nonstructural role of N in ribosome biogenesis. Substitution of lysine residues at positions 43 and 44 with glycine residues has been shown to destroy the pat7 motif and prevents the nuclear localization of N. In the current study, the NLS motif was modified to "PGGGNKK" to knock out the nuclear function of N using a full-length infectious cDNA clone. NLS-null mutant virus was obtained and compared with the wild-type virus for phenotypic changes in cells and pigs. The NLS-null virus was stable in cell culture and grew to a titer of 100-fold lower than wild-type virus. Pigs infected with the NLS-null virus exhibited a reduced severity of disease with milder viremia and higher neutralizing antibody and ELISA antibody titers than wild-type virus-infected pigs.

*Changhee Lee, Douglas C. Hodgins, Dongwan Yoo, University of Guelph, Guelph, Ontario, Canada. Jay G. Calvert, Siao-Kun Wan Welch, Rika Jolie, Pfizer Animal Health, Kalamazoo, Michigan 49001.

Strong selection pressure for reversion at the NLS locus was observed during viremia and persistence in pigs. The N protein nuclear localization may be associated with the virulence and pathogenesis of PRRSV *in vivo*.

2. MATERIALS AND METHODS

The NLS motif in the N protein was modified using a shuttle plasmid to substitute codons for lysine at 43 and 44 to glycine. The NLS-modified full-length cDNA clone was designated pCMV-S-P129-GG. Infectious virus was generated by direct transfection of the plasmid into MARC-145 cells using Lipofectin (Invitrogen) and designated P129-GG. N protein nuclear localization was determined by immunofluorescence of infected cells using N-specific monoclonal antibody SDOW17. Virus in the supernatant from transfected cells was expanded by three serial passages on MARC-145 cells, titrated by plaque assays, and used in this study.

Twenty-one piglets were obtained from a PRRSV-free swine herd at 5 weeks of age and were divided into three groups, seven piglets per group. Pigs received 5 x 10^4 pfu wild-type virus, P129-GG virus, or a placebo intranasally. Clinical signs (general condition, depression, loss of appetite, coughing, sneezing, and respiratory distress) were monitored, and rectal temperatures were measured daily in all pigs for the first one week after inoculation. Blood samples were taken on days 0, 4, 7, 10, 14, 21, and 28 post-infection for virus isolation and serology.

3. RESULTS AND DISCUSSION

The NLS motif of the N protein was modified from PGKKSKK to PGGGNKK in the full-length infectious cDNA clone. Cells were transfected with either the wild-type clone or the mutant clone. Both clones produced infectious progeny virus. CPE induced by the two viruses was similar (Fig. 1B, 1C), but plaques formed by the P129-GG NLS-null virus were smaller than wild-type plaques. No detectible translocation of N protein was identified in the nucleus of cells infected with P129-GG virus (Fig. 1F). Growth kinetics of both viruses were similar with maximum yields at 4–5 days postinfection, but P129-GG reached a peak titer approximately 2 logs lower than that of wild - type.

Groups of seven pigs at six weeks of age were infected intranasally with 5 x 10^4 pfu of P129-WT or P129-GG. All pigs became positive for virus in the serum at four days postinfection, and some animals remained viremic through day 10.

The mean virus titers were higher in the P129-WT group throughout the viremic period. The average duration of viremia was also higher in the P129-WT than in the P129-GG group, while the mock-infected pigs remained negative for PRRSV throughout the study. The appearance of anti-PRRSV antibody was monitored in sera using a commercial ELISA kit (IDEXX). Antibody was first detectable at 7 days postinoculation and continued to increase throughout the study. The mean S/P ratio in the P129-GG group remained consistently higher than that of the P129-WT group in spite of the lower level of viremia in the P129-GG group. Serum neutralization (SN) titers against PRRSV

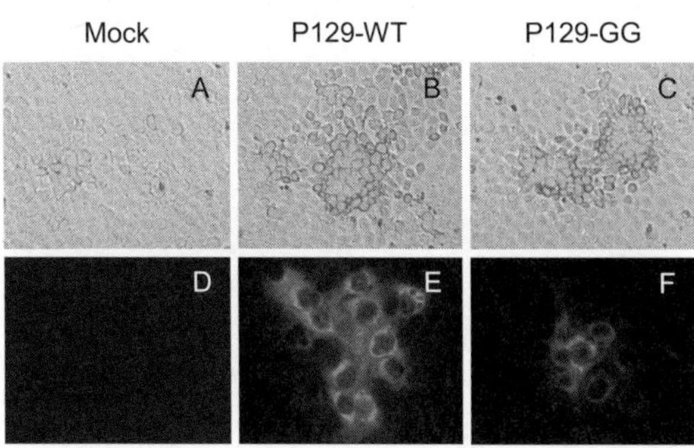

Figure 1. Infectivity of the NLS-null P129-GG mutant clone. Cells were transfected with 2 μg of DNA and incubated for 3 days. Cytopathic effects became visible (A, B, C) at 3 days post-transfection. For immunofluorescence (D, E, F), cells were stained with N-specific monoclonal antibody SDOW17 followed by goat anti-mouse antibody conjugated to Alexa green. Magnification 20×.

were also determined. The mean titer in the P129-GG group increased dramatically between 14 and 28 days. By the end of the study, the average neutralization titer in the P129-GG group was approximately five times higher than in the P129-WT group (Table 1).

One pig in the P129-GG group exhibited viremic and serologic patterns similar to those of animals in the P129-WT group, and a possible reversion of P129-GG was suspected. Viral plaques were prepared from serum of this pig and the N gene was amplified by RT-PCR. Of 10 plaques examined, only 1 plaque retained the original NLS sequence of P129-GG, and 9 plaques were all reverted to the wild-type sequence (Fig. 2). No other mutation was detected elsewhere in the N gene, suggesting strong and specific selection pressure at the NLS locus. N proteins from NLS revertants were found to regain the ability of translocation in the nucleus (Fig. 2).

In the present study, the biological significance of the N protein nuclear localization was studied for PRRS virus. The NLS-null P129-GG virus was generated using an infectious cDNA clone, demonstrating that NLS is nonessential for virus infectivity. The NLS-null virus grew to a titer 2 logs lower than the wild-type virus. In pigs, the NLS-null virus induced shorter duration viremia and lower virus titers in comparison to the wild-type infected pigs. Despite the reduced level of viremia, higher levels of ELISA and neutralizing antibody titers were observed in the NLS-null virus infected pigs. The ability of the NLS-null virus to induce higher level antibodies suggests a possible role of N in the host response modulation. Reversions at the NLS locus occurred in pigs during viremia, and N proteins from the revertants were found to localize in the nucleus. The data suggest that the PRRS virus N protein may play an important role in viral pathogenesis. This unique property of N may be associated with an evasion strategy of PRRS virus from the host defenses.

Table 1. Antibody response in pigs at 28 days postinoculation.

Virus	Mean values	
	ELISA titers (S/P ratio)	SN titers
P129-WT	1.76	5.3
P129-GG	2.31	24.4

virus	nucleotides	amino acid
P129-WT	. CCG GGC AAG AAA AGT AAG AAG .	. P G **K K S** K K .
P129-GG	. CCG GGA GGG GGA AAT AAG AAG .	. P G **G G N** K K .
P129-GG-1	. CCG GGA **A**GG **A**GA AAT AAG AAG .	. P G **R R** N K K .

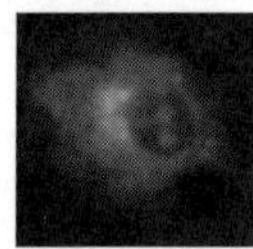

41-PGKKSKK

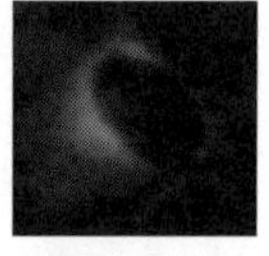

41-PGGGNKK

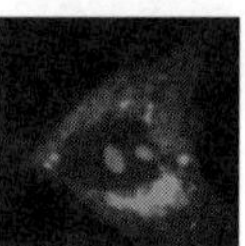

41-PGRRNKK

Figure 2. Reversion of P129-GG virus in pigs during viremia. Viral plaques were prepared from the sera of pig 45 infected with P129-GG and PCR-amplified and sequenced. Nucleotide and amino acid sequences in the NLS region are shown. Immunofluorescence shows the N protein nuclear localization of the revertant.

4. REFERENCES

1. Wootton, S., Rowland, R. R., and Yoo, D., 2002, Phosphorylation of the porcine reproductive and respiratory syndrome virus (PRRSV) nucleocapsid protein, *J. Virol.* **76**:10569-10576.
2. Wootton, S. K., and Yoo, D., 2003, Homo-oligomerization of the porcine reproductive and respiratory syndrome virus nucleocapsid protein and the role of disulfide linkages, *J. Virol.* **77**:4546-4557.
3. Lee, C., Calvert, J. G., Welch, S. W., and Yoo, D., 2005, A DNA-launched reverse genetics system for porcine reproductive and respiratory syndrome virus reveals that homodimerization of the nucleocapsid protein is essential for virus infectivity, *Virology* **331**:47-62.
4. Rowland, R. R., Kervin, R., Kuckleburg, C., Sperlich, A., and Benfield, D. A., 1999, The localization of porcine reproductive and respiratory syndrome virus nucleocapsid protein to the nucleolus of infected cells and identification of a potential nucleolar localization signal sequence, *Virus Res.* **64**:1-12.
5. Rowland, R. R., Schneider, P., Fang, Y., Wootton, S., Yoo, D., and Benfield, D. A., 2003, Peptide domains involved in the localization of the porcine reproductive and respiratory syndrome virus nucleocapsid protein to the nucleolus, *Virology* **316**:135-145.
6. Yoo, D., Wootton, S. K., Li, G., Song, C., and Rowland, R. R., 2003, Colocalization and interaction of the porcine arterivirus nucleocapsid protein with the small nucleolar RNA-associated protein fibrillarin, *J. Virol.* **77**:12173-12183.

SARS CORONAVIRUS ACCESSORY ORFs ENCODE LUXURY FUNCTIONS

Matthew B. Frieman, Boyd Yount, Amy C. Sims, Damon J. Deming, Thomas E. Morrison, Jennifer Sparks, Mark Denison, Mark Heise, and Ralph S. Baric*

1. INTRODUCTION

Severe acute respiratory syndrome (SARS) is a potentially fatal disease that was recognized in the Guangdong Province of China in the fall of 2002.[1,2] The disease quickly spread across Asia, Europe, and North America, and by the end of that outbreak more than 8000 people were infected resulting in about 800 deaths and economic losses in the tens of billions worldwide.[3] The disease is caused by a new human coronavirus (CoV), named the SARS-CoV, which is unlike any previous known coronavirus but classified among the group II coronaviruses like MHV.[4] Recent findings that antibodies against SARS-CoV–like virus were present in the human population prior to the outbreak suggest that the virus previously already circulated in humans.[5] Thus, resurgence of SARS from zoonotic sources remains a distinct possibility, making further understanding of pathogenic mechanisms essential.[6,7]

The SARS-CoV viral gene order is similar to other known coronaviruses with the first 2 open reading frames (ORFs) encoding the viral replicase and the downstream ORFs encoding structural proteins, S, E, M, and N. These downstream ORFs are interspaced with the accessory ORFs thought to be nonstructural proteins of unknown function (ORF3a/b, ORF6, ORF7a/b, ORF8a/b, and ORF9b).[8] The accessory ORFs likely influence the pathogenesis of the SARS-CoV, as the accessory ORFs of other coronaviruses contribute to *in vivo* pathogenesis but are not essential for growth *in vitro* (Figure 1).

Using an infectious clone of the SARS-CoV (icSARS), we tested the hypothesis that the accessory ORFs are not essential for *in vitro* replication but encode virulence determinants.[9] A set of SARS-CoV recombinant viruses lacking one or combinations of

* Matthew B. Frieman, Boyd Yount, Amy C. Sims, Damon J. Deming, Thomas E. Morrison, Mark Heise, Ralph S. Baric, University of North Carolina, Chapel Hill, North Carolina. Jennifer Sparks, Mark Denison, Vanderbilt University, Nashville, Tennessee.

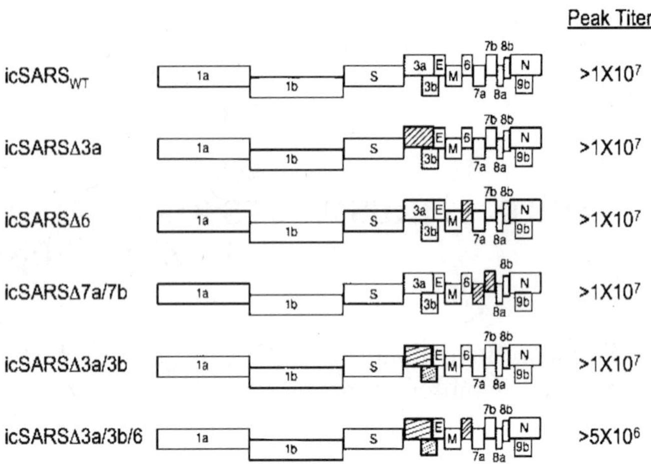

Figure 1. Schematic of SARS genome and mutant viruses. Deletion viruses were made with the icSARS CoV molecular cDNA clone. Cross-hatched boxes designate deleted ORFs. Shown on right is peak titers seen when grown in Vero cells. All mutants grow to wild-type–like titer except for Δ3a/3b/6 triple mutant virus, which is ½ log lower.

accessory ORFs (ORF3a, ORF3b, ORF6, ORF7a, ORF7b) or encoding zoonotic strain variations (full-length ORF8) were isolated by standard reverse genetic techniques. We then tested *in vitro* (e.g., growth, RNA synthesis, protein expression, CPE, and apoptosis) and *in vivo* phenotypes (mouse model) to determine accessory ORF function in replication and pathogenesis. All recombinant viruses were viable and plaque purified for future use.

These viruses were tested *in vitro* and *in vivo* for growth. We found that all viruses with any of the accessory ORFs deleted replicated to similar titers in Vero, MA104, and Caco cells, approaching10^7 pfu in ~24 hr (data not shown). We then tested the mutant viruses in a mouse model by intranasal inoculation. In this system, lungs were harvested 2 days postinfection and virus extracted from the tissue and analyzed by plaque assay. We found no significant differences between the wild-type Urbani strain, our icSARS strain, and any of the mutants deleted for the accessory ORFs (Figure 2). This lead us to postulate two conclusions. First, the accessory ORFs may not encode critical functions in viral pathogenesis. In other viruses that encode accessory ORFs, deletion of these genes allows for *in vitro* growth; however, some attenuate while others have little impact on *in vivo* pathogenesis. In SARS-CoV, it is possible that these accessory ORFs encode minor effects on *in vivo* pathogenesis. Such a result is surprising given data suggesting that ORF3a and ORF7 induce apoptosis and that ORF3a is a structural protein.[10] Alternatively,

	Lung Titer Greater than 10^6
Urbani	+
icSARS	+
Δ3a	+
Δ3a,b	+
Δ3a,b,6	+
Δ7a	+

Figure 2. Balb-C mice were infected intranasally with 1×10^5 virus. Lungs were dissected and virus titer assayed 2 dpi. Shown are the deletion strains tested in this model. A plus (+) denotes those viruses that grew to higher than 10^6 pfu/ml of mouse lung. ORF7a is ORF7a and ORF7b.

the current mouse model may not be sufficiently robust for dissecting out the role of the accessory ORFs in SARS-CoV pathogenesis. We have found that although SARS does infect mouse lungs, it fails to have any pathogenic effect on the animal, and virus is cleared from the infected mouse by day 5. New model systems, such as ferrets and human airway epithelial cultures, will need to be developed to investigate further roles of the accessory ORFs in pathogenesis.

2. IMMUNE RESPONSE

We tested whether one or more accessory ORFs of SARS might interfere with the host innate immunity and interferon signaling, resulting in increased pathogenicity *in vivo*. The innate immune response includes IFN signaling, cytokine activation, and anti-viral proteins and is essential for host clearance of invading viral pathogens. To test this hypothesis, we evaluated the induction of several interferon response genes upon SARS infection. Vero and 293 cells were transfected with constructs with promoters of normally induced antiviral genes, driving luciferase to assay expression. We found that Interferon (IFN) beta, NFkB, and p65 are not induced upon SARS infection, however they are highly upregulated upon Sendai virus infection. When the same assay was tested with the deletion mutants described above, identical results were obtained. We found no induction of IFN beta, NFkB, or p65 upon infection. NFkB should be induced upon viral infection from sensing of virus and activation by IKK; p56 should be induced by IRF3 signaling via sensing of viral replication products. We conclude that SARS is either blocking induction of the antiviral response of infected cells or functionally invisible to these response elements, which may be important in its initial survival when infecting a host. We also find that deletion of individual accessory ORFs does not diminish this modulation of the immune response.

3. CONCLUSIONS

The SARS CoV is a newly emerged virus that is highly pathogenic and evades the host innate immune system. Using the SARS-CoV molecular clone, we isolated a panel of recombinant viruses lacking several accessory ORFs and demonstrated that the replication of these mutants was similar to wild-type virus *in vitro* and *in vivo*. The direct manipulation of the genome of SARS will allow us to discern the function of the accessory ORFs contained in the virus as well and better understand the roles of the non-structural ORFs in the virus. These data suggest that ORF3a is a nonessential structural gene and that ORF3b, ORF6, and ORF7a/b are nonessential. Our data suggests that either the ORFs have no role in pathogenesis or the mouse model is not robust enough to identify virulence alleles in the SARS-CoV genome.

We have also investigated the immune response to SARS infection. Focusing on the innate immune response, we find a block in the induction of the early interferon pathway. IFN beta, NFkB, and p65 are all not induced upon infection. Further analysis will find where in the induction pathway this block is occurring.

Our data suggests a need for further development of new animal models for SARS that more readily recapitulate the human disease phenotype. Current work on the ferret model and non-human primates will aid in understanding the role of the accessory proteins in pathogenesis and the pathway that SARS takes to cause disease.

4. REFERENCES

1. Drosten, C., Gunther, S., Preiser, W., et al., 2003, Identification of a novel coronavirus in patients with severe acute respiratory syndrome, *N. Engl. J. Med.* **348**:1967-1976.
2. Ksiazek, T. G., Erdman, D., Goldsmith, C. S., et al., 2003, A novel coronavirus associated with severe acute respiratory syndrome, *N. Engl. J. Med.* **348**:1953-1966.
3. Han, Y., Geng, H., Feng, W., Tang, X., Ou, A., Lao, Y., Xu, Y., Lin, H., Liu, H., and Li, Y., 2003, A follow-up study of 69 discharged SARS patients, *J. Tradit. Chin. Med.* **23**:214-217.
4. Gorbalenya, A. E., Snijder, E. J., and Spaan, W. J., 2004, Severe acute respiratory syndrome coronavirus phylogeny: toward consensus, *J. Virol.* **78**:7863-7866.
5. Zheng, B. J., Wong, K. H., Zhou, J., Wong, K. L., Young, B. W., Lu, L. W., and Lee, S. S., 2004, SARS-related virus predating SARS outbreak, Hong Kong, *Emerg. Infect. Dis.* **10**:176-178.
6. Dowell, S. F., and Ho, M. S., 2004, Seasonality of infectious diseases and severe acute respiratory syndrome-what we don't know can hurt us, *Lancet Infect. Dis.* **4**:704-708.
7. Webster, R. G., 2004, Wet markets--a continuing source of severe acute respiratory syndrome and influenza? *Lancet* **363**:234-236.
8. Rota, P. A., Oberste, M. S., Monroe, S. S., et al., 2003, Characterization of a novel coronavirus associated with severe acute respiratory syndrome, *Science* **300**:1394-1399.
9. Yount, B., Curtis, K. M., Fritz, E. A., et al., 2003, Reverse genetics with a full-length infectious cDNA of severe acute respiratory syndrome coronavirus, *Proc. Natl. Acad. Sci. USA* **100**:12995-13000.
10. Ito, N., Mossel, E. C., Narayanan, K., Popov, V. L., Huang, C., Inoue, T., Peters, C. J., and Makino, S., 2005, Severe acute respiratory syndrome coronavirus 3a protein is a viral structural protein, *J. Virol.* **79**:3182-3186.

PRODUCTION AND CHARACTERIZATION OF MONOCLONAL ANTIBODIES AGAINST THE NUCLEOCAPSID PROTEIN OF SARS-CoV

Ying Fang, Andrew Pekosz, Lia Haynes, Eric A. Nelson, and Raymond R. R. Rowland*

1. INTRODUCTION

To elucidate the basic mechanism of nucleocapsid (N) protein function in the pathogenesis of the disease and in aid of developing diagnostic tests for SARS-CoV, we produced a panel of monoclonal antibodies (MAbs) against SARS-CoV N protein. We further tested their application to the detection of SARS-CoV infection using the methods of Western blotting, immunofluorescent staining (IFA), and ELISA.

2. MATERIALS AND METHODS

DNA encoding the N protein was amplified from cDNA prepared from the Urbani isolate (CDC) and cloned into the pBAD-TOPO protein expression vector (Invitrogen). The N protein was expressed in *E. coli* as a 6-His tagged fusion protein and purified using Ni-NTA columns (Qiagen). BALB/c mice were immunized with purified recombinant N protein at 2-week intervals for 8 weeks. Splenocytes from immunized mice were fused with NS-1 myeloma cells and cultured on 24-well plates. Cell culture supernatants from wells containing hybridoma colonies were initially screened by ELISA using recombinant N protein. Hybridoma cells from positive wells were subcloned, expanded, and retested.

For Western blot analysis, Vero cells (ATCC) were mock-infected or infected with SARS-CoV (Urbani strain). The cell lysate was separated on a 15% polyacrylamide gel, transferred to PVDF membranes, and then incubated with an anti-N MAb. After extensive washing, the membranes were incubated with a HRP conjugated goat anti-mouse IgG (Jackson Laboratories). After 60 minutes, the blot was washed extensively

* Ying Fang, Eric A. Nelson, South Dakota State University, Brookings, South Dakota. Andrew Pekosz, Washington University School of Medicine, St. Louis, Missouri. Lia Haynes, Centers for Disease Control and Prevention, Atlanta, Georgia. Raymond R.R. Rowland, Kansas State University, Manhattan, Kansas.

and visualized with ECL Plus Western blotting reagent (Amersham Biosciences) followed by exposure to autoradiographic film (Molecular Technologies).

For IFA, infected Vero cells were fixed with 2% paraformaldehyde at 24 h after infection and then permeabilized with PBS containing 3% normal goat sera and 0.2% saponin (blocking buffer) for 15 minutes. MAbs were diluted 1:500 in blocking buffer and incubated with cells for 1 h at room temperature. After extensive wash, the cells were incubated with a goat anti-mouse IgG antibody conjugated to AlexaFluor 594 (Molecular Probes) or FITC (Jackson Laboratories) and the nuclear stain ToPro-3 (Molecular Probes). After incubation, the cells were washed and mounted onto microscope slides. Samples were analyzed on a Zeiss LSM 510 confocal microscope.

For ELISA, detergent extracted, gamma-irradiated SARS-CoV or recombinant SARS N protein were coated onto 96-well plates (Dynex) in bicarbonate buffer (pH 9.6) overnight at 4^0C. The plates were blocked with 2% BSA for 1 h at 37^0C. After washing, 100 μl of serially diluted MAbs were added and incubated for 1 h at 37^0C, and then HRP-goat anti-mouse IgG/IgM was added and incubated for 45 minutes. Plates were developed using ABTS, and OD values were determined using an ELISA plate reader.

3. RESULTS AND DISCUSSION

Initial hybridoma screening by ELISA using recombinant N protein yielded three candidate MAbs, SA 87-A1, SA 46-4, and SA 41-48. The isotypes of these MAbs were determined using an IsoStrip Kit (Serotech) and results showed that SA 87-A1 was an IgG1, SA 46-4 was an IgG2a, and SA 41-48 was an IgM.

We further tested the reactivity of these MAbs by IFA on Vero cells expressing recombinant EGFP-N protein. The SARS-CoV (Ubani strain) N gene was cloned into pEGFP-C2 vector (Clontech), and Vero cells were transfected with plasmid pEGFP-C2-N. MAb SA 46-4 showed the highest sensitivity and specificity in recognizing the EGFP-N protein in Vero cells, but the other two MAbs showed weak fluorescence and high background (data not show). To further confirm the specificity of MAb SA 46-4, it was tested for cross-reactivity with other coronaviruses and arteriviruses. It did not cross-react with porcine transmissible gastroenteritis virus (TGEV), mouse hepatitis virus (MHV), or porcine reproductive and respiratory syndrome virus (PRRSV).

We further explored the sensitivity and specificity of these MAbs for detection of SARS-CoV infection using Western blotting, IFA, and ELISA. Protein specificity of MAbs SA 87-A1 and SA 46-4 was first studied by Western blot analysis of Vero cell lysates infected with SARS-CoV. The results in Fig. 1 showed that MAbs SA87-A1 and SA46-4 reacted with N from infected cells and not with proteins from mock-infected cells. These MAbs reacted with a viral protein, which migrated as a doublet close to 50 kDa, the predicted size for the N protein. The presence of N in two forms was also shown by Western blotting using *in vitro* expressed N protein (data not shown). The presence of N in two forms may reflect a post-translational modification. This result is consistent with the results reported by Leung et al.[1] In their experiments, Western blots prepared using lysates from SARS-CoV infected cells probed with sera from patients showed several N protein-specific bands. They indicated that the N protein cleavage site is at the C-terminal, and this type of cleavage is specifically found in every preparation of the crude extracts. Post-translational cleavage of N would be a unique property of SARS-CoV, which is a subject of further study in our laboratory.

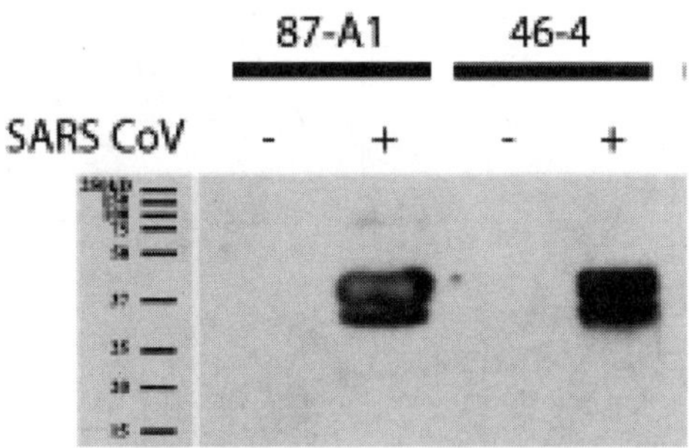

Figure 1. Detection of SARS-CoV N protein with Western blot. SARS-CoV viral lysates were electrophoresed with SDS-PAGE. After blotting on the PVDF membrane, proteins were detected by incubation with anti-N MAbs, followed by incubation with a horseradish peroxidase conjugated goat anti-mouse IgG. The blot was visualized with ECL Plus Western blotting reagent followed by exposure to autoradiographic film.

To further define MAb specificity, IFA was performed on Vero cells infected with SARS-CoV Urbani strain. The cells were fixed and incubated with MAbs SA87-A1 and SA46-4, and then double stained with AlexaFluor 594 (Molecular Probes) conjugated goat anti-mouse IgG and a nuclear stain ToPro-3 (Molecular Probes). As shown in Fig. 2, both MAbs SA46-4 and SA87-A1 specifically recognized the viral N protein, which is accumulated in the cytoplasm, in contrast with the ToPro-3 nuclear staining.

A commonly used method for the diagnosis of SARS-CoV infection is ELISA. To determine if these MAbs have any value for future development of serological tests, we tested MAbs SA46-4, SA87-A1, and SA41-48 in indirect ELISA using either *in vitro* expressed recombinant N protein or viral lysate as antigen. Antibodies were applied as twofold serial dilutions. Consistent with the other tests, MAb SA46-4 showed the highest sensitivity (Fig. 3).

Currently, although SARS-CoV appears to be under control, its future reemergence is still possible. Accurate and timely diagnosis of SARS infection is a critical step in preventing another global outbreak. Diagnostic reagents for SARS that are highly sensitive, specific, and convenient need to be developed. In this study, we produced a panel of anti-N MAbs, and MAb SA 46-4 consistently showed good sensitivity and specificity in all the tests conducted. Preliminary epitope mapping determined that MAb SA 46-4 recognizes an epitope located in the region between amino acids 50 to 200.

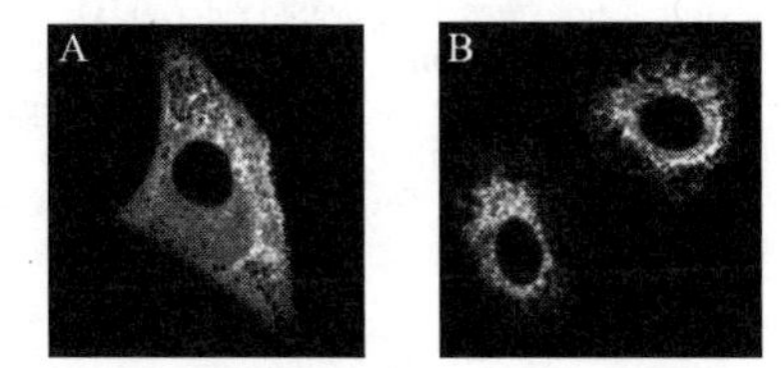

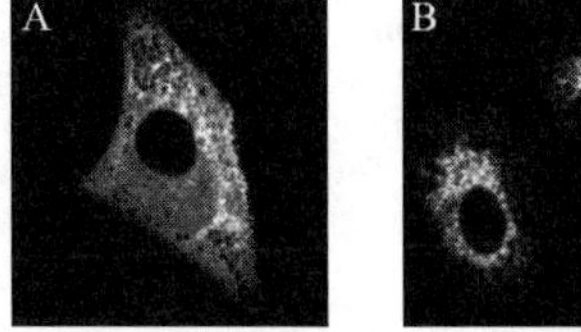

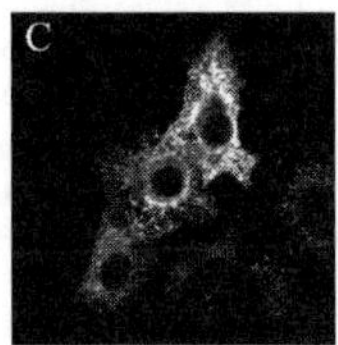

Figure 2. Confocal microscopy of SARS-CoV–infected Vero cells immunostained with anti-N MAbs SA 46-4 (A), SA 87-A1 (B), or SA 41-48 (C). Vero cells were parformaldehyde-fixed at 24 h postinfection. Fixed cells were incubated with anti-N MAbs and stained with AlexaFluor 594-labeled goat anti-mouse IgG (red). Cells were counterstained with the nuclear stain, ToPro-3 (data not shown).

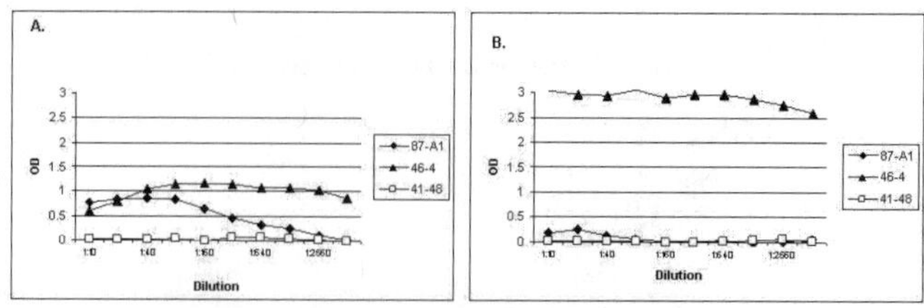

Figure 3. ELISA comparison of anti-N MAb reactivity to SARS-CoV (A) or to recombinant N protein (B). Detergent extracted, gamma-irradiated SARS-CoV or recombinant SARS N protein was coated onto 96-well Immunon-2 plates. Serially titrated anti-N MAbs were reacted for 1 h at room temperature and the bound MAbs were detected by peroxidase labeled goat anti-mouse IgG/IgM followed by color development with ABTS.

Other anti-N MAbs have been reported.[2,3] Some of them have been mapped to epitope sites on the N protein, and they recognize different epitopes than reported here.[2] Previous studies showed that none of the antigenic epitopes reacted with all test sera from SARS patients.[4] Therefore, a "MAb cocktail" combining currently available anti-N MAbs recognizing different antigenic epitopes could improve detection. Especially, because previous studies demonstrated that N protein detection exhibited a high positive rate between day 3 and day 5 after the onset of symptoms, these anti-N MAbs will be suitable reagents for early detection of SARS-CoV infection.[5] Furthermore, the N protein has been determined to be the most antigenic structural protein, expressed in high abundance in infected cells, and N protein sequences are relatively conserved among different strains of SARS-CoV.[1, 4, 6] These MAbs may have a substantial impact on the development of diagnostic tests for SARS.

4. REFERENCES

1. D. T. M. Leung, F. C. H. Tam, C. H. Ma, et al., Antibody response of patients with severe acute respiratory syndrome (SARS) targets the viral nucleocapsid, *J. Infect. Dis.* **190**, 379-386 (2004).
2. Y. He, Y. Zhou, H. Wu, Z. Kou, S. Liu, and S. Jiang, Mapping of antigenic sites on the nucleocapsid protein of the severe acute respiratory syndrome coronavirus, *J. Clin. Micro.* **42**, 5309-5314 (2004).
3. K. Ohnishi, M. Sakaguchi, T. Kaji, et al., Immunological detection of severe acute respiratory syndrome coronavirus by monoclonal antibodies, *Jpn. J. Infect. Dis.* **58**, 88-94 (2005).
4. J. Wang, J. Wen, J. Li, et al., Assessment of immunoreactive synthetic peptides from the structural proteins of severe acute respiratory syndrome coronavirus, *Clin. Chem.* **49**, 1989-1996 (2003).
5. B. Di, W. Hao, Y. Gao, et al., Monoclonal antibody-based antigen capture enzyme-linked immunosorbent assay reveals high sensitivity of the nucleocapsid protein in acute-phase sear of severe acute respiratory syndrome patients, *Clin. Diag. Lab. Immun.* **12**, 135-140 (2005).
6. S. Li, L. Lin, H. Wang, et al., The epitope study on the SARS-CoV nucleocapsid protein, *Genomics Proteomics Bioinformatics* **1**, 198-206 (2003).

MOUSE HEPATITIS CORONAVIRUS NUCLEOCAPSID PHOSPHORYLATION

Tiana C. White and Brenda G. Hogue*

1. INTRODUCTION

The nucleocapsid (N) protein of coronaviruses is a major structural component that packages the large viral RNA genome into a helical nucleocapsid within the mature virion. The N protein is multifunctional. The protein interacts with itself and with the viral M protein.[1,2,3] N has also been shown to bind viral RNA at both the packaging signal and the 5' leader sequence that is common to both viral genomic RNA and all subgenomic RNAs.[4,5,6] In addition to its role in virion structure and RNA binding, N has been implicated as playing a role in viral replication. Coronavirus replicons either expressing N or replicating in the presence of N protein supplied in *trans* showed enhanced activity over replicons transfected without the presence of N.[7] This evidence has led to the general opinion that N protein is involved in the transcription and/or replication complexes of the virus. Taken together, N is clearly a dynamic viral protein.

The N protein is phosphorylated, a feature conserved across the family. Mouse hepatitis virus A59 (MHV), a group II coronavirus, is being used as a model to study N protein phosphorylation. Data suggest that the N protein of BCV and MHV exists in at least two phosphorylated forms, indicated by differing molecular weights, during the viral life cycle.[8] Data also suggests that only one form is packaged into virions. Thus, phosphorylation of the N protein may play a role in viral assembly. Alternatively, different phosphorylated forms of the protein may perform distinct functions in assembly, replication, and/or transcription. To begin understanding the role that phosphorylation plays in any of the functions provided by N, we have begun identifying which amino acids are phosphorylated in the mature virion and in infected cells. This report focuses on preliminary identification of sites that are phosphorylated in the mature virion. Mass spectrometry is being used to identify which of the many predicted phosphorylation sites within the N protein are actually modified in the virion and in infected cells. Serine 389 and serine 424 have been preliminarily identified on the N protein from purified virions

* Arizona State University, Tempe, Arizona.

as sites that are phosphorylated. Complete mass spectrometry analysis will provide a basis for understanding the role of N protein phosphorylation.

2. MATERIALS AND METHODS

Mouse 17 clone 1 (17Cl1) cells were infected with MHV A59 at an MOI of 0.1. The extracellular media were collected and clarified by centrifugation. The supernatant containing the virions was precipitated with polyethylene glycol (PEG) 8000 in 0.4 M sodium chloride and was purified through a 20–60% sucrose gradient. Fractions containing virus particles were combined and concentrated by pelleting through a 30% sucrose cushion. Virions were resuspended in TMEN buffer. Purified virion proteins were analyzed in a 10% SDS-PAGE gel. The band containing the N protein was removed from the gel, destained, and incubated under reducing conditions with DTT prior to in-gel digestion with proteomics grade porcine pancreatic trypsin (Sigma). Resulting peptides were extracted and dried under vacuum. A portion of the peptide mixture was passed through a gallium (III) immobilized metal affinity column (IMAC) (Pierce) to enrich and isolate phosphopeptides. The IMAC elution was dried under vacuum prior to MALDI-TOF mass spectrometry analysis.

3. RESULTS

Identification of phosphorylation sites began by analyzing tryptic digestion products of N protein from mature virions by MALDI-TOF mass spectrometry. The FindMod tool on ExPASy (http://us.expasy.org/tools/findmod) was used to rapidly identify peaks reported by MALDI that corresponded to predicted peptide masses with an addition of approximately 80 Da, representing the addition of one phosphate group. Of the several putative masses identified by the software, only two of the peaks were detectable above the noise level that exhibited normal isotopic patterns (Fig. 1A). The first identified peak had a mass of 555.084 Da, corresponding to peptide -NVpSR- encompassing amino acids 422–425. The predicted mass for this peptide is 475.263 Da, the experimental mass detected by MALDI results from an additional 79.8217 Da, the equivalent of one phosphate group. The second peak that was identified had a mass of 1534.6747 Da, corresponded to peptide -DGGADVVpSPKPQRK- which includes amino acids 382–394. This peak was identified by the software as being 80.8023 Da larger than the predicted mass for that peptide.

To confirm that the identified peaks were actually phosphopeptides, total digestion products were bound to an immobilized metal affinity column under highly acidic conditions. Bound peptides were eluted under basic conditions after several wash steps. A mass at 556.242 eluted from the column that exhibited a normal isotopic pattern (Fig. 1B). This peak confirmed that the mass detected earlier at 555 Da was in fact a phosphorylated peptide.

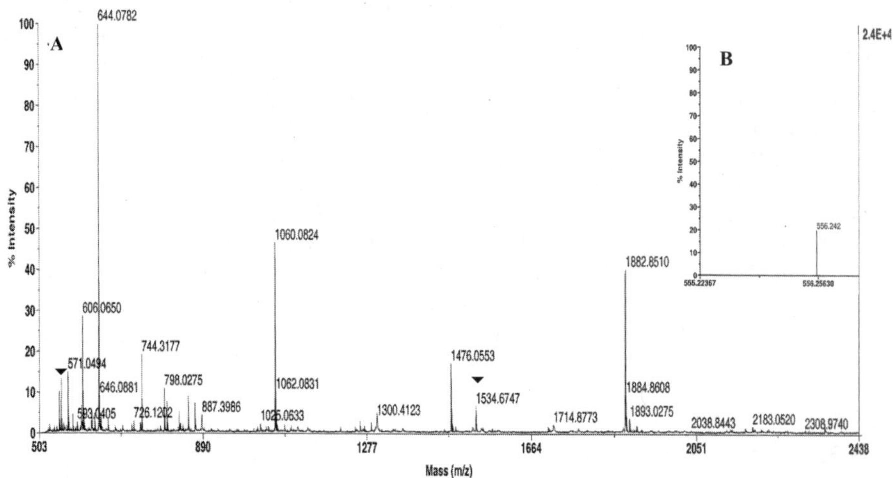

Figure 1. MALDI-TOF spectrum of peptides produced from a tryptic digest of N. The spectra obtained represent an average of 100 laser shots per spectrum. Samples were resuspended in α-cyano 4-hydroxycinnamic acid matrix and shot in positive ion mode with delayed extraction. (A) Arrows indicate masses that were identified as being approximately 80 Da larger than predicted tryptic peptide masses for N. (B) Spectrum obtained after peptides were eluted from gallium (III) IMAC column. A mass at 556.242 Da confirms the mass observed initially as being a phosphopeptide.

4. DISCUSSION

Two peaks, with masses 555.084 Da and 1534.578 Da, were identified as possible phosphopeptides from a tryptic digestion of the MHV A59 virion N protein. The peptides were calculated to be approximately 80 Da larger than the predicted peptide masses. The masses of these peaks preliminarily place the phosphorylation sites on the virion N protein at serine 424 and 389. However, when peptides were selected by retention on a gallium (III) IMAC column, only one of the sites, serine 424, was confirmed to be phosphorylated. Because the Voyager DE STR MALDI-TOF (Applied Biosystems) mass spectrometer used to obtain the mass information was not equipped with tandem capabilities, the sequence of the peptides could not be confirmed. Therefore, we preliminarily conclude that the serines at positions 424 and 389 are phosphorylated based on the combination of mass spectrometry data, software predictions, and past data illustrating that the N protein is most likely phosphorylated exclusively on serine residues.[9] Ongoing tandem mass spectrometry experiments will provide more conclusive identification of these sites.

All coronavirus nucleocapsid proteins are phosphorylated. How conserved the positioning of these sites are remains to be determined. To date, phosphorylation sites for two coronaviruses have been identified. Recombinant N protein from infectious bronchitis virus, a group III coronavirus, expressed in mammalian cells is phosphorylated at serines 190, 192, 379, and threonine 378.[12] Transmissible gastroenteritis virus N protein, a member of group I, isolated from infected cells is phosphorylated at serines 9, 156, 254, and 256.[13] The two preliminary sites on the group II MHV presented here are located in the carboxy end of the N protein within domain III. Two of the identified IBV sites map at the carboxy end of the protein, whereas none of the identified sites in the TGEV N, from either infected cells or mature virions, are present in the domain III carboxy region. Additionally, both IBV and TGEV have phosphorylated residues in domain II, which also contains the RNA binding domain of N.[6,14] It remains to be determined if additional sites will be identified for MHV in domain II.

Given that phosphorylation plays an important role in the regulation of both cellular and viral proteins, the role of N protein phosphorylation remains an intriguing, important area of study. There are precedents for phosphorylation involvement in both virus assembly and RNA transcription. Vesicular stomatitis virus uses phosphorylation as a way to control transcriptional activity.[10] The import of hepatitis B viral capsids into the nucleus of infected cells is dependent on phosphorylation of the HBV capsid protein.[11] Thus, identification of phosphorylation sites for both coronavirus intracellular and mature virion N proteins is well justified.

5. ACKNOWLEDGMENTS

This work was supported by Public Health Service grant AI54704, from the National Institute of Allergy and Infectious Diseases.

6. REFERENCES

1. L. Kho and P. Masters, Genetic evidence for a structural interaction between the carboxy termini of the membrane and nucleocapsid proteins of mouse hepatitis virus, *J. Virol.* **76**, 4987-4999 (2002).
2. D. Escors, J. Ortego, H. Laude, and L. Enjuanes, The membrane M protein carboxy terminus binds to transmissible gastroenteritis coronavirus core and contributes to core stability, *J. Virol.* **75**, 1312-1324 (2001).
3. K. Narayanan, K. H. Kim, and S. Makino, Characterization of N protein self-association in coronavirus ribonucleoprotein complexes, *Virus Res.* **98**, 131-140 (2003).
4. R. Cologna, J. F. Spagnolo, and B. G. Hogue, Identification of nucleocapsid binding sites within coronavirus-defective genomes, *Virology* **277**, 235-249 (2000).
5. R. Molenkamp and W. Spaan, Identification of a specific interaction between the coronavirus mouse hepatitis virus A59 nucleocapsid protein and packaging signal, *Virology* **239**, 78-86 (1997).
6. G. W. Nelson, S. A. Stohlman, and S. M. Tahara, High affinity interaction between nucleocapsid protein and leader/intergenic sequence of mouse hepatitis virus RNA, *J. Gen. Virol.* **81**, 181-188 (2000).
7. F. Almazán, C. Galán, and L. Enjuanes, The nucleoprotein is required for efficient coronavirus genome replication, *J. Virol.* **78**, 12683-12688 (2004).
8. B. G. Hogue, Bovine coronavirus nucleocapsid protein processing and assembly, *Adv. Exp. Med. Biol.* **380**, 259-263 (1995).
9. S. A. Stohlman and M. Lai, Phosphoproteins of murine hepatitis viruses, *J. Virol.* **32**, 672-675 (1979).
10. S. Barik and A. Banerjee, Sequential phosphorylation of the phosphoprotein of vesicular stomatitis virus by cellular and viral protein kinases is essential for transcription activation, *J. Virol.* **66**, 1109-1118 (1992).
11. B. Rabe, N. Pante, A. Helenius, and M. Kann, Nuclear import of hepatitis B virus capsids and release of the viral genome, *Proc. Natl. Acad. Sci. USA* **100**, 9849-9854 (2003).
12. H. Chen, et. al., Mass spectroscopic characterization of the coronavirus infectious bronchitis virus nucleoprotein and elucidation of the role of phosphorylation in RNA binding by using surface plasmon resonance, *J. Virol.* **79**, 1164-1179 (2005).
13. E. Calvo, D. Escors, J. A. López, J. M. González, A. Álvarez, E. Arza, and L. Enjuanes, Phosphorylation and subcellular localization of transmissible gastroenteritis virus nucleocapsid protein in infected cells, *J. Gen. Virol.* **86**, 2255-2267 (2005).
14. P. Masters, Localization of an RNA-binding domain in the nucleocapsid protein of the coronavirus mouse hepatitis virus, *Arch. Virol.* **125**, 141-160 (1992).

III. VIRAL ASSEMBLY AND RELEASE

GENETIC AND MOLECULAR BIOLOGICAL ANALYSIS OF PROTEIN-PROTEIN INTERACTIONS IN CORONAVIRUS ASSEMBLY

Paul S. Masters, Lili Kuo, Rong Ye, Kelley R. Hurst, Cheri A. Koetzner, and Bilan Hsue*

1. INTRODUCTION

Virions of coronaviruses (CoVs) are pleiomorphic, with a roughly spherical structure brought about by cooperation among a relatively small set of structural proteins and a membranous envelope acquired from the endoplasmic reticulum–Golgi intermediate compartment (ERGIC) (Fig. 1). Three integral membrane proteins reside in the envelope. The most salient of these is the spike glycoprotein (S), which mediates receptor attachment and fusion of the viral and host cell membranes. The membrane protein (M) is the most abundant virion component and gives the envelope its shape. The third constituent is the envelope protein (E), which, although minor in both size and quantity, plays a decisive role is envelope formation. In some group 2 CoVs, an additional protein, the hemagglutinin-esterase (HE), appears in the viral envelope. Finally, interior to the envelope, monomers of the nucleocapsid protein (N) wrap the genome into a helical structure.

A number of approaches have been taken to elucidate the network of interactions, among the canonical structural proteins S, M, E, and N, and the genomic RNA, that lead to assembly of virions. The earliest efforts employed the fractionation and reassociation of components of purified virions. These studies were followed by molecular genetic and co-immunoprecipitation analyses of expressed proteins or proteins from virus-infected cells. More recently, reverse-genetic techniques have become available. This chapter will briefly review the current understanding of CoV assembly, highlighting some recent results from our laboratory in the context of work that has been done by numerous other groups in this field.

From a large body of work extending over two decades, the main principle that has emerged is that M is the central organizer of CoV assembly. The M protein (~25 kDa)

*Paul S. Masters, Lili Kuo, Rong Ye, Kelley R. Hurst, Cheri A. Koetzner, New York State Department of Health, Albany, New York. Bilan Hsue, Stratagene, La Jolla, California.

has a small, amino-terminal ectodomain that is either O-glycosylated (group 2 CoVs) or N-glycosylated (groups 1 and 3 CoVs) (Fig. 1). This domain is followed by three transmembrane segments and a large carboxy-terminal endodomain.[1–4] For the group 1 CoV TGEV, it has been shown that roughly one-third of M protein assumes a topology in which part of the endodomain constitutes a fourth transmembrane segment, thereby positioning the carboxy terminus on the exterior of the virion[5]; however, this has not yet been demonstrated for other CoV family members.

The dominant role of M was, in part, deduced by the process of elimination. Early experiments with tunicamycin-treated MHV-infected cells showed that (noninfectious) virions could assemble without S protein.[6,7] This finding was also consistent with the properties of an S gene *ts* mutant, which failed to incorporate spikes into virions at the nonpermissive temperature.[8] Careful co-immunoprecipitation studies subsequently demonstrated that M selects both the S and HE proteins for assembly.[9,10] However, it was apparent that M could not act on its own: expression of M protein alone does not lead to formation of virion-like structures. In addition, M, expressed in the absence of other viral proteins, travels to the Golgi, whereas CoVs bud into the ERGIC.[11–15] This paradox was resolved by the landmark demonstration that co-expression of MHV M protein and a previously overlooked structural protein, E, resulted in the formation of virus-like particles (VLPs).[16,17] That just the M and E proteins are necessary and sufficient for the formation and release of VLPs has since been shown for CoVs of all three groups: BCoV[18] and SARS-CoV[19] (group 2), TGEV[18](group 1), and IBV[20,21](group 3). To date, the only apparent exception is a report that, for SARS-CoV, M protein and N protein were necessary and sufficient for VLP formation.[22] It remains to be seen whether this finding indicates a unique aspect of SARS-CoV virion morphogenesis, or whether it reflects a singular characteristic of the cell line or the expression system that was used.

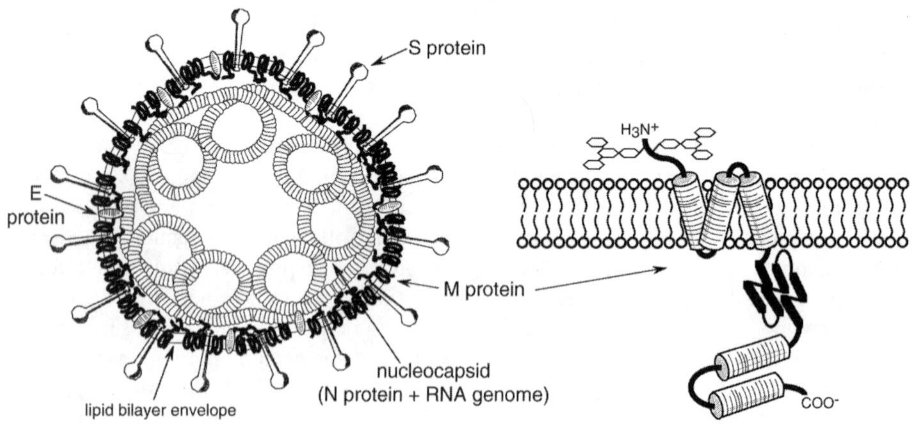

Figure 1. Structure of the CoV virion (left), and a model of the M protein (right).

2. INTERACTIONS OF S WITH M

Experiments with VLPs made possible the systematic manipulation of individual constituents of the virion envelope, leading to the first glimpses of the function of E protein and the partial localization of M-S intermolecular interactions.[23, 24] The S protein is a large (~180 kDa) type I transmembrane protein that assembles into trimers to form the distinctive CoV spikes. Although it is not required for VLP formation, S protein, if present, becomes incorporated into VLPs. The heavily N-glycosylated amino-terminal ectodomain of S, which makes up more than 95% of the mass of the molecule, has been found to be essentially inert in the assembly process. Construction of chimeric MHV-FIPV S proteins with swapped ectodomains showed that the 61-amino-acid transmembrane domain and endodomain determined the incorporation of the S protein into VLPs formed by the homologous M and E proteins.[24]

When this principle – the functional separation of the domains of S for receptor binding and for virion assembly – was extended to whole viruses, it allowed the development of host range-based selective systems for the reverse genetics of CoVs through targeted RNA recombination.[25,26] A mutant of MHV (designated fMHV) was constructed, in which the ectodomain of the MHV S protein was replaced by that of the FIPV S protein. This chimeric virus had the host cell-restricted growth pattern that was predicted from its precursors: it had simultaneously lost the ability to grow in murine cells and gained the ability to grow in feline cells. The use of fMHV as the recipient virus in targeted RNA recombination enabled us to efficiently carry out reverse genetics on MHV, by restoring the MHV S ectodomain (in conjunction with mutations of interest) and selecting for the reacquisition of the ability to grow in murine cells.[27] Selections of even greater stringency were subsequently made possible by the rearrangement of genes downstream of S, in fMHV.v2; this rearrangement effectively precluded the possibility of unwanted secondary crossover events during targeted RNA recombination.[28]

Among the many problems to which this system has been applied was the genetic dissection of the transmembrane domain and endodomain of the MHV S protein, in order to localize the determinants of S incorporation into virions.[29] We used two strategies for this investigation (Fig. 2). First, the S protein transmembrane and endodomains were attached to a heterologous ectodomain to produce a surrogate virion structural protein (named Hook), which could be mutated without consequence to viral infectivity. Second, significant mutations from Hook were transferred to the S protein (in the absence of Hook), to enable examination of their effects on viral phenotype. We found that assembly competence mapped to the endodomain of S, which was sufficient to target Hook for incorporation into virions. Further mutational analysis indicated a major role for the charge-rich carboxy-terminal region of the endodomain. Additionally, we found that the adjacent, membrane-proximal, cysteine-rich region of the endodomain is critical for cell-cell fusion during infection, thus confirming results previously obtained with S protein expression systems.[30, 31] A separate study[32] came to the same fundamental conclusion that virion incorporation was determined by the endodomain of S, but in the latter work the major role in assembly was ascribed to the cysteine-rich region of the endodomain. The differences among the detailed conclusions of these two studies may have been due to the relative importances of particular endodomain residues that were ablated in the differently constructed deletion mutants.

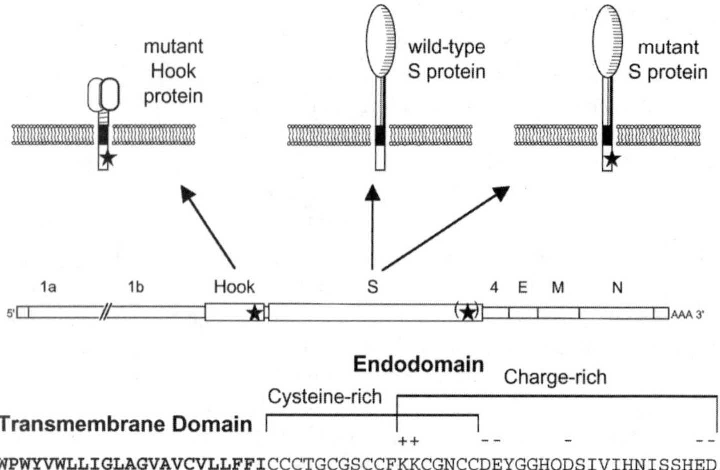

Figure 2. Genetic dissection of the determinants for incorporation of the S protein into virions.

3. INTERACTIONS OF N WITH M

We have also examined the association between the M protein and the N protein of MHV. This was initially accomplished through the construction of a highly impaired mutant, MΔ2, containing a two-amino-acid truncation of the M protein.[27] This mutant formed tiny plaques and grew to maximal titers that were three orders of magnitude lower than those of the wild type. Analysis of multiple second-site revertants of MΔ2 revealed a number of changes in either the M protein or the N protein that could individually compensate for the lesion in MΔ2 (Fig. 3). The latter set of suppressors provided the first genetic evidence for a structural interaction between the M and N proteins, and they allowed that interaction to be localized to the carboxy termini of both proteins.

The MHV N protein (~50 kDa) comprises three conserved domains that are separated by two highly divergent spacer regions.[33] Domains 1 and 2, which make up most of the molecule, are very basic, and the RNA-binding capability of N maps to domain 2.[34, 35] In contrast, domain 3, the carboxy-terminal 45 amino acids of N, has an excess of acidic over basic residues. To complement the results obtained with the MΔ2 mutant, we recently created a complete set of clustered charged-to-alanine mutants in domain 3 of N.[36] One of these mutants, CCA4, was extremely defective, thereby implicating a pair of aspartate residues (D440 and D441) as making a major contribution by N protein to the N-M interaction. Moreover, independent second-site reverting mutations of CCA4 were found to map in the carboxy-terminal region of either the N or the M protein (Fig. 3), thereby displaying genetic cross-talk reciprocal to that uncovered with the MΔ2 mutant. Indeed, one particular mutation in N domain 3 (Q437L) was isolated multiple times, either as a suppressor of the MΔ2 mutation or as a suppressor of the N CCA4 mutation. Additionally, we showed that the transfer of N protein domain 3 to a heterologous protein (GFP) was sufficient to allow incorporation of GFP into MHV virions.

It is not yet clear how this genetically defined N-M interaction is related to connections that have been uncovered by molecular biological and biochemical means. For TGEV, the interaction between N and M was assayed by binding of *in vitro*–translated M protein to immobilized nucleocapsid purified from virions.[37] Deletion mapping was used to localize this binding to a region of the TGEV M protein, the MHV counterpart of which partially overlaps with critical residues that we have identified in the MHV M protein by suppressor mapping. A key early study of MHV, which used biochemical procedures to fractionate the components of purified virions, found a temperature-dependent association between nonionic detergent-solubilized M protein and the viral nucleocapsid.[38] More recently, it was shown that N protein could be co-immunoprecipitated from MHV-infected cells by mAbs specific for M protein. Significantly, although N was shown to be intracellularly associated with all viral RNAs, both subgenomic and genomic,[39–41] the M protein bound only to those complexes of N molecules that were, in turn, bound to genomic RNA.[41] Such selectivity was determined to depend upon the presence of the genomic RNA packaging signal; this signal, if transferred to a heterologous RNA, was sufficient to allow its packaging into virions.[42] Surprisingly, further work with co-expressed MHV proteins and RNAs attributed the selection of packaging signal RNA to the M protein.[43] Thus, VLPs composed of M and E, but not N protein, were found to incorporate an RNA molecule only if it contained the MHV packaging signal. Although the N-M interaction that we have localized genetically appears to be independent of RNA, it is conceivable that the accessibility of N protein domain 3 is modulated by the binding of N to particular RNA substrates.

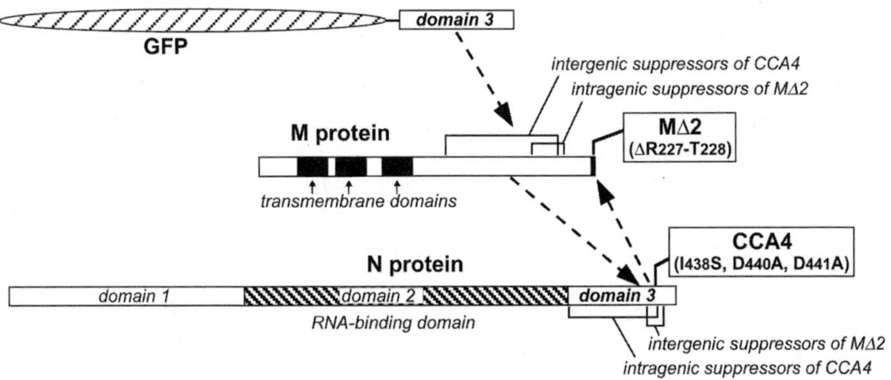

Figure 3. Intermolecular assembly interactions between the MHV N and M proteins revealed by genetic cross-talk and by transfer of domain 3 of the N protein to a heterologous protein (GFP).

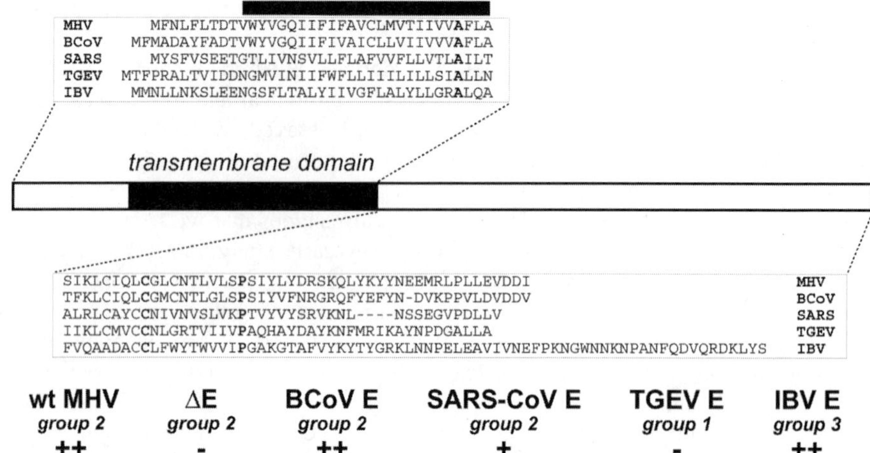

Figure 4. Alignment of the E proteins of various CoVs (top) and summary of the relative abilities of heterologous E proteins to functionally replace the E protein of MHV (bottom).

4. THE ROLE OF E PROTEIN

The CoV E protein is a small polypeptide (~10 kDa) that is only a minor constituent of virions. Nevertheless, it profoundly affects both VLP and virus assembly. E protein sequences diverge widely across the three CoV groups, but all CoV E proteins have the same architecture: a short hydrophilic amino terminus, followed by a large hydrophobic region, and a hydrophilic carboxy-terminal tail that constitutes one-half to two-thirds of the molecule (Fig. 4). Investigations with both the MHV and IBV E proteins are in agreement that E is an integral membrane protein and that its carboxy-terminal tail is cytoplasmic (corresponding to the interior of the virion).[20,44] Moreover, for IBV E, the carboxy-terminal tail alone can specify targeting to the budding compartment.[45] The disposition of the amino terminus is less clear, however. A lumenal (or virion-exterior) topology has been inferred for the IBV E protein amino terminus, based on its inaccessibility to antibodies at the cytoplasmic face of the Golgi membrane.[20] Such a single transit across the membrane would be consistent with the transmembrane oligomers of E predicted by molecular dynamics simulations.[46] Conversely, for MHV, the E protein amino terminus has been proposed to be buried within the membrane near the cytoplasmic face, based on the reactivity of an engineered amino-terminal epitope tag at the cytoplasmic face.[47] This orientation would require that the E protein hydrophobic domain form a hairpin looping back through the membrane, as envisioned in a recent biophysical analysis of the SARS-CoV E protein transmembrane domain.[48]

For MHV, we previously showed that particular clustered charged-to-alanine mutations constructed in the E gene rendered the virus defective in growth: assembled virions of one such mutant were found to have strikingly aberrant morphology, exhibiting pinched and elongated shapes that were rarely seen among wild-type virions.[49] This finding clearly demonstrated an important role for E in virion assembly, as shown earlier for

VLP assembly.[16,17] We were thus surprised to find that we could successfully generate a viable, albeit highly defective, MHV recombinant (ΔE) in which the E gene, as well as genes 4 and 5a, were entirely deleted from the viral genome.[50] This indicated that the MHV E protein is a critical, but not essential, participant in virion assembly. To more specifically focus on the E protein, we have very recently generated an additional recombinant virus (E-KO), in which E protein expression has been ablated by mutation of the initiation codon and placement of stop codons in all three reading frames. The E-KO mutant exhibits the same tiny plaque phenotype and extremely defective growth as does the ΔE mutant. This confirms that the phenotype observed for the ΔE mutant was a direct result of the E gene deletion.

Similarly, it has recently been found that knockout of SARS-CoV E protein expression results in a virus that is viable in tissue culture (Almazen, DeDiego, Alvarez, and Enjuanes, this volume). By contrast, for TGEV it has been shown by two distinct reverse genetic methods that if the E gene is knocked out, then no viable virus can be recovered; the resulting defect can only be rescued by E protein provided *in trans*.[51,52] This may indicate that basic morphogenic differences exist between the CoVs of group 2 (MHV and SARS-CoV) and group 1 (TGEV). Alternatively, it may suggest that E protein has multiple activities, one of which is essential for group 1 CoVs but unnecessary for group 2 CoVs.

To learn more about the constraints on E protein sequence, relative to the specificity of this protein's interaction with M protein, we investigated whether E proteins from different CoVs could functionally replace that of MHV. Toward this end, we exchanged the MHV E gene with that from viruses of each of the three CoV groups. In every case, exact ORF-for-ORF substitutions were made, so that each heterologous E gene was expressed in the same context as MHV E (i.e., as the second ORF in a message whose unique region is bicistronic). The results of this work revealed an unexpected flexibility in the sequence requirements of the E protein (Fig. 4). As predicted, the relatively closely related E protein of BCoV (group 2) could fully substitute for the MHV E protein. Replacement of MHV E with the more phylogenetically distant group 2 SARS-CoV E protein resulted in a virus with a slightly smaller plaque size than wild-type MHV. Very surprisingly, the group 3 IBV E protein, which is extremely divergent from MHV E in both size and sequence, was completely functional in MHV infection and assembly. This could indicate that E protein does not need to directly contact M protein in order to carry out its role in virion budding. By contrast, the E protein of TGEV (group 1) was not functional in MHV; the TGEV E substitution mutant had a phenotype indistinguishable from that of the ΔE mutant. These results lend further support to the notion that there are differences between the assembly mechanisms of group 1 and group 2 CoVs. We have been able to isolate multiple independent gain-of-function mutants from the TGEV E substitution recombinant, and we have found that these viruses have mutations clustering in two small regions of the TGEV E gene. Systematic analysis of these chimeric viruses should help to further elucidate the functions of E protein.

5. FUTURE QUESTIONS

A great deal remains to be learned about the rules governing CoV assembly. One particularly intriguing question raised by the work discussed above is: does the E protein need to directly physically interact with the M protein, or does E act at a distance? These two possibilities are not necessarily mutually exclusive. A direct E-M interaction is suggested by the observation that there are certain unallowed interspecies combinations of M and E with respect to VLP assembly[18] or virus assembly (see above). The close physical proximity of the two proteins is also supported by the demonstration that IBV E and M can be cross-linked to one another in infected or transfected cells.[21] Conversely, some results appear to argue that E acts independently of M. The individual expression of MHV or IBV E protein results in vesicles that are exported from cells.[20,53] It has also been found that the expression of MHV E protein alone leads to the formation of clusters of convoluted membranous structures highly similar to those seen in CoV-infected cells,[44] suggesting that one role of E is to induce membrane curvature in the ERGIC. The functional replacement of the MHV E protein by the highly divergent IBV E protein (see above) also suggests that a specific interaction with M is not necessary for viral assembly. Moreover, in multiple revertant searches, we have yet to find a suppressor of an E gene mutant that maps in M or in any gene other than E.[49] Similarly, we have never found intergenic suppressors of the MΔ2 mutant or the N CCA4 mutant that map in E.[27, 36] A mechanism for the independent action(s) of E in CoV assembly may be found in the recent demonstration that the SARS-CoV E protein is a cation-selective ion channel.[54]

A second pressing question arising from the roles of M protein discussed above is: what is the structural basis for the central position of M in the network of interactions that determine viral assembly? M associates with other monomers of M,[23] with the endodomain of S,[24, 29, 30] with domain 3 of N[36] and, possibly, with E and with genomic RNA.[43] We have noted that the viral M protein is extremely sensitive to mutations. This sensitivity would be consistent with the constraints imposed by M needing to maintain simultaneous contacts with multiple structural partners. On the other hand, M appears able to accommodate some radically altered versions of either the S endodomain[29,30] or N protein domain 3,[36] suggesting that M offers a variety of surfaces with which interacting polypeptides can establish alternative binding sites, if their primary interactions have been abolished by mutation. This versatility of M protein may be a component of the forces that drive CoV evolution, allowing the incorporation of altered or new proteins into virion envelopes. Such considerations clearly point to the necessity to obtain structural information about this crucial virion component.

We acknowledge a very productive collaboration with the laboratory of Peter Rottier on a number of the subjects discussed here. We are also grateful to Kathryn Holmes, David Brian, Carolyn Machamer, John Fleming, and Lawrence Sturman for the provision of viruses, clones, and antisera, and for very insightful discussions and advice. This work was supported by Public Health Service grants AI 39544, AI 45695, and AI 060755 from the National Institutes of Health.

6. REFERENCES

1. P. J. M. Rottier, in: *The Coronaviridae*, edited by S. G. Siddell (Plenum Press, New York, 1995), pp. 115-139.
2. J. Armstrong, H. Niemann, S. Smeekens, P. Rottier, and G. Warren, Sequence and topology of a model intracellular membrane protein, E1 glycoprotein, from a coronavirus, *Nature* **308**, 751-752 (1984).

3. P. Rottier, D. Brandenburg, J. Armstrong, B. van der Zeijst, and G. Warren, Assembly in vitro of a spanning membrane protein of the endoplasmic reticulum: the E1 glycoprotein of coronavirus mouse hepatitis virus A59, *Proc. Natl. Acad. Sci. USA* **81**, 1421-1425 (1984).
4. P. J. M. Rottier, G. W. Welling, S. Welling-Wester, H. G. M. Niesters, J. A. Lenstra, and B. A. M. Van der Zeijst, Predicted membrane topology of the coronavirus protein E1, *Biochemistry* **25**, 1335-1339 (1986).
5. C. Risco, I. M. Anton, C. Sune, A. M. Pedregosa, J. M. Martin-Alonso, F. Parra, J. L. Carrascosa, and L. Enjuanes, Membrane protein molecules of transmissible gastroenteritis coronavirus also expose the carboxy-terminal region on the external surface of the virion, *J. Virol.* **69**, 5269-5277 (1995).
6. K. V. Holmes, E. W. Dollar, and L. S. Sturman, Tunicamycin resistant glycosylation of a coronavirus glycoprotein: demonstration of a novel type of viral glycoprotein, *Virology* **115**, 334-344 (1981).
7. P. J. M. Rottier, M. C. Horzinek, and B. A. M. van der Zeijst, Viral protein synthesis in mouse hepatitis virus strain A59-infected cells: effects of tunicamycin, *J. Virol.* **40**, 350-357 (1981).
8. C. S. Ricard, C. A. Koetzner, L. S. Sturman, and P. S. Masters, A conditional-lethal murine coronavirus mutant that fails to incorporate the spike glycoprotein into assembled virions, *Virus Research* **39**, 261-276 (1995).
9. D.-J. E. Opstelten, M. J. B. Raamsman, K. Wolfs, M. C. Horzinek, and P. J. M. Rottier, Envelope glycoprotein interactions in coronavirus assembly, *J. Cell Biol.* **131**, 339-349 (1995).
10. V.-P. Nguyen and B. Hogue, Protein interactions during coronavirus assembly, *J. Virol.* **71**, 9278-9284 (1997).
11. J. Tooze, S. A. Tooze, and G. Warren, Replication of coronavirus MHV-A59 in Sac- cells: determination of the first site of budding of progeny virions, *Eur. J. Cell Biol.* **33**, 281-293 (1984).
12. P. J. M. Rottier and J. K. Rose, Coronavirus E1 protein expressed from cloned cDNA localizes in the Golgi region, *J. Virol.* **61**, 2042-2045 (1987).
13. C. E. Machamer and J. K. Rose, A specific transmembrane domain of a coronavirus E1 glycoprotein is required for its retention in the Golgi region, *J. Cell Biol.* **105**, 1205-1214 (1987).
14. C. E. Machamer, S. A. Mentone, J. K. Rose, and M. G. Farquhar, The E1 glycoprotein of an avian coronavirus is targeted to the cis Golgi complex, *Proc. Natl. Acad. Sci. USA* **87**, 6944-6948 (1990).
15. J. Klumperman, J. Krijnse Locker, A. Meijer, M. C. Horzinek, H. J. Geuze, and P. J. M. Rottier, Coronavirus M proteins accumulate in the Golgi complex beyond the site of virion budding, *J. Virol.* **68**, 6523-6534 (1994).
16. H. Vennema, G.-J. Godeke, J. W. A. Rossen, W. F. Voorhout, M. C. Horzinek, D.-J. E. Opstelten, and P. J. M. Rottier, Nucleocapsid-independent assembly of coronavirus-like particles by co-expression of viral envelope protein genes, *EMBO J.* **15**, 2020-2028 (1996).
17. E. C. W. Bos, W. Luytjes, H. van der Meulen, H. K. Koerten, and W. J. M. Spaan, The production of recombinant infectious DI-particles of a murine coronavirus in the absence of helper virus, *Virology* **218**, 52-60 (1996).
18. P. Baudoux, C. Carrat, L. Besnardeau, B. Charley, and H. Laude, Coronavirus pseudoparticles formed with recombinant M and E proteins induce alpha interferon synthesis by leukocytes, *J. Virol.* **72**, 8636-8643 (1998).
19. E. Mortola and P. Roy, Efficient assembly and release of SARS coronavirus-like particles by a heterologous expression system, *FEBS Lett.* **576**, 174-178 (2004).
20. E. Corse and C. E. Machamer, Infectious bronchitis virus E protein is targeted to the Golgi complex and directs release of virus-like particles, *J. Virol.* **74**, 4319-4326 (2000).
21. E. Corse and C. E. Machamer, The cytoplasmic tails of infectious bronchitis virus E and M proteins mediate their interaction, *Virology* **312**, 25-34 (2003).
22. Y. Huang, Z. Y. Yang, W. P. Kong, and G. J. Nabel, Generation of synthetic severe acute respiratory syndrome coronavirus pseudoparticles: implications for assembly and vaccine production, *J. Virol.* **78**, 12557-12565 (2004).
23. C. A. M. de Haan, H. Vennema, and P. J. M. Rottier, Assembly of the coronavirus envelope: homotypic interactions between the M proteins, *J. Virol.* **74**, 4967-4978 (2000).
24. G.-J. Godeke, C. A. de Haan, J. W. Rossen, H. Vennema, and P. J. M. Rottier, Assembly of spikes into coronavirus particles is mediated by the carboxy-terminal domain of the spike protein, *J. Virol.* **74**, 1566-1571 (2000).
25. L. Kuo, G.-J. Godeke, M. J. B. Raamsman, P. S. Masters, and P. J. M. Rottier, Retargeting of coronavirus by substitution of the spike glycoprotein ectodomain: crossing the host cell species barrier, *J. Virol.* **74**, 1393-1406 (2000).
26. P. S. Masters and P. J. M. Rottier, Coronavirus reverse genetics by targeted RNA recombination, *Curr. Topics Microbiol. Immunol.* **287**, 133-159 (2005).
27. L. Kuo and P. S. Masters, Genetic evidence for a structural interaction between the carboxy termini of the membrane and nucleocapsid proteins of mouse hepatitis virus, *J. Virol.* **76**, 4987-4999 (2002).

28. S. J. Goebel, B. Hsue, T. F. Dombrowski, and P. S. Masters, Characterization of the RNA components of a putative molecular switch in the 3' untranslated region of the murine coronavirus genome, *J. Virol.* **78**, 669-682 (2004).
29. R. Ye, C. Montalto-Morrison, and P. S. Masters, Genetic analysis of determinants for spike glycoprotein assembly into murine coronavirus virions: distinct roles for charge-rich and cysteine-rich regions of the endodomain, *J. Virol.* **78**, 9904-9917 (2004).
30. E. C. W. Bos, W. Luytjes, and W. J. M. Spaan, The function of the spike protein of mouse hepatitis virus strain A59 can be studied on virus-like particles: cleavage is not required for infectivity, *J. Virol.* **71**, 9427-9433 (1997).
31. K. W. Chang, Y. W. Sheng, and J. L. Gombold, Coronavirus-induced membrane fusion requires the cysteine-rich domain in the spike protein, *Virology* **269**, 212-224 (2000).
32. B. J. Bosch, C. A. M. de Haan, S. L. Smits, and P. J. M. Rottier, Spike protein assembly into the coronavirion: exploring the limits of itssequence requirements, *Virology* **334**, 306-318 (2005).
33. M. M. Parker and P. S. Masters, Sequence comparison of the N genes of five strains of the coronavirus mouse hepatitis virus suggests a three domain structure for the nucleocapsid protein, *Virology* **179**, 463-468 (1990).
34. P. S. Masters, Localization of an RNA-binding domain in the nucleocapsid protein of the coronavirus mouse hepatitis virus, *Arch. Virol.* **125**, 141-160 (1992).
35. G. W. Nelson and S. A. Stohlman, Localization of the RNA-binding domain of mouse hepatitis virus nucleocapsid protein, *J. Gen. Virol.* **74**, 1975-1979 (1993).
36. K. R. Hurst, L. Kuo, C. A. Koetzner, R. Ye, B. Hsue, and P. S. Masters, A major determinant for membrane protein interaction localizes to the carboxy-terminal domain of the mouse coronavirus nucleocapsid protein, *J. Virol.* **79**, in press (2005).
37. D. Escors, J. Ortego, H. Laude, and L. Enjuanes, The membrane M protein carboxy terminus binds to transmissible gastroenteritis coronavirus core and contributes to core stability, *J. Virol.* **75**, 1312-1324 (2001).
38. L. S. Sturman, K. V. Holmes, and J. Behnke, Isolation of coronavirus envelope glycoproteins and interaction with the viral nucleocapsid, *J. Virol.* **33**, 449-462 (1980).
39. R. S. Baric, G. W. Nelson, J. O. Fleming, R. J. Deans, J. G. Keck, N. Casteel, and S. A. Stohlman, Interactions between coronavirus nucleocapsid protein and viral RNAs: implications for viral transcription, *J. Virol.* **62**, 4280-4287 (1988).
40. R. Cologna, J. F. Spagnolo, and B. G. Hogue, Identification of nucleocapsid binding sites within coronavirus-defective genomes, *Virology* **277**, 235-249 (2000).
41. K. Narayanan, A. Maeda, J. Maeda, and S. Makino, Characterization of the coronavirus M protein and nucleocapsid interaction in infected cells, *J. Virol.* **74**, 8127-8134 (2000).
42. K. Narayanan and S. Makino, Cooperation of an RNA packaging signal and a viral envelope protein in coronavirus RNA packaging, *J. Virol.* **75**, 9059-9067 (2001).
43. K. Narayanan, C. J. Chen, J. Maeda, and S. Makino, Nucleocapsid-independent specific viral RNA packaging via viral envelope protein and viral RNA signal, *J. Virol.* **77**, 2922-2927 (2003).
44. M. J. B. Raamsman, J. Krijnse Locker, A. de Hooge, A. A. F. de Vries, G. Griffiths, H. Vennema, and P. J. M. Rottier, Characterization of the coronavirus mouse hepatitis virus strain A59 small membrane protein E, *J. Virol.* **74**, 2333-2342 (2000).
45. E. Corse and C. E. Machamer, The cytoplasmic tail of infectious bronchitis virus E protein directs Golgi targeting, *J. Virol.* **76**, 1273-1284 (2002).
46. J. Torres, J. Wang, K. Parthasarathy, and D. X. Liu, The transmembrane oligomers of coronavirus protein E, *Biophys. J.* **88**, 1283-1290 (2005).
47. J. Maeda, J. F. Repass, A. Maeda, and S. Makino, Membrane topology of coronavirus E protein, *Virology* **281**, 163-169 (2001).
48. E. Arbely, Z. Khattari, G. Brotons, M. Akkawi, T. Salditt, and I. T. Arkin, A highly unusual palindromic transmembrane helical hairpin formed by SARS coronavirus E protein, *J. Mol. Biol.* **341**, 769-779 (2004).
49. F. Fischer, C. F. Stegen, P. S. Masters, and W. A. Samsonoff, Analysis of constructed E gene mutants of mouse hepatitis virus confirms a pivotal role for E protein in coronavirus assembly, *J. Virol.* **72**, 7885-7894 (1998).
50. L. Kuo and P. S. Masters, The small envelope protein E is not essential for murine coronavirus replication, *J. Virol.* **77**, 4597-4608 (2003).
51. J. Ortego, D. Escors, H. Laude, and L. Enjuanes, Generation of a replication-competent, propagation-deficient virus vector based on the transmissible gastroenteritis coronavirus genome, *J. Virol.* **76**, 11518-11529 (2002).

52. K. M. Curtis, B. Yount, and R. S. Baric, Heterologous gene expression from transmissible gastroenteritis virus replicon particles, *J. Virol.* **76**, 1422-1434 (2002).
53. J. Maeda, A. Maeda, and S. Makino, Release of E protein in membrane vesicles from virus-infected cells and E protein-expressing cells, *Virology* **263**, 265-272 (1999).
54. L. Wilson, C. McKinlay, P. Gage, and G. Ewart, SARS coronavirus E protein forms cation-selective ion channels, *Virology* **330**, 322-331 (2004).

NEW INSIGHTS ON THE STRUCTURE AND MORPHOGENESIS OF BERNE VIRUS

Ana Garzón, Ana M. Maestre, Jaime Pignatelli, M. Teresa Rejas, and Dolores Rodríguez*

1. INTRODUCTION

Berne virus (BEV) is the protopype member of the torovirus genus and the only torovirus that can be grown in tissue culture. Torovirus genome consists of 6 ORFs, ORFs 1a and 1b, comprising the 5'-most two-thirds of the genome, and four additional ORFs encoding the structural proteins S, M, N, and HE. In BEV, ORF4, corresponding to the HE protein, is partially deleted, leaving only an 0.5-Kb fragment, that represents about one-third of the gene found in the bovine and porcine isolates. A distinct characteristic of the torovirus particles is the high morphological variability that they exhibit once they are released from infected cells. Spherical, oval, elongated, and kidney-shaped virions can be observed in the supernatant from infected cells after negative staining. The process of torovirus morphogenesis is still poorly understood and further studies are required to characterize this process at both morphological and molecular levels as well as to understand the structure of torovirus particles. All the information about the structure and morphogenesis of toroviruses comes from early studies performed by conventional electron microscopy examination of thin sections from cultured equine dermis cells infected with BEV,[1,2] as well as from intestinal tissue from calves infected with the bovine torovirus BRV.[3] These studies provided detailed descriptions of the different viral assemblies. However, in the past two decades new methods for structural analysis by electron microscopy have been developed to achieve an optimal preservation of the ultrastructure of cellular and viral components. Freeze-substitution is one of these high preservation methods that has been shown to greatly reduce artifactual changes in shape and size that can occur during fixation and dehydration of the sample when conventional treatments are used. Thus, in this work, BEV infected cells were fixed and treated by this technique to examine torovirus particles in the context of infected cells. To perform a thorough study of torovirus morphogenesis and to analyze virion structure, we need to identify viral components. For this purpose we have produced antibodies

* Centro Nacional de Biotecnologia, CSIC, Madrid, Spain

specific against BEV structural proteins. These antibodies were used to perform immunolabeling to detect viral proteins at the electron microscopy level and by immunofluorescence and confocal microscopy.

2. RESULTS AND DISCUSSION

2.1. Ultrastructure of Purified Virions

Viral particles released to the culture medium by BEV-infected Equine dermis (Ederm) cells were concentrated by centrifugation over a 20% sucrose cushion at 25000 rpm for 2 h. The pellet was resuspended in TEN buffer (10 mM Tris-HCl, pH 7.4, 1 mM EDTA, 150 mM NaCl) and layered over a 15–45% sucrose gradient that was centrifuged at 25000 rpm for 2 h. The fraction containing the virions was concentrated by ultracentrifugation. The characteristic polymorphism of torovirus particles was observed after negative staining of the purified preparation, and spherical, oval, elongated, kidney-shaped particles could be seen (Figure 1A, B). At higher magnification, surface projections or peplomers could be observed with more detail (Figure 1C, D).

2.2. Immunolabeling of Purified Virions

We have produced polyclonal antibodies against BEV-N protein by immunizing animals (mice, rats, and rabbits) with the protein produced as a recombinant product in the baculovirus expression system. We have also produced antibodies against the N and C termini of the M protein by immunizing animals with two synthetic peptides encompassing aminoacid residues 2 to 13 and 222 to 233, respectively. Specific reactivity of anti-N and anti-M antibodies was confirmed by Western blot using purified viral particles as antigen. In addition, monoclonal antibodies (mAb), G11 and AF3, were produced after immunization of mice with purified BEV virions. While the G11 mAb reacted in Western blot with the S protein, no reactivity against any viral protein was observed by this assay with the mAb AF3 (not shown).

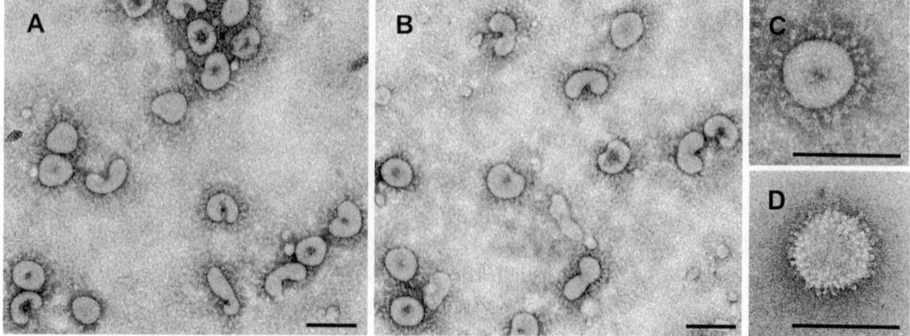

Figure 1. Polymorphism of BEV particles. Purified BEV virions adsorbed to collodion-carbon coated copper grids were negative stained for 1 min with 2% phosphotungstic acid (PTA) and examined by electron microscopy. (A) and (B) Low magnification fields showing particles with different shapes. (C) and (D) Isolated particles where surface peplomers can be clearly observed. Bars, 200 nm.

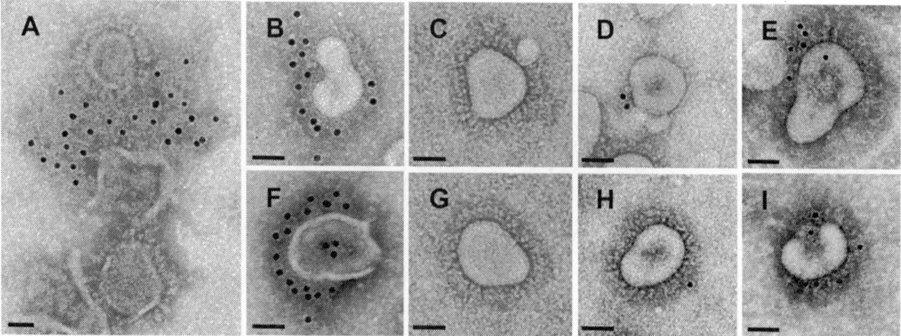

Figure 2. Immunolabeling of BEV particles. Purified BEV particles were incubated with different anti-BEV antibodies, followed by 10 nm gold-conjugated secondary antibodies and visualized by electron microscopy after negative staining with 2% PTA. (A) Anti-N polyclonal antibodies label the material released from virions partially disrupted by treatment at 4°C with 0.05% Tween 20 in 100 mM Tris-HCl-10 mM $MgCl_2$, pH 8.0. (B) and (F) Particles labeled with a polyclonal serum against the N terminus of the M protein. (C) and (G) Antibodies to the C-terminus of the M protein do not label the surface of viral particles. (D) and (H) Weak but specific labeling with anti-S mAb G11. (E) and (I) Surface labeling with mAb AF3. Bars, 200 nm.

To further characterize these antibodies, immunolabeling of purified viral particles was performed using gold-conjugated secondary immunoglobulins, and viral particles were visualized after negative staining with 2% PTA. As shown in Figure 2A, anti-N antibodies label the internal material being released from viral particles after mild detergent treatment, but there is no reactivity with intact virions (not shown). Antibodies to the N terminus of the M protein clearly decorated the surface of the virion (Fig. 2B, F), while those directed against the C-terminal end of the protein do not react with intact particles (Fig. 2C, G). This result is in agreement with the topological model proposed for this protein by Den Boon et al.[4] Using *in vitro* protein synthesis and protease treatment they proposed that the N-terminus of the M protein would be exposed on the surface of the virion while the C-terminus would be buried inside the particle, in contact with the nucleocapsid. The mAbs G11 and AF3 provide a weak but specific surface labeling, and both are able to neutralize viral infectivity (not shown).

2.3. Analysis of the Ultrastructure of BEV-Infected Cells After Freeze-Substitution

Because it has been shown that freeze-substitution after ultrarapid freezing significantly improves preservation of biological samples as compared with conventional embedding methods,[5, 6] we used this methodology to study BEV-infected cells. Ederm cells were infected with BEV at high multiplicity of infection, fixed at 10 and 24 hpi and treated for freeze-substitution as previously described.[7] As observed in Figure 3, the milder dehydration conditions used in this procedure prevent extraction of cellular components and structure collapse. In the cytoplasm of these infected cells, we observed rod-like viral particles of homogenous size (Fig. 3A, B), and in some of them the profile of the viral envelope can be clearly distinguished (Fig. 3B). Secretory vesicles containing few viral particles can be seen at 10 hpi (Fig. 3C), and they become enlarged by 24 hpi (Fig. 3D). Tubular structures of similar diameter to viral particles but of variable length and devoid of membrane can be observed in the cytoplasm enclosed within rough-endoplasmic reticulum cisternae (Fig. 3E), but also in the nucleus (not shown). Most

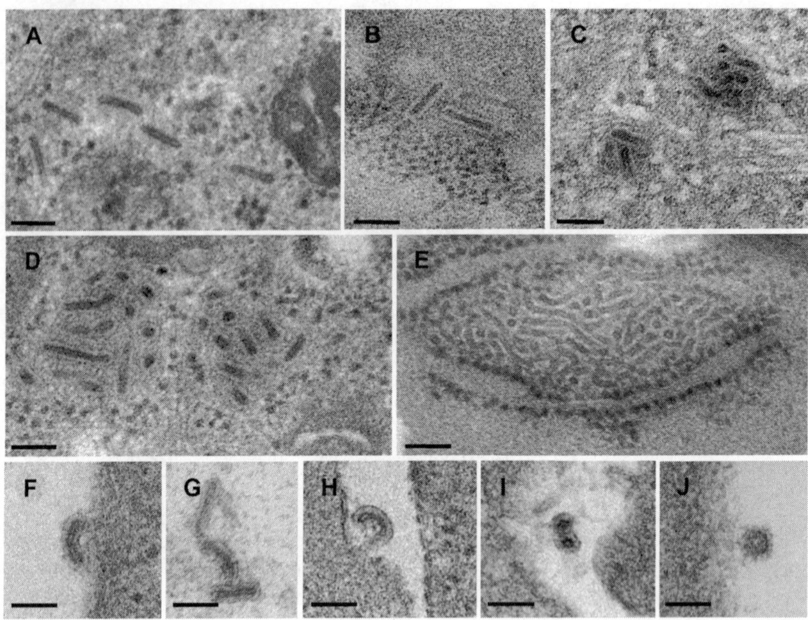

Figure 3. Thin sections of BEV-infected cells treated by freeze-substitution. Ederm cells infected with BEV at high multiplicity of infection were fixed *in situ* at 10 (A–C) or 24 hpi (D–J) with a mixture of 2% glutaraldehyde and 1% tannic acid in 0.4 M HEPES buffer (pH 7.5) for 1 h at RT. After washings, fixed cells were treated for freeze-substitution and embedding in epoxi-resin EML-812 as described in detail previously.[7] Ultrathin sections of the samples were stained with saturated uranyl acetate and lead citrate. (A) and (B) Rod-shaped particles of homogenous size. (C) and (D) Groups of particles enclosed within small (C) or large (D) secretory vesicles. (E) Tubular structures enclosed within a rough endoplasmic reticulum cisterna. (F)–(J) Extracellular viral particles, those in (F) and (G) show the same size and morphology as the intracellular rod-shaped particles seen in (A) and (B), although the profile of the viral envelope can be more clearly seen. In (H) and (I), elongated particles that are bent are observed in longitudinal (F) or cross - section (G). (J) Shows an example of a spherical particle surrounded by peplomers that are occasionally observed. Bars, 200 nm.

extracellular particles resemble the rod-shape particles seen in the cytoplasm (Fig. 3F, G), although some of them appear to be bent (longitudinally sectioned in Fig. 3H and cross-sectioned in Fig. 3I). Round particles with spikes on their surface can be occasionally seen (Fig. 3J). Our results are in agreement with those reported in previous studies,[1, 2, 3] and indicate that torovirus particles are elongated, with a rod-like appearance, resembling the morphology of the more recently described yellow head virus (YHV)[8] and gill associated virus (GAV),[9] both belonging to the *Roniviridae* family of the *Nidovirales* order. The different shapes adopted by torovirus particles outside the cells might be due to absence of a rigid structure that would impose a defined morphology.

2.4. Subcellular Localization of BEV Structural Proteins

We have examined the subcellular distribution of BEV proteins during the course of infection by confocal microscopy. At early times postinfection (6 h), the N protein is mainly localized in the perinuclear area (Fig. 4A), while at later times it is widely distributed all over the cytoplasm (Fig. 4F, K). Similarly, at 6 hpi the two anti-M antibodies, against C (Fig. 4B) and N (Fig. 4C) termini, show a perinuclear localization of the protein, and this distribution is maintained with both antibodies up to 12 hpi (Fig. 4G, H), however, the signal almost disappears at later times with the antibody to the C terminus (Fig. 4L), while the anti-N terminus labels the cell surface at this time post-infection (Fig. 4M). This result indicates that once the protein is incorporated in the virion the C terminus would be inaccessible to the antibodies while the N terminus remains exposed on the surface of the virion, and this is also in agreement with the topological model previously proposed for the M protein.[4] The anti-S mAbG11 shows a diffuse distribution of the protein throughout the cytoplasm at the different times post-infection (Fig. 4D, I, N), while the signal associated to mAb AF3 can be first observed at 10–12 hpi in the perinuclear area (compare Fig. 4E and 4J), and at later times the signal is localized in the cell surface (Fig. 4O).

The perinuclear signal observed with the anti-M antibodies and with mAb AF3 is reminiscent of the signal observed with the protein marker of the endoplasmic reticulum–Golgi intermediate compartment ERGIC-53, and thus we performed a double labeling with the anti-M-N terminal antibody and a mAb specific for ERGIC-53 in BEV-infected cells. The result shown in Figure 5 reveals a clear co-localization of the two proteins, indicating that M is incorporated in this membranous compartment, and suggesting that, as occurs in coronaviruses, budding of progeny viruses occurs in this membranous compartment.

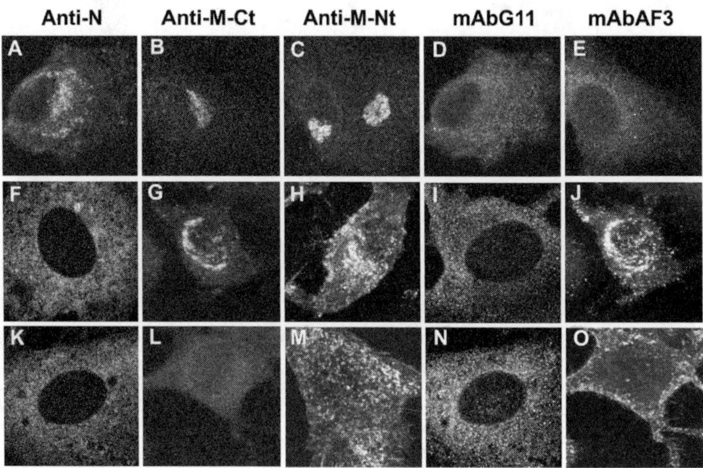

Figure 4. Intracellular distribution of BEV structural proteins. Ederm cells grown in coverslips were infected with BEV at high multiplicity and were fixed at 6 (A–E), 10 (F–J), and 16 (K–O) hpi. After fixing with 4% paraformaldehyde, cells were permeabilized in 0.1% Triton X100 and incubated with antibodies against the N protein (A–K), and against the C (B–L) and N (C–M) termini of M, and with the mAbs G11 (anti-S) and AF3 as indicated on the top of the figure, and cells were observed in a Bio-Rad Radiance 2000 confocal laser microscope.

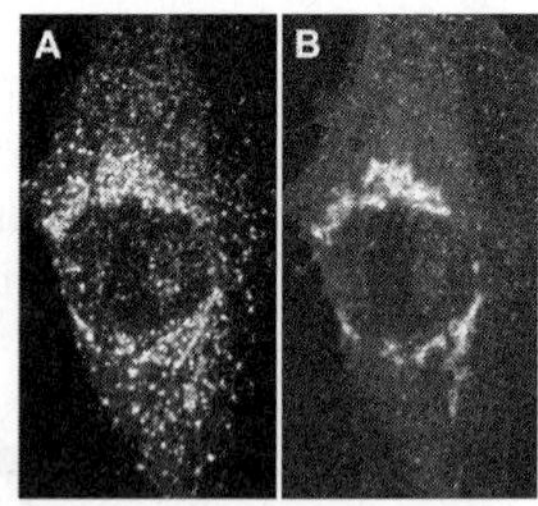

Figure 5. Co-localization of M protein with ERGIG-53. Ederm cells grown in coverslips were infected with BEV at high multiplicity. They were then fixed at 10 hpi. as described in legend to Figure 4, and simultaneously incubated with a mAb against ERGIC-53 (A) and a polyclonal serum against the N terminus of M protein (B).

3. ACKNOWLEDGMENTS

We are grateful to Raoul de Groot (University of Utrecht) for kindly providing BEV. We also thank Hans Peter Hauri (Biozentrum, University of Basel) for his anti-ERGIC-53 antibody, Sylvia Gutiérrez for her support with the confocal microscope, Milagros Guerra and Francisca Ocaña for their excellent technical assistance on sample preparation for electron microscopy, and Leonor Kremer for sharing her expertise in preparing monoclonal antibodies.

This work was supported by a grant from the Ministerio de Educación y Ciencia of Spain (BIO-2002-03739).

4. REFERENCES

1. M. Weiss, F. Steck, and M. C. Horzinek, Purification and partial characterization of a new enveloped RNA virus (Berne Virus), *J. Gen. Virol.* **64**, 1849-1858 (1983).
2. M. Weiss and M. C. Horzinek, Morphogenesis of Berne Virus (Proposed Family Toroviridae), *J. Gen. Virol.* **67**, 1305-1314 (1986).
3. J. A. Flagerland, J. F. L. Pohlenz, and G. N. Woode, A morphological study of the replicaton of Breda Virus (Proposed Family Toroviridae), *J. Gen. Virol.* **67**, 1293-1304 (1986).
4. J. A. Den Boon, E. J. Snider, J. Krinjse-Locker, M. C. Horzinek, and P. J. M. Rottier, Another triple-spanning envelope protein among intracellular budding RNA viruses: the torovirus E protein, *Virology* **182**, 655-663 (1991).
5. D. M. R Harvey, Freeze-substitution, *J. Microsc.* **127**, 209-221 (1982).
6. S. Hippe-Sanwald, The impact of freeze-substitution on biological electron microscopy. *Microsc. Res. Tech.* **24**, 400-422 (1993).
7. C. Risco, J. R. Rodríguez, C. López-Iglesias, J. L. Carrascosa, M. Esteban, and D. Rodríguez, Endoplasmic reticulum-Golgi intermediate compartment membranes and vimentin filaments participate in vacccinia virus assembly, *J. Virol.* **76**, 1839-1855 (2002).
8. C. Chantanachookin, S. Boonyaratpalin, J. Kasornchandra, D. Sataporn, U. Ekpanithanpong, K. Supamataya, S. Riurairatana, and T. W. Flegel, Histology and ultrastructure reveal a new granulosis-like virus in Penaeus monodon affected by yellow-head disease, *Dis. Aquat. Organ.* **17**, 145-157 (1993).
9. J. A. Cowley, C. M. Dimmock, C. Wongteerasupaya, V. Boonsaeng, S. Panyim, and P. J. Walker, Yellow head virus from Thailand and gill-associated virus from Australia are closely related but distinct prawn viruses, *Dis. Aquat. Organ.* **36**, 153-157 (1999).

ULTRASTRUCTURE OF SARS-CoV, FIPV, AND MHV REVEALED BY ELECTRON CRYOMICROSCOPY

Benjamin W. Neuman, Brian D. Adair, Craig Yoshioka, Joel D. Quispe, Ronald A. Milligan, Mark Yeager, and Michael J. Buchmeier*

1. INTRODUCTION

The current understanding of coronavirus ultrastructure relies heavily on transmission electron microscopy of negatively stained images. Such images typically show desiccated specimens and derive contrast from the accumulation of heavy metal negative stains, distorting the sample in the resulting image. Electron cryomicroscopy (cryo-EM) avoids some of the drawbacks of negative staining by imaging frozen specimens preserved in a fully hydrated state in vitreous ice. Cryo-EM images typically derive contrast solely from the density of the imaged sample and the surrounding ice matrix. A limited analysis of porcine transmissible gastroenteritis virus (TGEV) imaged by cryo-EM has been previously reported.[1] In this report we present a more detailed description of the supramolecular design of three coronaviruses: SARS coronavirus (SARS-CoV), feline infectious peritonitis virus (FIPV), and murine hepatitis virus (MHV).

Coronaviruses are usually classified as non-icosahedral, pleomorphic, enveloped viruses. Cryo-EM has revealed that other pleomorphic viruses have a roughly spherical appearance, studded with projections that correspond to oligomers of the attachment and fusion protein. Examples include influenza virus[2-4]; several retroviruses such as foamy virus,[5] human immunodeficiency virus,[6-10] murine leukemia virus,[11] and Rous sarcoma virus[12,13]; La Crosse virus[14,15]; Sendai virus[16]; and Pichinde, Tacaribe, and lymphocytic choriomeningitis viruses.[17] Based on single-particle image analysis of arenaviruses imaged by cryo-EM, we have proposed that pleomorphic arenavirus particles are constructed from overlapping paracrystalline lattices of proteins, and that these lattices span the viral membrane.[11] We hypothesized that coronaviruses may contain a similar supramolecular arrangement of proteins comprising a membrane-proximal scaffold. Here we used cryo-EM to examine the ultrastructure of a selection of coronaviruses representing two of the three proposed phylogenetic groups.

*The Scripps Research Institute, La Jolla, California.

2. PARTICLE CHARACTERISTICS

Particles of SARS-CoV, FIPV, and MHV were prepared from Vero-E6, AK-D, and DBT cells, respectively. MHV and SARS-CoV were also produced in cells cultured with tunicamycin, to form spike-depleted particles with low infectivity. For safety reasons, all particles were fixed with 10% (for SARS-CoV) or 1% (for FIPV and MHV) formalin in pH 6.5 HEPES-buffered physiological saline before imaging. All viruses were collected by sucrose gradient ultracentrifugation, and each remained highly infectious until fixed.

Each virus appeared approximately round in cryo-EM images, with a fringe of spikes protruding from the viral membrane and a region of lower density near the virion center (Fig. 1A–B). The average diameter of the membrane-enclosed part of each virus was similar, ranging from ~830 Å for SARS-CoV to ~960 Å for FIPV (Fig. 1C). The diameters of MHV and SARS-CoV virions were distributed more tightly than diameters of FIPV or spike-depleted, tunicamycin-grown MHV. The mean diameters of native and tunicamycin-grown MHV were similar.

Particles of SARS-CoV and MHV produced from tunicamycin-treated cells lacked the characteristic fringe of spikes, but were otherwise indistinguishable from particles grown under standard culture conditions (Fig. 2). Spike-depleted SARS-CoV particles appeared similar to spike-depleted MHV particles in negative stain, but were produced in lower yield, not suitable for effective cryo-EM imaging. Particles were imaged in several degrees of focus in order to emphasize different structural elements. Fine features such as the phospholipids headgroup densities of the viral membrane and individual nucleocapsid protein densities are revealed more clearly in images recorded relatively near to focus (Figs. 1A–B, 2A right). Images recorded farther from focus reveal spikes more clearly at the edge and center of each particle (Fig. 2A, left).

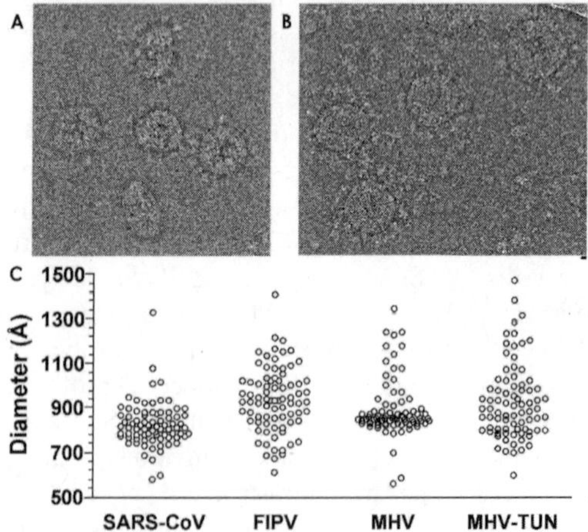

Figure 1. Cryo-EM images of formalin-fixed coronavirus particles in vitreous ice. Images are presented in "reversed contrast" with density depicted in white. Near-to-focus images of fields of SARS-CoV-TOR2 (A) and FIPV-Black (B) virions show pleomorphic enveloped particles with a slight electron-lucent hollow region near the center. Side projections of spikes are visible at the virion edge, and indistinct end-on projections of the spike are located nearer to the virion center. The scale bar below (B) represents 100 Å. The membrane-to-membrane diameter of eighty virions was measured from cryo-EM images of SARS-CoV, FIPV, MHV-OBLV60 (MHV), or tunicamycin-grown spike-depleted MHV-OBLV60 (MHV-TUN).

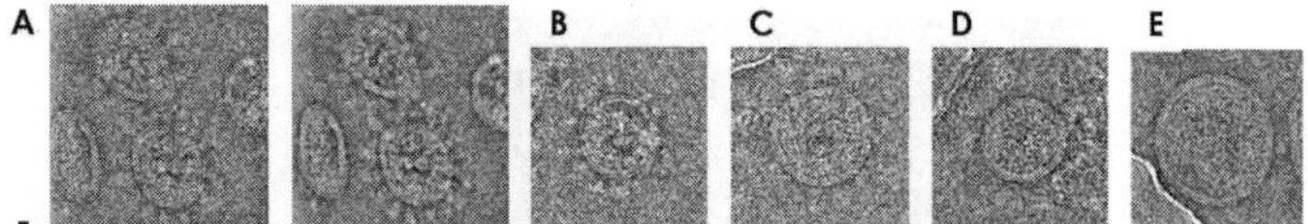

Figure 2. Cryo-EM images of formalin-fixed coronavirus particles in vitreous ice. Close-to-focus images of fields of SARS-CoV-TOR2 (A, 2 µm below true focus; B, 4.5 µm below focus; C, 2.5 µm below focus), FIPV-Black (C, 2.5 µm below focus), MHV-OBLV60 (D, 2.5 µm below focus), and tunicamycin-grown spike-depleted MHV-OBLV60 (E, 2.5 µm below focus).

3. VIRAL RIBONUCLEOPROTEIN

Preparations of each virus contained a small amount of material that was consistent with the appearance of coronavirus ribonucleoprotein (RNP).[18] A particularly interesting image of a SARS-CoV particle trapped in a partially uncoated state at the time of freezing (Fig. 3A–B) shows the spiral RNP partially uncoiled from an approximately round RNP core. The RNP proximal to the extruded membrane segment remains roughly spherical, and appears to be connected to the inner face of the membrane at the ruptured fringe (Fig. 3C).

4. STRUCTURE OF THE VIRION

The supramolecular architecture of SARS-CoV, FIPV, and MHV appears quite similar. Each virus is covered with spikes that extend ~200 Å from the peak density of the headgroups in the outer leaflet of the viral membrane. There appears to be a gap between adjacent end-projected spikes near the virion center (Fig. 2A right, for example). The arrangement of spike densities near the center of some particles approximates a rhombus, which would not be inconsistent with a paracrystalline organization of spikes as observed in the virions of pleomorphic arenavirus particles,[17] or a local hexagonal close-packing of structural proteins as observed in retroviral particles.[11] Coronavirus particles, as previously reported, appear pleomorphic, and deviate more sharply from a circular profile than other

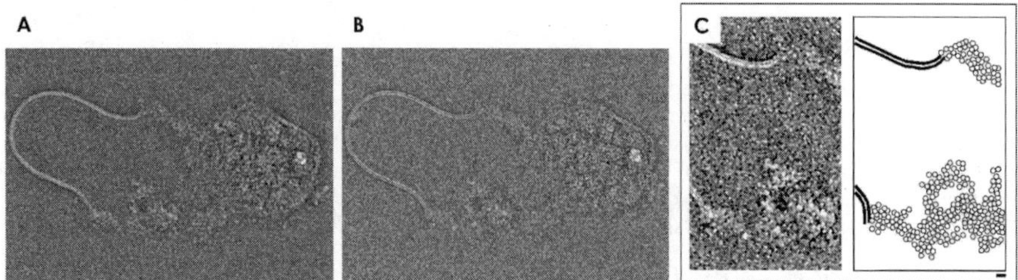

Figure 3. Ribonucleoprotein released from a spontaneously disrupted SARS-CoV particle. The viral ribonucleoprotein is shown at two levels of focus, (A, 4.5 µm below true focus; B, 2.0 µm below true focus). A third image, created by superimposing the low-resolution components of (A) with the high-resolution components of (B) shows the ribonucleoprotein more clearly (C, left). An interpretation of this image (C, right) depicts the viral membrane in bold lines and nucleoprotein molecules as circles. Scale bars denote 100 Å.

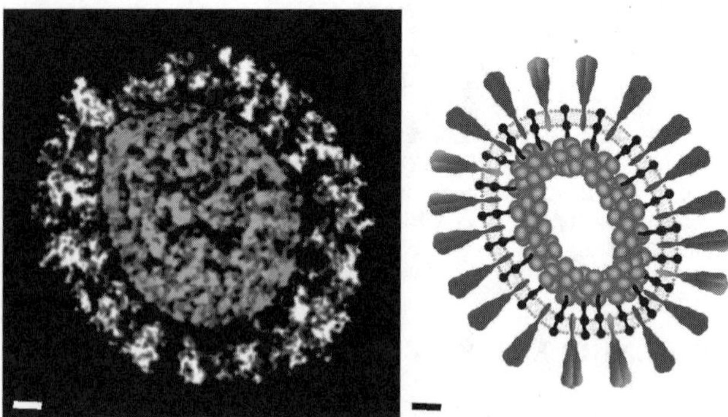

Figure 4. Image of a SARS-CoV particle extracted from the background (left panel) and a schematic cross-sectional representation of that part (right panel). Shaded spike proteins are shown at the surface of the virion, with black M proteins forming a connection to the shaded spiral ribonucleoprotein in the virion core. Scale bars denote 100 Å.

pleomorphic virions we have examined by cryo-EM. The observed variability in shape and size of the coronavirus particle would typically be considered inconsistent with icosahedral organization. The observation that the helical RNP is retained in a rough sphere through apparent interaction with proteins resident in the viral membrane is consistent with the spherical arrangement of the viral nucleocapsid proposed for TGEV.[19] However, further image analysis and biochemical experimentation will be required to determine the supramolecular organization of the virion.

5. ACKNOWLEDGMENTS

Some of the work described here was conducted at the National Resource for Automated Molecular Microscopy (NRAMM), which is supported by the National Institutes of Health though the National Center for Research Resources' P41 program (RR17573). This work was supported by NIH grants AI059799, AI025913, and NS41219, and by NIH/NIAID contract HHSN266200400058C.

6. REFERENCES

1. C. Risco, I. M. Anton, L. Enjuanes, and J. L. Carrascosa, The transmissible gastroenteritis coronavirus contains a spherical core shell consisting of M and N proteins, *J. Virol.* **70**, 4773-4777 (1996).
2. F. P. Booy, R. W. Ruigrok, and E. F. van Bruggen, Electron microscopy of influenza virus. A comparison of negatively stained and ice-embedded particles, *J. Mol. Biol.* **184**, 667-676 (1985).
3. Y. Fujiyoshi, N. P. Kume, K. Sakata, and S. B. Sato, Fine structure of influenza A virus observed by electron cryo-microscopy, *EMBO J.* **13**, 318-326 (1994).
4. T. Shangguan, D. P. Siegel, J. D. Lear, P. H. Axelsen, D. Alford, and J. Bentz, Morphological changes and fusogenic activity of influenza virus hemagglutinin, *Biophys. J.* **74**, 54-62 (1998).
5. T. Wilk, V. Geiselhart, M. Frech, S. D. Fuller, R. M. Flugel, and M. Lochelt, Specific interaction of a novel foamy virus Env leader protein with the N-terminal Gag domain, *J. Virol.* **75**, 7995-8007 (2001).

6. J. A. Briggs, T. Wilk, R. Welker, H. G. Krausslich, and S. D. Fuller, Structural organization of authentic, mature HIV-1 virions and cores, *EMBO J.* **22**, 1707-1715 (2003).
7. S. D. Fuller, T. Wilk, B. E. Gowen, H. G. Krausslich, and V. M. Vogt, Cryo-electron microscopy reveals ordered domains in the immature HIV-1 particle, *Curr. Biol.* **7**, 729-738 (1997).
8. T. Goto, T. Ashina, Y. Fujiyoshi, N. Kume, H. Yamagishi, and M. Nakai, Projection structures of human immunodeficiency virus type 1 (HIV-1) observed with high resolution electron cryo-microscopy, *J. Electron Microsc. (Tokyo)* **43**, 16-19 (1994).
9. M. V. Nermut, C. Grief, S. Hashmi, and D. J. Hockley, Further evidence of icosahedral symmetry in human and simian immunodeficiency virus, *AIDS Res. Hum. Retroviruses* **9**, 929-938 (1993).
10. T. Wilk, I. Gross, B. E. Gowen, T. Rutten, F. de Haas, R. Welker, H. G. Krausslich, P. Boulanger, and S. D. Fuller, Organization of immature human immunodeficiency virus type 1, *J. Virol.* **75**, 759-771 (2001).
11. M. Yeager, E. M. Wilson-Kubalek, S. G. Weiner, P. O. Brown, and A. Rein, Supramolecular organization of immature and mature murine leukemia virus revealed by electron cryo-microscopy: implications for retroviral assembly mechanisms, *Proc. Natl. Acad. Sci. USA* **95**, 7299-7304 (1998).
12. R. L. Kingston, N. H. Olson, and V. M. Vogt, The organization of mature Rous sarcoma virus as studied by cryoelectron microscopy, *J. Struct. Biol.* **136**, 67-80 (2001).
13. F. Yu, S. M. Joshi, Y. M. Ma, R. L. Kingston, M. N. Simon, and V. M. Vogt, Characterization of Rous sarcoma virus Gag particles assembled in vitro, *J. Virol.* **75**, 2753-2764 (2001).
14. Y. Talmon, B. V. Prasad, J. P. Clerx, G. J. Wang, W. Chiu, and M. J. Hewlett, Electron microscopy of vitrified-hydrated La Crosse virus, *J. Virol.* **61**, 2319-2321 (1987).
15. G. J. Wang, M. Hewlett, and W. Chiu, Structural variation of La Crosse virions under different chemical and physical conditions, *Virology* **184**, 455-459 (1991).
16. Y. Hosaka and T. Watabe, Cryoelectron microscopy of vitrified Sendai virions, *J. Virol. Methods.* **22**, 347-349 (1988).
17. B. W. Neuman, B. D. Adair, J. W. Burns, R. A. Milligan, M. J. Buchmeier, and M. Yeager, Complementarity in the supramolecular design of arenaviruses and retroviruses revealed by electron cryomicroscopy and image analysis, *J. Virol.* **79**, 3822-3830 (2005).
18. M. R. Macnaughton, H. A. Davies, and M. V. Nermut, Ribonucleoprotein-like structures from coronavirus particles, *J. Gen. Virol.* **39**, 545-549 (1978).
19. D. Escors, J. Ortego, H. Laude, and L. Enjuanes, The membrane M protein carboxy terminus binds to transmissible gastroenteritis coronavirus core and contributes to core stability, *J. Virol.* **75**, 1312-1324 (2001).

ROLE OF MOUSE HEPATITIS CORONAVIRUS ENVELOPE PROTEIN TRANSMEMBRANE DOMAIN

Ye Ye and Brenda G. Hogue*

1. INTRODUCTION

All coronaviruses contain only a few molecules of the small envelope (E) protein, but the protein plays an important role in virus assembly. Expression of E and M alone is sufficient for virus-like particle (VLP) assembly.[1-3] E protein containing vesicles are released from cells when E is expressed alone.[4] E protein may also be involved in determining the virus budding site at the endoplasmic reticulum–Golgi intermediate compartment membrane (ERGIC).[5] E protein is important for virion morphology and virus production.[6,7] Recently it was demonstrated that the E proteins of both MHV-CoV and SARS-CoV exhibit viroporin activity.[9-11] Coronavirus E proteins are small: 83 amino acids for the mouse hepatitis A59 (MHV-CoV A59) protein. All have a long hydrophobic domain (Fig. 1A). The transmembrane (TM), in addition to anchoring the protein in the membrane, must also be functionally important for the recently described viroporin activity. We used alanine scanning insertion mutagenesis to begin understanding the role and structural requirements of the E protein TM domain in virus assembly. Our work illustrates the importance of the TM domain and identifies potentially important residues within the domain that may affect the function of the protein.

2. MATERIALS AND METHODS

Alanine insertion mutants were created using a whole plasmid PCR protocol with a pair of primers containing the desired mutations. A plasmid containing the E gene was used as the template for PCR. After confirmation of the introduced mutations, the gene was subcloned into the G subclone of the MHV-CoV A59 infectious clone that is part of the seven cDNA fragment (A–G) system, kindly provided by Dr. Ralph Baric at the University of North Carolina at Chapel Hill.[12] All full-length cDNA clones were assembled and RNAs were transcribed using a protocol basically as described.[12, 13] After electroporation, mutant viruses were recovered by plaque purification. Multiple plaques

* Arizona State University, Tempe, Arizona.

were passaged on mouse L2 cells, and the presence of mutations were confirmed by reverse transcription and PCR amplification from total RNA from infected cells. The entirety of the E and M genes were sequenced directly from the PCR products. Viruses were passaged six times to determine growth and sequence stability. All viruses were analyzed for plaque size/morphology and growth kinetics relative to the wild-type virus.

3. RESULTS

Alanine scanning mutagenesis was used as part of our studies directed at understanding the functional role of the long TM in the MHV-CoV A59 E protein.[13] We chose this approach because insertion of alanine residues disrupts potential helix-helix interactions of residues in the membrane environment. It is a useful approach for mapping the approximate localization of functionally and structurally important parts of TM domains.[14] Eight individual alanine insertions were introduced across the TM domain (Fig. 1A).[13] The insertions were studied in the context of a full-length infectious clone. Individual viruses were designated as Ala 1–8. Viable mutant viruses were isolated for all of the mutants. Sequencing of the viral RNA confirmed that all, with one exception, retained the inserted alanine, and that no additional changes were present in the remainder of the E or within the M genes. One exception was Ala 5. Ala 5 was made two times. The virus that was recovered from the initial attempt, which we subsequently named Ala 5*, retained the inserted alanine, but also had changes in the adjacent resides on the COOH side of the insertion that resulted in deletion of residues 25 and 26 (methionine and valine) and addition of isoleucine (Fig. 1B). A second ala 5 mutant was generated that did retain the alanine insertion with no other changes. The latter was designated Ala 5.

Plaque purified viruses were passaged six times. The sequence of each mutant was again confirmed. All mutants retained the introduced alanines and no additional changes. The growth characteristics, including plaque size, morphology, and growth kinetics, of P6 of each mutant were analyzed. The viruses were grouped according to their growth properties relative to the wild-type virus (Fig. 1C). Ala 5*, 1 and 2 viruses exhibited growth characteristics similar to the wild-type virus, whereas the growth of the other mutant viruses was significantly reduced.[13] Ala 5* grew significantly better than Ala 5, suggesting that the changes observed adjacent to the introduced Ala 5* provide the virus with a growth advantage.

4. DISCUSSION

Most of the alanine insertions within the MHV-CoV E TM domain had a significant effect on the growth of the mutant viruses. The plaque size and growth of the mutant viruses were reduced. All together, the results indicate that the E TM domain is functionally important. The role of E in the virus life cycle is not fully understood. The protein is a minor component of the virion envelope, but it plays an important role in virus assembly.[6-8] To gain insight into the role of the E protein in the virus life cycle, we asked what effect disruption of the TM would have on the virus by inserting alanine residues at various positions across the domain. The TM domain is presumed to span the membrane as an alpha-helix. Insertion of a single residue in a TM helix results in displacement of residues on the amino-terminal side of the insertion relative to the COOH-side, thus disrupting the helix-helix packing interface of residues. To understand the potential effect of the insertion at position 5 (Ala 5 and Ala5*) for the two mutants described here, amino acids in the TM domain were displayed on an alpha-helical wheel (Fig. 2). Of particular note from this analysis is the positioning of the polar hydrophilic residues glutamine (Gln) 15, threonine (Thr) 27, and serine (Ser) 36, as well as cysteine (Cys) 27. All of these residues are predicted to be positioned on one side of the helix in the wild-type protein (Fig. 2, upper), however insertion of alanine at position 5 in the Ala 5 mutant (Fig. 1) is predicted to shift the relative positions of the Ser and Thr residues (Fig. 2, lower left). Threonine is predicted to be shifted to the opposite side of the helix (Fig. 2, lower left). The predicted positioning of these residues in the recovered Ala 5* virus, which exhibited a phenotype closer to that of the wild-type virus, restores the relative positioning of the polar residues (Fig. 2, lower right). Our data, taken with this analysis, suggest that the positioning of the polar residues may be important for the function of the E protein. Of particular interest is the potential role of the residues and their positioning for the recently described viroporin activity of coronavirus E proteins. Disruption of the positioning of key residues within the hydrophilic pore may impact the ion channel activity that is likely important for virus assembly.

A

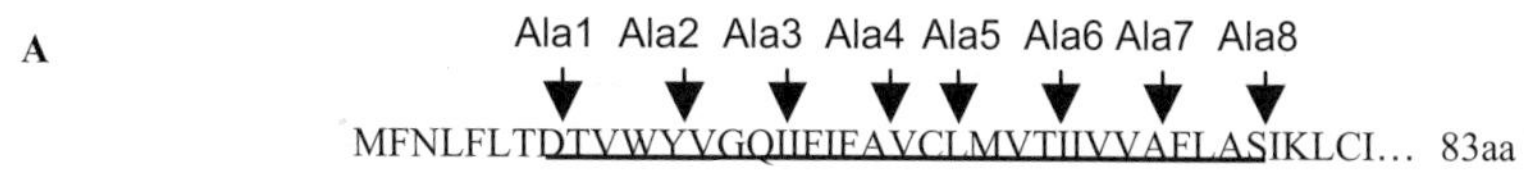

Transmembrane Domain

B.

Ala5 TVWYVGQIIFIFAVC**LAMV**TIIVVAFLASI
*Ala5** TVWYVGOIIFIFAVC**LAI**TIIVVAFLASI

C

Virus	Phenotype
WT	++++
Ala1, Ala2, Ala5*	+++
Ala7, Ala8	++
Ala3, Ala4, Ala5, Ala6	+

Figure 1. Alanine insertion mutants. (A) Amino acid sequence of wild-type MHV-CoV A59 E protein TM domain and positions of 8 alanine insertion mutations across TM domain. The putative TM domain is underlined. (B) Comparation of sequence results from RT-PCR of RNA extracted for cells infected with plaque purified Ala 5 and Ala 5* mutant viruses. Amino acids surrounding the ala insertion are bolded. (C) Viruses were grouped according to their phenotype relative to the wild-type virus based on plaque morphology and growth kinetics.

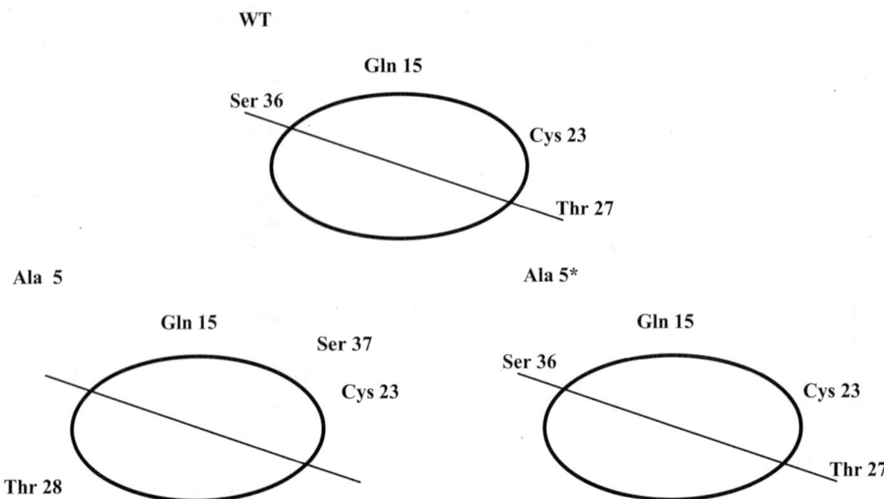

Figure 2. Simplified helical wheel diagram illustrating the relative positions of polar residues Gln 15, Cys 23, Thr 27, and Ser 36 in the wild-type and mutant Ala 5 and Ala 5* E proteins. Amino acid numbers are based on the full-length protein. Thr 27 and Ser 36 are shifted one position due to the insertion in Ala 5.

5. ACKNOWLEDGMENTS

This work was supported by Public Health Service grant AI54704, from the National Institute of Allergy and Infectious Diseases. We thank Ralph Baric for providing us with the MHV infectious clone.

6. REFERENCES

1. E. C. Bos, W. Luytjes, H. V. van der Meulen, H. K. Koerten, and W. J. Spaan, The production of recombinant infectious DI-particles of a murine coronavirus in the absence of helper virus, *Virology* **218**, 52-60 (1996).
2. H. Vennema, G. J. Godeke, J. W. Rossen, W. F. Voorhout, M. C. Horzinek, D. J. Opstelten, and P. J. Rottier, Nucleocapsid-independent assembly of coronavirus-like particles by co-expression of viral envelope protein genes, *EMBO J.* **15**, 2020-2028 (1996).
3. E. Corse and C. E. Machamer, Infectious bronchitis virus E protein is targeted to the Golgi complex and directs release of virus-like particles, *J. Virol.* **74**, 4319-4326 (2000).
4. J. Maeda, A. Maeda, and S. Makino, 1999, Release of coronavirus E protein in membranevesicles from virus-infected cells and E protein-expressing cells, *Virology* **263**, 265-272(1999).
5. M. J. Raamsman, J. K. Locker, A. de Hooge, A. A. de Vries, G. Griffiths, H. Vennema, and P. J. Rottier, Characterization of the coronavirus mouse hepatitis virus strain A59 small membrane protein E, *J. Virol.* **74**, 2333-23342 (2000).
6. F. Fischer, C. F. Stegen, P. S. Masters, and W. A. Samsonoff, Analysis of constructed E gene mutants of mouse hepatitis virus confirms a pivotal role for E protein in coronavirus assembly, *J. Virol.* **72**, 7885-7894 (1998).
7. J. Ortego, D. Escors, H. Laude, and L. Enjuanes, Generation of a replication-competent, propagation-deficient virus vector based on the transmissible gastroenteritis coronavirus genome, *J. Virol.* **76**, 11518-11529 (2002).
8. L. Kuo and P. S. Masters, The small envelope protein E is not essential for murine coronavirus replication, *J. Virol.* **77**, 4597-4608 (2003).
9. L. Wilson, C. McKinlay, P. Gage, and G. Ewart, SARS coronavirus E protein forms cation-selective ion channels, *Virology* **330**, 322-331 (2004).

10. Y. Liao, J. Lescar, J. P. Tam, and D. X. Liu, Expression of SARS-coronavirus envelope protein in *Escherichia coli* cells alters membrane permeability, *Biochem. Biophys. Res. Commun.* **325**, 374-380 (2004).
11. V. Madan, M. J. Garcia, M. A. Sanz, and L. Carrasco, Viroporin activity of murine hepatitis virus E protein, *FEBS Lett.* **579**, 3607-3612 (2005).
12. B. Yount, M. R. Denison, S. R. Weiss, and R. S. Baric, Systematic assembly of a full-length infectious cDNA of mouse hepatitis virus strain A50, *J. Virol.* **76**, 11065-11078. (2002).
13. Y. Ye and B. G. Hogue, Role of coronavirus envelope (E) protein transmembrane domain in virus assembly, Manuscript submitted (2005).
14. I. Mingarro, P. Whitley, M. A. Lemmon, and G. von Heijne, Ala-insertion scanning mutagenesis of the glycophorin A transmembrane helix: a rapid way to map helix-helix interactions in integral membrane proteins, *Protein Sci.* **5**, 1339-1341 (1996).

THE TRANSMEMBRANE DOMAIN OF THE INFECTIOUS BRONCHITIS VIRUS E PROTEIN IS REQUIRED FOR EFFICIENT VIRUS RELEASE

Carolyn E. Machamer and Soonjeon Youn*

1. INTRODUCTION

The envelope protein (E) of coronaviruses plays an important role in virus assembly, even though it is only incorporated at low levels into virions. Virus-like particles (VLPs) are produced when the membrane (M) protein and E protein are co-expressed, but not when M is expressed alone.[1,2] Thus, the E protein may help to induce membrane curvature at precise places within a scaffold made up of the M protein. Using coronavirus infectious clones, it was shown that the transmissible gastroenteritis virus E protein is essential for virus production,[3] and murine hepatitis virus lacking E protein is viable but extremely debilitated.[4]

The infectious bronchitis virus (IBV) E protein is a small protein that spans the membrane once, with its C-terminus in the cytoplasm.[5] We previously showed that the cytoplasmic tail of the IBV E protein mediated its targeting to Golgi membranes,[6] as well as its interaction with the IBV M protein.[7] Mutations in the cytoplasmic domain of IBV E reduced Golgi retention and blocked E-M association and production of VLPs. By contrast, complete replacement of the transmembrane domain of the IBV E protein with a heterologous membrane-spanning domain had no effect on the Golgi targeting of E, the association of M with E, or the production of VLPs.[6,7] We concluded that the sequence of the transmembrane domain of the protein was unimportant for its function.

Some enveloped viruses including influenza and human immunodeficiency virus encode small membrane proteins that form ion channels in infected cells.[8] The E protein of the severe acute respiratory syndrome (SARS) coronavirus has recently been shown to form a cation-specific ion channel in synthetic membranes.[9] This observation suggested that we reevaluate the IBV E mutant with a substituted transmembrane domain. Because the pore for ion movement forms from the transmembrane segments of ion channels, replacing the sequence would be expected to block channel function. After replacing the wild-type E protein sequence for that of the transmembrane-substituted E protein in an infectious clone for IBV, we recovered and characterized the

*Johns Hopkins University School of Medicine, Baltimore, Maryland 21205.

recombinant virus. Our results suggest that the amino acid sequence of the transmembrane domain is important but not essential for production of infectious virus.

2. METHODS

The cDNA encoding EG3[7] was used as a template to introduce the mutant E sequence into a molecular clone of the Beaudette strain of IBV.[10] *In vitro* ligation, transcription of RNA, electroporation, and recovery of virus in Vero cells were as previously described. Virus was plaque-purified and after amplification, RT-PCR sequencing confirmed the mutation. Two independent clones were used for the characterization of IBV-EG3. Purified virus was examined by negative stain electron microscopy on parlodian-coated grids stained with phosphotungstic acid. Thin section EM was performed after Epon embedding on osmium tetroxide stained samples as previously described.[11] IBV antibodies and immunoblotting have also been described.[5,12,13]

3. RESULTS

Because replacing the transmembrane domain of the IBV E protein had no effect on targeting or assembly of VLPs in transfected cells, we had an unprecedented opportunity to ask if the E transmembrane domain had another function in the context of a virus infection. Using a recently developed infectious clone for IBV,[10] we replaced the sequence of E with that for EG3,[7] a mutant IBV E that contains the transmembrane domain sequence of a heterologous membrane protein (VSV G). Infectious virus was recovered, indicating that the amino acid sequence of the E transmembrane domain was not essential for the virus replication cycle. We first characterized the mutant IBV-EG3 for its growth properties. In a single step growth curve, IBV-EG3 reached a peak titer later than the wild-type virus, and produced 10-fold less total infectious virus (Fig. 1A). In addition, there was a significant defect in release of infectious virus when the supernatants and cells were titered separately (Fig. 1B). At 14 h post-infection, cells infected with IBV-EG3 released about 200-fold less infectious virus into the supernatant compared to cells infected with IBV.

We showed that the EG3 protein interacted with IBV M as well as the wild-type E protein in infected cells, and was targeted and modified with palmitate normally (data not shown). We also found that entry and early stages of virus replication appeared normal. Due to the reduced release of infectious virus, we compared the virus particles produced in cells infected with IBV-EG3 to those from cells infected with wild-type IBV.

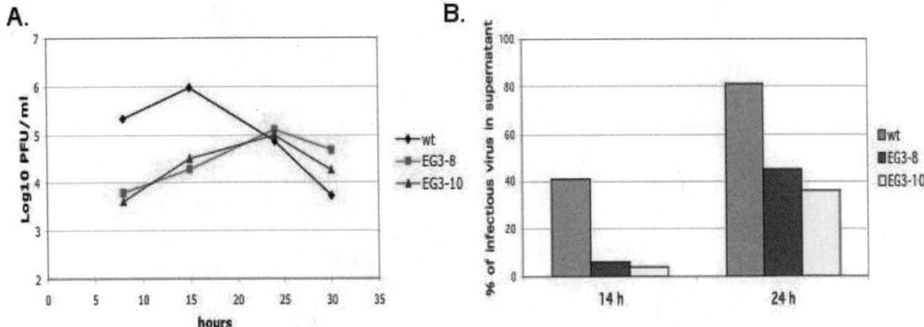

Figure 1. (A) Single-step growth curves for IBV and IBV-EG3. Vero cells were infected with wild-type IBV or two different clones of IBV-EG3 (8 and 10) at a multiplicity of infection of 4. Total virus was harvested at the times indicated by freezing the supernatant and cells together. Titers were determined by plaque assay on Vero cells. (B) The supernatants and cells were harvested separately from cells infected with IBV or IBV-EG3 at 14 h and 24 h postinfection. The percent release was calculated from the total infectious virus in cells plus supernatant at each time point.

Although infectious virus was significantly lower in supernatants of cells infected with IBV-EG3 relative to IBV, particle production was only decreased about 50%. By negative stain electron microscopy, most of the IBV-EG3 particles appeared defective, with absent or reduced spikes (Fig. 2). By contrast, wild-type IBV had the typical coronavirus appearance. When we examined the polypeptide content of purified particles, there was a significant reduction of S protein by immunoblotting (Fig. 3). Most of the S protein appeared to be cleaved from the particles, with only the internal cytoplasmic tail and transmembrane domain remaining.

Finally, we examined infected cells by thin section EM to determine if particles accumulated in cells infected with the mutant IBV-EG3. Budding profiles and virions inside Golgi membranes were observed for both viruses (Fig. 4). However, cells infected with IBV-EG3 contained more vacuoles filled with virions compared to those infected with wild-type IBV (Fig. 4). In some cases, the vacuoles in IBV-EG3 infected cells appeared to be autophagosomes, because degenerating organelles could also be found inside them.

4. DISCUSSION

The results presented here support the idea that the transmembrane domain of the coronavirus E protein plays an important role in a late stage of the virus replication cycle. Particles purified from the supernatants of IBV-EG3-infected cells lacked a full complement of spikes and contained what appeared to be a small, C-terminal proteolytic fragment of the spike protein. Thus, although the transmembrane domain of the IBV E protein is not required for formation of virus particles, it does appear to be required for their efficient release in an infectious form.

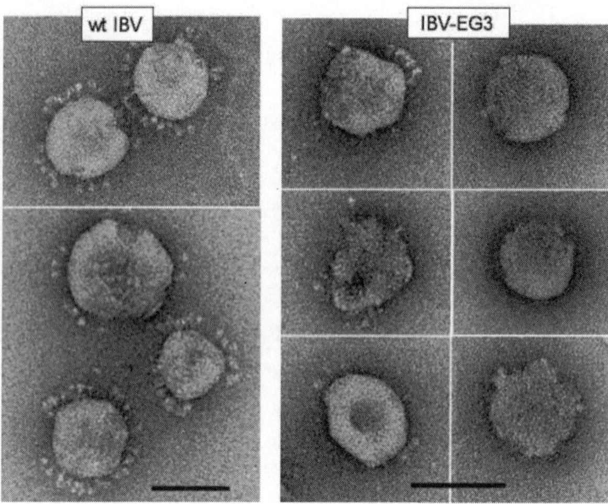

Figure 2. IBV-EG3 particles are defective. Negative staining was performed on particles purified on sucrose gradients from cells infected with IBV or IBV-EG3. Most of the IBV-EG3 particles lacked a full complement of spikes, whereas IBV had the normal coronavirus appearance. Bars, 100 nm.

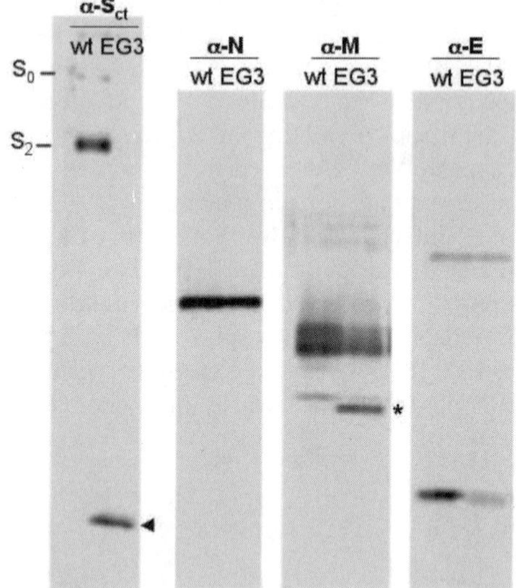

Figure 3. Biochemical analysis of purified particles. Purified particles were electrophoresed in 10% (S and N) or 15% (M and E) polyacrylamide-SDS gels, transferred to Immobilon membrane, and immunoblotted for IBV S, N, M or E proteins. IBV-EG3 lacked full-length S protein, and instead contained a small fragment that reacted with an antibody recognizing the C-terminus of IBV S (arrowhead). In addition, about 50% of the M protein lacked the N-terminal domain (asterisk).

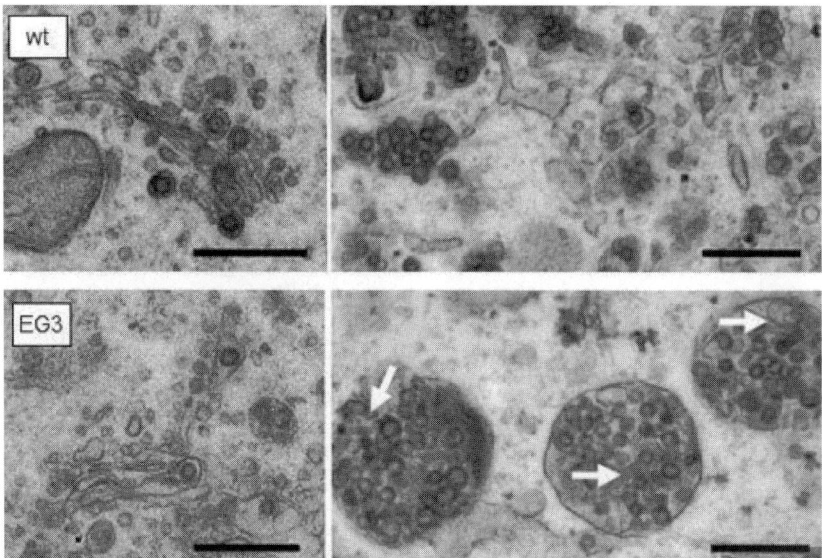

Figure 4. Virus-containing vacuoles accumulate in IBV-EG3-infected cells. Cells infected with wild-type IBV (top panels) or IBV-EG3 (bottom panels) were fixed at 14 h postinfection and embedded in Epon. Analysis of thin sections by electron microscopy showed budding virions in the Golgi region for both (left panels), as well as vacuoles containing virions (right panels). More vacuoles were observed in IBV-EG3 infected cells than in cells infected with wild-type IBV. In addition, some of the vacuoles in IBV-EG3 infected cells contained degenerating organelles (arrows), suggesting they were autophagosomes. Bars, 500 nm.

If the coronavirus E protein forms a cation-specific ion channel, we predict that the EG3 protein would lack such an activity. Because a recombinant IBV containing the chimeric E protein could be isolated, such an activity is not essential for the virus. However, the putative ion channel activity could be important for protecting virus-containing secretory vesicles from fusion with lysosomes. The exact mechanism of coronavirus release after budding into the endoplasmic reticulum–Golgi intermediate compartment is not known. Large vacuoles containing budded virions are observed in infected cells, presumably en route to the plasma membrane where fusion of the vacuole results in release of virions.[14] Perhaps the excess E protein produced in infected cells functions to modify membrane traffic pathways to enhance fusion of virus-containing vacuoles with the plasma membrane rather than with lysosomes. This could either be by modification of microenvironments via ion channel activity or by interaction with host cell membrane trafficking machinery directly. Fusion of vacuoles containing partially degraded virions with the plasma membrane would release these particles into the supernatant, explaining the biochemical and morphologic appearance of IBV-EG3. Regardless of whether the E protein forms an ion channel or interacts directly with membrane traffic machinery, the transmembrane domain of the E protein appears to play an important role in this second, nonstructural role in the virus replication cycle. Future experiments will be directed towards understanding the role of the E transmembrane domain in coronavirus release.

5. ACKNOWLEDGMENTS

We thank Lauren Wilson for stimulating discussions. This work was supported by National Institutes of Health grant R01 GM64647.

6. REFERENCES

1. Bos, E. C., Luytjes, W., van der Meulen, H. V., Koerten, H. K., and Spaan, W. J, 1996, The production of recombinant infectious DI-particles of a murine coronavirus in the absence of helper virus, *Virology* **218**:52.
2. Vennema, H., Godeke, G. J., Rossen, J. W., Voorhout, W. F., Horzinek, M. C., Opstelten, D. J., and Rottier, P. J., 1996, Nucleocapsid-independent assembly of coronavirus-like particles by co-expression of viral envelope protein genes, *EMBO J.* **15**:2020.
3. Ortego, J., Escors, D., Laude, H., and Enjuanes, L., 2002, Generation of a replication-competent, propagation-deficient virus vector based on the transmissible gastroenteritis coronavirus genome, *J. Virol.* **76**:11518.
4. Kuo, L., and Masters, P. S., 2003, The small envelope protein E is not essential for murine coronavirus replication, *J. Virol.* **77**:4597.
5. Corse, E., and Machamer, C. E., 2000, Infectious bronchitis virus E protein is targeted to the Golgi complex and directs release of virus-like particles, *J. Virol.* **74**:4319.
6. Corse, E., and Machamer, C. E., 2002, The cytoplasmic tail of infectious bronchitis virus E protein directs Golgi targeting, *J. Virol.* **76**:1273.
7. Corse, E., and Machamer, C. E., 2003, The cytoplasmic tails of infectious bronchitis virus E and M proteins mediate their interaction, *Virology* **312**:25.
8. Fischer, W.B., and Sansom, M.S., 2002, Viral ion channels: structure and function, *Biochim. Biophys. Acta* **1561**:27.
9. Wilson, L., McKinlay, C., Gage, P., and Ewart, G., 2004, SARS coronavirus E protein forms cation-selective ion channels, *Virology* **330**:322.
10. Youn, S., Leibowitz, J. L., and Collisson, E. W., 2005, In vitro assembled, recombinant infectious bronchitis viruses demonstrate that the 5a open reading frame is not essential for replication, *Virology* **332**:206.
11. Cluett, E. B., and Machamer, C. E., 1996, The envelope of vaccinia virus reveals an unusual phospholipid in Golgi complex membranes, *J. Cell. Sci.* **109**:2121.
12. Lontok, E., Corse, E., and Machamer, C. E., 2004, Intracellular targeting signals contribute to localization of coronavirus spike proteins near the virus assembly site, *J. Virol.* **78**:5913.
13. Machamer, C. E., and Rose, J. K., 1987, A specific transmembrane domain of a coronavirus E1 glycoprotein is required for its retention in the Golgi region, *J. Cell Biol.* **105:**1205.
14. Tooze, J., Tooze, S. A., and Fuller, S. D., 1987, Sorting of progeny coronavirus from condensed secretory proteins at the exit from the trans-Golgi network of AtT20 cells, *J. Cell Biol.* **105**:1215.

VIROPORIN ACTIVITY OF SARS-CoV E PROTEIN

Ying Liao, James P. Tam, and Ding X. Liu*

1. INTRODUCTION

Viroporins are integral membrane proteins encoded by viruses that contain a highly hydrophobic domain able to form an amphipathic α-helix and tend to oligomerize to form a hydrophilic pore after insertion into cellular membranes. They affect the vesicle system of host cells, glycoprotein trafficking, and membrane permeability, leading to the promotion of viral particle release.[1] In this study, we showed that SARS-CoV E protein could obviously enhance membrane permeability to hygromycin B (HB), a protein synthesis inhibitor, upon expression in mammalian cells. This activity was shown to be associated with the transmembrane domain of E protein.

2. MATERIALS AND METHODS

HB assay and construction of plasmids: HeLa cells were transfected with appropriate plasmids, pretreated with different concentrations of HB (Sigma), and labeled with [^{35}S] methionine/cysteine (Amersham). Cells were incubated in the presence or absence of HB, harvested and lysed. Total proteins were immunoprecipitated with appropriate antibodies, and analyzed by SDS 15% polyacrylamide gel electrophoresis.

Plasmids pFlagE and pFlagN was constructed by cloning an *Eco*RV- and *Eco*RI-digested PCR fragments into *Eco*RV- and *Eco*RI-digested pFlag vector. The Flag-tag was fused to the N-terminal end of the E protein. Mutations were introduced into the E gene by two rounds of PCR and confirmed by automated sequencing. The mutant constructs included in this study are summarized in Fig. 1.

3. RESULTS

3.1. Alteration of Membrane Permeability by the Expression of E Protein

To test if E protein can alter the membrane permeability of mammalian cells, the Flag-tagged E protein was expressed in HeLa cells and HB assay was performed.

*School of Biological Sciences, Nanyang Technological University, Singapore 637551, and Institute of Molecular and Cell Biology, Proteos, Singapore 138673189.

```
                1        10        20        30        40        50        76
E               MYSFVSEETGTLIVNSVLLFLAFVVFLLVTLAILTALRLCAYCCNIVNVSLVK.......
C40-A           ---------------------------------------A--------------------
C43-A           ------------------------------------------A-----------------
C44-A           -------------------------------------------A----------------
C40/44-A        ---------------------------------------A---A----------------
C40/43-A        ---------------------------------------A--A-----------------
C43/44-A        ------------------------------------------AA----------------
C40/43/44-A     ---------------------------------------A--AA----------------
Em1             -----------------EK-E---------------------------------------
Em2             ------------------------E-KE--------------------------------
Em3             -----------------EK-----E-KE--------------------------------
Em4             -----------------EKEKEKE------------------------------------
Em5             --------------E---------------------------------------------
Em6             -----------------EKEKEKE---------------A--AA----------------
```

Figure 1. Amino acid sequence of wild-type and mutant SARS-CoV E protein. The putative transmembrane domain is underlined.

Extracts prepared from cells without treatment with HB showed detection of the E protein and some other cellular proteins (lane 1). In cells treated with 1 and 2 mM of HB, no obvious detection of the E protein and other cellular proteins was obtained (Fig. 2, lanes 2 and 3). However, as a negative control, in cells transfected with the SARS-CoV N protein, a similar amount of the N protein was detected in cells both treated and untreated with HB (Fig. 2, lanes 4–6).

3.2. Mutational Analysis of the Three Cysteine Residues of E Protein

SARS-CoV E protein contains three cysteine residues at amino acid positions 40, 43, and 44, respectively. These residues are located 3–7 amino acids downstream of the transmembrane domain (Fig. 1). Mutations of these residues to alanine were made to generate seven mutants (C40-A, C43-A, C44-A, C43/44-A, C40/44-A, C40/43-A and C40/43/44-A) (Fig. 1). Western blot of cells expressing wild-type and most mutant constructs showed specific detection of three isoforms of E protein migrating at a range of molecular masses from 14 to 18 kDa under reducing conditions (Fig. 3a). These isoforms may be derived from post-translational modifications of E protein.

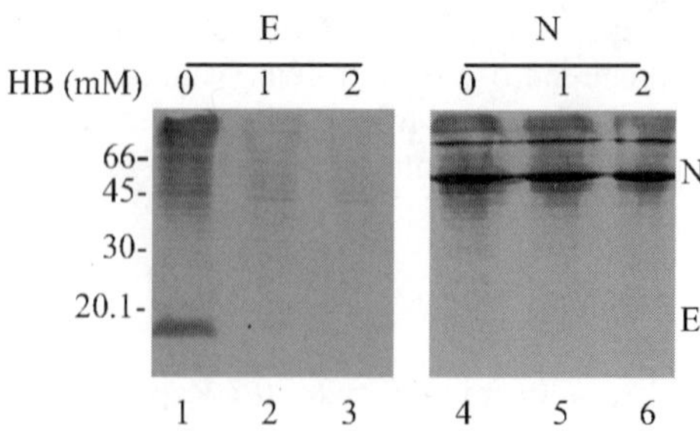

Figure 2. Modification of HeLa cells membrane permeability by SARS-CoV E protein.

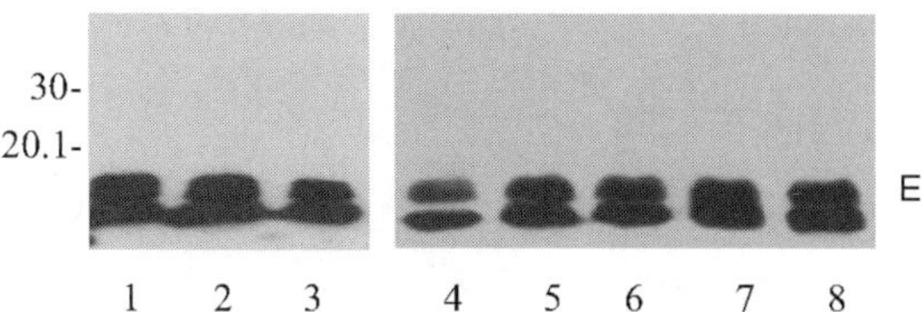

Figure 3a. Expression of wild-type (1), C40-A (2), C43-A (3), C44A (4), C43/44-A (5), C40/44-A (6), C40/43-A (7) and C40/43/44-A (8) in HeLa cells.

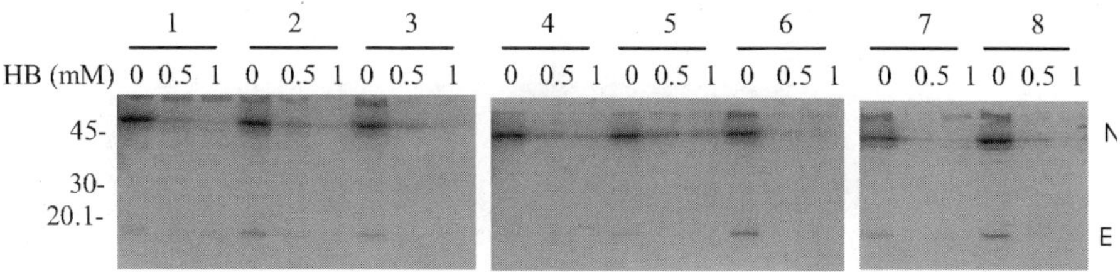

Figure 3b. Entry of Hygromycin B into HeLa cells expressing wild type and cysteine mutant E proteins. 1: E+N; 2: C40-A+N; 3: C43-A+N; 4:C44A+N; 5: C43/44-A+N; 6: C40/44-A+N; 7: C40/43-A+N; 8: C40/43/44-A+N.

In the membrane permeability assay shown in Fig. 3b, SARS-CoV N protein was co-transfected into HeLa cells together with wild-type and mutant E proteins. Expression of wild-type and mutant E protein showed that similar levels of inhibition of protein synthesis by HB were obtained (Fig. 3b).

3.3. Mutational Analysis of the Transmembrane Domain of E Protein

SARS-CoV E protein contains a long putative transmembrane domain of 29 amino acid residues.[2] This domain may be involved in the formation of ion channel by oligomerization.[3] Mutations of the putative transmembrane domain were carried out to study its functional roles. Four mutants, Em1, Em2, Em3, and Em4, were made by mutation of 3–7 leucine/valine residues to charged amino acid residues in the transmembrane domain (Fig. 1). Em5 was constructed based on the molecular simulation studies showing that N15 residue may be essential for oligomerization of the protein (Fig. 1). Em6 was made by combination of Em4 and C40/43/44-A (Fig. 1). Expression of these mutants showed the detection of polypeptides with apparent molecular masses ranging from 10 to 18 kDa (Fig. 4a). In the membrane permeability assay shown in Fig. 4b, cells expressing Em1, Em2, and Em5 exhibited a similar degree of inhibition of host protein synthesis as wild-type E protein (Fig 4b, lanes 1–9 and 16–18). In cells expressing Em3 and Em4, much less inhibition of protein synthesis by HB was observed compared with wild-type E protein (Fig. 4b, lanes 10–15). No inhibition of protein synthesis was observed in cells expressing Em6 (Fig. 4b, lanes 19–21).

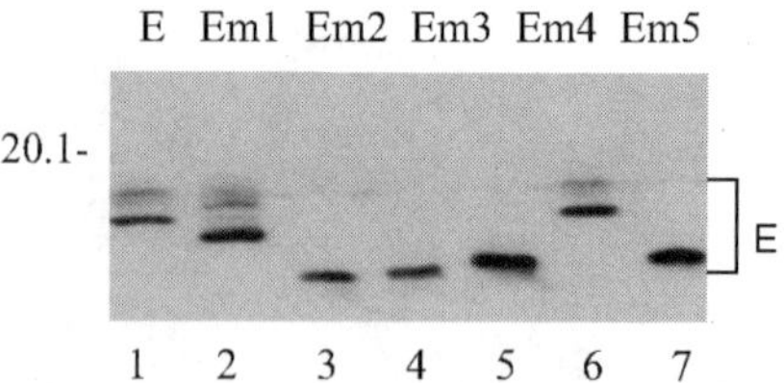

Figure 4a. Expression of the wild-type and transmembrane domain mutant E proteins in Hela cells.

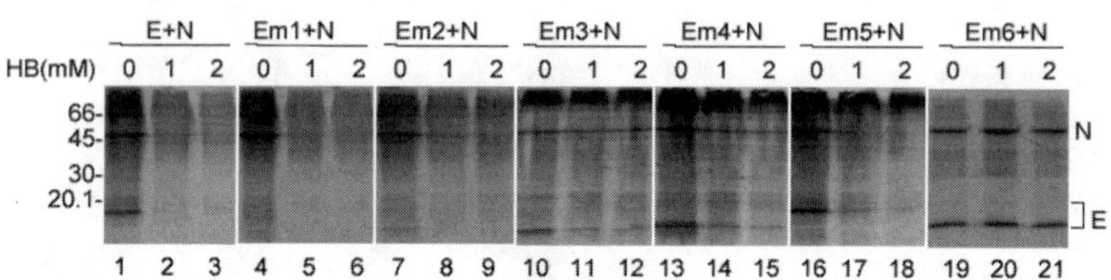

Figure 4b. Entry of Hygromycin B into Hela cells expressing wild-type and transmembrane domain mutant E proteins.

4. DISCUSSION AND CONCLUSION

In this study, we demonstrate that the membrane permeabilizing activity of SARS-CoV E protein is associated with the transmembrane domain. The C40 and C44 of the E protein were previously shown to play important roles in modification of membrane permeability in bacterial cells,[4] but no obvious effect was observed in mammalian cells in this study. Only after combining mutation of the transmembrane domain and three cysteine residues was the membrane-permeabilizing activity of E protein totally disrupted. It suggests that these cysteine residues may play certain roles in the membrane association and membrane-permeabilizing activities of the E protein. A recent report that showed SAR-CoV E protein could form cation-selective ion channels further demonstrated E protein is a viroporin.[5] It was also reported that MHV E protein could enhance membrane permeability in bacterial cells.[6] These results suggest that E protein from different coronaviruses is endowed with viroporin activity. Disruption of the function of viroporins would abrogate viral infectivity, rendering E protein suitable targets for the development of antiviral drugs.

5. REFERENCES

1. Ciampor, F., 2003, The ion channels coded by viruses, *Acta Microbiol. Immunol. Hung.* **50**:433-442.
2. Arbely, E., et al., 2004, A highly unusual palindromic transmembrane helical hairpin formed by SARS coronavirus E protein, *J. Mol. Biol.* **341**:769-779.
3. Torres, J., et al., 2005, The transmembrane oligomers of coronavirus protein E, *Biophys. J.* **88**:1283-1290.
4. Liao, Y., Lescar, J., Tam, J. P., and Liu, D. X., 2004, Expression of SARS-coronavirus envelope protein in *Escherichia coli* cells alters membrane permeability, *Biochem. Biophys. Res. Commun.* **325**:374-380.
5. Wilson, L., McKinlay, C., Gage, P., and Ewart, G., 2004, SARS coronavirus E protein forms cation-selective ion channels, *Virology* **330**:322-331.
6. Madan, V., et al., 2005, Viroporin activity of murine hepatitis virus E protein, *FEBS Lett.* **579**:3607-3612.

EFFICIENT TRANSDUCTION OF DENDRITIC CELLS USING CORONAVIRUS-BASED VECTORS

Klara K. Eriksson, Divine Makia, Reinhard Maier, Luisa Cervantes, Burkhard Ludewig, and Volker Thiel*

1. INTRODUCTION

Coronavirus-based vectors are currently considered a promising means to deliver multiple heterologous genes to specific target cells. During replication of the coronavirus RNA genome in the host cell cytoplasm, 6–8 subgenomic mRNAs encoding for structural and accessory proteins are produced. Most of these genes can be replaced by heterologous genes without affecting RNA replication.[1,2] This allows the insertion of more than 6 kb into coronavirus-based vectors.[1-4] Replication without a DNA intermediate in the host cell cytoplasm makes insertion of vector-derived sequences into the host cell genome unlikely. This, together with replacement of structural viral genes in the vector by heterologous sequences, makes these noninfectious vectors safe. An important consideration for viral vaccine vectors is the potential to efficiently deliver genetic material to specific target cells. Targeting of viral vaccine vectors to professional antigen-presenting cells, such as dendritic cells (DCs), is highly desirable in order to optimize vaccine efficacy.[5,6] The receptors of human coronavirus 229E and mouse hepatitis virus (MHV) are expressed on DCs,[7-9] indicating that vectors based on these viruses can be used to deliver genetic cargo efficiently to DCs via receptor-mediated transduction.[1] Therefore, recombinant MHV vectors in the context of a murine model can serve as a paradigm for the development and evaluation of coronavirus vaccine vectors suitable for *in vitro* and *in vivo* transduction of human DCs.

To investigate to what extent coronavirus vectors can induce antitumoral and antiviral humoral and cellular immune responses *in vivo*, we generated vectors based on mouse hepatitis virus to be used for studies of the immunological response to antigens expressed on murine dendritic cells. In these vectors, the structural genes encoding the viral envelope (E) and membrane (M) proteins have been deleted and replaced by sequences encoding a reporter protein (green fluorescent protein [GFP]) fused to an

*Kantonal Hospital St.Gallen, 9007 St.Gallen, Switzerland.

immunogenic epitope (e.g., LCMV gp33). We generated packaging cells stably expressing MHV E and M proteins and transfected MHV vector RNA into these by electroporation. Vector RNA-transfected packaging cells replicated and transcribed recombinant vector RNA as shown by reporter gene expression and syncytia formation in cell cultures. Vector RNA was packaged into viral-like particles as shown by transduction of dendritic cells after transfer of cell culture supernatants from vector transfected packaging cells. Further studies will address activation and antigen presentation of vector-transduced dendritic cells in a mouse model. Stable integration in packaging cells of the sequence encoding either MHV S, or of a chimeric S protein that will give vector particles the ability to transduce human dendritic cells, will further improve the applicability of these packaging cell lines for production of vectors for immunotherapeutical studies.

2. RESULTS

2.1. MHV Vector RNA Structure

We have previously established a reverse genetic system based on full-length cDNA copies of coronavirus genomes cloned and propagated in vaccinia virus.[10-12] Based on this system we have constructed a cDNA encoding a MHV prototype vector RNA using vaccinia virus-mediated homologous recombination. As illustrated in Figure 1, the vector RNA encodes the MHV 5'- and 3'- non-translated regions, the MHV replicase, spike and nucleocapsid genes. Furthermore, we inserted a reporter gene encoding a fusion protein comprised of GFP fused to an immunogenic epitope (e.g., LCMV gp33). The MHV vector RNA can be produced by *in vitro* transcription from genomic DNA of the recombinant vaccinia virus.

2.2. Generation of Packaging Cell Lines

In order to package MHV vector RNAs, we established a packaging cell line expressing MHV structural proteins E and M. These cells are based on 17 clone-1 cells that are susceptible to MHV infection.[13, 14] Genes encoding MHV E and M proteins were cloned downstream of a SV40 and CMV promoter, respectively, in a plasmid DNA conferring neomycin resistance. Using G418 selection we could obtain several stable cell clones that were analyzed for E and M protein expression. First, genomic DNA was

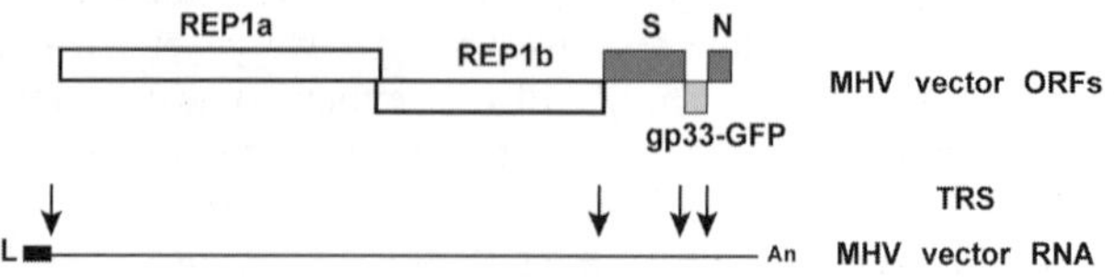

Figure 1. MHV vector RNA structure. The structure of the MHV vector RNA is illustrated. Open reading frames are indicated as boxes designated by encoded gene products. L, leader RNA; An, synthetic poly(A) tail. Transcription regulatory sequences (TRS) proceed each gene and are indicated as arrows.

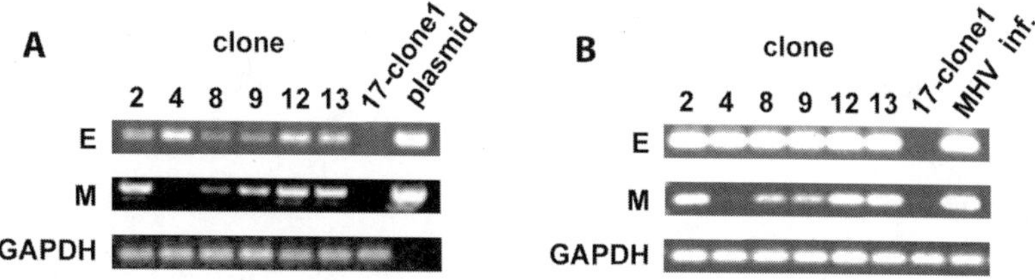

Figure 2. Analysis of packaging cell lines. Plasmids encoding MHV structural proteins E and M were transfected into 17-clone1 cells. Individual stable clones were selected and analyzed for E and M expression by PCR using genomic DNA as template (A) and RT-PCR using poly(A)-containing RNA as template (B).

isolated, and PCR was performed to verify integration of the E and M-encoding genes (Fig. 2a). Then, poly-A containing RNA was isolated as described[17] and oligo-dT-primed reverse transcription (RT) reactions followed by PCR were performed to verify expression of the E and M transcripts (Fig. 2b). Finally, E and M protein expression was confirmed by immunofluorescence analysis using E- and M-specific antisera (data not shown).

2.3. Generation of Vector RNA-Containing VLPs and Transduction of Murine DCs

After transfection of the MHV vector RNA into E and M protein-expressing packaging cells, the cells were monitored by fluorescence microscopy. After 24 h green fluorescent syncytia and plaques became apparent (Fig. 3b, left panel). To test whether VLPs have been formed and whether these particles are capable of transducing murine dendritic cells (DCs), we transferred tissue culture supernatant of MHV vector RNA-transfected packaging cells to murine DCs. After 12 h green fluorescent DCs were

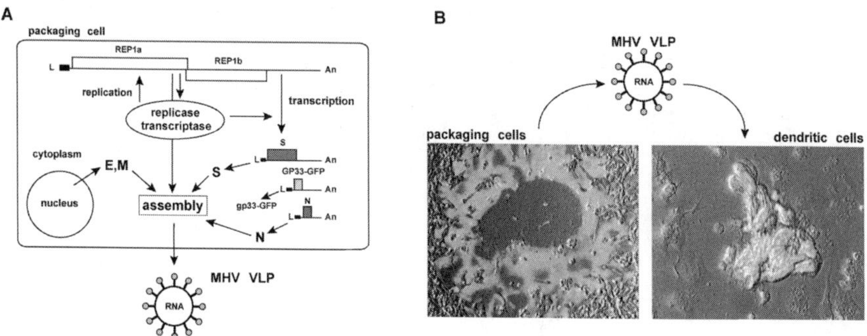

Figure 3. Generation of vector RNA-containing VLPs and transduction of murine DCs. The strategy for the production of MHV vector RNA-containing VLPs is illustrated (A). Green fluorescent syncytia and plaques were detected in vector RNA-transfected packaging cells (B, left panel). Bone marrow-derived murine DCs were cultured with supernatant from vector RNA-transfected packaging cells. Green fluorescent DCs became apparent within 24 h (B, right panel).

detectable (Fig. 3b, right panel). Therefore, we conclude that VLPs containing MHV vector RNAs have been formed in packaging cells and that these VLPs can be used to transduce murine DCs.

3. ACKNOWLEDGMENTS

This work was supported by the Gebert-Rüf Foundation, The Swiss Cancer League, and the UBS Optimus Foundation, Switzerland.

4. REFERENCES

1. V. Thiel, N. Karl, B. Schelle, P. Disterer, I. Klagge, and S. G. Siddell, Multigene RNA vector based on coronavirus transcription, *J. Virol.* **77**, 9790-9798 (2003).
2. V. Thiel, J. Herold, B. Schelle, and S. G. Siddell, Viral replicase gene products suffice for coronavirus discontinuous transcription, *J. Virol.* **75**, 6676-6681 (2001).
3. P. J. Bredenbeek and C. M. Rice, Animal RNA virus expression systems, *Semin Virol* **3**, 297-310 (1992).
4. L. Enjuanes, I. Sola, F. Almazan, et al., Coronavirus derived expression systems, *J. Biotechnol.* **88**, 183-204 (2001).
5. R. M. Steinman and M. Pope, Exploiting dendritic cells to improve vaccine efficacy, *J. Clin. Invest.* **109**, 1519-1526 (2002).
6. W. Lu, X. Wu, Y. Lu, W. Guo, and J. M. Andrieu, Therapeutic dendritic-cell vaccine for simian AIDS, *Nat. Med.* **9**, 27-32 (2003).
7. C. L. Yeager, R. A. Ashmun, R. K. Williams, C. B. Cardellichio, L. H. Shapiro, A. T. Look, and K. V. Holmes, Human aminopeptidase N is a receptor for human coronavirus 229E, *Nature* **357**, 420-422 (1992).
8. R. Thomas, L. S. Davis, and P. E. Lipsky, Isolation and characterization of human peripheral blood dendritic cells, *J. Immunol.* **150**, 821-34 (1993).
9. B. C. Turner, E. M. Hemmila, N. Beauchemin, and K. V. Holmes, Receptor-dependent coronavirus-infection of dendritic cells, *J. Virol.* **78**, 5486-5490 (2004).
10. V. Thiel, J. Herold, B. Schelle, and S. G. Siddell, Infectious RNA transcribed in vitro from a cDNA copy of the human coronavirus genome cloned in vaccinia virus, *J. Gen. Virol.* **82**, 1273-1281 (2001).
11. V. Thiel and S. G. Siddell, Reverse genetics of coronaviruses using vaccinia virus vectors, *Curr. Top. Microbiol. Immunol.* **287**, 199-227 (2005).
12. S. E. Coley, E. Lavy, S. G. Sawicki, L. Fu, B. Schelle, N. Karl, S. G. Siddell, and V. Thiel, Recombinant mouse hepatitis virus strain A59 from cloned, full-length cDNA replicates to high titers in vitro and is fully pathogenic in vivo, *J. Virol.* **79**, 3097-3106 (2005).

IV. VIRAL ENTRY

INSIGHTS FROM THE ASSOCIATION OF SARS-CoV S-PROTEIN WITH ITS RECEPTOR, ACE2

Wenhui Li, Hyeryun Choe, and Michael Farzan*

1. INTRODUCTION

Angiotensin-converting enzyme 2 (ACE2) is the cellular receptor of the coronavirus (SARS-CoV) that is the etiological agent of severe acute respiratory syndrome (SARS). Biochemical and functional studies of animal ACE2 and of the SARS-CoV spike (S) protein from SARS-CoV isolated from potential sources of SARS-CoV shed some light on the origin of the virus and perhaps the severity of disease caused in 2002–2003.

2. THE EMERGENCE OF SARS CORONAVIRUS

Severe acute respiratory syndrome (SARS) was first described in November 2002, when inhabitants of Guangdong Province, China, presented with an influenza-like illness that began with headache, myalgia, and fever, often followed by acute atypical pneumonia, respiratory failure, and death. The outbreak spread over Asia, and to Europe and North America. A total of 8,096 cases were recorded, of which 774 (9.6%) died.[1-5] The etiological agent of SARS was identified as a novel coronavirus, SARS-CoV.[6-10] This 2002–2003 SARS-CoV epidemic strain was successfully contained by conventional public health measures by July 2003.[11, 12]

SARS-CoV reemerged in Guangdong Province in the winter of 2003–2004, when it infected four individuals, all of whom recovered.[13-15] No subsequent human-to-human transmission was observed in these latter cases. The infections in 2002–2003 and 2003–2004 were unlikely to be the first instances of SARS-CoV transmission to humans; almost 2% (17 of 938) of serum samples collected in 2001 from one Hong Kong cohort recognized and neutralized SARS-CoV.[16] Additional SARS cases resulted from accidental laboratory infections in 2003 and 2004.[17, 18]

Exotic animals from the Guandong marketplace are likely to have been the immediate origin of SARS-CoV that infected humans in winters of both 2002–2003 and

*Wenhui Li, Michael Farzan, Harvard Medical School, Southborough, Massachusetts. Hyeryun Choe, Harvard Medical School, Boston, Massachusetts.

2003–2004. Marketplace Himalayan palm civets (*Paguma larvata*) and raccoon dogs (*Nyctereutes procyonoides*) harbored viruses highly similar to SARS-CoV.[19] Palm civets are of special interest because virus could be isolated from most marketplace civets, and SARS-CoV can persist in palm civets for weeks.[20] Moreover, the sporadic infections observed in 2003–2004 were associated with restaurants in which palm civet meat was prepared and consumed.[13, 14] Additionally, culling of palm civets dramatically reduced the number of infected animals in the Guandong marketplace and may be responsible for the absence of virus in humans after the winter of 2003–2004.[12, 21] Finally, functional studies of the viral receptor, described below, also support a critical role for palm civets in transmitting virus to humans.[22]

Although the palm civet is likely to have been the immediate source of virus found in humans, evidence suggests that they served as a conduit for virus from another reservoir or precursor host. For example, although SARS anti-sera and virus was overwhelmingly present in marketplace palm civets in Guangdong, the vast majority of civets on farms and in the wild were found free of infection.[23-25] Further, analysis of the rates of coding changes in the genomes of viruses isolated from palm civets suggest that the genome is not at equilibrium in the palm civet host.[14, 23] Recently, SARS-CoV-like viruses have been isolated from several bat species, predominately horseshoe bats (genus *Rhinolophus*).[26, 27] The genetic diversity of this virus in bat hosts, and the absence of overt disease, is consistent with a role for bats as a reservoir for SARS-CoV. However, as described below, substantial genetic changes in the spike (S) protein of bat SARS-CoV are likely necessary for this virus to infect humans.

3. CORONAVIRUS S PROTEINS AND THEIR RECEPTORS

Three distinct genetic and serological groups of coronaviruses have been defined.[28, 29] Coronaviruses from groups 1 and 2 are known to cause disease in humans.[30] Human coronavirus 229E (HCoV-229E), a group 1 virus, and human coronavirus OC43 (HCoV-OC43), a group 2 virus, cause mild upper respiratory infections that result in self-resolving common colds in otherwise healthy individuals.[29, 30] Human coronavirus NL63 (HCoV-NL63; also referred to as HCoV-NH and HCoV-NL) has recently been identified as a group 1 virus causing conjunctivitis, croup, and sometimes serious respiratory infections in children.[31-33] HCoV-NL63 is also notable for its use of the SARS-CoV cellular receptor ACE2 to infect cells.[34] Another group 2 coronavirus (HCoV-HKU1) was recently isolated from a 71-year old man with pneumonia.[35] SARS-CoV and SARS-CoV-like viruses found in animals also cluster with group 2 viruses, although they are outliers of group 2 and have been also described as group 4, or, more recently, group 2b viruses.[36-38]

Several coronavirus cell-surface receptors have been identified. Aminopeptidase N (APN, CD13) was shown to be the receptor for canine coronavirus, feline infectious peritonitis virus, HCoV-229E, porcine epidemic diarrhea virus, and transmissible gastroenteritis virus, all of which are group 1 coronaviruses.[39, 40] Members of the pleiotropic family of carcinoembryonic antigen-cell adhesion molecules (CEACAMs) were identified as receptors for the group 2 pathogen murine hepatitis virus,[41-43] whereas bovine group 2 coronaviruses bind to 9-*O*-acetylated sialic acids.[44] In 2003, ACE2 was identified as a functional cellular for SARS-CoV.[45] The role of ACE2 in HCoV-NL63 infection was demonstrated after isolation and characterization of this recently described group 1 coronavirus.[34]

4. ACE2, THE SARS-CoV RECEPTOR

ACE2 was identified as a functional receptor for SARS-CoV using a direct biochemical approach.[45] The S1 region of the SARS-CoV S protein was used to precipitate ACE2 from Vero E6 cells, an African green monkey kidney cell line previously shown to support efficient viral replication. Robust syncytia formed between HEK 293T cells expressing the S protein and those over-expressing ACE2. Transfection of cell lines with ACE2 rendered them permissive to infection with SARS-CoV and with retroviruses pseudotyped with S protein.[45, 46] Anti-ACE2 antisera, but not identically prepared anti-ACE1 sera, blocked replication of SARS-CoV, as did a soluble form of ACE2.

Many lines of evidence further implicate ACE2 as the principal receptor utilized *in vivo* by SARS-CoV. ACE2 is expressed in the lung and in the gastrointestinal tract, the major sites of replication of the virus.[47-50] The efficiency of infection in humans, mice, rats, and palm civets correlates with the ability of the ACE2 of each species to support viral replication.[20, 22, 51-53] ACE2 binds S protein specifically, with approximately 2 nM affinity.[54] Although many cell lines do not express ACE2, all cell lines shown to support efficient SARS-CoV infection express this receptor.[55,56] The ACE2-binding region of the S protein raises a protective neutralizing antibody response in mice, and anti-S-protein antibodies that block ACE2 association protect mice and hamsters against infection.[57-60] Finally, little or no viral replication is observed in ACE2$^{-/-}$ mice.[61] Additional factors may also contribute to the efficiency of infection. DC-SIGNR (L-SIGN, CD209L), DC-SIGN (CD209), L-SECTIN have been shown to enhance infection of ACE2-expressing cells,[62-65] these proteins do not appear to mediate efficient infection in the absence of ACE2.[63, 64]

ACE2 is a type I transmembrane protein with a single metalloprotease active site with a HEXXH zinc binding motif.[66, 67] The enzyme has been shown to cleave a variety of regulatory peptides *in vitro*, among them angiotensin I and II, des-Arg-bradykinin, kinetensin, and neurotensin.[66, 68] Some cleavage products have been shown to be potent vasodilators with antidiuretic effects. This finding suggests that ACE2 counterbalances the actions of ACE1, which mediates vasoconstriction.[69] Furthermore, targeted disruption of ACE2 in mice resulted in severe cardiac contractility defects.[70] The enzymatic activity of ACE2 does not contribute its ability to mediate fusion and viral entry, and small molecule inhibitors that block catalysis do not inhibit SARS-CoV infection.[22] However, ACE2 proteolysis has been implicated in SARS pathogenesis, and in acute respiratory distress syndrome (ARDS) caused by other viruses.[61, 71] These studies also demonstrated that SARS-CoV S protein can downregulate pulmonary ACE2, and that soluble ACE2 can protect mice from lung injury in a model of ARDS.

5. THE SARS-CoV RECEPTOR-BINDING DOMAIN

Discrete, independently folded, receptor-binding domains (RBDs) of the S proteins of several coronaviruses have been described.[72-77] The first 330 amino acids of the 769-residue S1 subunit of the murine hepatitis virus S protein is sufficient to bind its receptor,

CEACAM1.[72] A very different region of the S1 domain of HCoV-229E, between residues 407 and 547, is sufficient to associate with CD13.[73, 74] A 192-amino-acid fragment of the SARS-CoV S1 domain, residues 319–510, binds human ACE2 with greater efficiency than does the full-length S1 domain.[75-77] The RBDs of these coronaviruses are found in distinct regions of the primary structure of the S protein. This pattern may suggest that coronavirus S proteins are adapted for easy acquisition of novel binding domains, or for rapid shifts in receptor usage.

The crystal structure of the SARS-CoV RBD is consistent with this speculative possibility.[78] The RBD contains two subdomains: a core and an extended loop. The core is a five-stranded, anti-parallel β-sheet, with three short connecting α-helices. The loop, residues 424–494, termed the receptor-binding motif (RBM), is the only domain that contacts ACE2 directly. Although the RBD core domain is homologous with similar regions of other group 2 coronaviruses, the RBM is unique to SARS-CoV. Some evidence supports the suggestion that the RBM has been acquired from another coronavirus, perhaps a group 1 virus relative of HCoV-NL63. As indicated, HCoV-NL63 also enters cells through ACE2,[34] and its extended RBD region includes a stretch of residues with weak homology to the SARS-CoV RBM (unpublished observations).

Moreover, the recently described SARS-CoV-like viruses isolated from bats lack this stretch of residues, including those residues directly contacting ACE2.[26, 27, 78] The absence of these RBM residues is consistent with the inability of these viruses to grow on tissue culture cells permissive for SARS-CoV.[26, 27] If indeed bats are reservoir animals for a SARS-CoV predecessor, acquisition of this ACE2-binding region is likely to have been a critical event in the evolution of the virus. According to this scenario, the virus found in bats utilizes another receptor. A recombination event, perhaps with a group 1 virus similar to HCoV-NL63, occurring in bats, palm civets or another host, may have given rise to SARS-CoV.

6. INSIGHTS FROM ANIMAL ACE2 AND ANIMAL-DERIVED VIRAL ISOLATES

The ability of the ACE2 proteins of mice, rats, and palm civets to support SARS-CoV infection has been compared with that of human ACE2.[22, 51] Compared with cells expressing human receptor, SARS-CoV infection was less efficient in cells expressing murine ACE2. Infection was nearly absent in those expressing rat ACE2. Consistent with a role for palm civets in transmitting virus, palm civet ACE2 supported SARS-CoV infection as efficiently as human ACE2. These results correlated with affinity of each of these receptors for the S protein and its RBD.[22, 51] Chimeras between human and rat ACE2 receptors were used to identify the S-protein binding site on ACE2.[22] Mutation of four rat ACE2 residues (82–84 and 353) to their human equivalents converted rat ACE2 into an efficient SARS-CoV receptor. Residues 82–84 comprise a glycosylation site on the rat receptor that is not present on mouse, palm civet, or human receptor. Residue 353 is a histidine in mouse and rat receptors, and a lysine in palm civet and human ACE2. Strikingly, alteration of histidine 353 of mouse ACE2 to the human lysine results in a receptor that supports infection as efficiently as human ACE2 (Wenhui Li, unpublished observation). Alterations of additional residues along the first helix of human ACE2 (lysine 31 and tyrosine 41) to alanine interfered with S-protein-mediated infection and

RBD association. Collectively these data localize the S-protein-binding region to the membrane-distal lobe of the cleft that contains the catalytic site of ACE2.[22, 78]

Three S proteins of distinct origins have been compared for their ability to use human and palm civet ACE2.[22, 79, 80] The first, TOR2, was isolated during the 2002–2003 epidemic.[81] The second, denoted as GD03, was isolated from the sporadic infections in 2003–2004.[82] The third, SZ3, was obtained from palm civets.[19] Both SZ3 and, less expectedly, GD03 bound and utilized palm civet ACE2 much more efficiently than human ACE2.[22] In contrast, TOR2 utilized both receptors efficiently. The efficiency with which virus from both human outbreaks utilized palm civet receptor is consistent with recent transfer of SARS-CoV from palm civets to humans. The lower efficiency with which GD03 utilized human ACE2 compared with TOR2 may in part account for the mildness of symptoms, and absence of subsequent transmission observed during the 2003–2004 infections.[13, 14]

The differences in these three S-proteins were also reflected in the ability of their RBDs to bind human and palm civet ACE2. Two amino acids, residues 479 and 487, largely determined the much greater efficiency with which the TOR2 RBD bound human ACE2.[22, 79] Residue 479 is an asparagine or serine in all S proteins isolated from humans, either during the 2002–2003 epidemic or during 2003–2004 infections. However most sequences isolated from palm civets or raccoon dogs encode a lysine at this position. This lysine is incompatible with human ACE2, but palm civet ACE2 can efficiently bind S proteins expressing either lysine or asparagine, without an apparent preference for either.[22] Palm civets may therefore be an important intermediate in the transfer of SARS-CoV to humans, permitting the emergence of viruses that express a small, uncharged amino-acids at S-protein residue 479.

Residue 487 is also of interest. Residues 487 is a threonine in all of the more than 100 S protein sequences obtained during the 2002–2003 outbreak.[82] It is a serine in S proteins from viruses isolated during the mild 2003–2004 infections, and in all but one of the 20 or so S-proteins sequences obtained from palm civets and raccoon dogs. The relatively modest change of threonine in the TOR2 RBD to serine resulted in an approximately 20-fold decrease in binding to human ACE2.[22] A corresponding increase was observed when a threonine was introduced into the SZ3 RBD. A threonine at position 487 also substantially increased association with palm civet ACE2. Notably, the single palm-civet-derived S protein sequence that encoded a threonine at position 487 also encoded an asparagine at position 479 (Zhihong Hu, personal communication). The emergence of this rare combination of S-protein residues in palm civet-derived virus may have been necessary to generate a SARS-CoV that could efficiently transmit between humans. The infrequency of threonine 487 in animal-derived viruses may suggest that the receptor of the ultimate reservoir of SARS-CoV better utilizes a serine at this position.

The co-crystal of ACE2 with the SARS-CoV RBD clarifies these observations.[78] TOR2 S-protein asparagine 479, most commonly a lysine in palm civet virus, interacts with a network of residues that include lysine 31 of human ACE2. Palm civet and murine ACE2 express small, uncharged residues at this position, presumably better accommodating an S-protein lysine. S-protein residue 487, a threonine in all epidemic SARS-CoV isolates, directly contacts critical ACE2 lysine 353. Interaction of the threonine methyl group with lysine 353 provides a clear explanation for the decrease in affinity for human and palm civet ACE2 when this threonine is altered to serine.

7. CONCLUSIONS

Important questions remain. What receptor does bat SARS-CoV utilize? If bats are indeed a reservoir of SARS-CoV-like viruses, when and in which species did these viruses acquire an S protein capable of using palm civet and human ACE2? Did SARS-CoV gain the use of ACE2 through recombination, and if so, with what transcript? Are changes in the S protein that enhanced human-to-human transmission a probable consequence of incubation in palm civets and other animals, or a unique event unlikely to recur? What other changes in other viral proteins were necessary for SARS-CoV to transmit efficiently among humans? Our experience with SARS has taught us much about zoonotic transmission and coronaviral evolution, and there is yet more to learn.

8. REFERENCES

1. J. S. Peiris, S. T. Lai, L. L. Poon, Y. Guan, L. Y. Yam, W. Lim, J. Nicholls, W. K. Yee, W. W. Yan, M. T. Cheung, V. C. Cheng, K. H. Chan, D. N. Tsang, R. W. Yung, T. K. Ng, and K. Y. Yuen, Coronavirus as a possible cause of severe acute respiratory syndrome, *Lancet* **361**, 1319-1325 (2003).
2. I. T. Yu, Y. Li, T. W. Wong, W. Tam, A. T. Chan, J. H. Lee, D. Y. Leung, and T. Ho, Evidence of airborne transmission of the severe acute respiratory syndrome virus, *N. Engl. J. Med.* **350**, 1731-1739 (2004).
3. N. Zhong, Y. Ding, Y. Mao, Q. Wang, G. Wang, D. Wang, Y. Cong, Q. Li, Y. Liu, L. Ruan, B. Chen, X. Du, Y. Yang, Z. Zhang, X. Zhang, J. Lin, J. Zheng, Q. Zhu, D. Ni, X. Xi, G. Zeng, D. Ma, C. Wang, W. Wang, B. Wang, J. Wang, D. Liu, X. Li, X. Liu, J. Chen, R. Chen, F. Min, P. Yang, Y. Zhang, H. Luo, Z. Lang, Y. Hu, A. Ni, W. Cao, J. Lei, S. Wang, Y. Wang, X. Tong, W. Liu, M. Zhu, W. Chen, X. Xhen, L. Lin, Y. Luo, J. Zhong, W. Weng, S. Peng, Z. Pan, R. Wang, J. Zuo, B. Liu, N. Zhang, J. Zhang, B. Zhang, L. Chen, P. Zhou, L. Jiang, E. Chao, L. Guo, X. Tan, and J. Pan, Consensus for the management of severe acute respiratory syndrome, *Chin. Med. J. (Engl.)* **116**, 1603-1635 (2003).
4. N. Lee, D. Hui, A. Wu, P. Chan, P. Cameron, G. M. Joynt, A. Ahuja, M. Y. Yung, C. B. Leung, K. F. To, S. F. Lui, C. C. Szeto, S. Chung, and J. J. Sung, A major outbreak of severe acute respiratory syndrome in Hong Kong, *N. Engl. J. Med.* **348**, 1986-1994 (2003).
5. J. D. Cherry, The chronology of the 2002-2003 SARS mini pandemic, *Paediatr. Respir. Rev.* **5**, 262-269 (2004).
6. C. Drosten, S. Gunther, W. Preiser, S. van der Werf, H. R. Brodt, S. Becker, H. Rabenau, M. Panning, L. Kolesnikova, R. A. Fouchier, A. Berger, A. M. Burguiere, J. Cinatl, M. Eickmann, N. Escriou, K. Grywna, S. Kramme, J. C. Manuguerra, S. Muller, V. Rickerts, M. Sturmer; S. Vieth, H. D. Klenk, A. D. Osterhaus, H. Schmitz, and H. W. Doerr, Identification of a novel coronavirus in patients with severe acute respiratory syndrome, *N. Engl. J. Med.* **348**, 1967-1976 (2003).
7. R. A. Fouchier, T. Kuiken, M. Schutten, G. van Amerongen, G. J. van Doornum, B. G. van den Hoogen, M. Peiris, W. Lim, K. Stohr, and A. D. Osterhaus, Aetiology: Koch's postulates fulfilled for SARS virus, *Nature* **423**, 240 (2003).
8. T. G. Ksiazek, D. Erdman, C. S. Goldsmith, S. R. Zaki, T. Peret, S. Emery, S. Tong, C. Urbani, J. A. Comer, W. Lim, P. E. Rollin, S. F. Dowell, A. E. Ling, C. D. Humphrey, W. J. Shieh, J. Guarner, C. D. Paddock, P. Rota, B. Fields, J. DeRisi, J. Y. Yang, N. Cox, J. M. Hughes, J. W. LeDuc, W. J. Bellini, and L. J. Anderson, A novel coronavirus associated with severe acute respiratory syndrome, *N. Engl. J. Med.* **348**, 1953-1966 (2003).
9. T. Kuiken, R. A. Fouchier, M. Schutten, G. F. Rimmelzwaan, G. van Amerongen, D. van Riel, J. D. Laman, T. de Jong, G. van Doornum, W. Lim, A. E. Ling, P. K. Chan, J. S. Tam, M. C. Zambon, R. Gopal, C. Drosten, S. van der Werf, N. Escriou, J. C. Manuguerra, K. Stohr, J. S. Peiris, and A. D. Osterhaus, Newly discovered coronavirus as the primary cause of severe acute respiratory syndrome, *Lancet* **362**, 263-270 (2003).
10. N. S. Zhong, B. J. Zheng, Y. M. Li, Poon, Z. H. Xie, K. H. Chan, P. H. Li, S. Y. Tan, Q. Chang, J. P. Xie, X. Q. Liu, J. Xu, D. X. Li, K. Y. Yuen, Peiris, and Y. Guan, Epidemiology and cause of severe acute respiratory syndrome (SARS) in Guangdong, People's Republic of China, in February, 2003, *Lancet* **362**, 1353-1358 (2003).
11. J. S. Peiris, Y. Guan, and K. Y. Yuen, Severe acute respiratory syndrome, *Nat. Med.* **10**, S88-97 (2004).

12. N. Zhong, Management and prevention of SARS in China, *Philos. Trans. R. Soc. Lond. B Biol. Sci.* **359**, 1115-1116 (2004).
13. G. Liang, Q. Chen, J. Xu, Y. Liu, W. Lim, J. S. Peiris, L. J. Anderson, L. Ruan, H. Li, B. Kan, B. Di, P. Cheng, K. H. Chan, D. D. Erdman, S. Gu, X. Yan, W. Liang, D. Zhou, L. Haynes, S. Duan, X. Zhang, H. Zheng, Y. Gao, S. Tong, D. Li, L. Fang, P. Qin, and W. Xu, Laboratory diagnosis of four recent sporadic cases of community-acquired SARS, Guangdong Province, China, *Emerg. Infect. Dis.* **10**, 1774-1781 (2004).
14. H. D. Song, C. C. Tu, G. W. Zhang, S. Y. Wang, K. Zheng, L. C. Lei, Q. X. Chen, Y. W. Gao, H. Q. Zhou, H. Xiang, H. J. Zheng, S. W. Chern, F. Cheng, C. M. Pan, H. Xuan, S. J. Chen, H. M. Luo, D. H. Zhou, Y. F. Liu, J. F. He, P. Z. Qin, L. H. Li, Y. Q. Ren, W. J. Liang, Y. D. Yu, L. Anderson, M. Wang, R. H. Xu, X. W. Wu, H. Y. Zheng, J. D. Chen, G. Liang, Y. Gao, M. Liao, L. Fang, L. Y. Jiang, H. Li, F. Chen, B. Di, L. J. He, J. Y. Lin, S. Tong, X. Kong, L. Du, P. Hao, H. Tang, A. Bernini, X. J. Yu, O. Spiga, Z. M. Guo, H. Y. Pan, W. Z. He, J. C. Manuguerra, A. Fontanet, A. Danchin, N. Niccolai, Y. X. Li, C. I. Wu, and G. P. Zhao, Cross-host evolution of severe acute respiratory syndrome coronavirus in palm civet and human, *Proc. Natl. Acad. Sci. USA* **102**, 2430-2435 (2005).
15. F. Fleck, SARS virus returns to China as scientists race to find effective vaccine, *Bull. World Health Organ.* **82**, 152-153 (2004).
16. B. J. Zheng, K. H. Wong, J. Zhou, K. L. Wong, B. W. Young, L. W. Lu, and S. S. Lee, SARS-related virus predating SARS outbreak, Hong Kong, *Emerg. Infect. Dis.* **10**, 176-178 (2004).
17. D. Normile, Infectious diseases. Mounting lab accidents raise SARS fears, *Science* **304**, 659-661 (2004).
18. P. L. Lim, A. Kurup, G. Gopalakrishna, K. P. Chan, C. W. Wong, L. C. Ng, S. Y. Se-Thoe, L. Oon, X. Bai, L. W. Stanton, Y. Ruan, L. D. Miller, V. B. Vega, L. James, P. L. Ooi, C. S. Kai, S. J. Olsen, B. Ang, and Y. S. Leo, Laboratory-acquired severe acute respiratory syndrome, *N. Engl. J. Med.* **350**, 1740-1745 (2004).
19. Y. Guan, B. J. Zheng, Y. Q. He, X. L. Liu, Z. X. Zhuang, C. L. Cheung, S. W. Luo, P. H. Li, L. J. Zhang, Y. J. Guan, K. M. Butt, K. L. Wong, K. W. Chan, W. Lim, K. F. Shortridge, K. Y. Yuen, J. S. Peiris, and L. L. Poon, Isolation and characterization of viruses related to the SARS Coronavirus from animals in Southern China, *Science* **302**, 276-278 (2003).
20. D. Wu, C. Tu, C. Xin, H. Xuan, Q. Meng, Y. Liu, Y. Yu, Y. Guan, Y. Jiang, X. Yin, G. Crameri, M. Wang, C. Li, S. Liu, M. Liao, L. Feng, H. Xiang, J. Sun, J. Chen, Y. Sun, S. Gu, N. Liu, D. Fu, B. T. Eaton, L. F. Wang, and X. Kong, Civets are equally susceptible to experimental infection by two different severe acute respiratory syndrome coronavirus isolates, *J. Virol.* **79**, 2620-2625 (2005).
21. M. Wang, H. Q. Jing, H. F. Xu, X. G. Jiang, B. Kan, Q. Y. Liu, K. L. Wan, B. Y. Cui, H. Zheng, Z. G. Cui, M. Y. Yan, W. L. Liang, H. X. Wang, X. B. Qi, Z. J. Li, M. C. Li, K. Chen, E. M. Zhang, S. Y. Zhang, R. Hai, D. Z. Yu, and J. G. Xu, Surveillance on severe acute respiratory syndrome associated coronavirus in animals at a live animal market of Guangzhou in 2004, *Zhonghua Liu Xing Bing Xue Za Zhi* **26**, 84-87 (2005).
22. W. Li, C. Zhang, J. Sui, J. H. Kuhn, M. J. Moore, S. Luo, S. K. Wong, I. C. Huang, K. Xu, N. Vasilieva, A. Murakami, Y. He, W. A. Marasco, Y. Guan, H. Choe, and M. Farzan, Receptor and viral determinants of SARS-coronavirus adaptation to human ACE2, *EMBO J.* **24**, 1634 (2005).
23. B. Kan, M. Wang, H. Jing, H. Xu, X. Jiang, M. Yan, W. Liang, H. Zheng, K. Wan, Q. Liu, B. Cui, Y. Xu, E. Zhang, H. Wang, J. Ye, G. Li, M. Li, Z. Cui, X. Qi, K. Chen, L. Du, K. Gao, Y.-T. Zhao, X.-Z. Zou, Y.-J. Feng, Y.-F. Gao, R. Hai, D. Yu, Y. Guan, and J. Xu, Molecular evolution analysis and geographic investigation of severe acute respiratory syndrome coronavirus-like virus in palm civets at an animal market and on farms, *J. Virol.* **79**, 11892-11900 (2005).
24. C. Tu, G. Crameri, X. Kong, J. Chen, Y. Sun, M. Yu, H. Xiang, X. Xia, S. Liu, T. Ren, Y. Yu, B. T. Eaton, H. Xuan, and L. F. Wang, Antibodies to SARS coronavirus in civets, *Emerg. Infect. Dis.* **10**, 2244-2248 (2004).
25. L. L. Poon, D. K. Chu, K. H. Chan, O. K. Wong, T. M. Ellis, Y. H. Leung, S. K. Lau, P. C. Woo, K. Y. Suen, K. Y. Yuen, Y. Guan, and J. S. Peiris, Identification of a novel coronavirus in bats, *J. Virol.* **79**, 2001-2009 (2005).
26. W. Li, Z. Shi, M. Yu, W. Ren, C. Smith, J. H. Epstein, H. Wang, G. Crameri, Z. Hu, H. Zhang, J. Zhang, J. McEachern, H. Field, P. Daszak, B. T. Eaton, S. Zhang, and L. F. Wang, Bats are natural reservoirs of SARS-like coronaviruses, *Science* **310**, 676-679 (2005).
27. S. K. Lau, P. C. Woo, K. S. Li, Y. Huang, H. W. Tsoi, B. H. Wong, S. S. Wong, S. Y. Leung, K. H. Chan, and K. Y. Yuen, Severe acute respiratory syndrome coronavirus-like virus in Chinese horseshoe bats, *Proc. Natl. Acad. Sci. USA* **102**, 14040-14045 (2005).
28. J. M. Gonzalez, P. Gomez-Puertas, D. Cavanagh, A. 'E. Gorbalenya, and L. Enjuanes, A comparative sequence analysis to revise the current taxonomy of the family Coronaviridae, *Arch. Virol.* **148**, 2207-2235 (2003).

29. D. A. Brian and R. S. Baric, Coronavirus genome structure and replication, *Curr. Top. Microbiol. Immunol.* **287**, 1-30 (2005).
30. K. McIntosh, Coronaviruses in the limelight, *J. Infect. Dis.* **191**, 489-491 (2005).
31. F. Esper, C. Weibel, D. Ferguson, M. L. Landry, and J. S. Kahn, Evidence of a novel human coronavirus that is associated with respiratory tract disease in infants and young children, *J. Infect. Dis.* **191**, 492-498 (2005).
32. R. A. Fouchier, N. G. Hartwig, T. M. Bestebroer, B. Niemeyer, J. C. de Jong, J. H. Simon, and A. D. Osterhaus, A previously undescribed coronavirus associated with respiratory disease in humans, *Proc. Natl. Acad. Sci. USA* **101**, 6212-6216 (2004).
33. L. van der Hoek, K. Pyrc, M. F. Jebbink, W. Vermeulen-Oost, R. J. Berkhout, K. C. Wolthers, P. M. Wertheim-van Dillen, J. Kaandorp, J. Spaargaren, and B. Berkhout, Identification of a new human coronavirus, *Nat. Med.* **10**, 368-373 (2004).
34. H. Hofmann, K. Pyrc, L. van der Hoek, M. Geier, B. Berkhout, and S. Pohlmann, Human coronavirus NL63 employs the severe acute respiratory syndrome coronavirus receptor for cellular entry, *Proc. Natl. Acad. Sci. USA* **102**, 7988-7983 (2005).
35. P. C. Woo, S. K. Lau, C. M. Chu, K. H. Chan, H. W. Tsoi, Y. Huang, B. H. Wong, R. W. Poon, J. J. Cai, W. K. Luk, L. L. Poon, S. S. Wong, Y. Guan, J. S. Peiris, and K. Y. Yuen, Characterization and complete genome sequence of a novel coronavirus, coronavirus HKU1, from patients with pneumonia, *J. Virol.* **79**, 884-895 (2005).
36. A. E. Gorbalenya, E. J. Snijder, and W. J. Spaan, Severe acute respiratory syndrome coronavirus phylogeny: toward consensus, *J. Virol.* **78**, 7863-7866 (2004).
37. E. J. Snijder, P. J. Bredenbeek, J. C. Dobbe, V. Thiel, J. Ziebuhr, L. L. Poon, Y. Guan, M. Rozanov, W. J. Spaan, and A. E. Gorbalenya, Unique and conserved features of genome and proteome of SARS-coronavirus, an early split-off from the coronavirus Group 2 lineage, *J. Mol. Biol.* **331**, 991-1004 (2003).
38. A. J. Gibbs, M. J. Gibbs, and J. S. Armstrong, The phylogeny of SARS coronavirus, *Arch. Virol.* **149**, 621-624 (2004).
39. B. Delmas, J. Gelfi, R. L'Haridon, L. K. Vogel, H. Sjostrom, O. Noren, and H. Laude, Aminopeptidase N is a major receptor for the entero-pathogenic coronavirus TGEV, *Nature* **357**, 417-420 (1992).
40. C. L. Yeager, R. A. Ashmun, R. K. Williams, C. B. Cardellichio, L. H. Shapiro, A. T. Look, and K. V. Holmes, Human aminopeptidase N is a receptor for human coronavirus 229E, *Nature* **357**, 420-422 (1992).
41. G. S. Dveksler, C. W. Dieffenbach, C. B. Cardellichio, K. McCuaig, M. N. Pensiero, G. S. Jiang, N. Beauchemin, and K. V. Holmes, Several members of the mouse carcinoembryonic antigen-related glycoprotein family are functional receptors for the coronavirus mouse hepatitis virus-A59, *J. Virol.* **67**, 1-8 (1993).
42. G. S. Dveksler, M. N. Pensiero, C. B. Cardellichio, R. K. Williams, G. S. Jiang, K. V. Holmes, and C. W. Dieffenbach, Cloning of the mouse hepatitis virus (MHV) receptor: expression in human and hamster cell lines confers susceptibility to MHV, *J. Virol.* **65**, 6881-6891 (1991).
43. R. K. Williams, G. S. Jiang, and K. V. Holmes, Receptor for mouse hepatitis virus is a member of the carcinoembryonic antigen family of glycoproteins, *Proc. Natl. Acad. Sci. USA* **88**, 5533-5536 (1991).
44. B. Schultze and G. Herrler, Bovine coronavirus uses N-acetyl-9-O-acetylneuraminic acid as a receptor determinant to initiate the infection of cultured cells, *J. Gen. Virol.* **73**, 901-906 (1992).
45. W. Li, M. J. Moore, N. Vasilieva, J. Sui, S. K. Wong, M. A. Berne, M. Somasundaran, J. L. Sullivan, C. Luzeriaga, T. C. Greenough, H. Choe, and M. Farzan, Angiotensin-converting enzyme 2 is a functional receptor for the SARS coronavirus, *Nature* **426**, 450-454 (2003).
46. M. J. Moore, T. Dorfman, W. Li, S. K. Wong, Y. Li, J. H. Kuhn, J. Coderre, N. Vasilieva, Z. Han, T. C. Greenough, M. Farzan, and H. Choe, Retroviruses pseudotyped with the severe acute respiratory syndrome coronavirus spike protein efficiently infect cells expressing angiotensin-converting enzyme 2, *J. Virol.* **78**, 10628-10635 (2004).
47. P. K. Chan, K. F. To, A. W. Lo, J. L. Cheung, I. Chu, F. W. Au, J. H. Tong, J. S. Tam, J. J. Sung, and H. K. Ng, Persistent infection of SARS coronavirus in colonic cells in vitro, *J. Med. Virol.* **74**, 1-7 (2004).
48. Y. Ding, L. He, Q. Zhang, Z. Huang, X. Che, J. Hou, H. Wang, H. Shen, L. Qiu, Z. Li, J. Geng, J. Cai, H. Han, X. Li, W. Kang, D. Weng, P. Liang, and S. Jiang, Organ distribution of severe acute respiratory syndrome (SARS) associated coronavirus (SARS-CoV) in SARS patients: implications for pathogenesis and virus transmission pathways, *J. Pathol.* **203**, 622-630 (2004).
49. I. Hamming, W. Timens, M. L. Bulthuis, A. T. Lely, G. J. Navis, and H. van Goor, Tissue distribution of ACE2 protein, the functional receptor for SARS coronavirus. A first step in understanding SARS pathogenesis, *J. Pathol.* **203**, 631-637 (2004).
50. D. Harmer, M. Gilbert, R. Borman, and K. L. Clark, Quantitative mRNA expression profiling of ACE 2, a novel homologue of angiotensin converting enzyme, *FEBS Lett.* **532**, 107-110 (2002).

51. W. Li, T. C. Greenough, M. J. Moore, N. Vasilieva, M. Somasundaran, J. L. Sullivan, M. Farzan, and H. Choe, Efficient replication of severe acute respiratory syndrome coronavirus in mouse cells is limited by murine Angiotensin-converting enzyme 2, *J. Virol.* **78**, 11429-11433 (2004).
52. K. Subbarao, J. McAuliffe, L. Vogel, G. Fahle, S. Fischer, K. Tatti, M. Packard, W. J. Shieh, S. Zaki, and B. Murphy, Prior infection and passive transfer of neutralizing antibody prevent replication of severe acute respiratory syndrome coronavirus in the respiratory tract of mice, *J. Virol.* **78**, 3572-3577 (2004).
53. D. E. Wentworth, L. Gillim-Ross, N. Espina, and K. A. Bernard, Mice susceptible to SARS coronavirus, *Emerg. Infect. Dis.* **10**, 1293-1296 (2004).
54. J. Sui, W. Li, A. Murakami, A. Tamin, L. J. Matthews, S. K. Wong, M. J. Moore, A. St Clair Tallarico, M. Olurinde, H. Choe, L. J. Anderson, W. J. Bellini, M. Farzan, and W. A. Marasco, Potent neutralization of severe acute respiratory syndrome (SARS) coronavirus by a human mAb to S1 protein that blocks receptor association, *Proc. Natl. Acad. Sci. USA* **101**, 2536-2541 (2004).
55. H. Hofmann, M. Geier, A. Marzi, M. Krumbiegel, M. Peipp, G. H. Fey, T. Gramberg, and S. Pohlmann, Susceptibility to SARS coronavirus S protein-driven infection correlates with expression of angiotensin converting enzyme 2 and infection can be blocked by soluble receptor, *Biochem. Biophys. Res. Commun.* **319**, 1216-1221 (2004).
56. Y. Nie, P. Wang, X. Shi, G. Wang, J. Chen, A. Zheng, W. Wang, Z. Wang, X. Qu, M. Luo, L. Tan, X. Song, X. Yin, M. Ding, and H. Deng, Highly infectious SARS-CoV pseudotyped virus reveals the cell tropism and its correlation with receptor expression, *Biochem. Biophys. Res. Commun.* **321**, 994-1000 (2004).
57. T. C. Greenough, G. J. Babcock, A. Roberts, H. J. Hernandez, W. D. Thomas, Jr., J. A. Coccia, R. F. Graziano, M. Srinivasan, I. Lowy, R. W. Finberg, K. Subbarao, L. Vogel, M. Somasundaran, K. Luzuriaga, J. L. Sullivan, and D. M. Ambrosino, Development and characterization of a severe acute respiratory syndrome-associated coronavirus-neutralizing human monoclonal antibody that provides effective immunoprophylaxis in mice, *J. Infect. Dis.* **191**, 507-514 (2005).
58. J. Sui, W. Li, A. Roberts, L. J. Matthews, A. Murakami, L. Vogel, S. K. Wong, K. Subbarao, M. Farzan, and W. A. Marasco, Evaluation of human mAb 80R in immunoprophylaxis of SARS by an animal study, epitope mapping and analysis of spike variants, *J. Virol.* in press, 2005.
59. Y. He, H. Lu, P. Siddiqui, Y. Zhou, and S. Jiang, Receptor-binding domain of severe acute respiratory syndrome coronavirus spike protein contains multiple conformation-dependent epitopes that induce highly potent neutralizing antibodies, *J. Immunol.* **174**, 4908-4915 (2005).
60. Y. He, Y. Zhou, H. Wu, B. Luo, J. Chen, W. Li, and S. Jiang, Identification of immunodominant sites on the spike protein of severe acute respiratory syndrome (SARS) coronavirus: implication for developing SARS diagnostics and vaccines, *J. Immunol.* **173**, 4050-4057 (2004).
61. K. Kuba, Y. Imai, S. Rao, H. Gao, F. Guo, B. Guan, Y. Huan, P. Yang, Y. Zhang, W. Deng, L. Bao, B. Zhang, G. Liu, Z. Wang, M. Chappell, Y. Liu, D. Zheng, A. Leibbrandt, T. Wada, A. S. Slutsky, D. Liu, C. Qin, C. Jiang, and J. M. Penninger, A crucial role of angiotensin converting enzyme 2 (ACE2) in SARS coronavirus-induced lung injury, *Nat. Med.* **11**, 875-879 (2005).
62. T. Gramberg, H. Hofmann, P. Moller, P. F. Lalor, A. Marzi, M. Geier, M. Krumbiegel, T. Winkler, F. Kirchhoff, D. H. Adams, S. Becker, J. Munch, and S. Pohlmann, LSECtin interacts with filovirus glycoproteins and the spike protein of SARS coronavirus, *Virology* **340**, 224-236 2005.
63. S. A. Jeffers, S. M. Tusell, L. Gillim-Ross, E. M. Hemmila, J. E. Achenbach, G. J. Babcock, W. D. Thomas, Jr., L. B. Thackray, M. D. Young, R. J. Mason, D. M. Ambrosino, D. E. Wentworth, J. C. Demartini, and K. V. Holmes, CD209L (L-SIGN) is a receptor for severe acute respiratory syndrome coronavirus, *Proc. Natl. Acad. Sci. USA* **101**, 15748-15753 (2004).
64. A. Marzi, T. Gramberg, G. Simmons, P. Moller, A. J. Rennekamp, M. Krumbiegel, M. Geier, J. Eisemann, N. Turza, B. Saunier, A. Steinkasserer, S. Becker, P. Bates, H. Hofmann, and S. Pohlmann, DC-SIGN and DC-SIGNR interact with the glycoprotein of Marburg virus and the S protein of severe acute respiratory syndrome coronavirus, *J. Virol.* **78**, 12090-12095 (2004).
65. Z. Y. Yang, Y. Huang, L. Ganesh, K. Leung, W. P. Kong, O. Schwartz, K. Subbarao, and G. J. Nabel, pH-dependent entry of severe acute respiratory syndrome coronavirus is mediated by the spike glycoprotein and enhanced by dendritic cell transfer through DC-SIGN, *J. Virol.* **78** , 5642-5650 (2004).
66. M. Donoghue, F. Hsieh, E. Baronas, K. Godbout, M. Gosselin, N. Stagliano, M. Donovan, B. Woolf, K. Robison, R. Jeyaseelan, R. E. Breitbart, and S. Acton, A novel angiotensin-converting enzyme-related carboxypeptidase (ACE2) converts angiotensin I to angiotensin 1-9, *Circ. Res.* **87**, E1-9 (2000).
67. S. R. Tipnis, N. M. Hooper, R. Hyde, E. Karran, G. Christie, and A. J. Turner, A human homolog of angiotensin-converting enzyme. Cloning and functional expression as a captopril-insensitive carboxypeptidase, *J. Biol. Chem.* **275**, 33238-33243 (2000).

68. C. Vickers, P. Hales, V. Kaushik, L. Dick, J. Gavin, J. Tang, K. Godbout, T. Parsons, E. Baronas, F. Hsieh, S. Acton, M. Patane, A. Nichols, and P. Tummino, Hydrolysis of biological peptides by human angiotensin-converting enzyme-related carboxypeptidase, *J. Biol. Chem.* **277**, 14838-14843 (2002).
69. Y. Yagil and C. Yagil, Hypothesis: ACE2 modulates blood pressure in the mammalian organism, *Hypertension* **41**, 871-873 (2003).
70. M. A. Crackower, R. Sarao, G. Y. Oudit, C. Yagil, I. Kozieradzki, S. E. Scanga, A. J. Oliveira-dos-Santos, J. da Costa, L. Zhang, Y. Pei, J. Scholey, C. M. Ferrario, A. S. Manoukian, M. C. Chappell, P. H. Backx, Y. Yagil, and J. M. Penninger, Angiotensin-converting enzyme 2 is an essential regulator of heart function, *Nature* **417**, 822-828 (2002).
71. Y. Imai, K. Kuba, S. Rao, Y. Huan, F. Guo, B. Guan, P. Yang, R. Sarao, T. Wada, H. Leong-Poi, M. A. Crackower, A. Fukamizu, C. C. Hui, L. Hein, S. Uhlig, A. S. Slutsky, C. Jiang, and J. M. Penninger, Angiotensin-converting enzyme 2 protects from severe acute lung failure, *Nature* **436**, 112-116 (2005).
72. H. Kubo, Y. K. Yamada, and F. Taguchi, Localization of neutralizing epitopes and the receptor-binding site within the amino-terminal 330 amino acids of the murine coronavirus spike protein, *J. Virol.* **68**, 5403-5410 (1994).
73. A. Bonavia, B. D. Zelus, D. E. Wentworth, P. J. Talbot, and K. V. Holmes, Identification of a receptor-binding domain of the spike glycoprotein of human coronavirus HCoV-229E, *J. Virol.* **77**, 2530-2538, (2003).
74. J. J. Breslin, I. Mork, M. K. Smith, L. K. Vogel, E. M. Hemmila, A. Bonavia, P. J. Talbot, H. Sjostrom, O. Noren, and K. V. Holmes, Human coronavirus 229E: receptor binding domain and neutralization by soluble receptor at 37 degrees C, *J. Virol.* **77**, 4435-4438 (2003).
75. G. J. Babcock, D. J. Esshaki, W. D. Thomas, Jr., and D. M. Ambrosino, Amino acids 270 to 510 of the severe acute respiratory syndrome coronavirus spike protein are required for interaction with receptor, *J. Virol.* **78**, 4552-4560 (2004).
76. S. K. Wong, W. Li, M. J. Moore, H. Choe, and M. Farzan, A 193-amino acid fragment of the SARS coronavirus S protein efficiently binds angiotensin-converting enzyme 2, *J. Biol. Chem.* **279**, 3197-3201 (2004).
77. X. Xiao, S. Chakraborti, A. S. Dimitrov, K. Gramatikoff, and D. S. Dimitrov, The SARS-CoV S glycoprotein: expression and functional characterization, *Biochem. Biophys. Res. Commun.* **312**, 1159-1164 (2003).
78. F. Li, W. Li, M. Farzan, and S. C. Harrison, Structure of SARS coronavirus spike receptor-binding domain complexed with receptor, *Science* **309**, 1864-1868 (2005).
79. X. X. Qu, P. Hao, X. J. Song, S. M. Jiang, Y. X. Liu, P. G. Wang, X. Rao, H. D. Song, S. Y. Wang, Y. Zuo, A. H. Zheng, M. Luo, H. L. Wang, F. Deng, H. Z. Wang, Z. H. Hu, M. X. Ding, G. P. Zhao, and H. K. Deng, Identification of two critical amino acid residues of the severe acute respiratory syndrome coronavirus spike protein for its variation in zoonotic tropism transition via a double substitution strategy, *J. Biol. Chem.* **280**, 29588-29595 (2005).
80. Z. Y. Yang, H. C. Werner, W. P. Kong, K. Leung, E. Traggiai, A. Lanzavecchia, and G. J. Nabel, Evasion of antibody neutralization in emerging severe acute respiratory syndrome coronaviruses, *Proc. Natl. Acad. Sci. USA* **102**, 797-801 (2005).
81. M. A. Marra, S. J. Jones, C. R. Astell, R. A. Holt, A. Brooks-Wilson, Y. S. Butterfield, J. Khattra, J. K. Asano, S. A. Barber, S. Y. Chan, A. Cloutier, S. M. Coughlin, D. Freeman, N. Girn, O. L. Griffith, S.R. Leach, M. Mayo, H. McDonald, S. B. Montgomery, P. K. Pandoh, A. S. Petrescu, A. G. Robertson, J. E. Schein, A. Siddiqui, D. E. Smailus, J. M. Stott, G. S. Yang, F. Plummer, A. Andonov, H. Artsob, N. Bastien, K. Bernard, T. F. Booth, D. Bowness, M. Czub, M. Drebot, L. Fernando, R. Flick, M. Garbutt, M. Gray, A. Grolla, S. Jones, H. Feldmann, A. Meyers, A. Kabani, Y. Li, S. Normand, U. Stroher, G. A. Tipples, S. Tyler, R. Vogrig, D. Ward, B. Watson, R. C. Brunham, M. Krajden, M. Petric, D. M. Skowronski, C. Upton, and R. L. Roper, The genome sequence of the SARS-associated coronavirus, *Science* **300**, 1399-1404 (2003).
82. J. F. He, G. W. Peng, J. Min, D. W. Yu, W. J. Liang, S. Y. Zhang, R. H. Xu, H. Y. Zheng, X. W. Wu, J. Xu, Z. H. Wang, L. Fang, X. Zhang, H. Li, X. G. Yan, J. H. Lu, Z. H. Hu, J. C. Huang, and X. W. Wan, Molecular evolution of the SARS coronavirus during the course of the SARS epidemic in China, *Science* **303**, 1666-1669 (2004).

ATTACHMENT FACTOR AND RECEPTOR ENGAGEMENT OF SARS CORONAVIRUS AND HUMAN CORONAVIRUS NL63

Heike Hofmann, Andrea Marzi, Thomas Gramberg, Martina Geier, Krzysztof Pyrc, Lia van der Hoek, Ben Berkhout, and Stefan Pöhlmann*

1. INTRODUCTION

The cellular membrane constitutes a physical barrier against viral infection. Enveloped viruses developed specialized proteins to overcome this barrier. These proteins, which are often extensively glycosylated, are inserted into the viral membrane and mediate both recognition of target cells and fusion of the viral membrane with a host cell membrane. The latter process allows introduction of the viral genome and associated viral proteins into the host cell lumen and is therefore critical for establishment of productive infection.[1,2] Because of their important function, the viral membrane glycoproteins, which in the case of coronaviruses (CoV) are termed spike (S) proteins, are attractive targets for inhibitors and vaccines.

Enveloped viruses evolved two prototypes of glycoproteins to enter target cells, termed class I and class II fusion proteins.[3, 4] Class I fusion proteins are found in, e.g., retroviruses and paramyxoviruses, while, e.g., flaviviruses and alphaviruses encode class II fusion proteins. Both types of fusion proteins exhibit a distinct functional organization, which is reflected by their different spatial orientations. Thus, class I fusion proteins are oriented perpendicular to the cellular membrane and are visible as spikes in electron micrographs, while class II proteins are oriented horizontally relative to the cellular membrane and are well ordered on the virion surface. Viral class I fusion proteins are organized into a globular surface unit (SU), which interacts with cellular receptors, and a transmembrane unit (TM), which harbors highly conserved sequence elements required for membrane fusion.[3] Membrane fusion is initiated by binding of SU to cellular receptor(s) or by exposure of the glycoprotein to low pH, which triggers conformational changes in the glycoprotein that activate TM. TM-driven membrane fusion is initiated by insertion of a N-terminal fusion peptide into the target cell membrane, followed by

*Heike Hofmann, Andrea Marzi, Thomas Gramberg, Martina Geier, Stefan Pöhlmann, University of Erlangen-Nürnberg, Germany. Krzysztof Pyrc, Lia van der Hoek, Ben Berkhout, University of Amsterdam, The Netherlands.

conformational changes in TM during which two heptad repeats in the extracellular part of TM fold back onto each other and pull the viral and target cell membranes into close contact, which ultimately promotes membrane fusion. Despite the different functional organization of class II fusion proteins, which, e.g., harbor the fusion peptide in SU, membrane fusion driven by these proteins follows similar principles.[5]

The S-proteins of CoVs, which protrude from the viral membrane and provide virus particles with the typical corona like shape, exhibit the characteristics of class I fusion proteins.[1] Some S-proteins are cleaved between S1 and S2,[1] as is the case with human coronavirus OC43 (HCoV-OC43), and cleavage is possibly important for function. In contrast, cleavage of the spike proteins of HCoV-229 and other group I CoVs has not been observed and cleavage of mouse hepatitis virus (MHV) S appears to be cell-type dependent and not essential for function.[1] The S1 domains are adapted to bind to specific cellular receptors, and the spike-receptor interaction is the major determinant of viral cell tropism.[1] Because the S-proteins of CoVs from different groups are, in most cases, adapted to interact with different receptors, they show little sequence conservation. In contrast, the S2 domains share the same task, fusion of viral and cellular membranes, and therefore exhibit considerable sequence homology.[1]

SARS-CoV infects the lower respiratory tract with fatal outcome in about 10% of infected individuals.[6] In contrast, infection with HCoV-NL63 does not cause severe disease, but is often associated with bronchiolitis and cold like symptoms.[7, 8] Although SARS-CoV was, mainly due to air travel, spread into 29 different countries in 2003, the majority of cases were observed in Asia. HCoV-NL63 in turn seems to be a globally distributed pathogen, with HCoV-NL63 infections being reported in Europe, Japan, Canada, and Australia. Sequence analysis revealed that also the S-proteins of SARS-CoV and HCoV-NL63 exhibit features of class I fusion proteins. Thus, the S2 subunits contain heptad repeats, a transmembrane domain and a short intracellular domain,[1] elements found in the S2 subunits of all CoV S-proteins. The S1 subunit of SARS-CoV, which engages angiotensin converting enzyme 2 (ACE2) as a receptor for cellular entry,[9] shares little sequence homology with other CoV S-proteins. In contrast, the sequence of the S1 subunit of HCoV-NL63 is 56% identical to S1 of hCoV-229E,[8] which employs aminopeptidase N/CD13 for entry into target cells.

We thought to investigate the range of target cells susceptible to SARS-CoV-S and HCoV-NL63-S dependent infection as well as the interaction of the respective S-proteins with cellular membrane proteins and their recognition by sera from infected patients. For these analyses, we employed retroviral reporter viruses pseudotyped with the CoV-S-proteins. These viruses, so called pseudotypes, were generated by cotransfection of 293T cells with a plasmid encoding a retroviral genome, in which the *env* open reading frame was inactivated and in which a reporter gene was inserted, together with an expression plasmid for the CoV-S-protein to be studied. Such particles harbor the CoV S-protein in their membrane and enter target cells in a S-protein dependent manner. However, once membrane fusion is completed, all processes leading to viral gene expression are dependent on retroviral proteins. Efficiency of infection with pseudotyped viruses can be conveniently quantified because of expression of the reporter gene encoded by the proviral genome.

2. IDENTIFICATION OF TARGET CELLS SUSCEPTIBLE TO SARS-CoV-S DRIVEN INFECTION

We employed HIV-1 derived reporter viruses harboring the S-protein of the SARS-CoV Frankfurt strain to investigate the range of cells permissive to SARS-CoV-S dependent infection.[10] Pseudotypes bearing the G-protein of the amphotropic vesicular stomatitis virus (VSV) were used as positive control, while viruses without an envelope protein were employed as negative control. Infection of a panel of cell lines revealed that SARS-CoV-S mediates efficient entry into liver (Huh-7, Hep-G2) and kidney (293T) derived cell lines, and the hepatoma cell line Huh-7 was found to be permissive for SARS-CoV replication.[10] Entry into liver and kidney derived cell lines is in agreement with subsequent studies demonstrating infection of these organs in SARS patients.[11] In contrast, lymphoid cell lines were refractory to SARS-CoV-S driven infection, and similar observations were made by an independent study examining replication competent SARS-CoV,[12] suggesting that lymphoid cells might not support SARS-CoV spread *in vivo*. We next determined if SARS-CoV-S driven cellular entry depends on an acidic environment. To address this question, Huh-7 and 293T cells were preincubated with the lysosomotropic agents ammonium chloride and bafilomycin A1 and infected with SARS-CoV-S, VSV-G or murine leukemia virus (MLV) bearing pseudotypes. In agreement with published data, entry mediated by the MLV glycoprotein was not blocked by lysosomotropic agents, while entry driven by VSV-G was efficiently inhibited in a dose dependent manner.[10] Infectious entry of SARS-CoV-S bearing pseudotypes was also blocked, indicating that the SARS-CoV-S protein requires a low pH environment to unfold its fusogenic activity,[10] an observation confirmed by several independent studies.[13,14] Finally, we investigated if sera from SARS patients recognize the SARS-CoV-S protein and neutralize infection. Transient expression of SARS-CoV-S on 293T cells followed by staining of cells with sera from healthy patients or SARS patients revealed that SARS patient sera recognize the S-protein.[10] Sera from SARS patients but not control sera neutralized SARS-CoV-S dependent infection,[10] indicating that infected individuals mount a S-specific neutralizing antibody response.

3. ANALYSIS OF THE INTERACTION BETWEEN SARS-CoV-S AND ITS RECEPTOR ACE2

Studies by Li and colleagues revealed that SARS-CoV employs ACE2 for entry into target cells.[9] Specifically, the S1 subunit was shown to bind to ACE2, and fusion of SARS-CoV-S expressing cells with ACE2 but not control cells was demonstrated.[9] Inhibition analysis indicated that replication of SARS-CoV depends on ACE2, and amino acids 318–510 in SARS-CoV-S were shown to function as an independent receptor binding domain.[9,15] While these studies indicated that ACE2 is an important receptor for SARS-CoV, it was unclear if the virus engages receptors besides ACE2 for infection of target cells. Therefore, we investigated if expression of ACE2 correlates with permissiveness to SARS-CoV-S dependent infection. Analysis of mRNA expression in a panel of cell lines of known susceptibility to SARS-CoV-S mediated entry revealed that SARS-CoV-S bearing pseudotypes infected exclusively cell lines that express ACE2.[16]

These findings suggest that ACE2 is of paramount importance for SARS-CoV entry and most likely constitutes the only receptor for SARS-CoV — a finding substantiated by several subsequent studies.[12,17] The interaction between SARS-CoV-S and ACE2 is an attractive target for therapeutic intervention. Indeed, several studies described peptide inhibitors that mimic the second heptad repeat in the S2 subunit of SARS-CoV-S and block SARS-CoV-S infection in the low micromolar range.[1] We investigated if inhibitors based on the ACE2 ectodomain might also be effective. To this end, we incubated SARS-CoV-S and VSV-G bearing pseudotypes with concentrated soluble ACE2 ectodomain and analysed infection of target cells. The ACE2 ectodomain inhibited SARS-CoV-S in a potent and specific manner,[16] suggesting that polypeptides based on the ACE2 ectodomain could, at least in theory, be developed as therapeutics. Finally we asked if the cytoplasmic domain of ACE2, which harbors consensus sites for tyrosine kinases and casein kinase II motifs, is required for receptor function. However, ACE2 variants in which the cytoplasmic tail was stepwise deleted were fully capable of promoting SARS-CoV-S dependent entry into transiently transfected 293T cells,[16] indicating that the cytoplasmic domain of ACE2 might be dispensable for receptor function – at least under conditions of overexpression in already permissive cells.

4. ROLE OF CELLULAR LECTINS IN SARS-CoV INFECTION

Engagement of cellular receptors by viral glycoproteins is essential for virus entry into target cells. However, engagement of cellular factors other than the viral receptor can affect infection efficiency. Thus, binding to so called attachment factors can concentrate viruses on the surface of target cells, thereby increasing the chance of receptor engagement and subsequent infectious entry.[18] The lectin DC-SIGN is a universal pathogen attachment factor and promotes infection by a variety of viral and non viral pathogens.[1,19] Maybe most strikingly, DC-SIGN is expressed on dendritic cells (DCs) and facilitates the HIV interaction with these cells, which is believed to be important for viral dissemination.[19] DC-SIGN binds to high-mannose carbohydrates in the HIV-Env protein and facilitates both infection of the DC-SIGN expressing cells (in case they also express the HIV receptors CD4 and CCR5/CXCR4) and of adjacent susceptible cells.[19] The former process is termed infection in cis, while the latter process is termed infection in trans or transmission. Apart from DC-SIGN, the related protein DC-SIGNR, for DC-SIGNrelated (also termed L-SIGN, for Liver-SIGN), was also shown to function as a pathogen attachment factor.[20,21] In contrast to DC-SIGN, DC-SIGNR is expressed on sinusoidal endothelial cells in lymph node and liver[20,21] and might promote spread of HIV-1 and hepatotropic viruses.

Analysis of ACE2 positive cells expressing DC-SIGN, DC-SIGNR, or a control plasmid revealed that both lectins bind to the S1 unit of SARS-CoV-S and augment SARS-CoV-S dependent infection.[14,22] Importantly, however, expression of DC-SIGN and DC-SIGNR on nonpermissive cells did not allow for readily detectable SARS-CoV-S mediated infectious entry, and DC-SIGN positive immature DCs were refractory to infection,[14,22] indicating that DC-SIGN and DC-SIGNR function as SARS-CoV attachment factors and not as viral receptors. DC-SIGN expressing, nonpermissive cells and DCs transmitted SARS-CoV-S bearing pseudotypes and replication competent SARS-CoV to adjacent permissive cells,[14,22] indicating that DCs might promote SARS-CoV dissemination in infected individuals. Similarly, DC-SIGNR expression in the lung might promote SARS-CoV spread in this major target organ.[23]

5. IDENTIFICATION OF ACE2 AS A RECEPTOR FOR HCoV-NL63

5.1. Pseudotypes Bearing the S-Proteins of HCoV-NL63-S and SARS-CoV-S Exhibit a Comparable Cell Tropism

Monkey kidney cells were shown to be permissive for HCoV-NL63 infection,[7,8] however, the range of susceptible target cells had not been identified. In order to analyze the cellular tropism of HCoV-NL63, we employed lentiviral pseudotypes carrying the S-protein of HCoV-NL63 and included virions pseudotyped with SARS-CoV-S and HCoV-229E-S as controls. The latter S-protein is highly homologous to the NL63-S-protein and it has been suggested that both might employ the same cellular receptor for entry and might thus enter the same target cells.[8] However, we observed a striking difference in the cell tropism of 229E-S and NL63-S bearing pseudotypes (Table 1), as HOS, MRC-5 and feline FCWF cells were susceptible to 229E-S but not NL63-S-driven infection, whereas only NL63-S mediated entry into 293T kidney cells.[24] Interestingly, the cell tropism of NL63-S harboring pseudotypes matched that previously described for SARS-CoV-S bearing pseudovirions,[10,13,14] suggesting that both S-proteins might engage the same cellular factors for entry.[24] We next addressed if NL63-S bearing pseudotypes reflect the cell tropism of replication competent HCoV-NL63. Huh-7 cells were highly permissive to NL63-S and SARS-CoV-S driven infection[10,24] and we therefore expected HCoV-NL63 to replicate in these cells. Indeed, four to five days after inoculation with HCoV-NL63, a cytopathic effect was readily visible in the infected culture in comparison with mock-infected cells,[24] indicating that the hepatic cell tropism of NL63-S bearing pseudoparticles is reflected by replication-competent HCoV-NL63.

Table 1. Analysis of the cell tropism of CoV-S-pseudotypes. The indicated cell lines were infected with lentiviral pseudotypes carrying the S-proteins of hCoV-229E, -NL63, and SARS-CoV and reporter gene activities in cellular lysates quantified. Infection efficiency is shown as: (-), no infection , (+), low, (+++) high.

Cell type	Cell line	229E-S	NL63-S	SARS-S
T-lymphocyte	C81-66	-	-	-
	CEMx174	-	-	-
B-lymphocyte	BL41	-	-	-
Kidney	293T	-	+	+
Fibroblast	HFF	+	-	n.d.
	MRC-5	+++	-	-
Fibroblast (feline)	FCWF	+++	-	-
Glioblastoma	U373	-	-	-
Epithelial	Hela S3	-	-	-
	Hep-2	-	+	+
Osteosarcoma	HOS	+++	-	-
Liver	HepG2	+	+	+
	Huh7	+++	+++	+++

5.2. HCoV-NL63-S Dependent Entry Requires Low pH

Because NL63-S and SARS-CoV-S bearing pseudotypes exhibited a comparable cell tropism, we next asked if entry mediated by both S-proteins depends on low pH. To address this question, Huh-7 target cells were incubated with the lysosomotropic agents bafilomycin A1 or ammonium chloride and infected with pseudovirions bearing VSV-G, MLV, and the spike proteins of HCoV-NL63, HCoV-229E, and SARS-CoV (data not shown). As expected, VSV-G driven entry was inhibited by the lysosomotropic agents, while MLV glycoprotein dependent entry was not affected. Entry mediated by all CoV S-proteins examined was dependent on low pH, suggesting that SARS-CoV, HCoV-NL63 and HCoV-229E employ the same route of entry.

5.3. The S-Protein of HCoV-NL63 Engages ACE2 but Not CD13 for Cellular Entry

Feline CD13 (fCD13) serves as a receptor for all CoVs of the phylogenetic group I. Because HCoV-NL63-S is a group I virus, it was expected that NL63-S might also engage fCD13 for entry. In the HCoV-NL63 cell tropism experiments described above, however, we observed that the feline FCWF cells, which express fCD13, were refractory to NL63-S mediated infection. Similarly, 293T cells overexpressing fCD13 or human CD13 were permissive to 229E-S mediated infection, while expression of both feline or human CD13 did not augment infection driven by NL63-S,[24] indicating that CD13 is not involved in HCoV-NL63 entry. Because the cellular tropism of NL63-S and SARS-CoV-S bearing pseudotypes was identical, we next investigated if the SARS-CoV receptor ACE2 plays a role in NL63-S dependent entry. To address this question, we transiently expressed ACE2 on 293T cells and over-infected the cells with 229E-, NL63- or SARS-CoV-S bearing pseudotypes. Hereby, a significant enhancement of infection could be documented for NL63-S- and SARS-CoV-S-, but not for 229E-S-bearing pseudotypes.[24] The interaction of NL63-S and ACE2 was specific, as expression of the closely related ACE1 protein did allow for augmentation of infection.[24] Additionally, we were able to inhibit both NL63-S mediated entry and replication of HCoV-NL63 by an ACE2-specific antiserum, but not by antibodies directed against ACE1,[24] confirming the specificity of the interaction. Finally, we employed soluble NL63-S1 and SARS-CoV-S1 proteins to investigate if NL63-S directly interacts with ACE2. Both NL63-S and SARS-CoV-S bound efficiently to 293T cells expressing ACE2 but not control cells,[24] indicating that NL63-S directly contacts ACE2.

We next addressed the question whether ACE2 alone is sufficient to mediate HCoV-NL63 entry. For this, we transiently expressed ACE2 on nonpermissive Hela cells followed by infection with NL63- or SARS-S bearing pseudotypes. Whereas the presence of ACE2 allowed for efficient SARS-S-mediated entry, we observed only a slight enhancement of NL63-S dependent entry (data not shown). This finding can be explained in three ways: (i) SARS-S exhibits a higher affinity for ACE2 than NL63-S, (ii) a much higher amount of ACE2 has to be present on a target cell for efficient NL63-S mediated entry compared to SARS-CoV-S driven infection, or (iii) HCoV-NL63 requires a so far unidentified co-receptor which is not or only insufficiently present on Hela cells. Additionally, it is possible that a cellular factor involved in steps after receptor engagement might be critical for NL63-S mediated infection, but could be dispensable for SARS-S dependent cellular entry. Further experiments are required to decipher these differences between HCoV-NL63 and SARS-CoV entry into target cells.

5.4. Evidence That HCoV-NL63 Infection Is Common and Usually Acquired During Childhood

Initial PCR-based screening experiments suggested that HCoV-NL63 infection is relatively frequent,[7,8] but serological data were not included in these studies. Employing NL63-S bearing pseudoparticles, we investigated whether adults with or without respiratory tract illness exhibit a neutralizing antibody response against NL63-S. Interestingly, we found strong neutralizing activities against NL63-S, but not 229E-S in the sera of all adults tested.[24] Therefore, HCoV-NL63 infection seems to be frequent and more prevalent than infection with hCoV-229E. Furthermore, sera that neutralized NL63-S dependent infection did not necessarily block 229E-S mediated infectious entry,[24] suggesting that no cross-reactive antibodies are induced in infected individuals, despite the high amino acid similarity between both S-proteins. When investigating sera from children of different age groups, we observed that a neutralizing antibody response against NL63-S can first be detected at the age of 1.5 years and is found in all samples from donors aged at least 8 years.[24] In contrast, none of the sera investigated showed reactivity against SARS-CoV, indicating that NL63-S and SARS-CoV-S, despite using the same receptor for cellular entry, do not share determinants recognized by neutralizing antibodies.[24]

6. CONCLUDING REMARKS

We employed retroviral pseudotypes to analyze the S-proteins of SARS-CoV and HCoV-NL63. Our analysis, and its comparison with independent studies using replication competent CoVs, show that S-protein bearing pseudotypes adequately reflect cell tropism, receptor and attachment factor usage and route of entry of the CoVs from which the S-proteins were derived. We found that SARS-CoV-S bearing pseudotypes infect a relatively broad range of cells,[10] leading us to the conclusion that SARS-CoV might target organs other than the lung in infected patients. Indeed, subsequent studies examining tissues from SARS patients confirmed that SARS-CoV targets a variety of organs.[11] Entry of SARS-CoV-S pseudotypes[10,13,14] and replication competent virus[25] can be inhibited by compounds that impede acidification of the endosomal compartment, suggesting that low pH might trigger structural rearrangements in SARS-CoV-S pivotal to membrane fusion. However, a subsequent study revealed that an acidic environment is required for the activity of cellular cathepsin proteases, which cleave the SARS-CoV-S, and possibly the NL63-S-protein, and cleavage was found to be required for infectious cellular entry.[25] Thus, SARS-CoV-S dependent entry follows a novel principle, and offers new targets for therapeutic intervention.[25] Entry of SARS-CoV was strictly dependent on expression of ACE2,[16] indicating that ACE2 is the only cellular receptor for SARS-CoV. The ACE2 ectodomain was found to inhibit SARS-CoV-S dependent entry[16] and it might be possible to generate ACE2 derived inhibitors. Compounds based on the ACE2 ectodomain might be particularly promising, as it has been proposed that down-modulation of ACE2 during SARS-CoV infection is responsible for much of the SARS pathology, which can be prevented by application of the soluble ACE2 ectodomain.[26] While ACE2 promotes entry of SARS-CoV, viral entry is enhanced by S-protein binding to the lectins DC-SIGN and DC-SIGNR.[14,22] Our studies indicate that DC-SIGN and DC-

SIGNR augment infection but do not allow for infectious entry in the absence of ACE2.[10] However, another study employing replication competent SARS-CoV reported that DC-SIGNR functions as a viral receptor.[23] In any case, it will be interesting to analyze if polymorphisms in the DC-SIGNR neck region, which are frequently found, modulate the risk or outcome of SARS-CoV infection. SARS-CoV shares ACE2 as a receptor and DC-SIGN/DC-SIGNR as attachment factors with the group I HCoV-NL63.[24] The observation that HCoV-NL63 employs ACE2 for infection and consequently enters the same target cells as SARS-CoV, but does not induce severe disease, poses a variety of interesting questions. Are the accessory genes responsible of these differences in pathogenicity, nine of which are present in SARS-CoV compared with only one in NL63? Do SARS-CoV-S and NL63-S engage ACE2 differentially, and are potential differences associated with differences in pathogenicity? Is HCoV-NL63-S sensitive to ACE2 inhibitors? Can a chimeric NL63-S/229E-S protein be generated that induces neutralizing antibodies against both viruses and can be developed as a vaccine? Is ACE2 the only receptor for HCoV-NL63? While experiments with transiently transfected non permissive HeLa cells indeed suggest that HCoV-NL63 could use a coreceptor for entry (data not shown), a thorough comparative analysis of SARS-CoV and HCoV-NL63 infectious entry, including the establishment of animal models for HCoV-NL63 infection, is required to answer these questions.

7. REFERENCES

1. H. Hofmann and S. Pöhlmann, Cellular entry of the SARS coronavirus, *Trends Microbiol.* **12**, 466-472 (2004).
2. A. E. Smith and A. Helenius, How viruses enter animal cells, *Science* **304**, 237-242 (2004).
3. D. M. Eckert and P. S. Kim, Mechanisms of viral membrane fusion and its inhibition, *Annu. Rev. Biochem.* **70**, 777-810 (2001).
4. F. X. Heinz, K. Stiasny, and S. L. Allison, The entry machinery of flaviviruses, *Arch. Virol. Suppl.* (18), 133-137 (2004).
5. D. J. Schibli and W. Weissenhorn, Class I and class II viral fusion protein structures reveal similar principles in membrane fusion, *Mol. Membr. Biol.* **21**, 361-371 (2004).
6. J. S. Peiris, K. Y. Yuen, A. D. Osterhaus, and K. Stohr, The severe acute respiratory syndrome, *N. Engl. J. Med.* **349**, 2431-2441 (2003).
7. R. A. Fouchier, N. G. Hartwig, T. M. Bestebroer, B. Niemeyer, J. C. de Jong, J. H. Simon, and A. D. Osterhaus, A previously undescribed coronavirus associated with respiratory disease in humans, *Proc. Natl. Acad. Sci. USA* **101**, 6212-6216 (2004).
8. L. van der Hoek, K. Pyrc, M. F. Jebbink, W. Vermeulen-Oost, R. J. Berkhout, K. C. Wolthers, P. M. Wertheim-van Dillen, J. Kaandorp, J. Spaargaren, and B. Berkhout, Identification of a new human coronavirus, *Nat. Med.* **10**, 368-373 (2004).
9. W. Li, M. J. Moore, N. Vasilieva, J. Sui, S. K. Wong, M. A. Berne, M. Somasundaran, J. L. Sullivan, K. Luzuriaga, T. C. Greenough, H. Choe, and M. Farzan, Angiotensin-converting enzyme 2 is a functional receptor for the SARS coronavirus, *Nature* **426**, 450-454 (2003).
10. H. Hofmann, K. Hattermann, A. Marzi, T. Gramberg, M. Geier, M. Krumbiegel, S. Kuate, K. Uberla, M. Niedrig, and S. Pöhlmann, S protein of severe acute respiratory syndrome-associated coronavirus mediates entry into hepatoma cell lines and is targeted by neutralizing antibodies in infected patients, *J. Virol.* **78**, 6134-6142 (2004).
11. J. Gu, E. Gong, B. Zhang, J. Zheng, Z. Gao, Y. Zhong, W. Zou, J. Zhan, S. Wang, Z. Xie, H. Zhuang, B. Wu, H. Zhong, H. Shao, W. Fang, D. Gao, F. Pei, X. Li, Z. He, D. Xu, X. Shi, V. M. Anderson, and A. S. Leong, Multiple organ infection and the pathogenesis of SARS, *J. Exp. Med.* **202**, 415-424 (2005).
12. K. Hattermann, M. A. Muller, A. Nitsche, S. Wendt, M. O. Donoso, and M. Niedrig, Susceptibility of different eukaryotic cell lines to SARS-coronavirus, *Arch. Virol.* **150**, 1023-1031 (2005).

13. G. Simmons, J. D. Reeves, A. J. Rennekamp, S. M. Amberg, A. J. Piefer, and P. Bates, Characterization of severe acute respiratory syndrome-associated coronavirus (SARS-CoV) spike glycoprotein-mediated viral entry, *Proc. Natl. Acad. Sci. USA* **101**, 4240-4245 (2004).
14. Z. Y. Yang, Y. Huang, L. Ganesh, K. Leung, W. P. Kong, O. Schwartz, K. Subbarao, and G. J. Nabel, pH-dependent entry of severe acute respiratory syndrome coronavirus is mediated by the spike glycoprotein and enhanced by dendritic cell transfer through DC-SIGN, *J. Virol.* **78**, 5642-5650 (2004).
15. S. K. Wong, W. Li, M. J. Moore, H. Choe, and M. Farzan, A 193-amino acid fragment of the SARS coronavirus S protein efficiently binds angiotensin-converting enzyme 2, *J. Biol. Chem.* **279**, 3197-3201 (2004).
16. H. Hofmann, M. Geier, A. Marzi, M. Krumbiegel, M. Peipp, G. H. Fey, T. Gramberg, and S. Pöhlmann, Susceptibility to SARS coronavirus S protein-driven infection correlates with expression of angiotensin converting enzyme 2 and infection can be blocked by soluble receptor, *Biochem. Biophys. Res. Commun.* **319**, 1216-1221 (2004).
17. E. C. Mossel, C. Huang, K. Narayanan, S. Makino, R. B. Tesh, and C. J. Peters, Exogenous ACE2 expression allows refractory cell lines to support severe acute respiratory syndrome coronavirus replication, *J. Virol.* **79**, 3846-3850 (2005).
18. S. Ugolini, I. Mondor, and Q. J. Sattentau, HIV-1 attachment: another look, *Trends Microbiol.* **7**, 144-149 (1999).
19. Y. van Kooyk and T. B. Geijtenbeek, DC-SIGN: escape mechanism for pathogens, *Nat. Rev. Immunol.* **3**, 697-709 (2003).
20. A. A. Bashirova, T. B. Geijtenbeek, G. C. van Duijnhoven, S. J. van Vliet, J. B. Eilering, M. P. Martin, L. Wu, T. D. Martin, N. Viebig, P. A. Knolle, V. N. KewalRamani, Y. van Kooyk, and M. Carrington, A dendritic cell-specific intercellular adhesion molecule 3-grabbing nonintegrin (DC-SIGN)-related protein is highly expressed on human liver sinusoidal endothelial cells and promotes HIV-1 infection, *J. Exp. Med.* **193**, 671-678 (2001).
21. S. Pöhlmann, E. J. Soilleux, F. Baribaud, G. J. Leslie, L. S. Morris, J. Trowsdale, B. Lee, N. Coleman, and R. W. Doms, DC-SIGNR, a DC-SIGN homologue expressed in endothelial cells, binds to human and simian immunodeficiency viruses and activates infection in trans, *Proc. Natl. Acad. Sci. USA* **98**, 2670-2675 (2001).
22. A. Marzi, T. Gramberg, G. Simmons, P. Moller, A. J. Rennekamp, M. Krumbiegel, M. Geier, J. Eisemann, N. Turza, B. Saunier, A. Steinkasserer, S. Becker, P. Bates, H. Hofmann, and S. Pöhlmann, DC-SIGN and DC-SIGNR interact with the glycoprotein of Marburg virus and the S protein of severe acute respiratory syndrome coronavirus, *J. Virol.* **78**, 12090-12095 (2004).
23. S. A. Jeffers, S. M. Tusell, L. Gillim-Ross, E. M. Hemmila, J. E. Achenbach, G. J. Babcock, W. D. Thomas, Jr., L. B. Thackray, M. D. Young, R. J. Mason, D. M. Ambrosino, D. E. Wentworth, J. C. Demartini, and K. V. Holmes, CD209L (L-SIGN) is a receptor for severe acute respiratory syndrome coronavirus, *Proc. Natl. Acad. Sci. USA* **101**, 15748-15753 (2004).
24. H. Hofmann, K. Pyrc, H. L. van der, M. Geier, B. Berkhout, and S. Pöhlmann, Human coronavirus NL63 employs the severe acute respiratory syndrome coronavirus receptor for cellular entry, *Proc. Natl. Acad. Sci. USA* **102**, 7988-7993 (2005).
25. G. Simmons, D. N. Gosalia, A. J. Rennekamp, J. D. Reeves, S. L. Diamond, and P. Bates, Inhibitors of cathepsin L prevent severe acute respiratory syndrome coronavirus entry, *Proc. Natl. Acad. Sci. USA* **102**, 11876-11881 (2005).
26. K. Kuba, Y. Imai, S. Rao, H. Gao, F. Guo, B. Guan, Y. Huan, P. Yang, Y. Zhang, W. Deng, L. Bao, B. Zhang, G. Liu, Z. Wang, M. Chappell, Y. Liu, D. Zheng, A. Leibbrandt, T. Wada, A. S. Slutsky, D. Liu, C. Qin, C. Jiang, and J. M. Penninger, A crucial role of angiotensin converting enzyme 2 (ACE2) in SARS coronavirus-induced lung injury, *Nat. Med.* **11**, 875-879 (2005).

INTERACTIONS BETWEEN SARS CORONAVIRUS AND ITS RECEPTOR

Fang Li, Wenhui Li, Michael Farzan, and Stephen C. Harrison*

1. INTRODUCTION

The spike protein on the envelope of SARS-coronavirus (SARS-CoV) guides viral entry into cells by first binding to its cellular receptor and then fusing viral envelope and cellular membranes.[1] It consists of a large ecdotomain (S-e) (residues 12~1190), a trans-membrane anchor, and a short intracellular tail. S-e contains two regions, a receptor-binding region S1 and a membrane-fusion region S2. The S1 region contains a defined receptor-binding domain (RBD) (about residues 300~500).[2,5] SARS-CoV uses a zinc peptidase, ACE2, as its cellular receptor.[4] The crystal structure of ACE2 shows that it has a claw-like structure.[5] Ligand binding triggers an open-closed conformational change between its two lobes. The SARS-CoV RBD is sufficient for tight binding to ACE2, and thus it is the most important determinant of virus-receptor interactions, viral host range, and tropism. It is believed that a few residue changes on the RBD play a pivotal role in the cross-species transmission of SARS-CoV.[6,7] We have identified the boundaries of the RBD by limited proteolysis, purified the RBD, and determined its crystal structure in complex with ACE2 at 2.9 Å resolution. The structure reveals in atomic detail the specific and high-affinity interactions between the virus and its receptor. It sheds light on critical residue changes that dictate the species specificity of the virus.

2. RESULTS AND DISCUSSION

We constructed and expressed the SARS-CoV S-e in insect cells, purified it from the cell culture medium, and identified the S1/S2 boundary (after residue 667) by limited proteolysis of the purified S-e. We then constructed and expressed S1 in insect cells, purified it, and identified the N terminus of the RBD (before residue 306) by limited proteolysis of the purified S1. To obtain structural information on the RBD, we made a

* Fang Li, Stephen C. Harrison, Harvard Medical School and Children's Hospital, Boston, Massachusetts, 02115, and Howard Hughes Medical Institute (SCH). Wenhui Li, Michael Farzan, Harvard Medical School and New England Primate Research Center, Southborough, Massachusetts 01772.

series of constructs with the newly defined N-terminus of RBD but with different C-termini. We expressed and purified each of these fragments. Extensive crystallization trials of these fragments, by themselves or in complex with ACE2, did not yield useful crystals. By further proteolysis of one of these fragments (306–575), we generated a shorter fragment (306–527) that corresponds to the most stable version of the RBD. This RBD binds to ACE2 with high affinity, as shown by gel filtration experiments (Figure 1); it was subsequently co-crystallized with ACE2 in space group $P2_1$ (Figure 2).

The structure of the ACE2/RBD complex was determined by molecular replacement using ACE2 as the search model and refined to an R-factor of 22.1% (R free = 27.5%) at 2.9 Å resolution.[8] The final model of the complex contains the N-terminal peptidase domain of human ACE2 (residues 19–615) and the spike RBD of human SARS-CoV (residues 323–502, missing residues 376–381). The model also includes glycans N-linked to residues 53, 90, 322, 546 of ACE2 and to residue 330 of the RBD.

The RBD structure contains two subdomains – a core structure and an extended loop (Figure 3A). The core structure is a five-stranded antiparallel ß-sheet, with three short connecting α helices. The extended loop presents a gently curved surface to interact with the receptor. The base of this surface is a two-stranded antiparallel ß-sheet that cradles

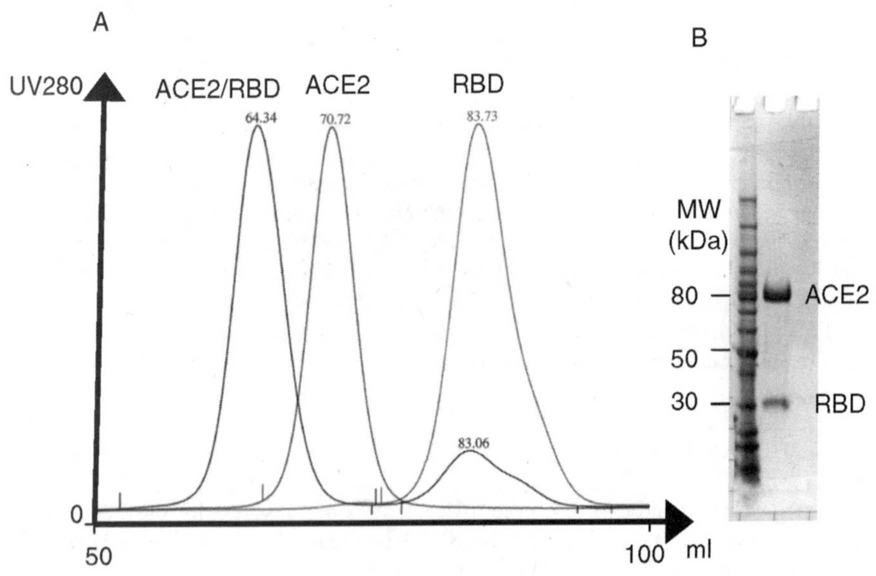

Figure 1. Interactions between SARS-CoV spike RBD and ACE2 in solution. (A) Gel filtration chromatography on Superdex 200 of RBD (right), ACE2 (middle), and ACE2/RBD complex (left; RBD is in excess). The elution volumes of each sample are indicated above the peaks. (B) Coomassie-blue stained reducing SDS-PAGE. The right lane is the ACE2/RBD complex collected from the left peak in (A).

the N-terminal helix of ACE2. One ridge of the surface contacts a loop of ACE2, while the other inserts between two ACE2 loops (Figure 4). Because this extended loop makes all the contacts with the receptor, we refer to it as the "receptor-binding motif," or RBM.

In the crystal there are two complexes in each asymmetric unit. The ACE2 molecule in one of the complexes is in the open conformation; the other is slightly closed. The RBD binds to the outer surface of the N-terminal lobe of ACE2, away from the peptidase active site (Figure 3B). Therefore, SARS-CoV binding is independent of ACE2 conformation and is unlikely to interfere with the peptidase activity of ACE2.

At least four features contribute to the specific and high-affinity binding between ACE2 and the RBD. First, the two proteins are perfectly complementary in shape. Second, the RBM is rich in tyrosine that has both a polar hydroxyl group and a hydrophobic aromatic ring, generating a combination of specific hydrogen-bond interactions and strong nonpolar contacts. Third, the RBM is reinforced by a disulfide bond. Fourth, the binding buries 1700 $Å^2$ at the interface. Thus, the interactions between the two proteins are both extensive and specific.

The structure reveals important residue changes at the binding interface that determine the species specificity of SARS-CoV. Previous genomic analysis and mutagenesis studies suggested possible roles for residues 479 and 487 in cross-species infection by SARS-CoV.[6,7] Detailed structural analysis sheds light on the significance of these residues in virus-receptor interactions (Figure 4). On most human SARS isolates, 479 is an asparagine, while on most civet SARS-like viral isolates, it is a lysine. Lys479 would have steric and electrostatic interference with residues on the N-terminal helix of human ACE2 such as His34. A K479N mutation would remove an unfavorable interaction at the interface and

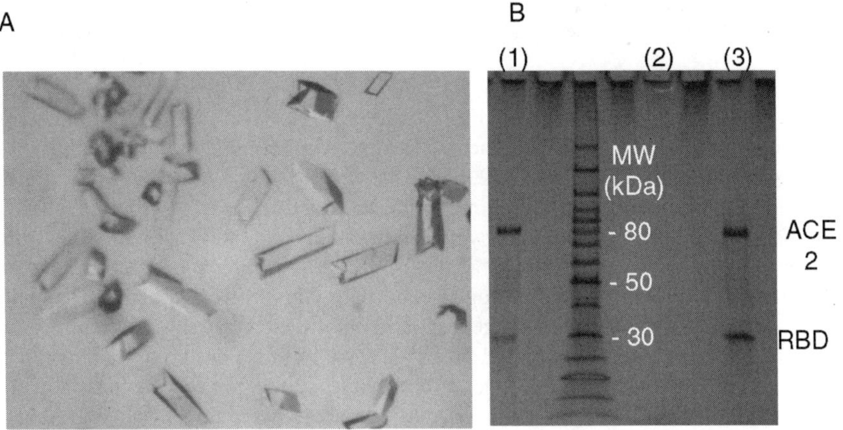

Figure 2. Crystallization of the ACE2/RBD complex. (A) Crystals of the ACE2/RBD complex were grown at room temperature from a mother liquor containing 24% PEG6000, 150 mM NaCl, 100 mM Tris pH 8.2, and 10% ethylene glycol. (B) Silver-stained reducing SDS-PAGE. Lane (1) is the ACE2/RBD complex purified by gel filtration chromatography as in Figure 1. Lane (2) is the crystal wash buffer from the last round. Lane (3) is dissolved crystal after several rounds of washes.

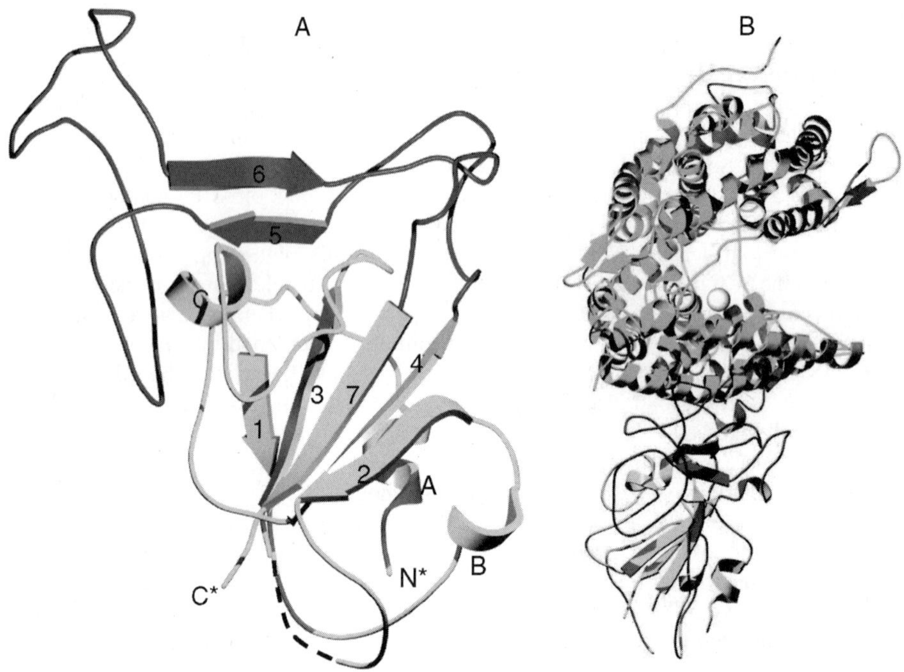

Figure 3. Crystal structure of the ACE2/RBD complex. (A) Structure of the RBD that contains two subdomains: a core structure (in cyan) and a receptor-binding motif (in red). (B) Structure of the ACE2/RBD complex. The RBD binds to the outer surface of the N-terminal lobe of ACE2 (in green). (See color plate).

enhance the binding affinity of the virus to its receptor. Hence the K479N mutation is a critical step for SARS to cross the species barrier to infect humans. In all human SARS-CoV sequences from the year 2002–2003 SARS epidemic, 487 is a threonine; in all civet SARS-like viral sequences, it is a serine. In the structure, the methyl group of Thr487 lies in a hydrophobic pocket bounded by the side chains of Tyr41 and Lys353 from the receptor. Lys353 on the receptor is at the center of a complex interaction network. It forms a main chain-main chain hydrogen bond with Gly488 from the virus, its charge is neutralized by Asp38 from the receptor, and its side chain is sandwiched between Tyr41 from the receptor and Tyr491 from the virus. Thus a serine at 487 would leave a hole in this tight hydrophobic pocket and decrease the binding affinity. Unlike the previous year, 2003–2004 saw no human to human transmission of SARS. Sequences from the second year have a serine at 487. It appears that the methyl group on the 487 side chain is a key factor in determining the severity of SARS and potentially viral transmissibility from human to human. Isolates from the 2002–2003 SARS epidemic all have a leucine at 472, but those from the second year have a proline. In the crystal structure, Leu472 forms a hydrophobic interaction with Met82 from the receptor (Figure 4). So L472P could be another attenuation mutation for SARS-CoV, besides T487S.

The crystal structure allows us to inspect and examine the evolutionary relationship between SARS-CoV and potential animal hosts. Rat ACE2 does not support SARS-CoV

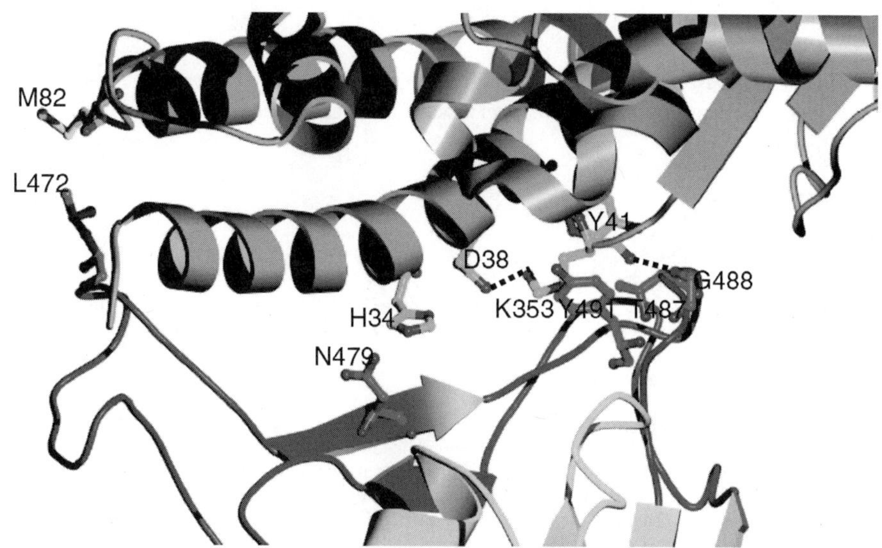

Figure 4. Residues located at the ACE2/RBD interface and important to the species specificity of SARS-CoV. Leu472 on the RBD has a hydrophobic interaction with Met82 on ACE2. L472P mutation may attenuate the virus. On rat ACE2, residue 82 is glycosylated, preventing the binding of SARS-CoV. A K479N mutation on the RBD is critical for SARS-CoV to jump from civets to humans. Thr487 on the RBD forms a hydrophobic interaction with Lys353 on ACE2. A S487T mutation on the RBD is important for SARS-CoV to transmit from human to human. On both rat ACE2 and mouse ACE2, residue 353 is a histidine, disfavoring the binding of SARS-CoV. (See color plate).

infection for two reasons. First, rat ACE2 has a histidine at 353, and it is thus unable to form the same interaction network at the interface as does Lys353. Second, rat ACE2 has an asparagine at 82, introducing a glycosylation site. A glycan at this position would have steric interference with the viral RBD. Mouse cells can be infected by SARS-CoV at low levels, probably because mouse ACE2 contains a histidine at 353 but does not have the glycan at 82. In fact, a single H353K mutation greatly enhances both binding affinity and viral infectivity.[9]

The structure provides insights into antiviral strategies. The RBD is sufficient to elicit neutralizing antibodies against the virus[10,11] and thus could be used in subunit vaccines. To date, at least two neutralizing antibodies are known to recognize epitopes on the base of the RBM.[11,12] Indeed, the structural properties of the RBM, including the relative flatness of the binding interface, the conservation in sequence, and the lack of glycosylation, suggest that immunization with the RBD could be a route to protective immunity.

In summary, the crystal structure of SARS-CoV spike RBD in complex with ACE2 has revealed detailed interactions between the virus and its receptor. Analysis of these interactions uncovers important aspects of the invasion mechanisms of SARS-CoV. It sheds light on the origination and severity of the SARS epidemic and can guide future antiviral studies. The approach we used to determine the crystal structure may be extended to study the interactions between other coronaviruses and their cellular receptors.

3. ACKNOWLEDGMENTS

This work was supported by NIH grants CA13202 (to S.C.H.) and AI061601 (to M.R.F.); S.C.H. is an investigator in the Howard Hughes Medical Institute.

4. REFERENCES

1. Lai, M. M. C., and Holmes, K. V., in: *Fields Virology*, D. M. Knipe, and P. M. Howley, eds. (Lippincott Williams & Wilkins, New York, 2001), Chap. 35.
2. Wong, S. K., Li, W., Moore, M. J., Choe, H., Farzan, M., 2004, A 193-amino acid fragment of the SARS coronavirus S protein efficiently binds angiotensin-converting enzyme 2, *J. Biol. Chem.* **279**:3197-3201.
3. Babcock, G. J., Esshaki, D. J., Thomas, W. D. Jr, Ambrosino, D. M., 2004, Amino acids 270 to 510 of the severe acute respiratory syndrome coronavirus spike protein are required for interaction with receptor, *J. Virol.* **78**:4552-4560.
4. Li, W., Moore, M. J., Vasilieva, N., Sui, J., Wong, S. K., et al., 2003, Angiotensin-converting enzyme 2 is a functional receptor for the SARS coronavirus, *Nature* **426**:450-454.
5. Towler, P., Staker, B., Prasad, S. G., Menon, S., Tang, J., et al., 2004, ACE2 X-ray structures reveal a large hinge-bending motion important for inhibitor binding and catalysis, *J. Biol. Chem.* **279**:17996-18007.
6. Song H. D., Tu, C. C., Zhang, G. W., Wang, S. Y., Zheng, K., et al., 2005, Cross-host evolution of severe acute respiratory syndrome coronavirus in palm civet and human, *Proc. Natl. Acad. Sci. USA* **102**:2430-2435.
7. Li, W., Zhang, C., Sui, J., Kuhn, J. H., Moore, M. J., et al., 2005, Receptor and viral determinants of SARS-coronavirus adaptation to human ACE2, *EMBO J.* **24**:1634-1643.
8. Li, F., Li, W., Farzan, M., and Harrison, S. C., 2005, Structure of SARS coronavirus spike receptor-binding domain complexed with receptor, *Science*. **309**:1864-1868.
9. Li, W., Greenough, T. C., Moore, M. J., Vasilieva, N., Somasundaran, M., et al., 2004, Efficient replication of severe acute respiratory syndrome coronavirus in mouse cells is limited by murine angiotensin-converting enzyme 2, *J. Virol.* **78**:11429-11433.
10. Sui, J., Li, W., Murakami, A., Tamin, A., Matthews, L. J., et al., 2004, Potent neutralization of severe acute respiratory syndrome (SARS) coronavirus by a human mAb to S1 protein that blocks receptor association, *Proc. Natl. Acad. Sci. USA* **101**:2536-2541.
11. van den Brink, E. N., Ter Meulen, J., Cox, F., Jongeneelen, M. A., Thijsse, A., et al., 2005, Molecular and biological characterization of human monoclonal antibodies binding to the spike and nucleocapsid proteins of severe acute respiratory syndrome coronavirus, *J. Virol.* **79**:1635-1644.
12. Sui, J., Li, W., Roberts, A., Matthews, L. J., Murakami, A., et al., 2005, Evaluation of human monoclonal antibody 80R for immunoprophylaxis of Severe Acute Respiratory Syndrome by an animal study, epitope mapping, and analysis of spike variants, *J. Virol.* **79**:5900-5906.

PROTEOLYSIS OF SARS-ASSOCIATED CORONAVIRUS SPIKE GLYCOPROTEIN

Graham Simmons, Andrew J. Rennekamp and Paul Bates*

1. INTRODUCTION

Severe acute respiratory syndrome-associated coronavirus (SARS-CoV) mediates attachment, receptor engagement and entry via its spike glycoprotein (S). S-dependent viral entry requires the presence of a primary receptor, angiotensin converting enzyme-2 (ACE2),[1] while the C-type lectins, DC-SIGN, DC-SIGNR, and LSECtin act as attachment factors, promoting binding to a subset of target cells.[2-5]

SARS-CoV S, like other coronavirus spike glycoproteins, contains two leucine/ isoleucine heptad repeats (HR1 and HR2) located in the C-terminal third of the glycoprotein.[6-8] By homology to other fusion proteins, such as influenza hemagglutinin (HA) and HIV gp160, HR1 and HR2 are thought to drive membrane fusion through their interaction together. HR1 and HR2 from each member of the trimeric spike pack together to form a highly stable structure known as the six-helix bundle. The events following receptor engagement and leading up to six-helix bundle formation are less defined.

Viral entry mediated by SARS-CoV S is exquisitely sensitive to compounds able to raise the pH of cellular endosomes.[2,9,10] Thus, a pH-dependent component appears to be required for cell-free viral infection mediated by S. However, S-dependent cell-to-cell fusion can occur at neutral pH,[1] and furthermore fusion is not enhanced by acidic pH.[9] Therefore, it appears unlikely that S has a direct requirement for acidic conditions in order to undergo the conformational rearrangements required for six-helix bundle formation, as is seen for other pH-dependent fusion proteins such as HA. In contrast, when cells expressing endogenous, low levels of ACE2 were used as targets for cell-to-cell fusion, treatment of the S glycoprotein expressing effectors with trypsin is necessary in order for efficient fusion to be observed.[9] Thus, we hypothesize that in order for efficient cell-free virus infection to occur, similar proteolytic activity is required, and that trypsin mimics the activity of a pH-dependent cellular protease, hence explaining the sensitivity to compounds able to raise endosomal pH.

* A.J. Rennekamp and P. Bates, University of Pennsylvania School of Medicine, Philadelphia, Pennsylvania 19104. G. Simmons, Blood Sytems Research Institute, San Francisco, CA 94118.

2. TRYPSIN - MEDIATED PROTEOLYSIS OF S PROTEIN

The precise requirement and role of glycoprotein processing for coronavirus entry and membrane fusion is not well defined. Many coronaviruses, such as mouse hepatitis virus (MHV), contain a furin cleavage site within S that yields S1 and S2 subunits.[10,11] This site, however, is not absolutely required for infection, although lack of cleavage lowers S-mediated cell-cell fusion.[11] In contrast, SARS-CoV S, both when overexpressed in cells, and on mature virions, is predominantly in a full-length, unprocessed form (Fig. 1A, left-hand lane). However, cell-to-cell fusion mediated by SARS-CoV S was enhanced by pretreatment of effector cells with trypsin.[9] Thus, the role of proteolysis of SARS-CoV S by trypsin-like proteases in cell-free virus infection was examined.

2.1. Trypsin Treatment of Cell-Free Virions

Lentiviral-based pseudovirions incorporating SARS-CoV S into their lipid coats[9] were produced in 293T cells [virions termed HIV(SARS-CoV S)]. Analysis revealed that following pretreatment with trypsin, a C-terminal fragment of approximately 100 kDa is detectable using polyclonal sera raised against the C-terminal extracellular portion of S (Fig. 1A, middle lane). A fragment of this size would be expected following processing at the predicted S1/S2 boundary.[1] However, rather than enhancing titers as might be predicted from processing of other viral fusion proteins such as influenza HA,[12] pre-treatment of HIV(SARS-CoV S) with trypsin led to a 95% reduction in infectivity (Fig. 1B). Mutagenesis suggested that trypsin-mediated cleavage of S did indeed occur at basic residues around the predicted S1/S2 boundary, although alteration of these sites did not dramatically alter infectivity titers (data not shown).

2.2. Trypsin-Mediated Enhancement of Infection by Cell-Bound Virions

We hypothesize that trypsin proteolysis can mimic the action of an endosomal protease. Thus, it is likely that ACE2 engagement at the cell surface occurs prior to proteolysis in the endosome. Therefore, the trypsin activation was attempted following attachment of virus to the cell surface, rather than in solution.

As previously demonstrated,[2,9,10] pretreatment of cells with agents able to raise the pH of endosomes, such as ammonium chloride, dramatically reduces infection mediated by SARS-CoV S (Fig. 1C). However, if HIV(SARS-CoV S) particles are first bound to the cell surface at 4°C, then trypsin activated, infection occurred even in the presence of ammonium chloride (Fig. 1C). Similar results were obtained using live, replication competent SARS-CoV virus infection of Vero E6 cells.[13] Thus, trypsin treatment of S not only enhances its ability to mediate membrane fusion, but also relieves it of a requirement for acid pH during the viral entry process.

S lacking basic residues at the S1/S2 boundary (arg667 and lys672, both mutated to ala) is no longer processed by trypsin to give a 100 - kDa C-terminal fragment (data not shown). However, these mutants can still be activated by trypsin cleavage at the cell surface to overcome the block to infection mediated by ammonium chloride (data not shown). Thus, it is unlikely that the proteolytic processing mediated by trypsin that leads to activation of infectivity occurs at the S1/S2 boundary, but rather at a distinct site.

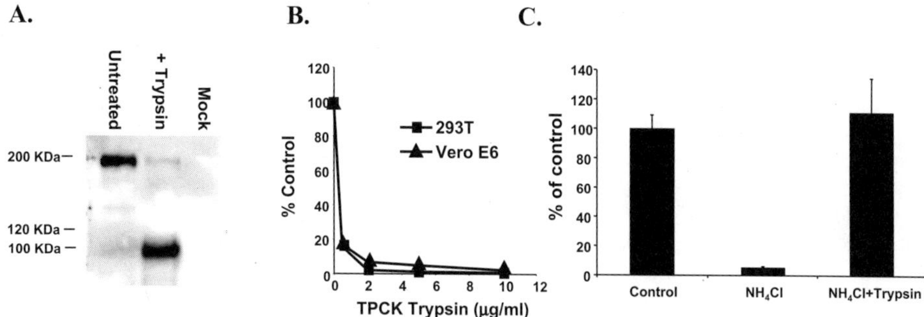

Figure 1. Effect of trypsin on SARS-CoV S. (A) Ultracentrifuge concentrated HIV(SARS-CoV S) pseudovirions were incubated with TPCK-trypsin (10 µg/ml) for 15 minutes at 25°C and analyzed by SDS-PAGE. (B) HIV(SARS-CoV S) encoding luciferase was incubated with trypsin, and used to challenge 293T or VeroE6 cells. After 48 hours, cells were analyzed for luciferase activity. Results are presented as a percentage of no trypsin control. (C) Vero E6 cells, pretreated with PBS (Control) or 20 mM ammonium chloride (NH_4Cl), were incubated with HIV(SARS-CoV S) at 4°C to allow binding, but not entry. Cells were incubated with serum-free medium at 37°C for 10 minutes and treated with trypsin (NH_4Cl+Trypsin). Cells were assayed as in (B). Results are presented as a percentage of no NH_4Cl, no trypsin control.

3. EFFECTS OF PROTEASE INHIBITORS ON SARS-CoV ENTRY

The ability of trypsin treatment to overcome ammonium chloride inhibition of S-mediated infection suggests that lysosomotropic agents may prevent proteolysis of S by pH-dependent, endosomal proteases. Thus, the effects of various protease inhibitors on HIV(SARS-CoV S), as well as live SARS-CoV, infection were examined in detail.

Leupeptin is an inhibitor of both serine and cysteine proteases, while aprotinin and E64c specifically inhibit serine and cysteine proteases, respectively. Pepstatin is an aspartate protease inhibitor. Entry of HIV pseudovirions mediated by SARS-CoV S was efficiently blocked by both leupeptin and E64c, with IC_{95}'s of 15.2 and 8.2 µM, respectively (Table 1). Infection mediated by VSV-G, a pH-dependent viral membrane fusion protein was not inhibited by either leupeptin nor E64c (Table 1). Likewise, pretreatment of target cells with either pepstatin or aprotinin had no effect on infection mediated by either of the tested envelopes (Table 1).

Infection of 293T cells transiently expressing ACE2 by replication-competent SARS-CoV was also inhibited by leupeptin.[13] Similarly to the ammonium chloride results (Fig. 1C), trypsin treatment of virions bound at the cell surface overcame the block to infection mediated by pretreatment with leupeptin.[13]

The specificity of endosomal cysteine proteases in mediating SARS-CoV infection was further examined using more specific inhibitors of cysteine proteases. Z-leu-leu-leu-fluoromethyl ketone (Z-lll-FMK), an inhibitor of both cathepsin B and L efficiently inhibited infection of HIV(SARS-CoV S), but not HIV(VSV-G). In contrast, CA-074, a selective inhibitor of cathepsin B, did not dramatically affect infection of either pseudovirus (Table 1). These results suggest that cathepsin L, but not cathepsin B, plays a critical role in SARS-CoV S–mediated entry.

Table 1. Inhibition of S-mediated entry by protease inhibitors.

Inhibitor	Target proteases	IC_{95} (μM)	
		HIV(SARS-CoV S)	HIV(VSV-G)
Leupeptin	Serine, cysteine	15.2	>200
Pepstatin	Aspartate	>200	>200
E64c	Cysteine	8.2	>200
Aprotinin	Serine	>200	>200
Z-lll-FMK[a]	Cathepsin B and L	3.5	>200
CA-074	Cathepsin B	>200	>200

[a] Z-lll-FMK, Z-leu-leu-leu—fluoromethyl ketone.

A panel of relatively specific inhibitors of cathepsin L activity, Z-Phe-Phe-CH_2F, Z-Phe-Tyr-CHO, Z-Phe-Tyr-(t-Bu)-CHN_2, and 1-naphthalenesulfonyl-Ile-Trp-CHO, were used to determine the role of cathepsin L in viral entry. Indeed, all four compounds potently inhibited HIV(SARS-CoV S) pseudotype infection (Fig. 2). In contrast, these inhibitors were found to have no effect on HIV(VSV-G) entry.[13] An inhibitor of cathepsin K (Boc-Phe-Leu-NHNH-CO-NHNH-Leu-Z) was significantly less effective at inhibiting SARS-CoV S-mediated entry, although at high concentrations some inhibition was noted (Fig. 2A). Whether this is due to cross inhibition of cathepsin L at high inhibitor concentrations, or suggestive of cathepsin K playing a minor role in S activation, is unclear.

4. *IN VITRO* PROTEOLYSIS OF S PROTEIN

In order to directly address the role of cathepsin L, a novel intervirion fusion assay was established in order to allow the study of S-mediated membrane fusion in a cell-free environment. Lentiviral pseudovirions can be prepared incorporating either SARS-CoV S, or its receptor, ACE2.[13] The two sets of particles are then mixed in the presence or absence of proteases in order to study the requirements for activation of S glycoprotein's membrane fusion potential. In order to quantify intervirion fusion, the ACE2 bearing virions package luciferase as a reporter gene, while the SARS-CoV S enveloped particles lack a reporter gene, but co-incorporate the envelope glycoprotein from subgroup A avian sarcoma and leukosis virus (ASLV-A env). Following mixing of the two viral populations, the particles are plated on HeLa cells that lack ACE2, but stably express the ASLV-A receptor, Tva. Thus, ASLV-A env will mediate infection of these cells by the dual ASLV-A/SARS-CoV S particles. Only if S has mediated intervirion fusion with ACE2 expressing particles will the cells be transduced by the luciferase reporter gene encoded by the ACE2 particles. Hence, luciferase activity acts as a direct measure of intervirion fusion. In addition, in order to prevent uptake of mixed virions into endosomes and subsequent activation of S by endogenous proteases, the target cells were pretreated with leupeptin. As a consequence, mixing of the two populations of virions led to no resultant luciferase activity in the HeLa cells (Fig. 2B). In contrast, if, after mixing the two populations of virions, samples were treated with either trypsin or recombinant cathepsin L, efficient transduction of target cells by the luciferase reporter gene was observed (Fig.2B). As predicted from the lack of inhibition by specific cathepsin B inhibitors, recombinant cathepsin B showed no enhancement of intervirion fusion (Fig. 2B).

Enhancement of intervirion fusion by cathepsin L was found to be pH-dependent, with the highest levels of fusion seen at a pH of 5, or below.[13] In contrast, little, or no intervirion fusion was observed with cathepsin L at neutral pH.[13] Conversely, trypsin most efficiently gave intervirion fusion at neutral pH (data not shown), arguing against pH having a direct effect on S. Thus, the apparent requirement for cathepsin L activity for efficient S-mediated entry explains the sensitivity to compounds such as ammonium chloride that raise endosomal pH.

Due to the requirement for binding to receptor-positive cells before treatment by trypsin (Fig. 1), we hypothesized that receptor interactions may be required prior to proteolysis. Indeed, in the intervirion assay, following mixing of HIV(SARS-CoV S/ASLV-A env) with HIV-luc(ACE2) virions, an incubation step at elevated temperatures was required in order for proteolysis to enhance fusion.[13] If virions were incubated at 4°C prior to trypsin treatment, no intervirion fusion was observed.[13] These results suggest that conformational changes induced by binding to ACE2 may indeed be required prior to proteolysis.

5. DISCUSSION

Overall, these experiments suggest a new paradigm for viral entry into target cells. Namely, that for SARS-CoV S, receptor-mediated conformational changes induce exposure of a cryptic cleavage site within the viral envelope glycoprotein. Cleavage at this site by pH-dependent cellular proteases is then necessary in order to fully activate the S protein's membrane fusion potential. Further characterization of this phenomena is likely to highlight steps in the activation of S that may yield targets for specific inhibitors of entry. Indeed, the finding that cathepsin L is an important activating protease for SARS infection suggests cellular proteases as a target for therapeutic intervention. The entry process described here for SARS-CoV S protein also raises the question whether

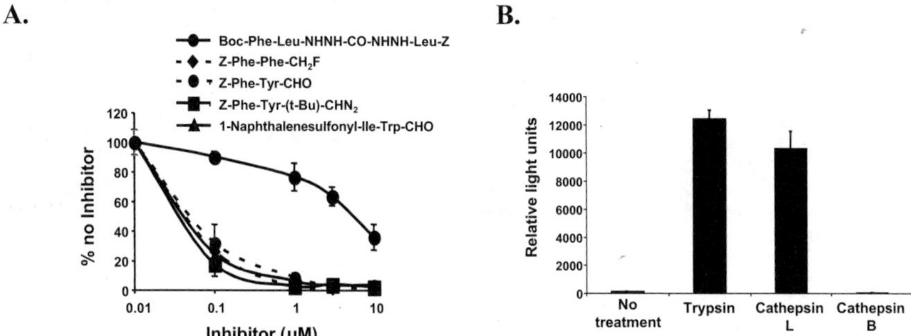

Figure 2. Role of cathepsin L. (A) Inhibition of HIV(SARS-CoV S) pseudovirions by cathepsin inhibitors. (B) Activation of intervirion fusion by protease treatment. HIV-luc(ACE2) and HIV(SARS-CoV S/ASLV-A env) particles were mixed, pre-incubated at 37°C and then treated with PBS, trypsin (10 μg/ml), or cathepsin L or B (2 μg/ml) for 10 minutes at 25°C. Proteases were inactivated by leupeptin (20 μg/ml), incubated at 37°C for 30 minutes to allow membrane fusion and plated on HeLa/Tva cells pretreated with leupeptin (20 μg/ml).

other classically defined pH-dependent viruses display this dependence due to a requirement for acidic protease activation and not pH-induced structural rearrangements as is commonly assumed.

This work was by the NIH Mid Atlantic Regional Center for Biodefense and Emerging Infectious Diseases Grant U54 AI057168 and NIH grants R01 AI43455, R21 AI059172, and R21 AI058701.

6. REFERENCES

1. W. Li, M. J. Moore, N. Vasilieva, J. Sui, S. K. Wong, M. A. Berne, M. Somasundaran, J. L. Sullivan, K. Luzuriaga, T. C. Greenough, H. Choe, and M. Farzan, Angiotensin-converting enzyme 2 is a functional receptor for the SARS coronavirus, *Nature* **426**, 450-454 (2003).
2. Z. Y. Yang, Y. Huang, L. Ganesh, K. Leung, W. P. Kong, O. Schwartz, K. Subbarao, and G. J. Nabel, pH-Dependent entry of SARS-CoV is mediated by the spike glycoprotein and enhanced by dendritic cell transfer through DC-SIGN, *J. Virol.* **78**, 5642-5650 (2004).
3. A. Marzi, T. Gramberg, G. Simmons, P. Moller, A. Rennekamp, M. Krumbiegel, M. Geier, J. Eisemann, N. Turza, B. Saunier, A. Steinkasserer, S. Becker, P. Bates, H. Hofmann, and S. Pohlmann, DC-SIGN and DC-SIGNR interact with Marburg virus and the S protein of SARS-CoV, *J. Virol.* **78**, 12090-12095 (2004).
4. S. Jeffers, S. Tusell, L. Gillim-Ross, E. Hemmila, J. Achenbach, G. Babcock, W. Thomas, Jr., L. Thackray, M. Young, R. Mason, D. Ambrosino, D. Wentworth, J. Demartini, and K. Holmes, CD209L (L-SIGN) is a receptor for SARS-CoV, *Proc. Natl. Acad. Sci. USA* **101**, 15748-15753 (2004).
5. T. Gramberg, H. Hofmann, P. Moller, P. F. Lalor, A. Marzi, M. Geier, M. Krumbiegel, T. Winkler, F. Kirchhoff, D. H. Adams, S. Becker, J. Munch, and S. Pohlmann, LSECtin interacts with filovirus glycoproteins and the spike protein of SARS coronavirus, *Virology* **340**, 224-236 (2005).
6. B. J. Bosch, B. E. Martina, R. Van Der Zee, J. Lepault, B. J. Haijema, C. Versluis, A. J. Heck, R. De Groot, A. D. Osterhaus, and P. J. Rottier, Severe acute respiratory syndrome coronavirus infection inhibition using spike protein heptad repeat-derived peptides, *Proc. Natl. Acad. Sci. USA* **101**, 8455-8460 (2004).
7. S. Liu, G. Xiao, Y. Chen, Y. He, J. Niu, C. R. Escalante, H. Xiong, J. Farmar, A. K. Debnath, P. Tien, and S. Jiang, Interaction between heptad repeat 1 and 2 regions in spike protein of SARS-CoV: implications for virus fusogenic mechanism and identification of fusion inhibitors, *Lancet* **363**, 938-947 (2004).
8. B. Tripet, M. W. Howard, M. Jobling, R. K. Holmes, K. V. Holmes, and R. S. Hodges, Structural characterization of the SARS-coronavirus spike S fusion protein core, *J. Biol. Chem.* **279**, 20836-20849 (2004).
9. G. Simmons, J. D. Reeves, A. J. Rennekamp, S. M. Amberg, A. J. Piefer, and P. Bates, Characterization of SARS-CoV spike glycoprotein-mediated viral entry, *Proc. Natl. Acad. Sci. USA* **101**, 4240-4245 (2004).
10. M. Frana, J. Behnke, L. Sturman, and K. Holmes, Proteolytic cleavage of the E2 glycoprotein of murine coronavirus: host-dependent differences in proteolytic cleavage and cell fusion, *J. Virol.* **56**, 912-920 (1985).
11. C. De Haan, K. Stadler, G. Godeke, B. Bosch, and P. Rottier, Cleavage inhibition of the murine coronavirus spike protein by a furin-like enzyme affects cell-cell but not virus-cell fusion, *J. Virol.* **78**, 6048-6054 (2004).
12. H. D. Klenk, R. Rott, M. Orlich, and J. Blodorn, Activation of influenza A viruses by trypsin treatment, *Virology* **68**, 426-439 (1975).
13. G. Simmons, D. N. Gosalia, A. J. Rennekamp, J. D. Reeves, S. L. Diamond, and P. Bates, Inhibitors of cathepsin L prevent severe acute respiratory syndrome coronavirus entry, *Proc. Natl. Acad. Sci. USA* **102**, 11876-11881 (2005).

FLUORESCENCE DEQUENCHING ASSAYS OF CORONAVIRUS FUSION

Victor C. Chu, Lisa J. McElroy, Beverley E. Bauman, and Gary R. Whittaker*

1. INTRODUCTION

For all enveloped viruses, a critical event during entry into cells is the fusion of the viral envelope with the membrane of the host cell.[1, 2] Our current understanding of viral fusion has been driven by fundamental problems first solved with influenza hemagglutinin (HA).[3] Whereas the trigger for HA-mediated fusion is the low pH of the endosome, other viruses (e.g., paramyxoviruses and most retroviruses) undergo a receptor-primed fusion with the plasma membrane at neutral pH.[1] In the case of coronaviruses, however, there is little consensus as to whether virus entry and fusion occur following endocytosis or at the plasma membrane.[4, 5]

The coronavirus spike protein (S) is a primary determinant of cell tropism and pathogenesis, being responsible (and apparently sufficient) for receptor binding and fusion.[6] The S protein is categorized as a class I fusion protein, based on the presence of characteristic heptad repeats[7-9]; as such it shows features of the fusion proteins of influenza virus (HA), retroviruses (Env), and paramyxoviruses (F), for which there is extensive characterization at structural and biophysical level.[10] Although class I fusion proteins share similar structural features, they can have quite different biological properties; i.e., they can be triggered for fusion by low pH or by receptor interaction. Receptor-induced conformational changes have been described for several coronaviruses, and the virus has generally been considered to exhibit a neutral or slightly alkaline pH optimum.[5] However, these fusion data are principally based on cell–cell fusion assays with S-expressing cells and may not recapitulate the fusion event that takes place during virus entry. Indeed, despite being considered to be pH-independent for fusion, there is increasing evidence that coronavirus entry is a low pH-dependent process as infection is sensitive to endosome neutralization.[11]

A powerful means of analyzing membrane fusion is the application of fluorescence assays. These techniques offer a number of advantages, including high sensitivity, relative ease in obtaining quantitative data, and the possibility of monitoring fusion by

*Cornell University, Ithaca, New York 14843.

either spectrofluorimetry or fluorescence microscopy.[12] In particular, an assay that has found wide application in studies of virus fusion is that based on relief of fluorescence self-quenching.[13, 14] Fusion of many different viruses has been studied with great effect using this technique; these include influenza virus, Sendai virus, vesicular stomatitis virus, and avian leukosis virus, among others. The assay involves the exogenous insertion of a fluorescent probe, typically octadecyl-rhodamine B chloride (R18), into the viral envelope by briefly incubating a virus suspension with an ethanolic solution (<1% v/v) of the probe. The concentration of the probe is such that it will cause efficient quenching of fluorescence when inserted into the lipid bilayer of the virus, yet when viruses fuse with nonlabeled target membrane the probe becomes diluted and its surface density decreases. A concomitant increase in fluorescence is observed, which increases proportionally with fusion progression, allowing kinetic and quantitative measurements of fusion to be made.[15] R18 is by far the most widely used probe used for FdQ studies. Other fluorescent probes with self-quenching properties, e.g., DiO, DiI, etc., may be used for fusion studies, but these have generally not found wide acceptance. Recently however, DiD has been used with notable impact in an analysis of influenza virus fusion by single-particle tracking,[16] and the use of alternative probes with different fluorescent properties (e.g., resistance to photobleaching, use in double-label experiments, etc.) may become more accepted in the future.

Avian infectious bronchitis virus (IBV) is a coronavirus that can be isolated, purified and labeled appropriately for molecular studies of virus fusion. Here, we examined coronavirus–cell fusion using fluorescence dequenching (FdQ) assays of octadecyl rhodamine (R18)-labeled viruses with host cells. We used a pathogenic strain of IBV (Massachusetts 41), in combination with primary chick kidney (CK) cells.

2. METHODS

2.1. Virus Purification

IBV (strain Massachusetts 41) was obtained from Dr. Benjamin Lucio-Martinez, Unit of Avian Health, Cornell University and propagated in 11-day-old embryonated chicken eggs. Virus was harvested from the allantoic fluid after 48 h of infection and purified on a sucrose gradient prior to labeling for FdQ studies.

2.2. Fluorescence Dequenching (FdQ) Fusion Assay

Fusion assays were based on fluorescence dequenching of octadecyl rhodamine (R18)-labeled virus.[13, 17] Typically, 100 μl of purified virus (2 mg/ml) was labeled by the addition of 1 μl of 1.7 mM octadecyl-rhodamine B chloride (R18) (Molecular Probes) and the mixture was incubated in the dark on a rotary shaker at room temperature for 60 min. Excess dye removed with a Sephadex G25 column (Pharmacia). Under such labeling and purification conditions there was no significant drop in virus infectivity (data not shown). Fifteen microliters of labeled virus (approximately 5 pfu/cell) was bound to 1.5×10^6 cells at 4°C for 1 h in binding buffer (RPMI1640 medium containing with 0.2% BSA, pH 6.8). Unbound virus was removed by washing with binding buffer and cells were resuspended in fusion buffer (5 mM HEPES, 5mM MES, 5 mM succinate, 150 mM

NaCl (HMSS) buffer, pH 7.0, 15 μM monensin) at 37°C. Fusion of IBV with the cell membrane was triggered by adding a pre-titrated amount of 250 mM HCl to obtain a final pH of between 5.0 and 7.0. FdQ was measured using a QM-6SE spectrofluorimeter (Photon Technology International), with excitation and emission wavelengths set to 560 nm and 590 nm respectively. Fusion efficiency was determined following addition of Triton X-100 (final concentration 1%) to obtain 100% dequenching.

2.3. Preparation of Primary Chick Kidney (CK) Cells

SPF White Leghorn Chicks (11–14 days of age) were placed in a CO_2 chamber for an appropriate amount of time such that the chick expires, but not long enough for large amounts of individual cell necrosis (typically 5 min). The chick was then placed on a clean surface. After rinsing down with water to dampen feathers (to reduce dust), the skin was opened. Using a new pair of sterile scissors, the abdomen muscle was opened and kidneys removed from each side of the chick. Kidney tissue was placed in 25–50 ml sterile PBS and the container shaken gently to remove clots and red blood cells. The supernatant containing the cells was removed by carefully decanting, and cells rinsed a second time with an equivalent volume of sterile PBS. 25 ml trypsin/EDTA was added and allowed to rinse/digest for approx. 5 min with a stir bar on a stir plate (or by hand-swirling). The trypsin/EDTA was decanted or aspirated and a further 25 ml more trypsin/EDTA added. This was then allowed to digest for 10–15 min with a stir bar on stir plate (or by hand-swirling). The supernatant was poured through sterile cheesecloth into sterile beaker and the trypsin/EDTA digest repeated 1–2 more times until all chunks of tissue were digested. Ten to 15 ml calf serum was then added to neutralize the trypsin/EDTA and the neutralized supernatant placed into a 50 ml Falcon tubes and centrifuged at 1,000 rpm for 2 min. Cells were resuspended in 25 ml M20 media and counted on a hemocytometer. Cells were adjusted to a concentration of 1–1.5 x 10^6/ml with M25 media containing 5% FBS.

3. RESULTS

R18-labeled virus was bound to the surface of CK cells at 4°C and shifted to 37°C in fusion buffer (pH 7.0) in the presence of monensin to prevent any entry from acidic endosomes. Even after a significant time period at 37°C (400 s), we saw little or no dequenching of virus signal that would indicate virus–cell fusion at neutral pH (Fig. 1). Upon addition of Triton X-100, extensive dequenching occurred showing that the virus binding had occurred and the virions were labeled appropriately. This indicated that the lack of dequenching was due to a lack of fusion activity at pH 7.0. As IBV appeared to be unable to fuse with cells at neutral pH, we wished to determine if coronavirus fusion was pH dependent. At pH 6.0 and above dequenching was negligible, however at pH 5.75 limited dequenching was apparent. At pH 5.5 and pH 5.25, high levels of dequenching were observable, which were maximal at pH 5.0. At pH 5.0, the overall extent of fusion typically reached between 40% and 60% of that in the presence of Triton X-100, with little or no appreciable lag time after pH change. Below pH 5.0, the fusion reaction was unstable and calibration was not possible (not shown).

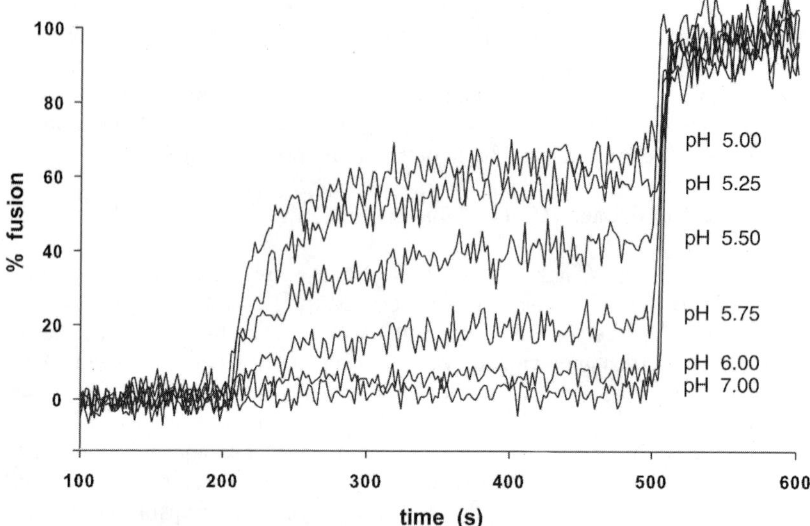

Figure 1. R18-labeled IBV (Massachusetts 41) was bound to CK cells at 4°C and samples added to a spectrofluorimeter cuvette in pH 7.0 buffer maintained at 37°C (t = 0). At t = 200 s, the pH was reduced to between 6.0 and 5.0, or was maintained at pH 7.0, and samples were monitored for fluorescence dequenching at 37°C, before addition of 1% Triton X-100 (final concentration) at t = 500 s to obtain complete (100%) dequenching.

To define a pH threshold for fusion, we calculated the initial rate of fusion between pH 7.0 and 5.0 (Fig. 2). Typically, we did not see an abrupt threshold for low-pH activated IBV fusion, as would be expected for influenza virus,[18] but a more gradual increase in fusion activity between pH 6.0 and 5.0. In our FdQ system, the half maximal pH ($pH_{1/2}$) at which IBV fusion occurred was approximately 5.6.

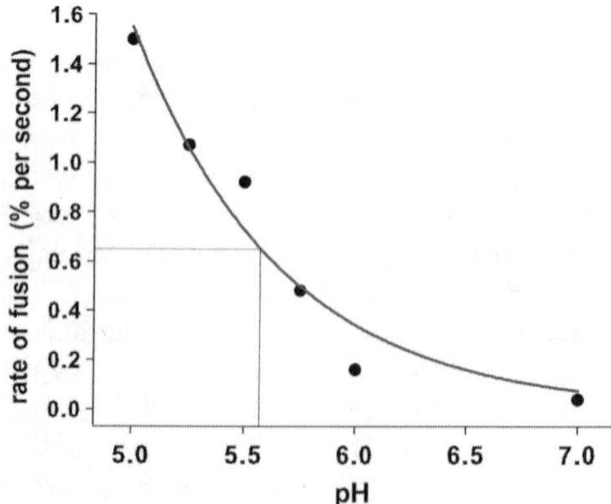

Figure 2. The initial rate of fusion (as obtained from data in Figure 1) was analyzed by 4-parameter, exponential decay and is plotted against pH. The pH at which the initial rate of IBV fusion was half maximal ($pH_{1/2}$) is shown.

4. DISCUSSION

Using an established assay of virus–cell fusion, we show here fusion of the coronavirus IBV with host cells does not occur at neutral pH, and that fusion activation is a low pH-dependent process. How then to rationalize the substantial data showing a neutral pH coronavirus fusion reaction (at least for cell–cell fusion), with our own data that clearly show activation of virus–cell fusion at pH 5.5? We consider one likely explanation is that the coronavirus S protein has a reversible fusion trigger, as is the case for VSV.[19] With VSV infection, syncytia can form even with a fusion protein that is clearly triggered by low pH.[20] In a similar fashion, a fraction of the IBV S protein that is expressed at the cell surface may transiently attain a fusion-competent state during maturation and delivery, allowing cell–cell fusion at the plasma membrane. While the pH of the Golgi is only mildly acidic, the pH of secretory vesicles can be as low at pH 5.5,[21] and would be low enough to activate fusion in such a model.

Biophysical measurements of viral fusion require relatively large amounts of pure virus preparations. Contamination of the virus preparation with inactive particles or cellular debris complicates the analysis by nonspecific dequenching of the probe. In order to achieve self-quenching, high concentrations of lipid probe need to be incorporated into the membrane, which may produce microcrystals that can also cause nonspecific dequenching. The major disadvantage of FdQ assays, therefore, is the nonspecific redistribution of the probe, which becomes a significant problem when the time of incubation is long.[22] We have performed extensive analysis of IBV and find no evidence for significant dequenching over the time course of our experiments (not shown).

The early events in coronavirus–cell interactions can be difficult to study, in part because of the tendency of S1 to detach from the virions.[23] To ensure that we were using intact virus, we analyzed the relative ratio of S1 to S2 of R18-labeled virions by ELISA assays, and saw no significant change after labeling (not shown). Overall, we consider that the fusion monitored by our FdQ studies is a bona fide receptor-mediated event.

One caveat with FdQ studies, such as these presented here, is that fusion is induced at the cell surface by artificially lowering the external pH. Under normal circumstances, fusion would occur following the drop in pH within the endosome. Although fluorimeter-based studies have been used to monitor viral fusion from endosomes,[13] it is important to remember that individual fusion events are asynchronous in endosomes and that fluorimeter assays are ensemble experiments; i.e., individual fusion events may be missed with this technique. One future application of fluorescence-based coronavirus fusion assays involves the use of single-particle tracking.[16] Use of this technique in our laboratory with R18-labeled virus has been hampered due to photobleaching and alternative probes are currently being investigated.

FdQ assays such as those described here are very powerful tools in the study of the fusion event occurring during coronavirus entry, especially when applied in combination with related cell biological and biochemical studies. Ultimately however, a complete understanding of the molecular events in virus fusion awaits the crystallization and structure determination of the intact coronavirus S protein.

5. REFERENCES

1. L. J. Earp, S. E. Delos, H. E. Park, and J. M. White, The many mechanisms of viral membrane fusion proteins, *Curr. Top. Microbiol. Immunol.* **285**, 25-66 (2004).
2. S. B. Sieczkarski and G. R. Whittaker, Viral entry. *Curr. Top. Microbiol. Immunol.* **285**, 1-23 (2005).
3. J. J. Skehel and D. C. Wiley, Receptor binding and membrane fusion in virus entry: the influenza hemagglutinin. *Annu. Rev. Biochem.* **69**, 531-69 (2000).
4. J. S. Peiris, Y. Guan, and K. Y. Yuen, Severe acute respiratory syndrome, *Nat. Med.* **10**, S88-97 (2004).
5. D. Cavanagh, in: *The Coronaviridae*, edited by S. G. Siddell (Plenum Press, New York, 1995), pp. 73-113.
6. T. M. Gallagher and M. J. Buchmeier, Coronavirus spike proteins in viral entry and pathogenesis, *Virology* **279**, 371-374 (2001).
7. P. Chambers, C. R. Pringle, and A. J. Easton, Heptad repeat sequences are located adjacent to hydrophobic regions in several types of virus fusion glycoproteins, *J. Gen. Virol.* **71**, 3075-3080 (1990).
8. B. J. Bosch, R. van der Zee, C. A. de Haan, and P. J. Rottier, The coronavirus spike protein is a class I virus fusion protein: structural and functional characterization of the fusion core complex, *J. Virol.* **77**, 8801-8811 (2003).
9. Y. Kliger and E. Y. Levanon, Cloaked similarity between HIV-1 and SARS-CoV suggests an anti-SARS strategy, *BMC Microbiol.* **3**, 20 (2003).
10. P. M. Colman and M. C. Lawrence, The structural biology of type I viral membrane fusion, *Nat. Rev. Mol. Cell. Biol.* **4**, 309-319 (2003).
11. H. Hofmann and S. Pohlmann, Cellular entry of the SARS coronavirus, *Trends Microbiol.* **12**, 466-472 (2004).
12. R. Blumenthal, S. A. Gallo, M. Viard, Y. Raviv, and A. Puri, Fluorescent lipid probes in the study of viral membrane fusion, *Chem. Phys. Lipids* **116**, 39-55 (2002).
13. T. Stegmann, H. W. M. Morselt, J. Scholma, and J. Wilschut, Fusion of influenza virus in an intracellular acidic compartment measured by fluorescence dequenching, *Biochem. Biophys. Acta* **904**, 165-170 (1987).
14. D. Hoekstra, T. de Boer, K. Klappe, and J. Wilschut, Fluorescence method for measuring the kinetics of fusion between biological membranes, *Biochemistry* **23**, 5675-5681 (1984).
15. A. Puri, M. J. Clague, C. Schoch, and R. Blumenthal, in: *Methods in Enzymology, Vol. 220: Membrane Fusion Techniques Part A*, edited by N. Duzgunes (Academic Press, San Diego, 1993), pp. 277-287.
16. M. Lakadamyali, M. J. Rust, H. P. Babcock, and X. Zhuang, Visualizing infection of individual influenza viruses, *Proc. Natl. Acad. Sci. USA* **100**, 9280-9285 (2003).
17. J. M. Gilbert, D. Mason, and J. M. White, Fusion of Rous sarcoma virus with host cells does not require exposure to low pH, *J. Virol.* **64**, 5106-5113 (1990).
18. D. Hoekstra and K. Klappe, in: *Methods in Enzymology, Vol. 220: Membrane Fusion Techniques Part A*, edited by N. Duzgunes (Academic Press, San Diego, 1993), pp. 261-276.
19. Y. Gaudin, Reversibility in fusion protein conformational changes. The intriguing case of rhabdovirus-induced membrane fusion, *Subcell. Biochem.* **34**, 379-408 (2000).
20. P. C. Roberts, T. Kipperman, and R. W. Compans, Vesicular stomatitis virus G protein acquires pH-independent fusion activity during transport in a polarized endometrial cell line, *J. Virol.* **73**, 10447-10457 (1999).
21. M. M. Wu, M. Grabe, S. Adams, R. Y. Tsien, H. P. Moore, and T. E. Machen. Mechanisms of pH regulation in the regulated secretory pathway, *J. Biol. Chem.* **276**, 33027-33035 (2001).
22. S. Ohki, T. D. Flanagan, and D. Hoekstra, Probe transfer with and without membrane fusion in a fluorescence fusion assay, *Biochemistry* **37**, 7496-7503 (1998).
23. K. V. Holmes and S. R. Compton, in: *The Coronaviridae*, edited by S. G. Siddell (Plenum Press, New York, 1995), pp. 55-71.

PORCINE ARTERIVIRUS ENTRY IN MACROPHAGES

Heparan sulfate–mediated attachment, sialoadhesin-mediated internalization, and a cell-specific factor mediating virus disassembly and genome release

Peter L. Delputte and H. J. Nauwynck*

1. INTRODUCTION

Porcine reproductive and respiratory syndrome (PRRS) was first described as a new disease in pig herds in North America late in the 1980s and beginning of the 1990s in Europe and is characterized by respiratory disease in young piglets and late-term reproductive failure.[1,2] The causative agent of the disease was identified in Europe in 1991 and in the US in 1992 as a virus, PRRS virus (PRRSV).[2-5]

PRRSV is classified with equine arteritis virus (EAV), lactate dehydrogenase-elevating virus (LDV), and simian hemorrhagic fever virus (SHFV) in the family *Arteriviridae*, which is grouped with the *Coronaviridae* and the *Roniviridae* in the order *Nidovirales.*[6-8] The virus is an enveloped particle with a diameter of 50 to 65 nm and contains a polyadenylated, positive-strand RNA genome.[9] The genome is about 15 kDa and encodes 9 open reading frames: ORF1a and 1b code for nonstructural proteins, ORF2a, 2b, 3, 4, 5 and 6 code for structural membrane proteins, and ORF7 codes for the nucleocapsid protein (N).[10-13]

Currently, PRRSV is present in most, if not all swine-producing areas of the world, including North and South America, Western and Eastern Europe, and Asia. As the virus is now enzootic in most countries, the number of acute disease outbreaks has diminished and infections are in general mild and subclinical.[14] From an economical point of view, however, the virus still causes major losses and is considered as the most important pig disease worldwide.

Characteristic for all members of the *Arteriviridae* is not only that they have a very narrow host tropism, but also that they share a marked *in vivo* tropism for cells of the monocyte/macrophage lineage.[14,15] Even in cell culture, arteriviruses have, with the exception of EAV, a very narrow cell specificity, only allowing replication in primary

---

*Ghent University, Merelbeke, Belgium.

macrophages from their respective hosts and in a limited number of cell lines upon adaptation of the virus.[16]

For PRRSV, it has been shown that *in vivo*, the virus infects a subpopulation of resident macrophages present throughout the body, and that alveolar macrophages are primary target cells for the virus.[17, 18] *In vitro*, porcine alveolar macrophages (PAM) have been shown to efficiently sustain virus replication.[5, 19, 20] Peripheral blood monocytes can also be infected *in vitro* at very low levels, but only when they have been cultivated for 24 h.[19, 21] Interestingly, peritoneal macrophages were shown to be refractory for PRRSV infection.[19] Besides primary macrophages of porcine origin, only the African green monkey kidney cells MA-104, and cells derived thereof, such as Marc-145 cells, can sustain *in vitro* virus replication.[22] Although PRRSV has thus both *in vivo* and *in vitro* a very restricted cell tropism, it can replicate in several cell lines upon transfection of the genomic RNA, indicating that the restricted cell tropism is very likely the result of the presence or absence of specific receptors on the membrane of macrophages and of other macrophage-specific factors.[23, 24]

In our laboratory, we have been studying for several years PRRSV entry in macrophages, and two receptors on macrophages have been identified: (1) heparan sulfate and (2) sialoadhesin. The glycosaminoglycan heparan sulfate was identified as a receptor because both heparan sulfate and heparin, an analogue of heparan sulfate, strongly reduce infection of macrophages with both European and American PRRSV strains when present during virus inoculation.[25, 26] Other glycosaminoglycans, such as chondroitin sulfate A and dermatan sulfates had no effect on PRRSV infection, indicating that the observed effect was specific. Also, treatment of macrophages with heparinase I, an enzyme which destroys heparan sulfate, reduced infection of macrophages. Using flow cytometry and labeled PRRSV, it was observed that heparin strongly reduced virus attachment to macrophages.[26] Heparan sulfate has also been proposed as receptor on Marc-145 cells, because heparin reduces PRRSV infection when present during infection, or when these cells are pretreated with heparinase I.[27]

Sialoadhesin, a macrophage-specific protein belonging to the sialic acid-binding immunoglobulin-like lectin (Siglec) family, was identified as a PRRSV receptor on macrophages because (a) a sialoadhesin-specific monoclonal antibody (mAb 41D3) is able to completely block infection of macrophages, (b) mAb 41D3 reduces PRRSV attachment to macrophages, (c) PRRSV co-localizes with sialoadhesin on the surface of macrophages, (d) both the *in vivo* and *in vitro* PRRSV susceptible porcine cells express sialoadhesin, and (e) nonsusceptible cells internalize PRRSV upon expression of a recombinant sialoadhesin.[28-30]

In this overview, our findings about the different steps of PRRSV entry in macrophages (attachment, internalization and genome release into the cytoplasm) will be discussed and these will be related to the findings of others on PRRSV entry in both macrophages and Marc-145 cells.

2. PRRSV ATTACHMENT: INITIATED BY AN INTERACTION WITH HEPARAN SULFATE AND FOLLOWED BY ATTACHMENT TO SIALOADHESIN

The role of heparan sulfate and sialoadhesin in PRRSV attachment to macrophages was first evaluated using flow cytometric attachment studies in the presence or absence of heparin and/or sialoadhesin-specific mAb 41D3, this to block respectively virus attachment to heparan sulfate, sialoadhesin, or both.[31] Both heparin and mab 41D3 reduced attachment up to a maximum reduction of respectively 83% and 50%. A combination of both heparin and mAb 41D3 could however completely block virus attachment. Clearly, from these results it can not only be concluded that both heparan sulfate and sialoadhesin mediate PRRSV attachment to the macrophage, but also that no other receptors are involved in this process. Furthermore, analysis of the attachment kinetics to macrophages showed that PRRSV first binds to heparan sulfate, followed by an interaction with sialoadhesin.[31]

3. PRRSV INTERNALIZATION IN MACROPHAGES IS A CLATHRIN-DEPENDENT PROCESS MEDIATED BY THE MACROPHAGE-SPECIFIC RECEPTOR SIALOADHESIN

Previously, it was shown that PRRSV internalization in macrophages is a clathrin-dependent process.[32] Recently, we identified the macrophage-specific receptor sialoadhesin as a receptor involved in PRRSV internalization, because expression of a recombinant sialoadhesin on the surface of cells which are nonpermissive for PRRSV infection, such as PK-15 or CHO cells, makes these cells capable of internalizing the virus.[28, 31] Furthermore, it was shown that the heparan sulfate receptor, which mediates virus attachment, is not necessary for internalization, as recombinant sialoadhesin internalizes PRRSV both in CHO cells expressing heparan sulfate and in CHO cells lacking heparan sulfate.[31] Because heparan sulfate is thus not essential for PRRSV attachment to sialoadhesin and subsequent sialoadhesin-mediated internalization, it is thought that the interaction with heparan sulfate serves to concentrate virus particles on the cell surface, which enhances virus binding to sialoadhesin, resulting in enhanced internalization and infection. Mouse and human sialoadhesins were previously described as a sialic acid binding lectin[33-35] and recently, we found that porcine sialoadhesin is also a sialic acid binding lectin.[36] We also observed that enzymatic removal of sialic acid from the PRRS virion surface almost completely blocked infection of macrophages, indicating that virus attachment to sialoadhesin is mediated by sialic acids present on one or more structural viral glycoproteins.[36]

PRRSV internalization in macrophages via sialoadhesin was shown to be a clathrin-mediated process[28,32] and although PRRSV internalization in Marc-145 cells is also clathrin-dependent,[37] our data indicate that sialoadhesin is not involved in PRRSV infection of Marc-145 cells and that another receptor is responsible for virus internalization in these cells. This conclusion is based on the fact that (a) sialoadhesin could not be detected on Marc-145 cells, neither using sialoadhesin-specific mAb 41D3, nor using a sialoadhesin specific polyclonal serum generated by DNA immunization in mice, (b) removal of sialic acid from the surface of PRRSV, which results in an almost complete block of PRRSV infection of macrophages, has no effect on infection of Marc-145 cells. Although an internalization receptor has not been identified on Marc-145 cells, it was suggested by preliminary studies that this receptor is a protein. Two monoclonal antibodies have been described which block PRRSV infection of Marc-145 cells and which recognize a yet unidentified 60–66 kDa Marc-145 cell surface protein.[38, 39]

4. VIRUS DISASSEMBLY AND GENOME RELEASE REQUIRES ENDOSOME ACIDIFICATION AND A CELL-SPECIFIC FACTOR

Previously, it was shown both for macrophages and Marc-145 cells that PRRSV is internalized via a clathrin-dependent process and that endosome acidification is essential for infectivity.[32,37] When endosome acidification was inhibited using lysosomotropic bases, no infected cells were detected using either confocal or electron microscopy up to several hours after inoculation. This was in contrast to a normal infection in which virus was disassembled after internalization.[32,37] From these results, it was concluded that an acidic pH is needed to trigger fusion between viral and endosomal membrane, which is a key element in virus disassembly and release of the viral genome in the cytoplasm.

Although we observed that PRRSV can be internalized in several cell types refractory for PRRSV infection upon expression of a recombinant sialoadhesin, a productive infection was not observed in these cells.[28,31] A comparison of the viral entry process in macrophages, Marc-145 cells and cells expressing recombinant sialoadhesin revealed that virus was disassembled upon internalization in macrophages and Marc-145 cells, while it remained present in endocytic vesicles in recombinant sialoadhesin expressing cells.[28] Apparently, productive infection not only requires a receptor such as sialoadhesin, which mediates virus internalization, but also a cell-specific factor mediating virus disassembly and genome release. Since it was previously shown that PRRSV also remains in endosomes upon internalization in macrophages and Marc-145 cells when the pH drop is blocked by drugs, we investigated if acidification of the extracellular medium could overcome this blockade in cells expressing recombinant sialoadhesin. Although this method was shown to be successful in the case of Marc-145 cells to counteract the effects of ammonium chloride,[37] it did not result in virus disassembly and productive infection in recombinant sialoadhesin expressing cells. So, while clearly endosome acidification is essential for infection, other cell-specific factors, such as fusion receptors or proteases, are also involved in PRRSV infection. Currently, experiments are being conducted to identify these co-factors involved in the post-internalization steps of PRRSV infection of macrophages.

Interestingly, two sets of monoclonal antibodies that reduce or block PRRSV infection of macrophages have been described.[40] One set recognizing a 220-kDa protein completely blocked infection. The nature of this protein has not been identified, but as suggested by Wissink et al. (2003)[40] and based on our recent results, this 220-kDa protein is likely to be the macrophage-specific protein sialoadhesin. Another set of monoclonal antibodies that recognizes a 150-kDa protein doublet was shown to strongly reduce, but not block PRRSV infection of macrophages. This protein is not identified, but could represent a receptor that is potentially involved in fusion between viral and endosomal membrane and subsequent release of the viral genome in the cytoplasm. Future studies using these monoclonal antibodies could clarify these issues.

5. ACKNOWLEDGMENTS

The authors would like to thank all the past and present members of our laboratory who have contributed to our understanding of the complexity of the entry process of this intriguing virus: Xiaobo Duan, Nathalie Vanderheijden, Gerald Misinzo, and Sarah Costers.

6. REFERENCES

1. K. D. Rossow, Porcine reproductive and respiratory syndrome, *Vet. Pathol.* **35**, 1 (1998).
2. J. E. Collins, D. A. Benfield, W. T. Christianson, et al., Isolation of swine infertility and respiratory syndrome virus (isolate ATCC VR-2332) in North America and experimental reproduction of the disease in gnotobiotic pigs, *J. Vet. Diagn. Invest.* **4**, 117 (1992).
3. J. J. Meulenberg, M. M. Hulst, E. J. de Meijer, P. L. Moonen, A. den Besten, E. P. de Kluyver, G. Wensvoort, and R. J. Moormann, Lelystad virus belongs to a new virus family, comprising lactate dehydrogenase-elevating virus, equine arteritis virus, and simian hemorrhagic fever virus, *Arch. Virol. Suppl.* **9**, 441 (1994).
4. C. Terpstra, G. Wensvoort, and J. M. Pol, Experimental reproduction of porcine epidemic abortion and respiratory syndrome (mystery swine disease) by infection with Lelystad virus: Koch's postulates fulfilled, *Vet. Q.* **13**, 131 (1991).
5. G. Wensvoort, C. Terpstra, J. M. Pol, et al., Mystery swine disease in The Netherlands: the isolation of Lelystad virus, *Vet. Q.* **13**, 121 (1991).
6. L. Enjuanes, W. J. Spaan, E. Snijder, and D. Cavanagh, in: *Virus Taxonomy. Classification and Nomenclature of Viruses*, edited by M. H. V. van Regenmortel, C. M. Fauquet, D. H. L. Bishop, et al. (Academia Press, San Diego, 2000), p. 827.
7. M. A. Mayo, A summary of taxonomic changes recently approved by ICTV, *Arch. Virol.* **147**, 1655 (2002).
8. D. Cavanagh, Nidovirales: a new order comprising Coronaviridae and Arteriviridae, *Arch. Virol.* **142**, 629 (1997).
9. J. J. Meulenberg, M. M. Hulst, E. J. de Meijer, P. L. Moonen, A. den Besten, E. P. de Kluyver, G. Wensvoort, and R. J. Moormann, Lelystad virus, the causative agent of porcine epidemic abortion and respiratory syndrome (PEARS), is related to LDV and EAV, *Virology* **192**, 62 (1993).
10. J. J. Meulenberg, A. Petersen-den Besten, E. P. De Kluyver, R. J. Moormann, W. M. Schaaper, and G. Wensvoort, Characterization of proteins encoded by ORFs 2 to 7 of Lelystad virus, *Virology* **206**, 155 (1995).
11. J. J. Meulenberg and A. Petersen-den Besten, Identification and characterization of a sixth structural protein of Lelystad virus: the glycoprotein GP2 encoded by ORF2 is incorporated in virus particles, *Virology* **225**, 44 (1996).
12. W. H. Wu, Y. Fang, R. Farwell, M. Steffen-Bien, R. R. Rowland, J. Christopher-Hennings, and E. A. Nelson, A 10-kDa structural protein of porcine reproductive and respiratory syndrome virus encoded by ORF2b, *Virology* **287**, 183 (2001).
13. S. Dea, C. A. Gagnon, H. Mardassi, B. Pirzadeh, and D. Rogan, Current knowledge on the structural proteins of porcine reproductive and respiratory syndrome (PRRS) virus: comparison of the North American and European isolates, *Arch. Virol.* **145**, 659 (2000).
14. E. J. Snijder and J. J. M. Meulenberg, in: *Fields Virology (4th ed.)*, edited by D. M. Knipe and P. M. Howley (Lippincott Williams & Wilkins, Philadelphia, 2001), p. 1205.
15. P. G. Plagemann and V. Moennig, Lactate dehydrogenase-elevating virus, equine arteritis virus, and simian hemorrhagic fever virus: a new group of positive-strand RNA viruses, *Adv. Virus Res.* **41**, 99 (1992).
16. E. J. Snijder and J. J. Meulenberg, The molecular biology of arteriviruses, *J. Gen. Virol.* **79**, 961 (1998).
17. J. P. Teifke, M. Dauber, D. Fichtner, M. Lenk, U. Polster, E. Weiland, and J. Beyer, Detection of European porcine reproductive and respiratory syndrome virus in porcine alveolar macrophages by two-colour immunofluorescence and in-situ hybridization-immunohistochemistry double labelling, *J. Comp. Pathol.* **124**, 238 (2001).
18. X. Duan, H. J. Nauwynck, and M. B. Pensaert, Virus quantification and identification of cellular targets in the lungs and lymphoid tissues of pigs at different time intervals after inoculation with porcine reproductive and respiratory syndrome virus (PRRSV), *Vet. Microbiol.* **56**, 9 (1997).
19. X. Duan, H. J. Nauwynck, and M. B. Pensaert, Effects of origin and state of differentiation and activation of monocytes/macrophages on their susceptibility to porcine reproductive and respiratory syndrome virus (PRRSV), *Arch. Virol.* **142**, 2483 (1997).
20. E. M. Bautista, S. M. Goyal, I. J. Yoon, H. S. Joo, and J. E. Collins, Comparison of porcine alveolar macrophages and CL 2621 for the detection of porcine reproductive and respiratory syndrome (PRRS) virus and anti-PRRS antibody, *J. Vet. Diagn. Invest.* **5**, 163 (1993).
21. I. L. Voicu, A. Silim, M. Morin, and M. A. Elazhary, Interaction of porcine reproductive and respiratory syndrome virus with swine monocytes, *Vet. Rec.* **134**, 422 (1994).
22. H. S. Kim, J. Kwang, I. J. Yoon, H. S. Joo, and M. L. Frey, Enhanced replication of porcine reproductive and respiratory syndrome (PRRS) virus in a homogeneous subpopulation of MA-104 cell line, *Arch. Virol.* **133**, 477 (1993).

23. L. C. Kreutz, Cellular membrane factors are the major determinants of porcine reproductive and respiratory syndrome virus tropism, *Virus Res.* **53**, 121 (1998).
24. J. J. Meulenberg, J. N. Bos-de Ruijter, R. van de Graaf, G. Wensvoort, and R. J. Moormann, Infectious transcripts from cloned genome-length cDNA of porcine reproductive and respiratory syndrome virus, *J. Virol.* **72**, 380 (1998).
25. N. Vanderheijden, P. Delputte, H. Nauwynck, and M. Pensaert, Effects of heparin on the entry of porcine reproductive and respiratory syndrome virus into alveolar macrophages, *Adv. Exp. Med. Biol.* **494**, 683 (2001).
26. P. L. Delputte, N. Vanderheijden, H. J. Nauwynck, and M. B. Pensaert, Involvement of the matrix protein in attachment of porcine reproductive and respiratory syndrome virus to a heparinlike receptor on porcine alveolar macrophages, *J. Virol.* **76**, 4312 (2002).
27. E. R. Jusa, Y. Inaba, M. Kouno, and O. Hirose, Effect of heparin on infection of cells by porcine reproductive and respiratory syndrome virus, *Am. J. Vet. Res.* **58**, 488 (1997).
28. N. Vanderheijden, P. L. Delputte, H. W. Favoreel, J. Vandekerckhove, J. Van Damme, P. A. van Woensel, and H. J. Nauwynck, Involvement of sialoadhesin in entry of porcine reproductive and respiratory syndrome virus into porcine alveolar macrophages, *J. Virol.* **77**, 8207 (2003).
29. X. Duan, H. J. Nauwynck, H. Favoreel, and M. B. Pensaert, Porcine reproductive and respiratory syndrome virus infection of alveolar macrophages can be blocked by monoclonal antibodies against cell surface antigens, *Adv. Exp. Med. Biol.* **440**, 81 (1998).
30. X. Duan, H. J. Nauwynck, H. W. Favoreel, and M. B. Pensaert, Identification of a putative receptor for porcine reproductive and respiratory syndrome virus on porcine alveolar macrophages, *J. Virol.* **72**, 4520 (1998).
31. P. L. Delputte, S. Costers, and H. J. Nauwynck, Analysis of porcine reproductive and respiratory syndrome virus attachment and internalization: distinctive roles for heparan sulfate and sialoadhesin, *J. Gen. Virol.* **86**, 1441 (2005).
32. H. J. Nauwynck, X. Duan, H. W. Favoreel, P. Van Oostveldt, and M. B. Pensaert, Entry of porcine reproductive and respiratory syndrome virus into porcine alveolar macrophages via receptor-mediated endocytosis, *J. Gen. Virol.* **80**, 297 (1999).
33. Y. Kumamoto, N. Higashi, K. Denda-Nagai, M. Tsuiji, K. Sato, P. R. Crocker, and T. Irimura, Identification of sialoadhesin as a dominant lymph node counter-receptor for mouse macrophage galactose-type C-type lectin 1, *J. Biol. Chem.* **279**, 49274 (2004).
34. A. Hartnell, J. Steel, H. Turley, M. Jones, D. G. Jackson, and P. R. Crocker, Characterization of human sialoadhesin, a sialic acid binding receptor expressed by resident and inflammatory macrophage populations, *Blood* **97**, 288 (2001).
35. P. R. Crocker, S. Kelm, C. Dubois, B. Martin, A. S. McWilliam, D. M. Shotton, J. C. Paulson, and S. Gordon, Purification and properties of sialoadhesin, a sialic acid-binding receptor of murine tissue macrophages, *EMBO J.* **10**, 1661 (1991).
36. P. L. Delputte and H. J. Nauwynck, Porcine arterivirus infection of alveolar macrophages is mediated by sialic acid on the virus, *J. Virol.* **78**, 8094 (2004).
37. L. C. Kreutz, and M. R. Ackermann, Porcine reproductive and respiratory syndrome virus enters cells through a low pH-dependent endocytic pathway, *Virus Res.* **42**, 137 (1996).
38. D. Therrien, Y. St-Pierre, and S. Dea, Preliminary characterization of protein binding factor for porcine reproductive and respiratory syndrome virus on the surface of permissive and non-permissive cells, *Arch. Virol.* **145**, 1099 (2000).
39. D. Therrien, and S. Dea, Monoclonal antibody directed against a membranous protein of MARC-145 cells blocks infection by PRRSV, *Adv. Exp. Med. Biol.* **494**, 395 (2001).
40. E. H. Wissink, H. A. van Wijk, J. M. Pol, G. J. Godeke, P. A. van Rijn, P. J. Rottier, and J. J. Meulenberg, Identification of porcine alveolar macrophage glycoproteins involved in infection of porcine respiratory and reproductive syndrome virus, *Arch. Virol.* **148**, 177 (2003).

ENHANCEMENT OF SARS-CoV INFECTION BY PROTEASES

Shutoku Matsuyama, Makoto Ujike, Koji Ishii, Shuetsu Fukushi, Shigeru Morikawa, Masato Tashiro, and Fumihiro Taguchi*

1. INTRODUCTION

The severe acute respiratory syndrome (SARS) is caused by a newly emergent coronavirus (SARS-CoV).[1, 2] This virus grows in a variety of tissues that express its receptor, but the mechanism of the severe respiratory illness is not well understood. SARS-CoV is supposed to enter cells via endosome, and its spike (S) protein, which is responsible for cell entry of this virus, is activated by a certain protease active only in acidic conditions in the endosome.[3] To see whether this is correct or not, we began to study the SARS-CoV entry mechanism. In the course of this study, we found that various proteases facilitated SARS-CoV entry from cell surface. This indicated that SARS-CoV has a potential to enter cells via two different pathways, endosomal and cell-surface pathways, depending upon the presence of proteases. Moreover, SARS-CoV entry from the cell surface mediated by proteases was a 100-fold more efficient infection than entry through endosomes. These results suggest that severe illness in the lung and intestine can be attributed to the proteases produced in these organs in inflammatory responses or physiological conditions.

2. MATERIALS AND METHODS

2.1. Viruses and Cells

The SARS-CoV Frankfurt 1 strain, kindly provided by Dr. J. Ziebuhr,[1] was propagated and titered using Vero E6 cells. Recombinant vaccinia virus harboring SARS-CoV S gene (DIs-S) was used to express S protein. This recombinant virus was made from highly attenuated vaccinia virus DIs.[4] Vesicular stomatitis virus (VSV) pseudotype bearing SARS-CoV S protein was produced as reported previously.[5]

---

* National Institute of Infectious Diseases, Tokyo, Japan.

2.2. Proteases

Various proteases were dissolved in phosphate-buffered saline, pH 7.2 (PBS), and used at the indicated concentration in DMEM containing 5% FCS. The proteases used in this study were trypsin (Sigma, St Louis, MO, T-8802), thermolysin (Sigma, P1512), chymotrypsin (Sigma, C-3142), dispase (Roche, 1 276 921, Branchburg, NJ), papain (Worthington Biochemicals, 53J6521, Freehold, NJ), proteinase K (Wako, Tokyo, Japan), collagenase (Sigma, C-5183) and elastase (Sigma, E-0258).

2.3. Western Blot

S protein expressed in Vero E6 cells was analyzed by Western blotting. Preparation of cell lysates, electrophoresis in SDS-polyacrylamide gel, and electrical transfer of the protein onto a transfer membrane were described previously.[6] S protein was detected with anti-S antibody, IMG-557 (IMGENEX, San Diego, CA, USA).

2.4. Real-Time PCR

SARS-CoV entry or replication in VeroE6 cells was examined by real-time PCR to detect the copy number of mRNA9. The primers for amplification were complementary to the leader sequence (forward) and N gene (reverse) of SARS-CoV. The reaction was performed using a LightCycler instrument (Roche).

3. RESULTS

3.1. Proteases Induce Syncytia Formation and S Protein Cleavage

VeroE6 cells were infected with the Frankfurt-1 strain of SARS-CoV at a multiplicity of infection (MOI) of 0.5, and infected cells were treated with trypsin (200 μg/ml) at room temperature (RT) for 5 min after 20 h incubation. Cell fusion was detected from approximately 2 h after trypsin treatment. Fusion was also found after treatment with thermolysin or dispase. Little or no fusion occurred following treatment with papain, chymotrypsin, proteinase K, or collagenase. S proteins in cells treated with proteases that induce fusion were cleaved approximately in the middle, and a fragment corresponding to S2 of ca. 100 kDa protein was detected. However, no S2 band was detected in SARS-CoV infected cells treated with proteases that failed to induce fusion. VeroE6 cells were infected with DIs-S that harbors SARS-CoV S gene at MOI of 1, and these cells were also treated with various proteases as described above. Trypsin, thermolysin, and dispase induced fusion of S protein expressing cells, while other proteases failed to induce substantial cell fusion (Fig. 1). The results obtained using VeroE6 cells expressing S protein by recombinant vaccinia virus were very similar to those observed in VeroE6 cells infected with SARS-CoV. These results showed that various proteases activate the fusion activity of the SARS-CoV S protein by inducing its cleavage. It was also revealed that SARS-CoV infection was extensively inhibited by treatment of cells with bafilomycin (1 mM), which perturbs endosomal pH (Fig. 2). Collectively, these results suggest that SARS-CoV takes an endosomal pathway for its entry and that S protein cleavage is important for fusion activity, which is in good agreement with the observations of a previous report.[3]

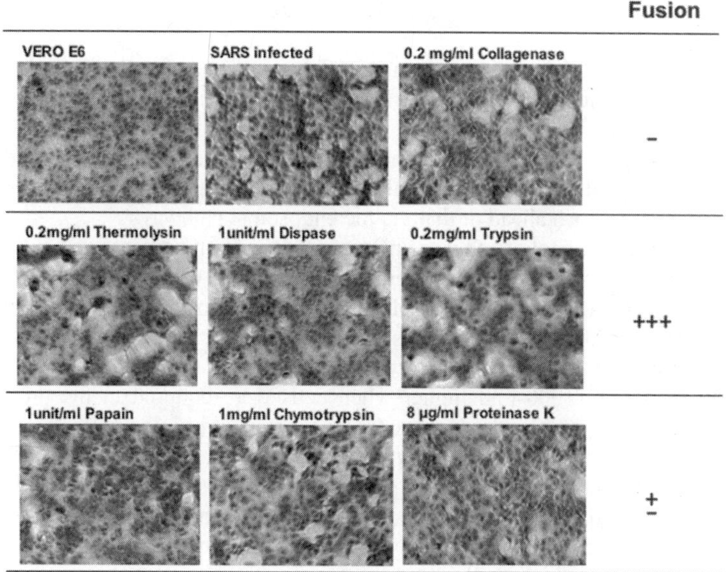

Figure 1. Fusion of SARS-CoV infected VeroE6 cells after treatment with various proteases.

3.2. Proteases Facilitate SARS-CoV Entry from the Cell Surface

The hypothesis proposed by Simmons *et al.*[3] stated that SARS-CoV is able to enter cells directly from their surface, if receptor-bound virus is treated with trypsin and other proteases that induce fusion. Treatment of VeroE6 cells with bafilomycin was shown to suppress SARS-CoV infection via the endosomal pathway to less than 1/100 (Fig. 2). Bafilomycin-treated cells were inoculated with SARS-CoV at an MOI of 1 and incubated on ice for 30 min. This allows virus binding to its receptor, but does not allow virus entry into cells. Cells were then treated with trypsin for 5 min at RT and incubated at 37°C for 6 h in the presence of bafilomycin. Virus entry was estimated by the newly synthesized mRNA9 measured quantitatively by real-time PCR. It was shown that trypsin with fusion-inducing activity extensively facilitated viral entry (Fig. 2). Thermolysin and dispase also facilitated entry into VeroE6 cells treated with bafilomycin. In contrast, two proteases that did not induce fusion, papain and collagenase, failed to do so. Pseudotype VSV bearing SARS-CoV S protein infection was also facilitated in bafilomycin-treated VeroE6 cells after treatment with proteases that induce fusion of SARS-CoV infected cells. Treatment of cells with trypsin before virus infection did not enhance viral entry (Fig. 2), indicating that the effects of trypsin on cells are not involved in this infection. Trypsin treatment of SARS-CoV prior to infection did not enhance infectivity, but reduced it by 10- to 100-fold (Fig. 3). These results demonstrate that SARS-CoV, when adsorbed onto the cell surface, fuses with the plasma membrane via the S protein after cleavage, suggesting a non-endosomal, direct entry of SARS-CoV into cells in the presence of proteases. Those findings also support the hypothesis drawn by Simmons *et al.*[3] that trypsin-like protease plays an important role in facilitating membrane fusion.

3.3. Various Proteases Enhance SARS-CoV Infection

Treatment with a high concentration of trypsin augmented virus entry or replication by approximately 10-fold during an early phase of the infection, from 3 to 6 h postinfection, compared with the standard infection. This implies that infection through the cell surface is approximately 10-fold more efficient than infection via the endosomal pathway. These data also imply that viral replication after entry via the cell surface proceeds approximately 1 h ahead of that via the endosomal pathway.

Because SARS-CoV replication was enhanced by trypsin treatment, we next assessed the efficiency of virus spread in the presence or absence of trypsin in a low MOI that mimics natural infection in humans. Ten pfu of virus were inoculated onto 10^5 confluent VeroE6 cells (MOI = 0.0001), and the cells were incubated at 37°C for 20 h in the media with or without various concentration of trypsin. The level of mRNA9 showed that virus replication was 100- to 1000-fold higher when cells were cultured in the presence of trypsin, compared to replication in the absence of trypsin. Viral infectivity also indicated that trypsin treatment enhanced viral growth by ca. 100-fold. This enhancement of viral replication observed in the presence of trypsin was also observed when infected VeroE6 cells were cultured in the presence of proteases, such as thermolysin and dispase, which induce fusion, but no enhancement was encountered when cultured in the presence of papain or collagenase, which fail to induce fusion. These observations suggest that proteases that facilitate SARS-CoV entry from the cell surface support efficient SARS-CoV infection. Thus, protease is likely to be responsible for the high multiplication of SARS-CoV in the target organs of SARS, such as the lungs, where various proteases are produced (e.g., by inflammatory cells), as well as in the intestines, where a number of proteases are physiologically secreted. Elastase is reported to be one of major proteases produced in inflammatory lungs.[7] Thus we examined whether elastase enhances SARS-CoV infection as do trypsin and thermolysin. Elastase enhanced SARS-CoV infection in cultured VeroE6 cells at low multiplicities of infection. This finding strongly suggests that SARS-CoV replication can be enhanced in the lungs by elastase. Thus, elastase is possibly a protease that is responsible for an acute severe illness caused by SARS-CoV.

4. DISCUSSION

SARS-CoV infection was evident in a number of organs, such as the liver, cerebrum, and kidneys, as well as in major target organs such as the lungs and intestines.[8, 9] In the latter organs, drastic tissue damage by SARS-CoV infection was observed, while the other organs were not so severely affected. Although the pathogenic mechanism of SARS has not been elucidated, the present study suggests that proteases secreted in major target organs play an important role in the high multiplication of virus in those organs, which could result in severe tissue damage. SARS-CoV may initially infect pneumocytes via an endosomal pathway. This would induce inflammation that generates a variety of proteases such as elastase. Once those proteases are present in the lungs, they may mediate a robust infection, which may result in enhanced replication. Although lung damage is reportedly mediated by cytokine storm,[8, 9] higher virus multiplication could also contribute to the cytokine storm by killing a large number of infected cells. Various proteases secreted in

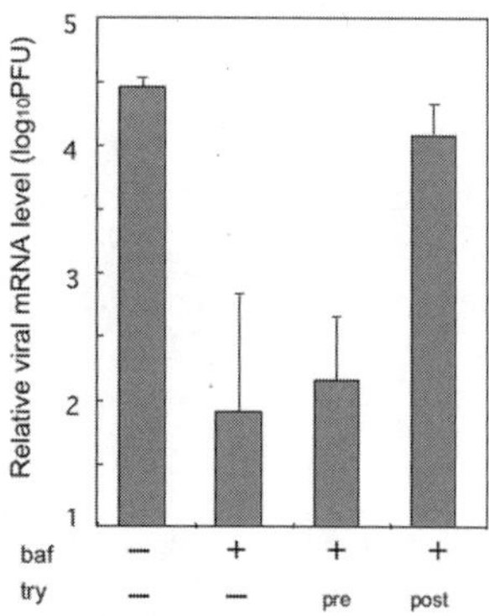

Figure 2. VeroE6 cells treated with bafilomycin were infected with SARS-CoV at 4°C for 30 min and then treated with trypsin at RT for 5 min (baf +, try post). Bafilomycin treated cells were treated with trypsin at RT for 5 min before SARS-CoV infection (baf +, try pre). Untreated cells (baf -, try -) or cells treated with bafilomycin alone (baf +, try -) were infected as controls. At 6 h postinfection, mRNA level was measured by real-time PCR and shown as log_{10}PFU.

another target organ, the small intestines, could also contribute to the high viral titers detected in these tissues, which, in turn, may result in diarrhea.[10]

The present studies suggest that co-infection of SARS-CoV with nonpathogenic respiratory agents, such as *Chlamydia* or mycoplasma, could result in severe lung disease as a consequence of protease production or induction by the non-SARS-CoV agents, as has been shown by the enhancement of disease caused by influenza virus co-infected with nonpathogenic bacteria.[11, 12] Studies are in progress to examine whether co-infection exacerbates pneumonia in mice infected with SARS-CoV.

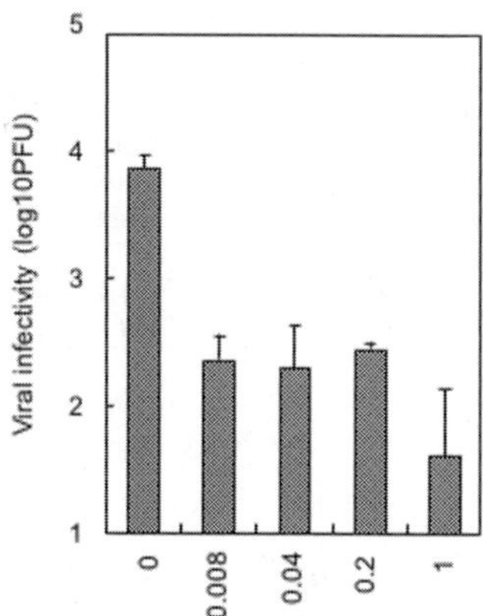

Figure 3. Treatment of SARS-CoV by trypsin. SARS-CoV was treated with various concentration of trypsin (from 0 to 1 mg/ml) in DMEM containing 5% FCS at RT for 30 min and infectivity was examined by plaque assay.

5. ACKNOWLEDGMENTS

We thank Miyuki Kawase for her excellent technical assistance throughout the experiments. This work was financially supported by a grant from the Ministry of Education, Culture, Sports, Science and Technology (16017308) and grant from the Ministry of Health, Labor and Welfare (H16-Shinkoh-9).

6. REFERENCES

1. T. G. Ksiazek, D. Erdman, C. Goldsmith, S. R. Zaki, T. Peret, S. Emery, S. Tong, C. Urbani, J. A. Comer, W. Lim, et al., A novel coronavirus associated with severe acute respiratory syndrome, *N. Engl. J. Med.* **348**, 1953-1966 (2003).
2. C. Drosten, S. Gunther, W. Preiser, S. Van Der Werf, H. R. Brodt, S. Becker, H. Rabenau, M. Panning, L. Kolesnikowa, R. A. Fouchier, et al., Identification of a novel coronavirus in patients with severe acute respiratory syndrome, *N. Engl. J. Med.* **348**, 1967-1976 (2003).
3. G. Simmons, J. D. Reeves, A. J. Rennekamp, S. M. Amberg, A. J. Piefer, and P. Bates, Characterization of severe acute respiratory syndrome-associated coronavirus (SARS-CoV) spike glycoprotein-mediated viral entry, *Proc. Natl. Acad. Sci. USA* **101**, 4240-4245 (2004).
4. K. Ishii, Y. Ueda, K. Matsuo, Y. Matsuura, T. Kitamura, K. Kato, Y. Izumi, K. Someya, T. Ohsu, M. Honda, and T. Miyamura, Structural analysis of vaccinia virus DIs strain: application as a new replication-deficient viral vector, *Virology* **302**, 433-444 (2002).
5. S. Fukushi, T. Mizutani, S. Saijo, S. Matsuyama, N. Miyajima, F. Taguchi, S. Itamura, I. Kurane, and S. Morikawa, Vesicular stomatitis virus pseudotyped with severe acute respiratory syndrome coronavirus spike protein, *J. Gen. Virol.* **86**, 2269-2274 (2005).
6. S. Matsuyama and F. Taguchi, Impaired entry of soluble receptor-resistant mutants of mouse hepatitits virus into cells expressing MHVR2 receptor, *Virology* **273**, 80-89 (2000).
7. K. Kawabata, T. Hagio, and S. Matsuoka, The role of neutrophil elastase in acute lung injury, *Eur. J. Pharmacol.* **451**, 1-10 (2002).
8. J. M. Nicholls, L. L. Poon, K. C. Lee, W. F. Ng, S. T. Lai, C. Y. Leung, C. M. Chu, P. K. Hui, K. L. Mak, W. Lim, et al., Lung pathology of fatal severe acute respiratory syndrome, *Lancet* **361**, 1773-1778 (2003).
9. G. M. Tse, K. F. To, P. K. Chan, A. W. Lo, K. C. Ng, A. Wu, N. Lee, H. K. Wong, S. M. Mac, K. F. Chan, et al., Pulmonary pathological features in coronavirus associated severe acute respiratory syndrome (SARS), *J. Clin. Pathol.* **57**, 260-265 (2004).
10. J. Zhan, W. Chen, C. Li, W. Wu, J. Li, S. Jiang, J. Wang, Z. Zeng, Z. Huang, and H. Huang, Digestive system manifestations in patients with severe acute respiratory syndrome, *Clin. Med. J. (Engl.)* **116**, 1265-1266 (2003).
11. M. Tashiro, P. Ciborowski, H.-D. Klenk, G. Pulverer, and R. Rott, Role of Staphylococcus protease in the development of influenza pneumonia, *Nature* **325**, 536-537 (1987).
12. N. Kishida, Y. Sakoda, M. Eto, Y. Sunaga, and H. Kida, Co-infection of Staphylococcus aureus or Haemophilus paragallinarum exacerbates H9N2 influenza A virus infection in chickens, *Arch. Virol.* **149**, 2095-2140 (2004).

INCREASED VIRAL TITERS AND SUBTLE CHANGES IN PLAQUE MORPHOLOGY UPON PASSAGE OF SARS-CoV IN CELLS FROM DIFFERENT SPECIES

Laura Gillim-Ross, Lindsay K. Heller, Emily R. Olivieri, and David E. Wentworth*

1. INTRODUCTION

Despite rapid advances in our knowledge of SARS-Coronavirus (SARS-CoV), the regions of the virus spike glycoprotein and host receptor that are important for virus entry remain to be completely elucidated. The tropism of SARS-CoV for cells derived from many diverse species was analyzed by virus titration and by the use of a multiplex RT-PCR assay that differentiates input virus from virus that has entered a cell and initiated replication.[1] Cells derived from monkey, human, and mink were shown to be productively infected by SARS-CoV. Interestingly, the level of virus produced varied dramatically (4 $\log_{10}$) between the susceptible cell lines. Human angiotensin-converting enzyme 2 (ACE2) was shown to have SARS-CoV receptor activity.[2] Using conserved oligonucleotide primers and RT-PCR to amplify ACE2, we determined that the SARS-CoV susceptible cells expressed ACE2 RNA. We hypothesized that passage of SARS-CoV in cell lines expressing different levels of ACE2, or ACE2 from novel species (e.g., mink), will lead to mutations in the spike glycoprotein gene, and/or in other genes that enhance virus replication. SARS-CoV/Urbani was passaged in susceptible cell lines derived from monkey, human, and mink. The titer of SARS-CoV in the supernatants increased upon passage, and viral plaques in VeroE6 cells showed subtle differences from wild-type SARS-CoV/Urbani. We are currently analyzing the sequence of the SARS-CoV quasi-species selected by passage; such analysis may identify regions of the spike protein that are critical for virus binding and/or fusion. The identification of regions within the viral spike glycoprotein critical for interaction with the receptor is important for the development of antivirals and/or immunogens for vaccine development.

*Laura Gillim-Ross, Lindsay Heller, New York State Department of Health, Albany, New York 12002. Emily Olivieri, David E. Wentworth, New York State Department of Health and State University of New York, Albany, New York 12002.

1.1. Background

In 2002–2003, an outbreak of severe acute respiratory syndrome (SARS) spread from Southern China to 28 other countries.[3] SARS infected at least 8,096 people, and led to 774 deaths.[4] A novel coronavirus (CoV), SARS-Coronavirus (SARS-CoV) was identified as the causative agent of the outbreak.[5-7] Although the origin/reservoir of SARS-CoV has not been identified, the identification of SARS-CoV-like viruses and anti-SARS-CoV antibodies, in several species in live animal markets and on farms in China, strongly suggests that the SARS outbreak resulted from zoonotic transmission(s).[8, 9] Specifically, SARS-CoV–like viruses and antibodies to SARS-CoV have been detected in Himalayan palm civets *(Paguma larvata)*.[8, 9] Rapid evolution of the SARS-CoV in palm civets and humans[10] suggests that the predecessor of SARS-CoV is endemic in another common source, and that palm civets probably played a role as an intermediate or amplifying host, leading to zoonosis.

CoVs belong to the order *Nidovirales*, family *Coronaviridae*. CoVs are a diverse group of enveloped, positive-sense RNA viruses. The CoV genome, 27–32 kilobases in length, is the largest of the known RNA viruses. CoV attachment, and fusion of the viral lipid envelope with host lipid membrane, is mediated by trimeric spike (S) glycoproteins that project from the virion.[11] Although CoVs infect a wide variety of species, including dogs, cats, cattle, mice, birds, and humans, the natural host range of each strain is typically limited to a single species.[11] Interaction of virus S with host cell receptors is a major determinant of the species specificity and tissue tropism of CoVs. Upon entry into target cells, CoVs initiate viral replication, utilizing a complex discontinuous RNA transcription mechanism to generate 3' co-terminal subgenomic RNAs, which share a short 5' leader sequence.[12]

Using a multiplex RT-PCR assay that detects SARS-CoV genomic and subgenomic RNA, we previously showed that cells derived from African green monkey kidney (VeroE6), human liver (Huh7), human kidney (HEK-293T), and mink lung (Mv1Lu) were all productively infected by SARS-CoV.[1] The titer of virus produced varied dramatically (4 $\log_{10}$) between the different susceptible cell lines.[1] At 24 h post-inoculation, VeroE6 cells released the highest titer of virus (5 x 10^7 $TCID_{50}$/ml), Huh7 cells (2 x 10^4 $TCID_{50}$/ml), HEK-293T (5 x 10^3 $TCID_{50}$/ml), and Mv1Lu (8 x 10^2 $TCID_{50}$/ml) released much lower titers of virus.[1]

We hypothesized that passage of SARS-CoV in cells with varying levels of receptor, or with species-specific differences in the receptors, will select for SARS-CoV with mutations in the S gene, and/or in other genes, that are advantageous to virus replication. Therefore, analysis of selected mutants will identify regions of S that are critical to entry of the SARS-CoV. In this study, we generated mutant SARS-CoVs by subjecting SARS-CoV/Urbani to selective pressure, primarily at the level of host cell receptors. Virus was serially passaged 12 times in cell lines that were derived from African green monkey, human, and mink. In addition, two human cell lines that express differing levels of receptor were used. To determine whether phenotypic changes had been selected, we analyzed the amount of virus released into the supernatant, as well as the plaque morphology of SARS-CoV variants, after passage in each of the cell lines.

2. RESULTS AND DISCUSSION

Human ACE2 and CD209L were recently identified as functional receptors for SARS-CoV.[2, 13] Although both are functional receptors, human ACE2 is a more efficient receptor for SARS-CoV than is CD209L. The cell lines that we identified as productively infected by SARS-CoV were assayed for the presence of ACE2. One-step RT-PCR (Qiagen) with oligonucleotide primers conserved between mouse and human ACE2 was used to amplify ACE2 RNA from 1 μg of total RNA in susceptible cells derived from various species. Sense primer (ACE2-902, 5'-CTTGGTGATATGTGGGGTAGA) and an anti-sense primer (ACE2-1548R, 5'-CGCTTCATCTCCCACCACTT) amplify a 646-base-pair fragment of ACE2 when RNA is expressed. SARS-CoV susceptible Huh7, VeroE6, HEK-293T, and Mv1Lu cells[1] were assayed for ACE2 transcript. ACE2 RNA was detected in all of the susceptible cell lines (Figure 1A). Qualitatively, the level of ACE2 differed among the various cell lines. VeroE6 had the strongest ACE2 amplicon, whereas Mv1Lu cells showed the weakest product (Figure 1A). Although ACE2 amplicons from Mv1Lu cells are weak, RT-PCR amplification with mink-specific primers generates a robust amplicon (Heller *et al.*, this volume). The level of protein expressed among these four cell lines was greatest in VeroE6, and lowest in HEK-293T (Olivieri and Wentworth, unpublished data). Of the cell lines susceptible to SARS-CoV, Mv1Lu cells, which were derived from mink lung, are most closely related to those of the Himalayan palm civet. The predicted amino acid sequence of mink ACE2 has 83% identity with the human ACE2 sequence and 88% identity with palm civet ACE2 (Heller *et al.*, this volume). Therefore, we analyzed the ability of mink ACE2 to function as a SARS-CoV receptor. Expression of mink ACE2 in normally non-permissive BHK-21 cells resulted in SARS-CoV infection (Heller *et al.*, this volume). Taken together, the data show that these four cell lines all express ACE2 that is a functional receptor for SARS-CoV.

Variation in the levels of virus production among the four cell lines may correspond to differing levels of ACE2 expression, or to species-specific variations in ACE2. VeroE6 cells produce high titers of SARS-CoV,[1] and they also express high levels of ACE2.[2] In contrast, HEK-293T cells produce significantly less SARS-CoV (4 $\log_{10}$ lower titer),[1] and they express very low levels of ACE2.[2] SARS-CoV/Urbani (previously passaged 4 times in VeroE6) was passaged in VeroE6, Huh7, HEK-293T, or Mv1Lu cells for a total of 12 passages. Two flasks of each cell line were inoculated with SARS-CoV/Urbani (Pass 4) at an MOI of 1.0 (replicates A and B). Cells were observed every 24 h for cytopathic effect, and viral supernatants were collected at 24 h, 48 h, or 72 h postinoculation. The viral supernatants were either immediately passed, or were titered in VeroE6 cells before subsequent passage. Cytopathic effect was observed only in VeroE6 cells throughout the experiment. Viral titers were determined for all passages by $TCID_{50}$ in VeroE6 cells, as previously described.[1] The titer of virus produced by VeroE6 cells showed a modest increase after passage (Figure 1B). In contrast, virus passaged in Huh7 cells increased in titer by ~2.5 $\log_{10}$ (Figure 1B). Passage of virus in HEK-293T cells resulted in a decrease in virus titer from 8×10^6 $TCID_{50}$/ml to 2×10^5 $TCID_{50}$/ml (Figure 1B). The replicates (A and B) of VeroE6, Huh7, and HEK-293T passage had very similar titers. In contrast, one of the replicates (B) of Mv1Lu passage showed a dramatic increase in viral titer upon passage. The titer increased from 1×10^3 $TCID_{50}$/ml (average of passages 1 and 2) to 3.7×10^7 $TCID_{50}$/ml (average of passages 11 and 12); the latter titer is similar to that produced by VeroE6 cells (Figure 1B). The other replicate Mv1Lu passage (A) never reached a titer exceeding 2×10^2 $TCID_{50}$/ml, and was not detectable by passage 7 (data not shown). Thus, the titer of replicate

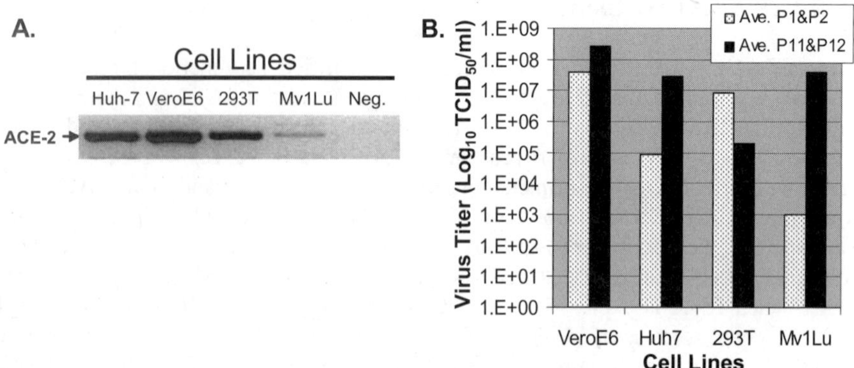

Figure 1. A. ACE2 RNA expression in cell lines susceptible to SARS-CoV. Negative image of ethidium bromide stained gel of RT-PCR amplicons from 1 μg of total RNA. Water was used as template for negative control reaction (Neg.). B. Passage of SARS-CoV in different cell lines results in changes in viral titer. SARS-CoV/Urbani was passaged 12 times in African green monkey kidney (VeroE6), human liver (Huh7), human kidney (293T), or mink lung (Mv1Lu) cell lines. The average of viral titer of passage 1 and 2 (P1 & P2) was compared with average titer of passage 11 and 12 (P11 & P12).

A decreased until the virus could not be sustained under the passage conditions. Alternatively, adaptation to mink ACE2 results in poor interaction with African green monkey ACE2, and consequently an artificially low $TCID_{50}$ in VeroE6 cells.

To determine whether the observed changes in viral titer correlated with changes in plaque phenotype, we analyzed viral supernatants obtained from the four cell lines at passage 12 (P12). We compared the plaque phenotype of the passaged variants with the parental wild-type virus (Urbani) using VeroE6 cell monolayers. VeroE6-P12 virus did not exhibit a significant change in plaque morphology when compared to Urbani. Both Urbani and VeroE6-P12 contained a population of subtly pleomorphic plaques. The average Urbani plaque was 0.7 mm, and the average VeroE6-P12 plaque was 1.0 mm in diameter. Huh7-P12, Mv1Lu-P12, and HEK-293T-P12 variants all showed increases in average plaque diameter to 1.2, 1.5, and 1.8 mm, respectively.

Changes observed in virus titer and in plaque morphology after 12 passages show that viral adaptation occurred as a result of the various selective pressures exerted by growth in primate, human, or mink cells. Species-specific differences in ACE2 expressed by these cells likely played a strong role in the selection process. In addition, differences in the level of human ACE2 expressed by Huh7 and HEK-293T likely selected for viruses with S glycoproteins that differ in their affinities for human ACE2, or that have enhanced abilities to use alternative receptors. For example, the largest increase in plaque size occurred after selection in HEK-293T cells, which express the lowest amount of ACE2, among the cell lines studied. We are currently analyzing the consensus nucleotide sequences of the quasi-species of the passaged viruses, to identify potential changes in S and/or other portions of the genome. We are also analyzing plaque-purified variants, because the plaque morphologies obtained suggest multiple genotypes exist in the passaged population of viruses. Comparison of the amino-acid sequence differences between human and mink ACE2 in the S binding domain, coupled with analysis of the adaptive changes occurring within S, will provide a better understanding of the SARS-CoV S-ACE2 interaction. Additionally, selection of SARS-CoV S variants that bind human ACE2 with higher affinity than does wild-type SARS-CoV will aid in the development of inhibitors that block viral entry. Lastly, the identification of changes in SARS-CoV that alter the tropism of the virus will provide additional information as to the mechanism(s) of zoonotic transmission of SARS-CoV.

3. ACKNOWLEDGMENTS

We thank Drs. William Bellini and Thomas Ksiazek for providing SARS-CoV/Urbani, and Dr. Aleem Siddiqui for providing Huh7 cells. The authors thank Noel Espina for his expert assistance, the Wadsworth Center Molecular Genetics Core for DNA sequencing, and the Wadsworth Center Tissue Culture Core Facility. L.G.-R. was supported by an appointment to the Emerging Infectious Diseases Fellowship program administered by the Association for Public Health Laboratories and funded by the CDC, and E.O. was supported by NIH/NIAID training grant T32AI05542901A1. This study was also funded in part by the Public Health Preparedness and Response to Bioterrorism cooperative agreement between the Department of Health and Human Services, and the Centers for Disease Control and Prevention, a research agreement with Diagnostic Hybrids Inc., and NIH/NIAID grants N01-AI-25490, and P01-AI-0595760.

4. REFERENCES

1. L. Gillim-Ross, J. Taylor, D. R. Scholl, J. Ridenour, P. S. Masters, and D. E. Wentworth, Discovery of novel human and animal cells infected by the severe acute respiratory syndrome coronavirus by replication-specific multiplex reverse transcription-PCR, *J. Clin. Microbiol.* **42**, 3196-3206 (2004).
2. W. Li, M. J. Moore, N. Vasilieva, et al., Angiotensin-converting enzyme 2 is a functional receptor for the SARS coronavirus, *Nature* **426**, 450-454 (2003).
3. N. S. Zhong and G. Q. Zeng, Our strategies for fighting Severe Acute Respiratory Syndrome (SARS), *Am. J. Respir. Crit. Care Med.* **168**, 7-9 (2003).
4. Summary of probable SARS cases with onset of illness from 1 November 2002 to 31 July 2003, *World Health Organization,* http://www.who.int/csr/sars/country/table2004_04_21/en/index.html, 4-21-2004.
5. C. Drosten, S. Gunther, W. Preiser, et al., Identification of a novel coronavirus in patients with severe acute respiratory syndrome, *N. Engl. J. Med.* **348**, 1967-1976 (2003).
6. T. G. Ksiazek, D. Erdman, C. S. Goldsmith, et al., A novel coronavirus associated with severe acute respiratory syndrome, *N. Engl. J. Med.* **348**, 1953-1966 (2003).
7. J. S. Peiris, S. T. Lai, L. L. Poon, et al., Coronavirus as a possible cause of severe acute respiratory syndrome, *Lancet* **361**, 1319-1325 (2003).
8. Y. Guan, B. J. Zheng, Y. Q. He, et al., Isolation and characterization of viruses related to the SARS coronavirus from animals in southern China, *Science* **302**, 276-278 (2003).
9. C. Tu, G. Crameri, X. Kong, et al., Antibodies to SARS coronavirus in civets, *Emerg. Infect. Dis.* **10**, 2244-2248 (2004).
10. H. D. Song, C. C. Tu, G. W. Zhang, et al., Cross-host evolution of severe acute respiratory syndrome coronavirus in palm civet and human, *Proc. Natl. Acad. Sci. USA* **102**, 2430-2435 (2005).
11. K. V. Holmes, in: *Fields Virology, Vol. 1*, edited by D. M. Knipe, P. M. Howley, D. E. Griffin, R. A. Lamb, M. A. Martin, and B. Roizman (Lippincott Williams & Wilkins, Philadelphia, 2001), pp. 1187-1203.
12. S. G. Sawicki and D. L. Sawicki, Coronavirus transcription: a perspective, *Curr. Top. Microbiol. Immunol.* **287**, 31-55 (2005).
13. S. A. Jeffers, S. M. Tusell, L. Gillim-Ross, et al., CD209L (L-SIGN) is a receptor for severe acute respiratory syndrome coronavirus, *Proc. Natl. Acad. Sci. USA* **101**, 15748-15753 (2004).

HUMAN CORONAVIRUS 229E CAN USE CD209L (L-SIGN) TO ENTER CELLS

Scott A. Jeffers, Erin M. Hemmila, and Kathryn V. Holmes*

1. INTRODUCTION

The primary receptor for SARS-CoV is ACE2, a metallopeptidase expressed on membranes of renal and cardiovascular tissues as well as the gastrointestinal tract.[1-3] Marzi *et al.*, showed that retroviral pseudotypes bearing the SARS-CoV glycoprotein could also bind to DC-SIGN and CD209L (also called L-SIGN or DC-SIGNR) on cell membranes, but not use these C-type lectins to enter cells.[4] We showed that CD209L, which is expressed on sinusoidal endothelial cells of the liver, endothelial cells of the lymph nodes, Peyer's patches, capillaries in the villous lamina propria of the terminal ileum, and in type II alveolar cells and endothelial cells in the lung, has receptor activity for SARS-CoV.[5] Briefly, Chinese hamster ovary (CHO) cells, which are refractory to binding of SARS-CoV spike glycoprotein and entry of SARS-CoV, were transduced with a human lung cDNA library in a retroviral vector. Cells that bound soluble SARS-CoV spike glycoprotein were detected by flow cytometry and inoculated with SARS-CoV under BSL-3 conditions. Virus entry and viral subgenomic RNA and protein synthesis were detected by RT-PCR and immunofluorescence. Less than 1% of the cells expressed SARS-CoV nucleocapsid protein 24 hours after inoculation. CD209L cDNA was cloned from subcloned cells positive for both binding of SARS-CoV spike and expression of SARS-CoV subgenomic RNA and N protein. Transfection of CHO cells with cDNA encoding CD209L made these cells as susceptible to SARS-CoV as the retrovirus transduced cells.

DC-SIGN on dendritic cells has been shown to act as an attachment factor for Ebola virus, HIV-1, and other enveloped viruses.[6] The dendritic cells, which are professional antigen-presenting cells, can present virus to macrophages or $CD4^+$ T-cells that express the appropriate receptor (CD4 and co-receptors for HIV-1). Thus DC-SIGN is said to mediate infection *in trans*.[7]

Our data suggest that CD209L may act as an alternative weak receptor for SARS-CoV. Coronaviruses bud from the ER-Golgi intermediate compartment (ERGIC) and are believed to be released from cells by exocytosis.[8, 9] The maturation of virus in the ERGIC

*University of Colorado Health Sciences Center at Fitzsimons, Aurora, Colorado.

may limit the amount of trimming of *N*-linked glycans on spike by endoglycosidases in the Golgi. We postulate that glycans on the coronavirus S proteins may be predominantly composed of the high mannose type rather than the more complex glycans on the spikes of enveloped viruses that bud from the plasma membrane. SARS-CoV may bind to CD209L on the plasma membrane through high mannose glycans on S, which may then mediate relatively inefficient virus entry. We have begun to study the roles of C-type lectins in entry of other coronaviruses such as human coronavirus (HCoV)-229E in addition to SARS-CoV.

2. RESULTS AND DISCUSSION

HCoV-229E uses human aminopeptidase N (hAPN) as its principal receptor[10] and can also use feline APN as an alternative receptor.[11] To determine whether the human coronavirus, HCoV-229E, could also use CD209L as a receptor, we used flow cytometry to compare binding of purified HCoV-229E virions to cells expressing hAPN (MRC5), CD209L (CHO/CD209L) or neither receptor protein (CHO) (Table 1). HCoV-229E virions bound equally well to MRC5 cells via hAPN and CHO/CD209L cells via CD209L.

Virus entry and expression of viral proteins were compared in CHO cells, CHO cells expressing CD209L, and mouse cells expressing hAPN. In cells that did not express hAPN or CD209L no viral antigen was detected at any time. However, in cells that expressed hAPN or CD209L viral antigens were observed 8 hours after virus inoculation. Twenty-four and 48 hours after inoculation, the hAPN-expressing cells were all positive for viral antigen. In contrast, fewer CD209L-expressing cells expressed viral antigens at 24 hours; however, by 48 hours after inoculation CD209L-expressing CHO cells were all positive for viral antigens. The pattern of antigen staining in the CD209L cells was granular with patches of bright cytoplasmic fluorescence; whereas staining in hAPN expressing cells was more uniform and bright throughout the cytoplasm. Seventy-two hours after inoculation the hAPN-expressing cells had high levels of cytopathic effect (CPE, large areas of rounded cells), whereas, the CD209L-expressing CHO cells showed no CPE and had formed a confluent monolayer of antigen positive cells. Thus, HCoV-229E entered cells that expressed CD209L, viral antigens were detected by immunofluorescence assay up to 72 hours after inoculation, and virus infection spread from cell-to-cell in the culture. These data suggest that CD209L can act as a weak receptor for entry of HCoV-229E, but has less HCoV-229E receptor activity than hAPN. No infectious

Table 1. Binding of HCoV-229E to cell lines.

Cells	% Cells that bind virus*
CHO	<5
MRC5	97
CHO/CD209L	96

* Virus binding was detected by flow cytometry with monoclonal antibody 511H6 directed against the viral spike glycoprotein.

virus was released into the medium of CD209L-expressing CHO cells, as determined by plaque assay. These data suggest that CD209L allows entry of HCoV-229E into some cells leading to synthesis of viral RNA and proteins. Possibly, CD209L on cell membranes traps newly formed virus preventing release of virus into the medium, while still allowing cell-to-cell spread of infection.

In order to further examine the interaction of CD209L with HCoV-229E spike, soluble CD209L with a 6-His tag was expressed in bacteria and purified to near homogeneity using mannose affinity chromatography and nickel affinity chromatography. Cell surfaces are replete with high-mannose glycans. To determine if purified soluble CD209L could bind to high mannose glycans on cells, trypsinized cells that express CD209L, hAPN, or neither receptor were incubated with 200 μg of soluble CD209L for 1 hour at 4°C. FACS analysis with anti-CD209L antibody showed that soluble CD209L bound to CHO cells expressing CD209L, MRC5 cells expressing hAPN, and control CHO cells that expressed neither receptor (Table 2).

To examine the effects of soluble CD209L on binding of HCoV-229E to cell lines, various concentrations of the purified soluble CD209L protein were incubated with purified HCoV-229E virions for 1 hour at 4°C. The virions were then incubated with trypsinized CHO cells expressing CD209L, MRC5 cells expressing hAPN, or control CHO cells expressing neither receptor. Virus binding was detected by flow cytometry with monoclonal anti-229E spike antibody 511H6. Table 3 shows that lower levels of soluble CD209L (50 μg/ml) significantly blocked binding of HCoV-229E virions to hAPN or CD209L in cell membranes. However, larger amounts of soluble CD209L (200 μg/ml) increased binding of virions to CD209L and hAPN-expressing cells. Interestingly, virions incubated with 200 μg/ml of soluble CD209L bound to CHO cells that do not express either hAPN or CD209L. The soluble CD209L protein apparently acted as a bridge between the virions and the cellular surface. It is possible that binding of CD209L treated virions to CHO cells occurred when the carbohydrate recognition domain of soluble tetrameric CD209L bound to glycans on the surface of the virion, to glycans on soluble tetrameric CD209L and to glycans on the cell surface. Perhaps other soluble C-type lectins in the lung, such as surfactant protein D (SP-D), could also act as a bridge to bind virions to cells. For example HCoV-229E or SARS-CoV combined with a soluble C-type lectin might be able to enter cells that do not express a specific viral receptor such as hAPN or ACE2.

Table 2. Soluble CD209L binding to cell lines.

Cell line	[sCD209L] μg/ml	% Cells that bind antibody to CD209L*
CHO	0	<5
CHO	200	93
MRC5	0	7
MRC5	200	94
CHO/CD209L	0	97
CHO/CD209L	200	99

* Detected by flow cytometry with goat polyclonal antibody to the C terminal domain of CD209L.

Table 3. Effects of soluble CD209L on binding of HCoV-229E virions to cell lines.

Cell line	[sCD209L] μg/ml	% Cells that bind virus*
CHO	0	31
CHO	50	<5
CHO	200	87
MRC5	0	92
MRC5	50	22
MRC5	200	81
CHO/CD209L	0	94
CHO/CD209L	50	59
CHO/CD209L	200	96

* Virus binding was detected by flow cytometry with monoclonal antibody 511H6 directed against the viral spike glycoprotein.

Surfactant protein D in the lungs a C-type lectin that contains a carbohydrate recognition domain similar to that of CD209L[12] is an important innate host defense mechanism that binds to high-mannose glycans on the surfaces of bacteria and viruses. Bacteria and viruses coated by SP-D form aggregates and are phagocytosed by macrophages. Because soluble CD209L can bind to HCoV-229E virions and enhance binding of virions to receptor negative cells in culture, it is possible that SP-D could also bind to HCoV-229E and mediate the attachment and eventual entry of virus into macrophages and type II alveolar cells in the lung.

The National Center for Biotechnology Information database lists 47 sequences for CD209L (also called L-SIGN or DC-SIGNR) from humans, apes, and mice. Many mammalian species have homologues of C-type lectins such as CD209L, SP-D, and mannose binding lectin. These anchored or soluble C-type lectins may act as a bridge to allow for the initial crossing of the species barrier by viruses. If the only block to infection is receptor usage, it is possible that an enveloped virus that normally infects a non-human species could use human CD209L as a weak receptor to enter a small number of human type II alveolar cells or macrophages. Virus mutants that could use a different receptor in the lung more efficiently might develop and these would have a selective advantage for growth in human lung. The selected virus mutants might then be able to spread from human to human using the new receptor.

Further study of CD209L interacting with HCoV-229E in mouse or hamster models will help to elucidate the significance of this receptor in coronavirus infection *in vivo*.

3. ACKNOWLEDGMENTS

We are grateful to Dr. T. Miura for insightful discussion and Dr. P. Talbot for the generous gift of the 511H6 hybridoma. This research was supported by NIH grant AI59576. S.A.J. was supported by NIH postdoctoral training grant 1 T32 AI07587.

4. REFERENCES

1. W. Li, M. J. Moore, et al., Angiotensin-converting enzyme 2 is a functional receptor for the SARS coronavirus, *Nature* **426**, 450-454 (2003).
2. S. K. Wong, W. Li, et al., A 193-amino acid fragment of the SARS coronavirus S protein efficiently binds angiotensin-converting enzyme 2, *J. Biol. Chem.* **279**, 3197-3201 (2004).
3. D. Harmer, M. Gilbert, et al., Quantitative mRNA expression profiling of ACE 2, a novel homologue of angiotensin converting enzyme, *FEBS Lett.* **532**, 107-110 (2002).
4. A. Marzi, T. Gramberg, et al., DC-SIGN and DC-SIGNR interact with the glycoprotein of Marburg virus and the S protein of severe acute respiratory syndrome coronavirus, *J. Virol.* **78**, 12090-12095 (2004).
5. S. A. Jeffers, S. M. Tusell, et al., Cd2091 (L-Sign) is a receptor for severe acute respiratory syndrome coronavirus, *Proc. Natl. Acad. Sci. USA* **101**, 15748-15753 (2004).
6. R. W. Doms and D. Trono, The plasma membrane as a combat zone in the HIV battlefield, *Genes Dev.* **14**, 2677-2688 (2000).
7. M. T. Yu Kimata, M. Cella, et al., Capture and transfer of simian immunodeficiency virus by macaque dendritic cells is enhanced by DC-SIGN, *J. Virol.* **76**, 11827-11836 (2002).
8. E. Lontok, E. Corse, et al., Intracellular targeting signals contribute to localization of coronavirus spike proteins near the virus assembly site, *J. Virol.* **78**, 5913-5922 (2004).
9. A. G. Bost, E. Prentice, et al., Mouse hepatitis virus replicase protein complexes are translocated to sites of M protein accumulation in the ERGIC at late times of infection, *Virology* **285**, 21-29 (2001).
10. C. L. Yeager, R. A. Ashmun, et al., Human aminopeptidase N is a receptor for human coronavirus 229E, *Nature* **357**, 420-422 (1992).
11. D. B. Tresnan, R. Levis, et al., Feline aminopeptidase N serves as a receptor for feline, canine, porcine, and human coronaviruses in serogroup I, *J. Virol.* **70**, 8669-8674 (1996).
12. J. K. van de Wetering, L. M. van Golde, et al., Collectins: players of the innate immune system, *Eur. J. Biochem.* **271**, 1229-1249 (2004).

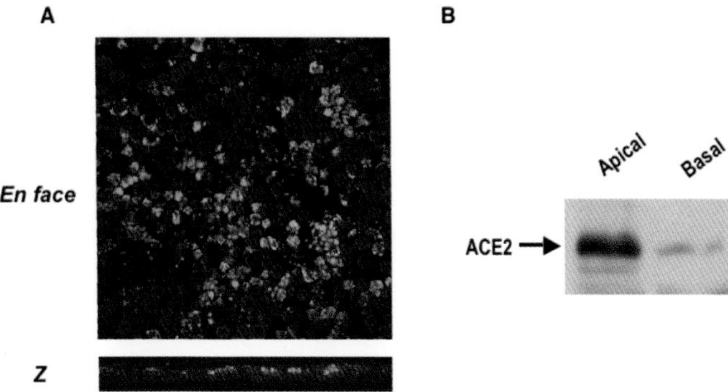

Figure 1. ACE2 is expressed in human airway epithelia. (A) ACE2 protein location in polarized human airway epithelia was determined using immunofluorescence staining for ACE2 (green) and the nucleus (ethidium bromide, red). Confocal fluorescence photomicroscopic images are presented *en face* (top) and from vertical sections in the z axis (bottom). (B) ACE2 protein location in polarized human airway epithelia was determined by selective apical or basolateral biotinylation, immunoprecipitation of biotinylated surface proteins, and immunoblat analysis for ACE2. (See page 480).

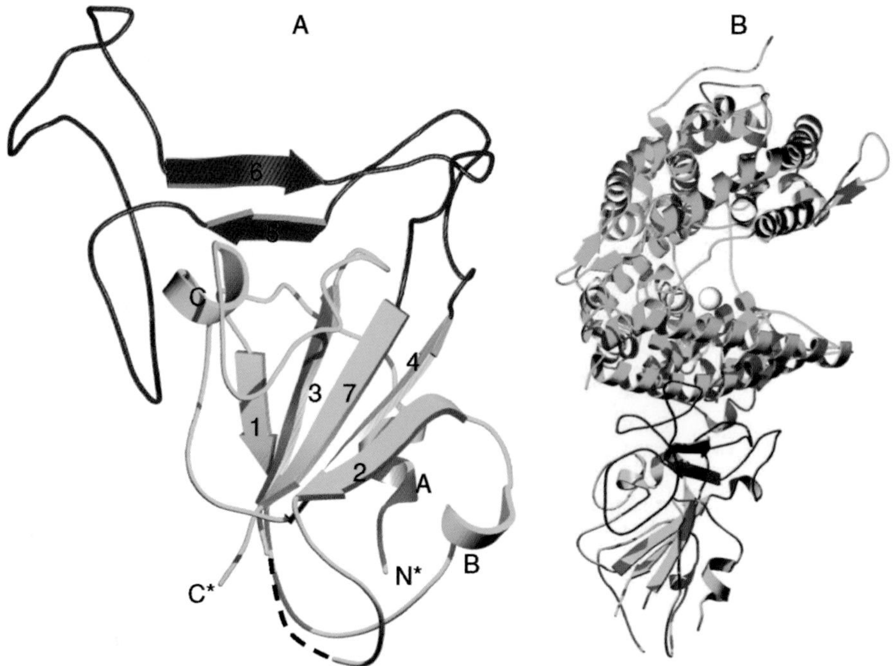

Figure 3. Crystal structure of the ACE2/RBD complex. (A) Structure of the RBD that contains two subdomains: a core structure (in cyan) and a receptor-binding motif (in red). (B) Structure of the ACE2/RBD complex. The RBD binds to the outer surface of the N-terminal lobe of ACE2 (in green).

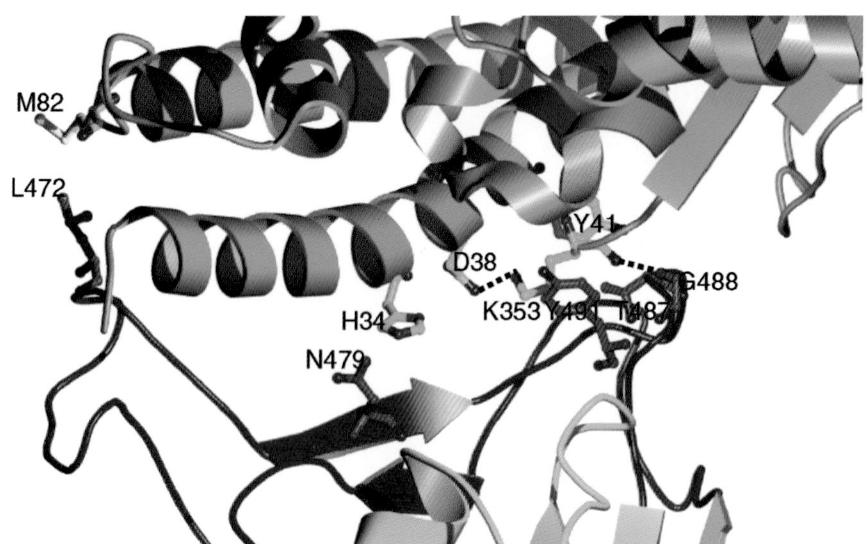

Figure 4. Residues located at the ACE2/RBD interface and important to the species specificity of SARS-CoV. Leu472 on the RBD has a hydrophobic interaction with Met82 on ACE2. L472P mutation may attenuate the virus. On rat ACE2, residue 82 is glycosylated, preventing the binding of SARS-CoV. A K479N mutation on the RBD is critical for SARS-CoV to jump from civets to humans. Thr487 on the RBD forms a hydrophobic interaction with Lys353 on ACE2. A S487T mutation on the RBD is important for SARS-CoV to transmit from human to human. On both rat ACE2 and mouse ACE2, residue 353 is a histidine, disfavoring the binding of SARS-CoV. (See pages 232 and 233).

INTRACELLULAR TRANSPORT OF THE S PROTEINS OF CORONAVIRUSES

Christel Schwegmann-Weßels, Xiaofeng Ren, and Georg Herrler*

1. INTRODUCTION

Coronaviruses mature by a budding process at intracellular membranes. For two of the viral membrane proteins, M and E, it has been shown that they are intracellularly retained. Upon single expression, the M proteins of transmissible gastroenteritis virus (TGEV) and avian infectious bronchitis virus are localized in the cis-Golgi network or cis-Golgi complex.[1, 2] The small membrane protein E transiently resides in a pre-Golgi compartment[3] before it progresses to the Golgi apparatus.[4, 5] The S protein of TGEV is retained intracellularly.[6] Retention is mediated by a tyrosine-based signal within the cytoplasmic tail. In contrast, the S protein of SARS-CoV lacks a tyrosine-residue in the corresponding tail portion, and in fact, it is transported to the cell surface.[6, 7]

We analyzed the protein expression of TGEV S protein and SARS-CoV S protein in two different expression systems. In the pTM1 vector, gene expression is under the control of a T7 promoter.[8] This expression system requires the use of cells expressing the T7 RNA-polymerase, e.g., BSR-T7/5 cells. To exclude the possibility that retention of the S protein of TGEV is affected by the expression system, we compared pTM1-driven expression with expression by a plasmid under the control of a CMV promoter. As coronavirus S proteins cannot be expressed by standard plasmid vectors containing a CMV promoter, we used the vector pCG1 (kindly provided by Dr. Cattaneo), which contains a rabbit β-globin intron. This plasmid vector allows the expression of the S protein in different cell lines independent of T7 expression.

2. METHODS

The pTM1 plasmids were constructed as described previously.[6] For the construction of the pCG1 plasmids, the 5'-end (first 1200 nucleotides) of the TGEV S protein gene was amplified from the plasmid TGEVS-pTM1 by PCR using oligonucleotides a and b (see Table 1, Fig. 1). Primer a contained an BamHI restriction site. After digestion of TGEVS-pTM1 with restriction enzymes XhoI and PstI, the resulting fragment of about

*Institut für Virologie, Tierärztliche Hochschule Hannover, 30559 Hannover, Germany.

Table 1. Primers used for plasmid constructions.

	5' – sequence – 3'
a	TTTGGATCCCACACCATGAAAAAACTATTTGTGGTTTTGG
b	ACAGTACCGTGGTCCATCAGTTAC
c	GTTAACCAGAATGCTCAAGCATTAA
d	GGCCTCTAGATTATGTGTAATGTAATTTGACACCCTTGAG

3400 nucleotides and the 1200 bp PCR product were ligated in the pCG1 vector via restriction sites BamHI and PstI. The new plasmid was designated TGEVS-pCG1 (Fig. 1). For the generation of mutant Y1440A-pCG1, the respective pTM1 construct was digested with SpeI (restriction site at position 3834) and PstI. The resulting fragment was inserted into TGEVS-pCG1 via these restriction sites (Fig. 1). The 3'-end of the SARS-CoV S protein was amplified from the plasmid pcDNA-spike (kindly provided by Dr. Deng) by PCR using oligonucleotides c and d (see Table 1, Fig. 1). Primer d contained an XbaI restriction site. After incubation of pcDNA-spike with BamHI and EcoRV, the resulting 5'-end of the gene was ligated with the PCR product into the pCG1 vector and designated SARSS-pCG1 (Fig. 1).

Immunofluorescence analysis was performed as described previously.[6] For the detection of the SARS-CoV S protein, a rabbit antiserum directed against SARS-CoV was used (kindly provided by Dr. Eickmann).

3. RESULTS

The transport of the coronavirus S proteins was analyzed using plasmid vectors to avoid overexpression. As expression vectors that depend on nuclear transcription were very inefficient, all constructs were first cloned into the pTM1 vector under the control of the T7 promoter. The constructs were transiently expressed in BSRT7/5 cells (kindly provided by Dr. Conzelmann) that stably express the T7 RNA-polymerase. Figure 2 shows that the TGEV S protein was only detectable intracellularly by immunofluorescence

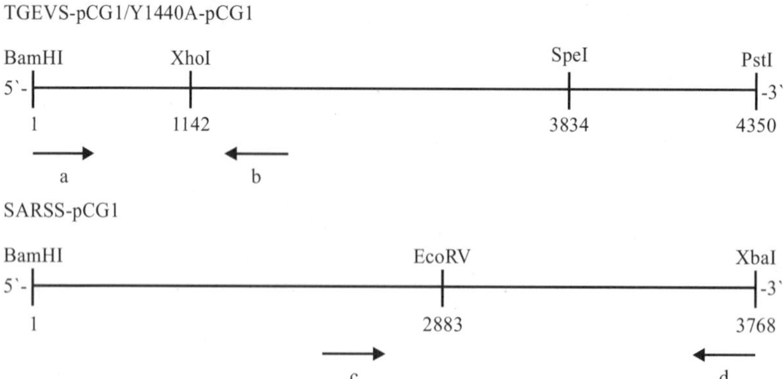

Figure 1. Schematic drawing of the pCG1-constructs made for this study. The primer binding sites are indicated with arrows. The important restriction sites and their position are indicated.

microscopy. The replacement of the tyrosine at amino acid position 1440 by alanine (Y1440A) resulted in a protein, which is transported to the cell surface (Fig. 2). In contrast to the TGEV S protein, the SARS-CoV S protein is expressed at the cell surface (Fig. 2). As expression with the pTM1 vector is only possible in cells expressing T7 polymerase, it was not possible to express these constructs in other cells. For this reason we cloned the respective S genes into the pCG1 vector, which depends on nuclear transcription but in contrast to standard vectors, contains a β-globin intron. With this vector it was possible to express the coronavirus S proteins in BHK21 cells. By fluorescence microscopy, the TGEV S protein was readily detectable inside cells (Fig. 2). On the cell surface we found occasionally some faint fluorescent patches, which might represent small amounts of S protein. This surface fluorescence can be explained by the stronger expression obtained with the pCG1 vector in BHK21 cells. The Y1440A mutant and the SARS-CoV S protein were efficiently transported to the cell surface (Fig. 2). Taken together, our results indicate that the transport of the analyzed coronavirus S proteins was similar in BSR-T7/5 cells and in BHK21 cells. The protein expression was much stronger in the latter cells transfected with the pCG1 vector, i.e., the percentage of expressing cells and intensity of expression per cell was higher in BHK21 cells than in BSR-T7/5 cells. The intracellular localization of the TGEV S protein was similar in BSR-T7/5 and BHK21 cells, with most of the antigen concentrated on one side of the nucleus. These findings demonstrate that the tyrosine-based signal in the TGEV S protein acts as a retention signal irrespective of the vector and cell line used for expression.

4. CONCLUSION

Coronaviruses mature by a budding process at the cis-Golgi network/endoplasmic reticulum-Golgi intermediate compartment.[9] The two coronavirus envelope proteins M and E are known to be intracellularly retained.[1, 2] For the coronavirus S proteins, others found some surface expression by using efficient expression vectors like vaccinia virus or baculovirus.[10, 11, 12] Even with vaccinia virus, the majority of the S protein was found intracellularly. The weak surface expression can be explained by an overexpression of the S protein resulting in saturation of the retention machinery. By using the pCG1 vector in BHK21 cells, we observed the same phenomenon. Because of the strong expression, small amounts of S protein could be seen at the cell surface in fluorescence microscopy. The majority of the TGEV S protein is localized intracellularly in a region that may represent the endoplasmic reticulum–Golgi intermediate compartment. For optimal virus production, it appears reasonable that the S protein is retained at the site of virus budding. The transport behavior of the mutant Y1440A indicates that the tyrosine plays an important role for the localization of the TGEV S protein. In our experiments, the SARS-CoV S protein is transported to the cell surface independent from the vector/cell line used. Other groups also reported surface expression of this protein.[7] Thus, with the pCG1 vector that we used in BHK21 cells and with the pCAGGS vector Simmons and coworkers used in 293T cells, efficient surface expression of SARS-CoV S protein is detectable. As the pCG1 vector contains a CMV promoter and the pCAGGS vector a chicken β-actin promoter, the kind of promoter does not appear to be responsible for the efficient expression. With other vectors containing a CMV promoter, e.g., pcDNA3.1, the S protein expression via the nucleus was very inefficient. As coronaviruses replicate in the cytoplasm, the S gene may contain cryptic splice sites or other sequence elements that are detrimental for mRNA processing in the nucleus. As the pCG1 and pCAGGS vectors contain a β-globin intron, this intron appears to prevent the degradation of the viral mRNA.

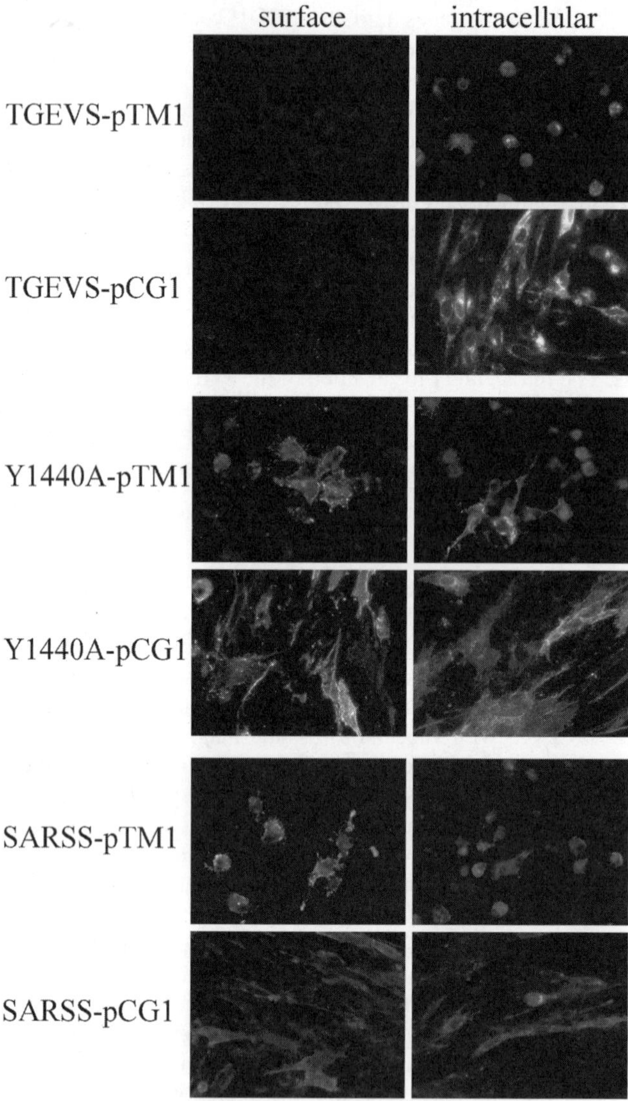

Figure 2. Surface and intracellular immunofluorescence analysis of parental TGEV S protein (TGEVS-pTM1, TGEVS-pCG1), the TGEV S mutant (Y1440A-pTM1, Y1440A-pGC1; the amino acid exchange and position is indicated), and parental SARS-CoV S protein (SARSS-pTM1, SARSS-pCG1). BSR-T7/5 cells (for pTM1 constructs) and BHK21 cells (for pCG1 constructs) were transfected with the genes indicated. Cells were analyzed for surface and intracellular expression of the proteins at 24 h post-transfection by immunofluorescence microscopy.

In future studies, we want to analyze the S protein expression in different cell lines to see if there is a difference in protein transport and localization between different cell lines. By using cellular compartment markers, the TGEV S protein localization will be determined.

5. ACKNOWLEDGMENTS

This work was supported by grants from the Sino-German Center for Rresearch Promotion, from Deutsche Forschungsgemeinschaft (He1168/12-1 and SFB621), and from the European Community (No. 511064). We thank Drs. Cattaneo, Conzelmann, Deng, Eickmann, and Enjuanes for providing cells, DNA, and antibodies. Parts of Figure 2 were taken from Schwegmann-Weβels et al., *J. Biol. Chem.* 279, 43661-43666, with permission of the copyright owner.

6. REFERENCES

1. C. E. Machamer, S. A. Mentone, J. K. Rose, and M. G. Farquhar, The E1 glycoprotein of an avian coronavirus is targeted to the cis Golgi complex, *Proc. Natl. Acad. Sci. USA* **87**, 6944-6948 (1990).
2. J. Klumperman, J. K. Locker, A. Meijer, M. C. Horzinek, H. J. Geuze, and P. J. Rottier, Coronavirus M proteins accumulate in the Golgi complex beyond the site of virion budding, *J. Virol.* **68**, 6523-6534 (1994).
3. K. P. Lim and D. X. Liu, The missing link in coronavirus assembly. Retention of the avian coronavirus infectious bronchitis virus envelope protein in the pre-Golgi compartments and physical interaction between the envelope and membrane proteins, *J. Biol. Chem.* **276**, 17515-17523 (2001).
4. E. Corse and C. E. Machamer, Infectious bronchitis virus E protein is targeted to the Golgi complex and directs release of virus-like particles, *J. Virol.* **74**, 4319-4326 (2000).
5. E. Corse and C. E. Machamer, The cytoplasmic tail of infectious bronchitis virus E protein directs Golgi targeting, *J. Virol.* **76**, 1273-1284 (2002).
6. C. Schwegmann-Weβels, M. Al Falah, D. Escors, Z. Wang, G. Zimmer, H. Deng, L. Enjuanes, H. Y. Naim, and G. Herrler, A novel sorting signal for intracellular localization is present in the S protein of a porcine coronavirus but absent from severe acute respiratory syndrome-associated coronavirus, *J. Biol. Chem.* **279**, 43661-43666 (2004).
7. G. Simmons, J. D. Reeves, A. J. Rennekamp, S. M. Amberg, A. J. Piefer, and P. Bates, Characterization of severe acute respiratory syndrome-associated coronavirus (SARS-CoV) spike glycoprotein-mediated viral entry, *Proc. Natl. Acad. Sci. USA* **101**, 4240-4245 (2004).
8. B. Moss, O. Elroy-Stein, T. Mizukami, W. A. Alexander, and T. R. Fuerst, Product review. New mammalian expression vectors, *Nature* **348**, 91-92 (1990).
9. J. Tooze, S. Tooze, and G. Warren, Replication of coronavirus MHV-A59 in sac- cells: determination of the first site of budding of progeny virions, *Eur. J. Cell Biol.* **33**, 281-293 (1984).
10. M. Godet, D. Rasschaert, and H. Laude, Processing and antigenicity of entire and anchor-free spike glycoprotein S of coronavirus TGEV expressed by recombinant baculovirus, *Virology* **185**, 732-740 (1991).
11. D. J. Pulford and P. Britton, Intracellular processing of the porcine coronavirus transmissible gastroenteritis virus spike protein expressed by recombinant vaccinia virus, *Virology* **182**, 765-773 (1991).
12. H. Vennema, L. Heijnen, A. Zijderveld, M. C. Horzinek, and W. J. Spaan, Intracellular transport of recombinant coronavirus spike proteins: implications for virus assembly, *J. Virol.* **64**, 339-346 (1990).

ANALYSIS OF SARS-CoV RECEPTOR ACTIVITY OF ACE2 ORTHOLOGS

Emily R. Olivieri, Lindsey K. Heller, Laura Gillim-Ross, and David E. Wentworth*

1. INTRODUCTION

An outbreak of severe acute respiratory syndrome that killed 774 of the 8,096 people infected in 2002–2003 was caused by a coronavirus (SARS-CoV).[1] Zoonotic transmission of SARS-CoV was likely responsible for the outbreak, and the virus infects multiple species. Human angiotensin-converting enzyme 2 (ACE2) is a functional receptor for SARS-CoV.[2] Coronavirus spike-receptor interactions are major determinants of species specificity, and transfection of viral genomic RNA or expression of receptors in nonpermissive cell lines usually results in productive infection. Our hypothesis is that species-specific differences in ACE2 are important in SARS-CoV infection, and analysis of ACE2 orthologs will permit the identification of regions of the receptor that are critical for virus entry. We analyzed cell lines that were derived from numerous species for their susceptibility to SARS-CoV. Cell lines derived from human, monkey, and mink were permissive to SARS-CoV and ACE2 RNA transcripts were detected in all of these cell lines. ACE2 RNA was also detected in nonpermissive dog (*Canis familiaris*) and chicken (*Gallus gallus*) cells. We used regions conserved between human ACE2 (hACE2) and mouse ACE2 to amplify regions of ACE2 from diverse species by RT-PCR and 3'RACE. Sequence analysis demonstrated that dog ACE2 had 87% nucleotide identity and 81% amino acid identity with hACE2, and chicken ACE2 has 79% nucleotide identity and 62% amino acid identity with hACE2. The ACE2 open reading frames from dog and chicken cDNA clones were subcloned into eukaryotic expression to analyze their function as SARS-CoV receptors.

* Emily R. Olivieri, David E. Wentworth, New York State Department of Health and State University of New York, Albany, New York 12002. Laura Gillim-Ross, Lindsay K. Heller, New York State Department of Health, Albany, New York 12002.

1.1. Background

Most coronaviruses (CoVs) have a very narrow host range.[3] SARS-CoV, however, was isolated from a number of animals naturally, including several Himalayan palm civets, a raccoon dog,[4] and a pig.[5] SARS-CoV does not appear to be enzootic in these animals, and the natural reservoir of the virus has not been identified. SARS-CoV also infects multiple species experimentally, including nonhuman primates, ferrets, cats, mice, and hamsters.[6-8]

SARS-CoV is an enveloped, single-stranded virus with a positive-sense 29.7-kb mRNA-like genome. CoVs have type 1 membrane glycoproteins, called spike, that project from their surface. Spike (S) interacts with cellular receptors and mediates binding and subsequent fusion of the viral envelope with the host cell membrane.[3] S-receptor interactions are a major determinant of species specificity, and of pathogenesis of CoVs.[3] Human ACE2, a zinc-containing carboxypeptidase, was identified as an efficient receptor for SARS-CoV.[2] CD209L was also recently shown to be a functional receptor for SARS-CoV, but it was a less efficient receptor than hACE2.[9] Although some regions of S and ACE2 that are important in virus binding were recently identified,[10] much remains to be learned about binding of SARS-CoV, and its fusion with the host cell membrane.

The goals of this study were to determine whether ACE2 transcript is produced by permissive and nonpermissive cells derived from diverse species and to use species-specific differences in ACE2 to identify amino acids or other post-translational modifications critical for SARS-CoV binding and/or fusion.

2. RESULTS AND DISCUSSION

Cell lines derived from monkey (VeroE6, pRhMk, and pCMK), human (HRT-18, HEK293T, and Huh-7), mink (Mv1Lu), dog (MDCK), cat (CRFK), hamster (BHK-21), and chicken (CEF) were analyzed for susceptibility to SARS-CoV.[11] VeroE6, pRhMK, pCMK, HEK293T, Huh-7, and Mv1Lu were susceptible to SARS-CoV, whereas MDCK and CEF were not permissive to SARS-CoV.[11] To determine whether ACE2 mRNA was expressed by permissive and non-permissive cells, we designed primers for RT-PCR based on consensus sequence alignments between mouse ACE2 and hACE2. One-step RT-PCR (Gillim-Ross, *et al.*, this volume), was used to amplify a region of ACE2 transcript corresponding to nucleotides 902-1548 of hACE2. ACE2 RNA was detected in all of the permissive cells derived from monkey, human, and mink, and it was not detected in CRFK cells (Figure 1A). Dog (MDCK) and chicken (CEF) cells expressed ACE2 RNA, but are not permissive to SARS-CoV (Figure 1A). This data suggests that dog ACE2 and chicken ACE2 may have amino acid or other differences, which inhibit their function as SARS-CoV receptors. Alternatively, SARS-CoV infection may be inhibited after binding and/or fusion of the virus. To identify regions of ACE2 that are unique in the nonpermissive species, we amplified overlapping fragments of the predicted open reading frames (ORFs) from dog and chicken ACE2 by RT-PCR, using consensus primers. The nucleotide sequence of these amplicons was used to generate oligonucleotide primers specific for the dog and chicken ACE2 ORFs. The phylogenic relationship of the ACE2 ORFs from human, Himalayan palm civet, mink, cow, mouse, rat, pig, dog, and chicken was analyzed (Figure 1B).

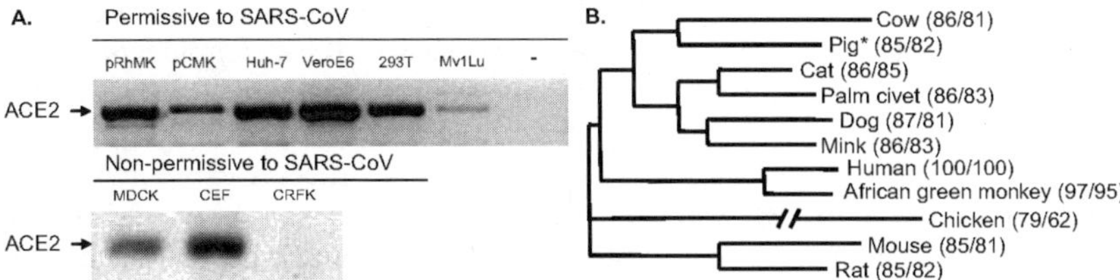

Figure 1. A. ACE2 RNA is expressed by cell lines from diverse species. ACE2 RNA was amplified by RT-PCR from 1 μg of total RNA template. Amplicons were visualized by ethidium bromide staining after agarose gel electrophoresis and a negative image is shown. Water was used as template for negative control (-) reaction. B. Amino acid relationship of ACE2 orthologs. The deduced amino acid sequences of ACE2 ORFs from various species were analyzed using Vector NTI Advance9.1 (Invitrogen). The percent nucleotide/amino acid identity as compared with human ACE2 is shown in parenthesis.

The predicted ORF for hACE2 encodes an 805-amino-acid protein. Among mammalian species, the predicted ACE2 protein had strong amino acid conservation, which ranged from 95% (African green monkey) to 81% (mouse) identity with hACE2 (Figure 1B). Chicken ACE2 showed the greatest divergence from hACE2, and had only 62% amino acid identity. Comparison of hACE2 with dog or chicken ACE2 showed 145 or 305 amino acid differences, respectively. Intriguingly, dog ACE2 showed 90% amino acid identity and 92% similarity with mink ACE2, which is a functional SARS-CoV receptor (Heller *et al.*, this volume). Thus, the limited amino acid differences between dog and mink ACE2 block SARS-CoV entry, or MDCK cells are resistant to SARS-CoV because of inhibition of at another stage of virus replication. To investigate this further, we tested another dog cell line (A72) and nonpermissive BHK-21 cells, transfected with a dog ACE2 expression construct, for their susceptibility to SARS-CoV. Our preliminary results suggest that dog ACE2 also functions as a SARS-CoV receptor (data not shown).

We used the crystal structure of hACE2[12] to analyze differences between efficient and inefficient SARS-CoV receptors. Amino acid differences between human and dog, or human and chicken ACE2 were mapped on the crystal structure of hACE2 using Cn3D version 4.1.[13] We used multiple sequence alignment of efficient and poor receptors, combined with difference mapping on the hACE2 structure, to identify specific residues or post-translational modifications within ACE2 that may be critical for SARS-CoV entry. Lysine 353 of hACE2 was recently shown to be important in the binding of an S1-Ig fusion protein (SARS-CoV S1 domain fused to the Fc domain of human IgG1).[10] This residue is a histidine in mouse and rat ACE2, which are less efficient SARS-CoV receptors.[10] Our data indicate that human, monkey, palm civet, mink, cow, pig, dog, cat, and chicken ACE2 all have K353. Yet, some of these molecules (e.g., chicken ACE2) appear to be poor or nonfunctional SARS-CoV receptors. Therefore, additional residues and/or post-translational modifications are influencing species specificity of SARS-CoV. Li *et al.*, also showed that K31, Y41, MYP 82-84, D355, and R357 affect binding.[10] Our data suggest that additional residues such as T20, H34, S113, T122, D136, H228, E232, E233, A246, E329, V339, G354, P426, N432, K465, M474, N572, D629, and WND 635-637 may have a role in ACE2 receptor activity (see also Heller *et al.*, this volume). We also identified differences in putative N-linked glycosylation sites at amino acids 216–218, 280–282, and 299–301 that could influence SARS-CoV entry. Chicken ACE2 has a potential glycosylation site at amino acids 280–282. In contrast, efficient SARS-CoV receptors such as, human, monkey, palm civet, and mink ACE2 don't contain a putative glycosylation site at 280–282. Differences in glycosylation can have dramatic effects on virus receptor activity and were shown to be species-specific determinants of Group1 CoV receptor CD13.[14] Additionally, glycosylation of rat ACE2 at N84 decreases binding of SARS-CoV S1-Ig fusion protein.[10]

The data demonstrate that cells derived from diverse species express ACE2 RNA transcripts, and some ACE2 RNA positive cell lines were not permissive to SARS-CoV. Multiple sequence alignment of ACE2 orthologs, combined with mapping the amino acid substitutions on the crystal structure of hACE2, was used to identify 24 residues that may influence SARS-CoV receptor activity. Site-directed mutagenesis of ACE2 expression constructs is being used to determine how specific amino acids, or post-translational modifications, influence binding and/or fusion of SARS-CoV.

3. ACKNOWLEDGMENTS

We would like to thank Noel Espina for his technical assistance and the Wadsworth Center Molecular Genetics Core for DNA sequencing. E.O. was supported by NIH/NIAID grant T32AI05542901A1, and L.G.-R. was supported by the EID Fellowship Program administered by the APHL and funded by the CDC. This work was also supported by N01-AI-25490, and P01-AI-05957601 from the NIH/NIAID.

4. REFERENCES

1. Summary of probable SARS cases with onset of illness from 1 November 2002 to 31 July 2003, *World Health Organization,* http://www.who.int/csr/sars/country/table2004_04_21/en/index.html, 2004.
2. W. H. Li, M. J. Moore, N. Vasilieva, et al., Angiotensin-converting enzyme 2 is a functional receptor for the SARS coronavirus, *Nature* **426**, 450-454 (2003).
3. K. V. Holmes, in: *Fields Virology, Vol. 1*, edited by D. M. Knipe, P. M. Howley, D. E. Griffin, R. A. Lamb, M. A. Martin, and B. Roizman (Lippincott Williams & Wilkins, Philadelphia, 2001), pp. 1187-1203.
4. Y. Guan, B. J. Zheng, Y. Q. He, et al., Isolation and characterization of viruses related to the SARS coronavirus from animals in southern China, *Science* **302**, 276-278 (2003).
5. W. Chen, M. Yan, L. Yang, et al., SARS-associated coronavirus transmitted from human to pig, *Emerg. Infect. Dis.* **11**, 446-448 (2005).
6. D. E. Wentworth, L. Gillim-Ross, N. Espina, and K. A. Bernard, Mice susceptible to SARS coronavirus, *Emerg. Infect. Dis.* **10**, 1293-1296 (2004).
7. K. Subbarao, J. McAuliffe, L. Vogel, et al., Prior infection and passive transfer of neutralizing antibody prevent replication of severe acute respiratory syndrome coronavirus in the respiratory tract of mice, *J. Virol.* **78**, 3572-3577 (2004).
8. A. Roberts, L. Vogel, J. Guarner, et al., Severe acute respiratory syndrome coronavirus infection of golden Syrian hamsters, *J. Virol.* **79**, 503-511 (2005).
9. S. A. Jeffers, S. M. Tusell, L. Gillim-Ross, et al., CD209L (L-SIGN) is a receptor for severe acute respiratory syndrome coronavirus, *Proc. Natl. Acad. Sci. USA* **101**, 15748-15753 (2004).
10. W. Li, C. Zhang, J. Sui, et al., Receptor and viral determinants of SARS-coronavirus adaptation to human ACE2, *EMBO J.* **24**, 1634-1643 (2005).
11. L. Gillim-Ross, J. Taylor, D. R. Scholl, et al., Discovery of novel human and animal cells infected by the severe acute respiratory syndrome coronavirus by replication-specific multiplex reverse transcription-PCR, *J. Clin. Microbiol.* **42**, 3196-3206 (2004).
12. P. Towler, B. Staker, S. G. Prasad, et al., ACE2 X-ray structures reveal a large hinge-bending motion important for inhibitor binding and catalysis, *J. Biol. Chem.* **279**, 17996-18007 (2004).
13. Cn3D 4.1, *National Center for Biotechnology Information,* http://www.ncbi.nlm.nih.gov, 2005.
14. D. E. Wentworth and K. V. Holmes, Molecular determinants of species specificity in the coronavirus receptor aminopeptidase N (CD13): influence of N-linked glycosylation, *J. Virol.* **75**, 9741-9752 (2001).

INTERACTION BETWEEN THE SPIKE PROTEIN OF HUMAN CORONAVIRUS NL63 AND ITS CELLULAR RECEPTOR ACE2

Stefan Pöhlmann, Thomas Gramberg, Anja Wegele, Krzysztof Pyrc, Lia van der Hoek, Ben Berkhout, and Heike Hofmann*

1. INTRODUCTION

Coronavirus (CoV) infection of humans has so far not been associated with severe disease. However, the discovery of the severe acute respiratory syndrome (SARS) CoV revealed that highly pathogenic human CoVs (hCoVs) can evolve. As the characterization of new hCoVs is therefore an important task, we studied the cellular entry of hCoV-NL63, which was recently isolated from patients with lower respiratory tract illness.[1]

Entry of CoVs into target cells is determined by the major viral envelope glycoprotein termed "spike" (S), which provides the virions with their characteristic corona-like shape.[2,3] The main function of the S-protein in CoV entry is the binding to a host cell receptor followed by fusion of viral and cellular membranes.[4,5] The domains in S that are required for membrane fusion locate to the C-terminal half of the protein (S2 subunit). Receptor engagement is conferred by the N-terminal S1 subunit; consequently, the S-protein of a given CoV can determine its cell tropism.[6] Although the S-proteins of animal and human CoVs exhibit the same functional organization, particularly the S1 subunits differ in amino acid sequence, resulting in interaction with specific cellular receptors. Within group I CoVs, hCoV-NL63 is phylogenetically highly linked to hCoV-229E, and especially the S-proteins of both viruses share a high sequence homology. The S-protein of hCoV-229E is known to employ the human aminopeptidase N (hAPN, also called CD13, herein referred to as hAPN/CD13) for infection of target cells[7]; therefore, it was speculated that NL63 might use the same receptor for cellular entry.[1]

The S-protein is sufficient to mediate CoV entry into receptor-positive target cells and can be incorporated into heterologous viral particles. Thus, CoV S-proteins can be expressed in the envelope of lentiviral particles, and these pseudotyped viruses ("pseudotypes") proved to be a useful experimental system to analyze SARS-CoV-S

*Stefan Pöhlmann, Thomas Gramberg, Anja Wegele, Heike Hofmann, University of Erlangen-Nürnberg, Germany. Krzysztof Pyrc, Lia van der Hoek, Ben Berkhout, University of Amsterdam, The Netherlands.

mediated cellular entry.[8–11] For production of pseudotypes, envelope-defective lentiviral genomes encoding a reporter gene and a plasmid encoding a CoV S-protein are expressed in cells, which then secrete lentiviral particles harbouring the S-protein in their envelope. These particles infect susceptible cells in a CoV-S dependent manner, and entry efficiency can be quantified by determination of the reporter gene activity.

2. IDENTIFICATION OF ANGIOTENSIN-CONVERTING ENZYME 2 (ACE2) AS A RECEPTOR FOR hCoV-NL63

2.1. The S-proteins of hCoV-NL63 and -229E Interact with Different Receptors

Replication competent hCoV-NL63 has been cultured on tertiary monkey kidney cells,[1,12] but permissive human cell lines had not been identified so far. Using lentiviral pseudotypes, we therefore analyzed a panel of human cell lines for susceptibility to hCoV-NL63 S-mediated entry (chapter 4.2). Interestingly, the cell tropism measured for NL63-S was congruent with that observed for SARS-CoV,[8,10,11] but differed significantly from that measured for 229E-S, suggesting that both glycoproteins might interact with different receptors despite their high amino acid identity. Furthermore, we were able to show that hCoV-NL63 does not engage feline aminopeptidase N (fAPN) in contrast to all CoVs of class I investigated so far.[13] However, cells expressing the SARS-CoV receptor protein ACE2[14,15] were susceptible to NL63-S driven infection. This was unexpected as NL63-S has no striking homology to either the whole S1 subunit of SARS-CoV or the already identified ACE2 interaction domain in SARS-CoV-S,[16] suggesting that both proteins either form a common three-dimensional structure that allows ACE2 engagement in a similar fashion or that both S-proteins evolved different strategies to target ACE2. The interaction between NL63-S and ACE2 was specific, as the closely related ACE1 protein did not react with NL63-S, and on the other hand, ACE2 was not able to confer 229E-S-mediated infection, suggesting that ACE2 is not a functional equivalent of fAPN in class I CoV entry.

2.2. Direct Interaction Between ACE2 and NL63-S

In order to investigate whether NL63-S directly contacts ACE2, we employed a FACS-based binding assay. For this, the NL63-S1 region was fused to the constant chain of human immunoglobulin (IgG), expressed in 293T cells and concentrated from the culture supernatant. Then, NL63-S-IgG was incubated with cells expressing either hAPN/CD13 or ACE2 followed by staining with specific antibodies directed against the respective receptor protein to quantify receptor expression levels and an anti-human Fc antibody to detect the bound IgG fusion protein. Hereby, we demonstrated that NL63-S in contrast to 229E-S does not react with hAPN/CD13, but like SARS-CoV-S directly binds to ACE2. Finally, we analyzed the NL63-S interaction with ACE2 by employing ACE1 or ACE2 specific polyclonal antisera. Only the ACE2 serum interfered with infection by NL63-S-bearing lentiviral pseudotypes and replication-competent hCoV-NL63, thus confirming that ACE2 is a receptor for hCoV-NL63.

3. MAPPING OF THE ACE2-INTERACTION DOMAIN IN NL63-S

3.1. Role of the Unique Domain in NL63-S

As mentioned above, the S-proteins of hCoV-229E and hCoV-NL63 share an overall amino acid identity of more than 50%. However, hCoV-NL63 S harbors a 178 amino acid N-terminal extension that is not present in 229E-S or any other known protein and that is therefore designated "unique domain".[1] In order to investigate if the unique region is involved in ACE2 binding, we first analyzed binding of the isolated domain to ACE2 expressing cells in a FACS based binding assay. However, the unique domain alone did not show any interaction with ACE2, indicating that it does not serve as an independent receptor binding domain. Because it is possible that the unique domain might confer ACE2-binding together with other sequences in the S1-domain of hCoV-NL63 and possibly hCoV-229E, we constructed a chimeric mutant comprising the N-terminal 178 amino acids of NL63-S fused to the S1 subunit of 229E. This mutant, however, showed no ACE2-interaction, but bound to hAPN/CD13 as efficient as the wildtype protein, indicating that the unique domain does not interfere with hAPN/CD13 recognition and does not allow binding to ACE2. When the unique domain was removed from NL63-S, the remaining protein still bound ACE2 and showed no affinity for hAPN/CD13, confirming that the unique region is dispensable for ACE2 binding. In summary, these observations indicate that amino acids in the highly conserved S1 regions of NL63- and 229E-S confer specificity for the interaction with ACE2 and hAPN/CD13, respectively.

3.2. Analysis of hCoV-NL63-S1 Deletion Mutants and Chimeric hCoVNL63-229E-S Variants

In order to map which region in the NL63 S1-protein is responsible for targeting ACE2, we analyzed a panel of N-terminal S1-deletion mutants. By this, we were able to narrow down the ACE2 interaction domain in NL63-S to amino acids 232 to 741. In parallel, we constructed chimeric S1-proteins comprising defined regions of NL63-S fused to complementary domains within 229E-S and analyzed them for hAPN/CD13 and ACE2 interactions. We found that several sequence elements in the center and C-terminus of the proteins can impact receptor binding, suggesting that some of these mutations might interfere with the integrity of the possibly complex three-dimensional structure of the proteins and thus with their capacity to recognize receptors. Our observations are in agreement with a model suggesting that the central region in the hCoV229E-S and possibly NL63-S proteins might determine the correct folding or orientation of a C-terminal receptor binding domain, as has been suggested previously for hCoV-229E-S.[17,18] Taken together, a detailed point mutagenesis of NL63-S1 will be required to identify residues with a critical function in ACE2 interaction.

4. DISCUSSION AND OUTLOOK

We were able to show that the SARS-CoV receptor ACE2 is used by the recently identified hCoV-NL63 for entry into target cells. Simultaneously, the same observation was reported by Smith and colleagues (M.K. Smith, et al., chapter 4.13). The interaction of NL63-S with ACE2 was unexpected, as NL63-S and SARS-CoV-S share no significant amino acid homology. In contrast, NL63-S is highly related to the glycoprotein of hCoV-229E, which binds hAPN/CD13, and the hAPN/CD13 interaction domain is well conserved in NL63-S. The most striking difference between the S-proteins of hCoV-NL63 and -229E is a 178 amino acid extension that is exclusively present in NL63-S. This unique domain, however, is dispensable for ACE2-interaction; thus, amino acids in the highly conserved central portions and C-termini within the S-proteins of hCoV-229E and hCoV-NL63 determine the recognition of their respective receptors. Therefore, detailed point mutagenesis in combination with the determination of the three-dimensional structure of both S-proteins is required to identify amino acids that mediate receptor binding. These data in turn will help to develop specific small molecule inhibitors against NL63-S–mediated infection.

5. REFERENCES

1. van der Hoek, L., et al., 2004, Identification of a new human coronavirus, *Nat. Med.* **10**:368.
2. Holmes, K. V., 2001, The Coronaviridae, in: *Fields Virology*, D. Knipe, ed., Lippincott Wiliams & Wilkins, Philadelphia, pp. 1187-1203.
3. Stadler, K., et al., 2003, SARS-beginning to understand a new virus, *Nat. Rev. Microbiol.* **1**:209.
4. Gallagher, T. M., and Buchmeier, M. J., 2001, Coronavirus spike proteins in viral entry and pathogenesis, *Virology* **279**:371.
5. Hofmann, H., and Pöhlmann, S., 2004, Cellular entry of the SARS coronavirus, *Trends Microbiol.* **12**:466.
6. Kuo, L., et al., 2000, Retargeting of coronavirus by substitution of the spike glycoprotein ectodomain: crossing the host cell species barrier, *J. Virol.* **74**:1393.
7. Yeager, C. L., et al., 1992, Human aminopeptidase N is a receptor for human coronavirus 229E, *Nature* **357**:420.
8. Hofmann, H., et al., 2004, S-protein of severe acute respiratory syndrome-associated coronavirus mediates entry into hepatoma cell lines and is targeted by neutralizing antibodies in infected patients, *J. Virol.* **78**:6134.
9. Moore, M. J., et al., 2004, Retroviruses pseudotyped with the severe acute respiratory syndrome coronavirus spike protein efficiently infect cells expressing angiotensin-converting enzyme 2, *J. Virol.* **78**, 10628.
10. Simmons, G., et al., 2004, Characterization of severe acute respiratory syndrome-associated coronavirus (SARS-CoV) spike glycoprotein-mediated viral entry, *Proc. Natl. Acad. Sci. USA* **101**:4240.
11. Yang, Z. Y., et al., 2004, pH-dependent entry of severe acute respiratory syndrome coronavirus is mediated by the spike glycoprotein and enhanced by dendritic cell transfer through DC-SIGN, *J. Virol.* **78**:5642.
12. Fouchier, R. A., et al., 2004, A previously undescribed coronavirus associated with respiratory disease in humans, *Proc. Natl. Acad. Sci. USA* **101**:6212.
13. Tresnan, D. B., and Holmes, K. V., 1998, Feline aminopeptidase N is a receptor for all group I coronaviruses, *Adv. Exp. Med. Biol.* **440**:69.
14. Li, W., et al., 2003, Angiotensin-converting enzyme 2 is a functional receptor for the SARS coronavirus, *Nature* **426**:450.
15. Wang, P., et al., 2004, Expression cloning of functional receptor used by SARS coronavirus, *Biochem. Biophys. Res. Commun.* **315**:439.
16. Wong, S. K., et al., A 193-amino acid fragment of the SARS coronavirus S protein efficiently binds angiotensin-converting enzyme 2, 2004, *J. Biol. Chem.* **279**:3197.
17. Bonavia, A., et al., 2003, Identification of a receptor-binding domain of the spike glycoprotein of human coronavirus HCoV-229E, *J. Virol.* **77**:2530.
18. Breslin, J. J., et al., 2003, Human coronavirus 229E: receptor binding domain and neutralization by soluble receptor at 37 degrees, *J. Virol.* **77**:4435.

HUMAN ANGIOTENSIN-CONVERTING ENZYME 2 (ACE2) IS A RECEPTOR FOR HUMAN RESPIRATORY CORONAVIRUS NL63

M. K. Smith, Sonia Tusell, Emily A. Travanty, Ben Berkhout, Lia van der Hoek, and Kathryn V. Holmes*

1. INTRODUCTION

Until 2003, only two human coronaviruses, HCoV-OC43 (group 2) and HCoV-229E (group 1), were well adapted to growth in tissue culture. These HCoVs have generally been associated with mild upper respiratory disease, although HCoV-229E can cause pneumonia in immunocompromised patients.[5] In contrast, animal coronaviruses (CoVs) have been associated with a wide variety of diseases, many of them severe, in multiple animal and avian species. From the animal coronaviruses, a great deal about CoV biology and pathogenesis is known. The variety of animal hosts and diseases among CoVs are due in large part to differences in the spike glycoprotein.[1,13,6,8,10] CoV spikes are large type 1 membrane glycoproteins that are the major determinants of receptor specificity as well as virulence.

The recent discoveries of HCoV-NL63 (group 1) and its association with a variety of respiratory tract infections, including pneumonia and croup in young children,[15] and HKU1 (group 2), isolated from a patient with pneumonia,[16] shows that coronaviruses are emerging as important pathogens of the human lower respiratory tract.

The previously known group 1 CoVs use aminopeptidase N (APN) glycoprotein from their respective host species as their principal receptor.[11] In addition, feline APN (fAPN) is a receptor for all of these group 1 coronaviruses (Table 1).[12] HCoV-NL63 is most closely related to the group 1 coronaviruses. Indeed, the spike glycoprotein of HCoV-NL63 shares 55% amino acid identity with the spike of HCoV-229E. We therefore tested whether this new virus would also utilize hAPN as its cellular receptor.

* M. K. Smith, Sonia Tusell, Emily A. Travanty, Kathryn V. Holmes, University of Colorado Health Sciences Center, Aurora, Colorado 80045. Ben Berkhout, Lia van der Hoek, University of Amsterdam, The Netherlands.

Table 1. Group 1 coronavirus receptors.[11]

Virus	hAPN[a]	pAPN[b]	cAPN[c]	fAPN[d]
HCoV-229E	+	-	-	+
PRCoV	-	+	-	+
TGEV	-	+	-	+
CCoV	-	-	+	+
FCoV	-	-	-	+

[a] Human aminopeptidase N.
[b] Porcine aminopeptidase N.
[c] Canine aminopeptidase N.
[d] Feline aminopeptidase N.

2. METHODS AND RESULTS

2.1. NL63 Infection of Cell Lines

LLC-MK2 cells, Vero E6 cells, and a human lung fibroblast-like cell line, MRC-5 (ATCC # CCL-171), were inoculated with NL63 and then observed for signs of CPE for 96 hours postinfection. Mild and transient CPE, consisting mostly of rounding and the appearance of vacuoles, was observed at 36 hours postinfection in all three cell types. HCoV-NL63 viral antigens were detected by immunofluorescence with a cross-reactive polyclonal goat antibody raised against HCoV-229E virions followed by fluorescein isothionate (FITC)-conjugated donkey anti-goat IgG as previously described.[2] Viral antigen was detectable from 24 to 96 hours postinfection, with a maximum of about 25% of cells positive for antigen at 72 hours postinfection.

We inoculated LLC-MK2 and Vero E6 cells, which are permissive for HCoV-NL63 infection,[14] as well as NIH-3T3 cells stably expressing hAPN (NIH-3T3/hAPN), BHK-21 cells stably expressing fAPN (BHK/fAPN), which are permissive for HCOV-229E infection, with either HCoV-NL63 or HCoV-229E at an MOI of 0.07. HCoV-229E and HCoV-NL63 viral antigens were both detected in susceptible cells with immunofluorescence. Viral antigens were detected in LLC-MK2 and Vero E6 cells inoculated with HCoV-NL63. No infection was detected in LLC-MK2 or Vero E6 cells inoculated with HCoV-229E. No viral antigens were detected in NIH-3T3/ hAPN or BHK/fAPN cells inoculated with HCoV-NL63, but robust infection was apparent in these cells after inoculation with HCoV-229E (Table 2).

The observations that both SARS-CoV and HCoV-NL63 grow in LLC-MK2 and Vero E6 cells led us to test the hypothesis that HCoV-NL63 uses the same receptor as SARS-CoV, angiotensin-converting enzyme 2 (ACE2).[9] BHK-21 cells were transiently transfected with a plasmid encoding full-length human ACE2 (hACE2, kindly provided by Michael Farzan), and inoculated 48 hours after transfection with HCoV-NL63 or HCoV-229E. HCoV-NL63 antigen was detected in hACE2-transfected BHK-21 cells, but not in non-transfected controls. No HCoV-229E antigen was detected in either transfected or hACE2-negative control BHK-21 cells. ACE2-transfected cells were permissive for HCoV-NL63 replication, with peak yields of 10^5 $TCID_{50}$/ml as assayed on LLC-MK2 cells.

To determine which receptor HCoV-NL63 uses to infect MRC-5 cells, we tested MRC-5s for surface expression of hACE2 by flow cytometry. MRC-5 cells expressed a small amount of ACE2, relative to LLC-MK2 and Vero E6 cells (Table 3).

Table 2. Susceptibility of cultured cells to HCoV-NL63 and HCoV-229E infection.

	Virus	
Cell type	HCoV-229E	HCoV-NL63
LLC-MK2	-	+
VERO E6	-	+
MRC-5	+	+
BHK	-	-
BHK/hACE2	-	+
BHK/fAPN	+	-

3. DISCUSSION

In this paper, we show that infectious HCoV-NL63 used human ACE2 as an efficient receptor and that this group 1 coronavirus cannot use feline APN, the receptor for all other group 1 coronaviruses tested, or human APN, the principal receptor for HCoV-229E. Hoffmann et al., also showed that human ACE2 served as a receptor for HCoV-NL63 using retrovirus pseudotyped with HCoV-NL63 spike glycoprotein.[4]

The finding that HCoV-NL63 uses ACE2 as a receptor is surprising in that HCoV-NL63 is most closely related to HCoV-229E, and its genome structure is most similar to that of PEDV in group 1, while SARS-CoV, which also uses human ACE2 as a receptor, is a distantly related member of group 2 coronaviruses. Although the group 1 spikes studied to date share no more than 48% amino acid sequence identity, all but HCoV-NL63 are able to use fAPN as a receptor.

Interestingly, although the spike glycoprotein of HCoV-NL63 shares only 25% amino acid sequence identity with that of SARS-CoV (Urbani strain), the two viruses use the same receptor. It is possible that the two viruses have evolved independently to bind to ACE2. The finding that HCoV-NL63 and SARS-CoV utilize the same human receptor glycoprotein has important implications for coronavirus evolution. The genomes of all group 2 coronaviruses except SARS-CoV contain a gene encoding an HE glycoprotein that binds to O-acetylated sialic acids,[7] apparently derived by recombination of the mRNA encoding HE from influenza C with the genome of an ancestral group 2 coronavirus.[7] Group 1 coronaviruses including HCoV-NL63 lack this HE gene. Perhaps both HCoV-NL63 and SARS-CoV descended from a common ancestral coronavirus that used ACE2 as its receptor, and this lineage split from the ancestral group 2 coronaviruses before the HE gene was acquired.

Understanding HCoV-NL63 S glycoprotein and how its interactions with its receptor, hACE2, will elicudate how pathogenic coronaviruses emerge and may suggest new strategies for prevention and therapy.

Table 3. Expression of ACE2 on Vero E6, LLC-MK2, and MRC-5 cells.

	Percent cells gated	
Cell type	Control antibody	Anti-ACE2
Vero E6	3.1	77.6
LLC-MK2	17.7	70.8
MRC-5	7.7	27.0

4. ACKNOWLEDGMENTS

The authors are grateful for helpful discussions with D. Wentworth, T. Miura, S. Jeffers, L. Thackray, and B. Turner. This research was supported by NIH grant RO1 A1 26075.

5. REFERENCES

1. M. L. Ballesteros, C. M. Sanchez, and L. Enjuanes, Two amino acid changes at the N-terminus of transmissible gastroenteritis coronavirus spike protein result in the loss of enteric tropism, *Virology* **227**, 378-388 (1997).
2. A. Bonavia, B. D. Zelus, D. E. Wentworth, P. J. Talbot, and K. V. Holmes, 3 A.D. Identification of a receptor-binding domain of the spike glycoprotein of human coronavirus HCoV-229E, *J. Virol.* **77**, 2530-2538 (2003).
3. R. A. Fouchier, N. G. Hartwig, T. M. Bestebroer, B. Niemeyer, J. C. De Jong, J. H. Simon, and A. D. Osterhaus, A previously undescribed coronavirus associated with respiratory disease in humans, *Proc. Natl. Acad. Sci. USA* **101**, 6212-6216 (2004).
4. H. Hofmann, K. Pyrc, L. van der Hoek, M. Geier, B. Berkhout, and S. Pohlmann, Human coronavirus NL63 employs the severe acute respiratory syndrome coronavirus receptor for cellular entry, *Proc. Natl. Acad. Sci. USA* **102**, 7988-7993 (2005).
5. F. Pene, A. Merlat, A. Vabret, F. Rozenberg, A. Buzyn, F. Dreyfus, A. Cariou, F. Freymuth, and P. Lebon, Coronavirus 229E-related pneumonia in immunocompromised patients, *Clin. Infect. Dis.* **37**, 929-932 (2003).
6. J. D. Rempel, S. J. Murray, J. Meisner, and M. J. Buchmeier, Mouse hepatitis virus neurovirulence: evidence of a linkage between S glycoprotein expression and immunopathology, *Virology* **318**, 45-54 (2004).
7. S. L. Smits, G. J. Gerwig, A. L. van Vliet, et al., Nidovirus sialate-O-acetylesterases: evolution and substrate specificity of coronaviral and toroviral receptor-destroying enzymes, *J. Biol. Chem.* **280**, 6933-6941 (2005).
8. H. D. Song, C. C. Tu, G. W. Zhang, et al., Cross-host evolution of severe acute respiratory syndrome coronavirus in palm civet and human, *Proc. Natl. Acad. Sci. USA* **102**, 2430-2435 (2005).
9. J. Sui, W. Li, A. Murakami, et al., Potent neutralization of severe acute respiratory syndrome (SARS) coronavirus by a human mAb to S1 protein that blocks receptor association, *Proc. Natl. Acad. Sci. USA* **101**, 2536-2541 (2004).
10. L. B. Thackray and K. V. Holmes, Amino acid substitutions and an insertion in the spike glycoprotein extend the host range of the murine coronavirus MHV-A59, *Virology* **324**, 510-524 (2004).
11. D. B. Tresnan and K. V. Holmes, Feline aminopeptidase N is a receptor for all group I coronaviruses, *Adv. Exp. Med. Biol.* **440**, 69-75 (1998).
12. D. B. Tresnan, R. Levis, and K. V. Holmes, Feline aminopeptidase N serves as a receptor for feline, canine, porcine, and human coronaviruses in serogroup I, *J. Virol.* **70**, 8669-8674 (1996).
13. J. C. Tsai, L. de Groot, J. D. Pinon, K. T. Iacono, J. J. Phillips, S. H. Seo, E. Lavi, and S. R. Weiss, Amino acid substitutions within the heptad repeat domain 1 of murine coronavirus spike protein restrict viral antigen spread in the central nervous system, *Virology* **312**, 369-380 (2003).
14. L. van der Hoek, K. Pyrc, M. F. Jebbink, et al., Identification of a new human coronavirus, *Nat. Med.* **10**, 368-373 (2004).
15. L. van der Hoek, K. Sure, G. Ihorst, et al., Croup is associated with the novel coronavirus NL63, *PLoS. Med.* **2**, e240 (2005).
16. P. C. Woo, S. K. Lau, C. M. Chu, et al., Characterization and complete genome sequence of a novel coronavirus, coronavirus HKU1, from patients with pneumonia, *J. Virol.* **79**, 884-895 (2005).

MOLECULAR INTERACTIONS OF GROUP 1 CORONAVIRUSES WITH FELINE APN

Sonia M. Tusell and Kathryn V. Holmes*

1. INTRODUCTION

Most coronaviruses in phylogenetic group 1 can cause disease in only one animal species. Within group 1, porcine transmissible gastroenteritis virus (TGEV), feline infectious peritonitis virus (FIPV), human coronavirus 229E (HCoV-229E), and canine coronavirus (CCoV) use aminopeptidase N (APN) from their natural host for entry and infection of cells.[1-4] APN is a highly conserved type II transmembrane glycoprotein in mammals (70–80% at the amino acid level). Remarkably, although each of these group 1 coronaviruses uses APN of its normal host for entry, all of them can also use feline APN (fAPN) as a receptor for entry and infection in cell culture.[2] Previous studies used chimeras between APN proteins of different species to identify domains in APN that are required for coronavirus receptor activity.[4-7] Studies on receptor specificities of chimeras between human and feline APN or pig and human APN suggest that the spike glycoproteins of TGEV, FIPV, and HCoV-229E interact with two discontinous regions within APN.[6] Also, species-specific N-linked glycosylations in APN can affect receptor activity for HCoV-229E.[8] In *in vivo* studies, transgenic mice expressing human APN (hAPN) were resistant to infection with HCoV-229E, but cells harvested from the transgenic mice were susceptible to HCoV-229E.[9] hAPN transgenic mice in a Stat-1 knockout background (*hAPN* $^{+/+}$*Stat* $^{-/-}$) were susceptible to HCoV-229E, which was adapted for growth in cells from these double transgenic mice.[10] These studies suggest that other host factors in addition to the receptor are needed for infection *in vivo*.

In this study, we used chimeras between mouse APN (mAPN) and fAPN to identify domains of fAPN that are necessary for entry by group 1 coronaviruses. Baby hamster kidney (BHK-21) cells, which are resistant to infection by TGEV, FIPV, CCoV, and HCoV-229E, were transfected with cDNAs encoding wild-type mAPN, fAPN, or mouse-feline APN (MF) chimeras. The transfected cells were inoculated with TGEV clone E, FIPV 79-1146, HCoV-229E, or CCoV 1-71 virus strains. Virus entry and infection was demonstrated by immunofluorescence with antiviral antibodies.

*University of Colorado Health Sciences Center, Aurora, Colorado 80045.

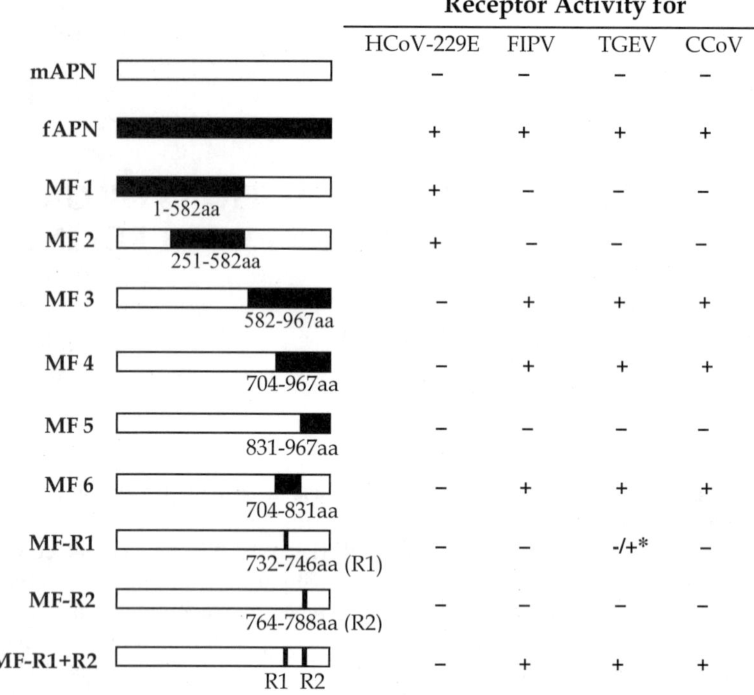

Figure 1. Coronavirus receptor activities of mouse APN (mAPN), feline APN (fAPN), and mouse-feline APN (MF) glycoproteins. All constructs were cloned into pCDNA3.1 TOPO/D V5 his and transiently expressed in BHK-21 cells. Forty-eight hours after transfection, cells were inoculated with each of the coronaviruses, and 10 or 24 hours after virus inoculation, cells were fixed and immunolabeled with antibodies against viral antigens. *Very few cells stained positive for TGEV antigens in BHK-21 cells transfected with MF-R1.

2. RESULTS AND DISCUSSION

BHK-21 cells transfected with mAPN cDNA remained resistant to infection with FIPV, TGEV, CCoV, and HCoV-229E. Chimera MF1, which consists of amino acids 1 to 582 of fAPN in a mAPN backbone, promoted entry of HCoV-229E only, while MF3 promoted entry of FIPV, TGEV, and CCoV, but not HCoV-229E. Construct MF2 consisting of a smaller region of fAPN (amino acids 251–582) in a mAPN backbone functioned as a receptor only for HCoV-229E. The receptor activities of MF1 and MF3 indicated that amino acids 582 to 967 of fAPN were necessary and sufficient for entry of the cat, pig, and dog coronaviruses, but not for HCoV-229E. In contrast, amino acids closer to the N terminus of the protein (251–582aa) of fAPN were required for entry of HCoV-229E. Additional chimeras, MF 4, 5, and 6 were constructed to identify a smaller region in fAPN that would be sufficient for FIPV, TGEV, and CCoV entry. Importantly, MF6 identified a sequence of 127 amino acids in fAPN (aa 704–831) that when substituted into mAPN was sufficient for FIPV, TGEV, and CCoV receptor activity. These data agree with previously published conclusions based on other chimeric APN proteins.[4, 6, 7] Further mutational analysis identified several amino acid residues in aa

704–831 of fAPN that when introduced into mAPN were sufficient to promote entry of the pig, dog, and cat viruses. These residues were located in two discontinuous segments corresponding to amino acids 732–746 (R1) and amino acids 764–788 (R2). Interestingly, TGEV was able to enter BHK-21 cells transfected with the MF-R1 construct, although virus entry was very inefficient relative to TGEV receptor activity of fAPN. However, MF-R1 had no receptor activity for FIPV and CCoV. MF-R2 had no receptor activity for any of these viruses. Efficient receptor activity for these three group 1 coronaviruses was only detected when R1 and R2 from fAPN were substituted together into mAPN.

In summary, amino acid residues in fAPN that are important for entry of FIPV, TGEV, and CCoV were localized to two discrete regions within the C terminal region of fAPN, whereas HCoV-229E entry required an N-terminal domain of fAPN. Without a crystal structure for APN, it is unclear whether these functionally important regions are adjacent in the three-dimensional structure of the receptor glycoprotein. These observed differences in fAPN receptor utilization correlate well with differences in the spike glycoproteins of these viruses, as the cat, pig, and dog virus spike glycoproteins are more closely related to each other at the amino acid level than to the HCoV-229E spike glycoprotein. Characterization of the molecular interactions between the spikes of these group 1 coronaviruses and their APN receptors will identify residues that affect the host ranges of these viruses and provide insight into the evolution of group 1 coronaviruses.

3. ACKNOWLEDGMENTS

This research was supported by NIH grant R01-AI-26075. Sonia Tusell was supported by NIH grant AI 52066.

4. REFERENCES

1. B. Delmas, J. Gelfi, R. L'Haridon, et al., Aminopeptidase N is a major receptor for the entero-pathogenic coronavirus TGEV, *Nature* **357**, 417-420 (1992).
2. D. B. Tresnan, R. Levis, and K. V. Holmes, Feline aminopeptidase N serves as a receptor for feline, canine, porcine, and human coronaviruses in serogroup I, *J. Virol.* **70**, 8669-8674 (1996).
3. C. L. Yeager, R. A. Ashmun, R. K. Williams, et al., Human aminopeptidase N is a receptor for human coronavirus 229E, *Nature* **357**, 420-422 (1992).
4. L. Benbacer, E. Kut, L. Besnardeau, et al., Interspecies aminopeptidase-N chimeras reveal species-specific receptor recognition by canine coronavirus, feline infectious peritonitis virus, and transmissible gastroenteritis virus, *J. Virol.* **71**, 734-737 (1997).
5. A. F. Kolb, A. Hegyi, J. Maile, et al., Molecular analysis of the coronavirus-receptor function of aminopeptidase N, *Adv. Exp. Med. Biol.* **440**, 61-67 (1998).
6. A. Hegyi and A. F. Kolb, Characterization of determinants involved in the feline infectious peritonitis virus receptor function of feline aminopeptidase N, *J. Gen. Virol.* **79**, 1387-1391 (1998).
7. A. F. Kolb, A. Hegyi, and S. G. Siddell, Identification of residues critical for the human coronavirus 229E receptor function of human aminopeptidase N, *J. Gen. Virol.* **78**, 2795-2802 (1997).
8. D. E. Wentworth and K. V. Holmes, Molecular determinants of species specificity in coronavirus receptor aminopeptidase N (CD13): Influence of N-linked glycosylation, *J. Virol.* **75**, 9741-9752 (2001).
9. D. E. Wentworth, D. B. Tresnan, B. C. Turner, et al., Cells of human aminopeptidase N (CD13) transgenic mice are infected by human coronavirus-229E in vitro, but not in vivo, *Virology* **335**, 185-197 (2005).
10. C. Lassnig, C. M. Sanchez, M. Egerbacher, et al., Development of a transgenic mouse model susceptible to human coronavirus 229E, *Proc. Natl. Acad. Sci. USA* **102**, 8275-8280 (2005).

PSEUDOTYPED VESICULAR STOMATITIS VIRUS FOR FUNCTIONAL ANALYSIS OF SARS CORONAVIRUS SPIKE PROTEIN

Shuetsu Fukushi, Tetsuya Mizutani, Masayuki Saijo, Shutoku Matsuyama, Fumihiro Taguchi, Ichiro Kurane, and Shigeru Morikawa*

1. INTRODUCTION

The entry of severe acute respiratory syndrome–associated coronavirus (SARS-CoV) into susceptible cells is mediated by binding of the viral spike (S) protein to its receptor molecule, angiotensin-converting enzyme 2 (ACE2). A pseudotyping system with vesicular stomatitis virus (VSV) particles [the VSVΔG* system, in which the VSV-G gene is replaced by the green fluorescent protein (GFP) gene] was reported to produce pseudotyped VSV incorporating envelope glycoproteins from RNA viruses.[1,2] This system is useful for studies of viral envelope glycoproteins due to the ability to grow at high titers in a variety of cell lines. Infection of target cells with pseudotyped VSV can be detected readily as GFP-positive cells within 16 hours postinfection (hpi) because of the high level of GFP expression in the VSVΔG* system.[2] Thus, pseudotyping of SARS-CoV-S protein using the VSVΔG* system may have advantages for studying the function of SARS-CoV-S protein as well as for developing a rapid system for examining neutralizing antibodies specific for SARS-CoV infection. In this report, we describe a rapid detection system for SARS-CoV-S protein–bearing VSV pseudotype infection. The effects of ACE2-binding peptides on SARS-CoV-S–mediated infection were investigated using this system.

2. MATERIALS AND METHODS

Plasmids: cDNAs of the full-length or a truncated version of SARS-CoV-S protein lacking the C-terminal 19 amino acids were cloned into the mammalian expression vector,

* National Institute of Infectious Diseases, Tokyo 208-0011, Japan.

pKS336[3] yielding the plasmids, pKS-SARS-S and pKS-SARS-St19, respectively. The plasmid, pKS-SARS-St19rev, carried the same cDNA as pKS-SARS-St19 but in the reverse orientation in pKS336, and was used as a negative control for experiments regarding pseudotype production.

Preparation of VSV pseudotype: At 24 h after transfection of 293T cells with pKS-SARS-S, pKS-SARS-St19, or pKS-SARS-St19rev, the cells were infected with VSVΔG* in which the G gene was replaced by the GFP gene.[1] After 24 h, culture supernatants were collected and stored at –80°C until use. Vero E6 cells grown on 24-well glass slides were inoculated with pseudoviruses. Infection by pseudotype virus was detected by monitoring GFP expression under a fluorescence microscope, and the number of GFP-expressing cells was counted using ImageJ software (http://rsb.info.nih.gov/ij/). For inhibition assays, VSV pseudotypes were incubated with serially diluted inhibitors for 1 h at 37°C, and the mixtures were then inoculated onto Vero E6 cells.

Inhibitors: Angiotensin I, angiotensin II, and desArg^9-bradykinin were purchased from Sigma. An ACE2 inhibitor, DX600, was purchased from Phoenix Pharmaceuticals.

3. RESULTS AND DISCUSSION

3.1. Production of SARS-CoV-S–Bearing VSV Pseudotype

To generate VSV pseudotyped with full-length SARS-CoV-S protein, the expression plasmid pKS-SARS-S was transfected into 293T cells, followed by infection with VSVΔG*. When the culture supernatants of the infected 293T cells were inoculated onto Vero E6 cells, a cell line commonly used for SARS-CoV propagation, only small numbers of GFP-expressing cells were observed (data not shown). These observations indicated that VSV pseudotype bearing the full-length SARS-CoV-S protein was not highly infectious. Next, we generated VSV pseudotyped with SARS-CoV-S protein in which the C-terminal 19 amino acids were truncated using the plasmid, pKS-SARS-St19. The plasmid, pKS-SARS-St19rev, was used as a negative control. 293T cells transfected with either pKS-SARS-St19 or pKS-SARS-St19rev were infected with VSVΔG*. After 24 h, the culture supernatants of infected cells were collected and inoculated onto Vero E6 cells. As shown in Figure 1, the number of infectious units (IU) of pseudotyped VSV (5.0×10^5/ml), referred to as VSV-SARS-St19, obtained from 293T cells transfected with pKS-SARS-St19 was significantly higher than that of the negative control. As partial deletion of the cytoplasmic domain of the SARS-CoV-S protein allowed efficient incorporation into VSV particles and led to pseudotype generation at high titer, the intact cytoplasmic domain of SARS-CoV-S protein may interrupt proper assembly of the pseudotype particles.

3.2. Time Course Analysis of GFP Expression in VSV Pseudotype–Infected Cells

Infection by retrovirus-based pseudotypes is usually measured at 48 hpi, while infection of VSV-based pseudotypes can be detected at 16 h.[2] Interestingly, GFP expression in Vero E6 cells was detected clearly at 7 h after inoculation with VSV-SARS-St19. Time course analysis of the number of GFP-positive cells indicated that it was possible to quantify VSV-SARS-St19 infection at 7 hpi (Fig. 1B). Therefore, in subsequent analyses, we counted the number of GFP-positive cells infected with VSV-SARS-St19 at 7 hpi.

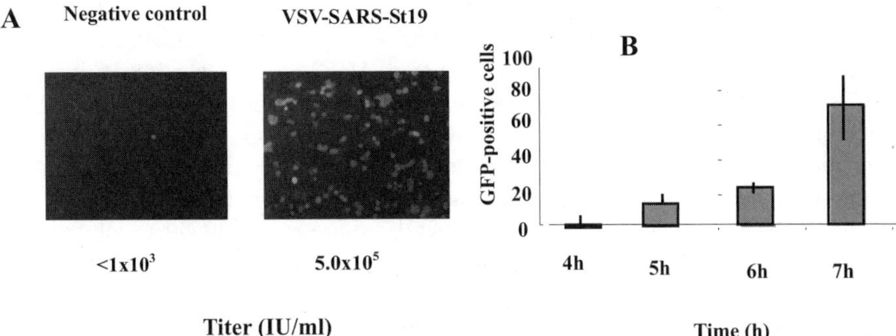

Figure 1. Infectivity of VSV pseudotypes. (A) VSV-SARS-St19 or negative control was inoculated onto Vero E6 cell monolayers. GFP expression was examined by fluorescence microscopy. (B) Cells were photographed under a fluorescence microscope at various time points after inoculation. The numbers of GFP-expressing cells in the photographs are shown.

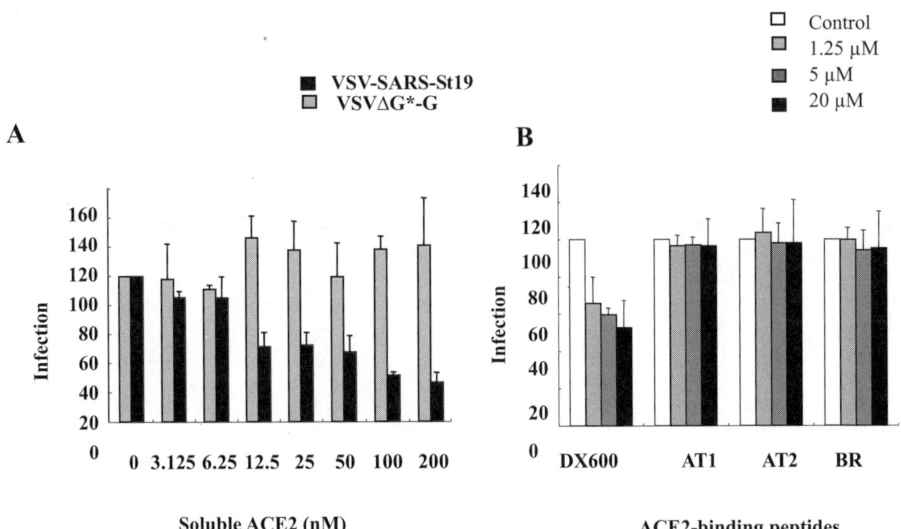

Figure 2. Inhibition of VSV-SARS-St19 infection. (A) VSV-SARS-St19 or VSVΔG*-G was pre-incubated with serially diluted soluble ACE2 followed by inoculation onto Vero E6 cells. (B) VSV-SARS-St19 was pre-incubated with DX600, angiotensin I (AT1), angiotensin II (AT2), or desArg^9-bradykinin (BR) followed by inoculation onto Vero E6 cells. Infectivity of the pseudotypes was examined using the methods described in Figure 1.

3.3. Inhibition of VSV-SARS-St19 Infection

VSV-SARS-St19 infection of Vero E6 cells was neutralized by anti-SARS-CoV antibody (data not shown). Furthermore, a recombinant human ACE2 ectodomain protein, soluble ACE2,[4] strongly affected VSV-SARS-St19 infection but did not affect infection of VSV-G–bearing pseudotype (VSVΔG*-G; Fig. 2A). These results indicated that VSV-SARS-St19 infection is mediated by SARS-CoV-S protein in an ACE2-dependent manner. We then investigated whether a known ACE2-specific peptide inhibitor can compete with ACE2-mediated pseudotype virus infection. As shown in Figure 2B, pretreatment of Vero E6 cells with DX600, which has been shown to inhibit ACE2 enzymatic activity,[5] inhibited VSV-SARS-St19 infection, while pretreatment with ACE2 peptide substrates, angiotensin I, angiotensin II, or desArg^9-bradykinin, did not. Higher concentrations (>1.25 μM) of DX600 were required for 30–50% inhibition of VSV-SARS-St19 infection, indicating that this inhibition was weak (Fig. 2B). Enzymatic activity is not required for ACE2 protein to act as a SARS-CoV receptor.[6] However, our results indicated that DX600 partially influenced the function of ACE2 as a SARS-CoV receptor. Further investigations, including inhibition studies with live SARS-CoV, are necessary to elucidate the efficacy of DX600. Our results suggested that ACE2-binding peptides can be used as specific inhibitors of SARS-CoV-S–mediated infection. Based on the results of neutralization experiments using anti-SARS-CoV antibody and soluble ACE2, we concluded that VSV-SARS-St19 infection of target cells is mediated by SARS-CoV-S protein. The assay system described here will be useful not only for developing a safe and rapid method to detect neutralizing antibodies to SARS-CoV but also for screening for inhibitors of SARS-CoV-S–mediated infection.

4. ACKNOWLEDGMENTS

This work was supported in part by a grant-in-aid from the Ministry of Health, Labor, and Welfare of Japan and the Japan Health Science Foundation, Tokyo, Japan.

5. REFERENCES

1. Matsuura, Y., Tani, H. Suzuki, K. et al., 2001, Characterization of pseudotype VSV possessing HCV envelope proteins, *Virology* **286**:263.
2. Ogino, M., Ebihara, H., Lee, B. H., et al., 2003, Use of vesicular stomatitis virus pseudotypes bearing hantaan or seoul virus envelope proteins in a rapid and safe neutralization test, *Clin. Diagn. Lab. Immunol.* **10**:154.
3. Saijo, M., Qing, T., Niikura, M., et al., 2002, Immunofluorescence technique using HeLa cells expressing recombinant nucleoprotein for detection of immunoglobulin G antibodies to Crimean-Congo hemorrhagic fever virus, *J. Clin. Microbiol.* **40**:372.
4. Fukushi, S., Mizutani, T., Saijo, M. et al., 2005, Vesicular stomatitis virus pseudotyped with severe acute respiratory syndrome coronavirus spike protein, *J. Gen. Virol.* **86**:2269.
5. Huang, L., Sexton, D.J., Skogerson, K., et al., 2003, Novel peptide inhibitors of angiotensin-converting enzyme 2. *J. Biol .Chem.* **278**:15532.
6. Li, W., Moore, M. J. Vasilieva, N., et al., 2003, Angiotensin-converting enzyme 2 is a functional receptor for the SARS coronavirus, *Nature* **426**:450.

SUBCELLULAR LOCALIZATION OF SARS-CoV STRUCTURAL PROTEINS

Lisa A. Lopez, Ariel Jones, William D. Arndt, and Brenda G. Hogue*

1. INTRODUCTION

Coronaviruses are enveloped viruses that assemble at intracellular membranes of the endoplasmic reticulum–Golgi intermediate compartment (ERGIC) in infected cells.[1] S and M are the main components of the viral envelope. The E protein is a minor component of the envelope, but plays an important role in virus assembly.[2] Coronavirus envelope formation is nucleocapsid independent. Expression of only the E and M proteins is sufficient for the formation of virus-like-particles (VLPs) of many coronaviruses.[3-5] The N protein is a multifunctional phosphoprotein that encapsidates the viral genome and plays a role in virus assembly.[6, 7] N also appears to be involved in viral RNA replication and/or transcription.[8, 9]

The aim in this study was to provide a comprehensive view of the subcellular localization of the main SARS-CoV structural proteins. The S, M, E, and N genes were expressed in BHK-21 cells, and localization of the proteins was analyzed by indirect immunofluorescence microscopy. The proteins were co-analyzed with specific organelle markers for the endoplasmic reticulum (ER) and Golgi. Additionally, the M and E proteins were co-localized with a cellular marker for the ERGIC. The N protein remained cytoplasmic. The S, M, and E proteins were found to concentrate to the Golgi region, although some S appeared to also be transported to the cell surface.

2. MATERIALS AND METHODS

SARS-CoV M, E, and S genes were subcloned into the pCAGGS expression vector under the control of the chicken beta actin promoter.[10] SARS-CoV M and E genes were cloned into pCAGGS with an HA tag on the amino terminus. BHK-21 cells were grown on Lab-Tek chamber slides (Nunc Inc.) and were transfected with pCAGGS DNAs using Lipofectamine (Invitrogen Life Technologies). Cells were fixed 18 h post-transfection in 100% methanol for 15 min at -20ºC and blocked overnight in PBS containing 0.2%

*Arizona State University, Tempe, Arizona.

gelatin. The N gene was cloned into a pcDNA vector (Invitrogen Life Technologies) under the control of the T7 promoter and expressed with vaccinia vTF7-3 that expresses T7 RNA polymerase.[11] BHK-21 cells were infected with vTF7-3 1 h prior to transfection of pcDNA-SARS-CoV N and fixed as described above at 3 h after transfection. Indirect immunofluorescence was done using primary antibodies (α-S, CDC; α-N, ViroStat; α-HA, Santa Cruz; α-Calnexin, StressGen; α-Giantin, Convance; α-ERGIC-53, Alexis Corp). Cells were dual labeled with secondary antibodies conjugated to FITC (Santa Cruz) or Alexa Fluor-594 (Molecular Probes). After washing, nuclei were stained with 4,6-diamidino-2-phenylindole (DAPI) and mounted with ProLong Gold Antifade reagent (Molecular Probes). Images were collected on an epifluorescence Nikon inverted microscope (Nikon Inc.) using MetaMorph imaging software (Universal Imaging Corporation). Images were processed using Adobe Photoshop.

3. RESULTS AND DISCUSSION

To determine the subcellular localization sites for SARS-CoV E, M, S, and N proteins, the genes were expressed in BHK-21 cells. Proteins were co-localized against the ER, ERGIC, and Golgi marker proteins, using calnexin, ERGIC-53, and giantin, respectively.

SARS-CoV M co-localized with the ERGIC/Golgi markers with no overlap with the ER marker calnexin (Fig. 1, upper left). There was no overlap between the M or E proteins and ER calnexin. Localization of the M protein in the Golgi is consistent with other coronaviruses.[1] Our observation of M in the ERGIC agrees with a recent report which also noted an overlap of SARS-CoV M with the ERGIC in addition to localization in the Golgi.[12]

SARS-CoV E also localized to the Golgi (Fig. 1, upper right). Instead of a more punctuate pattern characteristic of ER, E exhibited a compact appearance that clearly overlapped the Golgi marker, giantin. Our lab has shown that mouse hepatitis A59 (MHV-CoV A59) E protein localizes in the perinuclear region that overlaps the ER (Lopez and Hogue unpublished data). Our results suggest that SARS and MHV E proteins localize differently. SARS-CoV E localization appears to be similar to avian infectious bronchitis virus (IBV) E, which localizes to the Golgi.[5]

SARS-CoV S protein co-localized with the ER and Golgi markers and was detected in compartments along the secretory pathway (Fig. 1, lower panel). This is typical of proteins that are transported to the cell surface. However the SARS-CoV S protein appeared to concentrate primarily in the Golgi region. The focused localization of S when expressed alone suggests there may be some functional significance for the presence of the protein near the site of SARS-CoV budding and assembly. Our results are consistent with the recent report demonstrating that SARS-CoV S contains a novel dibasic motif that retains the protein in the ERGIC.[13]

The N protein remained in the cytoplasm (Fig. 1, middle panel). The protein did not co-localize with either the ER or Golgi markers. Other coronavirus N proteins have been localized to both the cytoplasm and the nucleolus.[14] Several reports indicate that SARS-CoV N is transported to the nucleus, whereas another did not observe nuclear localization.[15-18] Consistent with the latter, we were unable to detect SARS-CoV N in the nucleus. More comprehensive studies are clearly required to fully resolve and understand the trafficking of the N protein.

Our data show that while there are obvious similarities in the localization of SARS-CoV structural proteins and those of other coronaviruses, there are apparent differences. Ultimately, it will be important to compare our data with staining patterns seen in SARS-CoV infected cells to get a complete picture of where these structural proteins localize in the context of the other viral proteins. Understanding viral protein trafficking and identifying which protein(s) directs the site of virus assembly will help us develop anti-viral drug platforms and possible vaccines to combat this important pathogen.

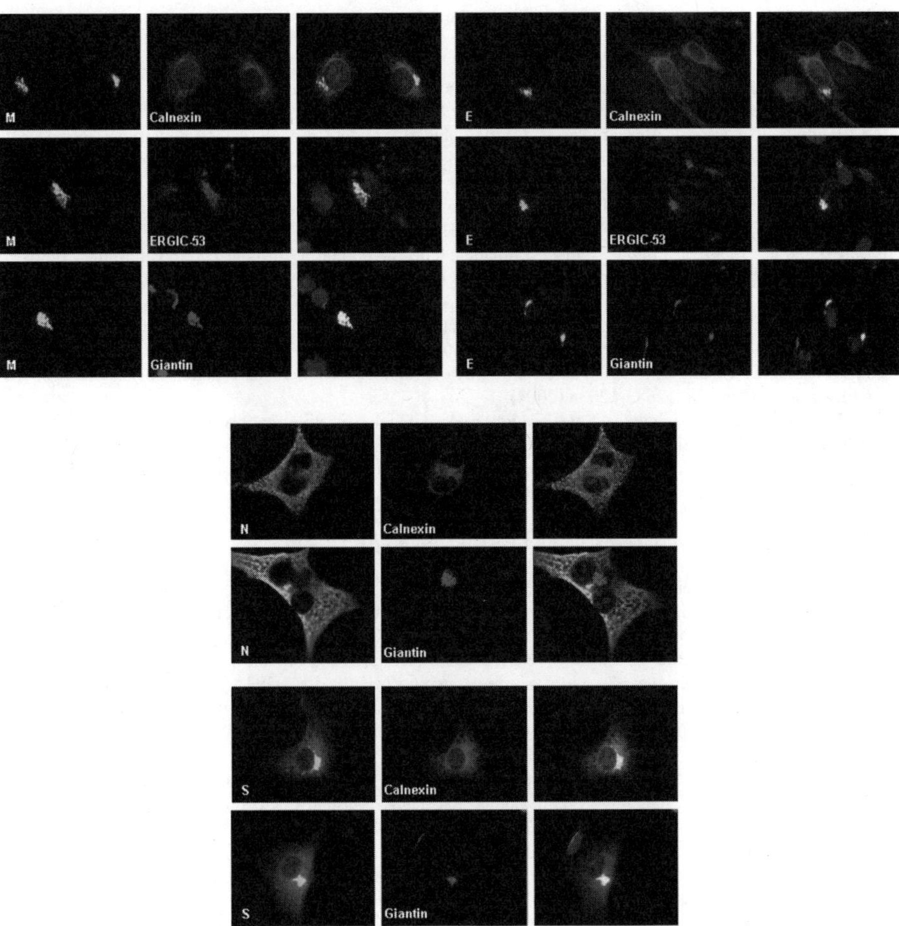

Figure 1. Subcellular localization of HA-tagged SARS-CoV M and E proteins, and localization of N and S proteins transiently expressed in BHK-21 cells. Singly expressed M, E, N, and S proteins were co-localized with cellular marker proteins. Cells were fixed and dual labeled for viral proteins and the corresponding marker proteins. Merged images are shown in the far right column (100 × magnification).

4. ACKNOWLEDGMENTS

This work was supported by Public Health Service grant AI54704 from the National Institute of Allergy and Infectious Diseases. We thank Jarrod Lauer and Tiana White for help in generating the M, E, and N subclones.

5. REFERENCES

1. K. V. Holmes and M. M. Lai, in: *Fields Virology, 4th edition*, edited by B. N. Fields, D. N. Knipe, and P. M. Howley (Lippincott-Raven Publishers, Philadelphia, 2001).
2. L. Kuo and P. S. Masters, The small envelope protein E is not essential for murine coronavirus replication, *J. Virol.* **77**, 4597-4608 (2003).
3. H. Vennema, G. J. Godeke, J. W. Rossen, W. F. Voorhout, M. C. Horzinek, D. J. Opstelten, and P. J. Rottier, Nucleocapsid-independent assembly of coronavirus-like particles by co-expression of viral envelope protein genes, *EMBO J.* **15**, 2020-2028 (2000).
4. P. Baudoux, C. Carrat, L. Besnardeau, B. Charley, and H. Laude, Coronavirus pseudoparticles formed with recombinant M and E proteins induce alpha interferon synthesis by leukocytes, *J. Virol.* **72**, 8636-8643 (1998).
5. E. Corse and C. E. Machamer, Infectious bronchitis virus E protein is targeted to the Golgi complex and directs release of virus-like particles, *J. Virol.* **74**, 4319-4326 (2000).
6. D. Escors, J. Ortego, H. Laude, and L. Enjuanes, The membrane M protein carboxy terminus binds to transmissible gastroenteritis coronavirus core and contributes to core stability, *J. Virol.* **75**, 1312-1324 (2001).
7. K. Narayanan, K. H. Kim, and S. Makino, Characterization of N protein self-associations in coronavirus ribonucleoprotein complexes, *Virus Res.* **98**, 131-140 (2003).
8. F. Almazan, C. Galan, and L. Enjuanes, The nucleoprotein is required for efficient coronavirus genome replication, *J. Virol.* **78**, 12683-12688 (2004).
9. B. Schelle, N. Karl, B. Ludewig, S. G. Siddell, and V. Thiel, Selective replication of coronavirus genomes that express nucleocapsid protein, *J. Virol.* **79**, 6620-6630 (2005).
10. H. Niwa, K. Yamamura, and J. Miyazaki, Efficient selection for high-expression transfectants with a novel eukaryotic vector, *Gene* **108**, 193-199 (1991).
11. T. R. Fuerst, E. G. Niles, F. W. Studier, and B. Moss, Eukaryotic transient-expression system based on recombinant vaccinia virus that synthesizes bacteriophage T7 RNA polymerase, *Proc. Natl. Acad. Sci. USA* **83**, 8122-8126 (1986).
12. B. Nal, C. Chan, F. Kien, L. Siu, J. Tse, K. Chu, J. Kam, I. Staropoli, B. Crescenzo-Chaigne, N. Escriou, S. vander Werf, K.-Y. Yuen, and R. Altmeyer, Differential maturation and subcellular localization of severe acute respiratory syndrome coronavirus surface proteins S, M and E, *J. Gen. Virol.* **86**, 1423-1434 (2005).
13. E. Lontok, E. Corse, and C. E. Machamer, Intracellular targeting signals contribute to localization of coronavirus spike proteins near the virus assembly site, *J. Virol.* **78**, 5913-5922 (2004).
14. T. Wurm, H. Chen, T. Hodgson, P. Britton, G. Brooks, and J. A. Hiscox, Localization to the nucleolus is a common feature of coronavirus nucleoproteins, and the protein may disrupt host cell division, *J. Virol.* **75**, 9345-9356 (2001).
15. M. A. Chang, Y. T. Lu, S. T. Ho, C. C, Wu, T. Y. Wei, C. J. Chen, Y. T. Hsu, P. C. Chu, C. H. Chen, J. M. Chu, Y. L. Jan, C. C. Hung, C. C. Fan, and Y. C. Yang, Antibody detection of SARS-CoV spike and nucleocapsid protein, *Biochem. Biophys. Res. Commun.* **314**, 931-936 (2004).
16. R. Zeng, H. Q. Ruan, X. S. Jiang, H. Zhou, L. Shi, L. Zhang, Q. H. Sheng, Q. Tu, Q. C. Xia, and J. R. Wu, Proteomic analysis of SARS associated coronavirus using two-dimensional liquid chromatography mass spectrometry and one-dimensional sodium dodecyl sulfate-polyacrylamide gel electrophoresis followed by mass spectrometric analysis, *J. Proteome Res.* **3**, 549-555 (2004).
17. M. Surjit, R. Kumar, R. N. Mishra, M. K. Reddy, V. T. Chow, and S. K. Lal, The severe acute respiratory syndrome coronavirus nucleocapsid protein is phosphorylated and localizes in the cytoplasm by 14-3-3-mediated translocation, *J. Virol.* **79**, 11476-11486 (2005).
18. R. R. Rowland, V. Chauhan, Y. Fang, A. Pekosz, M. Kerrigan, and M. D. Burton, Intracellular localization of the severe acute respiratory syndrome coronavirus nucleocapsid protein: absence of nucleolar accumulation during infection and after expression as a recombinant protein in Vero cells, *J. Virol.* **79**, 11507-11512 (2005).

SPIKE GENE DETERMINANTS OF MOUSE HEPATITIS VIRUS HOST RANGE EXPANSION

Willie C. McRoy and Ralph S. Baric*

1. INTRODUCTION

The emergence of new viruses is a poorly understood process, but one demanding increased attention in the wake of HIV, hantaviruses, avian influenza virus, and the SARS-CoV. We are using mouse hepatitis virus (MHV) as a model to explore potential mechanisms that mediate coronavirus cross-species transmission. These previously published models include a persistent infection system[1, 2] and a mixed infection system.[3] Both systems resulted in MHV variants with extended host range. Our current efforts involve characterizing the genetic determinants of the expanded host range phenotype and receptor usage of these variants as compared with the parental viruses.

2. RESULTS AND DISCUSSION

The persistent infection model produced variant V51 with an expanded host range that includes human and hamster cell lines, cell lines that the parental A59 strain is unable to productively infect. Sequencing of the Spike (S) gene revealed the presence of 13 amino acid mutations, seven in the S1 cleavage subunit and six in the S2 subunit. The mixed infection model resulted in variant C4, again with extended host range to human, hamster, and primate cell lines. Sequencing revealed 17 amino acid mutations in the S gene when compared with the A59 and JHM strains that initiated the co-infection. Ten changes are located in the S1 subunit while the remaining seven are in the S2 subunit. As a result of the co-infection, the C4 S gene is a chimera of the A59 and JHM spike genes, with the amino-terminal 85% of S derived from JHM and the remainder from A59. Interestingly, there are only two overlap mutations between our two model systems and a single overlap mutation with an additional persistent infection model characterized by Schickli et al.[4] The first overlap involves a codon deletion in C4 versus an amino acid change in V51 at position 939, while the second manifests as an amino acid change at position 949 in both V51 and C4 resulting in different amino acids.

*University of North Carolina at Chapel Hill, Chapel Hill, North Carolina 27599.

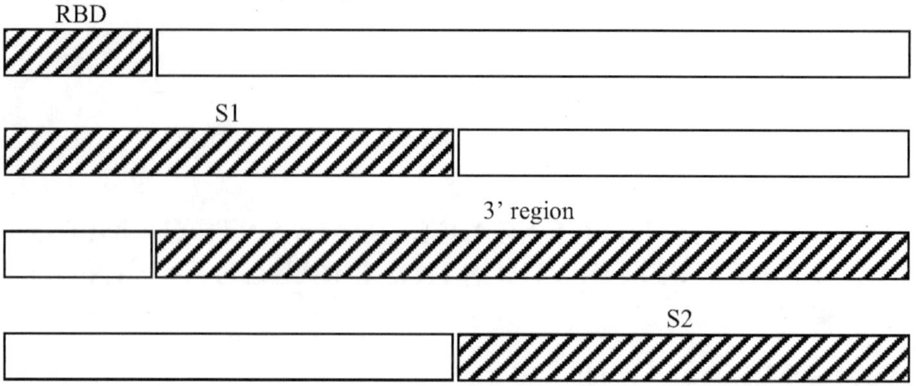

Figure 1. S gene recombinants produced for this study. The hatched regions represent S regions derived from host range isolates, while the unhatched regions are contributed by A59 from the targeted RNA recombination system, or a chimeric S gene derived from A59 and JHM for C4. From top to bottom, the recombinants depicted are: RBD recombinant, S1 recombinant, 3' region recombinant, and S2 recombinant.

To further characterize the impact of the identified mutations on the host range phenotype, we utilized the targeted RNA recombination system[5] and No See'm restriction site technology[6] to allow for precise insertion of S gene pieces from host range variants into the A59 genetic background. Isolated recombinants contained either the full length V51 or C4 S glycoproteins (Full S Recombinant) or specific domains of the molecules incorporated into the appropriate parental S gene. Recombinants include: (1) the receptor binding domain consisting of the amino terminal 330 residues (RBD recombinant), (2) the S1 cleavage subunit (S1 recombinant), (3) all residues but the receptor binding domain (3' region recombinant), and (4) the S2 cleavage subunit (S2 recombinant) (Figure 1). As the co-infection model resulted in a chimeric S gene (as noted above), recombinants containing pieces from C4 were paired with S gene regions from a prototype chimera S gene made with JHM and A59. All recombinants are viable, although the C4 3' region recombinant required multiple passages before CPE and appreciable titers could be observed. It has not been evaluated experimentally to date.

Growth curve analysis on human cell lines HepG2 and MCF7 indicates that the determinants for host range in both model systems are encoded in the S2 region, with the phenotype presenting most clearly in the V51 recombinants. Recombinants containing the S2 region alone from V51 or C4 are capable of productively replicating in HepG2 or MCF7 cell lines, respectively. Table 1 lists 48-hour titers supporting this observation. Additional cell lines from a wider array of species will be examined to determine if this trend holds or is unique to expansion into these human cell lines. Analysis of the location of the mutations in V51 S2 places two mutations at the 5' terminus of the recognized HR1 region[7] and two within the previously identified PEP3 region[8, 9] (a putative fusion domain). S2 mutations in C4 are not as tightly clustered as in V51, however one mutation is located in PEP3 (the overlap deletion versus residue change discussed earlier) and a second is located in HR1 (the common position change also discussed earlier). This unusual clustering of mutations is being more closely examined to determine its impact on host range expansion.

Table 1. Titers on HepG2 cells (V51 recombinants) and MCF7 cells (C4 recombinants).

Recombinant virus	48 hour titer (pfu/mL)	
	V51	C4
Full S gene recombinant	$1.7x10^6$	$3.45x10^4$
RBD recombinant	$6.5x10^2$	ND[1]
S1 recombinant	$6.5x10^2$	$1.5x10^2$
3' Region recombinant	$1.35x10^7$	Data not available
S2 recombinant	$5.35x10^6$	$1.3x10^4$
A59 Control	$2.5x10^2$	ND

[1] Indicates that no plaques were detected.

Receptor usage of C4, V51, and various recombinants was examined utilizing swine ST cells (refractory to infection by MHV and variants in their natural state) that express CEACAM1a, the well characterized high affinity receptor for MHV. The original V51 isolate, the Full S recombinant, the S2 recombinant, and the 3' region recombinant are all unable to productively infect ST-CEACAM1a cells. The RBD and S1 recombinants are capable of a productive infection, with 24-hour titers similar to that of A59 and are sensitive to CC1 blockade (a monoclonal antibody specific for the MHV binding domain on CEACAM1a). The situation is somewhat reversed on murine DBT cells, which express CEACAM1a along with CEACAM1b and CEACAM2. On DBT cells, the RBD and S1 recombinants are hampered by CC1 in a fashion similar to A59 while S2 and 3' region recombinants are fairly resistant to CC1 blockade. The C4 Full S recombinant and original C4 isolate are unable to productively infect ST-CEACAM1a cells, suggesting they may have lost the ability to use this receptor to initiate infection. These same viruses are resistant to blockade on DBT cells. Data on C4 derived domain recombinants are unavailable at this time. Table 2 summarizes these results.

Table 2. CEACAM1a receptor usage on ST-CEACAM1a and DBT cells in the presence of blocking antibody CC1 (24-hour titer, pfu/mL).

		ST-CEACAM1a		**DBT**	
	Virus	**Untreated**	**CC1 treated**	**Untreated**	**CC1 treated**
V51	A59 Control	$3.85x10^5$	ND[1]	$5.25x10^6$	ND
	Original isolate	$1x10^2$	ND	Data not available	
	Full S recombinant	$1x10^2$	ND	$1.39x10^7$	$5.35x10^5$
	S1 recombinant	$3.05x10^5$	ND	$3.65x10^7$	$2x10^2$
	S2 recombinant	$2x10^2$	ND	$4.15x10^7$	$6.3x10^5$
	RBD recombinant	$1.11x10^5$	ND	$4.54x10^7$	$5.5x10^2$
	3' Region recombinant	$7x10^2$	ND	$2.03x10^7$	$8.65x10^5$
C4	A59 Control	$1.4x10^7$	ND	$1x10^7$	$4.5x10^4$
	Full S recombinant	ND	ND	$2x10^7$	$5.1x10^7$
	Original isolate	$3.0x10^3$	ND	$1.5x10^7$	$4.7x10^7$
	JHM control	$4.05x10^6$	ND	Data not available	

[1] Indicates that no plaques were detected.

Although a definitive group of mutations leading to MHV host range expansion has not been examined, the data presented here expand traditional concepts of MHV S protein biology. Previous work has identified the receptor binding domain as an important determinant in MHV host range expansion[10, 11] as well as virulence and tissue tropism. We have demonstrated that the receptor binding domain does not appear to be a genetic factor in the expansion process into the human cell lines tested as part of this study. All host range determinants are encoded in the S2 region of the protein, suggesting the possibility for additional functions of S2 such as receptor binding (to an as yet unidentified receptor) and/or initiation of viral entry through mechanisms not traditionally associated with coronaviruses (processes such as endocytosis or phagocytosis).

3. REFERENCES

1. R. S. Baric, et al., Persistent infection promotes cross-species transmissibility of mouse hepatitis virus, *J. Virol.* **73**, 638-649 (1999).
2. W. Chen and R. S. Baric, Molecular anatomy of mouse hepatitis virus persistence: coevolution of increased host cell resistance and virus virulence, *J. Virol.* **70**, 3947-3960 (1996).
3. R. S. Baric, et al., Episodic evolution mediates interspecies transfer of a murine coronavirus, *J. Virol.* **71**, 1946-1955 (1997).
4. J. H. Schickli, et al., The murine coronavirus mouse hepatitis virus strain A59 from persistently infected murine cells exhibits an extended host range, *J. Virol.* **71**, 9499-9507 (1997).
5. L. Kuo, et al., Retargeting of coronavirus by substitution of the spike glycoprotein ectodomain: crossing the host cell species barrier, *J. Virol.* **74**, 1393-1406 (2000).
6. B. Yount, et al., Systematic assembly of a full-length infectious cDNA of mouse hepatitis virus strain A59, *J. Virol.* **76**, 11065-11078 (2002).
7. R. J. de Groot, et al., Evidence for a coiled-coil structure in the spike proteins of coronaviruses, *J. Mol. Biol.* **196**, 963-966 (1987).
8. Z. Luo and S. R. Weiss, Roles in cell-to-cell fusion of two conserved hydrophobic regions in the murine coronavirus spike protein, *Virology* **244**, 483-494 (1998).
9. P. Chambers, C. R. Pringle, and A. J. Easton, 1990, Heptad repeat sequences are located adjacent to hydrophobic regions in several types of virus fusion glycoproteins, *J. Gen. Virol.* **71**, 3075-3080 (1990).
10. J. H. Schickli, et al., The N-terminal region of the murine coronavirus spike glycoprotein is associated with the extended host range of viruses from persistently infected murine cells, *J. Virol.* **78**, 9073-9083 (2004).
11. L. B. Thackray and K. V. Holmes, Amino acid substitutions and an insertion in the spike glycoprotein extend the host range of the murine coronavirus MHV-A59, *Virology* **324**, 510-524 (2004).

VIRION-LIPOSOME INTERACTIONS IDENTIFY A CHOLESTEROL-INDEPENDENT CORONAVIRUS ENTRY STAGE

Joseph A. Boscarino, Jeffrey M. Goletz, and Thomas M. Gallagher*

1. INTRODUCTION

Entry of enveloped viruses depends on several cellular components, including protein or carbohydrate receptors and oftentimes co-receptors that both bind viruses to cells and catalyze the initial stages of viral surface protein refolding.[1] Endocytosis is a common prerequisite to successful entry; both the acidic pH and the proteases of the endosome can create further structural changes in the viral proteins mediating cell receptor binding and virus-cell membrane fusion.[2] Finally, an appropriate lipid environment, often abundant in sterols, is also frequently necessary to achieve facile fusion of viral and cellular membranes.[3] In numerous seminal investigations, each of these steps has been dissected by blocking virus entry with mutant transgenes or with drugs that alter receptors or endosome or lipid environments. Although these are powerful approaches, the findings can be complicated when the transgenes or drugs create untoward pleiotropic changes in cellular functions. A complementary reductionist approach involves *in vitro* virus entry in which enveloped viruses are bound to and then fused into synthetic liposomes.[4] In these test-tube reactions, receptors, pH, proteases and liposome bilayer compositions can be precisely defined and freely altered in ways not achievable in living cells. Such systems are useful adjuncts to understanding the biochemistry of enveloped virus entry.

In vitro virus-liposome binding assays have been recently developed for the murine coronaviruses.[5,6] In these studies, purified MHV particles acquired hydrophobic character by incubation with soluble MHV receptors or by exposures to elevated pH, thus causing a small proportion of the virus population to associate with liposomes. We aimed to advance these studies by creating liposomes that more closely reflect authentic MHV-susceptible cells. To this end, we created synthetic liposomes containing nickel-nitriloacetic acid (NiNTA) adducts on a fraction of lipid head groups. Soluble MHV receptors containing engineered polyhistidine tags could then be attached to the liposomes, thereby creating artificial targets for virus binding. These liposomes with

*Loyola University Medical Center, Maywood, Illinois 60153.

attached receptors have been useful in identifying the requirements for irreversible association of MHV to target membranes during entry.

2. RESULTS

We obtained lipids from Avanti Polar Lipids, Inc., including the novel synthetic product DOGS-NTA-Ni, which contains a chelated Ni^{+2}-NTA head group that can noncovalently complex with polyhistidine tags. Lipids in various formulations [typically 69:30:1 mole ratio phosphatidylcholine (PC), cholesterol (chol), DOGS-NiNTA] were suspended from a dried state into HEPES-buffered saline and then extruded through polycarbonate filters to create 100-nm-diameter liposomes. These liposomes were then mixed with baculovirus-expressed soluble MHV receptors containing C-terminal 6x-histidine tags.[7] These soluble receptors are designated soluble Carcino-Embryonic-Antigen, or sCEA6his. To remove unbound receptors, we adjusted the mixtures to contain 50% w/v sucrose, then overlaid stepwise with 40% – 30% – 10% w/v sucrose solutions and floated the liposomes to the 10–30% sucrose interface by ultracentrifugation. The liposome:sCEA6his complexes were collected and stored at 4°C. Estimates of the sCEA-6xhis concentrations by titration and immunoblotting indicated that our typical preparations had densities of ~50 sCEA per 100-nm liposome.

Metabolically-radiolabeled [^{35}S] MHV particles were purified by equilibrium density gradient sedimentation. Typically, we incubated ~50,000 cpm of [^{35}S] MHV (10^5 pfu) for 1 h at 4°C with 10 mM phospholipid liposomes, in a volume of 0.2-ml of HEPES-buffered saline. Mixtures were then brought to 50% sucrose and subjected to a second round of ultracentrifugation to float the liposomes. By identifying the [^{35}S]-containing gradient fractions, we determined that MHV binding to liposomes was absolutely dependent on NiNTA lipids and on sCEA6his (Fig. 1). Notably, a control sCEAFc lacking the 6xhis tag bound to the [^{35}S] virions, but failed to link virions to liposomes (Fig. 1), even after incubation at 37°C.

To determine whether the connections between [^{35}S] MHV and liposomes were reversible, we exploited the common knowledge that noncovalent NiNTA:6his linkages are disrupted by EDTA or imidazole. To this end, we harvested [^{35}S] MHV-liposome complexes from gradient fractions and added EDTA (10 mM final concentration) for 30 min at 4°C. Reflotation in fresh sucrose gradients revealed that the EDTA exposures separated the virions, leaving the [^{35}S] near the bottom of the ultracentrifuge tubes (Fig. 2). However, MHV-liposome complexes incubated for 30 min at 37°C were insensitive to dissociation by EDTA (Fig. 2). This discovery allowed us to conclude that physiologic temperature is required to establish EDTA-resistant linkages between virus and its target lipid bilayer.

MHV entry requires cholesterol in target lipid bilayers.[8] To determine whether the irreversible virus-liposome linkages required cholesterol, we prepared liposomes lacking and containing 30mol% cholesterol. These two liposome formulations were indistinguishable in their irreversible, 37°C temperature-dependent capture of [^{35}S] MHV particles (Fig. 2).

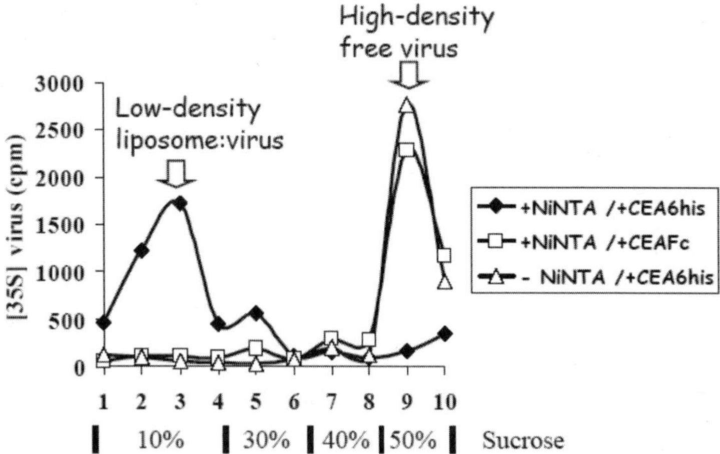

Figure 1. Virus-liposome interactions require both sCEA6his and NiNTA lipids. Liposomes with or without incorporated NiNTA lipids were incubated with either sCEA6his or sCEAFc, then with purified [^{35}S] MHV strain A59 for 1 h at 4°C before floating liposomes in sucrose gradients. The abundance of [^{35}S] in 1/10 of each of the gradient fractions were determined by scintillation counting.

3. DISCUSSION

Coronavirus particles are distinguished by their prominent spike protein trimers. These glycoproteins bind cell receptors and mediate virus-cell membrane fusions. The fusion reaction pathway is of the "class I" type[9], which involves a temperature-dependent refolding into helical bundles. In a productive process, this refolding is preceded by

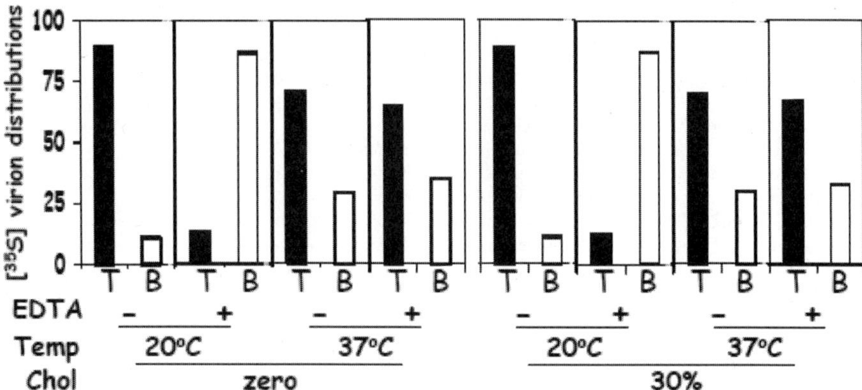

Figure 2. Stability of virus-liposome complexes against dissociation by EDTA. [^{35}S] MHV-liposome complexes, purified by floatation in sucrose gradients, were held at 20°C or 37°C prior to EDTA exposure. Samples were then subjected to a second round of floatation, and the proportion of [^{35}S] in the top (T) and bottom (B) of each gradient were determined by scintillation counting. The [^{35}S] in T and B represent virus-liposome complexes and free virus, respectively. EDTA released viruses from liposomes at 20°C, but 37°C temperatures created EDTA-stable virus-liposome complexes.

insertion of hydrophobic spike peptides, so-called fusion peptides, into target membranes. We are currently hypothesizing that the stabilized association of MHV with synthetic liposomes results from the insertion of their fusion peptides into the liposome membranes. If this hypothesis is borne out, then we can suggest that the intercalation of fusion peptides is strongly temperature-dependent but does not require cholesterol in the target membranes.

In assessing whether bona-fide fusion of virus and liposome membranes occurs, our principal strategy has been to monitor fluorescence changes as pyrene lipid probes redistribute from limiting membranes into the larger area of fused vesicles.[4] Despite numerous attempts, we have not yet documented MHV:liposome coalescence by these methods. It is entirely possible that components required to complete the MHV membrane fusion process are missing from our *in vitro* assays. Fusion may require exposure to acids or proteases, may require proteinaceous co-factors in addition to the principal CEA receptors, and likely also requires specific sterol-containing lipid environments. We are continuing our investigations to identify the conditions that may allow completion of the spike-mediated class I fusion pathway.

4. ACKNOWLEDGMENTS

We thank Dr. Fumihiro Taguchi (National Institute of Infectious Diseases, Tokyo, Japan) for providing baculovirus recombinants encoding sCEA6xhis. This work was funded by PHS grant NIH AI60030.

5. REFERENCES

1. Chen, B., Vogan, E. M., Gong, H., Skehel, J. J., Wiley, D. C., and Harrison, S. C., 2005, Structure of an unliganded simian immunodeficiency virus gp120 core, *Nature* **433**:834-841.
2. Simmons, G., Gosalia, D. N., Rennekamp, A. J., Reeves, J. D., Diamond, S. L., and Bates, P., 2005, Inhibitors of cathepsin L prevent severe acute respiratory syndrome coronavirus entry, *Proc. Natl. Acad. Sci. USA* **102**:11876-11881.
3. Rawat, S. S., Viard, M., Gallo, S. A., Rein, A., Blumenthal, R., and Puri, A., 2003, Modulation of entry of enveloped viruses by cholesterol and sphingolipids, *Mol. Membr. Biol.* **20**:243-254.
4. Smit, J. M., Waarts, B. L., Bittman, R., and Wilschut, J., 2003, Liposomes as target membranes in the study of virus receptor interaction and membrane fusion, *Methods Enzymol.* **372**:374-392.
5. Tsai, J. C., Zelus, B. D., Holmes, K. V., and Weiss, S. R., 2003, The N-terminal domain of the murine coronavirus spike glycoprotein determines the CEACAM1 receptor specificity of the virus strain, *J. Virol.* **77**:841-850.
6. Zelus, B. D., Schickli, J. H., Blau, D. M., Weiss, S. R., and Holmes, K. V., 2003, Conformational changes in the spike glycoprotein of murine coronavirus are induced at 37 degrees C either by soluble murine CEACAM1 receptors or by pH 8, *J. Virol.* **77**:830-840.
7. Taguchi, F., and Matsuyama, S., 2002, Soluble receptor potentiates receptor-independent infection by murine coronavirus, *J. Virol.* **76**:950-958.
8. Thorp, E. B., and Gallagher, T. M., 2004, Requirements for CEACAMs and cholesterol during murine coronavirus cell entry, *J. Virol.* **78**:2682-2692.
9. Bosch, B. J., van der Zee, R., deHaan, C. A., and Rottier, P. J., 2003, The coronavirus spike protein is a class I virus fusion protein: structural and functional characterization of the fusion core complex, *J. Virol.* **77**:8801-8811.
10. Sainz, B., Rausch, J. M., Gallaher, W. R., Garry, R. F., and Wimley, W. C., 2005, Identification and characterization of the putative fusion peptide of the severe acute respiratory syndrome-associated coronavirus spike protein, *J. Virol.* **79**:7195-7206.

AVIAN INFECTIOUS BRONCHITIS VIRUS ENTERS CELLS VIA THE ENDOCYTIC PATHWAY

Victor C. Chu, Lisa J. McElroy, A Damon Ferguson, Beverley E. Bauman, and Gary R. Whittaker*

1. INTRODUCTION

Avian infectious bronchitis virus (IBV) is a group III coronavirus that has a major economic impact in the poultry industry. Infected layers often have a large drop in egg production due to impaired ovary and oviduct functions. Clinical manifestations of IBV-infected chickens include respiratory, renal, and reproductive diseases and are often coupled with secondary infections such as airsaculitis and oviduct salpingitis.[1] Although the etiology, pathogenesis, and diagnosis have been described since the 1930s, the molecular mechanism of viral entry remains elusive.

Viruses utilize various mechanisms to gain entry into their host cells. Enveloped viruses need to shed the viral envelope in order to release their genomes into target cells and initiate further replication.[2] The process of uncoating typically follows two distinct mechanisms: pH-dependent fusion in the endosome or pH-independent fusion at the cell surface.[3] Many enveloped viruses (e.g., influenza viruses) utilize various endocytic pathways to travel deep into the cell cytoplasm in order to smoothly bypass the cortical cytoskeleton near the cell periphery.[4] The endocytic pathway also provides a suitable acidic environment for the pH-dependent membrane fusion to deliver the viral genome into the cell.[5, 6] In contrast, paramyxoviruses (e.g., Sendai viruses) are well established to enter target cells via pH-independent fusion with the plasma membrane.[7] For coronaviruses, contrasting results have been reported regarding the role of endosomes and low pH activation, and consequently, there is no general consensus regarding the entry mechanism.[8-11]

A major characteristic of IBV infected cells is the formation of large syncytia, which has become a key piece of evidence supporting fusion with the cell plasma membrane at neutral pH during viral entry.[12] In this model, pH-neutral syncytia formation is thought to be a representative model of coronavirus fusion during entry. However, electron microscopy from Patterson *et al.* clearly shows IBV entering chorioallantoic membrane (CAM) and chick kidney cells via "viroplexis" through cellular engulfment into

*Cornell University, Ithaca, New York 14843.

cytoplasmic vacuoles.[8] In addition, Li and Cavanagh demonstrated that IBV strain Beaudette (IBV/Bdtt) infection can be reduced by as much as 95% with the lysosomotropic agent ammonium chloride.[11] Recently, we established a tissue culture system using IBV/Bdtt viruses to infect baby hamster kidney (BHK-21) cells, using anti-S1 monoclonal antibody labeling to study coronavirus entry requirements. Our goal was to investigate the molecular mechanism of IBV/Bdtt entry. We found that IBV infection was sensitive to endocytosis inhibitors including monensin and chlorpromazine in a dose-dependent manner. We therefore conclude that despite syncytia formation during late times of infection, IBV strain Beaudette utilizes the endocytic pathway and low pH-induced fusion to gain entry into BHK-21 cells.

2. RESULTS AND DISCUSSION

2.1. Effect of Monensin and Chlorpromazine on IBV Strain Beaudette Infection

To examine the route of entry during IBV/Bdtt virus infection, we made use of a pH neutralizing agent monensin. As an ionophore, monensin neutralizes the pH gradient of endocytic vesicles by exchanging luminal protons (H^+) with potassium ions (K^+),[13] hence it inhibits low pH dependent virus-cell fusion. IBV strain Beaudette was propagated and cultured from the allantoic fluid of 10-day-old specific pathogen free (SPF) embryonated chicken eggs. Sendai virus, strain Cantell, (ATCC) was also propagated as a control virus. A standard IBV/Bdtt or Sendai/Cantell infection was achieved by inoculating a monolayer of BHK-21 cells (ATCC) with 5 MOI of each virus in 200 µl of RPMI1640 binding media (Cellgro) containing 0.2% bovine serum albumin (Cellgro), 10 mM HEPES (Cellgro) at pH 7.3 in a 24-well tissue culture treated plate. Virus was adsorbed in a 37°C and 5% CO_2-free incubator for 60 minutes with gentle rocking. Then viral inoculum was replaced with DMEM infection media (Cellgro) containing 2% fetal bovine serum (Cellgro), 1% penicillin and streptomycin (Cellgro), and 10 mM HEPES (Cellgro), and incubation was resumed in a 37°C and CO_2 incubator for additional 7 hours.

The BHK-21 cell monolayer was fixed with ice-cold methanol (Sigma) for 2 minutes at 8 hours postinoculation. IBV infection was identified using the anti-S1 monoclonal antibody 15:88.[14] A chicken polyclonal anti-Sendai virus antibody (USBiological) was used to identify Sendai virus infection. Secondary antibodies were Alexa 488-conjugated goat anti-mouse or anti-chicken (Molecular Probes). Cells were observed under a Nikon Eclipse E600 fluorescence microscope and images collected using a SPOT RT camera and SPOT 3.5 software.

For monensin treated samples, BHK-21 cell monolayers were pretreated with 15 µM monensin (CalBiochem) at 37°C for 10 minutes prior to virus inoculation. Fifteen µM monensin was added to both RPMI1640 binding media and DMEM infection media to prevent low pH dependent fusion activation during IBV/Bdtt infection. In contrast to Sendai virus infection, which was unaffected by the monensin treatment, Fig. 1A shows that IBV/Bdtt infection in BHK-21 cells was sensitive to monensin treatment at a concentration of 15 µM. The infection could be restored when monensin was added 2 hours post–viral inoculation (data not shown). This indicated that monensin had no indirect effects on viral replication and specifically affected virus entry.

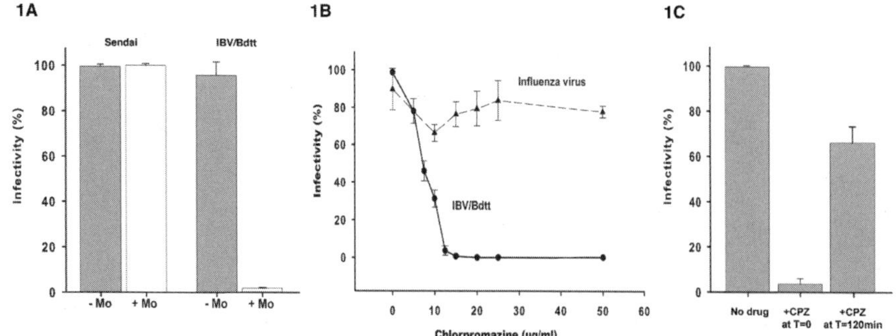

Figure 1. A. Effect of monensin (Mo) treatment during entry of Sendai virus and IBV/Bddt. B. Effect of chlorpromazine treatment during entry of influenza A/WSN/33 (dashed line) or IBV/Bdtt (solid line). C. Chlorpromazine (CPZ) delay treatment assay. Greater than 200 cells were counted in each sample. Each experimental condition was performed in triplicate. Statistical analysis was performed in SigmaPlot 9.0 software, and error bars represent the standard deviation of the mean.

To narrow down the specific route of IBV/Bdtt entry, we studied the effects of chlorpromazine during IBV entry. Chlorpromazine induces clathrin lattice assembly around endosomes while inhibiting adaptin-2 protein (AP-2) binding to the cellular membrane, hence preventing clathrin-coated pit formation and clathrin-mediated endocytosis.[15] Infection of BHK-21 cells by IBV/Bdtt or influenza virus, strain A/WSN/33, (ATCC) was performed essentially as described above. Chlorpromazine (Calbiochem) was diluted in RPMI1640 binding media to a concentration of 0–50 µg/ml, and drug treatment followed by virus infection was performed. Figure 1B shows that IBV/Bdtt infectivity is inhibited by chlorpromazine treatment in a dose-dependent manner. However, the control influenza virus remained unaffected by chlorpromazine treatment since influenza viruses can enter cells through both clathrin-dependent and-independent endocytic pathways.[16] To exclude any nonspecific effect of chlorpromazine on virus replication, chlorpromazine was added 2 hours post–virus inoculation to bypass its effect on IBV/Bdtt entry. IBV/Bdtt infection was restored to as much as 60% of the nontreated control suggesting that chlorpromazine specifically inhibited IBV/Bdtt entry through a clathrin-mediated endocytic pathway (Fig. 1C).

The use of drugs such as monensin and chlorpromazine to screen for IBV entry pathways served as an effective tool during our initial investigation. However, more specialized assays and screening methods targeting individual endocytic compartments or cellular pathway are required to draw a definitive conclusion regarding to the molecular mechanism during IBV entry to its target cell. Co-localization studies between specific cellular compartments (e.g., clathrin coated pits, caveolin, early, late, or recycling endosomes) and infective virions using a panel of antibody markers can help reveal more precise location at each stage of IBV entry. It has also been shown that cholesterol can function as a co-factor for fusion of mouse hepatitis virus (MHV)[17]; therefore, lipid rafts may also play a role in coronavirus entry. Finally, dominant negative proteins and small interfering RNAs may serve as effective tools to investigate the entry requirements for IBV, which will be addressed in future experiments.

3. ACKNOWLEDGMENTS

We would like to thank Dr. Benjamin Lucio-Martinez and Dr. Syed Naqi for their provision of viruses and antibodies and Xiangjie Sun for experimental support. This work was supported by NIH grant R03 AI060946.

4. REFERENCES

1. D. Cavanagh and S. Naqi, in: *Diseases of Poultry,* edited by Y.M. Saif (Blackwell, 2003), pp. 101-120.
2. A. E. Smith and A. Helenius, How viruses enter animal cells, *Science* **304**, 237-242 (2004).
3. L. J. Earp, S. E. Delos, H. E. Park, and J. M. White, The many mechanisms of viral membrane fusion proteins, *Curr. Top. Microbiol. Immunol.* **285**, 25-66 (2005).
4. V. C. Chu and G. R. Whittaker, Influenza virus entry and infection require host cell N-linked glycoprotein, *Proc. Natl. Acad. Sci. USA* **101**, 18153-18158 (2004).
5. L. V. Chernomordik, E. Leikina, V. Frolov, P. Bronk, and J. Zimmerberg, An early stage of membrane fusion mediated by the low pH conformation of influenza hemagglutinin depends upon membrane lipids, *J. Cell Biol.* **136**, 81-93 (1997).
6. S. B. Sieczkarski and G. R.Whittaker, Dissecting virus entry via endocytosis, *J. Gen. Virol.* **83**, 1535-1545 (2002).
7. M. C. Hsu, A. Scheid, and P. W. Choppin, Enhancement of membrane-fusing activity of Sendai virus by exposure of the virus to basic pH is correlated with a conformational change in the fusion protein, *Proc. Natl. Acad. Sci. USA* **79**, 5862-5866 (1982).
8. S. Patterson and R. W. Bingham, Electron microscope observations on the entry of avian infectious bronchitis virus into susceptible cells, *Arch. Virol.* **52**, 191-200 (1976).
9. G. Simmons, J. D. Reeves, A. J. Rennekamp, S. M. Amberg, A. J. Piefer, and P. Bates, Characterization of severe acute respiratory syndrome-associated coronavirus (SARS-CoV) spike glycoprotein-mediated viral entry, *Proc. Natl. Acad. Sci. USA* **101**, 4240-4245 (2004).
10. L. S. Sturman, C. S. Ricard, and K. V. Holmes, Conformational change of the coronavirus peplomer glycoprotein at pH 8.0 and 37 degrees C correlates with virus aggregation and virus-induced cell fusion, *J. Virol.* **64**, 3042-3050 (1990).
11. D. Li and D. Cavanagh, Coronavirus IBV-induced membrane fusion occurs at near-neutral pH, *Arch. Virol.* **122**, 307-316 (1992).
12. D. J. Alexander and M. S. Collins, Effect of pH on the growth and cytopathogenicity of avian infectious bronchitis virus in chick kidney cells, *Arch. Virol.* **49**, 339-348 (1975).
13. J. Malecki, A. Wiedlocha, J. Wesche, and S. Olsnes, Vesicle transmembrane potential is required for translocation to the cytosol of externally added FGF-1, *EMBO J.* **21**, 4480-90 (2002).
14. K. Karaca, S. Naqi, and J. Gelb, Jr., Production and characterization of monoclonal antibodies to three infectious bronchitis virus serotypes, *Avian Dis.* **36**, 903-915 (1992).
15. L. H. Wang, K. G. Rothberg, and R. G. Anderson, Mis-assembly of clathrin lattices on endosomes reveals a regulatory switch for coated pit formation, *J. Cell. Biol.* **123**, 1107-1117 (1993).
16. S. B. Sieczkarski and G. R. Whittaker, Influenza virus can enter and infect cells in the absence of clathrin-mediated endocytosis, *J. Virol.* **76**, 10455-10464 (2002).
17. E. B. Thorp and T. M. Gallagher, Requirements for CEACAMs and cholesterol during murine coronavirus cell entry, *J. Virol.* **78**, 2682-2692 (2004).

THREE-DIMENSIONAL RECONSTRUCTION OF THE NUCLEOLUS USING META-CONFOCAL MICROSCOPY IN CELLS EXPRESSING THE CORONAVIRUS NUCLEOPROTEIN

Jae-Hwan You, Mark L. Reed, Brian K. Dove, and Julian A. Hiscox*

1. INTRODUCTION

The coronavirus nucleoprotein (N protein) is one of the most abundantly expressed viral proteins in an infected cell, with the principal function of binding the viral RNA genome to form the ribonucleocapsid structure (RNP) and forming the viral core. N protein also has roles in viral replication, transcription, and translation as well as modulating cellular processes. We and others have shown that some coronavirus and arterivirus N proteins can localize to a dynamic subnuclear structure called the nucleolus and interact with nucleolar proteins.[1-3] The nucleolus is involved in ribosome subunit biogenesis, RNA processing, cell cycle control, and acts as a sensor for cell stress.[4] Morphologically the nucleolus can be divided into an inner fibrillar center (FC), a middle dense fibrillar component (DFC), and an outer granular component (GC). A directed proteomic analysis followed by subsequent bioinformatic analysis revealed that the nucleolus is composed of at least 400 proteins.

Coronavirus N proteins have the potential to be phosphorylated at multiple serine residues. However, mass spectroscopic analysis of both the avian infectious bronchitis virus (IBV)[5] and porcine transmissible gastroenteritis virus (TGEV)[6] N proteins have shown that phosphorylation occurs at only three or four residues. In the case of IBV N protein, these map to predicted casein kinase II sites.[5] Based on amino acid sequence comparisons, three conserved regions have been identified in the murine coronavirus, mouse hepatitis virus (MHV) N protein.[7] In general, other coronavirus N proteins would appear to follow this pattern.

We investigated the three-dimensional structure of the nucleolus and sub-nuclear bodies within cells expressing IBV and severe acute respiratory syndrome coronavirus (SARS-CoV) N protein. In many cases, viral proteins localize to discrete regions of the nucleolus and their specific localization can inform as to what effect they may be having

*University of Leeds, Leeds LS2 9JT, United Kingdom.

on the host cell. For example, proteins that localize to, and disrupt the GC, can affect cellular transcription. Therefore, we use coronavirus N proteins as a model to test our working hypothesis that disruption of nucleolar proteins and/or alterations in nucleolar architecture can perturb cellular functions.

2. MATERIALS AND METHODS

2.1. Expression Constructs

SARS-CoV N protein was N-terminally tagged with ECFP, creating pECFP-SARS-CoV-N. Briefly, the SARS-CoV N gene was amplified (from a clone containing the SARS-CoV N gene-kindly provided by Dr. Maria Zambon) using gene-specific primers to the 5' and 3' 20 nucleotides of the SARS-CoV N gene, but incorporating 5' restriction enzyme sites (in the case of the forward primer this was *Bsp*EI and in the reverse primer *Bam*HI) and then TOPO cloned into pCR2.1 (Invitrogen). The insert was subcloned into pECFPC1 (Clontech) such that SARS-CoV N protein would be C-terminal and in frame to ECFP. EGFP was added N-terminally to IBV N protein, creating pEGFP-IBV-N. The cloning strategy was identical to that described for SARS-CoV N gene, except the forward primer restriction site was *Bam*HI and the reverse primer restriction site was *Eco*RI. The gene was ligated into pEGFPC2 (Clontech), such that IBV N protein would be C-terminal and in frame to EGFP. All clones were verified by sequencing and expression of fusion proteins by Western blot (data not shown).

2.2. Meta-Confocal Microscopy

Confocal sections of fixed samples were captured on an LSM510 META microscope (Carl Zeiss Ltd.) equipped with a 63x, NA 1.4, oil immersion lens. Pinholes were set to allow optical sections of 1 μm to be acquired. In singly transfected cells, ECFP was excited with the 458-nm argon laser line running at 10%, and emission was collected through a BP435-485 emission filter. EGFP was excited with the 488-nm argon laser line running at 2%, and emission was collected through a LP505 filter. Propidium iodide (PI) was excited with the helium:neon 543-nm laser line in all cases, and emission was collected through a LP560 filter. Due to excitation of the EGFP molecule by the 458-nm argon laser line, co-transfected samples were linearly unmixed using the META detector. Lambda plots of EGFP and ECFP were generated from singly transfected reference samples excited with the 458-nm argon laser line and collected with the META detector between 461 and 536-nm, in 10.7-nm increments. These lambda plots were then utilized to separate, or unmix, overlapping emission signal from co-transfected samples. Z-sections of cells expressing EGFP, counterstained with PI, were generated by a two-step methodology. Firstly, serial confocal sections of EGFP were acquired with the META detector. PI was then collected as described using the same z-settings. Z-steps were collected 0.5 μm apart to allow over sampling of the data. The two sets of z-stacks were then pseudo-coloured and merged using the 'copy' facility within the LSM510 META software. Three-dimensional reconstruction and orthogonal views were also generated in the LSM510 META software.

3. RESULTS AND DISCUSSION

3.1. Expression of IBV N Protein and SARS-CoV Protein

Studying the nucleolar localisation of proteins can be problematic. Previous reports raised the possibility that charged proteins could migrate through cells postfixation and become localized to the nucleus and nucleolus.[8] Although in this instance the specific example of VP22 has been challenged, the possibility arises that localization of coronavirus N proteins to discrete subcellular structures could be an artifact of fixation conditions. In addition, the successful detection of nucleolar proteins using antibodies can be related to the concentration of the protein within the nucleolus, in that the nucleolus, because of the high protein concentration, is not always amenable to antibody staining.[9]

To address these concerns and also to investigate the subcellular localization of SARS-CoV N protein, we generated vectors that expressed fluorescent tagged SARS-CoV and IBV N proteins and determined the subcellular localization of these proteins first by live cell imagery, followed by fixation and confocal microscopy. SARS-CoV N gene was cloned downstream of ECFP (from vector pECFPC1, Clontech), creating vector pECFP-SARSCoV-N, and IBV N gene was cloned downstream of EGFP (from vector pEGFPC2, Clontech), creating pEGFP-IBV-N, and when expressed in cells, resulted in fluorescent fusion proteins ECFP-SARS-CoV-N and EGFP-IBV-N, respectively.

Cos-7 cells were transfected with pECFPC1, pEGFPC2, pEGFP-IBV-N or pECFP-SARSCoV-N (the former two as controls), and imaged 24 hr later by live cell imaging (Fig. 1a) or co-transfected with both pEGFP-IBV-N and pECFP-SARSCoV-N, fixed 24 hr post-transfection and analyzed by META-confocal microscopy (which unmixes ECFP from EGFP (Fig. 1b).

Live cell imaging data indicated that both EGFP and ECFP, when expressed as individual proteins, had no distinct distribution pattern and were present in both the cytoplasm and nucleus but not nucleolus (Fig. 1A). However, EGFP tagged IBV N

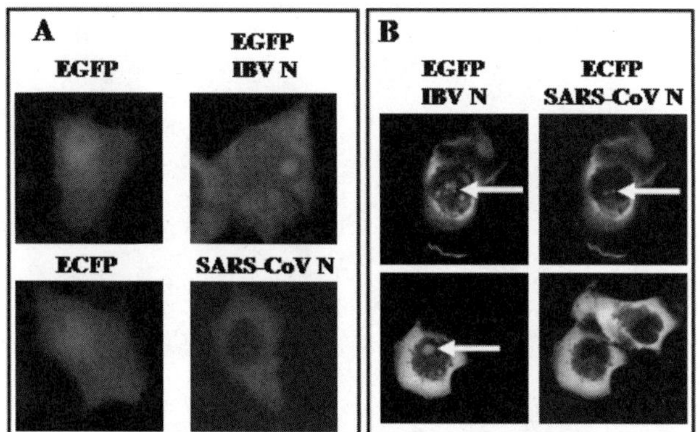

Figure 1. (A). Live cell imaging of cells expressing EGFP, ECFP, EGFP-IBV-N, and ECFP-SARS-CoV N protein. (B) META-confocal image of the same cells co-expressing EGFP-IBV-N (indicated) and ECFP-SARS-CoV N (indicated) protein; subnuclear structures are indicated by an arrow in these cells.

protein localized to both the cytoplasm and nucleus while ECFP tagged SARS-CoV N protein localized to the cytoplasm only. As ECFP tagged SARS-CoV N has a molecular weight lower than the size exclusion limit for the nuclear pore complex, the lack of any SARS-CoV N protein within the nucleus suggests this protein contains a cytoplasmic retention signal. Because these images were taken from live cells, the localization of IBV N protein to the nucleolus could not have been due to an artifact of fixation.

Confocal microscopy data (Fig. 1B) reflected the localization patterns observed using live cell imaging. In contrast to EGFP-tagged IBV N protein, ECFP-tagged SARS-CoV N protein localized to the cytoplasm and, in a minority of cells, to what appeared to be a nuclear body (arrowed). Based upon morphology this structure cannot be identified but it does not have the appearance of the nucleolus.

3.2. Three-Dimensional Reconstruction of the Nucleolus in IBV N–Expressing Cells

To investigate whether IBV N protein localized to a specific part of the nucleolus, Cos-7 cells were transfected with pEGFP-IBV-N, fixed 24 hr post-transfection, stained with PI to visualize the nucleus and nucleolus, then sectioned by confocal microscopy (Fig. 2).

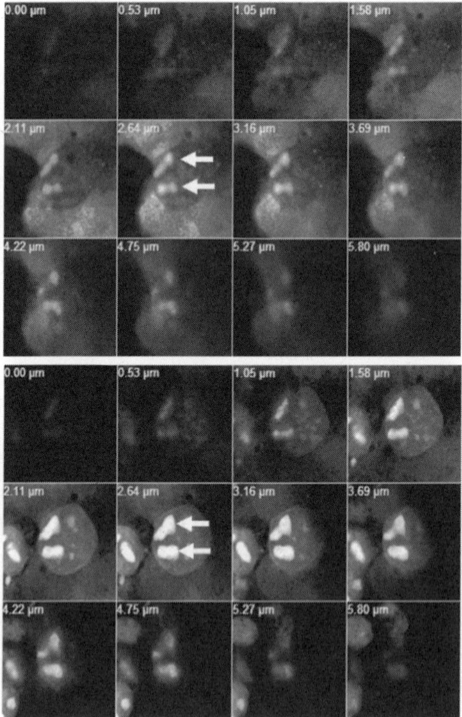

Figure 2. META-confocal image showing sections (indicated in the top left-hand corner of each image) through a Cos-7 cells expressing IBV N protein. The upper panel shows the distribution of IBV N protein and the lower panel shows the same sections but showing the signal from PI, which highlights the nucleus and nucleolus.

The data indicated that IBV N protein localized to a discrete area of the nucleolus. For example, compare the distribution of IBV N protein to the PI stained nucleolus in optical section 2.64 μm (Fig. 2, arrows), N protein would appear to occupy less nucleolar volume. To investigate this further, we utilized these Z-sections to construct a three-dimensional representation focusing specially on the nuclear region of the cell (Fig. 3). As can be observed, the volume taken up by IBV N protein in the nucleolus (Fig. 3A) is less than total nucleolar volume (Fig. 3B). From this data we hypothesize that IBV N protein localizes to the DFC but not the GC.

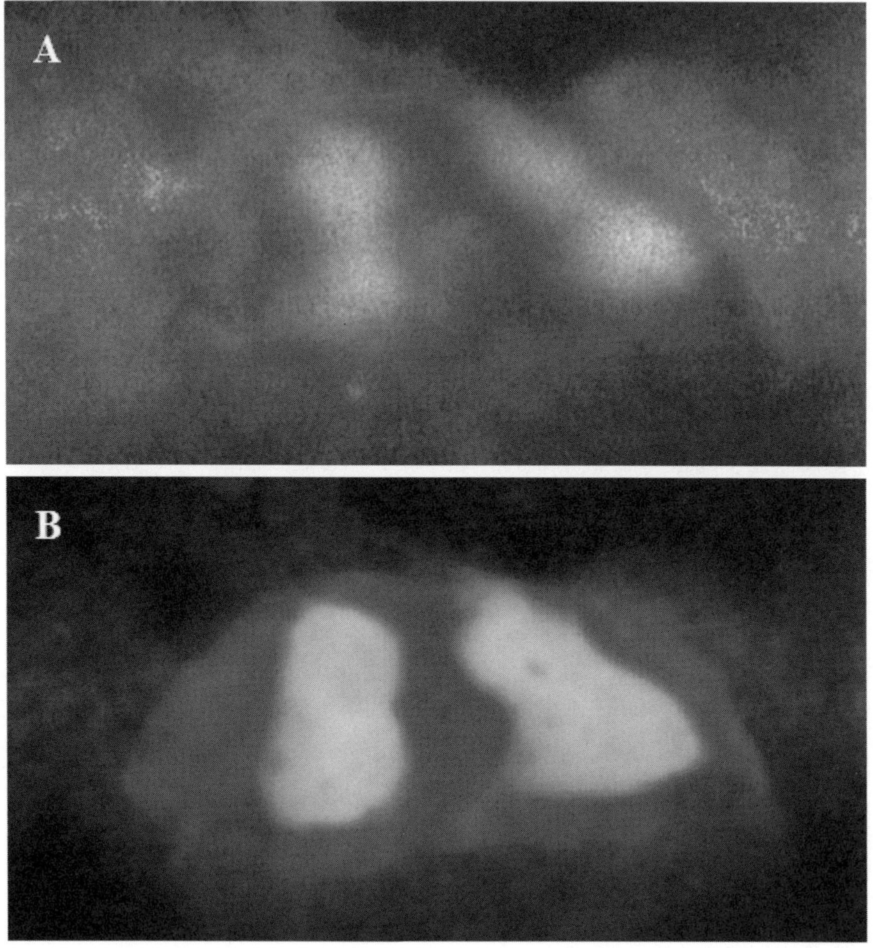

Figure 3. Three-dimensional reconstruction of a nucleus showing the distribution of IBV N protein (A) and the total nucleolar volume (B). Images were reconstructed from the data shown in Fig. 2.

In conclusion, our data demonstrate that IBV N protein localizes to the cytoplasm and nucleolus and is not an artifact of fixation conditions. In contrast, SARS-CoV N protein remains localized in the cytoplasm and does appear to cross the nuclear pore complex, despite being below the size exclusion limit for entry into the nucleus. We hypothesize that SARS-CoV N protein contains a dominant cytoplasmic retention motif. META-confocal analysis and three-dimensional reconstructions of cells expressing IBV N protein revealed that N protein does not localize throughout the nucleolus and may be confined to the DFC.

4. ACKNOWLEDGMENTS

This work was funded by the BBSRC, project grant number BBSB03416 and studentship BBSSP200310434 to J.A.H. The confocal microscope facility in the Astbury Centre for Structural Molecular Biology was funded by the Wellcome Trust and SRIF, and we would like to thank Gareth Howell for his help in using this facility.

5. REFERENCES

1. J. A. Hiscox, The interaction of animal cytoplasmic RNA viruses with the nucleus to facilitate replication, *Virus Res.* **95**, 13-22 (2003).
2. J. A. Hiscox, The nucleolus--a gateway to viral infection? *Arch. Virol.* **147**, 1077-1089 (2002).
3. R. R. Rowland and D. Yoo, Nucleolar-cytoplasmic shuttling of PRRSV nucleocapsid protein: a simple case of molecular mimicry or the complex regulation by nuclear import, nucleolar localization and nuclear export signal sequences, *Virus Res.* **95**, 23-33 (2003).
4. A. K. Leung, J. S. Andersen, M. Mann, and A. I. Lamond, Bioinformatic analysis of the nucleolus, *Biochem. J.* **376**, 553-569 (2003).
5. H. Chen, A. Gill, B. K. Dove, S. R. Emmett, C. F. Kemp, M. A. Ritchie, M. Dee, and J. A. Hiscox, Mass spectroscopic characterization of the coronavirus infectious bronchitis virus nucleoprotein and elucidation of the role of phosphorylation in RNA binding by using surface plasmon resonance, *J. Virol.* **79**, 1164-1179 (2005).
6. E. Calvo, D. Escors, J. A. Lopez, J. A. Gonzalez, A. Alvarez, E. Arza, and L. Enjuanes, Phosphorylation and subcellular localization of transmissible gastroenteritis virus nucleocapsid protein in infected cells, *J. Gen. Virol.* **86**, 2255-2267 (2005).
7. M. M. Parker and P. S. Masters, Sequence comparison of the N genes of five strains of the coronavirus mouse hepatitis virus suggests a three domain structure for the nucleocapsid protein, *Virology* **179**, 463-468 (1990).
8. M. Lundberg and M. Johansson, Positively charged DNA-binding proteins cause apparent cell membrane translocation, *Biochem. Biophys. Res. Commun.* **291**, 367-371 (2002).
9. E. V. Sheval, M. A. Polzikov, M. O. Olson, and O. V. Zatsepina, A higher concentration of an antigen within the nucleolus may prevent its proper recognition by specific antibodies, *Eur. J. Histochem.* **49**, 117-123 (2005).

DISSECTION OF THE FUSION MACHINE OF SARS-CORONAVIRUS

Megan W. Howard, Brian Tripet, Michael G. Jobling, Randall K. Holmes, Kathryn V. Holmes, and Robert S. Hodges*

1. INTRODUCTION

Infection of target cells by enveloped viruses occurs through two pathways. Upon binding to a receptor on the cell surface, the virus enters the cell either through the plasma membrane or via endocytosis through a pH-dependent fusion event with the endosomal membrane.[1] Coronaviruses (CoVs) enter host cells using both pathways[1, 2] through the interaction of the S (spike) glycoprotein with the target receptor on the cell surface. Different CoVs use different receptors to bind to cells and enter either at the plasma membrane or via endosomes. Previous work from our lab has identified murine CEACAM1a as a receptor for MHV.[4] SDS-PAGE and liposome flotation analysis showed that either soluble receptor or basic pH (pH 8.0) at 37°C triggers the S protein to undergo a conformational change that exposes trypsin cleavage sites and a hydrophobic region of the protein that can associate with membranes.[3]

The S protein of CoVs is a type-I viral fusion glycoprotein (VFG),[5] related to the fusion glycoproteins of many other enveloped viruses (including gp41 of HIV-1, gp2 of Ebola, HA of Influenza, and F of SV5).[6, 7] Type-I VFGs contain several conserved domains; two heptad repeats (HR1 and HR2 toward the N- and C-termini of the fusion domain of the protein, respectively), a hydrophobic fusion peptide (FP), and a trans-membrane domain (TM). Some of these proteins also contain a membrane-proximal domain (TMP) rich in aromatic amino acids.[7]

Binding of the S protein of CoVs to a specific receptor induces conformational changes in the S protein, changing it from an inactive pre-fusion state to a fusion-active state. This allows a previously hidden hydrophobic region of the protein, the fusion peptide, to interact with the host membrane. Another conformational change occurs that changes the trimeric S protein into a post-fusion conformation, containing a structure known as a 6-helix-bundle (6HB) which draws the host and viral membranes close together, facilitating membrane fusion and entry. The 6HB is formed through the

---

*University of Colorado Health Sciences Center, Aurora, Colorado 80045.

antiparallel association of the C-terminal HR with the N-terminal HRs in a 'trimer of dimers' conformation.[7]

Although CoV S proteins are type-I VFGs, they have several unusual features. Most VFGs must be cleaved by a serine protease before the protein is able to mediate membrane fusion. However some CoVs (e.g., HCoV-229E) function without a known serine protease activation event. The FP of CoV S proteins is not at the N terminus of the fusion domain, as seen with most type-I VFGs. Instead, the CoV FP is internal, preceded by ~230 aa's. A long interhelical domain of ~140 aa's links the two HRs. Thus S proteins of CoVs are variant type-I VFGs.

2. RESULTS AND DISCUSSION

To investigate the SARS-CoV S protein and its mechanism of fusion, we used bioinformatics to identify regions likely to be involved in membrane fusion. Using STABLECOIL,[8] a program based on empirical data on coiled coils, we identified portions of the S protein likely to fold into coiled coils. We identified two regions, both in the S2 domain: a N-terminal (HRN, aa882-1011), and a C-terminal (HRC, aa1151-1185) region, separated by a domain of ~140 aa residues (Fig 1).[9] The HRN region could be subdivided into three parts based on the heptad *a* and *d* positions. The presence of these alternative heptad 'registers' may indicate that the HRN region can fold into two structurally different coiled coils. Two of these 'registers' overlap, indicating that hydrophobic residues are present both in the interfacial *e* and *g* positions and in the intrafacial *a* and *d* positions. This 'double register' may be involved in HRC binding and association with HRN in the post-fusion state.

We synthesized different regions of the HRN and HRC regions in overlapping 35-aa peptides. Analysis of these peptides for helical structure and stability using circular dichroism (CD) spectroscopy and temperature denaturation curves showed that the HRN peptides were random coil. However addition of 50% trifluoreoethanol (TFE), a helix-inducing solvent, showed that they could adopt a helical structure, thus they do have intrinsic helical character. The HRC peptide showed helical structure, however it was of low helicity and stability (50% unfolded at 33°C). Mixing HRN and HRC peptides showed that the HRN10 peptide, corresponding to aa 916–950 of SARS S, could interact with the HRC peptide as the stability of the complex rose 24° to 57°C indicating a stabilizing interaction. The CD spectra of the complex indicated an alpha-helical structure. We examined the interaction of the HRN10 peptide with truncated HRC peptides. The temperature denaturation curves were similar for the full HRC and the truncated HRC4 (aa 1151–1185), indicating that the binding region of the HRN10 peptide (aa 916–950) on HRC is localized to aa 1151–1185.

Sedimentation equilibrium (SE) was used to determine the oligomeric states of the HRN (aa 882–973) and HRC peptides individually. The SE data indicated that HRC, aa 1151–1185, associates as a trimer. Surprisingly the HRN peptide associated in a tetrameric state. These results suggest that though the HRN is a trimer in the complete protein, truncation of the native sequence causes it to adopt a tetrameric configuration. The SE data on the HRN10/HRC complex fits best to a model with a 3:3 molar ratio of HRN10:HRC, indicating formation of a 6HB.

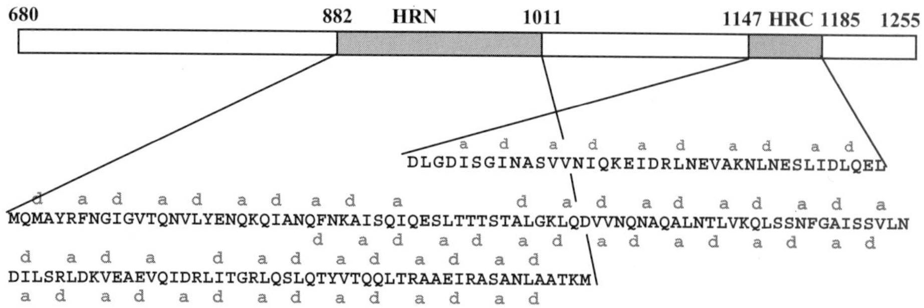

Figure 1. SARS-CoV S2 domain.

We investigated the orientation of the helicies in the HRN10/HRC complex by introducing disulfide bridges. Engineering a gly-gly-cys sequence onto the N- or C-termini of HRC and the N-termini of the HRN10 peptide allowed us to covalently link the peptides together, locking them into either a highly stable soluble form (the correct orientation) or a less stable form composed of head-to-tail aggregates (the incorrect orientation). Temperature denaturation studies showed that the antiparallel orientation was favored by an increase in stability (by 19°C) and solubility (>5 mg/mL). Surprisingly, the CD spectra of the covalently linked antiparallel complex ($-19200^{\circ}*cm^2*dmol^{-1}$ at 222 nm rather than $-34{,}500^{\circ}*cm^2*dmol^{-1}$) showed that it either contained unfolded residues at one end, or one of the peptides exists in a partially helical form. The crystal structure of the 6HB of SARS S was recently published, and confirmed that the 6HB is not 100% helical and contains extended regions at both ends of HRC.[11, 15]

3. CONCLUSIONS

We characterized the HRN and HRC regions in the ectodomain of the SARS-CoV S protein, a variant type-I VFG, and identified similarities to the MHV fusion core[16] and other type-I VFG fusion cores.[6, 7] Analysis of the HRN region showed that the region that interacts with HRC is aa 916–950. Peptides of HRN, aa 916–950, and HRC, aa 1151–1185, can form a 6HB, which appears similar to the interactions between the HRC and HRN of the SV5 F protein[6] that involves both helical and extended regions in HRC. The identification of the 6HB structure formed by the HRN and HRC peptides of the SARS-CoV fusion core suggests that the HRC peptide may inhibit viral fusion and entry, as observed with similar VFGs.[10-15] Recent publications show that the HRC peptide can inhibit infection with SARS-CoV.[10-15]

4. ACKNOWLEDGMENTS

We thank Mark Genest, Jennifer Labrecque, Larissa Thackray, and Mark Young for help with peptide synthesis, cloning, and purification. CD and analytical ultracentrifugation measurements were performed at the Biophysics Core Facility, University of Colorado Health Sciences Center. This work was supported by NIH grants RO1 A148717 (to

R.S.H.), RO1 AI31940 (to R.K.H.), RO1 A125231 (to K.V.H.), and T32 AI52066 (for M.W.H.).

5. REFERENCES

1. L. J. Earp, S. E. Delos, H. E. Park, and J. M. White, The many mechanisms of viral membrane fusion proteins, *Curr. Top. Microbiol. Immunol.* **285**, 25-66 (2004).
2. T. C. Nash and M. J. Buchmeier, Entry of mouse hepatitis virus into cells by endosomal and nonendosomal pathways, *Virology* **233**, 1-8 (1997).
3. B. D. Zelus, J. H. Schickli, D. M. Blau, S. R Weiss, and K. V. Holmes, Conformational changes in the spike glycoprotein of murine coronavirus are induced at 37 degrees C either by soluble murine CEACAM1 receptors or by pH 8, *J. Virol.* **77**, 830-840 (2003).
4. G. S. Dveksler, C. W. Dieffenbach, C. B. Cardellichio, K. McCuaig, M. N. Pensiero, G. S. Jiang, N. Beauchemin, and K. V. Holmes, Several members of the mouse carcinoembryonic antigen-related glycoprotein family are functional receptors for the coronavirus mouse hepatitis virus-A59, *J. Virol.* **67**, 1-8 (1993).
5. B. J. Bosch, R. van der Zee, C. A. de Haan, and P. J. Rottier, The coronavirus spike protein is a class I virus fusion protein: structural and functional characterization of the fusion core complex, *J. Virol.* **77**, 8801-8811 (2003).
6. K. A. Baker, R. E. Dutch, R A. Lamb, and T. S. Jardetzky, Structural basis for paramyxovirus-mediated membrane fusion, *Mol. Cell* **3**, 309-319 (1999).
7. P. M. Colman and M. C. Lawrence, The structural biology of type I viral membrane fusion, *Nat. Rev. Mol. Cell Biol.* **4**, 309-319 (2003).
8. B. Tripet and R. S. Hodges, in *Peptides: The Wave of the Future*, edited by M. Lebl and R. A. Houghten (American Peptide Society, San Diego, 2001), pp. 365-366.
9. B. Tripet, M. W. Howard, M. Jobling, R. K. Holmes, K. V. Holmes, and R. S. Hodges, Structural characterization of the SARS-coronavirus spike S fusion protein core, *J. Biol. Chem.* **279**, 20836-20849 (2004).
10. B. J. Bosch, B. E. Martina, R. van der Zee, J. Lepault, B. J. Haijema, C. Versluis, A. J. Heck, R. De Groot, A. D. Osterhaus, and P. J. Rottier, Severe acute respiratory syndrome coronavirus (SARS-CoV) infection inhibition using spike protein heptad repeat-derived peptides, *Proc. Natl. Acad. Sci. USA* **101**, 8455-8460 (2004).
11. Y. Xu, Z. Lou, Y. Liu, H. Pang, P. Tien, G. F. Gao, and Z. Rao, Crystal structure of severe acute respiratory syndrome coronavirus spike protein fusion core, *J. Biol. Chem.* **279**, 49414-49419 (2004).
12. S. Liu, G. Xiao, Y. Chen, Y. He, J. Niu, C. R. Escalante, H. Xiong, J. Farmar, A. K. Debnath, P. Tien, and S. Jiang, Interaction between heptad repeat 1 and 2 regions in spike protein of SARS-associated coronavirus: implications for virus fusogenic mechanism and identification of fusion inhibitors, *Lancet* **363**, 938-947 (2004).
13. Y. Xu, J. Zhu, Y. Liu, Z. Lou, F. Yuan, Y. Liu, D. K. Cole, L. Ni, N. Su, L. Qin, X. Li, Z. Bai Z., J. I. Bell, H. Pang, P. Tien, G. F. Gao, and Z. Rao, Characterization of the heptad repeat regions, HR1 and HR2, and design of a fusion core structure model of the spike protein from severe acute respiratory syndrome (SARS) coronavirus, *Biochemistry* **43**, 14064-14074 (2004).
14. K. Yuan, L. Yi, J. Chen, X. Qu, T. Qing, X. Rao, P. Jiang, J. Hu, Z. Xiong, Y. Nie, X. Shi, W. Wang, C. Ling, X. Yin, K. Fan, L. Lai, M. Ding, and H. Deng, Suppression of SARS-CoV entry by peptides corresponding to heptad regions on spike glycoprotein, *Biochem. Biophys. Res. Commun.* **319**, 746-752 (2004).
15. P. Ingallinella, E. Bianchi, M. Finotto, G. Cantoni, D. M. Eckert, V. M. Supekar, C. Bruckmann, A. Carfi, and A. Pessi, Structural characterization of the fusion-active complex of severe acute respiratory syndrome (SARS) coronavirus, *Proc. Natl. Acad. Sci. USA* **101**, 8709-8714 (2004).
16. Y. Xu, Z. Bai, L. Qin, X. Li, G. Gao, and Z. Rao, Crystallization and preliminary crystallographic analysis of the fusion core of the spike protein of the murine coronavirus mouse hepatitis virus (MHV), *Acta Crystallogr. D* **60**, 2013-2015 (2004).

CHARACTERIZATION OF PERSISTENT SARS-CoV INFECTION IN VERO E6 CELLS

Tetsuya Mizutani, Shuetsu Fukushi, Masayuki Saijo, Ichiro Kurane, and Shigeru Morikawa*

1. INTRODUCTION

Severe acute respiratory syndrome (SARS) is a newly discovered infectious disease caused by a novel coronavirus, SARS coronavirus (SARS-CoV).[1,2] Understanding the molecular mechanisms of the pathogenicity of SARS-CoV is a rational approach for the prevention of SARS. As the gene organization of SARS-CoV is similar to those of other coronaviruses, previous scientific data regarding coronaviruses can help in understanding the virological features of SARS-CoV. A human intestinal cell line, LoVo, was shown to permit SARS-CoV infection, resulting in the establishment of persistent infection.[3] However, the mechanism of persistence has yet to be clarified. The monkey kidney cell line, Vero E6, is often used in SARS-CoV research because of the high degree of sensitivity of these cells to the virus. This cell line expresses the viral receptor ACE-2[4] at high levels, and SARS-CoV infection of Vero E6 causes cytopathic effects within 24 h.[5–7] Recently, we showed that establishment of SARS-CoV persistently infected infected Vero E6 cells requires activation of JNK and Akt signaling pathways.[8]

2. RESULTS

2.1. Importance of JNK and PI3K/Akt Signaling Pathways for Establishment of Persistent Infection

Recently, we reported that both Akt and JNK signaling pathways are important for the establishment of persistent SARS-CoV infection in Vero E6 cells.[8] An inhibitor of

* National Institute of Infectious Diseases, Tokyo 208-0011, Japan.

p38 MAPK, SB203580, and an inhibitor of MEK, PD98059, did not affect the establishment of persistent infection, whereas no surviving cells were observed after treatment with the JNK inhibitor, SP600125, or the PI3K/Akt inhibitor, LY294002.

2.2. Downregulation of ACE-2 Viral Receptor by SARS-CoV Infection

Previously, we showed that ACE-2 was not detected in a persistently infected cell line on Western blotting analysis.[8] ACE-2 expression was also shown to be reduced in the acute phase of SARS-CoV infection. These results suggested that virus particles produced by persistently infected cells could not infect other cells due to a lack of the receptor, resulting in a decrease in the number of virus-infected cells.

2.3. Difference in Migration of Bcl-xL in a Persistently Infected Cell Line

The anti-apoptotic protein, Bcl-2, is activated in cells persistently infected with viruses.[9,10] In the present study, we established a persistently SARS-CoV-infected cell line after passage 6. However, this cell line did not show significant activation of Bcl-2 (data not shown). On the other hand, Bcl-xL, which is also an anti-apoptotic protein, showed a different migration pattern in the persistently infected cell line as compared with mock and acutely infected cells (Fig. 1). Previous studies[11–13] indicated that the fast-migrating Bcl-xL band is unphosphorylated Bcl-xL, which has been shown to have anti-apoptotic roles. The anti-Bcl-xL antibody (Cell Signaling Co. Ltd.) recognizes both phosphorylated and unphosphorylated Bcl-xL. The slowly migrating band shown in Fig. 1 may be an inactivated form of Bcl-xL. Thus, Bcl-xL may be involved in maintenance of persistent infection.

3. DISCUSSION

Previously, we concluded that a population of cells produced from parental Vero E6 cells had the potential to support persistent infection, and that acute infection caused by a major population of seed virus was necessary for persistent infection.[8]

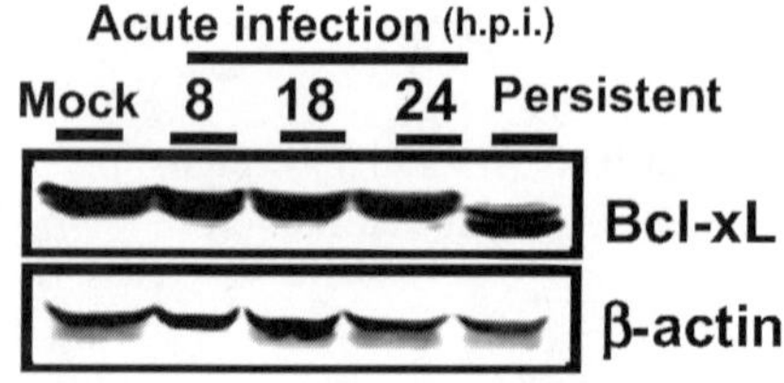

Figure 1. Activation of Bcl-xL in persistently infected cells.

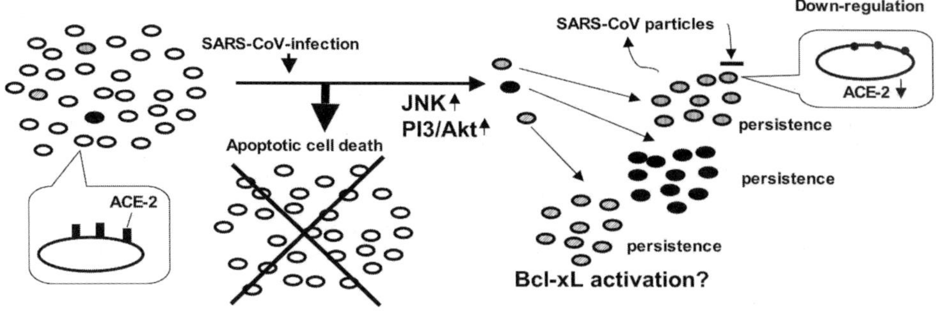

Figure 2. Model of establishing persistent infection of SARS-CoV.

The anti-apoptotic protein, Bcl-2, is capable of blocking apoptosis caused by RNA virus infection.[14–16] Bcl-2 plays a key role in the death or survival of virus-infected cells. Moreover, some studies indicated that Bcl-2 determines the establishment of persistence. A persistent strain of Sindbis virus induces upregulation of Bcl-2, whereas a virulent strain induces an increase in Bax.[10] On the other hand, the present study of one cell line persistently infected with SARS-CoV suggested accumulation of a form of Bcl-xL that was different in size of from that in parental Vero E6 cells, but significant activation of Bcl-2 was not observed. Bcl-xL is known to be phosphorylated at one site, Ser62, by treatment with taxol and 2-ME.[13] Further studies are necessary to confirm the dephosphorylation of Bcl-xL at the site in the persistently infected cells.

Here, we reported a possible mechanism of the establishment of persistent SARS-CoV infection in Vero E6 cells (Fig. 2). Although the majority of cells died due to apoptosis after SARS-CoV infection, activation of JNK and PI3K/Akt signaling pathways aided a minor population of cells with the potential to support persistent infection to establish persistence. One strategy for cell survival on viral infection is activation of Bcl-xL.

4. ACKNOWLEDGMENTS

We thank Ms. M. Ogata (National Institute of Infectious Diseases, Japan) for her assistance. This work was supported in part by a grant-in-aid from the Ministry of Health, Labor, and Welfare of Japan and the Japan Health Science Foundation, Tokyo, Japan.

5. REFERENCES

1. Rota, P. A., Oberste, M. S., Monroe, S. S., et al., 2003, Characterization of a novel coronavirus associated with severe acute respiratory syndrome, *Science* **300**:1394-1399.
2. Marra, M. A., Jones, S. J., Astell, C. R., et al., 2003, The genome sequence of the SARS-associated coronavirus, *Science* **300**:1399-1404.
3. Chan P. K., To, K. F., Lo, A. W., Cheung, J. L., Chu, I., Au, F. W., Tong, J. H., Tam, J. S., Sung, J. J., and Ng, H. K., 2004, Persistent infection of SARS coronavirus in colonic cells in vitro, *J. Med. Virol.* **74**:1-7.
4. Li, W., Moore, M. J., Vasilieva, N., et al., 2003, Angiotensin-converting enzyme 2 is a functional receptor for the SARS coronavirus, *Nature* **426**:450-454.
5. Mizutani, T., Fukushi, S., Saijo, M., Kurane, I., and Morikawa, S., 2004a, Phosphorylation of p38 MAPK and its downstream targets in SARS coronavirus-infected cells, *Biochem. Biophys. Res. Commun.* **319**:1228-1234.
6. Mizutani, T., Fukushi, S., Saijo, M., Kurane, I., and Morikawa, S., 2004b, Importance of Akt signaling pathway for apoptosis in SARS-CoV-infected Vero E6 cells, *Virology* **327**:169-174.
7. Mizutani, T., Fukushi, S., Murakami, M., Hirano, T., Saijo, M., Kurane, I., and Morikawa, S., 2004c. Tyrosine dephosphorylation of STAT3 in SARS coronavirus-infected Vero E6 cells, *FEBS Lett.* **577**:187-192.
8. Mizutani, T., Fukushi, S., Saijo, M., Kurane, I., and Morikawa, S., 2005, JNK and PI3K/Akt signaling pathways are required for establishing persistent SARS-CoV-infection in Vero E6 cells, *Biochim. Biophys. Acta.* **1741**: 4-10.
9. Liao C. L., Lin, Y. L., Shen, S. C., Shen, J. Y., Su, H. L., Huang, Y. L., Ma, S. H., Sun, Y. C., Chen, K. P. and Chen, L. K. 1998, Antiapoptic but not antiviral function of human bcl-2 assists establishment of Japanese encephalitis virus persistence in cultured cells. J Virol. **72**:9844-9854.
10. Appel, E., Katzoff, A., Ben-Moshe, T., Kazimirsky, G., Kobiler, D., Lustig, S., and Brodie, C., 2000, Differential regulation of Bcl-2 and Bax expression in cells infected with virulent and nonvirulent strains of sindbis virus, *Virology* **276**:238-242.
11. Poruchynsky, M. S., Wang, E. E., Rudin, C. M., Blagosklonny, M. V., and Fozo, T., 1998, Bcl-xL is phosphorylated in malignant cells following microtubule disruption, *Cancer Res.* **58**:3331–3338.
12. Kharbanda, S., Saxena, S., Yoshida, K., et al., 2000, Translocation of SAPK/JNK to mitochondria and interaction with Bcl-x(L) in response to DNA damage, *J. Biol. Chem.* **275**:322-327.
13. Basu, A., and Haldar, S., 2003, Identification of a novel Bcl-xL phosphorylation site regulating the sensitivity of taxol- or 2-methoxyestradiol-induced apoptosis, *FEBS Lett.* **538**:41-47.
14. Hinshaw, V. S., Olsen, C. W., Dybdahl-Sissoko, N., and Evans, D., 1994, Apoptosis: a mechanism of cell killing by influenza A and B viruses, *J. Virol.* **68**:3667-3673.
15. Ubol, S., Tucker, P. C., Griffin, D. E, and Hardwick, J. M., 1994, Neurovirulent strains of alphavirus induce apoptosis in *bcl-2*-expressing cells: role of a single amino acid change in the E2 glycoprotein, *Proc. Natl. Acad. Sci. USA* **91**:5202-5206.
16. Pekosz, A., Phillips, J., Pleasure, D., Merry, D., and Gonzalez-Scarano, F., 1996, Induction of apoptosis by La Crosse virus infection and role of neuronal differentiation and human *bcl-2* expression in its prevention, *J. Virol.* **70**:5329-5335.

RECEPTOR-INDEPENDENT SPREAD OF A NEUROTROPIC MURINE CORONAVIRUS MHV-JHMV IN MIXED NEURAL CULTURE

Keiko Nakagaki, Kazuhide Nakagaki, and Fumihiro Taguchi*

1. INTRODUCTION

Highly neurovirulent mouse hepatitis virus (MHV) JHMV strain multiplies in a variety of brain cells, although the expression of its receptor, carcinoembryonic antigen cell adhesion molecule 1 (CEACAM 1, MHVR), is expressed only in microglia/macrophages in the brain.[1] The present study was undertaken to clarify the mechanism of extensive JHMV infection of the brain by using neural cells isolated from mice.

2. MATERIALS AND METHODS

2.1. Virus, Neural Cell Culture, and Antibody

We used two strains of MHV: MHV-JHM cl-2 strain (designated wt JHMV), known to be a neurotropic MHV, and srr7 (soluble receptor resistant mutant), derived from wt JHMV. Srr7 has a single amino acid mutation in the S protein (L1114F).[2] Primary mixed neural cell cultures were established from the forebrains of 1- to 3-day-old neonate mice as described previously with minor modifications.[3] Primary antibodies used for immunocytochemistry or flow cytometry (FACS) to identify each cell type were as follows; anti-glial fibrillary acidic protein (GFAP) polyclonal antibody for astrocytes, O4 monoclonal antibodies (MAbs) for oligodendrocytes, and MAP-2 for neuron. Binding of the Griffonia simplicifolia lectin (GS-lectin) was used for microglia/macrophages identification. To detect MHV-specific antigen in cells, mouse anti-MHV MAbs were used.[4] MAb CC1to detect MHVR and to block the infection was kindly provided by Kay V. Holmes.[5]

---

* Keiko Nakagaki, Fumihiro Taguchi, National Institute of Infectious Diseases, Tokyo 208-0011, Japan. Kazuhide Nakagaki, Nippon Veterinary and Animal Science University, Tokyo 180-8602, Japan.

3. RESULTS

3.1. Infection of wt JHMV or srr-7 in Cultured Neural Cells and the Blockade of Initial Infection by Anti-MHVR MAb CC1

Virus-infected fused cells were sporadically detected in neural cell culture at 8 h after wt JHMV infection and the infection rapidly spread regardless of cell type. Twenty-four hours postinfection (PI), the majority of cells were included in syncytia and contained viral antigens. Unlike wt JHMV, the viral antigen in srr7-infected cells were detected in only a small proportion of club-like cells at 12 hour, PI, and syncytium formation was not observed as long as 24 hour, PI. The cells infected with srr7 were immunocytochemically identified as microglia/macrophages because of GS-lectin-positivity. Because the virus infection was blocked by the pretreatment of neural cells with CC1, we concluded that initial infection of those two viruses was mediated by MHVR.

3.2. Identification of MHVR-positive Cells and of Cells That Bound JHMV by FACS

Neural cells isolated from the cerebrum of neonatal mice were directly examined for GS-lectin positivity as well as for the presence of MHVR. MHVR-positive cells stained with CC1 accounted for 2.1% of the entire neural cell population, while GS-lectin-positive cells were 5.1%. Approximately 78% of the MHVR-positive cells were revealed to be GS-lectin positive, showing that the major population expressing MHVR were microglia/macrophages. In contrast, 48% of GS-lectin-positive cells were MHVR-positive, indicating that half of the microglia/macrophages express MHVR and the other half does not. Virus-bound cells accounted for only 1% of all cells isolated from brain, but more than 83% of those cells were GS-lectin-positive cells. This suggests that most of the cells bound with JHMV are microglia/ macrophage.

3.3. Virus Infection in Microglia/Macrophages-Enriched Culture

In GS-lectin-positive microglia/macrophages-enriched cultures, wt and srr7 similarly induced syncytia in approximately one-half of cultured cells. The finding indicates that both viruses spread and efficiently induce syncytium formation in microglia/macrophage cultures.

3.4. Effects of CC1 Treatment on Virus Spread from Initially Infected Cells

The results presented above are suggestive that the wt JHMV spreads from primarily infected microglia/macrophages to MHVR-negative cells in mixed neural cell culture, while srr7 infection is limited to MHVR-positive microglia/macrophages. To see whether this is a case or not, we have examined whether CC1 prevents wt spread from initially infected cells. Neural cells were infected with wt JHMV, and 1 hour later CC1 was added in the culture. There was no substantial difference in the proportion of wt JHM antigen-positive syncytium formation, when cultured in the presence or absence of CC1. This indicates that wt virus spread from initially infected cells is not mediated with MHVR. We have also done a similar experiment using either wt or srr7 in microglia/macrophage-enriched cultures. CC1 prevented the spread of srr7, while it

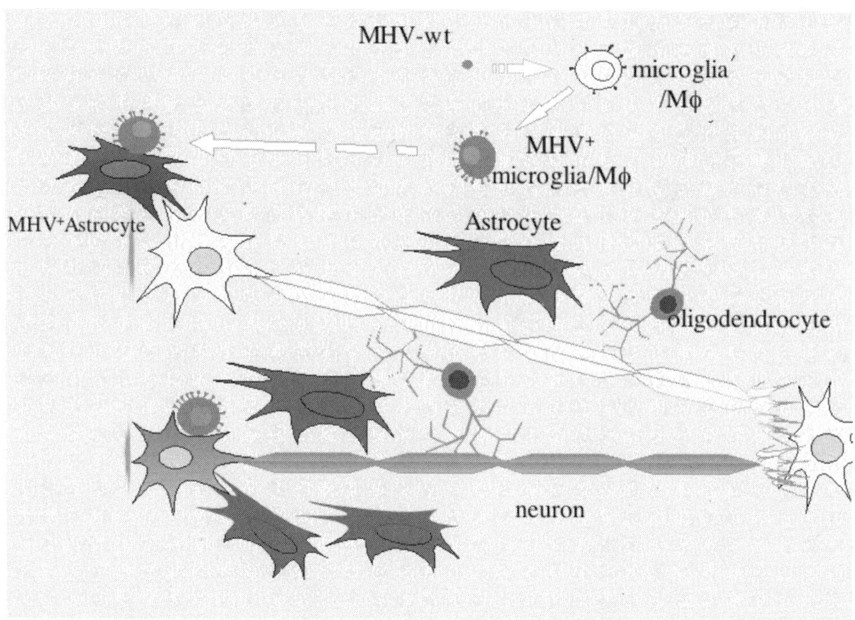

Figure 1. Schematic representation of mechanisms of MHVR-independent infection caused by wt JHMV in neural cells. The virus spreads from initially infected microglia/macrophages (m ϕ) to receptor-negative cells.

failed to prevent the wt virus spread in a similar way observed in srr7 infection, though it reduced wt syncytium formation to a certain extent. These results clearly showed that wt virus spread from initially infected cells to a variety of cells in an MHVR-independent fashion, while srr7 infection was solely MHVR-dependent.

4. CONCLUSIONS

The present study indicates that microglia/macrophages are the major population expressing MHVR in the brain, and they are the initial targets for MHV. Wt JHMV spreads from initially infected microglia/macrophages, which is MHVR-dependent manner, to a variety of cells such as astrocytes, oligodendrocytes or neuron in an MHVR-independent fashion as postulated by Gallagher and Buchmeier[6] (Fig. 1).

5. ACKNOWLEDGMENTS

We greatly thank Dr. Kay V. Holmes for MAb specific for MHVR (CC1). This work was partly supported by a grant ("Urgent Research on the Diagnosis and Test Techniques for SARS") from the Ministry of Education, Science, Sports and Culture of Japan.

6. REFERENCES

1. Gallagher, T. M., and Buchmeier, M. J., 2001, Coronavirus spike proteins in viral entry and pathogenesis, *Virology* **279**:371-374.
2. Hirayama, M., Silberberg, D. H., Lisak, R. P., and Pleasure, D., 1983, Long-term culture of oligodendrocytes isolated from rat corps callosum by percoll density gradient. Lysis by polyclonal antigalactocerebroside serum, *J. Neuropathol. Exp. Neurol.* **42**:16-28.
3. Kubo, H., Takase, Y-S., and Taguchi, F., 1993, Neutralization and fusion inhibition activities of monoclonal antibodies specific for S1 subunit of the spike protein of neurovirulent murine coronavirus HMV cl-2 variant, *J. Gen. Virol.* **74**:1421-1425.
4. Ramarkrishna, C., Bergmann, C. C., Holmes, K. V., and Stohlman, S., 2004, Expression of the mouse hepatitis virus receptor by central nervous system microglia, *J. Virol.* **78**:7828-7832.
5. Saeki, K., Ohtsuka, N., and Taguchi, F., 1997, Identification of spike protein residues of murine coronavirus responsible for receptor-binding activity by use of soluble receptor-resistant mutants, *J. Virol.* **71**:9024-9031.
6. Williams, R. K., Jiang, G. S., Snyder, S. W., Frana, M. F., and Holmes, K. V., 1990, Purification of the 110-kilodalton glycoprotein receptor for mouse hepatitis virus (MHV)-A59 from mouse liver and identification of a nonfunctional, homologous protein in MHV-resistant SJL/J mice, *J. Virol.* **64**:3817-3823.

RECEPTOR-INDEPENDENT INFECTION OF MOUSE HEPATITIS VIRUS: ANALYSIS BY SPINOCULATION

Rie Watanabe, Kazumitsu Suzuki, and Fumihiro Taguchi*

1. INTRODUCTION

Cell entry of mouse hepatitis virus (MHV) is mediated by the interaction of its spike (S) protein and cellular receptor carcinoembryonic antigen adhesion molecule 1 (CEACAM1, MHVR). However, a highly neurotropic MHV, wild-type (wt) JHMV is known to spread to receptor-negative BHK cells from firstly infected receptor-positive DBT cells (MHVR-independent infection).[1] Although the mechanism of this infection is still unclear, it is hypothesized that the S protein attached on cell surface is activated for fusion by natural dissociation of S1 from S2, which was revealed in wt JHMV but not mutants derived from it,[2] with or without MHVR-independent infection activity, respectively.[3,4] Wild–type JHMV fails to infect BHK cells by a standard infection procedure, because the virus is not able to attach cells without MHVR. However, the S protein expressed on cell surface infected with wt can attach onto MHVR-negative cells by overlaying these cells, which induces fusion/infection of MHVR deficient cells (infected-cell overlay test). If MHVR-independent infection occurs as expected above, then we will be able to make wt JHMV infect cells by attaching virions onto cells without MHVR. To test this possibility, we employed the spinoculation method, which has been shown to facilitate the binding of viruses onto cells[5].

Two strains of JHMV, cl-2 and its soluble receptor resistant mutant, srr7,[6] were used for spinoculation. There is only one single amino-acid substitution in the S2 subunit of srr7 as compared with cl-2. They were spinoculated onto BHK cells, which facilitated the attachment of these viruses to the same extent. It also facilitated the infection of cl-2 virus but not that of srr7, being in good agreement with the result obtained by infected-cell overlay test. Furthermore, dissociation of S1 from S2 was confirmed in cells expressing cl-2 S protein but not srr7 S. These results clearly support the proposed hypothesis for the mechanism of MHVR-independent infection.

* National Institute of Infectious Diseases, Tokyo 208-0011, Japan.

2. MATERIALS AND METHODS

2.1. Cells, Viruses, and Spinoculation

MHVR-positive DBT and MHVR-negative BHK cells were used as target cells. Wild-type JHMV cl-2 and its mutant srr7 were used.[6] Spinoculation was performed as previously described[5] with a slight modification. Cells in 24-well plates were infected with viruses in 300 μl of medium containing 1 μg/ ml of concanavalin A and were centrifuged at 1750 × g for 2 hours at 4°C. After 14 hours incubation at 37°C, cells were fixed, stained with crystal violet to count the number of syncytium.

2.2. Expression of S Proteins

Cells infected with vTF7.3[7] were transfected with S protein expression plasmids by electroporation. Twelve hours after transfection, the culture supernatants and cells were separately harvested. The S protein in culture supernatants was collected using anti-S monoclonal antibodies (MAbs).[8] S proteins in the supernatants and in cell lysates were analyzed by Western blot with MAbs.[9]

3. RESULTS AND DISCUSSION

3.1. Centrifugation Mediates the Infection of cl-2 to Receptor-Deficient Cells

Cl-2 and srr7 were spinoculated onto MHVR-positive DBT or negative BHK cells, and their infections were monitored by syncytium formation (Table 1). The centrifugation had small effects on virus infection (23-fold increase compared with no-centrifuged plate), when DBT cells were infected with cl-2. In contrast, cl-2 infection of receptor-deficient BHK cells was extensively (ca. 800-fold) increased by centrifugation, while virus hardly infected BHK cells without centrifugation (2.75 syncytia per well). On the other hand, no syncytium formation was observed in BHK cells spinoculated with JHMV srr7, while an increase in infection similar to cl-2 was observed in DBT cells. We then examined whether spinoculation increased the attachment of the viruses or not. Real-time PCR showed that attachment of both cl-2 and srr7 was increased significantly by spinoculation in both DBT and BHK cells (data not shown). These data suggest that spinouclation increased attachment to MHVR-negative cells regardless of viruses used, however, only cl-2 infected those cells as shown in Table 1. This result is in good agreement with the MHVR-independent infection observed by the infected-cell overlay test, suggesting that localization of cl-2 S protein in close proximity is an important condition for MHVR-independent infection.

Table 1. Infection of cl-2 and srr7 on DBT and BHK cells by spinoculation.

Viruses	Number of syncytium formed / well[a)]			
	DBT		BHK	
	spin(+)	spin(-)	spin(+)	spin(-)
JHMV cl-2	145000	6120	2200	2.75
JHMV srr7	47800	12800	ND[b)]	ND[b)]

[a)] 10^5 pfu/well of viruses were inoculated onto target cells and infection was monitored by syncytium formation. Three independent experiments were done,and mean values are shown.

[b)] ND, not detected.

3.2. S1 Subunit of cl-2, but Not srr7 S1, Is Dissociated Easily from S2

S proteins of both viruses were expressed on BHK cells and analyzed by Western blotting for their S1 dissociation from S2 (Fig. 1). Cl-2 S1 was released into culture supernatants of cells expressing S protein (lane 1) compared with the cells expressing srr7 S protein (lane 2). Because there was no significant difference on expression and cleavage of S protein between two strains, these results suggest that S1 of cl-2 is more releasable than srr7 S1. This result indicates the correlation between S1 dissociation and MHVR-independent infection activity.

4. CONCLUSIONS

To verify the proposed mechanism of MHVR-independent infection, we forced viruses attachment onto MHVR-negative BHK cells by spinning cells together with inoculated viruses (spinoculation). Cl-2 with MHVR-independent infection activity successfully infected BHK cells, whereas srr7 without this activity failed to infect. Furthermore, the S1 of cl-2 was removed from S2 in a naturally occurring event, but not S1 of srr7. These findings support a mechanism of MHVR-independent infection proposed from infected cell overlay test that the S protein of cl-2, in which S1 is easily

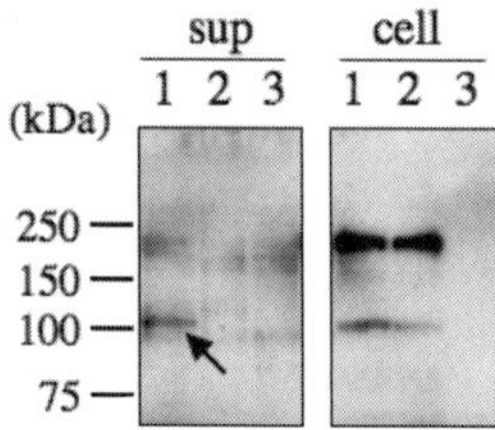

Figure 1. Dissociation of S1 subunit of cl-2 S protein. S protein of cl-2 (lane 1) and srr7 (lane 2) were expressed by T7 vacciniavirus expression system in BHK cells. Released (left panel) and intracellular (right panel) S were detected using S1-specific MAbs 11F. Lane 3 shows samples derived from mock-transfected cells.

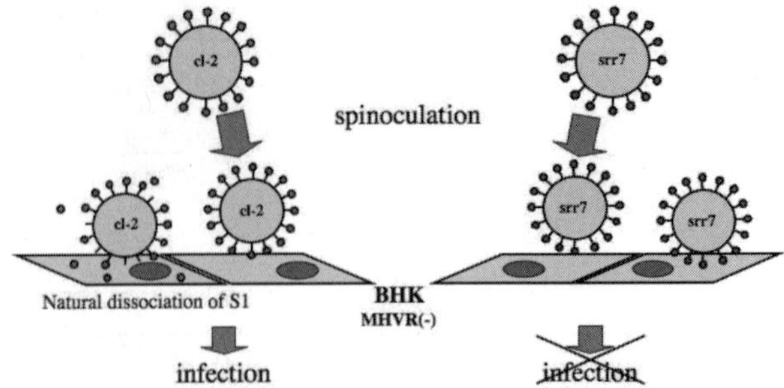

Figure 2. Schematic diagram for mechanism of MHVR-independent infection of cl-2 mediated by spinoculation.

removed from S2 without MHVR binding, mediates viral-cell membrane fusion, if it is placed onto or close to target cell membrane.

5. ACKNOWLEDGMENTS

We thank Shutoku Matsuyama and Keiko Nakagaki for valuable discussions and comments.

6. REFERENCES

1. Gallagher, T. M., Buchmeier, M. J., and Perlman, S., 1992, Cell receptor-independent infection by a neurotropic murine coronavirus, *Virology* **191**:517-522.
2. Gallagher, T. M., and Buchmeier, M. J., 2001, Coronavirus pike proteins in viral entry and pathogenesis, *Virology* **279**:371-374.
3. Krueger, D. K., Kelly, S. M., Lewicki, D. N., Ruffolo, R., and Gallagher, T. M., 2001, Variations in disparate regions of the murine coronavirus spike protein impact the initiation of membrane fusion, *J. Virol.* **75**:2792-2802.
4. Matsuyama, S., and Taguchi, F., 2002, Receptor-induced conformational changes of murine coronavirus spike protein, *J. Virol.* **76**:11819-11826.
5. O'Doherty, U., Swiggard, W. J., and Malim, M. H., 2000, Human immunodeficiency virus type 1 spinoculation enhances infection through virus binding, *J. Virol.* **74**:10074-10080.
6. Saeki, K., Ohtsuka, N., and Taguchi, F., 1997, Identification of spike protein residues of murine coronavirus responsible for receptor-binding activity by use of soluble receptor-resistant mutants, *J. Virol.* **71**:9024-9031.
7. Fuerst, T. R., Niles, E. G., Studier, F. W., and Moss, B., 1986, Eukaryotic transient-expression system based on recombinant vaccinia virus that synthesizes bacteriophage T7 RNA polymerase, *Proc. Natl. Acad. Sci. USA* **83**:8122-8126.
8. Kubo, H., Takase, Y-S., and Taguchi, F., 1993, Neutralization and fusion inhibition activities of monoclonal antibodies specific for S1 subunit of the spike protein of neurovirulent murine coronavirus HMV cl-2 variant, *J. Gen. Virol.* **74**:1421-1425.
9. Routledge, E., Stauber, R., Pfleiderer, M., and Siddel, S. G., 1991, Analysis of murine coronavirus surface glycoprotein functions by using monoclonal antibodies, *J. Virol.* **65**:254-262.

SARS-CoV, BUT NOT HCoV-NL63, UTILIZES CATHEPSINS TO INFECT CELLS

Viral entry

I-Chueh Huang, Berend Jan Bosch, Wenhui Li, Michael Farzan, Peter M. Rottier, and Hyeryun Choe*

1. INTRODUCTION

Three genetic and serologic groups of coronaviruses have been described. Group 1 human coronavirus HCoV-229E utilizes aminopeptidase N (APN; CD13) as its cellular receptor,[1, 2] whereas SARS-CoV, a group 2 virus, uses angiotensin-converting enzyme 2 (ACE2).[3, 4] A recently identified novel group 1 coronavirus, HCoV-NL63, utilizes the SARS-CoV receptor ACE2,[5] despite its close similarity to other group 1 coronaviruses.

Some coronavirus S proteins are cleaved into 2 domains by a furin-like protease in virus-producing cells. The resulting S1 domain mediates receptor binding,[6, 7] and the C-terminal S2 domain mediates fusion between viral and cellular membranes. This producer-cell processing of fusion protein is essential for the infection of HIV-1 and influenza virus.[8] Although the S1 and S2 domains can be identified by their similarity with other S proteins, SARS-CoV and HCoV-NL63 S proteins are not processed in the producer cells.

Cathepsins are a diverse group of endosomal and lysosomal proteases. The role of cathepsins in reovirus infection is well established.[9-12] After receptor-mediated endocytosis, degradation of outer capsid protein σ3 by cathepsins is essential for reovirus infection. Recently, it has been demonstrated that infection mediated by the GP protein of the Zaire Ebola virus depends on cathepsin B.[13] Here we show that cathepsins play an important role in SARS-CoV infection. In contrast, HCoV-NL63 infection is not dependent on cathepsin activities. Thus variations in cellular proteases can serve as an additional determinant of viral tropism.

* I-Chueh Huang, Hyeryun Choe, Harvard Medical School, Boston, Massachusetts 02115. Berend Jan Bosch, Peter M. Rottier, Utrecht University, Utrecht, The Netherlands. Wenhui Li, Michael Farzan, Harvard Medical School, Southborough, Massachusetts 01772.

2. RESULTS

Inhibitors for several classes of proteases were assayed for their ability to modulate SARS-CoV infection. HIV-1 is included as a control. Murine leukemia viruses carrying the gene for green fluorescent protein (GFP) were pseudotyped with the SARS-CoV S protein or HIV-1 gp160 (SARS/MLV and HIV-1/MLV, respectively). 293T cells expressing ACE2 or CD4/CXCR4 were incubated with these pseudotyped viruses in the presence of the aspartic protease inhibitor pepstatin A, serine protease inhibitor AEBSF, cysteine protease inhibitor E64d, and metalloprotease inhibitor phosphoramidon. As shown in Fig. 1A, only E64d, a general inhibitor of cysteine proteases, blocked SARS/MLV infection. Because SARS-CoV infection is sensitive to NH_4Cl, an inhibitor of lysosomal acidification,[14] and roles for the lysosomal cysteine protease cathepsins in reovirus infection have been described,[9-12] the ability of cathepsin inhibitors to block SARS/MLV infection was assessed. As shown in Fig. 1B, cathepsin L inhibitor (Z-FY(t-Bu)-DMK) potently blocked SARS/MLV infection. Cathepsin B inhibitor (CA-074 methyl ester) also showed consistent but less significant inhibition of infection. HIV-1/MLV infection was rather enhanced by these inhibitors, consistent with other reports indicating that lysosomal degradation interferes with productive HIV-1 infection.

The ability of cathepsin inhibitors to block infection was then examined using replication-competent SARS-CoV or HCoV-NL63, which also utilizes ACE2 as its receptor. Sindbis virus and vesicular stomatitis virus (VSV) were included as controls. Infection was assessed within 8 and 24 hours of incubation with SARS-CoV and HCoV-NL63, respectively, thereby minimizing any potential effects of inhibitors on post-entry steps in viral replication. As shown in Fig. 2, SARS-CoV infection was effectively blocked by cathepsin L inhibitor and less significantly by cathepsin B inhibitor. Neither of the cathepsin inhibitors showed significant suppression of HCoV-NL63 replication.

This difference between HCoV-NL63 and SARS-CoV was confirmed using pseudotyped viruses. Inhibitors of cathepsin B, L, K, and S—the only available specific cathepsin inhibitors–were assessed. None of these inhibitors had a detectable effect on the infection of MLV pseudotyped with the HCoV-NL63 S protein (NL63/MLV), or with the VSV G protein (VSV-G/MLV), whereas cathepsin L and S inhibitors efficiently blocked SARS/MLV entry (Fig. 3A). None of the protease inhibitors tested in Fig. 1A had any effect on NL63/MLV infection (data not shown), nor did E64d, which again potently inhibited SARS/MLV (Fig. 3B).

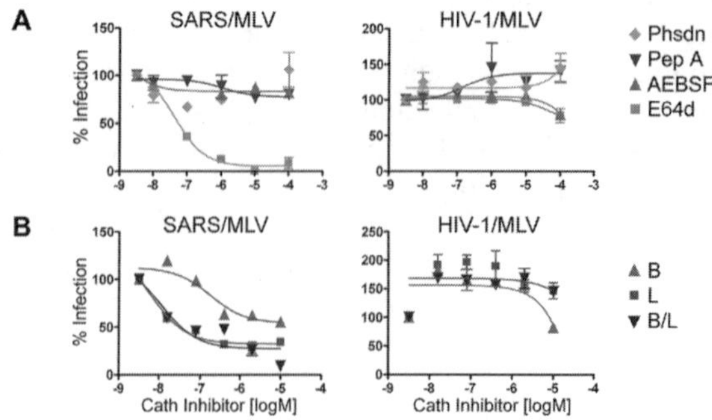

Figure 1. SARS coronavirus S-protein-mediated (SARS/MLV) entry is blocked by cathepsin L inhibitor.

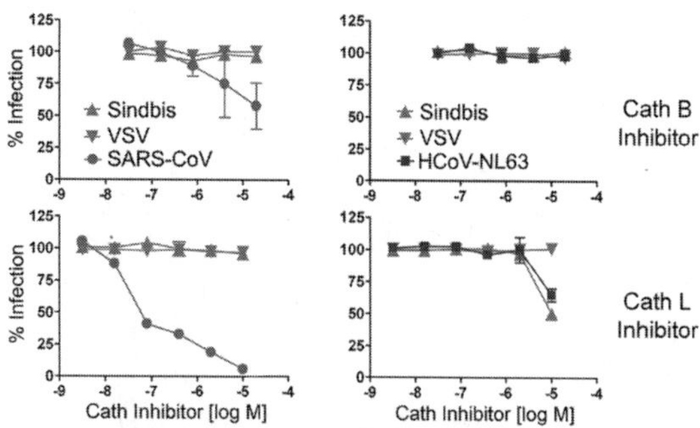

Figure 2. Cathepsin L inhibitor suppresses the infection of SARS-CoV, but not that of HCoV-NL63.

Because cathepsin inhibitors can cross-react, the roles of specific cathepsins were studied by introducing exogenous cathepsins into 293T cells. To ensure comparable ACE2 expression levels in cells transfected with various cathepsins, varying amounts of ACE2-expressing plasmid were used in transfection together with fixed amount of cathepsin plasmids. Also, ACE2 cell-surface expression was assessed by flow cytometry in each experiment. As shown in Fig. 4A, cathepsin L markedly increased infection of SARS/MLV but had no effect on NL63/MLV or VSV-G/MLV. Cathepsin S also modestly enhanced SARS/MLV infection, but, surprisingly, reduced NL63/MLV infection. In parallel, these cathepsins were immunoprecipitated from aliquots of the same cells that were metabolically labeled, and analyzed by SDS-PAGE. Figure 4B shows that exogenous cathepsin L expression was lower than that of cathepsin B or S, despite its greater effect on SARS-CoV infection. Similarly, as shown in Fig. 4C, introduction of cathepsin L into mouse embryonic fibroblasts derived from mice lacking cathepsin L resulted in enhanced infection by SARS/MLV and MLV pseudotyped with Marburg virus or Ebola virus GP proteins, while no enhancement of infection was observed with NL63/MLV or VSV-G/MLV. Collectively, these data show that cathepsins L and S contribute to SARS-CoV infection but not to that of HCoV-NL63.

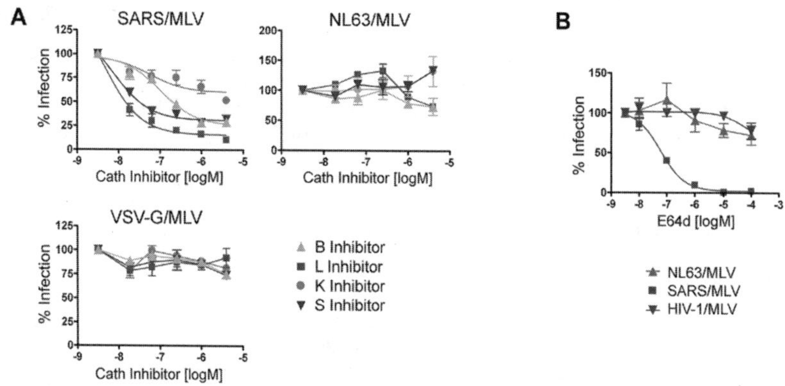

Figure 3. HCoV-NL63 S-protein-mediated (NL63/MLV) entry is not affected by cathepsin inhibitors.

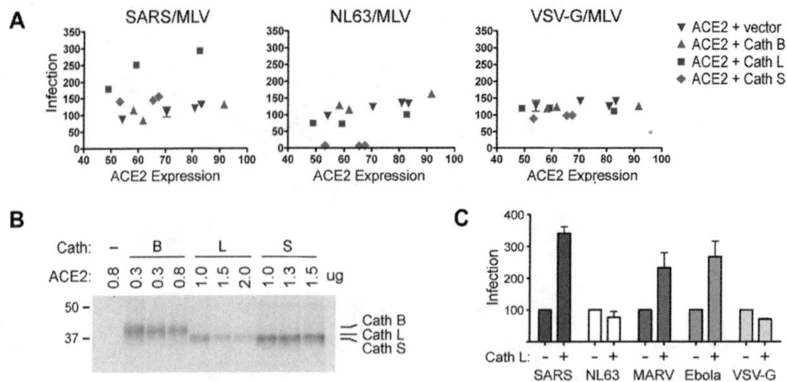

Figure 4. Exogenous cathepsin L enhances the entry of SARS/MLV, but not that of NL63/MLV.

3. REFERENCES

1. C. L. Yeager, R. A. Ashmun, R. K. Williams, C. B. Cardellichio, L. H. Shapiro, S. T. Look, and K. V. Holmes, Human aminopeptidase N is a receptor for human coronavirus 229E, *Nature* **357**:420-422 (1992).
2. D. B. Tresnan and K. V. Holmes, Feline aminopeptidase N is a receptor for all group I coronaviruses, *Adv. Exp. Med. Biol.* **440**, 69-75 (1998).
3. W. Li, M. J. Moore, N. Vasilieva, J. Sui, S. K. Wong, M. A. Berne, M. Somasundaran, J. L. Sullivan, C. Luzeriaga, T. C. Greenough, H. Choe, and M. Farzan, Angiotensin-converting enzyme 2 is a functional receptor for the SARS coronavirus, *Nature* **426**, 450-454 (2003).
4. P. Wang, J. Chen, A. Zheng, Y. Nie, X. Shi, W. Wang, G. Wang, M. Luo, H. Liu, L. Tan, X. Song, Z. Wang, X. Yin, X. Qu, X. Wang, T. Qing, M. Ding, and H. Deng, Expression cloning of functional receptor used by SARS coronavirus, *Biochem. Biophys. Res. Commun.* **315**, 439-444 (2004).
5. H. Hofmann, K. Pyrc, L. van der Hoek, M. Geier, B. Berkhout, and S. Pohlmann, Human coronavirus NL63 employs the severe acute respiratory syndrome coronavirus receptor for cellular entry, *Proc. Natl. Acad. Sci. USA* **16**, 7988-7993 (2005).
6. S. K. Wong, W. Li, M. J. Moore, H. Choe, and M. Farzan, A 193-amino acid fragment of the SARS coronavirus S protein efficiently binds angiotensin-converting enzyme 2, *J. Biol. Chem.* **279**, 3197-3201 (2004).
7. A. Bonavia, B. D. Zelus, D. E. Wentworth, P. J. Talbot, and K. V. Holmes, Identification of a receptor-binding domain of the spike glycoprotein of human coronavirus HCoV-229E, *J. Virol.* **77**, 2530-2538 (2003).
8. D. S. Dimitrov, Virus entry: molecular mechanisms and biomedical applications, *Nat. Rev. Microbiol.* **2**, 109-122 (2004).
9. J. W. Golden, J. Linke, S. Schmechel, K. Thoemke, and L. A. Schiff, Addition of exogenous protease facilitates reovirus infection in many restrictive cells, *J. Virol.* **76**, 7430-7443 (2002).
10. D. H. Ebert, J. Deussing, C. Peters, and T. S. Dermody, Cathepsin L and cathepsin B mediate reovirus disassembly in murine fibroblast cells, *J. Biol. Chem.* **277**, 24609-24617 (2002).
11. J. Jane-Valbuena, L. A. Breun, L. A. Schiff, and M. L. Nibert, Sites and determinants of early cleavages in the proteolytic processing pathway of reovirus surface protein sigma 3, *J. Virol.* **76**, 5184-5197 (2002).
12. G. S. Baer, D. H. Ebert, C. J. Chung, A. H. Erickson, and T. S. Dermody, Mutant cells selected during persistent reovirus infection do not express mature cathepsin L and do not support reovirus disassembly, *J. Virol.* **73**, 9532-9543 (1999).
13. K. Chandran, N. J. Sullivan, U. Felbor, S. P. Whelan, and J. M. Cunningham, Endosomal proteolysis of the Ebola virus glycoprotein is necessary for infection, *Science* **308**, 1643-1645 (2005).
14. G. Simmons, J. D. Reeves, A. J. Rennekamp, S. M. Amberg, A. J. Piefer, and P. Bates, Inhibitors of cathepsin L prevent severe acute respiratory syndrome coronavirus entry, *Proc. Natl. Acad. Sci. USA* **101**, 4240-4245 (2004).

V. PATHOGENESIS OF NON-HUMAN CORONAVIRUSES

CORONAVIRUS IMMUNITY
From T cells to B cells

Cornelia C. Bergmann, Chandran Ramakrishna, J. M. Gonzales, S. I. Tschen, and Stephan A. Stohlman*

1. INTRODUCTION

Coronavirus infections affect a variety of organs including the respiratory and enteric tracts as well as the central nervous system (CNS). Infections of the CNS provide a unique challenge to the host. Rapid responses are vital to control the pathogen yet also affect the communication network between highly specialized cells that control all host cognitive and vital functions.[1-3] CNS resident cells express few, if any, major histocompatibility complex (MHC) molecules in the quiescent state, thus minimizing the potential to activate T cells. Immunological activity in the CNS is also restrained by the absence of a dedicated lymphatic drainage system and constitutive secretion of neurotrophins and TGF-β.[3-5] Lastly, tight junctions between endothelial cells associated with the blood-brain barrier (BBB) and the limited expression of adhesion molecules limit large molecules, e.g., antibodies, as well as T cells from entering the CNS.[2, 5] Although a small number of activated/memory T cells randomly patrol the CNS in the absence of 'danger' signals, they disappear in the absence of antigen recognition.[2] This quiescent steady state contrasts dramatically with the vigorous inflammatory responses induced following many CNS infections.[3, 4] Infections of the murine CNS by neurotropic coronaviruses provide excellent model systems that illustrate the interactions of innate immune responses with adaptive host effector mechanisms and the control of virus replication at the cost of immune pathology.

Infection by the neurotropic mouse hepatitis virus (MHV) strain JHMV produces an acute demyelinating encephalomyelitis in mice.[6-8] Survivors have no detectable infectious virus by approx. 2 weeks postinfection (p.i.), yet viral antigen and more prominently RNA remain detectable exclusively within the CNS up to 2 years p.i. Despite exploitation of various T cell–mediated functions to control acute virus infection in distinct cell types, host regulatory mechanisms, presumably designed to protect CNS integrity, contribute to the failure to eliminate virus. An enigma has been the inability to isolate persisting virus

*Cleveland Clinic Foundation, Cleveland, OH 44106

from CNS explants or from immunosuppressed mice perfused to remove neutralizing antibody (Ab) prior to explant. Although survivors exhibit little or no clinical abnormalities, histological examination shows ongoing primary CNS demyelination, similar to the pathological changes associated with multiple sclerosis. Clues that virus may persist in a replication competent form were provided by mice genetically impaired in anti-viral Ab production.[9,10] Despite initial effective clearance of infectious virus, virus recrudesced in the absence of anti-viral Ab. The majority of data discussed in this review pertains to a monoclonal Ab neutralization JHMV escape mutant, designated 2.2v-1.[11] Microglia, astrocytes, and oligodendroglia are primary targets of infection. Neurons are only rarely infected, sparing mice from death due to neuronal dysfunction. Survival and demyelination associated with chronic infection is thus significantly enhanced. Furthermore, hepatitis, potentially interfering with CNS specific immune responses, is extremely rare after intracerebral (i.c.) infection.

2. LINK BETWEEN INNATE AND ADAPTIVE RESPONSES

JHMV CNS infection induces rapid, coordinated expression of chemokines, matrix metalloproteinases (MMPs), the tissue inhibitor of MMPs (TIMP-1), and pro-inflammatory cytokines.[12-18] MMP activation is associated with several physiological processes, including the influx of inflammatory cells into tissues, activation of cytokines, and CNS pathology.[19] Neutrophils, macrophages, and natural killer (NK) cells are the initial inflammatory cells recruited into the MHV-infected CNS.[20, 21] Secretion of pre-packaged MMP-9 by neutrophils contributes to a loss of BBB integrity and facilitates subsequent entry of inflammatory cells into the infected CNS.[20] JHMV infection itself only induces a subset of MMPs in CNS resident cells, namely MMP-3 and MMP-12.[18,22] However, their roles in cell migration or pathology are unclear. A potential mechanism for counteracting MMP-mediated CNS pathology may reside in sustained elevation of mRNA encoding the MMP inhibitor, TIMP-1.[18, 19]

The earliest chemokines induced following infection are CXCL10 and CCL3.[12,23] CXCL10 overexpression in the CNS using a recombinant MHV revealed that excessive NK cell accumulation in immunodeficient RAG1$^{-/-}$ mice may contribute to virus clearance.[24] By contrast, despite early NK cell recruitment, there is little evidence of a direct anti-viral role to combat wildtype virus. Their potential to secrete IFN-γ may, however, facilitate antigen presentation via upregulation of MHC class I and class II molecules. Increased CCL3 expression appears to link the innate and adaptive immune responses by stimulating recruitment, activation, and maturation of dendritic cells and T cells.[23, 25] Accumulation of macrophages, comprising the largest fraction of innate infiltrates, is enhanced by CCL5.[12,26,27] Cytokines rapidly induced within the MHV infected CNS include IL-1α, IL-1β, IL-6, and IL-12.[13-16] These cytokines vary in expression levels, but not overall pattern between distinct MHV variants.[15] TNF-α, IL-12, and IL-1β are induced in resident CNS cells in response to viral infection.[13,16] The absence of TNF-α neither alters MHV replication *in vivo* nor CNS pathology.[28, 29]

The innate CNS inflammatory response induced by MHV infection is succeeded by a prominent adaptive response, which peaks at 7–10 days p.i.[20,21,30,31] Although JHMV replication is undetectable at peripheral sites, virus-specific T cells are detected in the cervical lymph nodes (CLN) and spleen, prior to the CNS.[31] Initial virus replication in ependymal cells lining the ventricles[32] is likely to facilitate the peripheral activation of adaptive immune responses by antigen drainage into the CLN via the cerebral spinal fluid.[4, 5] The early detection of cells with a dendritic cell–like phenotype in the CNS parenchyma and CLN alternatively supports acquisition of viral antigens in the CNS by antigen presenting cells, followed by migration to CLN.[25] Activation of adaptive immune components alters both the composition of CNS infiltrating cells and chemokine expression. Although CXCL9, CCL2, CCL3, CCL4, CCL5, and CCL7 are expressed during acute infection, CXCL10 is the most prominent and sustained chemokine.[12] In addition to attracting NK cells,[24] CXCL9 and CXCL10 recruit activated T cells[33,34] and potentially plasmablasts,[35] via the CXCR3 receptor. Increasing T-cell accumulation coincides with a decline in neutrophils and NK cells; however, macrophages persist in the CNS.[21] The T cell–mediated reduction in CNS viral burden results in a decline in CXCL9, CCL2, CCL3, and CCL7[12] as well as IL-1α, IL-1β, IL-6, IL-12, and IFN-β mRNA.[13] By contrast, the T-cell chemoattractants CXCL10 and CCL5 remain elevated[12] correlating with increased T-cell recruitment and IFN-γ expression.[13,30] Peak IFN-γ mRNA coincided with peak T-cell infiltration and is functionally evident by maximal expression of both MHC class I and II on microglia.[13,21,36] In the absence of IFN-γ, MHC class I expression is reduced and MHC class II remains undetectable on microglia and most macrophages.[21,36] IFN-γ is thus crucial in facilitating interactions between immune effectors and CNS resident cells.

3. T CELL–MEDIATED IMMUNE CONTROL

T cells, most prominently the $CD8^+$ subset, provide the most critical anti-viral functions.[7] $CD4^+$ T cells provide crucial accessory function by enhancing virus-specific $CD8^+$ T-cell expansion and maintaining $CD8^+$ T-cell viability within the CNS.[37,38] The distinct localization of T-cell subsets early during inflammation constitutes an enigma. $CD4^+$ T cells cross the BBB and accumulate around blood vessels.[37] By contrast, $CD8^+$ T cells migrate into the parenchyma potentially guided by sites of virus replication. The differential ability of T-cell subsets to traffic through the infected tissue is linked to TIMP-1 expression by $CD4^+$ but not $CD8^+$ T cells.[22] These findings implicate the novel concept that migration into the CNS parenchyma is not only controlled by proteases that promote migration but also by protease inhibitors potentially stalling migration.

Although the majority of early T-cell infiltrates are memory T cells specific for irrelevant antigens, these are replaced by virus-specific T cells.[39] During peak T-cell accumulation the majority of both $CD8^+$ and $CD4^+$ T cells within the CNS are virus-specific.[7,30] Virus-specific $CD8^+$ T cells accumulate to 10-fold higher frequencies in the CNS compared with the periphery.[30,31] This high frequency correlates with virus specific *ex vivo* cytolysis and efficient control of viral replication.[10,30,40,41] Susceptibility of astrocytes, microglia/macrophages, and oligodendroglia to distinct T-cell effector function is cell type specific.[42,43] Replication in astrocytes and microglia, but not oligodendrocytes is controlled via perforin mediated cytolysis.[42] By contrast, IFN-γ controls replication in oligodendrocytes, but is insufficient for virus elimination in astrocytes and microglia.[43] The absence of the Fas/FasL pathway does not alter virus clearance or pathology.[44] Distinct anti-viral efficiencies of T-cell effector mechanisms were confirmed in infected, immunodeficient recipients of $CD8^+$ T cells lacking IFN-γ or perforin.[21,36] Overall a more prominent role of IFN-γ compared with perforin is evident by enhanced virus control and reduced mortality in the sole presence of either function alone.[21,36,42,43] Importantly, infection of mice with an IFN-γ signaling defect selectively in oligodendroglia support a direct role for IFN-γ rather than a secondary effect in controlling oligodendroglial infection.[45] The basis for oligodendroglial resistance to perforin-mediated cytolysis is unclear, especially as MHC class I is up regulated on oligodendroglia during JHMV infection.[46] Irrespectively, perforin-mediated control of virus replication in astrocytes and macrophage/microglia implicates MHC class I–mediated cytolysis.

After infectious virus is eliminated, inflammatory cells, viral antigen, and viral mRNA persist. The inability to achieve sterile immunity suggests viral evasion from, or loss of, T-cell function. Indeed, responsiveness by virus-specific $CD8^+$ T cells is lost at the cytolytic level concomitant with virus clearance.[30,47] Loss of function is independent of either demyelination or antigen load[10,41] and unlikely due to anergy, as $CD8^+$ T cells are not impaired in IFN-γ secretion.[30,47] The loss of $CD8^+$ T cell–mediated cytolysis during resolution of primary MHV infection and throughout persistence further contrasted with retention of cytolytic function in reactivated memory cells after neurotropic influenza virus challenge.[48] In an analogous study, MHV-specific memory $CD8^+$ T cells from the inflamed CNS of previously immunized mice exhibit increased IFN-γ and granzyme B production, as well as enhanced cytolysis at a single cell level compared with naïve mice after challenge.[47] Enhanced effector function resulted in more effective virus control. Importantly, reactivated memory $CD8^+$ T cells retained cytolytic function coincident with increased granzyme B levels compared with primary $CD8^+$ T cells.[47] Loss of virus-specific cytolytic function thus appears to reflect distinct differentiation states of primary compared to memory $CD8^+$ T cells rather than an intrinsic property of the inflamed CNS environment.

Despite their significant decline following clearance of infectious virus, persisting T cells are a hallmark of viral persistence and ongoing demyelination.[6,7] The percentages of virus-specific T cells within the $CD8^+$ compartment remain remarkably stable throughout infection[10,30,47] suggesting indiscriminate homeostatic retention or turnover. A role for viral persistence and/or continuing pathology in maintaining T-cell retention is supported by the complete loss of T cells from the CNS after infection with a neurotropic MHV not associated with persistence or myelin loss.[40] A virus driven component was also suggested by selection of $CD8^+$ T-cell populations with limited T-cell receptor specificities during persistence compared with the acute infection.[49] Finally, virus-induced TNF-α secretion by $CD8^+$ T cells during both acute infection and persistence is low,[47] suggesting that T-cell retention within the CNS may be due to decreased secretion of apoptosis inducing factors. The contribution of local homeostatic proliferation or ongoing recruitment to T-cell maintenance in the CNS remains unclear. Preliminary evidence suggests that IL-15, which regulates antigen-independent homeostasis of memory cells in lymphoid organs,[50] is not required (Bergmann, unpublished). Memory cells traffic poorly into the CNS[51] and activated T cells recruited in response to acute infection are only retained within the CNS upon cognate antigen recognition.[2,39] These recent observations support both very limited local turnover and peripheral recruitment.

4. HUMORAL IMMUNITY CONTROLS PERSISTENCE

A protective role of T cells during persistence is disputed by viral recrudescence in mice devoid of B cells. Mice lacking humoral immunity mount a normal inflammatory response during acute infection and control CNS infectious virus with kinetics similar to immunocompetent mice.[9, 10] However, unlike recovery of wild-type mice, mice unable to secrete Ab exhibit increased mortality associated with the reemergence of infectious virus within the CNS. By contrast, the MHV A59 strain, which infects both the liver and CNS, fails to reactivate in the liver in the absence of humoral immunity.[52] Whether this is due to the absence of viral persistence in liver, or reflects a fundamental difference in immune control in these two organs is unclear. Preexisting virus neutralizing Ab provides protection against MHV-induced CNS,[6] presumably by limiting virus replication after transport into the CNS parenchyma while BBB integrity is compromised.[18] Similarly, virus recrudescence in the CNS of Ab deficient mice can be prevented by transfer of polyclonal Ab at the time infectious virus is initially cleared.[9] Dissection of the specificities and mechanisms of Ab-mediated protection revealed that only spike (S) protein specific neutralizing IgG prevented recrudescence.[53] Non-neutralizing Ab specific for S, matrix, or nucleocapsid proteins had no anti-viral effect,[53] distinct from the apparent protective role for non-neutralizing Ab prior to infection.[6] Despite the initial promise of protection, antiviral control waned as transferred Ab decayed in B cell–deficient recipients.[53] These data suggested that virus is maintained in a replication competent form within the CNS and that sustained intrathecal Ab is crucial to maintain virus at undetectable levels during persistence.

Accumulation of virus-specific Ab secreting cells (ASC) within the CNS confirmed intrathecal Ab synthesis.[54] Unlike T cells, virus-specific IgG ASC accumulate prominently in the CNS approximately 1 week after clearance of infectious virus. Although virus specific ASC are barely detectable in the CNS during the virus clearance phase numerous heterologous ASC are already present, implicating nonspecific recruitment.[54] The delayed peak in virus-specific IgG ASC in the CNS compared with CLN, suggests maturation in secondary lymphoid organ germinal centers precedes migration into the CNS. Consistent with the meager presence of virus-specific ASC in either the CNS or lymphoid ogans during acute infection,[54] virus-specific serum Ab, including neutralizing Ab, is undetectable prior to the complete elimination of infectious virus. However, sustained serum Ab levels after clearance of infectious virus supports a role in controlling persistence. Virus-specific ASC are retained in the CNS at high frequencies for at least 3 months p.i. implicating ASC-specific survival factors in the CNS during viral persistence. Despite their progressive decline, virus-specific ASC are maintained at higher levels than virus-specific T cells. The CNS as a survival niche for ASC has previously been observed following other virus-induced CNS infections.[4]

5. BYSTANDER RECRUITMENT AND PATHOLOGY

Numerous approaches have been applied to assess the roles of individual effector molecules in the demyelinating process; however, a unifying mechanism has not emerged.[8] Despite their protective role in controlling acute virus replication, T cells are prominent mediators of demyelination.[6, 8, 34, 55] Immunocompromised SCID or RAG1$^{-/-}$ mice develop little if any demyelination, despite uncontrolled virus replication. Mice deficient for IFN-γ, perforin, inducible nitric oxide synthase, $CD4^+$, $CD8^+$, or B cells all develop demyelinating disease.[6,42,43,56] Similarly, immunodeficient recipients of $CD4^+$ T cells, $CD8^+$ T cells, or γδ T cells are all susceptible to immunopathology.[21,29,36,55,57,58] Interpretation is further confounded by distinct abilities of transferred populations to control virus. The common theme emerging is that recruitment of most immune cell populations as a single entity or overexpression of a specific chemokines can lead to demyelination.[59] A separate issue still unresolved is the role of bystander cells. Bystander cells of irrelevant specificity are commonly recruited by chemokines. Their activation within the CNS environment after microbial infection may influence subsequent CNS inflammation and/or enhance pathogenesis. Enhanced CNS pathology via activation of bystander $CD8^+$ T cells has recently been demonstrated after JHMV infection of LCMV-specific TCR/Rag $2^{-/-}$ mice.[60] By contrast, investigation of bystander effects in mice with a wild-type TCR repertoire demonstrated that although JHMV-induced encephalitis recruits heterologous memory T cells, they are not activated in the process.[39] The vast majority of CNS infiltrating $CD8^+$ T cells express the $CD44^{hi}$, $CD62L^{-/lo}$, $CD11a^{hi}$, and CD49d (VLA-4) activation/memory phenotypic markers.[30,39] However, $CD43^{hi}$, $CD127^{-/lo}$ expression discriminates virus-specific $CD8^+$ T cells within the CNS from cells specific for irrelevant antigens, which retain a $CD43^{int}$, $CD127^+$ phenotype. Bystander cells further did not acquire expression of *ex vivo* cytolytic activity, suggesting that bystander $CD8^+$ T-cell recruitment alone is unlikely to directly result in immune pathology in the absence of cognate or cross-reactive antigen.[39] These results are consistent with the requirement for an antigen-driven component in contributing to $CD8^+$ T cell–mediated CNS damage observed in other models.[61,62]

6. CONCLUSIONS

Several novel concepts have emerged from MHV-induced CNS infection. T-cell subsets appear to have differential abilities to migrate within the CNS. Cross-talk is indicated by enhanced $CD8^+$ T-cell function and survival in the presence of $CD4^+$ T cells. Distinct cellular targets of infection have differential susceptibility to T-cell effector function. The accumulation and maintenance of virus-specific ASC in the CNS, coupled with reactivation of infectious virus in the absence of antibody, indicates that antibody secretion within the CNS and not T-cell immunity is absolutely critical for the control of MHV CNS persistence. Lastly, retention of both T cells and ASC in the CNS during persistence suggests myelin loss is associated with an ongoing immune response, sustained by low-level oligodendroglial infection.

7. ACKNOWLEDGMENTS

This research was supported by the National Institutes of Health Grants NS18146 and AI47249.

8. REFERENCES

1. Z. Fabry, C. S. Raine, and M. N. Hart, Nervous tissue as an immune compartment: the dialect of the immune response in the CNS, *Immunol. Today* **15**, 218-224 (1994).
2. W. F. Hickey, Basic principles of immunological surveillance of the normal central nervous system, *Glia* **36**, 118-124 (2001).
3. R. Dorries, The role of T-cell-mediated mechanisms in virus infections of the nervous system, *Curr. Top. Microbiol. Immunol.* **253**, 219-245 (2001).
4. D. E. Griffin, Immune responses to RNA-virus infections of the CNS, *Nat. Rev. Immunol.* **3**, 493-502 (2003).
5. R. M. Ransohoff, P. Kivisakk, and G. Kidd, Three or more routes for leukocyte migration into the central nervous system, *Nat. Rev. Immunol.* **3**, 569-581 (2003).
6. S. Stohlman, C. Bergmann, and S. Perlman, in: *Persistent Viral Infection*, edited by R. Ahmed and I. Chen (John Wiley, New York, 1999), pp. 537-558.
7. N. W. Marten, S. A. Stohlman, and C. C. Bergmann, MHV infection of the CNS: mechanisms of immune-mediated control, *Viral Immunol.* **14**, 1-18 (2001).
8. M. J. Buchmeier and T. E. Lane, Viral-induced neurodegenerative disease, *Curr. Opin. Microbiol.* **2**, 398-402 (1999).
9. M. T. Lin, D. R. Hinton, N. W. Marten, C. C. Bergmann, and S. A. Stohlman, Antibody prevents virus reactivation within the central nervous system, *J. Immunol.* **162**, 7358-7368 (1999).
10. C. Ramakrishna, S. A. Stohlman, R. Atkinson, M. Schlomchik, and C. C. Bergmann, Mechanisms of central nervous system viral persistence: critical role of antibody and B cells, *J. Immunol.* **168**, 1204-1211 (2002).
11. J. O. Fleming, M. Trousdale, F. El-Zaatari, S. A. Stohlman, and L. P. Weiner, Pathogenicity of antigenic variants of murine coronavirus JHM selected with monoclonal antibodies, *J. Virol.* **58**, 869-875 (1986).
12. T. E. Lane, V. C. Asensio, N. Yu,. A. D., Paoletti, I. L.Campbell, and M. J. Buchmeier, Dynamic regulation of alpha- and beta-chemokine expression in the central nervous system during mouse hepatitis virus-induced demyelinating disease, *J. Immunol.* **160**, 970-978 (1998).
13. B. Parra, D. R. Hinton, M. T. Lin, D. J. Cua, and S. A. Stohlman, Kinetics of cytokine mRNA expression in the CNS following lethal and sublethal coronavirus-induced encephalomyelitis, *Virology* **233**, 260-270 (1997).
14. B. D. Pearce, M. V. Hobbs, T. S. McGraw, and M. J. Buchmeier, Cytokine induction during T-cell-mediated clearance of mouse hepatitis virus from neurons in vivo, *J. Virol.* **68**, 5483-5495 (1994).
15. J. D. Rempel, S. J. Murray, J. Meisner, and M. J. Buchmeier, Differential regulation of innate and adaptive immune responses in viral encephalitis, *Virology* **318**, 381-392 (2004).
16. J. D. Rempel, L. A. Quina, P. K. Blakelu-Gonzales, M. J. Buchmeier, and D. L. Gruol, Viral induction of central nervous system innate immune responses, *J. Virol.* **79**, 4369-4381 (2005).
17. T. E. Lane and M. J. Buchmeier, in: *Universes in Delicate Balance: Chemokines and the Nervous System*, edited by R. Ransohoff, et al. (Elsevier Press, New York, 2002), pp. 191-202.
18. J. Zhou, S. Stohlman, R. Atkinson, D. Hinton, and N. Marten, Matrix metalloproteinase expression correlates with virulence following neurotropic mouse hepatitis virus infection, *J. Virol.* **76**, 7373-7384 (2002).
19. V. W. Yong, C. Power, P. Forsyth, and D. R. Edwards, Metalloproteinases in biology and pathology of the nervous system, *Nature* **2**, 502-511 (2001).
20. J. Zhou, S. Stohlman, D. R. Hinton, and N. Marten, Neutrophils modulate inflammation during viral induced encephalitis, *J. Immunol.* **170**, 3331- 3336 (2003).
21. C. C. Bergmann, B. Parra, D. Hinton, C. Ramakrishna, M. Morrison, and S. A. Stohlman, Perforin mediated effector function within the CNS requires IFN-γ mediated MHC upregulation, *J. Immunol.* **170**, 3204-3213 (2003).
22. J. Zhou, N. W. Marten, C. C. Bergmann, W. B. Macklin, D. R. Hinton, and S. A. Stohlman, Expression of matrix metalloproteinases and their tissue inhibitor during viral encephalitis, *J. Virol.* **79**, 4764-4773 (2005).
23. M. J. Trifilo, C. C. Bergmann, W. A. Kuziel, and T. E. Lane, CC chemokine ligand 3 (CCL3) regulates $CD8^+$-T-cell effector function and migration following viral infection, *J. Virol.* **77**, 4004-4014 (2003).
24. M. J. Trifilo, C. Montalto-Morrison, L. N. Stiles, K. R. Hurst, J. L. Hardison, J. E. Manning, P. S. Masters, and T. E. Lane, CXC chemokine ligand 10 controls viral infection in the central nervous system: Evidence for a role in innate immune response through recruitment and activation of natural killer cells, *J. Virol.* **78**, 585-594 (2004).

25. M. J. Trifilo and T. E. Lane, The CC chemokine ligand 3 regulates CD11c+CD11b+CD8alpha-dendritic cell maturation and activation following viral infection of the central nervous system: implications or a role in T cell activation, *Virology* **327**, 8-15 (2004).
26. T. E. Lane, M. T. Liu, B. P. Chen, V. C. Asensio, R. M. Samawi, A. D. Paoletti, I. L. Campbell, S. L. Kunkel, H. S. Fox, and M. J. Buchmeier, A central role for CD4(+) T cells and RANTES in virus-induced central nervous system inflammation and demyelination, *J. Virol.* **74**, 1415-1424 (2000).
27. W. G. Glass, M. J. Hickey, J. L. Hardison, M. T. Liu, J. E. Manning, and T. E. Lane. Antibody targeting of the CC chemokine ligand 5 (CCL5) results in diminished leukocyte infiltration into the central nervous system and reduced neurologic disease in a viral model of multiple sclerosis, *J. Immunol.* **172**, 4018-4025 (2004).
28. S. A. Stohlman, D. R. Hinton, D. Cua, E. Dimacali, J. Sensintaffar, S. Tahara, F. Hofman, and Q. Yao, Tumor necrosis factor expression during mouse hepatitis virus induced demyelination, *J. Virol.* **69**, 5898-5903 (1995).
29. L. Pewe and S. Perlman, Cutting edge: CD8 T cell-mediated demyelination is IFN-gamma dependent in mice infected with a neurotropic coronavirus, *J. Immunol.* **168**, 1547-1551 (2002).
30. C. C. Bergmann, J. Altman, D. Hinton, and S. A. Stohlman, Inverted immunodominance and Impaired cytolytic function of CD8+ T cells during viral persistence in the CNS, *J. Immunol.* **163**, 3379-3387 (1999).
31. N. Marten, S. A. Stohlman, Z. Zhou, and C. C. Bergmann, Kinetics of virus specific CD8+ T cell expansion and trafficking following central nervous system infection, *J. Virol.* **77**, 2775-2778 (2003).
32. F. I. Wang, D. R. Hinton, W. Gilmore, M. D. Trousdale, and J. O. Fleming, Sequential infection of glial cells by the murine hepatitis virus JHM strain (MHV-4) leads to a characteristic distribution of demyelination, *Lab. Invest.* **66**, 744-754 (1992).
33. M. T. Liu, B. P. Chen, P. Oertel , M. J. Buchmeier, D. Armstrong , T. A. Hamilton, and T. E. Lane, The T cell chemoattractant IFN-inducible protein 10 (IP-10) is essential in host defense against viral-induced neurologic disease, *J. Immunol.* **165**, 2327-2330 (2000).
34. M. T. Liu, H. S. Keirstead, and T. E. Lane, Naturalization of the chemokine CXCL10 reduces inflammatory cell invasion and demyelination and improves neurological function in a viral model of multiple sclerosis, *J. Immunol.* **167**, 4091-4097 (2001).
35. A. E. Hauser, G. F. Debes, S. Arce, G. Cassese, A. Hamann, A. Radbruch, and R. A. Manz, Chemotactic responsiveness toward ligands for CXCR3 and CXCR4 is regulated on plasma blasts during the time course of a memory immune response, *J. Immunol.* **169**, 1277-1282 (2002).
36. C. C. Bergmann, B. Parra, D. R. Hinton, C. Ramakrishna, K. C. Dowdell, and S. Stohlman, Perforin and interferon gamma mediated control of coronavirus central nervous system infection by CD8 T cells in the absence of CD4 T cells, *J. Virol.* **78**, 1739-1750 (2004).
37. S. A. Stohlman, C. C. Bergmann, M. T. Lin, D. J. Cua, and D. R. Hinton, CTL effector function within the CNS requires CD4+ T cells, *J. Immunol.* **160**, 2896-2904 (1998).
38. J. Zhou, D. R. Hinton, S. A. Stohlman, and N. Marten, Maintenance of CD8+ T cells during acute viral infection of the central nervous system requires CD4+ T cells but not interleukin-2, *Virol. Immunol.* **18**, 162-169 (2005).
39. A. M. Chen, N. Khanna, S. A. Stohlman, and C. C. Bergmann, Virus-specific and bystander CD8 T cells recruited during virus-induced encephalomyelitis, *J. Virol.* **79**, 4700-4708 (2005).
40. N. Marten, S. Stohlman, and C. C. Bergmann, Role of Viral Persistence in Retaining CD8+ T cells within the central nervous system, *J. Virol.* **74**, 7903-7910 (2000).
41. N. W. Marten, S. A. Stohlman, R. Atkinson, D. A. Hinton, and C. C. Bergmann, Contributions of $CD8^+$ T cells and viral spread to demyelinating disease, *J. Immunol.* **164**, 4080-4088 (2000).
42. M. Lin, S. Stohlman, and D. Hinton, Mouse hepatitis virus is cleared from the central nervous system of mice lacking perforin-mediated cytolysis, *J. Virol.* **71**, 383-391 (1997).
43. B. Parra, D. R. Hinton, N. Marten, C. C. Bergmann, M. Lin, C. Yang, and S. A. Stohlman, Gamma interferon is required for viral clearance from central nervous system oligodendroglia, *J. Immunol.* **162**, 1641-1647 (1999).
44. B. Parra, M. Lin, S. Stohlman, C. Bergmann, R. Atkinson, and D. Hinton, Contributions of Fas-Fas ligand interactions to the pathogenesis of mouse hepatitis virus in the central nervous system, *J. Virol.* **74**, 2447-2450 (2000).
45. J. M. Gonzalez, C. C. Bergmann, B. Fuss, D. R. Hinton, W. B. Macklin, and S. A. Stohlman, Expression of a dominant negative IFN-γ receptor on mouse oligodendrocytes, *Glia* **51**, 22-34 (2005).
46. J. M. Redwine, M. J. Buchmeier, and C. F. Evans, In vivo expression of major histocompatibility complex molecules on oligodendrocytes and neurons during viral infection, *Am. J. Pathol.* **159**, 1219-1224 (2001).
47. C. Ramakrishna, S. Stohlman, R. Atkinson, D. H. Hinton, and C. C. Bergmann, Differential regulation of primary and secondary CD8+ T cells in the CNS, *J. Immunol.* **173**, 6265-6273 (2004).

48. S. Hawke, P. G. Stevenson, S. Freeman, and C. R. M. Bangham, Long term persistence of activated cytotoxic T lymphocytes after viral infection of the central nervous system, *J. Exp. Med.* **187**, 1575-1582 (1998).
49. N. Marten, S. Stohlman, W. Smith-Begolka, S. Miller, M. Dimicali, Q. Yao, S. Stohl, J. Goverman, and C. Bergmann, Selection of CD8+ T cells with highly focused specificity during viral persistence in the central nervous system, *J. Immunol.* **162**, 3905-3914 (1999).
50. D. Masopust and R. Ahmed, Reflections on CD8 T-cell activation and memory, *Immunol. Res.* **29**, 151-160 (2004).
51. L. Lefrancois and D. Masopust, T cell immunity in lymphoid and non-lymphoid tissues, *Curr. Opin. Immunol.* **14**, 503-508 (2002).
52. A. E. Matthews, S. R. Weiss, M. J. Shlomchik, L. G. Hannum, J. L. Gombold, and J. Y. Paterson, Antibody is required for clearance of infectious murine hepatitis virus A59 from the central nervous system, but not the liver, *J. Immunol.* **167**, 5254-5263 (2001).
53. C. Ramakrishna, C. Bergmann, R. Atkinson, and S. Stohlman, Control of central nervous system viral persistence by neutralizing antibody, *J. Virol.* **77**, 4670-4678 (2003).
54. S. I. Tschen, C. Bergmann, C. Ramakrishna, R. Atkinson, and S. Stohlman, Recruitment kinetics of antibody secreting cells within the CNS following viral encephalomyelitis, *J. Immunol.* **168**, 2922-2929 (2002).
55. G. F. Wu, A. A. Dandekar, L. Pewe, and S. Perlman, CD4 and CD8 T cells have redundant but not identical roles in virus-induced demyelination, *J. Immunol.* **165**, 2278-2286 (2000).
56. G. F. Wu, L. Pewe, and S. Perlman, Coronavirus-induced demyelination occurs in the absence of inducible nitric oxide synthase, *J. Virol.* **74**, 7683-7686 (2000).
57. L. Pewe, J. Haring, and S. Perlman, CD4 T-cell-mediated demyelination is increased in the absence of gamma interferon in mice infected with mouse hepatitis virus, *J. Virol.* **76**, 7329-7333 (2002).
58. A. A. Dandekar, K. O'Malley, and S. Perlman, Important roles for gamma interferon and NKG2D in gammadelta T-cell-induced demyelination in T-cell receptor beta-deficient mice infected with a coronavirus, *J. Virol.* **79**, 9388-9396 (2005).
59. T. S. Kim and S. Perlman, Viral expression of CCL2 is sufficient to induce demyelination in RAG1-/- mice infected with a neurotropic coronavirus, *J. Virol.* **79**, 7113-7120 (2005).
60. J. S. Haring, L. L. Pewe, and S. Perlman, Bystander CD8^{+} T cell-mediated demyelination after viral infection of the central nervous system, *J. Immunol.* **169**, 1550-1555 (2002).
61. J. Cabarrocas, J. Bauer, E. Piaggio, R. Liblau, and H. Lassman, Effective and selective immune surveillance of the brain by MHC class I-restricted cytotoxic T lymphocytes, *Eur. J. Immunol.* **33**, 1174-1182 (2003).
62. D. B. McGavern and P. Truong, Rebuilding an immune-mediated central nervous system disease: weighing the pathogenicity of antigen-specific versus bystander T cells, *J. Immunol.* **173**, 4779-4790 (2004).

RAT CORONAVIRUS INFECTION OF PRIMARY RAT ALVEOLAR EPITHELIAL CELLS

Tanya A. Miura, Jieru Wang, Robert J. Mason, and Kathryn V. Holmes*

1. INTRODUCTION

There are five human coronaviruses (HCoV) that cause respiratory disease. HCoV-229E and HCoV-OC43 cause upper respiratory tract infections and rarely cause lower respiratory tract disease in immunocompromised patients. In 2002–2003, a novel coronavirus, SARS-CoV, caused a pandemic of severe acute respiratory distress syndrome. Since the SARS epidemic, two additional HCoVs have been identified, HCoV-NL-63 and HCoV-HKU1. HCoV-NL-63 has been isolated from pediatric cases of respiratory disease[1,2] and HCoV-HKU1 has been isolated from two adult patients with pneumonia.[3] Coronaviruses also cause respiratory diseases in other species, including porcine respiratory coronavirus (PRCoV),[4] respiratory bovine coronavirus (BCoV-Resp),[5] and canine respiratory coronavirus (CRCoV).[6] Two strains of rat coronavirus (RCoV) cause respiratory disease in rats. Sialodacryoadenitis virus (RCoV-SDAV) and Parker's rat coronavirus (RCoV-P) have both been isolated from infected rat lungs. We are studying the pathogenesis of RCoVs as a model for coronavirus respiratory disease in the natural host.

RCoV-SDAV was initially isolated from the salivary glands of rats with sialodacryoadenitis.[7] In addition to pathogenesis in the salivary and lacrimal glands, RCoV-SDAV can cause chronic eye disease, reproductive disorders, and mild acute disease in the respiratory tract.[7-10] RCoV-SDAV–infected adult rats have lesions and inflammation in the upper respiratory tract. However, lesions in the lung are very mild. Infection of suckling rats by RCoV-SDAV results in more severe respiratory disease. RCoV-P was initially isolated from lungs of asymptomatic rats.[11] In contrast with RCoV-SDAV, RCoV-P is strictly pneumotropic. Inoculation of adult rats with RCoV-P causes asymptomatic infection of the upper and lower respiratory tract with focal interstitial pneumonia, while infection of neonates results in lethal interstitial pneumonia.[11,12] The cellular receptor(s) used by RCoV-SDAV and RCoV-P have not been identified. RCoV-P expresses an enzymatically active hemagglutinin esterase (HE) protein on the viral

* Tanya A. Miura, Kathryn V. Holmes, University of Colorado Health Sciences Center, Aurora, Colorado 80010. Jieru Wang, Robert J. Mason, National Jewish Medical and Research Center, Denver, Colorado 80206.

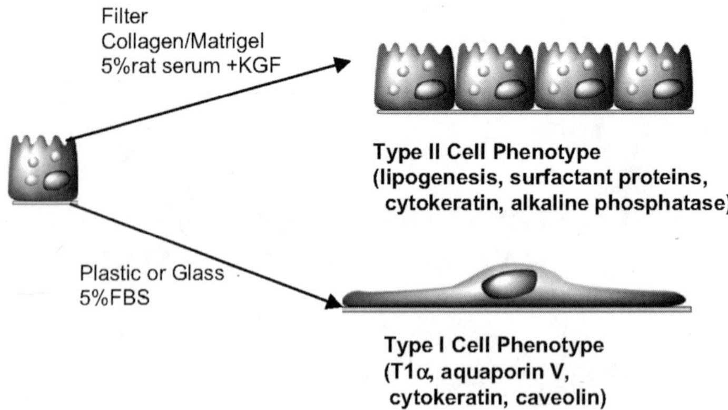

Figure 1. Primary rat alveolar epithelial cells can be cultured to maintain a type II cell phenotype or trans-differentiate into a type I cell phenotype.

envelope, whereas our laboratory isolate of RCoV-SDAV does not express HE.[13] The role of the HE protein in RCoV infection and pathogenesis has not been studied.

Here we report infection of primary cultures of differentiated rat alveolar epithelial cells by RCoV-P and RCoV-SDAV. Rat type II alveolar epithelial cells can be isolated and grown under culture conditions that either maintain a type II phenotype or trans-differentiate into a type I cell phenotype (Figure 1).[14,15] Type II alveolar cells are dividing, cuboidal cells that produce surfactant and regenerate the lung epithelium after injury. Type I alveolar cells are nondividing cells with a flattened morphology that mediate gas exchange and fluid homeostasis in the lung. Specific markers for differentiating type I and type II alveolar cells include surfactant proteins for type II cells and T1α, aquaporin V, and caveolin for type I cells.

Respiratory coronaviruses have been studied in bronchial epithelial cells but not in alveolar epithelial cells. Cultures of differentiated rat pneumocytes can be used to evaluate the cellular tropism of viral infection, viral cytopathic effects on alveolar cells, and viral modulation of cytokines and chemokines that control the immune response and tissue damage in the host. Inoculation with RCoV-SDAV results in more rapid and extensive cell fusion of type I cells than RCoV-P. We are evaluating determinants of differential cell fusion by RCoV strains. In conclusion, the ability of coronavirus to infect both alveolar type I and type II cells has implications in the pathogenesis of severe lung disease, such as that caused by SARS-CoV.

2. METHODS AND RESULTS

2.1. RCoV Infection of Primary Rat Alveolar Type II and Type I Cells

We used freshly isolated type II cells to evaluate infection of highly differentiated pneumocytes by RCoV-SDAV and RCoV-P. Rat alveolar type II cells were isolated and inoculated in suspension with RCoV-SDAV or RCoV-P. Cells were fixed at 72 hpi and

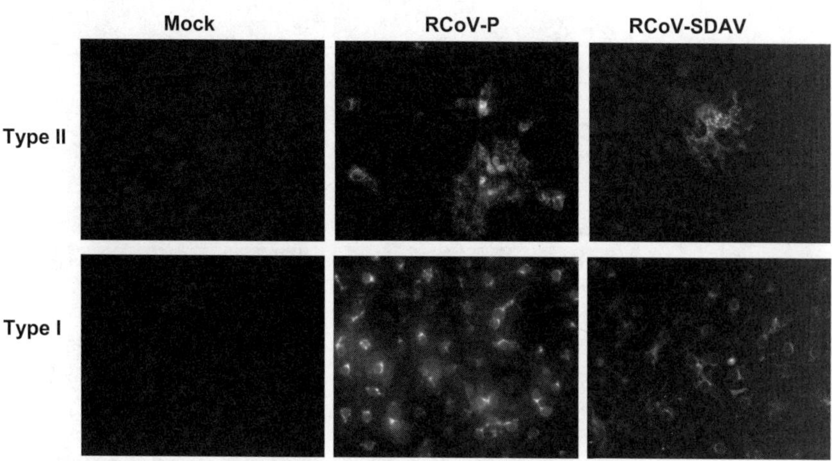

Figure 2. Primary rat alveolar epithelial cells are susceptible to infection by RCoV-P and RCoV-SDAV.

viral antigen was detected by immunofluorescence. To evaluate infection of type I phenotype cells, alveolar epithelial cells were cultured for 6 days under conditions that result in trans-differentiation into a type I cell phenotype (Figure 1) prior to virus inoculation. Viral antigen was detected in type II and type I cells that had been inoculated with RCoV-SDAV or RCoV-P but not in mock-inoculated cells (Figure 2). Infectious virus was detected in media from infected cultures (data not shown). Thus both RCoV-SDAV and RCoV-P can infect primary cultures of rat alveolar type II and type I cells in vitro.

We studied a time course of RCoV infection of primary cultures of rat alveolar type I cells. Cells were incubated for 6 days to fully differentiate into a type I cell phenotype, then inoculated with RCoV-SDAV or RCoV-P. Cells and medium were collected at various times postinoculation. The titer of infectious virus in culture supernatant was determined by plaque assay on L2P41.a cells. Viral antigen was present in RCoV-SDAV and RCoV-P infected cells by 10 hours postinoculation (hpi). Infection by RCoV-SDAV resulted in the formation of syncytia by 12 hpi, which were very large by 27 hpi (Figure 3) and were peeling off the coverslip by 51 hpi. In contrast, the RCoV-P infected cells did not form syncytia until 27 hpi, and syncytia were much smaller than those in the RCoV-SDAV-infected cells. Both RCoV-SDAV and RCoV-P grew to similar titers (1.5×10^4-3×10^4 pfu/mL) in primary type I cell cultures.

2.2. Sequence of the RCoV Spike Glycoprotein

The spike (S) glycoprotein of coronaviruses mediates cell to cell fusion. We therefore sequenced the spike glycoprotein of RCoV-P. We found 70 nucleotide differences in the sequence of the S gene of RCoV-P as compared with the published sequence of the RCoV-SDAV S (NCBI #AF207551). These mutations resulted in 22 amino acid changes in the S1

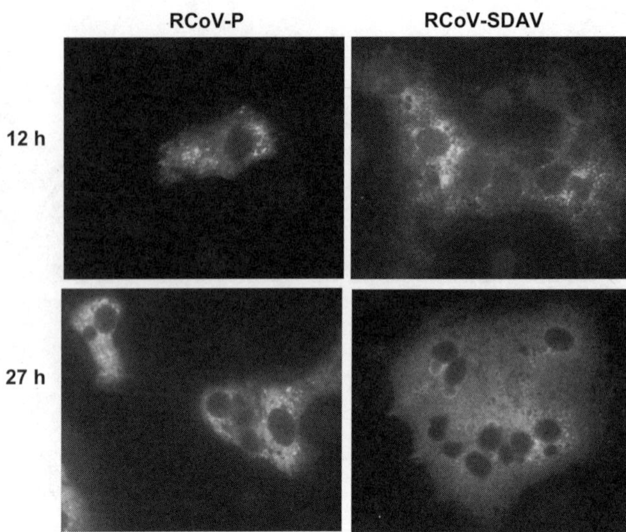

Figure 3. Infection of primary alveolar type I cells by RCoV-SDAV results in larger syncytia than infection by RCoV-P at 12h and 27h post-inoculation.

region that might affect receptor binding and 9 amino acid changes in the S2 region that might affect cell fusion.

The functions of the S glycoprotein of murine coronavirus MHV have been studied extensively.[16] The S proteins of MHV strains are 79–90% identical in amino acid sequence with RCoV-P and RCoV-SDAV. Four of the amino acids that were changed in RCoV-P as compared with RCoV-SDAV are shared with the non-fusogenic MHV-2 (NCBI #AF201929). These 4 amino acids may be important for the non-fusogenic phenotype of MHV-2 and the delayed fusogenic phenotype of RCoV-P. In particular, amino acid 750, which is located near the S1/S2 cleavage signal sequence, may be important in S1/S2 cleavage and the subsequent ability of spike to mediate cell-cell fusion. Hingley et al. previously showed that substitution of the MHV-A59 cleavage signal with the corresponding amino acids of MHV-2 resulted in less efficient cleavage of S1/S2 and a delayed fusion phenotype.[17] The spike glycoprotein of RCoV-SDAV grown in L2P41.a cells is completely cleaved into 90-kDa S1 and S2 proteins, while RCoV-P spike is only partially cleaved.[13] Like MHV-2, RCoV-P has a threonine at amino acid 750, while RCoV-SDAV has an isoleucine. We hypothesize that changing the isoleucine 750 to threonine in the RCoV-SDAV spike would result in decreased efficiency of spike cleavage and cell fusion. Alternatively, other amino acid changes in the spike of RCoV-P may contribute to the delayed fusion phenotype seen in RCoV-P-infected cells. The specific amino acids in spike that are important in the delayed cell to cell fusion phenotype of RCoV-P will be studied further.

3. DISCUSSION

The discovery of 3 human respiratory coronaviruses since 2003 increases the need to study the pathogenesis of coronaviruses in the lung. We are studying rat coronavirus infection of primary rat alveolar epithelial cells as a model for coronavirus pathogenesis in the lung of a natural host.

We have found that rat coronaviruses infect both type I and type II alveolar epithelial cell phenotypes. ACE-2, the principal receptor for SARS-CoV, is expressed in both type I and type II pneumocytes.[18] CD209L, which has been identified as an alternative receptor for SARS-CoV,[19] is also expressed in type I and type II pneumocytes. SARS-CoV RNA and antigen have been detected in alveolar epithelial cells of fatal SARS cases[20,21] and in type I pneumocytes in infected macaques.[22] Type I pneumocytes make up the majority of the epithelial surface area of the lung. Damage to type I cells interferes with gas exchange and fluid homeostasis. Injury to type I cells causes type II cells to dedifferentiate, proliferate, migrate, and trans-differentiate into type I cells to repair the damaged lung. A virus that can infect both type I and type II cells, like RCoV, could cause severe lung disease by preventing repair of the damaged epithelium. Type II cells are essential for the secretion of surfactant, which stabilizes the alveoli and prevents acute respiratory distress syndrome, and surfactant proteins that are important mediators of innate immunity. Alveolar epithelial cells elicit innate and acquired immune responses in the lung by secretion of specific cytokines and chemokines. The effects of virus infection on the immune response will be addressed in future studies.

4. ACKNOWLEDGMENTS

This work was supported by NIH P01-AI59576 and R01-AI25231. The authors thank Christian Pontillo and Karen Edeen for excellent technical assistance.

5. REFERENCES

1. L. van der Hoek, K. Pyrc, M. F. Jebbink, W. Vermeulen-Oost, R. J. Berkhout, K. C. Wolthers, P. M. Wertheim-van Dillen, J. Kaandorp, J. Spaargaren, and B. Berkhout, Identification of a new human coronavirus, *Nat. Med.* **10**, 368-373 (2004).
2. R. A. Fouchier, N. G. Hartwig, T. M. Bestebroer, B. Niemeyer, J. C. de Jong, J. H. Simon, and A. D. Osterhaus, A previously undescribed coronavirus associated with respiratory disease in humans, *Proc. Natl. Acad. Sci. USA* **101**, 6212-6216 (2004).
3. P. C. Woo, S. K. Lau, C. M. Chu, K. H. Chan, H. W. Tsoi, Y. Huang, B. H. Wong, R. W. Poon, J. J. Cai, W. K. Luk, et al., Characterization and complete genome sequence of a novel coronavirus, coronavirus HKU1, from patients with pneumonia, *J. Virol.* **79**, 884-895 (2005).
4. D. O'Toole, I. Brown, A. Bridges, and S. F. Cartwright, Pathogenicity of experimental infection with 'pneumotropic' porcine coronavirus, *Res. Vet. Sci.* **47**, 23-29 (1989).
5. J. Storz, X. Lin, C. W. Purdy, V. N. Chouljenko, K. G. Kousoulas, F. M. Enright, W. C. Gilmore, R. E. Briggs, and R. W. Loan, Coronavirus and Pasteurella infections in bovine shipping fever pneumonia and Evans' criteria for causation, *J. Clin. Microbiol.* **38**, 3291-3298 (2000).
6. K. Erles, C. Toomey, H. W. Brooks, and J. Brownlie, Detection of a group 2 coronavirus in dogs with canine infectious respiratory disease, *Virology* **310**, 216-223 (2003).
7. P. N. Bhatt, D. H. Percy, and A. M. Jonas, Characterization of the virus of sialodacryoadenitis of rats: a member of the coronavirus group, *J. Infect. Dis.* **126**, 123-130 (1972).
8. Y. L. Lai, R. O. Jacoby, P. N. Bhatt, and A. M. Jonas, Keratoconjunctivitis associated with sialodacryoadenitis in rats, *Invest. Ophthalmol.* **15**, 538-541 (1976).

9. K. Utsumi, Y. Yokota, T. Ishikawa, K. Ohnishi, and K. Fujiwara, in: *Coronaviruses and Their Diseases*, edited by D. Cavanagh and T. D. K. Brown, (Plenum Press, New York, 1990), pp. 525-532.
10. Z. W. Wojcinski and D. H. Percy, Sialodacryoadenitis virus-associated lesions in the lower respiratory tract of rats, *Vet. Pathol.* **23**, 278-286 (1986).
11. J. C. Parker, S. S. Cross, and W. P. Rowe, Rat coronavirus (RCV): a prevalent, naturally occurring pneumotropic virus of rats, *Arch. Gesamte. Virusforsch.* **31**, 293-302 (1970).
12. P. N. Bhatt and R. O. Jacoby, Experimental infection of adult axenic rats with Parker's rat coronavirus, *Arch. Virol.* **54**, 345-352 (1977).
13. S. Gagneten, C. A. Scanga, G. S. Dveksler, N. Beauchemin, D. Percy, and K. V. Holmes, Attachment glycoproteins and receptor specificity of rat coronaviruses, *Lab. Anim. Sci.* **46**, 159-166 (1996).
14. L. G. Dobbs and R. J. Mason, Pulmonary alveolar type II cells isolated from rats. Release of phosphatidylcholine in response to beta-adrenergic stimulation, *J. Clin. Invest.* **63**, 378-387 (1979).
15. R. J. Mason, M. C. Lewis, K. E. Edeen, K. McCormick-Shannon, L. D. Nielsen, and J. M. Shannon, Maintenance of surfactant protein A and D secretion by rat alveolar type II cells in vitro, *Am. J. Physiol. Lung Cell. Mol. Physiol.* **282**, L249-258 (2002).
16. T. M. Gallagher and M. J. Buchmeier, Coronavirus spike proteins in viral entry and pathogenesis, *Virology* **279**, 371-374 (2001).
17. S. T. Hingley, I. Leparc-Goffart, S. H. Seo, J. C. Tsai, and S. R. Weiss, The virulence of mouse hepatitis virus strain A59 is not dependent on efficient spike protein cleavage and cell-to-cell fusion, *J. Neurovirol.* **8**, 400-410 (2002).
18. I. Hamming, W. Timens, M. L. Bulthuis, A. T. Lely, G. J. Navis, and H. van Goor, Tissue distribution of ACE2 protein, the functional receptor for SARS coronavirus. A first step in understanding SARS pathogenesis, *J. Pathol.* **203**, 631-637 (2004).
19. S. A. Jeffers, S. M. Tusell, L. Gillim-Ross, E. M. Hemmila, J. E. Achenbach, G. J. Babcock, W. D. Thomas, Jr., L. B. Thackray, M. D. Young, R. J. Mason, et al., CD209L (L-SIGN) is a receptor for severe acute respiratory syndrome coronavirus, *Proc. Natl. Acad. Sci. USA* **101**, 15748-15753 (2004).
20. K. F. To, J. H. Tong, P. K. Chan, F. W. Au, S. S. Chim, K. C. Chan, J. L. Cheung, E. Y. Liu, G. M. Tse, A. W. Lo, et al., Tissue and cellular tropism of the coronavirus associated with severe acute respiratory syndrome: an in-situ hybridization study of fatal cases, *J. Pathol.* **202**, 157-163 (2004).
21. K. F. To and A. W. Lo, Exploring the pathogenesis of severe acute respiratory syndrome (SARS): the tissue distribution of the coronavirus (SARS-CoV) and its putative receptor, angiotensin-converting enzyme 2 (ACE2), *J. Pathol.* **203**, 740-743 (2004).
22. B. L. Haagmans, T. Kuiken, B. E. Martina, R. A. Fouchier, G. F. Rimmelzwaan, G. van Amerongen, D. van Riel, T. de Jong, S. Itamura, K. H. Chan, et al., Pegylated interferon-alpha protects type 1 pneumocytes against SARS coronavirus infection in macaques, *Nat. Med.* **10**, 290-293 (2004).

INFECTIOUS BRONCHITIS CORONAVIRUS INDUCES CELL-CYCLE PERTURBATIONS

Brian K. Dove, Katrina Bicknell, Gavin Brooks, Sally Harrison, and Julian A. Hiscox*

1. INTRODUCTION

The term 'cell cycle' is a generic description comprising the various stages that actively replicating cells undergo to proliferate. The cell cycle of replicating mammalian cells can be divided into five distinct phases[1]: gap 1 and gap 2 (G_1 and G_2) where cells undergo RNA and protein synthesis, synthesis (S) phase where cellular DNA replication occurs, and mitosis (M) phase followed by cytokinesis (cell division) (Fig. 1). Cells not undergoing replication and within a quiescent state are described as being in G_0 phase.

Viruses from a diverse range of families have been shown to be able to perturb the cell cycle of infected cells,[2-6] including the coronavirus mouse hepatitis virus, which induces a G_1 arrest.[7] Although the primary site of coronavirus replication is the cytoplasm, localization and interactions of coronavirus proteins with nuclear and sub-nuclear structures and proteins have been reported.[8-11] Therefore, we hypothesize that avian infectious bronchitis virus (IBV), a group 3 member of the coronavirus family, would induce cell-cycle perturbations as a consequence of virus infection.

To investigate this, we utilized dual-label flow cytometric analysis to accurately gate cells in the G_0/G_1, S and G_2/M phases of the cell cycle.[12,13] Prototypic single-color flow cytometric analysis of cycling cells uses propidium iodide (PI) to stain the total cellular DNA content of individual cells. This allows measurement of the percentage of cells in the G_0/G_1 (2N DNA content), G_2/M (4N DNA content), and S (intermediate DNA content) phases within a cell population. However, application of a dual-label approach, by the addition of thymidine analogue bromodeoxyuridine (BrdU) incorporated into DNA during cellular DNA synthesis allows increased discrimination and measurement of cell populations into the G_0/G_1, S, and G_2/M phases.

* Brian K. Dove, Sally Harrison, Julian A. Hiscox, University of Leeds, Leeds LS2 9JT United Kingdom. Katernia Bicknell, Gavin Brooks, University of Reading, Reading RG6 6AJ, United Kingdom.

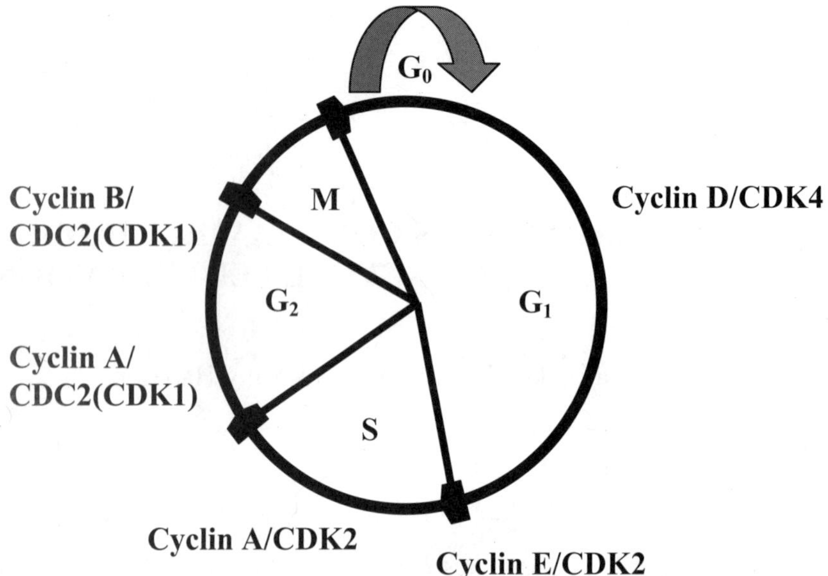

Figure 1. The five cellular phases that compr. the mammalian cell cycle. Positive control is regulated by the formation of specific cyclin/cyclin-dependent kinase (CDK) complexes, required at certain stages of the cell cycle for progression to occur.

2. MATERIALS AND METHODS

2.1. IBV Infections

Actively replicating asynchronous Vero cells (an African green monkey kidney-derived kidney cell line) were infected with either IBV B-US strain at an MOI of 1, ultraviolet (UV) inactivated IBV B-US, or mock infected. Cells were analyzed at the appropriate time postinfection (pi) by dual-label flow cytometric analysis.

2.2. Dual-Labeled Flow Cytometric Analysis

Two-color flow cytometric analysis was used to accurately determine the cell-cycle profile of both mock and infected cell populations. BrdU was added to cell medium within each flask 30 minutes prior to fixing the cells. BrdU-labeled DNA was detected by addition of mouse-anti-BrdU antibody followed by anti-mouse FITC antibody. The PI stain was then applied and the cell populations were analyzed for PI staining and BrdU incorporation using a FACS Calibur analyzer (Becton Dickinson) and the percentage of cells in the G_0/G_1, S, or G_2/M phases in each sample gated using CellQuest software (Becton Dickinson).

2.3. Cyclin-Dependent Kinase Assay

Cyclin-dependent kinase 2 (CDK2) and CDC2 (CDK1) associated complexes were immunoprecipitated from mock and IBV-infected Vero cells at 0, 8, 16, and 24 hr post-infection. Complexes were immunoprecipitated from total protein extracts using anti-CDK2 polyclonal antibody or anti-cyclin B1 monoclonal antibody, respectively. Associated kinase activities were measured using histone H1 as substrate. Phosphorylation of histone H1 was analyzed using a Molecular Dynamics PhosphorImager SI and quantified using the ImageQuant software package (Molecular Dynamics).

2.4. Western Blot Analysis of Cellular Proteins

Total cellular protein, extracted from mock and IBV-infected Vero cells at various times pi, was separated on a 10% Novex Bis-Tris polyacryamide pre-cast gel in MES SDS running buffer (Invitrogen). Western blotting was performed using ECL (Amersham/Pharmacia) as described in the manufacturer's instructions.

3. RESULTS AND DISCUSSION

To determine any IBV-infection induced cell-cycle perturbations, asynchronously replicating Vero cells were infected with both IBV and UV-inactivated IBV as well as mock-infected. Dual-label flow cytometric analysis determined that at 24hr pi, there was a significant increase ($p < 0.001$, n-3) in the number of cells in the G_2/M phase of the cell cycle compared with both mock-infected cells and cells infected with UV-inactivated virus. There was also a significant decrease in the number of cells in the G_0/G_1 phase of the cell cycle in infected cells compared with the controls ($p < 0.001$, n-3). While not discussed in this chapter, a similar G_2/M phase arrest was also observed in IBV-infected BHK cells, another IBV-permissible cell line.

Studies also focused on identifying any IBV-induced perturbations and interactions with specific cell-cycle factors. Both CDK2 and CDC2 kinase assays were performed to determine any alteration in the level of kinase activity of cyclinB/CDC2 complexes and of both cyclin A/CDK2 and E/CDK2 complexes in IBV infected cells. Figure 3 shows that while kinase activity fluctuated over time in both mock and infected cells, at 24hr pi CDK2 activity was reduced in IBV-infected Vero cells compared with mock-infected. Therefore, infection with IBV resulted in a reduction of either/or cyclin E/CDK2 and cyclin A/CDK2 activity. This reduction in activity in infected cells may be due to a reduced number of cells cycling through the cell cycle as a consequence of the virus-induced block. Western blot analysis was also performed on a variety of both positive and negative cellular cell

Table 1. IBV-infected Vero cell cycle profile.

Infectious state	Cell-cycle phase percentages		
	G_0/G_1	S	G_2/M
Mock infected	69.9 +/-0.7	21.3 +/-1.0	8.8 +/-1.7
U.V. inactive IBV	69.2 +/-1.0	22.7 +/-0.6	8.0 +/-0.5
IBV infected	59.6 +/-1.2*	17.8 +/-2.1	22.3 +/-1.8*

Specific cell-cycle stage percentages were calculated from BrdU/PI dual-label dot blots (Figure 2) +/- standard deviation. *, significantly different from corresponding mock infected and UV-inactivated cell controls ($p < 0.001$; n-3).

cycle regulatory factors. Analysis of mock-infected and IBV-infected Vero cell lysates for cyclins D1, D2, A, E, and B determined that at 16 and 24 hr a 10-and 16-fold reduction, respectively, of both cyclin D1 and D2 expression in IBV-infected cells, compared with mock infected cells, was detectable (data not shown). A reduction in cyclin D expression was also reported by Chen et al.[7] Analysis of cyclin A, E, and B expression, while undergoing slight perturbations, demonstrated no significant change in the level of expression between mock-infected and IBV-infected Vero cells. Western blot analysis of the negative cellular regulatory protein p21 and tumor suppressor protein p53 also demonstrated no significant alteration in the levels of p21 and p53 expressed between mock and infected cells (data not shown).

Our data indicate that IBV infection induces a G_2/M phase arrest or delay within infected cells. While Chen et al.[7] reported that MHV induced a G_0/G_1 phase arrest in cells, the difference in cell-cycle arrest state between IBV and MHV could simply be due to the inherent differences between the viruses or the type of cell-cycle analysis used. Currently, work is focused on the development of a model to explain how IBV induces a G_2/M phase arrest and to determine what physiological advantage a G_2/M phase arrest confers to IBV infection. Preliminary work indicates both an increase of viral protein production and progeny virus output in cells G_2/M synchronized compared with G_0/G_1 synchronized cells and asynchronously replicating cells (data not shown). Therefore, identification of the mechanism or mechanisms IBV manipulates to induce a cell-cycle arrest could lead to further insight into the mechanism of IBV replication.

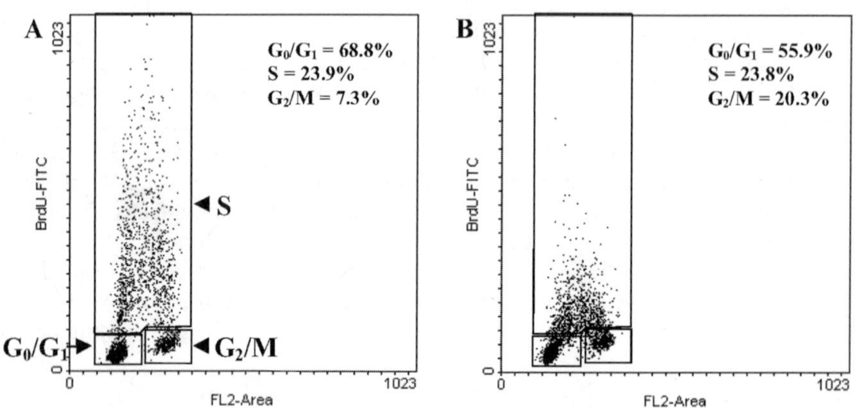

Figure 2. Representative BrdU-labeled (BrdU-FITC) / PI-stained (FL2-area) dot blot cell-cycle profiles of mock-infected (A) and IBV-infected (B) Vero cells at 20 hr pi. Utilizing a dual-label approach allows accurate gating of a cell population into the G_0/G_1, S, and G_2/M phases of the cell cycle.

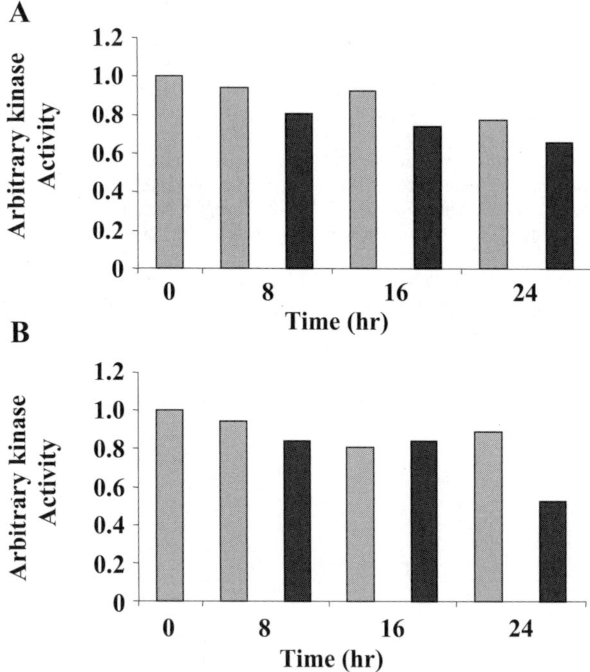

Figure 3. The level of histone H1 kinase activity from CDC2- (A) and CDK2- (B) associated complexes immunoprecipitated from total cellular protein extracted from mock-infected and IBV-infected Vero cells at 0, 8, 16, and 24 hr postinfection was quantified to calculate arbitrary kinase activity in mock-infected (gray) and IBV-infected cells (black). IBV infection results in a reduction of CDK2 kinase activity.

4. ACKNOWLEDGMENTS

This work was funded by the BBSRC (grant number BBS/B/03416) to J.A.H. and G.B. We would like to acknowledge the help and expertise of Graham Botley of the University of Leeds and Carmen Coxon and Jane Harper of the University of Reading with the flow cytometry. The flow cytometer at the University of Leeds was provided by a grant from Yorkshire Cancer Research.

5. REFERENCES

1. J. V. Harper and G. Brooks, The mammalian cell cycle: an overview, *Meth. Mol. Biol.* **296**, 113-153 (2005).
2. K. L. Tyler, P. Clarke, R. L. DeBiasi, D. Kominsky, and G. J. Poggioli, Reoviruses and the host cell, *Trends Microbiol.* **9**, 560-564 (2001).
3. A. Op De Beeck and P. Caillet-Fauquet, Viruses and the cell cycle, *Prog. Cell Cycle Res.* **3**, 1-19 (1997).
4. E. K. Flemington, Herpesvirus lytic replication and the cell cycle: arresting new developments, *J. Virol.* **75**, 4475-4481 (2001).
5. D. Naniche, S. I. Reed, and M. B. Oldstone, Cell cycle arrest during measles virus infection: a G0-like block leads to suppression of retinoblastoma protein expression, *J. Virol.* **73**, 1894-1901 (1999).
6. B. Groschel and F. Bushman, Cell cycle arrest in G2/M promotes early steps of infection by human immunodeficiency virus, *J. Virol.* **79**, 5695-5704 (2005).

7. C. J. Chen and S. Makino, Murine coronavirus replication induces cell cycle arrest in G0/G1 phase, *J. Virol.* **78**, 5658-5669 (2004).
8. J. A. Hiscox, T. Wurm, L. Wilson, P. Britton, D. Cavanagh, and G. Brooks, The coronavirus infectious bronchitis virus nucleoprotein localizes to the nucleolus, *J. Virol.* **75**, 506-512 (2001).
9. T. Wurm, H. Chen, T. Hodgson, P. Britton, G. Brooks, and J. A. Hiscox, Localization to the nucleolus is a common feature of coronavirus nucleoproteins, and the protein may disrupt host cell division, *J. Virol.* **75**, 9345-9356 (2001).
10. H. Chen, T. Wurm, P. Britton, G. Brooks, and J. A. Hiscox, Interaction of the coronavirus nucleoprotein with nucleolar antigens and the host cell, *J. Virol.* **76**, 5233-5250 (2002).
11. X. Yuan, Z. Yao, Y. Shan, B. Chen, Z. Yang, J. Wu, Z. Zhao, J. Chen, and Y. Cong, Nucleolar localization of non-structural protein 3b, a protein specifically encoded by the severe acute respiratory syndrome coronavirus, *Virus Res.* (2005).
12. F. Dolbeare, H. Gratzner, M. G. Pallavicini, and J. W. Gray, Flow cytometric measurement of total DNA content and incorporated bromodeoxyuridine, *Proc. Natl. Acad. Sci. USA* **80**, 5573-5577 (1983).
13. R. Nunez, DNA measurement and cell cycle analysis by flow cytometry, *Curr. Issues Mol. Biol.* **3**, 67-70 (2001).

GENES 3 AND 5 OF INFECTIOUS BRONCHITIS VIRUS ARE ACCESSORY PROTEIN GENES

Paul Britton, Rosa Casais, Teri Hodgson, Marc Davis, and Dave Cavanagh*

1. INTRODUCTION

Avian infectious bronchitis virus (IBV), a group 3 member of the genus *Coronavirus*, is a highly infectious pathogen of domestic fowl that replicates primarily in the respiratory tract but also in epithelial cells of the gut, kidney, and oviduct.[1-3] Interspersed amongst the IBV structural protein genes are two genes, 3 and 5 (Fig. 1),[4] whose role is unknown.[5-7] Gene 3 is functionally tricistronic,[8] expressing three proteins, 3a, 3b, and 3c, the latter being the structural E protein of IBV.[9] Expression studies have indicated that translation of the E protein is initiated as a result of ribosomes binding to a structure formed by the preceding 3a and 3b sequences.[10,11] Gene 5 is functionally bicistronic and expresses two proteins, 5a and 5b, which are expressed in IBV-infected cells.[12] To investigate the requirement for the 3a, 3b, 5a, and 5b proteins for replication, we have used our reverse genetic system[13-17] to produce isogenic recombinant IBVs (rIBVs), after site-specific mutagenesis of the appropriate sequences, with specific modifications in genes 3 and 5.

2. MATERIALS AND METHODS

2.1. Modification of IBV cDNAs by PCR Mutagenesis

Overlapping PCR mutagenesis was used to scramble the initiation codons of 3a, 3b, 5b, delete the 3ab coding sequences and introduce a *Kpn*I restriction endonuclease upstream of the gene 5 TAS. The scrambled gene 5 TASs and scrambled 5a initiation codon were introduced using adapters to replace the 45 bp *Kpn*I-*Spe*I fragment. The modified sequences comprising two scrambled initiation codons ScAUG3ab and ScAUG5ab were generated from sequences containing a singly scrambled ATG. The modified cDNAs are shown in Fig. 2.

---

* Institute for Animal Health, Compton Laboratory, Newbury, Berkshire RG20 7NN, United Kingdom.

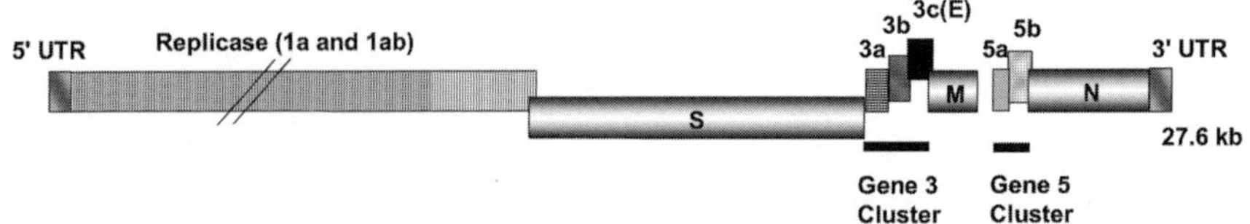

Figure 1. Schematic diagram of the IBV genome indicating the positions of genes 3 and 5.

2.2. Generation of rVVs with Modified IBV Full-Length cDNAs

Our IBV reverse genetics system is based on the use of vaccinia virus (VV) as a vector for the IBV full-length cDNA.[14] Recombinant VVs (rVV) containing the gene 3 and 5 modified cDNA sequences were generated by transient dominant selection (TDS)[18] using the *Eco gpt* (GPT) gene as the transient selectable marker.[13, 16] The modified IBV cDNAs were inserted into the Beaudette sequences in vNotI/IBV_{FL} as a result of homologous recombination and selection of rVVs expressing GPT in the presence of mycophenolic acid (MPA). MPA-sensitive vaccinia viruses, potentially containing the modified IBV cDNAs, were then generated from the MPA-resistant vaccinia viruses after the spontaneous loss of the GPT gene by three rounds of plaque purification in the absence of MPA.[13, 16] Two rVVs, representing each modification, identified by PCR amplification and sequence analysis, were isolated after two independent TDSs.

2.3. Recovery of Recombinant IBVs

Recombinant IBVs, containing each of the modified gene 3 and 5 sequences, were recovered from DNA isolated from the rVVs as shown in Fig. 3 and described in Refs.[13–16.] Recombinant IBVs were characterized and used for subsequent experiments after three passages in CK cells. Two independent clones of each rIBV were rescued from each of the two rVV DNAs, except for rIBVs ScAUG3b and ScAUG3ab, for which only one rIBV was recovered.

2.4. Growth Kinetics of rIBV

The growth kinetics of the rIBVs were analyzed on chick kidney (CK) cells, and the amounts of progeny virus produced, at specific time points, were determined by plaque titration in CK cells and compared with those produced from Beau-R.

3. RESULTS AND DISCUSSION

IBV and the coronaviruses isolated from other avian species, turkey,[19] pheasant,[20] peafowl (accession no. AY641576) and partridge (accession no. AY646283), all contain a tricistronic gene 3 and a bicistronic gene 5, the latter located between the M and N genes. The conservation of the gene 3 and 5 sequences in IBV and IBV-like viruses, isolated from other avian species, indicate they may play a role in the virus replication cycle. In order to determine whether the 3a, 3b, 5a, and 5b proteins are required for the replication of IBV, we have used a number of alternative ways to oblate the expression

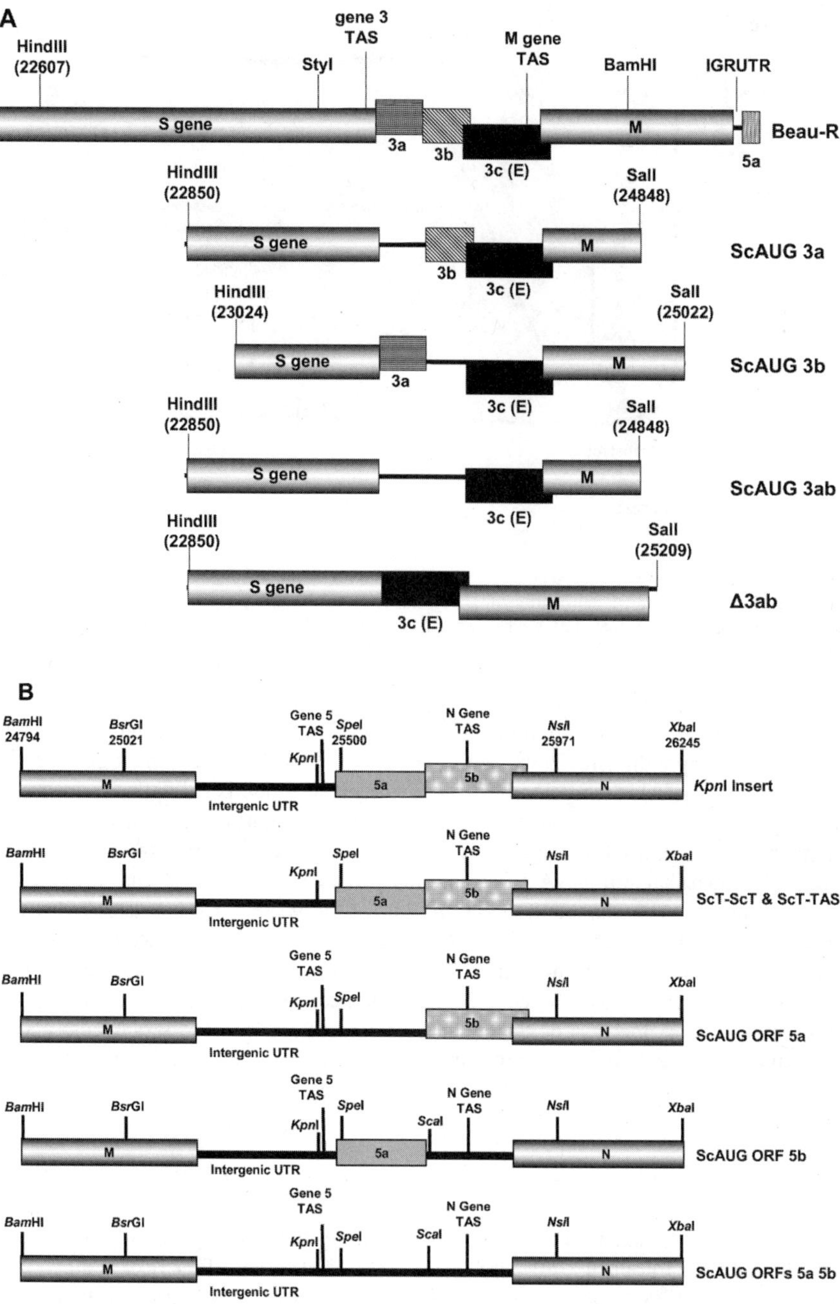

Figure 2. Summary of the modified (A) gene 3 and (B) gene 5 sequences. The positions of the IBV genes are shown with the horizontal black lines indicating that the coding sequences are retained but that translation of the gene product is lost. ScAUG-scrambled initiation codon. ScT-scrambled transcription associated sequence.

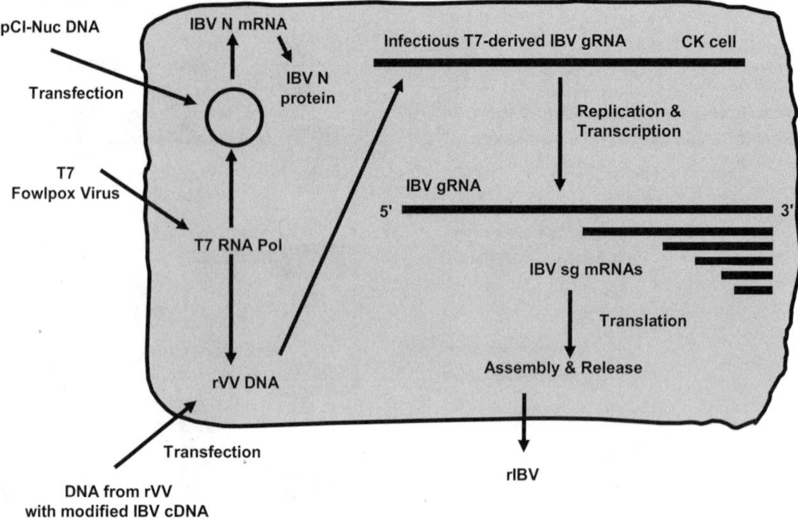

Figure 3. Schematic diagram representing the recovery of rIBV from DNA isolated from a rVV containing a full-length IBV cDNA under the control of a T7 promoter. A plasmid expressing the IBV nucleoprotein is required for successful rescue of IBV. The infectious IBV RNA is generated using T7 RNA polymerase expressed from fowlpox virus.

of these gene products. We modified the IBV genome corresponding to gene 3 by scrambling the 3a and 3b initiation codons, either singly or together, and by deleting the sequence corresponding to 3a3b. We have shown that 3a is no longer produced after scrambling of the AUG or deletion of the sequence (Unpublished data, Hodgson *et al.*). We modified gene 5 by scrambling the 5a and 5b initiation codons, either singly or together, and by scrambling the sg mRNA 5 TAS preventing expression of the sg mRNA. We have shown that sg mRNA 5 is no longer produced after scrambling of the TAS and that 5b is no longer produced after scrambling of the AUG.[16] Comparison of the growth kinetics of the rIBV with Beau-R, on CK cells, showed that there were no differences (Fig. 4), demonstrating that neither the IBV 3a, 3b, 5a, nor 5b proteins are essential for replication *per se*; they can be considered to be accessory proteins. We have rescued a rIBV that lacks expression of 3a and 3b, after deletion of their sequences, and lacks expression of 5a and 5b after scrambling of the gene 5 TAS indicating that both sets of gene products are dispensable *in vitro*.

4. ACKNOWLEDGMENTS

This work was supported by the Biotechnology and Biological Sciences Research Council (BBSRC) grant no. 201/15836 and the Department of Environment, Food and Rural Affairs (DEFRA) project code OD0712. T. Hodgson was supported by a graduate training award from the BBSRC.

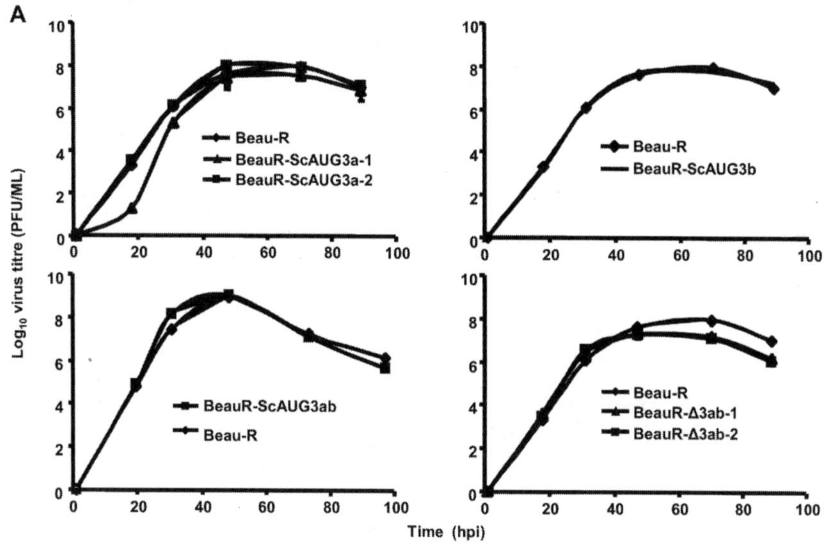

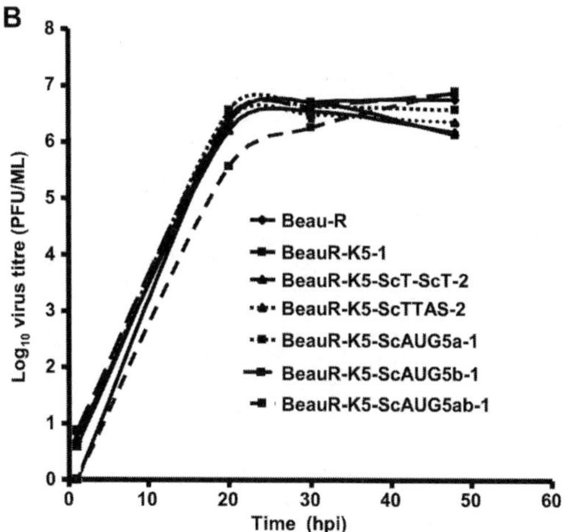

Figure 4. Comparison of the multistep growth kinetics of the (A) gene 3 and (B) gene 5 modified rIBVs on CK cells.

5. REFERENCES

1. J. K. Cook, J. Chesher, W. Baxendale, N. Greenwood, M. B. Huggins, and S. J. Orbell, Protection of chickens against renal damage caused by a nephropathogenic infectious bronchitis virus, *Avian Pathol.* **30**, 423-426 (2001).
2. D. Cavanagh, A nomenclature for avian coronavirus isolates and the question of species status, *Avian Pathol.* **30**, 109-115 (2001).
3. D. Cavanagh and S. Naqi, in: *Diseases of Poultry*, edited by Y. M. Saif, H. J. Barnes, J. R. Glisson, A. M. Fadly, L. R. McDougald, and D. E. Swayne (Iowa State University Press, Ames, 2003), pp. 101-119.

4. M. E. G. Boursnell, T. D. K. Brown, I. J. Foulds, P. F. Green, F. M. Tomley, and M. M. Binns, Completion of the sequence of the genome of the coronavirus avian infectious bronchitis virus, *J. Gen. Virol.* **68**, 57-77 (1987).
5. T. D. K. Brown and I. Brierly, in: *The Coronaviridae*, edited by S. G. Siddell (Plenum Press, New York, 1995), pp. 191-217.
6. M. M. Lai and D. Cavanagh, The molecular biology of coronaviruses, *Adv. Virus Res*. **48**, 1-100 (1997).
7. W. Luytjes, in: *The Coronaviridae*, edited by S. G. Siddell (Plenum Press, New York, 1995), pp. 33-54.
8. D. X. Liu, D. Cavanagh, P. Green, and S. C. Inglis, A polycistronic mRNA specified by the coronavirus infectious bronchitis virus, *Virology* **184**, 531-544 (1991).
9. A. R. Smith, M. E. G. Boursnell, M. M. Binns, T. D. K. Brown, and S. C. Inglis, Identification of a new membrane associated polypeptide specified by the coronavirus infectious bronchitis virus, *J. Gen. Virol.* **71**, 3-11 (1990).
10. D. X. Liu and S. C. Inglis, Internal entry of ribosomes on a tricistronic mRNA encoded by infectious bronchitis virus, *J. Virol.* **66**, 6143-6154 (1992).
11. S. Y. Le, N. Sonenberg, and J. V. Maizel, Jr., Distinct structural elements and internal entry of ribosomes in mRNA3 encoded by infectious bronchitis virus, *Virology* **198**, 405-411 (1994).
12. D. X. Liu and S. C. Inglis, Identification of two new polypeptides encoded by mRNA5 of the coronavirus infectious bronchitis virus, *Virology* **186**, 342-347 (1992).
13. P. Britton, S. Evans, B. Dove, M. Davies, R. Casais, and D. Cavanagh, Generation of a recombinant avian coronavirus infectious bronchitis virus using transient dominant selection, *J. Virol. Methods* **123**, 203-211 (2005).
14. R. Casais, V. Thiel, S. G. Siddell, D. Cavanagh, and P. Britton, Reverse genetics system for the avian coronavirus infectious bronchitis virus, *J. Virol.* **75**, 12359-12369 (2001).
15. R. Casais, B. Dove, D. Cavanagh, and P. Britton, Recombinant avian infectious bronchitis virus expressing a heterologous spike gene demonstrates that the spike protein is a determinant of cell tropism, *J. Virol.* **77**, 9084-9089 (2003).
16. R. Casais, M. Davies, D. Cavanagh, and P. Britton, Gene 5 of the avian coronavirus infectious bronchitis virus is not essential for replication, *J. Virol.* **79**, 8065-8078 (2005).
17. T. Hodgson, R. Casais, B. Dove, P. Britton, and D. Cavanagh, Recombinant infectious bronchitis coronavirus Beaudette with the spike protein gene of the pathogenic M41 strain remains attenuated but induces protective immunity, *J. Virol.* **78**, 13804-13811 (2004).
18. F. G. Falkner and B. Moss, Transient dominant selection of recombinant vaccinia viruses, *J. Virol.* **64**, 3108-3111 (1990).
19. D. Cavanagh, K. Mawditt, M. Sharma, S. E. Drury, H. L. Ainsworth, P. Britton, and R. E. Gough, Detection of a coronavirus from turkey poults in Europe genetically related to infectious bronchitis virus of chickens, *Avian Pathol.* **30**, 355-368 (2001).
20. D. Cavanagh, K. Mawditt, D. d. B. Welchman, P. Britton, and R. E. Gough, Coronaviruses from pheasants (*Phasianus colchicus*) are genetically closely related to coronaviruses of domestic fowl (infectious bronchitis virus) and turkeys, *Avian Pathol.* **31**, 81-93 (2002).

NKG2D SIGNALING AND HOST DEFENSE AFTER MOUSE HEPATITIS VIRUS INFECTION OF THE CENTRAL NERVOUS SYSTEM

Kevin B. Walsh, Melissa B. Lodoen, Lewis L. Lanier, and Thomas E. Lane*

1. INTRODUCTION

Natural killer (NK) cells and $CD8^+$ T cells exhibit potent anti-viral effector responses after infection with a wide variety of viruses. Through secretion of cytokines as well as cytolytic activity, these cells are capable of muting viral replication and reducing the amount of virus present within infected tissue. In recent years, additional activating and inhibitory receptors have been discovered that control both NK cell and $CD8^+$ T-cell activation in response to stress, transformation, and/or infection. In this study, the expression and functional role of the activating receptor NKG2D and its corresponding ligands were determined in viral infection of the central nervous system (CNS). Our data highlight a previously unappreciated role for NKG2D ligand-induced signaling in host defense after viral infection of the CNS by enhancing T-cell effector function and regulating immune-cell trafficking.

2. THE NKG2D RECEPTOR AND ITS LIGANDS

NKG2D is a type II transmembrane-anchored glycoprotein that is a member of the C-type lectin superfamily.[1] NKG2D receptor ligation by cell-surface glycoprotein ligands that are structurally related to MHC class I molecules results in activation of lymphocytes such as natural killer (NK) cells, $CD8^+$ T cells, and $\gamma\delta$-TcR^+ T cells.[2,3] The known mouse NKG2D ligands include the retinoic acid early inducible-1 (RAE-1) proteins (RAE-1 α, β, γ, δ, and ε)[4], minor histocompatibility antigen H60,[4] and murine UL16-binding protein-like transcript-1 (MULT1) glycoprotein,[5] which are expressed by cells undergoing cellular stress such as viral infection or transformation.

*Kevin B. Walsh, Thomas E. Lane, University of California, Irvine, California 92697-3900. Melissa B. Lodoen, Lewis L. Lanier, University of California, San Francisco, California 94143-0414.

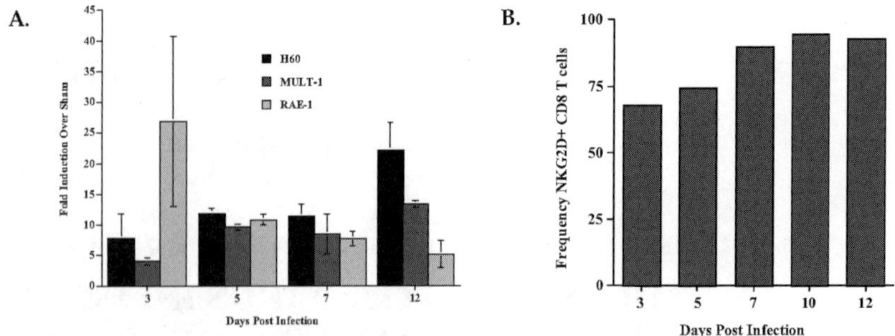

Figure 1. (A) Increased expression of NKG2D ligand mRNA transcripts within the brain after infection with MHV as determined by quantitative PCR. Ligand transcripts were not detected in the brains of sham-infected mice. (B) Flow cytometric analysis demonstrating that the majority of $CD8^+$ T cells infiltrating into the brain express the NKG2D receptor after MHV infection. Data presented represent the average frequency (percentage) of CD8+ T cells present within the brain on which the NKG2D receptor was detected. Data shown represent two separate experiments with a minimum of three mice per time point.

3. MHV INFECTION OF THE CNS

Mouse hepatitis virus (MHV) is a positive-strand RNA virus that is a member of the *Coronaviridae* family.[6] Instillation of MHV into the central nervous system (CNS) of susceptible mice results in an acute encephalomyelitis, followed by a chronic demyelinating disease.[7] Both $CD4^+$ and $CD8^+$ T cells are important in reducing viral burden through secretion of anti-viral cytokines, such as interferon gamma.[8, 9] In addition, cytolytic activity by virus-specific $CD8^+$ T cells is also necessary for viral clearance.[10]

4. NKG2D RECEPTOR AND LIGAND EXPRESSION

NKG2D ligand mRNA transcripts were upregulated within the brains at days 3, 5, 7, and 12 after intracranial (i.c.) infection of BALB/c mice with MHV (Figure 1A). Expression of NKG2D receptor was detected on NK cells entering the CNS early after MHV infection. NKG2D receptor was also detected on $CD8^+$ T cells and the frequency of $CD8^+$ T cells expressing the receptor increased over time and ultimately peaked at day 10 p.i. (Figure 1B). These data indicate that NKG2D ligand mRNA and receptor are expressed during the course of MHV infection of the CNS.

5. NKG2D NEUTRALIZATION IN IMMUNOCOMPETENT MICE

NKG2D neutralization in MHV-infected BALB/c mice resulted in a dramatic decrease in survival compared with mice treated with a control antibody (Figure 2A). Mice treated with anti-NKG2D had increased viral titers at 7 and 12 days p.i., when compared with control mice. Moreover, anti-NKG2D treatment did result in a reduction in NK cells, T cells, and macrophages. IFN-γ secretion was also reduced in anti-NKG2D treated mice and

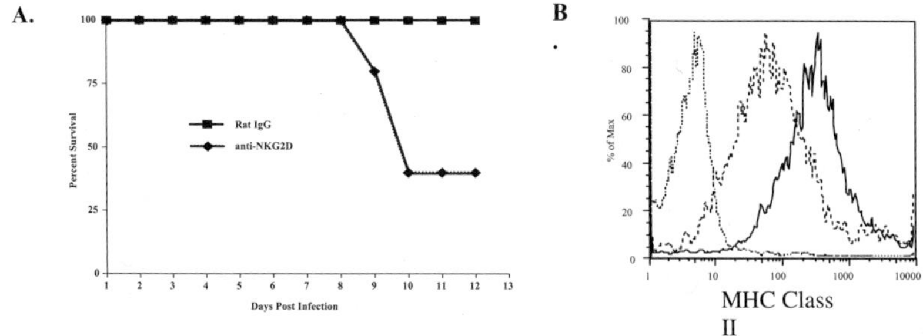

Figure 2. (A) Survival of BALB/c mice infected i.c. with MHV and treated with 100 μg of rat IgG (control) or anti-NKG2D at 2, 5, and 10 days p.i. Data presented represent two separate experiments with a minimum of 10 mice per experimental group. (B) MHC class II expression on microglia (CD45^{low} F480+) at 12 days p.i. in MHV-infected BALB/c mice treated with rat IgG (solid line) or anti-NKG2D (dashed line). Staining from sham-infected mice is indicated by the dotted line. Shown are representative histograms.

this was reflected by a decrease in MHC class II staining of microglia (Figure 2B). These data demonstrate that NKG2D signaling is required for generation of an efficient immune response in the CNS of immunocompetent mice infected with MHV.

6. CONCLUSIONS

This study demonstrated that after MHV infection, NKG2D ligands are up-regulated within the CNS and infiltrating NK cells and CD8$^+$ T cells express the NKG2D receptor. NKG2D neutralization in BALB/c mice resulted in reduced survival and diminished immune cell infiltration accompanied by decreased IFN-γ secretion that correlated with increased viral titers in the CNS. In addition, anti-NKG2D neutralization hampered the host's ability to mount an efficient immune response, demonstrating that NKG2D signaling is important in this model of viral-induced CNS disease.

7. ACKNOWLEDGMENTS

This work was supported by NIH grants CA89189 and CA095137 to L.L.L. T.E.L. was supported by NIH grants NS41249 and NS18146. L.L.L. is an American Cancer Society Research Professor.

8. REFERENCES

1. D. H. Raulet, Roles of the NKG2D immunoreceptor and its ligands, *Nat. Rev. Immunol.* **3**, 781-790 (2003).

2. A. Diefenbach, A. M. Jamieson, S. D. Liu, N. Shastri, and D. H. Raulet, Ligands for the murine NKG2D receptor: expression by tumor cells and activation of NK cells and macrophages, *Nat. Immunol.* **1**, 119-126 (2000).
3. A. M. Jamieson, A. Diefenbach, C. W. McMahon, N. Xiong, J. R. Carlyle, and D. H. Raulet, The role of the NKG2D immunoreceptor in immune cell activation and natural killing, *Immunity* **17**, 19-29 (2002).
4. A. Cerwenka, A. B. Bakker, T. McClanahan, J. Wagner, J. Wu, J. H. Phillips, and L. L. Lanier, Retinoic acid early inducible genes define a ligand family for the activating NKG2D receptor in mice, *Immunity* **12**, 721-727 (2000).
5. L. N. Carayannopoulos, O. V. Naidenko, D. H. Fremont, and W. M. Yokoyama, Cutting edge: murine UL16-binding protein-like transcript 1: a newly described transcript encoding a high-affinity ligand for murine NKG2D, *J. Immunol.* **169**, 4079-4083 (2002).
6. S. R. Perlman, T. E. Lane, and M. J. Buchmeier, in: *Effects of Microbes on the Immune System*, edited by M. W. Cunningham and R. S. Fujinami (Lippincott Williams & Wilkins, Philadelphia, 1999), pp. 331-348.
7. F. S. Cheever, J. B. Daniels, A. M. Pappenheimer, and O. T. Bailey, A murine virus (JHM) causing disseminated encephalomyelitis with extensive destruction of myelin, *J. Exp. Med.* **90**, 181-194 (1949).
8. J. S. Williamson and S. A. Stohlman, Effective clearance of mouse hepatitis virus from the central nervous system requires both CD4+ and CD8+ T cells, *J. Virol.* **64**, 4589-4592 (1990).
9. B. Parra, D. R. Hinton, N. W. Marten, C. C. Bergmann, M. T. Lin, C. S. Yang, and S. A. Stohlman, IFN-gamma is required for viral clearance from central nervous system oligodendroglia, *J. Immunol.* **162**, 1641-1647 (1999).
10. M. T. Lin, S. A. Stohlman, and D. R. Hinton, Mouse hepatitis virus is cleared from the central nervous systems of mice lacking perforin-mediated cytolysis, *J. Virol.* **71**, 383-391 (1997).

MURINE HEPATITIS VIRUS STRAIN 1 AS A MODEL FOR SEVERE ACUTE RESPIRATORY DISTRESS SYNDROME (SARS)

Nadine DeAlbuquerque, Ehtesham Baig, Max Xuezhong, Itay Shalev, M. James Phillips, Marlena Habal, Julian Leibowitz, Ian McGilvray, Jagdish Butany, Eleanor Fish, and Gary Levy*

1. INTRODUCTION

Severe acute respiratory syndrome (SARS) is a novel infectious disorder that was first diagnosed in China in November 2002.[1,2] SARS was documented in approximately 8,000 persons globally with more than 700 deaths. In Canada, there were 375 probable and suspect cases between March and July 2003 with 44 deaths, reflecting a mortality rate of 11%. Spread of SARS was shown to be by airborne droplets and results in acute pulmonary inflammation and epithelial damage.[3] It has now been determined that a novel coronavirus, SARS-CoV, is the etiologic agent in SARS. Based on phylogenetic sequence analysis, it best fits within group 2 coronaviruses, which include the mouse hepatitis viruses (MHV).[4,5]

As for most infections, SARS varies considerably in terms of its clinical severity. This variation is almost certainly due to population-based diversity in the genes controlling the immune response. Clearance of mouse hepatitis virus coincides with a robust innate immune response, including increased numbers of CD8 T cells. Disease and death do not correlate with high viral titers, and it has been suggested that disease reflects alteration in host innate immune response. Furthermore, host production and response to type 1 interferons (IFN) is a key determinant of outcome in MHV-infected mice.[6] However, IFNs and other cytokines regulate in a coordinate manner both inflammation and the Th1/Th2 character of the specific immune response. An imbalance in timing and proportions of cellular responses to inflammatory cytokines after viral infection can lead to chronic disease or death. Although a number of models for SARS have been proposed including SARS-CoV infection of mice, of cats and ferrets, and SARS-CoV infection of non-human primates, none of the models produce lung pathology similar to that seen in

* Nadine DeAlbuquerque, Ehtasham Baig, Max Xuezhong, Itay Shalev, M. James Phillips, Marlena Habal, Ian McGilvray, Jagdish Butany, Eleanor Fish, Gary Levy, University of Toronto, Canada. Julian Leibowitz, Texas A & M University System, College Station, Texas.

humans and serve only as models where agents including neutralizing antibodies, putative vaccines, or anti-virals can be studied for effect on viral replication.[7,8]

MHV-1 was first isolated in 1950, and mice infected by MHV-1 had massive hepatic necrosis on autopsy. On further analysis, it was realized that MHV-1 infected mice were co-infected with *Eperythrozoon coccoides*. When mice were infected with MHV-1 devoid of this bacterium, only mild hepatitis was seen and all mice survived. As described below, the MHV-1 mouse model established in our laboratory offers the potential to provide insights into the pathogenesis of MHV-induced lung injury and the contribution of both the virus and host immune response.

2. MATERIALS AND METHODS

Mice: Female Balb/cJ, A/J, and C3H mice, 6–8 weeks of age, were purchased from Jackson Laboratories and housed in the animal facility of the Toronto General Research and were fed with standard laboratory chow diet and water *ad libitum*.

Virus: MHV-1, MHV-A59, MHV-JHM, MHV-S, and MHV-3 was originally obtained from American Type Culture Collection (ATCC) and plaque purified on monolayers of DBT cells and titered on L2 cells using a standard plaque assay.[9]

Tissue processing: Lungs, spleens, livers, kidneys, small intestines, hearts, and brains were harvested from mice and samples snap frozen and stored in a -80°C freezer or fixed with 10% formalin for further analysis. For detection of fgl2 and fibrin, a standard immunohistochemical system was employed as previously described.

Cytokine assays: Serum cytokine levels were assayed using commercial cytometric bead array kits (BD Biosciences) for IL-6, IL-10, IL-12p70, IFN-γ, TNF-α, MCP-1. Samples were analyzed in triplicate using a BD FACS Calibur flow cytometer.

3. RESULTS

Balb/cJ mice infected with 10^5 pfu of MHV-1 intranasally developed severe pulmonary disease characterized by congestion, pulmonary infiltrates, hyaline membranes, and hemorrhage. In addition to diffuse pulmonary infiltrates, focal deposition of fibrin was seen around small arterial blood vessels and in alveolar spaces with entrapment of platelets. Changes were noted as early as 3 days p.i. and progressed to day 28. Clinically these mice became lethargic, with rapid respiration, but all of the mice survived and pulmonary pathology resolved within 21 days of infection. Balb/cJ mice infected intranasally with MHV-JHM developed neither liver or lung pathology. Although MHV-3 and MHV-A59 produced pulmonary lesions, these were milder than those generated by MHV-1 and did not have the characteristics of lesions caused by SARS. MHV-A59 and MHV-3 infected mice all developed severe hepatic necrosis and died of liver failure by day 10 and thus these strains of MHV do not represent relevant models of SARS (Table 1).

Table 1. Effect of different strains of MHV on lung pathology of Balb/cJ mice.

	MHV strain				
Features of SARS	MHV-1	MHV-3	MHV A59	MHV- S	MHV JHM
Congestion	Marked		X		
Edema			Few patches		
Hyaline membrane	X				
Interstitial thickening	X	X	X	X	Minimal
Airways	Inflammation, bronchopneumonia				
Pattern	Hemorrhage particularly in anterior portion		Changes in subpleural regions, perhaps giant		

These data suggest that MHV-1 induces a pathology most similar to human SARS. Therefore, we infected 3 inbred strains of mice (C57Bl/6J, C3H/HeJ, and A/J mice) that have previously been known to show varying degrees of susceptibility to other strains of MHV. Although C57Bl/6J developed acute pulmonary disease, these mice all survived and the pulmonary lesions resolved by day 21. C3H/HeJ mice showed an intermediate pattern of resistance/susceptibility with 40% of mice dying by day 28. Surviving C3H mice developed pulmonary fibrosis and bronchial hyperplasia (Figure 2). A/J mice all died within 7–10 days post–MHV-1 infection of severe pulmonary disease. Lungs showed 100% consolidated pneumonitis with hyaline membranes, fibrin deposition, and lymphocytic and macrophage infiltration (Figure 1A). We examined lung tissue from both susceptible A/J and resistant C57Bl/6J mice for presence of fgl2 mRNA transcripts by real-time PCR, protein, and fibrin, as fgl2 is known to cause thrombosis. Shown in Figure 1B, fgl2 protein was expressed by inflammatory cells and type 1 pneumocytes in juxtaposition with deposits of fibrin. Lungs from Balb/cJ and C57Bl/6 mice had neither fgl2 or fibrin deposits. Electron micrographs of lung showed virions mostly localized to type 1 pneumocytes and pulmonary macrophages (data not shown). By plaque assay, virus was detected in the lung by 12 hours, reaching maximal levels by 48–72 hours and in A/J mice persisted at high levels until death of animals. Virus was also detected in the lungs of C57Bl/6J mice by 12 hours p.i., reaching maximal levels by day 4 but disappearing by days 7–10 p.i.

Serum was collected from A/J and C57Bl/6J mice pre and post MHV-1 infection and cytokines measured (Table 2). Gene expression levels for the different IFNαs and IFN-ß were assessed in lung tissues from MHV-1 infected C57Bl/6Jand A/J mice, using quantitative real-time PCR. In contrast to the late (36 hr postinfection) and low levels of induction of IFN-αs and IFN-ß in the A/J mice, we observed a robust and sustained IFN-α and IFN-ß gene induction by 12 hr postinfection in the C57BL/6 mice (data not shown). The data suggest a correlation between the kinetics and extent of an IFN response and disease severity.

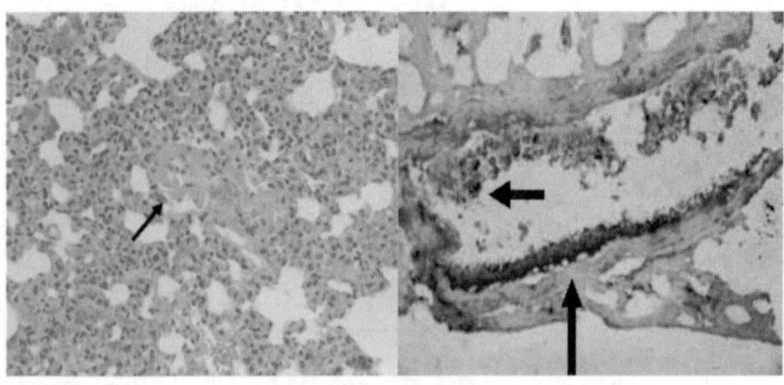

Figure 1. Left: SARS-like lung pathology in A/J mice infected with coronavirus MHV-1, day 7, postinfection. Severe interstitial pneumonitis with hyaline membranes (arrow). Most bronchi remain widely open but the alveolar spaces are completely consolidated. Right: Co-localization of fgl2 and fibrin detected by double immunochemistry staining. Widespread fibrin deposition near fgl2 expression, especially in microvasculature (arrows) of the lung.

Table 2. Serum cytokine profiles in MHV-1 infected mice (pg/mL).

	Day 0 p.i.		Day 6 p.i.	
Cytokines	AJ	C57BL6/J	AJ	C57BL6/J
IL-10	26.7±3	24.6±4	164±22	38±12
IL-6	53.8±4.9	48.6±5	546±23	215±20
IL-12p70	306±30.6	298±29	642.5±36	449.3±21
IFN-γ	14.7±1.4	13.9±2	555±62	236±34
TNF-α	25.2±2.5	18±2	163±22	68±12
MCP-1	166±16	159±16	7400±400	225±40

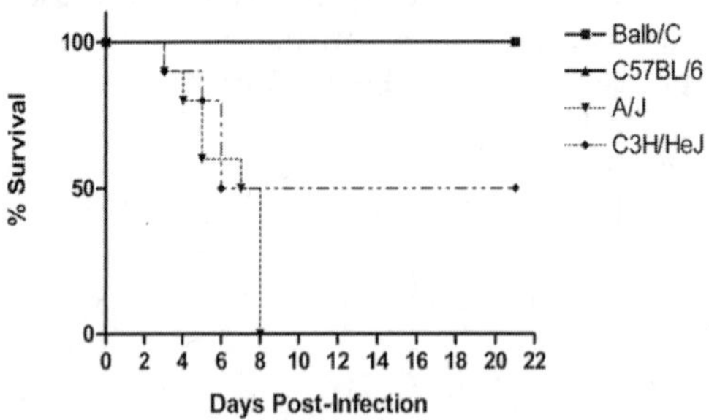

Figure 2. Mortality in MHV-1 infected (1 x 10^5 pfu) intranasally (n-10 per group).

4. DISCUSSION

Existing models for SARS in rodents and non-human primates fail to produce lung pathology or mortality similar to that seen in humans. Thus, these models only serve to assess the effects of agents including neutralizing antibodies, putative vaccines, or anti-virals on viral replication. The MHV-1 model described in this report produces a clinical syndrome in mice that serves as a model for SARS. First, we found that MHV-1 produces strain-dependent disease. Interestingly, A/J mice usually resistant to MHV-3 and MHV-A59 infection all died after intranasal infection with MHV-1. In contrast, although Balb/cJ and C57Bl/6 mice developed pulmonary disease, these animals cleared virus by day 14 and survived. C3H mice developed an intermediate pattern of susceptibility with a 40% mortality. Viral titers were higher in susceptible A/J mice and serum cytokines and chemokines were markedly elevated in these mice in comparison with resistant animals. These findings are consistent with the elevated levels of IFNγ, TNFα, IL-12p70, and IL-8 (CXCL8) detected in sera from SARS patients.[10] These elevated cytokines and chemokines could contribute to the immunopathology of SARS. The fact that corticosteroids ameliorated disease in some SARS patients is consistent with this.

We previously reported that MHV-3 induces fgl2, an inflammatory immune coagulant, which results in fibrin deposition and hepatic necrosis. In the MHV-1 model of SARS, both fgl2 mRNA transcripts and protein were also seen in association with deposits of fibrin in diseased lungs from A/J mice suggesting that this inflammatory mediator may be contributing to the pathogenesis of SARS as well. These results are compatible with what has been reported in humans with SARS CoV.

Susceptible animals failed to generate a robust type 1 interferon response, which, in addition to their anti-viral effects are known to inhibit inflammatory cytokines. IFN alfacon treatment of SARS patients accelerated resolution of inflammation, possibly contributing to increased survival consistent with the above observations. Thus collectively, these data support the concept that the pathogenesis of SARS reflects an altered innate immune response with marked inflammation. The molecular mechanism for these findings is not presently known.

This model offers the potential to conduct additional studies that will provide insights into the pathogenesis of coronavirus-induced lung injury and the contribution of both the virus and host immune response. We anticipate that data generated from these studies will provide novel insights into the pathogenesis of this serious human disease and provide avenues for therapy.

5. ACKNOWLEDGMENTS

This work was supported by the Canadian Institutes for Health Research (grant no. 37780).

6. REFERENCES

1. Lee, N., Hui, D., Wu, A., et al., 2003, A major outbreak of severe acute respiratory syndrome in Hong Kong, *N. Engl. J. Med.* **348**:1986-1994.

2. Poutanen, S. M., Low, D. E., Henry, B., et al., 2003, Identification of severe acute respiratory syndrome in Canada, *N. Engl. J. Med.* **348**:1995-2005.
3. Donnelly, C., Ghani, A., and Leung, G., 2003, Epidemiological determinants of spread of causal agent of severe acute respiratory syndrome in Hong Kong, *Lancet* **361**:1761-1766.
4. Drosten, C., Gunther, S., Preiser, W., et al., 2003 Identification of a novel coronavirus in patients with severe acute respiratory syndrome, *N. Engl. J. Med.* **348**:1967-1976.
5. Gorbalenya, A. E., Snijder, E. J., and Spaan, W. J. M., 2004, Severe acute respiratory syndrome coronavirus phylogeny: toward consensus, *J. Virol.* **78**:7863-7866.
6. Matsuyama, S., Henmi, S., Ichihara, N., Sone, S., Kikuchi, T., Ariga, T., and Taguchi, F., 2000, Protective effects of murine recombinant interferon-beta administered by intravenous, intramuscular or subcutaneous route on mouse hepatitis virus infection, *Antiviral Res.* **47**:131-137.
7. Subbarao, K., McAuliffe, J., Vogel, L., Fahle, G., Fischer, S., Tatti, K., Packard, M., Shieh, W. J., Zaki, S., and Murphy, B., 2004, Prior infection and passive transfer of neutralizing antibody prevent replication of severe acute respiratory syndrome coronavirus in the respiratory tract of mice, *Virology* **78**:3572-3577.
8. Martina, B. E., Haagmans, B. L., Kuiken, T., et al., 2003, Virology: SARS virus infection of cats and ferrets, *Nature* **425**:915.
9. Pachuk, C. J., Bredenbeek, P. J., Zoltick, P. W., et al., 1989, Molecular cloning of the gene encoding the putative polymerase of mouse hepatitis coronavirus, strain A59, *Virology* **171**:141-148.
10. Ward, S. E., Loutfy, M. R., Blatt, L. M., Siminovitch, K. A., Chen, J., Hinek, A., Wolff, B., Pham, D. H., Deif, H., LaMere, E. A., Kain, K. C., Farcas, G. A., Ferguson, P., Latchford, M., Levy, G., Fung, L., Dennis, J. W., Lai, E. K., and Fish, E. N., 2005, Dynamic changes in clinical features and cytokine/chemokine responses in SARS patients treated with interferon alfacon-1 plus corticosteroids, *Antivir. Ther.* **10**:263-275.

PERSISTENT CORONAVIRUS INFECTION OF PROGENITOR OLIGODENDROCYTES

Yin Liu and Xuming Zhang*

1. INTRODUCTION

Mouse hepatitis virus (MHV) is a prototype of murine coronavirus. It can infect rodents and causes enteritis, hepatitis, and central nervous system (CNS) diseases. Infection of mouse CNS with neurovirulent MHV strains usually results in acute encephalitis followed by demyelination.[1,2] If the majority of the virus can be cleared from the CNS, encephalitis then resolves; if mice survive the acute phase, demyelination develops. Although acute demyelination can be detected histologically as early as 6 days postinfection (p.i.), extensive demyelination is often not seen until 4 weeks p.i.[3] However, infectious virus can no longer be isolated from the CNS at this time, although viral RNAs continue to persist in the CNS for more than one year, during which time period demyelination is concomitantly detectable.[4,5] The correlation between viral RNA persistence and demyelination in the CNS suggests that viral persistence may be a prerequisite for the development of CNS demyelination. However, virtually nothing is known as to how viral persistence contributes to demyelination.

Previous studies attempted to establish an *in vitro* system of glial or fibroblast cell culture for viral persistence.[6,7] Unfortunately, the persistent infection established in these cells is productive, i.e., generation of infectious viruses with significant virus titers. This type of persistence does not reflect on the infection of animal CNS. Recently we established a persistent MHV infection in a progenitor rat oligodendrocyte. We showed that MHV RNAs were continuously detected in infected cells of more than 20 passages. However, no infectious virus could be isolated from these cells. This phenomenon resembles the persistent, nonproductive infection in animal CNS. To understand the molecular basis of viral persistence in host cells, we analyzed the gene expression profiles of the persistently infected cells by using DNA microarray technology and RT-PCR. We found that the expression of a substantial number of cellular genes was altered by viral persistence. Interestingly, although persistently infected progenitor cells could be induced to differentiate into mature oligodendrocytes, the number of dendrites and level of myelin basic protein were markedly reduced in persistent cells. This finding indicates

* University of Arkansas for Medical Sciences, Little Rock, Arkansas 72205.

that MHV persistence has an inhibitory effect on oligodendrocyte differentiation and dendrite outgrowth and provides the first direct evidence linking viral persistence to demyelination.

2. MATERIALS AND METHODS

2.1. Cell, Virus, and Reagents

The CG (central glial)-4 cell is a permanent, undifferentiated type 2 oligodendrocyte/astrocyte progenitor cell that was established during a primary neural cell culture derived from the brain of newborn Sprague-Dawley rat pups (1–3 days postnatal).[8] CG-4 cell culture was maintained as described previously.[9] MHV strain JHM was obtained from Michael Lai's laboratory. It was propagated in mouse astrocytoma cell line DBT cells and was used throughout this study. Virus titers were determined by plaque assay as described previously.[9]

2.2. Reverse Transcription–Polymerase Chain Reaction (RT-PCR)

The intracellular RNAs were reverse-transcribed into cDNAs by using a random hexomer oligonucleotide primer (Invitrogen, Inc.), and the cDNAs were amplified by PCR using gene specific primers as described previously.[10] The following gene-specific primer pairs were used in PCR: 5'BamN (5'-TAG GGA TCC ATG TCT TTT GTT CCT-3') and 3'EcoN515 (5'-TAG GAA TTC GGC AGA GGT CCT AG-3') for viral nucleocapsid (N) gene; 5'-cmyc (5'-TTT CTC GAG GCC ACG ATG CCC CTC AAC GTG AGC TTC-3') and 3'-cmyc (5'-TTT GAA TTC CCA GAG TCG CTG CTG GTG GTG GGC-3') for c-myc gene; 5'-sox (5'-TTT CTC GAG ATG GTG CAG CAG GCC GAG AGC-3') and 3'-sox (5'-TTG AAT TCC ATA CGT GAA CAC CAG GTC GGA-3') for Sox11 gene; 5'-bcl2 (5'-TTT CTC GAG GCC ACC ATG GCG CAC GAT GGG AGA ACA-3') and 3'-bcl2 (5'-TTT GAA TTC CCT TGT GGC CCA GAT AGG CAC CCA-3') for Bcl-2 gene; 5'mb-actin (5'-ACC AAC TGG GAC GAT ATG GAG AAT A-3') and 3'mb-actin (5'-TAC GAC CAG AGG CAT ACA GGG ACA-3') for β-actin, which was used as an internal control.

2.3. DNA Microarray Analysis

For DNA microarray analysis, mRNAs were extracted from persistent- or mock-infected CG-4 cells at passage 20 p.i. using the Qiagen RNAeasy kit according to the manufacturer's instructions. The purity and quantity of the RNAs were determined by spectrophotometry. The levels of individual mRNA species were determined by microarray using the Affymetrix Oligo Gene Chip (U34), which detects approximately 7,000 known genes and 1,000 EST clusters. The DNA microarray analysis was carried out at the University of Iowa DNA Core facility. A 2-fold or greater difference between the test (persistently infected CG-4 cells) and the control (mock-infected CG-4 cells) was considered a significant change while any genes that are absence (below detectable level) in both test and control cells were excluded from the analysis.

2.4. Western Blot Analysis

Western blot analysis was carried out as described.[9] The antibodies used in this study include a polyclonal rabbit antibody specific to Bcl-2 (0.2 μg/ml) (Cell Signals, Inc.), a monoclonal antibody (mAb) to rat myelin basic protein (MBP) (1 μg/ml) (Chemicon Internation, Inc.), and a mAb to β-actin (1:5000) (Sigma).

2.5. Plasmid Construction, DNA Transfection, and Selection of Stable Transfectants

The Bcl-2 gene was kindly provided by Marie Hardwick (Johns Hopkins University) and was subcloned into pcDNA3, resulting in pcDNA3/Bcl-2. CG-4 cells were transfected with pcDNA3/Bcl-2 DNA with FuGene-6 transfection reagent (Roche), and selected for stable expression of Bcl-2 with G418 and by Western blot analysis.

2.6. Assay for Cell Viability

Trypan blue was used for staining dead cells following MHV infection or mock-infection. The dead cells were counted in 3 independent experiments.

2.7. Immunofluorescence Staining

The immunocytochemistry method used in this study was previously described.[9] Stained cells were observed under a fluorescent microscope (Olympus IX70) and photographs were taken with an attached digital camera (MagnaFire).

3. RESULTS AND DISCUSSION

3.1. Establishment of Persistent, Nonproductive MHV Infection in Rat Progenitor Oligodendrocytes

In a recent study, we reported that a brief treatment of CG-4 cells with fetal bovine serum rescued cells from MHV killing.[10] To extend these observations, we collected cells at various passages and determined the presence of viral genomic RNA by RT-PCR and the infectious virus by plaque assay. The viral genomic RNA at the N gene locus was consistently detected by RT-PCR throughout all 20 passages. The identity of the RT-PCR fragments was confirmed by DNA sequencing. However, no infectious virus could be detected from these cells beyond passage 3. This result demonstrates that viral RNA was able to persist in CG-4 cells without the production of infectious virus. This phenomenon resembles MHV persistence in the animal CNS. The establishment of MHV RNA persistence in glial cells provides a useful system for studying the molecular mechanisms of viral persistence and demyelination.

Table 1. Genes that are upregulated in MHV-persistent CG-4 cells.

Category	Gene name	Acces. no.	Fold(↑)	Function
MARK/ phosphatase	Dual specificity phosphatase(MKP3)	X94185	29.9	Inhibits FGF/RAS/ MAPK pathway
Cell cycle/ apoptosis	Bcl-2	L14680	26	Anti-apoptotic, cell survival
Tanscription /oncogene	c-Myc	Y00396	9.8	Oncogenic, cell survival/apoptosis
Neural/glial cell-specific	Sry-related HMG-box protein Sox11	AJ004858	3.2	Inhibits oligo.matur./ myelin expression

3.2. Viral RNA Persistence Altered the Gene Expression Profile of Progenitor Oligodendrocytes

To determine whether viral RNA persistence in progenitor oligodendrocytes alters the cellular gene expression, we used DNA microarray technology to compare the mRNA levels in persistently infected CG-4 cells with those of mock-infected CG-4 cells. Overall, approximately 350 genes were significantly upregulated, whereas only about 30 genes were downregulated in persistently infected CG-4 cells as compared with those of mock-infected CG-4 cells at passage 20. Most notable among the differentially expressed genes are those that are involved in signal transduction, cell cycle, cell survival and death, and differentiation. For examples, the expression of cellular oncogenes Bcl-2 and c-myc increased by 26 and 9.6 fold, respectively; the Sry-related HMG-box protein Sox11 was increased by 3.2 fold (Table 1). To confirm the DNA microarray results, we selected a few genes for analysis with RT-PCR. Indeed, a significant increase of the Sox11, c-myc, and Bcl-2 mRNAs was detected at passage 20 in persistently infected CG-4 cells as compared to those in mock-infected CG-4 cells (Figure 1). These results suggest that persistent infection might have altered many biological properties of CG-4 cells, such as cell proliferation and differentiation. Significantly, because Sox11 is a transcription factor, a negative regulator of transcription of many myelin-associated proteins, the increase of Sox11 expression might have a negative consequence on myelination by persistently infected oligodendrocytes.

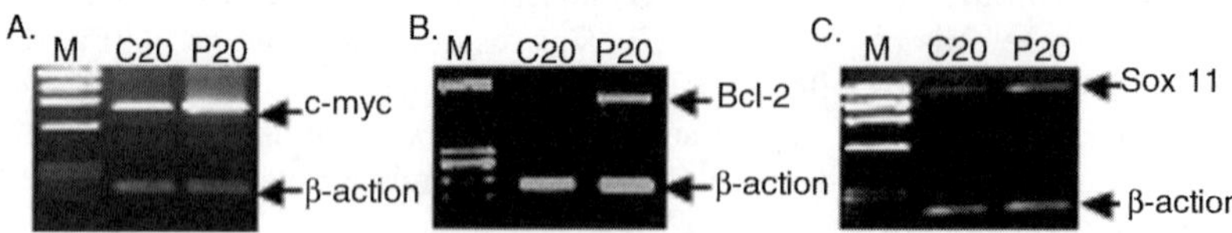

Figure 1. RT-PCR detection of mRNAs for c-myc (A), Bcl-2 (B), and Sox11 (C) genes in MHV persistently infected CG-4 cells. Intracellular RNAs were isolated from mock-infected cells as a control (lane C) or from MHV-persistently infected CG-4 cells at passage 20 (lane P20) and were detected by RT-PCR with gene specific primers. B-actin gene was used as an internal control. M, molecular size marker.

3.3. Role of Bcl-2 in the Establishment and Maintenance of Viral RNA Persistence in CG-4 Cells

To determine whether the change in host cellular gene expression also plays a role in the establishment of viral persistence in CG-4 cells, we chose the Bcl-2 gene as an example. Because Bcl-2 is an anti-apoptotic protein of the mitochondria, its increased expression by viral persistence must have an effect on cell survival. To test this hypothesis, we stably expressed the Bcl-2 gene in CG-4 cells. Overexpression of Bcl-2 in stable expressing CG-4 cells was confirmed by Western blot analysis (data not shown). When cells were infected with MHV, the extent of apoptosis in Bcl-2– expressing cells was significantly lower than in cells expressing the vector alone. This result suggests that upregulation of Bcl-2 by viral persistence may have beneficial effect on cell survival, regardless of whether the upregulation of Bcl-2 is the result of host response to viral infection or is the direct effect of virus infection. It is conceivable that the survival of host cells is also essential for the virus to persist. This finding may also provide a potential mechanism for the coexistence of the host and the parasite and represent a close relationship between the host cell and the persistence of viral RNA.

To determine whether and to what extent surviving cells still harbored virus, cells from various passages were collected and intracellular RNAs were isolated. The presence of the viral genome was determined by RT-PCR with primers specific to MHV N gene, and the production of infectious virus was measured by plaque assay in DBT cells. Indeed, the N gene region of the infecting viral genome was detected in all 10 passages, indicating the establishment of persistent infection. In contrast, infectious virus could be isolated only in the first 2 passages. These results suggest that cellular anti-apoptotic gene Bcl-2 may play a role in the establishment and maintenance of MHV persistence in CNS cells. By extrapolating the findings from Bcl-2, it is attempting to suggest that numerous cellular genes that are related to cell cycle, growth, proliferation, and survival and that are altered by viral persistence may play a vital role in the process of viral RNA persistence in CNS cells.

3.4. Effect of MHV Persistence on Differentiation of Progenitor CG-4 Cells

Because the destruction of the myelin sheath surrounding neuron axons in the CNS is the hallmark of demyelination in persistently MHV-infected animals and in multiple sclerosis patients, we further determined the ability of MHV-persistently-infected progenitor CG-4 cells to differentiate into mature oligodendrocytes and to form dendrites. The myelin basic protein, a marker expressed during the late stage of differentiation, was used to identify the morphology of the mature oligodendrocyte. Both persistently infected and mock-infected CG-4 cells were cultured under the condition that allows the cells to differentiate into mature oligodendrocyte. Although both cells could differentiate into mature oligodendrocytes, the oligodendrocytes differentiated from MHV-persistently infected CG-4 cells at both passage 20 and passage 30 had either fewer dendrites or less branched dendrites as compared with those derived from mock-infected CG-4 (data not shown). This result indicates that MHV persistence has an inhibitory effect on oligodendrocyte differentiation and dendrite formation. Consistent with this result is the finding that the expression of the myelin-basic protein was also reduced in persistently infected oligodendrocytes (data not shown). This study provides the first direct evidence that links MHV RNA persistence to malfunction of affected oligodendrocytes in dendrite (and possibly *in vivo* myelin) formation.

4. ACKNOWLEDGMENTS

This work was supported by grants from the National Institutes of Health (AI 47188 and NS 47499). We thank Paul Drew (UAMS) for kindly providing the CG-4 cells, Stephen Stohlman (Keck School of Medicine, University of Southern California, Los Angeles) for the N monoclonal antibody, and Marie Hardwick (Johns Hopkins University School of Medicine, Baltimore) for the Bcl-2 gene.

5. REFERENCES

1. Weiner, L. P., 1973, Pathogenesis of demyelination induced by a mouse hepatitis virus (JHM virus), *Arch. Neurol.* **28**:67-74.
2. Fleury, H. J., Sheppard, R. D., Bornstein, M. B., and Raine, C. S., 1980, Further ultrastructural observations of virus morphogenesis and myelin pathology in JHM virus encephalomyelitis, *Neuropathol. Appl. Neurobiol.* **6**:165-179.
3. Lavi, E., Gilden, D. H., Highkin, M. K., and Weiss, S. R., 1984, Persistence of mouse hepatitis virus A59 RNA in a slow virus demyelinating infection in mice as detected by in situ hybridization, *J. Virol.* **51**:563-566.
4. Knobler, R. L., Lampert, P. W., and Oldstone, M. B., 1982, Virus persistence and recurring demyelination produced by a temperature-sensitive mutant of MHV-4, *Nature* **298**:279-280.
5. Bergmann, C., Dimacali, E., Stohl, S., Wei, W., Lai, M. M. C., Tahara, S., and Marten, N., 1998, Variability of persisting MHV RNA sequences constituting immune and replication-relevant domains, *Virology* **244**:563-572.
6. Sawicki, S. G., Lu, J. H., and Holmes, K. V., 1995, Persistent infection of cultured cells with mouse hepatitis virus (MHV) results from the epigenetic expression of the MHV receptor, *J. Virol.* **69**:5535-5543.
7. Chen, W., and Baric, R. S., 1996, Molecular anatomy of mouse hepatitis virus persistence:coevolution of increased host cell resistance and virus virulence, *J. Virol.* **70**:3947-3960.
8. Louis, J. C., Magal, E., Muir, D., Manthorpe, M., and Varon, S., 1992, CG-4, a new bipotential glial cell line from rat brain, is capable of differentiating in vitro into either mature oligodendrocytes or type-2 astrocytes, *J. Neurosci. Res.* **31**:193-204.
9. Liu, Y., Cai, Y., and Zhang, X., 2003, Induction of caspase-dependent apoptosis in cultured rat oligodendrocytes by murine coronavirus is mediated during cell entry and does not require virus replication, *J. Virol.* **77**:11952-11963.
10. Liu, Y., and Zhang, X., 2005, Expression of cellular oncogene Bcl-xL prevents coronavirus-induced cell death and converts acute infection to persistent infection in progenitor rat oligodendrocytes, *J. Virol.* **79**:47-56.

CD8+ T-CELL PRIMING DURING A CENTRAL NERVOUS SYSTEM INFECTION WITH MOUSE HEPATITIS VIRUS

Katherine C. MacNamara and Susan R. Weiss*

1. INTRODUCTION

Infection with mouse hepatitis virus (MHV) provides an animal model with which to study central nervous system (CNS) diseases including both encephalitis and demyelination. Different strains of MHV induce disease with varying degrees of severity. The A59 strain induces acute encephalitis during the first week of infection and a strong CD8+ T-cell response is observed in the brain coinciding with virus clearance. Despite efficient clearance of infectious virus, demyelination is evident four weeks postinfection (p.i.). The JHM strain (also referred to as MHV-4 or JHM.SD in the literature)[1] induces lethal encephalomyelitis within the first week of infection and virus is typically not cleared. In this study, we investigate the CD8+ T-cell responses induced during infections with A59 and JHM.

Virus specific CD8+ T cells play a protective role against MHV strain A59 and are essential for clearance of infectious virus from the central nervous system (CNS). We have previously found that only early transfer, prior to 3 days postinfection (p.i.) with RA59-gfp/gp33, of gp33-specific CD8+ T cells (obtained from P14 transgenic mice) resulted in accumulation of activated epitope-specific CD8+ T cells within the brain.[2] We observed that P14 splenocytes did not accumulate in the brains of RA59-gfp/gp33 infected mice when the transfers were performed on day 3 or 5 p.i. In order to determine if this was due to a defect in trafficking or priming during the infection, we examined the expansion of transferred CFSE-labeled gp33-specific CD8+ T cells in the draining cervical lymph nodes following infection with RA59-gfp/gp33. In addition, we sought to determine why activated, virus-specific CD8+ T cells are detected at very low levels in the spleen and brain after infection with RJHM.

*University of Pennsylvania, Philadelphia, Pennsylvania 19104.

2. MATERIALS AND METHODS

2.1. Mice and Viruses

Four-week-old male mice were used in all experiments; B6 or B6-LY5.2/Cr (CD45.1) mice were obtained from the National Cancer Institute. P14 mice[3] were bred at the University of Pennsylvania. Recombinant MHV strain A59 expressing enhanced green fluorescent protein (EGFP) or expressing the gp33 epitope as fused to EGFP are described elsewhere.[4] Recombinant A59 (RA59), recombinant JHM (RJHM) and the recombinant chimeric virus expressing the JHM spike with A59 background genes (SJHM/RA59) have been described elsewhere.[5,6].

2.2. Isolation of Mononuclear Cells for Adoptive Transfer

Spleens were removed from P14 mice and suspensions were prepared by homogenizing in a nylon bag (64 μm diameter) in RPMI 1640 medium supplemented with 1% fetal calf serum. Red blood cells were lysed with 0.83% ammonium chloride and the lymphocyte suspension was washed twice in 1 x PBS and resuspended in 1 x PBS for transfer. Cells (at a concentration of 5 x 10^7 cells/ml) were labeled with 1 μl of 5 mM CFSE/ml. The total number of cells transferred was 2 x 10^7 cells in 0.5 ml.

2.3. Isolation of Mononuclear Cells from Brains, Spleens, or Lymph Nodes

Mice were perfused with 10 ml 1 x PBS and organs removed. Brain lymphocytes were isolated as previously described.[7,8] Cells were harvested from spleens and lymph nodes as described above. Intracellular IFN-γ was assayed as previously described.

2.4. Demyelination

Demyelination was analyzed in at least 10 sections of spinal cord from each animal and five to eight mice were examined in each of two separate experiments. Percent demyelination was calculated by counting quadrants of cross-sectioned spinal cord that was stained with the myelin specific dye, luxol fast blue. A neuropathologist examined the spinal cords to determine the severity score which was from 0 to 5 with 5 being the most severe demyelination.[2]

3. RESULTS

3.1. Early Transfer Required for Protection and Accumulation of gp33-Specific CD8+ T cells in the Brain

Transfer of naïve, gp33-specific CD8+ T cells one day prior to infection with RA59-gfp/gp33 protected against acute encephalitis and, importantly, virus spread to the spinal cord was markedly reduced. This correlated with a dramatic reduction in the quantity and severity of demyelination seen 28 days p.i. However, mice that received adoptive transfers of gp33-specific CD8+ T cells on days 3 or 5 p.i. were not protected from acute

Table 1. Percent and severity of demyelination on day 28 p.i. is reduced by early transfer.

Day of transfer	Percent demyelination[a]	Severity of demyelination[b]
No transfer	30.53 +/- 3.5	2.35
Transfer (-1)	10.34 +/- 2.63	0.3
Transfer (+3)	24.2 +/- 5.7	2.0
Transfer (+5)	19.88 +/- 3.5	1.4

[a] Percent demyelination was calculated by counting quadrants of cross-sectioned spinal cord that was stained with the myelin spscific dye, luxol fast blue (see "Materials and Methods").
[b] The severity of demyelination was observed and assessed by a neuropathologist. The scale was from 0 to 5 (see "Materials and Methods"). Data was presented in a different format in Ref. x.

disease, which was assessed by virus replication, viral antigen spread and encephalitis. Importantly, only the mice receiving the early transfer that were protected from acute disease had significantly reduced chronic demyelination as observed on day 28 p.i. (Table 1).

We observed that the early transfer, performed one day prior to infection, resulted in protection from acute and chronic disease. In addition, we observed that the transferred (CD45.2 positive) cells were activated and secreted IFN-γ in response to gp33 peptide and accumulated to high percentages within the brains by day 7 p.i. However, the transferred cells did not accumulate in the brain on day 7 p.i. when the transfers were performed on days 3 or 5 p.i. Thus, we examined the brain-derived mononuclear cells from transfer recipients at later time points, days 10 and 12 p.i. As is evident from the data shown in Table 2 when transfers were performed on days 3 or 5 p.i. significantly fewer transferred cells accumulated at the site of infection as compared to the transfer recipients that received the transfer prior to infection. Whereas nearly half of the CD8+ T cells were the transferred cells in the early transfer recipients on day 10 p.i., only about 10.0% and less than 1.0% of the CD8+ T cells were the transferred cells in the day 3 and day 5 transfer recipients, respectively. Furthermore, on day 12 p.i. the total numbers of both CD8+ T cells as well as the transferred CD45.2-positive cells decreased as compared to day 10 p.i. We concluded that the cells transferred on days 3 or 5 p.i. were defective in their activation and/or ability to traffic into the CNS. However, when transfers were performed in RAG-/-, we observed the accumulation of gp33-specific IFN-γ-secreting cells in the brain and the later the transfer was performed the higher the percentage of gp33-specific cells observed in the brain. RAG-/- do not contain endogenous T cells capable of lytic activity, thus, it is assumed that antigen presentation is prolonged. Thus, we predicted that when we transferred P14 splenocytes into B6 mice on days 3 or 5 the cells were not activated or recruited into the brain due to a lack of antigen presentation at that time point.

Table 2. Total brain-derived CD8+ T cells and transferred, CD45.2 positive, cells 7 days post transfer.

Day of transfer	C57Bl/6		RAG-/-
	CD8+ (% of total)	CD45.2+	gp33-specific IFN-γ+
No transfer	1×10^5 (14%)	--	--
Transfer (-1)	1×10^5 (18%)	++++	++
Transfer (+3)	6×10^4 (11%)	+	+++
Transfer (+5)	5.5×10^4 (8%)	+/-	++++

[a] Cells harvested 7 days post transfer. Some of these data were presented in a different format in Ref.2.

3.2. Duration of Antigen Presentation During Infection of the CNS

In order to determine whether there was a block in the ability for the gp33-specific CD8+ T cells to traffic into the brain or if there was a defect in priming of the transferred cells, we developed an adoptive transfer model to trace the expansion of transferred cells. P14 splenocytes were labeled with CFSE and transferred prior to infection or on day 3 post infection.

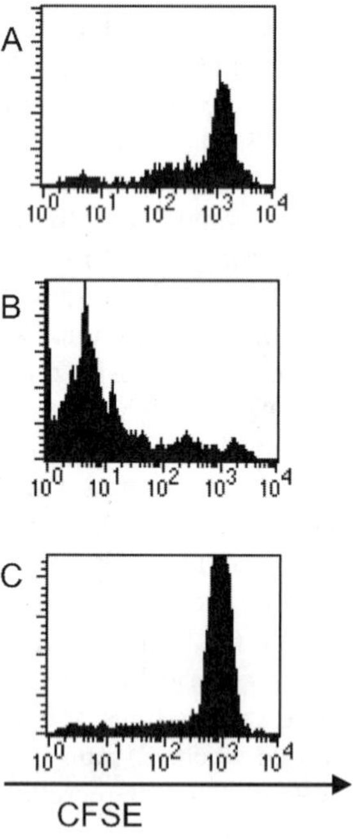

Figure 1. Proliferation of transferred P14 splenocytes. P14 splenocytes (CD45.2) were labeled with CFSE and then transferred via tail-vein injection to B6-LY5.2/Cr (CD45.1) recipients. Histograms represent CD8+, CD45.2+ cells harvested from cervical lymph nodes 72 hours post-transfer. Panels represent cells harvested from an uninfected mouse (A), a mouse infected with RA59-gfp/gp33 that received the transfer one day prior to infection (B), and a mouse that was infected with RA59-gfp/gp33 that received the adoptive transfer on day 3 postinfection (C).

3.3. RJHM Elicits a Weak CD8+ T-Cell Response

As previously been reported in the literature,[9] we observed that RJHM elicits a surprisingly weak CD8+ T-cell response and a poor epitope specific IFN-γ response. In order to rule out the possibility that RJHM causes destruction of the brain parenchyma that prevents recruitement of CD8+ T cells, animals were inoculated intranasally (i.n.) with their LD_{50} dose of 100 or 1000 pfu of RJHM or SJHM/RA59 (recombinant A59 expressing the JHM spike in place of the A59 spike), respectively. Intranasal infection results in a slower course of disease with most animals dying after the first week of infection allowed the analysis of recruitment of T cells into the brain. Mice were sacrificed at 7 days p.i. in order to analyze the epitope specific CD8+ T-cell response at the site of infection. Whereas 21.9% of the cells isolated from the brains of SJHM/RA59 infected animals were CD8+, only 0.8% of the cells isolated from the brains of RJHM infected animals were CD8+. Furthermore, the specific IFN-γ response to the two epitopes within the spike protein was much higher in the SJHM/RA59 infected animals (Fig. 2). This also indicates that the low CD8+ T-cell response is not due to the RJHM spike.

In order to determine if RJHM was capable of suppressing the immune response we coinfected mice with RJHM and RA59. Interestingly, the coinfected animals had a strong CD8+ T cell response to the subdominant S598 epitope that is present in both viruses, however, there was still no response to S510, the immunodominant epitope that is only present in RJHM.

4. CONCLUSIONS

In this study we define the window of antigen presentation during a CNS infection with RA59-egfp/gp33 to be within the first 72 hours of infection. When splenocytes were transferred on day 3 p.i., they were not activated to proliferate and, thus, did not accumulate within the brain. This provides more evidence that in order for CD8+ T cells to traffic to the site of infection they must undergo several rounds of division in the lymphoid organs. Furthermore, consistent with the idea that antigen presenting cells are destroyed by cytotoxic T lymphocytes as limiting determinant of immune activation we observed that transfers performed on days 3 and 5 p.i. in RA59-gfp/gp33 infected RAG-/- did result in the accumulation of virus-specific CD8+ T cells within the brain 7 days post transfer (Table 2).

Table 3. RJHM does not suppress the CD8+ T-cell response.

Virus	Total CD8+ T cells[a]	S510-specific	S598
RA59	$\sim 1 \times 10^6$	–	++
RJHM	$\sim 1 \times 10^4$	-/+	-/+
SJHM/RA59	$\sim 2 \times 10^6$	+++	+++
RA59 and RJHM	$\sim 2 \times 10^6$	–	++

[a] Cells per brain.

The neurotropic strains of MHV, A59, and JHM induce different courses of disease with JHM resulting in lethal encephalitis within the first week of infection. In addition, JHM induces a strong innate immune response but almost no CD8+ T- cells are found in the brains of infected animals. Following coinfection with JHM and A59, we observed that JHM is not capable of suppressing the CD8+ T-cell response but fails to elicit a CD8+ T-cell response. The weak CD8+ T-cell response induced during infections with JHM is not spike-determined as a chimeric virus expressing the JHM spike with the background genes derived from A59 results in a strong CD8+ T-cell response to both the S510 and S598 epitopes.

5. ACKNOWLEDGMENTS

We thank Peter T. Nelson for analyzing and scoring demyelination. This work was supported by NIH grant no.AI-47800.

6. REFERENCES

1. Ontiveros, E., Kim, T. S., et al., 2003, Enhanced virulence mediated by the murine coronavirus, mouse hepatitis virus strain JHM, is associated with a glycine at residue 310 of the spike glycoprotein. *J. Virol.* 77:10260-10269.
2. MacNamara, K. C., Chua, M. M., et al., 2005, Increased epitope-specific CD8+ T cells prevent murine coronavirus spread to the spinal cord and subsequent demyelination, *J. Virol.* **79**:3370-3381.
3. Brandle, D., Burki, K., et al., 1991, Involvement of both T cell receptor V alpha and V beta variable region domains and alpha chain junctional region in viral antigen recognition, *Eur. J. Immunol.* **21**: 2195-2202.
4. DasSarma, J., Scheen, E., et al., 2002, Enhanced green fluorescent protein expression may be used to monitor murine coronavirus spread in vitro and in the mouse central nervous system, *J. Neurovirol.* **8**:1-11.
5. Phillips, J. J., Chua, M. M., et al., 1999, Pathogenesis of chimeric MHV4/MHV-A59 recombinant viruses: the murine coronavirus spike protein is a major determinant of neurovirulence, *J. Virol.* **73**:7752-7760.
6. Macnamara, K. C., Chua, M. M., et al., 2005, Contributions of the viral genetic background and a single amino acid substitution in an immunodominant CD8+ T-cell epitope to murine coronavirus neurovirulence, *J. Virol.* **79**:9108-9118.
7. Murali-Krishna, K., Altman, J. D., et al., 1998, Counting antigen-specific CD8 T cells: a reevaluation of bystander activation during viral infection, *Immunity* **8**:177-187.
8. Phillips, J. J., Chua, M. M., et al., 2002, Murine coronavirus spike glycoprotein mediates degree of viral spread, inflammation, and virus-induced immunopathology in the central nervous system, *Virology* **301**:109-120.
9. Rempel, J. D., Murray, S. J., et al., 2004, Differential regulation of innate and adaptive immune responses in viral encephalitis, *Virology* **318**:381-392.

ANTIBODY-MEDIATED VIRUS CLEARANCE FROM NEURONS OF RATS INFECTED WITH HEMAGGLUTINATING ENCEPHALOMYELITIS VIRUS

Norio Hirano, Hideharu Taira, Shigehiro Sato, Tsutomu Hashikawa, and Koujiro Tohyama*

1. INTRODUCTION

Swine hemagglutinating hencephalomyelitis virus (HEV) causes vomiting and wasting disease or encephalitis in piglets. In our experimental studies of rats[1] and mice,[2] HEV-67N spread trans-synaptically from peripheral nerve to the central nervous system (CNS) and infected neurons but not any glial cells. These neurotropic properties of HEV are similar to those of rabies virus, indicating that HEV might be a good experimental model for the investigation of control of neurotropic viral infections including rabies. To examine the effectiveness of antiserum treatment, rats were inoculated with HEV following antiserum treatment and analyzed virologically.

2. MATERIALS AND METHODS

Plaque-purified HEV-67N strain was propagated and assayed for infectivity in SK-K cells as described previously.[3] Specific pathogen free of 6-week-old Wistar male rats were inoculated into the right hindleg by subcutaneous (s.c.) route ($1x10^6$ pfu). Five rats were used in each group. Antiserum was prepared to inoculate into infected rats 3 times at weekly intervals intraperitoneally (i.p.). On day 7 after the last inoculation, blood was collected from rats. For antiserum treatment, rats were administrated i.p. with 1 ml of rat antiserum (HI titer; > 1:1000).

The same experiment was made to confirm preventing fatal infection from rats by treating with antiserum heated at 56°C for 30 minutes. In addition, antiserum treatment was delivered by intravenous inoculation to compare the effectiveness of different routes.

*Norio Hirano, Hideharu Taira, Iwate University, Morioka 020-8550, Japan. Shigehiro Sato, Koujiro Tohyama, Iwate Medical University, Morioka 020-8505, Japan. Tsutomu Hashikawa, RIKEN, Wako 351-0198, Japan.

3. RESULTS

To prevent the virus spread into the CNS of rats infected with HEV by intracerebral (i.c.) or intraspinal (i.s.) inoculation, antiserum was administered before/after HEV inoculation. Among rats pretreated with antiserum 24 hr before i.s. inoculation, only 3 of 5 rats survived. Even with pretreatment 24 hr before i.c. inoculation, all rats infected by i.c. route died of encephalitis.

To measure virus growth in the spinal cord and brain of rats infected after s.c. inoculation, 3 rats per day were examined to detect virus in CNS. As shown in Figure 1, virus was first detected in spinal cord on day 2, and in brain on day 3. On day 4, the brain titers became higher than those of spinal cord, reaching 10^6 to 10^7 pfu/0.2 g.

As shown in Table 1, antiserum treatment at 0, 24, 48, 72 hr postinoculation (p.i.) prevented fatal infection in all rats after s.c. inoculation. At 96 hr p.i., all rats showed flapping ears as clinical signs. After antiserum treatment at 96 hr p.i., 3 of 5 rats survived.

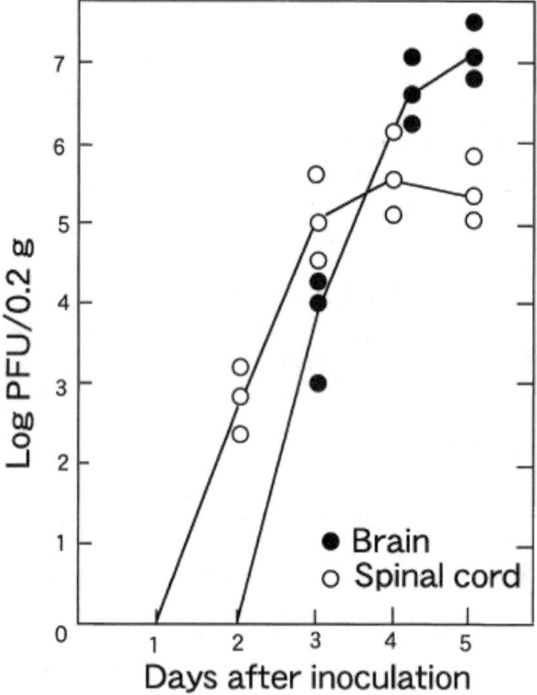

Figure 1. Virus growth in the spinal cord and brain.

Table 1. Antiserum treatment of rats after HEV inoculation into hind leg.

Antiserum treatment	CNS/tested	Dead/tested
0 hr	0/5	0/5
24	0/5	0/5
48	0/5	0/5
72	2/5	0/5
96	5/5	2/5
120	5/5	5/5
Nontreated	5/5	5/5

At 120 hr p.i., all rats developed CNS signs and died even with antiserum treatment. All nontreated rats died showing CNS signs within 7 days. The virus was detected in the brain of dead rats but not of the survivors. Approximately the same results were obtained in rats treated with heated antiserum as described above.

After antiserum treatment by both i.p., and i.v. routes, infected rats did not show any clinical signs and survived. The virus was not detected in brains of all of these animals on day 4 and 7 after treatment. Nontreated rats developed CNS signs and died within 7 days.

4. DISCUSSION

Antiserum treatment failed to prevent fatal infection from rats inoculated by i.c. and i.s. routes. However, rats inoculated s.c. were rescued by antiserum given at 0, 24, 48, and 72 hr p.i. By treatment at 96 hr p.i., 3 of 5 rats with CNS signs were protected. As shown in Figure 1, the virus already replicated in neurons of the brain at 72 to 96 hr p.i. These findings suggest that antibody mediated clearance of HEV was established in neurons of the brain of rats infected. Our previous studies demonstrated that HEV spreads trans-synaptically from peripheral nerve to CNS of rats.[4] Antibody might inhibit the spread of HEV in synaptic pathways via axons. Although this mechanism is unclear at present, antibody treatment might be useful tool for preventing neurotropic virus infection in the CNS of animals and humans.

5. REFERENCES

1. N. Hirano, T, Haga, F. Sada, and K. Tohyama, 2001. Susceptibility of rats of different ages to inoculation of swime haemagglutinating encephalomyelitis virus (coronavirus) by various routes, *J. Comp. Pathol.* **125**, 8-14.
2. N. Hirano, R. Nomura, T. Tawara, and K. Tohyama, 2004. Neurotropism of swine haemagglutinating encephalomyelitis virus (Coronavirus) in mice depending upon host age and route of infection, *J. Comp. Pathol.* **150**, 58-65.

3. N. Hirano, K. Ono, T. Takasawa, T, Murakami, and S. Haga, 1990. Replication and plaque formation of swine hemagglutinating encephalomyelitis virus (67N) in swine cell line, SK-K culture, *J. Virol. Methods*, **27**, 91-100.
4. N. Hirano, K. Tohyama, and H. Taira, 1998. Spread of swine hemagglutinating encephalomyelitis virus from peripheral nerve to the CNS in rats, *Adv. Exp. Med. Biol.* **440**, 601-607.

DEVELOPING BIOINFORMATIC RESOURCES FOR CORONAVIRUSES

Susan C. Baker, Dalia Jukneliene, Anjan Purkayastha, Eric E. Snyder, Oswald R. Crasta, Michael J. Czar, Joao C. Setubal, and Bruno W. Sobral*

1. INTRODUCTION

Virginia Bioinformatics Institute (VBI) and its partners have been awarded a 5-year contract from NIH-NIAID to establish a national Bioinformatics Resource Center (BRC) to facilitate research on microbial pathogens. As part of this initiative, VBI is developing the PathoSystems Resource Integration Center (PATRIC), a multi-organism relational database to support infectious disease research, especially as it affects biodefense and research on emerging infectious diseases (http://patric.vbi.vt.edu). We expect PATRIC to be used as a computational resource to gain insight into mechanisms of microbial pathogenesis and to hasten the development of improved vaccines, diagnostics, and therapeutics. The database will contain high-quality curated data: sequence annotations from published whole and partial genomes; relevant experimental data; metabolic pathway data; taxonomic data; literature citations; and a suite of visualization and analysis tools. Research experts and members of the scientific community will be closely involved at each step of the curation/annotation process. VBI is curating information on a set of eight different pathogen classes that include both bacteria and viruses. Included in this set is the genus *Coronavirus* (family *Coronaviridae*). At present we have archived the annotations of the 153 coronavirus species. These include both whole-genome (130) and partial-genome (23) annotations. This sequence archive represents the initial step in our efforts to curate data on *Coronavirus* species. We welcome active participation by the *Coronavirus* research community in developing PATRIC as a useful computational resource for infectious disease research.

* Susan C. Baker, Dalia Jukneliene, Loyola University Chicago Stritch School of Medicine, Maywood, Illinois. Anjan Purkayastha, Eric E. Snyder, Oswald R. Crasta, Michael J. Czar, Joao C. Setubal, Bruno W. Sobral, Virginia Polytechnic Institute and State University, Blacksburg, Virginia.

2. INTEGRATION OF ORGANISM INFORMATION

To facilitate the large-scale annotation/curation project that we have undertaken, we have built an annotation pipeline and associated curation tool interface. The annotation pipeline is composed of gene-prediction programs, similarity search algorithms, and protein structure and function prediction programs. The results of these programs and searches assembled by the annotation pipeline are used to propose biological features that are also stored in the curation database that uses the Genomics Unified Schema (GUS). The scenario for user interaction with the tools is presented in Figure 1. During the manual curation/annotation process, the curation tool interface retrieves the results of the automated annotation process [along with the proposed biological features] and presents them to a curator. Curators review the computational evidence in light of their collective expertise and accept proposed features or edit/remove them.

3. REFERENCE GENOMES

PATRIC genomes are organized into categories based on phylogenetic relationships. The simplest of these PATRIC categories consists of a relatively small number of sequenced genomes from a bacterial or viral family or genus. For the purposes of defining minimal, non-redundant set of genes characteristic of the category, one genome (usually the best-known or best-characterized) is identified as the "reference genome"; the remaining members of the class are called "associated genomes." For example, the Tor2 and Urbani isolates were the first two SARS coronavirus genomes to be sequenced and therefore were named as reference genomes. Efforts are underway to coordinate our system of reference and associated genomes with the RefSeqs from NCBI.[1]

For each organism category, a "reference gene set" is constructed consisting of a single representative of each orthologous group and is built by progressive identification of unique genes from the category's genomes. The reference genome has the highest precedence and therefore contributes its entire gene complement to the reference gene set. The reference set is then compared at the protein level to the first associated genome and vice versa. Genes from the associated genome identified as orthologs according to the "bidirectional best hit" test are annotated as such. This allows high-value, manually curated information from the corresponding reference genes to be automatically linked to the associated genes, provided minimal similarity criteria based on automated sequence analysis are satisfied. However, because the orthologous genes from the reference genome are already present in the reference gene set, only genes that fail the orthology test are added to the reference set. These genes are presumed to be novel and characteristic of the associated genome. This process is repeated for the remaining associated genomes.

4. PATRIC'S GENOME ANALYSIS PIPELINE (GAP)

The GAP is an automated system for annotating prokaryotic and viral genomes. It consists of two conceptual units, the Genomic Sequence Analysis Pipeline (GSAP) and Protein Analysis Pipeline (PAP) and is configured using GAPML, an XML-based pipeline description language. Submission of a genomic sequence to the database triggers

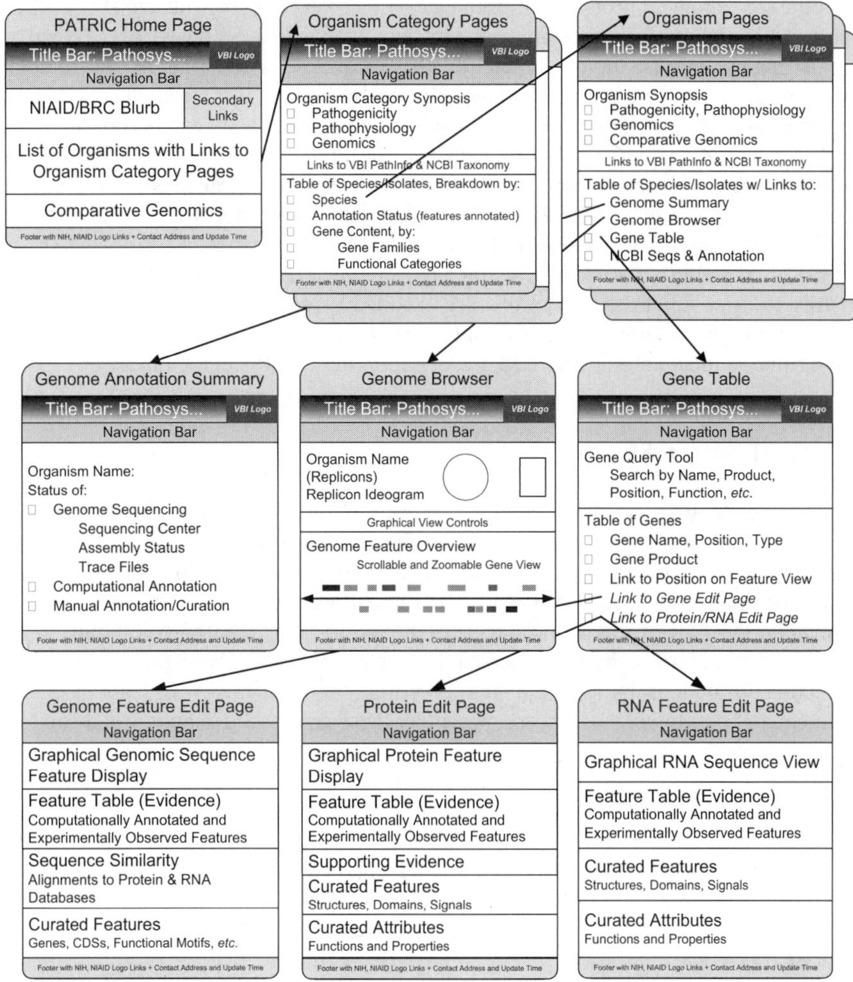

Figure 1. PATRIC browser / curation tool.

pipeline execution. Analysis begins in the GSAP with programs to identify tRNA, rRNA, and protein-coding genes. The programs tRNAscanSE, BLASTN, Glimmer, and GeneMark, respectively, make the gene predictions. The sequence is processed by the "putative gene interval" (PGI) parser to segment the genome into fragments containing a single gene. This breaks the genome into a manageable size for similarity searches and simplifies interpretation of their results. Because noncoding sequence is included within PGIs, genomic features such as putative RNA secondary structures, transcription regulatory sequences, and other features are annotated and queued for curatorial review. Curators make the final call on the predicted gene coordinates and translation and review the other results prior to submission to the GUS database. The translations are then passed to the PAP where it is first classified with respect to the Reference Protein Set, a collection of canonical proteins for each category of PATRIC organisms. If the protein is found to be similar to an existing entry in this database based on BLASTP search and very stringent cutoff, it is linked to that sequence and inherits its annotation. Inappropriate links can be broken at any point by curators. If no match is found, the protein is added to the Reference Set. Characterization continues with similarity searches to other databases such as SwissProt, GenPept and PIR. The sequence is then analyzed to identify physical characteristics such as signal peptides, transmembrane segments, secondary structure and PROSITE motifs. The final step involves characterization of functional domains using Pfam and TIGRfam, HMM libraries, SCOP, SMART, and BLOCKS. These programs/databases not only predict features but are also used by the curators to infer functions. Functions are encoded using terms from Genome Ontology (GO) and Enzyme Commission (EC) numbers. Features and functional assignments are then written to the database where they are used to infer pathway membership.

5. FUTURE DIRECTIONS

The information presented above reflects our immediate plans for basic genome annotation. This lays the foundation for our future work, which will include the analysis of metabolic and regulatory pathways and comparative genomics. In addition, we plan to relate this information to RNA and protein expression as data becomes available. Ultimately, the goal of this work is to help the biomedical research community leverage genomic information to better understand the physiology of these organisms and their interaction with their human and animal hosts. In time, this will lead to improved treatment and prophylaxis of disease caused by these potentially deadly organisms.

6. ACKNOWLEDGMENT

This project is funded by NIAID / NIH contract HHSN26620040035C to Bruno Sobral.

7. REFERENCES

1. Bao, Y., Federhen, S., Leipe, D., Pham, V., Resenchuk, S., Rozanov, M., Tatusov, R., and Tatusova, T., 2004, National Center for biotechnology information viral genomes project, *J. Virol.* **78**:7291-7298.

AUTOANTIBODIES EXACERBATE THE SEVERITY OF MHV-INDUCED ENCEPHALITIS

Renaud Burrer, Matthias G. von Herrath, Tom Wolfe, Julia D. Rempel, Antonio Iglesias, and Michael J. Buchmeier*

1. INTRODUCTION

The relationship between autoimmunity and infections by viruses and bacteria is complex. Infectious agents can initiate autoimmune responses by mechanisms such as molecular mimicry or bystander activation (reviewed in Ref.1). In addition, pathogens might also provoke relapses or worsen preexisting autoimmune pathologies. For example, common infections augment both the risk and the severity of relapses in multiple sclerosis (MS) patients.[2]

To study the effects of a viral infection on a preexisting autoimmune background, we have combined the neurotropic strain of mouse hepatitis virus (MHV) and mice that express auto-antibodies to a CNS antigen. MHV A59 provokes acute encephalitis, which is followed in a proportion of the surviving mice by a demyelinating disease that shares several features with MS.[3,4] Litzenburger and colleagues have generated transgenic mice (referred here as anti-MOG Ig mice) that constitutively express auto-antibodies specific of the myelin oligodendrocyte glycoprotein (MOG). These mice show an exacerbated version of experimental autoimmune encephalomyelitis (EAE), but do not develop any spontaneous disease.[5]

2. ANTIBODIES TO MOG AUGMENT THE SEVERITY OF A CNS INFECTION

Intracranial injection of a sublethal dose of MHV A59 resulted in an exacerbation of the clinical disease in mice with MOG-specific antibodies compared with controls (Figure 1, A). The mortality was increased in anti-MOG transgenic mice compared with controls, with 18% vs. 73% survival at day 21, respectively ($p<0.001$, Fisher's log-rank

*Renaud Burrer, Michael J. Buchmeier, The Scripps Research Institute, La Jolla, California. Matthias G. von Herrath, Tom Wolfe, La Jolla Institute of Allergy and Immunology, La Jolla, California. Julia Rempel, University of Manitoba, Winnipeg, Manitoba, Canada. Antonio Iglesias, Max Planck Institute for Neurobiology, München, Germany.

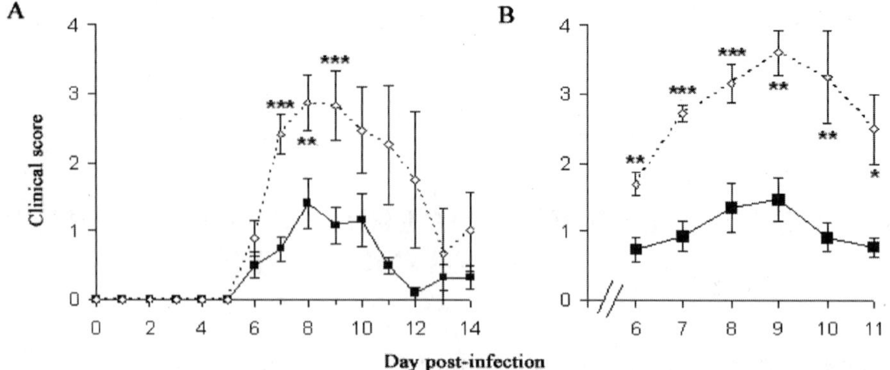

Figure 1. A. C57Bl/6 (solid line, n=15) or transgenic anti-MOG Ig knock-in mice (dashed lines, n=17) mice were infected 10 PFU MHV A59 I.C. Clinical signs of neurological disease were evaluated daily (0, no disease. 1, ruffled fur. 2, hunched posture. 3, lethargy. 4, moribund). B. C57Bl/6 mice received 500 μl of either normal mouse serum (solid line, n=13) or serum from anti-MOG Ig mice (dashed lines, n=16) at the time of infection with 10 pfu MHV A59 I.C. (Mann-Whitney test, **, p<0.01, ***, p<0.001).

survival test). In addition to the increased clinical scores and mortality, clinical signs were also detected 2 days earlier in the autoantibody transgenic mice than in the controls when a dose of 100 or 1000 pfu was used (data not shown). Injection of UV-inactivated virus (1000 pfu-equivalent) did not trigger any clinical signs in anti-MOG Ig mice, indicating that breaking the blood-brain barrier or the presence of viral antigens was not sufficient to trigger the pathogenicity of the autoantibodies, and that viral replication was required.

We found no significant difference between the viral titers in the CNS of control and MOG-Ig transgenic mice (Table 1), showing that the exacerbated disease could not be attributed to a difference in viral replication. Transfer of a single dose of serum from anti-MOG Ig mice to C57Bl/6 at the time of infection was sufficient to reproduce the clinical disease that was observed in the transgenic animals (Figure 1B), and resulted in a comparable increase of the mortality (13% survival at day 21 in recipients of anti-MOG Ig serum vs. 85% in controls, p<0.01). These results confirm that the anti-MOG autoantibodies account for the exacerbation of the virally-induced CNS disease in our model.

Table 1. Viral titers in brains of infected controls and mice with autoantibodies, as determined by plaque assay.

	Day postinfection		
	3	5	6
C57Bl/6	$6.5*10^4 \pm 5.8*10^4$ [a]	$5.9*10^5 \pm 1.5*10^5$	$2.2*10^7 \pm 7.5*10^6$
Anti-MOG Ig	$1.1*10^5 \pm 4.3*10^4$	$6.6*10^5 \pm 1.7*10^5$	$2.3*10^7 \pm 9.6*10^6$

[a] PFU/g, average±SE, n=3 to 6.

Table 2. Demyelination score in lesions in the brains of anti-MOG Ig and WT mice at day 9 postinfection.

	n	Demyelination score	Range
C57Bl/6	5	1.30±0.53[a]	0-3
Anti-MOG Ig	4	1.25±0.72	0-3

[a] Slides were examined in a blinded fashion, and given a score on a 0 to 4 scale. Mean±SE.

3. MHV-INDUCED EARLY DEMYELINATION IS NOT INCREASED BY ANTI-MOG AUTO-ANTIBODIES

In the MHV A59 model, demyelination can be detected as early as day 7 post-infection in the brains of a subset of mice. As anti-MOG antibodies have been shown to exacerbate demyelination in EAE[5,6] or in MS,[6] we have examined luxol fast blue–stained brain sections obtained from anti-MOG Ig and control mice. Preliminary results do not show any significant difference between the number or importance of demyelinating lesions in the brains of transgenic and control animals (Table 2).

The mechanisms responsible for the augmented virus-induced pathology in mice with CNS auto-antibodies are currently under investigation. We are also trying to determine if the ability to trigger the auto-aggressivity of the auto-antibodies is limited to MHV, or a more general feature of viral infections of the CNS.

Our study illustrates the additive effects of the concomitant presence of autoantibodies to a CNS antigen and a viral CNS infection, each of which, by themselves, is relatively harmless or leads to milder disease (Figure 1), respectively. These findings support the concept that infectious and autoimmune components can act in synergy leading to enhanced disease.

4. ACKNOWLEDGMENTS

This work was supported by NIH grants U19 AI51973, P01 AI058105, DK51091, AI44451, and JDRF 1-2002-726 to M.G.V.H., AI25913 and AI43103 to M.J.B., and Fellowships from the Canadian and U.S. National Multiple Sclerosis societies to J.D.R.

5. REFERENCES

1. Von Herrath, M., Fujinami, R. S., and Whitton, J. L., 2003, Viruses as triggers of autoimmunity – making the barren field fertile, *Nat. Rev. Microbiol.* **1**:151-157.
2. Buljevac, D., Flach, H. Z., Hop, W. C., Hijdra, D., Laman, J. D., Savelkoul, H. F., van Der Meche, F. G., van Doorn, P. A., and Hintzen, R. Q., 2002, Prospective study on the relationship between infections and multiple sclerosis exacerbations, *Brain* **125**:952-960.
3. Lane, T. E. and Buchmeier, M. J., 1997, Murine coronavirus infection: a paradigm for virus-induced demyelinating disease, *Trends Microbiol.* **5**:9-14.
4. Matthews, A. E., Weiss, S. R., and Paterson, Y., 2002, Murine hepatitis virus–a model for virus-induced CNS demyelination, *J. Neurovirol.* **8**:76-85.

5. Litzenburger, T., Fassler, R., Bauer, J., Lassmann, H., Linington, C., Wekerle, H., and Iglesias, A., 1998, B lymphocytes producing demyelinating autoantibodies: development and function in gene-targeted transgenic mice, *J. Exp. Med.* **188**:169-180.
6. Genain, C. P., Cannella, B., Hauser, S. L., and Raine, C. S., 1999, Identification of autoantibodies associated with myelin damage in multiple sclerosis, *Nat. Med.* **5**:170-175.

ANALYSIS OF THE N PROTEIN IN FELINE CORONAVIRUS STRAINS IN ITALY

Mara Battilani, Ambra Foschi, Alessandra Scagliarini, Sara Ciulli, Santino Prosperi, and Luigi Morganti*

1. INTRODUCTION

Feline coronaviruses (FCoVs) are responsible for an asymptomic or mild enteric infection but also cause a progressive, fatal immune-mediated disease, feline infectious peritonitis (FIP). The structural proteins of FCoVs include the spike (S), the membrane (M), and, the most representative, the nucleocapsid protein (N).

Coronavirus N proteins vary from 377 to 455 amino acids in length, are highly basic, have a high serine content (7–11%), and are potential targets for phosphorylation. Antigenic studies have shown that the N protein is one of the immunodominant antigens in the CoV family.[1] The cellular immune response against the N protein of some animal coronaviruses can enhance recovery from the virus infection. Immunization with a cell lysate using a recombinant baculovirus-expressing feline infectious peritonitis virus (FIPV) nucleocapsid protein was effective in preventing the progression of FIP.[2]

To investigate the antigenic role of the N protein, we carried out a computational analysis of the N protein of FCoV strains detected in healthy and diseased cats on the basis of the primary amino acid sequences.

2. MATERIALS AND METHODS

The N gene of FCoV strains detected in healthy and diseased cats coming from the same shelter (Table 1) was amplified by RT-PCR, and the amplicons were sequenced. Deduced amino acids sequences were aligned by ClustalW, and the alignment was visualized using the Genedoc program and Bioedit V5.06.

The PONDR program (Molecular Kinetics, Indianapolis, IN, USA, http://www. pondr .com) with the VL3-BA neuronetwork feedback predictor was used to predict the order/disorder regions of the N proteins of FCoV strains.

*University of Bologna, Italy.

Table 1. Details of viral strains analyzed.

Healthy cats	Diseased cats	Clinic form
352C	419 brain	Dry form
352N	419 kidney	Dry form
352S	420 lymph nodes	Wet form
368	420 small intestine	Wet form
	420 liver	Wet form

N protein phosphorylation site predictions were made with DISPHOS (DISorder-enhanced PHOSphorylation predictor, http://www.ist.temple.edu/DISPHOS). Only residues with a prediction > 0.5 are considered to be phosphorylated.

Computational analysis of the antigenic sites was carried out using the method of Kolaskar and Tongaonkar.[3]

Prediction of the immunodominant helper T-lymphocyte antigenic sites from amino acid sequence data, was carried out using the AMPHI algorithm; the identification of the antigenic sites that interact with mouse MHC II haplotype d was carried out using the SETTE algorithm.[4] Both methods were available in the Protean program of the DNASTAR multiple program package (Lasergene Inc., USA).

3. RESULTS AND DISCUSSION

The predicted amino acid sequences of the N gene are formed by 377 residues (the virulent strains 419 and 420) or 376 residues (the avirulent strains 352N, C, S and 368) as a consequence of the deletion of residue 205.

The alignment of the N sequences of virulent and avirulent strains showed that the major part of the mutations are located between aa 180-230 (Fig. 1). Analysis by the PONDR program suggested that a significant part of the N protein is disordered, but the virulent strains seem to be slightly more disordered in the region between aa 130–240, where the major numbers of mutations are located (Fig. 2).

Recent studies have found that unfolded protein or uncostructed protein regions are involved in molecular interactions such as receptor/ligand, protein/RNA, protein/protein. Disordered protein regions play an important role in cell signaling pathways.

To predict the phosphorylation sites in our sequences, we used the Web tool DISPHOS, which uses disorder information to detect phosphorylation sites. The virulent strains showed a major number of phosporylated serines with respect to avirulent strains, and all phosphorylation sites fall in the disordered regions.

Phosphorylated N protein bound to viral RNA with a higher binding affinity than non-viral RNA,[5] suggesting that phosphorylation of the N protein determined the recognition of viral RNA. Because the virulent strains showed more phosphorylation sites, we speculated that these viruses are more efficacious in assembly and packaging during viral replication.

```
            180        200         *       220         *       240
              QQSNNQNTNVEDTIVAVLQKLGVTDKQRSRSKSRERSSSNSRDTTPKNANKHTWKKTAGKGEVT
420_liver     ................................................................
420_lymph     ................................................................
420_int.      ................................................................
419_kidn.     ................................................................
419_brain     ................................................................
352           .S.....N..........R....AE-..P.....D.GN..NK....R..............D..
368           .S.....N..........R....AE-..P.....D.GN..NK....R..............D..
```

Figure 1. Alignment of predicted peptide sequence from the N gene. Nucleotide sequences from N PCR product were translated into amino acid sequences and aligned using ClustalW. Residues identical to the consensus are indicated by dots and gaps are indicated as dashes.

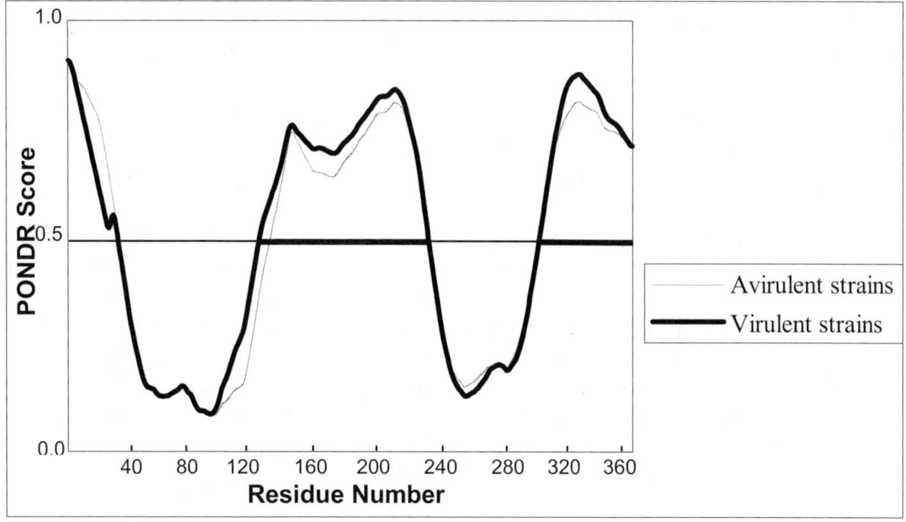

Figure 2. Order/disorder analysis of FCoVs N protein by the VL3-BA predictor of the PONDR program.

Antigenic analysis of the N protein was carried out by applying several methods. Our analysis demonstrated a substantial difference between virulent and avirulent strains in the mapping of antigenic sites. Using the method of Kolaskar and Tongaonkar, we predicted 13 antigenic sequences in the N protein of avirulent strains and 12 immunodominant sites in the virulent strains.

Antigenic analysis of N proteins carried out using SETTE and AMPHI algorithms showed different putative epitopes recognized by helper T cells and peptide antigenic sites confirming a possible involvement of the nucleocaspid protein in the protective immune response. All avirulent strains from healthy cats showed two additional motifs for IAd haplotypes detected using the Sette algorithm which are not present in other strains (aa 176–181; aa 261–266). On the basis of these results, we feel that virulent strains could elude immune surveillance, removing the immunodominant motifs from the N protein sequences.

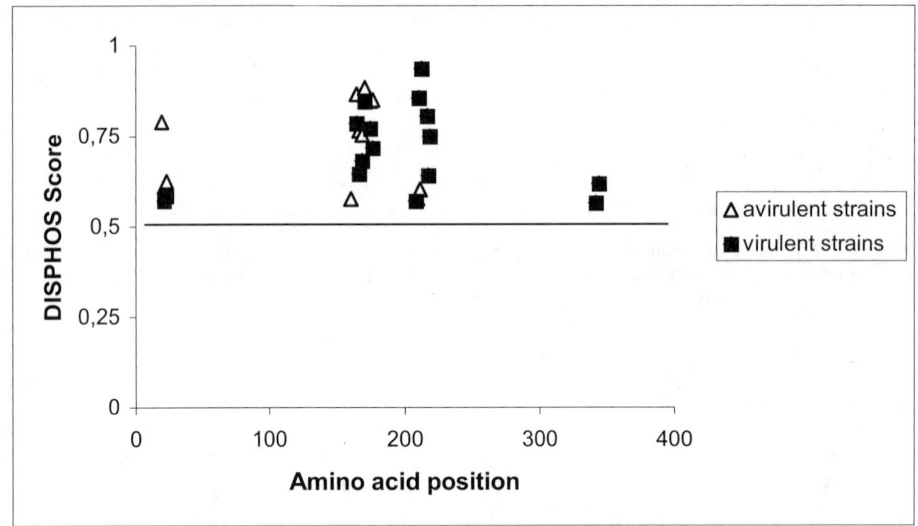

Figure 3. Prediction of phosphorylation sites.

It is interesting to compare the antigenic sites on N proteins with ordered/disordered regions in the amino acid sequence. The additional immunodominant site identified in the avirulent strains using the method of Kolaskar and Tongaonkar (sequence 12 TRKSCSK aa 336–342) lies in the disordered region at the border between the ordered and the disordered regions. The additional antigenic site (aa 176–181), detected using the Sette alghoritm, also falls in the disordered regions. As disordered regions are involved in molecular interaction, having the property of high specificity with modest binding affinity, the combined approach ordered/disordered regions and computational analysis of antigenic sites could be useful in the prediction of the antigenicity of the proteins.

4. REFERENCES

1. Lecomte, J., Cainelli-Gebara, V., Mercier, V., Mansour, S., Talbot, P. J., Lussier, G., and Oth, D., 1987, Protection from mouse hepatitis virus type 3-induced acute disease by an anti-nucleoprotein monoclonal antibody, *Arch. Virol.* **97**:123.
2. Hohdatsu, T., Yamato, H., Ohkawa, T., Kaneko, M., Motokawa, K., Kusuhara, H., Kaneshima, T., Arai, S., and Koyama, H., 2003, Vaccine efficacy of a cell lysate with recombinant baculovirus-expressed feline infectious peritonitis (FIP) virus nucleocapsid protein against progression of FIP. *Vet. Microbiol.* **97**:31.
3. Kolaskar, A. S., and Tongaonkar, P. C., 1990, A semi-empirical method for prediction of antigenic determinants on protein antigens, *FEBS Lett.* **276**:172.
4. Sette, A., Buus, S., Appella, E., Smith, J. A., Chesnut, R., Miles, C., Colon, S. M., and Grey, H. M., 1989, Prediction of major histocompatibility complex binding regions of protein antigens by sequence pattern analysis, *Proc. Natl. Acad. Sci. USA* **86**:3296.
5. Chen, H., Gill, A., Dove, B. K., Emmett, S. R., Kemp, C. F., Ritchie, M. A., Dee, M., and Hiscox, J. A., 2005, Mass spectroscopic characterization of the coronavirus infectious bronchitis virus nucleoprotein and elucidation of the role of phosphorylation in RNA binding by using surface plasmon resonance, *J. Virol.* **79**:1164.

DIFFERENTIAL INDUCTION OF PROINFLAMMATORY CYTOKINES IN PRIMARY MOUSE ASTROCYTES AND MICROGLIA BY CORONAVIRUS INFECTION

Dongdong Yu and Xuming Zhang*

1. INTRODUCTION

Mouse hepatitis virus (MHV) can infect rodents and cause digestive and central nervous system (CNS) diseases. The severity of the diseases is influenced by both viral and host factors. The strain JHM is highly neurovirulent while A59 is low neurovirulent, although both strains cause encephalitis and demyelination. By contrast, strain MHV-2 is non-neurovirulent, causing only mild meningitis but no encephalitis and demyelination.[1] Studies have shown that the viral spike protein is the major determinant for neurovirulence. Recombinant A59/JHM-S, which contains the JHM spike gene in the A59 genomic background, exhibited a neurovirulent phenotype in C57BL/6 mice similar to that of JHM[2] while A59/MHV2-S (Penn-98-1), which contains the spike of MHV-2, caused acute encephalitis but did not cause demyelination.[3] Thus, the spike can modulate the viral pathogenic phenotype. The host immune system also plays a critical role in the onset and progression of the CNS disease. There is clear evidence showing that MHV-induced CNS diseases are often accompanied by lymphocyte infiltration in the CNS. Some studies using immunodeficient mice showed that the demyelination induced by JHM is largely immune-mediated, whereas others using RAG1-/- mice found that the immune system is not absolutely required for the demyelination when mice are infected with A59.[4,5] Therefore, the precise role of individual components of the immune system in the demyelination process is not known.

The role of proinflammatory cytokines in the pathogenesis of CNS diseases has been studied both *in vitro* and *in vivo*. Indeed, significant upregulation of proinflammatory cytokines and chemokines has been observed during the course of CNS disease of mice infected with JHM and A59. Infection of primary mouse astrocytes and microglia with A59 significantly induced proinflammatory cytokines, especially TNF-α and IL-6, while infection with MHV-2 did not.[6] Thus, the ability of the virus strains to induce cytokines

*University of Arkansas for Medical Sciences, Little Rock, Arkansas 72205.

correlates with their neuropathogenic phenotypes, suggesting a role for proinflammatory cytokines in the disease process. However, removing TNF-α from mice through depletion with antibodies did not block or reduce demyelination induced by JHM,[7] thus negating the role for TNF-α in the demyelination process.

To further clarify this controversial issue and to determine whether the neurovirulent determinant spike protein is responsible for the induction of the cytokines in glial cells, we used Penn-98-1 in comparison with A59 and MHV-2 to determine proinflammatory cytokines in primary astrocytes and microglia. We found that a significant level of TNF-α and IL-6 was induced by A59 but not by MHV-2. Unexpectedly, Penn-98-1 induced TNF-α and IL-6 at a much higher level than that by A59. Our results thus suggest that the spike is not responsible for the induction of the proinflammatory cytokines in glial cells and that these proinflammatory cytokines might be associated with acute encephalitis but not with demyelination seen in A59-infected animals.

2. MATERIALS AND METHODS

2.1. Viruses, Cells, and Reagents

The following viruses were used: MHV-A59 and MHV-2 (kindly provided by Michael Lai, USC), and the recombinant virus Penn-98-1 (kindly provided by Ehud Lavi, University of Pennsylvania). Penn-98-1 contains the genome of A59 with the replacement of S gene by MHV-2.[3] All viruses were propagated in DBT cells, and were purified by ultracentrifugation through a 30% (wt/vol) sucrose cushion at 27,000 rpm for 3 h at 4°C (Beckman). The virus pellets were resuspended in phosphate-buffered saline (PBS). For preparation of mouse primary astrocytes and microglia, neonatal C57BL/6 mice were sacrificed. Astrocytes and microglia were isolated from the mouse brains by using a technique exploiting the differential adherence characteristics of astrocytes and microglia.[8] The purity of both astrocyte and microglia cell preparations was verified by immunofluorescence staining with antibodies specific to glial fibrillary acidic protein (GFAP) (Dako cytomation) and CD11b (Biosource), respectively. This isolation procedure routinely yielded cell populations with a purity of >95%. The MAPK p38 inhibitor SB203580 and the JNK inhibitor SP600125 were purchased from Biosource and were dissolved in DMSO.

2.2. Cytometric Beads Array (CBA) Assay and Western Blot Analysis

To quantify the cytokines produced and secreted into the culture medium, 50 μl of virus-infected or mock-infected culture medium were subjected to CBA assay according to the manufacturer's instruction (BD Sciences). The mouse inflammatory cytokine beads array kit includes 6 proinflammatory cytokines (TNF-α, IL-6, IL-10, IL-12p70, MCP-1, IFN-γ). To determine JNK activation, Western blot was carried out according to the manufacturer's instruction (Cell Signaling Technology Inc., CA).

3. RESULTS AND DISCUSSION

3.1. Cytokine Response to MHV Infection in Primary Astrocytes and Microglia

We determined the cytokine proteins in primary mouse astrocytes and microglia following infection with A59, MHV-2, and Penn-98-1 for 24 h, and compared them with those of mock-infected cells. We found that MCP-1 was significantly induced to a similar level by all 3 viruses, whereas no induction of IL-12p70, IFN-γ, and IL-10 was detected in cells infected with the 3 viruses. TNF-α and IL-6 were significantly higher in A59-infected than in MHV-2-infected cells. However, to our surprise, both TNF-α and IL-6 were much higher in Penn-98-1-infected cells than in A59-infected cells (Fig. 1).

Since the only difference between A59 and Penn-98-1 is the spike protein, the result suggests that the spike is not responsible for the induction of TNF-α and IL-6. Moreover, since Penn-98-1 does not cause demyelination in mice,[3] our finding suggests that the proinflammatory cytokines such as TNF-α and IL-6 might not be sufficient for causing demyelination.

3.2. Involvement of MAPK Pathways in the Regulation of Cytokine Induction in Primary Astrocytes by MHV Infection

In general, cytokines are regulated via MAPK pathways. To determine whether and which specific MAPK pathway(s) is involved in regulation of TNF-α and IL-6 induction by MHV infection, we used the p38 inhibitor SB203580 (40 μM) and JNK inhibitor SP600125 (20 μM) to treat astrocytes 1 h prior to infection, and determined the cytokines in the medium by CBA. The result showed that the p38 inhibitor significantly inhibited the induction of IL-6 but not of TNF-α, while the JNK inhibitor blocked the induction of both IL-6 and TNF-α. This result suggests that the p38 MAPK pathway is involved in regulation of IL-6 induction whereas the JNK MAPK pathway is involved in both IL-6 and TNF-α induction by MHV infection. To further determine the biological relevance of JNK activation to MHV pathogenesis, we compared the activation of JNK phosphorylation by the 3 MHV strains with different pathogenic phenotypes. Astrocytes were infected with A59, MHV2, Penn-98-1 at 5 m.o.i. or mock infected. Cells were collected at 24 h p.i., and Western blot analysis was carried out to detect JNK phosphorylation. Results showed that A59 and Penn-98-1, but not MHV2, induced JNK phosphorylation (data not shown). Because A59 and Penn-98-1 can induce TNF-α and IL-6 and cause acute encephalitis in mice whereas MHV-2 cannot, our findings suggest that the induction of the proinflammatory cytokines by A59 and Penn-98-1 may contribute to the acute encephalitis by the 2 viruses.

3.3. Viral Replication Is Required for the Induction of TNF-α and for the Activation of the JNK Signaling Pathway

To determine whether MHV replication was required for the induction of TNF-α and IL-6, we used UV light-irradiated or live viruses to infect primary astrocytes. At 24 h p.i., TNF-α and IL-6 in the culture medium were determined by CBA. We found that there was no induction of TNF-α in cells infected with the UV-irradiated virus. In contrast, UV-irradiated virus still induced IL-6 to a level similar to that induced by the live virus

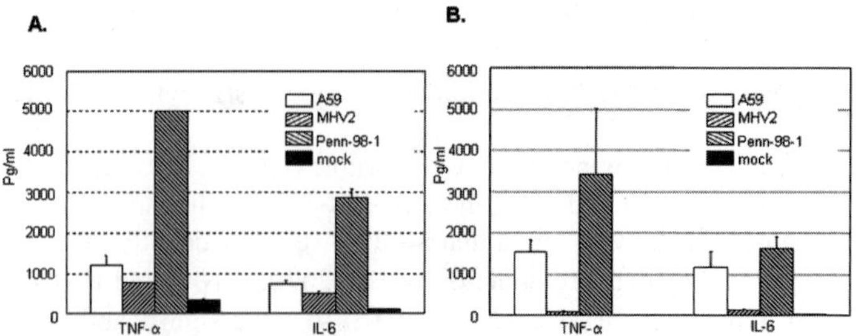

Figure 1. TNF-α and IL-6 production in primary astrocytes (A) and microglia (B) by MHV infection. Data indicate the means of the results from three independent experiments in the same pool of cells and the standard deviation of the means, which are representative of at least 3 pools of cells (9 independent experiments).

(data not shown). Taken together, these results demonstrate that MHV replication is required for TNF-α induction. We further determined if virus replication is required for the JNK pathway activation. c-Jun is the down-stream transcription factor of JNK that phosphorylates c-Jun. Our results showed that c-Jun was phosporylated by infection with A59 and Penn-98-1 but not by infection with MHV-2, and that c-Jun phosphorylation required viral replication (data not shown). These results thus correlate the activation of the JNK pathway with the induction of TNF-α by MHV infection.

4. ACKNOWLEDGMENTS

This work was supported by grants from the National Institutes of Health (AI 47188 and NS47499).

5. REFERENCES

1. Das Sarma, J., Fu, L., Hingley, S. T., Lai, M. M. C., and Lavi, E., 2001, Sequence analysis of the S gene of recombinant MHV-2/A59 coronaviruses reveals three candidate mutations associated with demyelination and hepatitis, *J. Neurovirol.* **7**:432-436.
2. Navas, S., and Weiss, S. R., 2003, Murine coronavirus-induced hepatitis: JHM genetic background eliminates A59 spike-determined hepatotropism, *J. Virol.* **77**:4972-4978.
3. Das Sarma, J., Fu, L., Tsai, J. C., Weiss, S. R., and Lavi, E., 2000, Demyelination determinants map to the spike glycoprotein gene of coronavirus mouse hepatitis virus, *J. Virol.* **74**:9206-9213.
4. Houtman, J. J., and Fleming, J. O., 1996, Dissociation of demyelination and viral clearance in congenitally immunodeficient mice infected with murine coronavirus JHM, *J. Neurovirol.* **2**:101-110.
5. Matthews, A. E., Lavi, E., Weiss, S. R., and Paterson, Y., 2002, Neither B cells nor T cells are required for CNS demyelination in mice persistently infected with MHV-A59, *J. Neurovirol.* **8**:257-264.
6. Li, Y., Fu, L., Gonzales, D. M., and Lavi, E., 2004, Coronavirus neurovirulence correlates with the ability of the virus to induce proinflammatory cytokine signals from astrocytes and microglia, *J. Virol.* **78**:3398-3406.
7. Stohlman, S. A., Hinton, D. R., Cua, D., Dimacali, E., Sensintaffar, J., Hofman, F. M., Tahara, S. M., and Yao, Q., 1995, Tumor necrosis factor expression during mouse hepatitis virus-induced demyelinating encephalomyelitis, *J. Virol.* **69**:5898-5903.
8. Esen, N., Tanga, F. Y., DeLeo, J. A., Kielian, T., 2004, Toll-like receptor 2 (TLR2) mediates astrocyte activation in response to the Gram-positive bacterium Staphylococcus aureus, *J. Neurochem.* **88**:746-758.

PREFERENTIAL INFECTION OF MATURE DENDRITIC CELLS BY THE JHM STRAIN OF MOUSE HEPATITIS VIRUS

Haixia Zhou and Stanley Perlman*

1. INTRODUCTION

The JHM strain of mouse hepatitis virus (MHV-JHM) causes acute encephalitis and acute and chronic demyelinating diseases in mice.[1] After infection, viruses are largely cleared by T cells; however, demyelination is induced as a consequence of this process. CD4 and CD8 T-cell responses, as well as an anti-viral antibody response are induced in infected mice, and they are critical to control virus replication and recrudescence.[2,3] Although an adaptive immune response is induced in infected mice, the anti-MHV CD8 T-cell response is reduced in mice infected with a virulent strain of MHV (named MHV-4 or JHM.SD[4]) when compared with mice infected with the A59 strain.[5] This might indicate suboptimal DC function during infection with JHM.SD. Because of the critical role DCs play in the host immune response to viral pathogens, viruses have developed strategies to depress the function of these cells.

Dendritic cells are readily infected with A59 *in vitro* and form syncytia. However, the antigen-specific T-cell response is very robust *in vivo,* suggesting that dendritic cells are still functional in this setting.[5,6] Previous work has focused on the effect of MHV infection on bone marrow (BM)-derived DCs after culture *in vitro*. These cultures usually include a mixture of mature and immature DCs. In general, only myeloid DCs are present in these cultures. In order to resolve the apparent contradiction between the *in vitro* and *in vivo* results, we assessed the extent to which mature and immature DCs are infected by MHV.

2. RESULTS

BM-derived DCs were prepared from B6 mice and cultured *in vitro* as previously described.[7] After 6–7 days in culture, cells were infected with the JHM strain of MHV. After 7 hr, extensive syncytia formation was observed. BM-derived cultures include cells

*University of Iowa, Iowa City, Iowa 52242.

that are positive and others that are negative for the DC marker, CD11c. By confocal microscopy, we showed that virus antigen was present in CD11c$^+$ cells. Infection was productive with increases in virus titers observed by 8 hr p.i. (data not shown). Thus, JHM, like A59, readily infects DCs *in vitro.*

Immature DCs are critical for antigen uptake whereas mature DCs present antigen to T cells and orchestrate the innate and adaptive immune responses. One explanation for the apparent contradiction between the *in vivo* and *in vitro* results described above is that MHV preferentially infects either mature or immature DCs. For this purpose, we infected unfractionated BM-derived DCs with a recombinant JHM that expresses GFP (rJHM.GFP) and analyzed cells by FACS at 9 hr p.i. In this virus, GFP was inserted by targeted recombination into gene 4 of the virus because this gene is not necessary for growth in tissue culture cells or mice.[8] The majority of infected CD11c$^+$ cells exhibited high expression of MHC class I and II antigen (data not shown) and of the costimulatory molecule, CD86 (Figure 1), consistent with a mature phenotype. An attenuated strain of JHM, J2.2-V-1,[9] and the A59 strain also preferentially infected mature DCs. However, compared with JHM, A59 infected immature DCs to a greater extent (Figure 1). Further, CD11c$^-$ precursor cells present in the DC culture were about 10 times more susceptible to infection with A59 compared with JHM (Figure 2).

In order to further determine if JHM directly infects mature DCs or JHM infects immature DCs and induces their subsequent maturation, we separated CD86^{hi} and CD86^{lo} DCs by flow cytometric sorting prior to infection with rJHM.GFP. As shown in Figure 3A, CD86^{hi} cells were about 5–10 times more susceptible to JHM infection than were immature DCs. We also quantified the proportion of GFP$^+$ cells, including syncytia, by fluorescent microscopy. Again, the percentage of GFP$^+$ cells was approximately 5–10 times higher in CD86^{hi} than in CD86^{lo} DCs (data not shown). We also found that CD86^{hi} DCs produced 10 times more JHM than CD86 ^{lo} DCs at the peak of the infection (Figure 3B).

MHV receptor CEACAM1a is expressed on DCs,[10] and infection of DCs *in vitro* is CEACAM1a-dependent.[6] One explanation for the differential infection of mature and immature DCs is that CEACAM1a is expressed at higher levels on CD86^{hi} cells. However, using anti-CEACAM1a mAb (kindly provided by Dr. K Holmes), we observed equivalent expression of CEACAM-1a on CD86^{hi} and CD86^{lo}CD11c$^+$DCs by FACS analysis (data not shown).

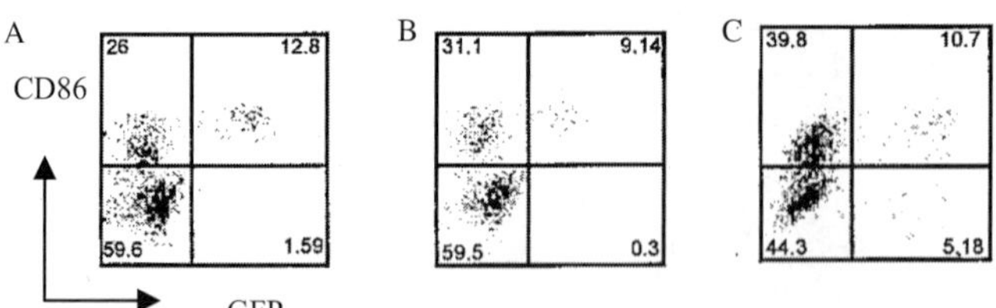

Figure 1. rJHM.GFP, rJ2.2-V-1.GFP, or rA59.GFP-infected DCs are mostly CD86^{high}. Shown are samples after gating on CD11c$^+$ cells. The percentage of infected cells was determined by GFP expression. Cells were harvested 9 hr after infection with rJHM.GFP (A) or rJ2.2-V-1.GFP (B) at an m.o.i of 10, or 7 hr after infection with rA59.GFP (C).

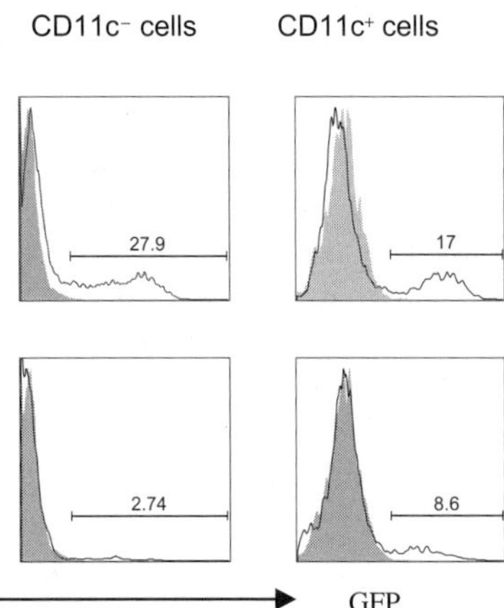

Figure 2. CD11c⁻ cells are more susceptible to infection with A59 than JHM. DCs were infected with either rA59.GFP top or rJHM.GFP bottom at an m.o.i of 10. Cells were harvested 7 hr after infection with rA59.GFP, and 9 hr after infection with rJHM.GFP.

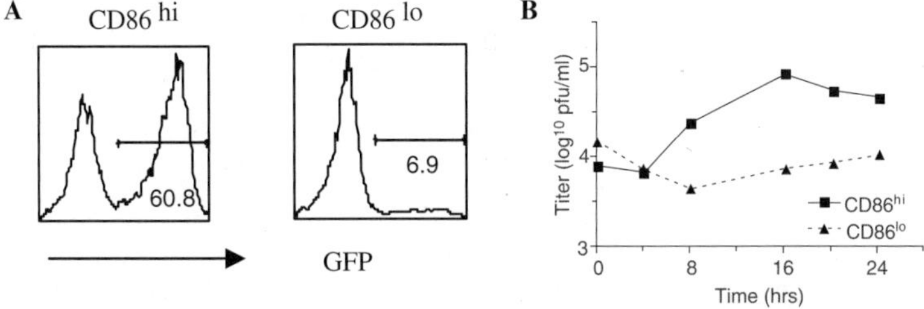

Figure 3. rJHM.GFP preferentially infects mature DCs. A. CD86^{hi} or CD86^{lo} DCs were separated using a flow cytometer prior to infection with rJHM.GFP (m.o.i. of 10). Cells were harvested at 9 hr p.i., and the percentage of GFP^{+} cells in each population was assessed by FACS. B. CD86^{hi} or CD86^{lo} DCs were infected with JHM at an m.o.i. of 10 and samples were harvested for titers at the indicated times.

3. DISCUSSION

Our results show that JHM infects cultured BM-derived DCs, with extensive syncytia formation. However, we could not detect a significant number of infected DCs in infected mice (data not shown). Mature DCs are the major population susceptible to JHM infection. However, the majority of DCs in naïve or infected animals are immature. This preferential infection of mature DCs may explain the low percentage of infected DCs that we observed in infected animals. Some other viruses also preferentially infect mature DCs. For example, mature blood-derived human DCs were infected at a higher level by RSV than were immature cells.[11] Although human CMV preferentially infects immature blood-borne DCs, recent results show that the virus has a tropism for mature Langerhans cells.[12]

Although MHV receptor CEACAM1a is expressed on DCs,[6] JHM does not infect cultured DCs efficiently: at an m.o.i. of 100, only 70% of mature DCs are infected with the virus (data not shown). Furthermore, immature DCs are more refractory to infection, even though levels of receptor are similar on the two types of cells. The essential role of

CEACAM1a in MHV infection of DCs was shown in a previous study, in which the presence of anti-CEACAM1a antibody blocked the process completely.[6] However, JHM may require a second host factor for efficient infection of some cells, as previous studies showed that JHM infected some cell lines less efficiently than A59.[13] Our data also show that A59 more readily infects CD11c$^-$ precursor cells and CD86^{lo} DCs than does JHM. These results suggest that JHM, compared with A59, might be more dependent upon the presence of this putative second factor for infection. This factor might be expressed at low levels on CD86^{hi} DCs, and even lower level on CD11c$^-$ cells and CD86^{lo} DCs.

Collectively, our results suggest that JHM entry into DCs or uncoating after entry might be dependent on a cofactor; endosomal proteases have been implicated in SARS-CoV entry, and MHV entry might also require a similar enzyme.

4. REFERENCES

1. Stohlman, S. A., Bergmann, C. C., and Perlman, S., 1998, Mouse hepatitis virus, in: *Persistent Viral Infections*, R. Ahmed, and I. Chen, eds., John Wiley & Sons, Ltd., New York, pp. 537-557.
2. Stohlman, S. A., Bergmann, C. C., van der Veen, R. C., and Hinton, D. R., Mouse hepatitis virus-specific cytotoxic T lymphocytes protect from lethal infection without eliminating virus from the central nervous system, *J. Virol.* **69**, 684-694 (1995).
3. Ramakrishna, C., Stohlman, S. A., Atkinson, R. D., Shlomchik, M. J., and Bergmann, C. C., Mechanisms of central nervous system viral persistence: the critical role of antibody and B cells, *J. Immunol.* **168**, 1204-1211 (2002).
4. Ontiveros, E., Kim, T. S., Gallagher, T. M., and Perlman, S., Enhanced virulence mediated by the murine coronavirus, mouse hepatitis virus Strain JHM, Is associated with a glycine at residue 310 of the spike glycoprotein, *J. Virol.* **77**, 10260-10269 (2003).
5. Rempel, J. D., Murray, S. J., Meisner, J., and Buchmeier, M. J., Differential regulation of innate and adaptive immune responses in viral encephalitis, *Virology* **318**, 381-392 (2004).
6. Turner, B. C., Hemmila, E. M., Beauchemin, N., and Holmes, K. V., Receptor-dependent coronavirus infection of dendritic cells, *J. Virol.* **78**, 5486-5490 (2004).
7. Inaba, K., et al., Generation of large numbers of dendritic cells from mouse bone marrow cultures supplemented with granulocyte/macrophage colony-stimulating factor, *J. Exp. Med.* **176**, 1693-1702 (1992).
8. Ontiveros, E., Kuo, L., Masters, P. S., and Perlman, S., Inactivation of expression of gene 4 of mouse hepatitis virus strain JHM does not affect virulence in the murine CNS, *Virology* **290**, 230-238 (2001).
9. Fleming, J. O., Trousdale, M. D., El-Zaatari, F., Stohlman, S. A., and Weiner, L. P., Pathogenicity of antigenic variants of murine coronavirus JHM selected with monoclonal antibodies, *J. Virol.* **58**, 869-875 (1986).
10. Kammerer, R., Stober, D., Singer, B. B., Obrink, B., and Reimann, J., Carcinoembryonic antigen-related cell adhesion molecule 1 on murine dendritic cells is a potent regulator of T cell stimulation, *J. Immunol.* **166**, 6537-6544 (2001).
11. Bartz, H., et al., Respiratory syncytial virus decreases the capacity of myeloid dendritic cells to induce interferon-gamma in naive T cells, *Immunology* **109**, 49-57 (2003).
12. Hertel, L., Lacaille, V. G., Strobl, H., Mellins, E. D., and Mocarski, E. S., Susceptibility of immature and mature Langerhans cell-type dendritic cells to infection and immunomodulation by human cytomegalovirus, *J. Virol.* **77**, 7563-7574 (2003).
13. Yokomori, K., Asanaka, M., Stohlman, S. A., and Lai, M. M. C., A spike protein-dependent cellular factor other than the viral receptor is required for mouse hepatitis virus entry, *Virology* **196**, 45-56 (1993).

ROLE OF THE REPLICASE GENE OF MURINE CORONAVIRUS JHM STRAIN IN HEPATITIS

Sonia Navas-Martín, Maarten Brom, and Susan R. Weiss*

1. INTRODUCTION

Mouse hepatitis virus (MHV) is the prototype of group II coronaviruses. Various strains of MHV induce different patterns of pathogenesis. MHV-JHM is a highly neurovirulent strain that causes severe acute encephalitis and chronic demyelination, but not hepatitis. MHV-A59 strain is dualtropic, causing mild to moderate hepatitis, as well as acute meningoencephalitis and chronic demyelination in C57BL/6 mice. Using a combination of targeted RNA recombination to precisely manipulate the coronavirus genome, and *in vivo* approaches (the mouse model), we have previously reported that the coronavirus spike protein is a major determinant of pathogenesis.[1, 2] Interestingly, we have also found that expression of the "hepatotropic" A59 spike glycoprotein within the background of the "neurotropic" JHM strain does not reproduce the A59 hepatotropic phenotype.[3] Thus, our studies demonstrated that genes other than the spike play a role in coronavirus tropism and virulence. These results prompted us to further investigate which genes may account for the lack of hepatotropism of the JHM strain. We have started to assess the role of the JHM replicase gene in pathogenesis.

Here, we have generated a recombinant chimeric JHM-A59 virus, in which the whole JHM replicase gene was introduced into the A59 background (JHMrep-RA59). We have performed *in vitro* replication kinetics analysis, and *in vivo* studies in order to compare JHMrep-RA59 with RJHM and RA59 (recombinant wild types). *In vitro* studies demonstrate that the JHMrep-RA59 virus replicates with similar kinetics to RA59. *In vivo* studies demonstrate that the presence of the 3' third of the A59 genome is sufficient to confer on a recombinant virus the ability to induce hepatitis; thus the replicase gene of JHM strain does not account for the non-hepatotropic phenotype of JHM. Our results suggest that genes other than spike and replicase must play a role in the tropism of JHM.

*Sonia Navas-Martín, Drexel University College of Medicine, Philadelphia, Pennsylvania 19102. Maarten Brom, Susan R. Weiss, University of Pennsylvania School of Medicine, Philadelphia, Pennsylvania 19104.

2. METHODS

2.1. Cells, Plasmids, and Viruses

Murine fibroblast (L2 and 17Cl.1) cells and *Felis catus* whole-fetus (FCWF) cells were maintained in Dulbecco's minimal essential medium (DMEM) and supplemented with 10% fetal bovine serum (FBS), 1% antibiotic-antimycotic (penicillin-streptomycin-amphotericin B), and 10 mM HEPES buffer solution. The helper virus fMHV-B3b as well as the plasmid pJHM were provided by Stanley Perlman (University of Iowa). pMH54 plasmid and helper virus fMHV were provided by Paul Masters (Wadsworth Center for Laboratories and Research, New York State Department of Health, Albany, NY).

2.2. Targeted RNA Recombination

All isogenic recombinant viruses were generated as previously described,[4,5] using fMHV and pMH54 (for A59 background viruses) and fMHV-JHM 3Bb and pJHM (for JHM background viruses). Isogenic chimeric A59 viruses expressing the replicase gene of the JHM strain were constructed using fMHV-JHM 3Bb as parental virus and pMH54 as donor plasmid (Figure 1). JHM and A59 wild-type recombinant viruses were generated using fMHV-JHM 3Bb / pJHM, and fMHV / pMH54, respectively. Recombinant viruses were selected by their ability to infect murine cells.

2.3. Viral Load and Histopathology in Liver

To study virulence, C57BL/6 mice were infected intracranially (IC) with 10-fold serial dilutions of virus, five mice per dilution. Fifty percent lethal dose (LD50) values were calculated as previously described.[6] In order to assess viral load in the liver, mice were inoculated intrahepatically (IH) with 500 plaque forming units (pfu) of virus as described previously.[2] Livers were harvested from infected and mock mice on day 5 postinfection (p.i.). A piece of the liver was fixed overnight with 10% buffered formalin, and the rest of the liver was used for virus titration. Formalin-fixed liver was embedded

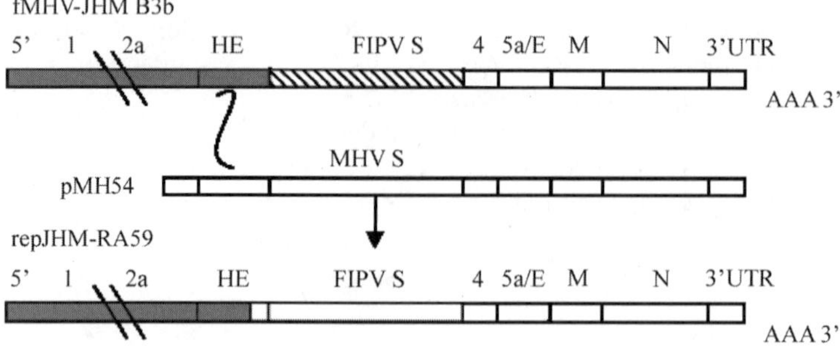

Figure 1. Schematic of targeted RNA recombination.

in paraffin, sectioned and stained with hematoxylin and eosin (H&E), and used for pathologic evaluation. Viral load (pfu/g liver) was determined by plaque assay.[2]

3. RESULTS AND DISCUSSION

MHV-JHM is a highly neurovirulent strain that does not cause hepatitis in infected mice. This is in contrast with other MHV strains, such as A59, that induce mild to moderate hepatitis.[7,8] Using isogenic recombinant viruses in the A59 background that differ only in the spike gene, we have previously demonstrated that the spike glycoprotein of murine coronavirus is a major determinant of hepatitis[2] and CNS disease.[1,9] However, we have also shown that recombinant viruses in the JHM background that express the spike of the hepatotropic A59 strain (SA59-RJHM) replicate poorly in the liver and do not induce hepatitis, while they do cause encephalitis to an intermediate level, between that of A59 and JHM (Ref. 3 and Iacono et al., mss in preparation). These results were surprising considering the role of the spike glycoprotein in pathogenesis and demonstrate that the JHM genetic background eliminates the ability of the A59 spike to mediate hepatotropism. Therefore, a major question arises from these studies, that is what JHM genes determine its non-hepatotropic phenotype?

Here, we wanted to assess whether the replicase gene of the neurotropic murine coronavirus JHM strain (MHV-JHM), determines its lack of hepatotropism. Using targeted RNA recombination, we have generated chimeric recombinant A59 viruses that express the replicase gene of the non-hepatotropic JHM strain (repJHM-RA59) (Figure 1). We have generated 2 independent chimeric repJHM-RA59 viruses, together with recombinant RA59 and RJHM wild-type viruses. Independent recombinants from each genotype exhibited the same phenotypes *in vitro* as well as *in vivo*. The data shown here represent results obtained from one recombinant virus per construct (RA59, RJHM and repJHM-RA59).

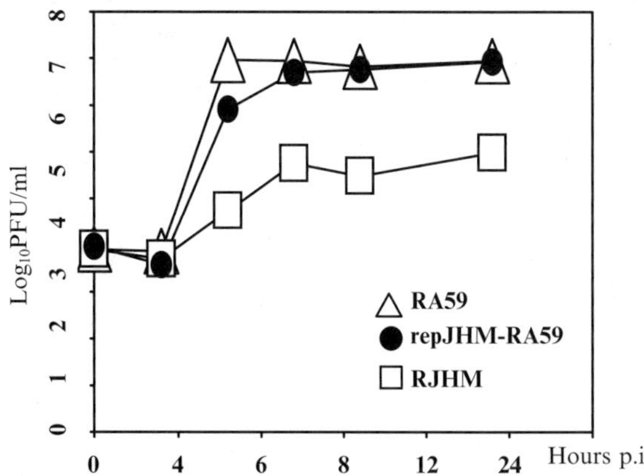

Figure 2. Growth kinetics of MHV recombinant viruses in L2 cells. Cells were infected at 37°C at an MOI of 1. Titers of released virus in the cultures were measured by plaque assay.

It is well-known that A59 and JHM viruses exhibit very different replication kinetics *in vitro*. MHV-A59 replicates to a higher titer, whereas JHM replicates with slower kinetics and to a lower final titer, and displays higher levels of fusion and cytotoxicity.[10] We have previously reported that RA59 and RJHM recombinant viruses generated by targeted RNA recombination mimic both *in vitro* as well as *in vivo* phenotypes of wild type A59 and JHM viruses.[3] Here, we found that recombinant A59 viruses expressing the replicase gene of JHM, exhibit the same replication kinetics pattern as RA59, whereas RJHM replicates with slower kinetics and to a lower final titer than RA59 and repJHM-RA59 (Figure 2). This suggests that, at least in the cells tested, structural genes rather than the replicase have a major role in JHM replication kinetics.

We next evaluated the virulence of repJHM-RA59 compared to wild type recombinants RA59 and RJHM. Interestingly, recombinant A59 viruses expressing the replicase gene of JHM exhibited the same range of virulence values as RA59 ($\log_{10}$ LD_{50} 3.6–3.8), in contrast with the highly neurotropic RJHM (Table 1). Although our study did not address the question of whether virulence factors are encoded by the replicase gene of JHM virus, it demonstrates that chimeric A59 virus expressing the replicase gene of JHM is as virulent as A59 wild-type in mice, and that the virulence is mostly determined by the 3' end of the viral genome, suggesting that structural genes have a major role in the highly neurovirulent JHM phenotype.

Finally, we evaluated viral load and hepatitis at day 5 p.i., that is the peak of replication in the liver (Figure 3, Table 2). We did not observe differences in viral load and histopathology in the livers of mice infected with repJHM-RA59 compared with RA59 viruses. repJHM-RA59 as well as RA59 infected mice exhibited mild to moderate hepatitis, in a range that we have previously observed for A59 strain.[2, 3] Viral load correlated with hepatitis in infected mice (data not shown). In contrast, RJHM induces none to minimal changes in the liver and replicates poorly, as expected.

In summary, we have generated recombinant A59 viruses expressing the replicase gene of the neurotropic JHM strain (repJHM-RA59) and compared their *in vitro* replication kinetics as well as their ability to replicate and induce hepatitis in mice. Our data demonstrate that (1) repJHM-RA59 viruses replicate *in vitro* to the same extent (load and kinetics) as RA59; (2) repJHM-RA59 viruses are as virulent in mice as A59 wild type; (3) repJHM-RA59 viruses induce hepatitis similarly to RA59.

Our data suggest that the lack of hepatotropism of the JHM strain is not determined by the replicase gene; rather the 3' end of the genome may have a major role in JHM phenotype in the liver. The role of the replicase gene of coronaviruses in pathogenesis

Table 1. Virulence of recombinant viruses after intracranial inoculation.

Virus	$\log_{10}$ (LD_{50}) i.c.[a]
RA59	3.6-3.8
RJHM	0.8
repJHM-RA59	3.6-3.8

[a] Intracranial virulence expressed as $\log_{10}$ (LD_{50}).

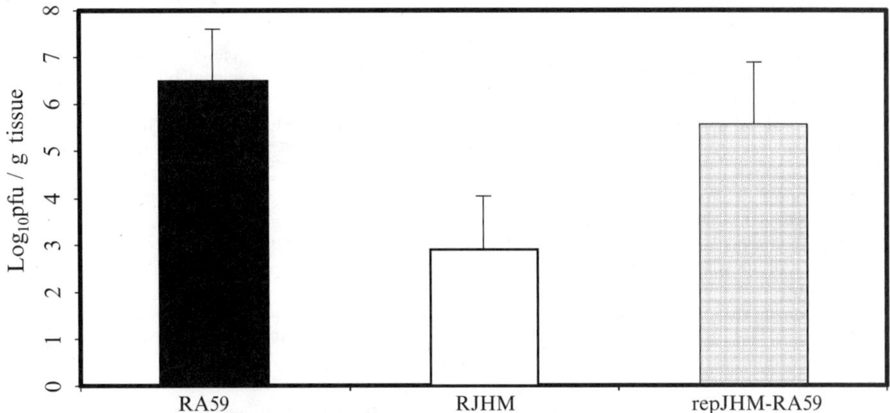

Figure 3. Viral load in the liver at day 5 p.i. (limit of detection 200 pfu/g).

remains poorly understood. The replicase gene is approximately 20 kb, and encodes a protein complex of up to 16 viral subunits that together with a number of cellular proteins form the replicase complex.[11] It should be pointed out that the data generated using these chimeras do not rule out the possibility that replicase proteins influence pathogenesis. In particular, Sperry et al.[12] have identified mutations in ORF 1b (p59-Our data suggest that the lack of hepatotropism of the JHM strain is not determined ns14) and 2a that attenuate virus replication and virulence in mice but do not affect *in vitro* replication. Their results suggest that proteins of the replicase complex (as well as nonstructural proteins such as 2a) serve roles in pathogenesis distinct from functions in virus replication.

Overall, our data demonstrates that structural genes of JHM may play a major role in coronavirus replication kinetics *in vitro*, and that the replicase gene of JHM may not determine the lack of hepatotropism of JHM strain.

4. ACKNOWLEDGMENTS

This work was supported by NIH grants AI-17418 and AI60021 (formerly NS-21954) to S.R.W. S.N.M. is supported by internal funds of Drexel University College of Medicine.

Table 2. Viral induced histopathology in the liver.

Virus	None	Minimal	Mild	Moderate	Severe
RA59			20%	60%	20%
RJHM	40%	60%			
repJHM-RA59			40%	60%	

Results are shown as percentage of mice exhibiting minimal, mild, moderate or severe hepatitis.

5. REFERENCES

1. J. J. Phillips, M. M. Chua, E. Lavi, and S. R. Weiss, Pathogenesis of chimeric MHV4/MHV-A59 recombinant viruses: the murine coronavirus spike protein is a major determinant of neurovirulence, *J. Virol.* **73**, 7752-7760 (1999).
2. S. Navas, S. H. Seo, M. M. Chua, J. D. Sarma, E. Lavi, S. T. Hingley, and S. R. Weiss, Murine coronavirus spike protein determines the ability of the virus to replicate in the liver and cause hepatitis, *J. Virol.* **75**, 2452-2457 (2001).
3. S. Navas and S. R. Weiss, Murine coronavirus-induced hepatitis: JHM genetic background eliminates A59 spike-determined hepatotropism, *J. Virol.* **77**, 4972-4978 (2003).
4. L. Kuo, G. J. Godeke, M. J. Raamsman, P. S. Masters, and P. J. Rottier, Retargeting of coronavirus by substitution of the spike glycoprotein ectodomain: crossing the host cell species barrier, *J. Virol.* **74**, 1393-1406 (2000).
5. E. Ontiveros, L. Kuo, P. S. Masters, and S. Perlman, Inactivation of expression of gene 4 of mouse hepatitis virus strain JHM does not affect virulence in the murine CNS, *Virology* **289**, 230-238 (2001).
6. L. J. Reed and H. Muench, A simple method of estimating fifty per cent points, *Am. J. Hygiene* **27**, 493-497 (1938).
7. J. Haring and S. Perlman, Mouse hepatitis virus, *Curr. Opin. Microbiol.* **4**, 462-446 (2001).
8. A. E. Matthews, S. R. Weiss, and Y. Paterson, Murine hepatitis virus--a model for virus-induced CNS demyelination, *J. Neurovirol.* **8**, 76-85 (2002).
9. J. Das Sarma, L. Fu, J. C. Tsai, S. R. Weiss, and E. Lavi, Demyelination determinants map to the spike glycoprotein gene of coronavirus mouse hepatitis virus, *J. Virol.* **74**, 9206-9213 (2000).
10. T. M. Gallagher and M. J. Buchmeier, Coronavirus spike proteins in viral entry and pathogenesis, *Virology* **279**, 371-374 (2001).
11. J. Ziebuhr, The coronavirus replicase, *Curr. Top. Microbiol. Immunol.* **287**, 57-94 (2005).
12. S. Sperry, L. Kazi, R. Graham, R. Baric, S. Weiss, and M. Denison, Single amino acid substitutions in non-structural ORF1b-nsp14 and ORF2a 30kDa proteins of the murine coronavirus MHV-A59 are attenuating in mice, *J. Virol.* **79**, 3391-3400 (2005).

IDENTIFICATION OF THE RECEPTOR FOR FGL2 AND IMPLICATIONS FOR SUSCEPTIBILITY TO MOUSE HEPATITIS VIRUS (MHV-3)-INDUCED FULMINANT HEPATITIS

Hao Liu, Li Zhang, Myron Cybulsky, Reg Gorczynski, Jennifer Crookshank, Justin Manuel, David Grant, and Gary Levy*

1. INTRODUCTION

After MHV-3 infection, susceptible mice develop microvascular disturbances, resulting in intravascular thrombosis and cell necrosis, which correlates with macrophage and endothelial cell production of a unique membrane-associated procoagulant, fibrinogen like protein2 (mFGL2). That FGL2 accounted for pathogenesis was shown by the fact that both neutralizing antibody to FGL2 ameliorated the pathological process and FGL2 knockout mice are resistant to liver disease and have increased survival.[1] First cloned from cytotoxic T lymphocytes, FGL2 was classified as a fibrinogen superfamily member.[2] The procoagulant activity has been localized to the linear N terminal domain of mFGL2. $CD4^+$ $CD25^+$ T cells have recently been shown to secrete a soluble form of FGL2 (sFGL2), which has now been proposed to have immunomodulatory activity.[3] We recently reported that sFGL2 inhibits T-cell proliferation and dendritic cell maturation.[4] That FGL2 exists as both soluble and membrane forms is not unique and has been described for other inflammatory molecules including tissue factor, CD16, and CD38. The purpose of this study is to identify the receptor(s) for sFGL2 and investigate the role of sFGL2 receptor interaction in MHV pathogenesis.

2. MATERIALS AND METHODS

Recombinant protein production: CHO-K1 cells were transfected with a Pires-neo3–Fc-mFGL2 plasmid in which the mouse IgG2a Fc tag was mutated to prevent the binding to FcγR and complement.

* Multi Organ Transplant Program, University Health Network, University of Toronto, Toronto, Canada.

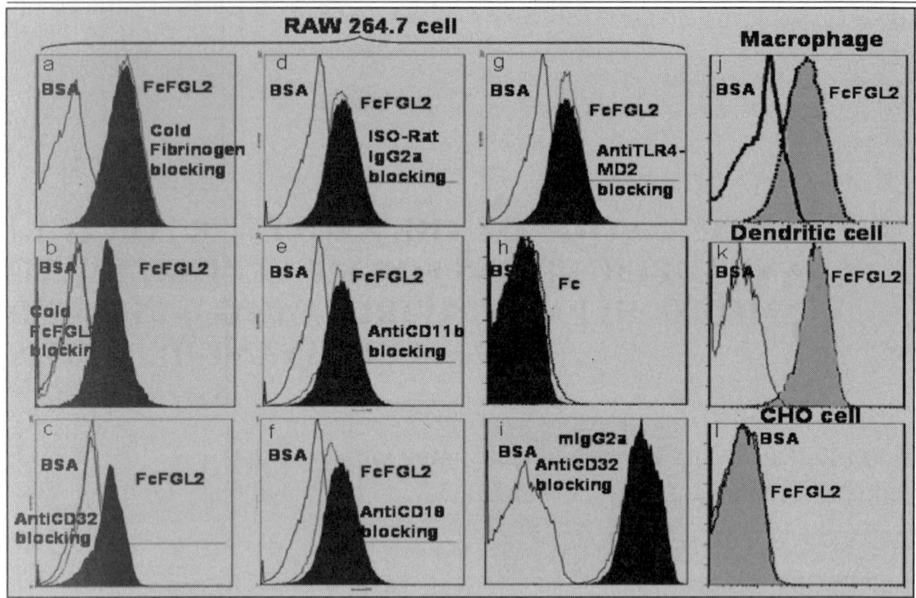

Figure 1. FcFGL2 binding with different cells can be specifically blocked by cold FGL2 and anti-mouse CD16/32.

Receptor positive cells identification: Biotinylated fusion FGL2 was used to study binding to RAW264.7, CHO-K1, peritoneal macrophages, bone marrow derived dendritic cells, and LPS stimulated spleen cells. To test the interaction specificity, 100-fold excess of cold BSA, fibrinogen, and FGL2 were added to compete the FGL2 probe binding.

Strain differences of FcγRIIB allotype: Spleen cells from C57BL6/J and AJ mice were harvested and stimulated with LPS for 48 hours, followed by staining with 2.4G2-FITC, Ly17.1/2-FITC, and anti-CD19-pc5.

3. RESULTS

We first showed that sFGL2 binds primarily to antigen presenting cells including macrophages, dendritic cells and LPS stimulated B cells (Figure 1). Preliminary data indicates that the FGL2 receptor might be one of the low-affinity receptors for immunoglobulin G (FcγR), a receptor family known to link the innate and acquired immunity (Figure 2). Furthermore, sFGL2 bound to B cells from susceptible C57BL/6J mice but not B cells from resistant A/J mice (Figure 3).

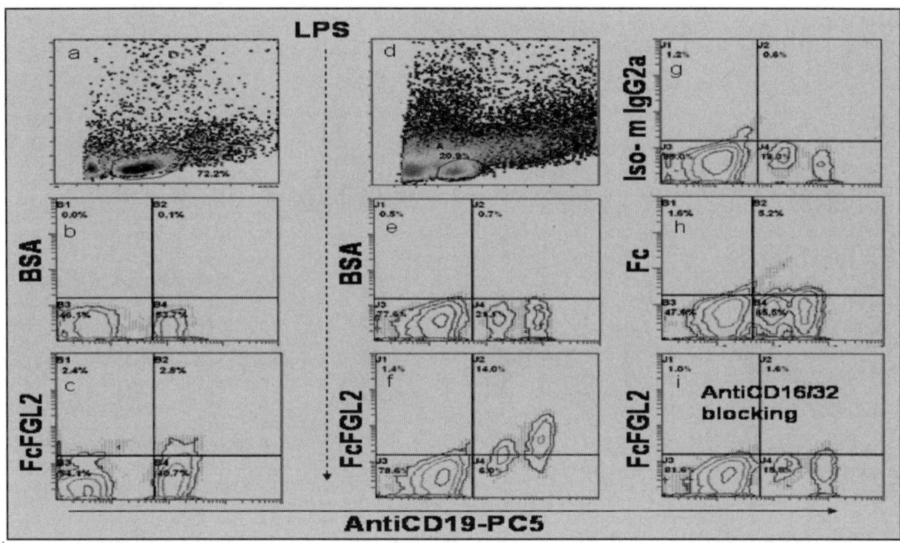

Figure 2. B cell from C57BL/6J binding with FcFGL2 after LPS stimulation can be blocked specifically by anti-CD16/32.

4. DISCUSSION

Here we have identified the inhibitory FcγRIIB as a receptor for sFGL2. As members of the triggering receptors expressed by myeloid cells (TREM), Fcγ receptors consist of both activating (FcγRI/CD64, FcγRIII/CD16 and FcγRIV/CD16-2) and inhibitory (Fcγ-RIIB/CD32) members. There are 2 allotypes of the Fcγ receptors, Ly17.1 and Ly17.2, which are associated with a two amino acid polymorphism in the second extracellular domain.[5] Uniquely, B lymphocytes express only CD32 and NK cells express only CD16 on their surfaces, while most immune cells express all three receptors.[6] The expression level of those receptors can be regulated by Th1/2 cytokines controlling the balance of immune responses.[7]

Viruses have evolved various strategies to counteract host immunity and thus, escape host immune surveillance. For example, coronavirus spike protein (S) displays Fcγ receptor activity. The anti-mouse FcγRII/III monoclonal antibody, 2.4 G2, has been shown to immunoprecipitate S protein of MHV, and by amino acid sequence analysis it was shown that the S protein and the FcγRs share homology.[8] Therefore, it was hypothesized that nonspecific antibody binding with the S protein will prevent antibody-dependent cellular cytotoxicity. This is not unique to MHV, and herpes simplex virus also can induce Fc receptor activity as a means of escaping immune surveillance.[9]

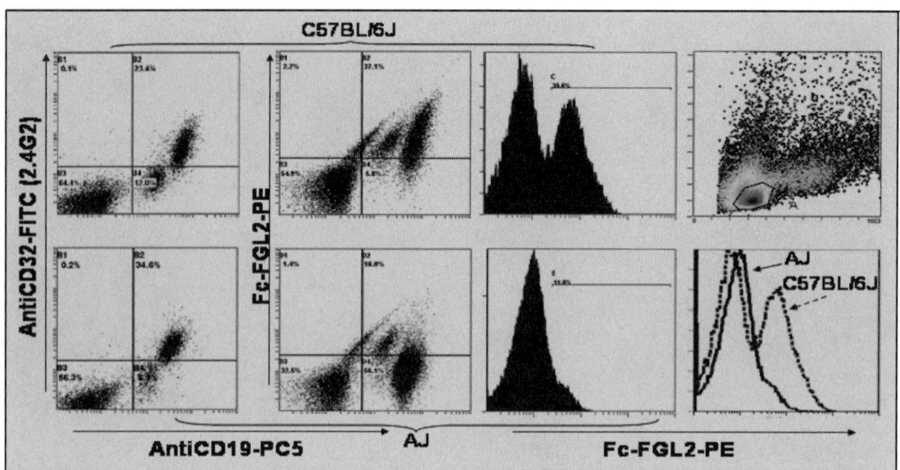

Figure 3. Recombinant Fc-FGL2 bind B cells from B6 but not A/J mice after LPS stimulation.

The potential biological effects of sFGL2 binding to the inhibitory low-affinity FcγRIIB remains unknown at present; however, it may result in inactivation of DC and macrophages and provide yet another mechanism by which MHV escapes immune surveillance. In susceptible mice, the interaction of sFGL2 with the activating CD16 may induce the activation of macrophages, NK cells, neutrophils, and platelets and contribute to disease. In addition, sFGL2 by binding with the S protein, may enhance viral replication. Finally, binding of sFGL2 to B cells may lead to apoptosis resulting in inhibition of neutralizing antibody production. Consistent with this concept sFGL2 does not bind to B cells from A/J mice known to be resistant to MHV, whereas it binds avidly to B cells from susceptible C57BL/6J mice.

5. ACKNOWLEDGMENTS

This work was supported by grants from the Canadian Institute of Health Research (grant no. 37780). Hao Liu is the recipient of a CIHR Regenerative Medicine Fellowship.

6. REFERENCES

1. Marsden, P. A., Ning, Q., Fung, L. S., Luo, X., Chen, Y., Mendicino, M., Ghanekar, A., Scott, J. A., Miller, T., Chan, C. W., Chan, M. W., He, W., Gorczynski, R. M., Grant, D. R., Clark, D. A., Phillips, M. J., Levy, G. A., 2003, The Fgl2/fibroleukin prothrombinase contributes to immunologically mediated thrombosis in experimental and human viral hepatitis, *J. Clin. Invest.* **112**:58-66.
2. Marazzi, S., Blum, S., Hartmann, R., Gundersen, D., Schreyer, M., Argraves, S., von Fliedner, V., Pytela, R., and Ruegg, C., 1998, Characterization of human fibroleukin, a fibrinogen-like protein secreted by T lymphocytes, *J. Immunol.* **161**:138-147.
3. Herman, A. E., Freeman, G. J., Mathis, D., and Benoist, C., 2004, CD4+CD25+ T regulatory cells dependent on ICOS promote regulation of effector cells in the prediabetic lesion, *J. Exp. Med.* **199**:1479-1489.
4. Chan, C. W., Kay, L. S., Khadaroo, R. G., Chan, M. W., Lakatoo, S., Young, K. J., Zhang, L., Gorczynski, R. M., Cattral, M., Rotstein, O., and Levy, G. A., 2003, Soluble fibrinogen-like protein 2/fibroleukin exhibits immunosuppressive properties: suppressing T cell proliferation and inhibiting maturation of bone marrow-derived dendritic cells, *J. Immunol.* **170**:4036-4044.

5. Schiller, C., Janssen-Graalfs, I., Baumann, U., Schwerter-Strumpf, K., Izui, S., Takai, T., Schmidt, R. E., and Gessner, J. E., 2000, Mouse FcgammaRII is a negative regulator of FcgammaRIII in IgG immune complex-triggered inflammation but not in autoantibody-induced hemolysis, *Eur. J. Immunol.* **30**:481-490.
6. Ravetch, J. V., and Kinet, J. P., 1991, Fc receptors, *Annu. Rev. Immunol.* **9**:457-492.
7. Pricop, L., Redecha, P., Teillaud, J. L., Frey, J., Fridman, W. H., Sautes-Fridman, C., and Salmon, J. E., 2001, Differential modulation of stimulatory and inhibitory Fc gamma receptors on human monocytes by Th1 and Th2 cytokines, *J. Immunol.* **166**:531-537.
8. Oleszak, E. L., Perlman, S., and Leibowitz, J. L., 1992, MHV S peplomer protein expressed by a recombinant vaccinia virus vector exhibits IgG Fc-receptor activity, *Virology* **186**:122-132.
9. Johnson, D. C., Frame, M. C., Ligas, M. W., Cross, A. M., and Stow, N. D., 1988, Herpes simplex virus immunoglobulin G Fc receptor activity depends on a complex of two viral glycoproteins, gE and gI, *J. Virol.* **62**:1347-1354.

VI. PATHOGENESIS OF ARTERIVIRUSES AND TOROVIRUSES

EQUINE VIRAL ARTERITIS

N. James MacLachlan and Udeni B. Balasuriya*

1. INTRODUCTION

Equine viral arteritis (EVA) is a contagious disease of horses caused by equine arteritis virus (EAV). EVA was first described perhaps 200 years or more ago. Horsemen also recognized long ago that otherwise healthy stallions could transmit the disease to susceptible mares at breeding, and that these "carrier" stallions could be a source of infection for many years. EAV was first isolated by Doll *et al.* during an outbreak of respiratory disease and abortion on a Standardbred breeding farm in Bucyrus, Ohio, in 1953.[1]

EAV infection of equines (horses, donkeys, and mules) occurs throughout the world, although the incidence of both EAV infection as well as clinical EVA varies markedly between countries and amongst horses of different breeds. The vast majority of EAV infections are inapparent or subclinical, but occasional outbreaks of EVA occur that are characterized by any combination of influenza-like illness in adult horses, abortion in pregnant mares, and fatal interstitial pneumonia in very young foals. International concern over EVA increased markedly following an extensive outbreak of the disease in Kentucky Throughbreds in 1984, and several other outbreaks have since been reported from North America and Europe. Similarly, EAV infection of horses has recently been identified in countries like Australia, New Zealand, and South Africa that were previously thought to be largely or completely free of the virus. This apparent global dissemination of EAV and rising incidence of EVA likely reflects the rapid national and international movement of horses for competition and breeding, as well as heightened diagnostic scrutiny as a consequence of increasing concern over the potential importance of EAV infection.[1,2]

2. EQUINE ARTERITIS VIRUS

EAV has previously been proposed to be an alphavirus, flavivirus, and a non-arthropod-transmitted togavirus.[3] It relatively recently was designated the prototype arterivirus in the family *Togaviridae* based on virion morphology.[4] More recent studies

*N. J. MacLachlan, School of Veterinary Medicine, University of California, Davis, California 95616. U. B. Balasuriya, University of Kentucky, Lexington, Kentucky 40509.

identified the distinctive nested set of 3' co-terminal mRNA that is generated during replication of EAV, which subsequently was shown to also be a feature of the replication of other arteriviruses as well members of the *Coronaviridae* and *Toroviridae.* Thus, these virus families now are grouped in the order *Nidovirales,* and EAV is the prototype virus in the family *Arteriviridae* (genus *Arterivirus*), a grouping that also includes porcine reproductive and respiratory syndrome virus, simian hemorrhagic fever virus, and lactate dehydrogenase–elevating virus of mice.[5]

The EAV virion is an enveloped, spherical 50–65 nm particle with an icosahedral core that contains a single-stranded, positive-sense RNA molecule of ~12.7 kilobases. The EAV genome includes a 5' leader sequence and nine open reading frames (ORFs). The two most 5'-proximal ORFs (1a and 1b) occupy approximately three-quarters of the genome and encode two replicase polyproteins that are extensively processed after translation. Although the early literature was highly confusing, it now is clearly established that the EAV virion includes six envelope proteins (E, GP2b, GP3, GP4, GP5, and M) and a nucleocapsid protein (N), which respectively are encoded by ORFs 2a, 2b, 3–7 that are located at the 3' proximal quarter of the genome. The greatest sequence variation in the ORFs encoding structural EAV proteins occurs in those (ORFs 3 and 5, respectively) that encode GP3 and GP5. GP5 expresses the major neutralization determinants of EAV and although there is considerable variation in the sequence of the GP5 protein of field strains of the virus, there is only one known serotype of EAV and all strains evaluated thus far are neutralized by polyclonal antiserum raised against the virulent Bucyrus strain. However, field strains of EAV are frequently distinguished on the basis of their neutralization phenotype with polyclonal antisera and monoclonal antibodies and, similarly, geographically and temporally distinct strains of EAV differ in the severity of the clinical disease they induce and in their abortigenic potential. Furthermore, although strains of EAV from North America and Europe share as much as 85% nucleotide identity, these viruses generally segregate into clusters reflective of their geographical origins following phylogenetic analysis.[6-9]

3. EPIDEMIOLOGY OF EQUINE ARTERITIS VIRUS INFECTION OF HORSES

EAV is spread by both the respiratory and venereal routes, and the persistently infected carrier stallion is the essential natural reservoir of the virus.[1, 2] Although the EAV carrier state in convalescent stallions had been recognized for many years, pioneering work to characterize this state was done by Drs. Peter Timoney and William McCollum in the course of their investigations of the 1984 outbreak of EVA in Kentucky. In subsequent investigations these investigators confirmed that the carrier state occurred in some 30–50% of exposed stallions and persisted for variable periods (short-term [< 3 months], intermediate [3 – 7 months], and long-term [> 7 months]). They also showed that EAV is confined to the reproductive tract during persistence, and that persistent infection does not occur in mares or geldings. The pathogenesis of the EAV carrier state remains poorly characterized, but it clearly is testosterone dependent as carrier stallions that are castrated but supplemented with testosterone continue to shed EAV in their semen whereas those that are not supplemented with testosterone cease to shed the virus. Not only is the carrier stallion the essential natural reservoir of EAV but genetic and antigenic variation is generated in the course of persistence, thus an increasingly diverse population of related viral variants (so-called quasispecies) is present in the semen of individual stallions.[10, 11] Outbreaks of EVA occur when one of these variants is transmitted to a susceptible cohort, which typically is a mare bred to the stallion although the virus also can be spread from carrier stallion by fomites (including semen-contaminated bedding). EAV can be efficiently transmitted by aerosol in populations of susceptible horses. Thus, the virus rapidly can spread amongst contact horses as occurred during an extensive outbreak of EVA that occurred in elite young racing Thoroughbred horses in the central U.S. in 1991 for example. The outbreak began at the Arlington track in Chicago and then spread to the tracks at Churchill Downs, Prairie Meadow, and Ak-Sar-Ben and ultimately affected more than 200 horses. In marked contrast to the quasispecies evolution of EAV in the reproductive tract of carrier stallions, there is minimal genetic change in EAV during outbreaks of EVA and the virus strains that cause individual outbreaks are genetically distinct.[12]

The seroprevalence of EAV infection varies not only between countries but also amongst horses of different breed and age, with especially marked disparity between the prevalence of infection of Standardbred (up to 85%) and Thoroughbred horses (< 5%) in the U.S.[13] EAV infection also is common in many European Warmblood breeds. There is no evidence of any breed-specific variation in susceptibility to EAV infection or in establishment of the carrier state, thus the number of actively shedding carrier stallions likely determines the prevalence of EAV infection in individual horse breeds. The virulence of the strains of EAV associated with individual horse breeds may not, however, be constant and those shed by carrier Standardbred stallions are often very highly attenuated and cause minimal if any disease in susceptible horses (regardless of breed).

4. EQUINE VIRAL ARTERITIS

Outbreaks of EVA recently have been reported from a number of European countries, Canada, and the U.S.[2] Outbreaks are often precipitated by the importation of carrier stallions, as in the first recorded outbreak of EVA in the United Kingdom, which followed the importation of an Anglo-Arab stallion from Poland. The clinical manifestations of EAV infection of horses vary markedly but most infections are inapparent.[1, 2] Outbreaks of clinical EVA are characterized by one more of the following: abortion of pregnant mares; fulminant infection of neonates leading to severe interstitial pneumonia or enteritis; systemic illness in adult horses with any combination of leukopenia and pyrexia, respiratory signs with nasal and ocular discharge, peripheral edema, hives, and persistent infection of stallions. The clinical signs observed in natural cases of EVA vary considerably among individual horses and between outbreaks, and depend on factors such as the age and physical condition of the horse(s), challenge dose and route of infection, strain of virus and environmental conditions. Although there is only one serotype of EAV the clinical disease produced by different virus strains ranges from severe, lethal infection caused by the horse-adapted Bucyrus strain to clinically inapparent infection.[14-16] Very young, old, debilitated or immunosuppressed horses are predisposed to severe EVA. Regardless of the infecting virus strain, the vast majority of naturally infected horses recover uneventfully from EVA. With the notable exceptions of abortion and fulminant respiratory disease in foals, mortality rarely if ever occurs in natural outbreaks of EVA. The highly virulent horse-adapted Bucyrus strain of EAV (that causes high mortality in healthy adult horses) is not representative of field strains of the virus and is best regarded as a laboratory strain; however, standard texts often describe the disease caused by this virus.

The clinical manifestations of EVA reflect vascular injury with increased permeability and leakage of fluid. EAV replicates in macrophages and endothelial cells within the lungs following aerosol respiratory infection, from where it rapidly is disseminated throughout the body.[14, 17] The relative roles and importance of direct virus-mediated endothelial cell injury versus virus-induced macrophage-derived vasoactive and inflammatory cytokines in the pathogenesis of EAV-induced vascular injury are not yet clearly defined, however it is abundantly clear that strains of EAV of different virulence to horses differ in both their cytopathogenicity to endothelial cells as well as their ability to induce proinflammatory cytokines.

5. SUMMARY

EVA is an important if uncommon disease of horses. Potential economic losses attributable to EVA include direct losses from abortion, pneumonia in neonates, and febrile disease in performance horses. Indirect losses are those associated with national and international trade/animal movement regulations, particularly those pertaining to persistently infected carrier stallions and their semen. However, EAV infection and EVA are readily prevented through serological and virological screening of horses, coupled with sound management practices that include appropriate quarantine and strategic vaccination.

6. REFERENCES

1. P. J. Timoney and W. H. McCollum, Equine viral arteritis, *Vet. Clin. North Am. Equine Pract.* **9**, 295-309 (1993).
2. U. B. Balasuriya and N. J. MacLachlan, in: *Infectious Diseases of the Horse*, edited by D. Sellon and M. Long (Elsevier), in press.
3. M.C. Horzinek, *Non-arthropod-borne togaviruses* (Academic Press, London, 1981).
4. F. G. Westaway, M. A. Brinton, S. Y. Gaidamovich, M. C. Horzinek, L. Igarashi, L. Kaarianen, D. K. Lvov, J. S. Porterfield, P. K. Russell, and D. W. Trent, Togaviridae, *Intervirology* **24**, 125-139 (1985).
5. E. J. Snijder and J. M. Meulenberg, The molecular biology of arteriviruses, *J. Gen. Virol.* **79**, 1-17 (1998).
6. U. B. Balasuriya, P. J. Timoney, W. H. McCollum, and N. J. MacLachlan, Phylogenetic analysis of open reading frame 5 of field isolates of equine arteritis virus and identification of conserved and nonconserved regions in the G_L envelope glycoprotein, *Virology* **214**, 690-697 (1995).
7. U. B. Balasuriya, H. W. Heidner, J. F. Hedges, J. C. Williams, N. L. Davis, R. E. Johnston, and N. J. MacLachlan, Expression of the two major envelope glycoproteins of equine arteritis virus as a heterodimer is necessary for induction of neutralizing antibodies in mice immunized with recombinant Venezuelan equine encephalitis virus replicon particles, *J. Virol.* **74**, 10623-10630 (2001).
8. U. B. Balasuriya, J. C. Dobbe, H. W. Heidner, V. L. Smalley, A. Navarette, E. J. Snijder, and N. J. MacLachlan, Characterization of the neutralization determinants of equine arteritis virus using recombinant chimeric viruses and site-specific mutagenesis of an infectious cDNA clone, *Virology* **321**, 235-246 (2004).
9. U. B. Balasuriya and N. J. Maclachlan, The immune response to equine arteritis virus: potential lessons for other arteriviruses, *Vet. Immunol. Immunopathol.* **102**, 107-129 (2004).
10. J. F. Hedges, U. B. Balasuriya, P. J. Timoney, W. H. McCollum, and N. J. MacLachlan, Genetic divergence with emergence of phenotypic variants of equine arteritis virus during persistent infection of stallions, *J. Virol.* **73**, 3672-3681 (1999).
11. U. B. Balasuriya, J. F. Hedges, V. L. Smalley, A. Navarrette, W. H. McCollum, P. J. Timoney, E. J. Snijder, and N. J. MacLachlan, Genetic characterization of equine arteritis virus during persistent infection of stallions, *J. Gen. Virol.* **85**, 379-390 (2004).
12. U. B. Balasuriya, J. F. Hedges, P. J. Timoney, W. H. McCollum, and N. J. MacLachlan, Genetic stability of equine arteritis virus during horizontal and vertical transmission in an outbreak of equine viral arteritis, *J. Gen. Virol.* **80**, 1949-1958 (1999).

13. P. J. Hullinger, I. A. Gardner, S. K. Hietala, G. L. Ferraro, and N. J. MacLachlan, Seroprevalence of antibodies against equine arteritis virus in horses residing in the United States and imported horses, *J. Am. Vet. Med. Assoc.* **219**, 946-949 (2001).
14. N. J. MacLachlan, U. B. Balasuriya, P. V. Rossitto, P. A. Hullinger, J. F. Patton, and W. D. Wilson, Fatal experimental equine arteritis virus infection of a pregnant mare: cellular tropism as determined by immunohistochemical staining, *J. Vet. Diagn. Invest.* **8**, 367-374 (1996).
15. J. F. Patton, U. B. Balasuriya, J. F. Hedges, T. M. Schweidler, P. J. Hullinger, and N. J. MacLachlan, Phylogenetic characterization of a highly attenuated strain of equine arteritis virus from the semen of a persistently infected standardbred stallion, *Arch. Virol.* **144**, 817-827 (1999).
16. B. F. Moore, U. B. Balasuriya, J. F. Hedges, and N. J. MacLachlan, Growth characteristics of a highly virulent, a moderately virulent, and an avirulent strain of equine arteritis virus in primary equine endothelial cells are predictive of their virulence to horses, *Virology,* **298**, 39-44 (2002).
17. B. F. Moore, U. B. Balasuriya, J. L. Watson, C. M. Bosio, R. J. MacKay, and N. J. MacLachlan, Virulent and avirulent strains of equine arteritis virus induce different quantities of TNF-alpha and other proinflammatory cytokines in alveolar and blood-derived equine macrophages, *Virology*, **314**, 662-670 (2003).

USE OF A PRRSV INFECTIOUS CLONE TO EVALUATE *IN VITRO* QUASISPECIES EVOLUTION

Susan K. Schommer and Steven B. Kleiboeker*

1. INTRODUCTION

Genetic, phenotypic, and antigenic heterogeneity of porcine reproductive and respiratory syndrome virus (PRRSV) has been well described.[1] Genetic diversity is a hallmark of RNA viruses. Virus populations *in vivo*, referred to as quasispecies, are comprised of a heterogeneous mix of related variants that are randomly generated as a result of errors by the viral RNA-dependent RNA polymerase. This diversity is considered to be an important mechanism of virus persistence and pathogenesis in many virus systems and provides a mechanism to rapidly respond to changes in the host environment.

Initially PRRSV genetic variation was studied using consensus genome sequences. There have been several papers published on PRRSV variation during experimental infection of pigs,[2–5] however only one started with a biologically cloned virus.[4] Our laboratory is currently using a North American infectious clone to investigate the mechanisms of PRRSV persistence and pathogenesis. One purpose of this study was to evaluate the stability of the infectious clone for its manipulation and subsequent use. Use of the infectious clone also allows us to begin with a single DNA sequence, providing a well-defined starting point for studying PRRSV evolution. The other goal of this study was to investigate PRRSV quasispecies evolution in an environment that excludes the immunologic pressure that is present in the previous studies which involved experimental infection in swine.

Four regions of the genome were selected for analysis in this study: Nsp2, ORF3, ORF5, and ORF6. The Nsp2 protein is the most variable region among arteriviruses,[6, 7] and has been implicated in having a role in the humoral immune response.[8] The ORF3 protein has the greatest percentage of amino acid changes between the modified live vaccine (Ingelvac) and its parent strain, VR-2332, the isolate from which the infectious clone used in this study was derived. ORF5 has been the focus of previous quasispecies investigations[2–5] and its corresponding protein has been associated with virus neutralization.[9] The most conserved region of the PRRSV genome across all North

* University of Missouri, Columbia, Missouri 65211.

American isolates is ORF6[10, 11] and in this study its analysis serves as a basis for comparison in order to safeguard against identifying variation due to errors in methodology or bias as true genetic diversity.

2. MATERIALS AND METHODS

The North American PRRSV infectious clone derived from VR-2332[12] was obtained, used to transform *Escherichia coli* Top10 strain (Invitrogen), and reisolated 3 times to obtain a pure colony of transformed bacteria. Plasmid DNA was isolated from this clone and linearized using Acl1. Linearized plasmid was used for in vitro transcription with T7 RNA polymerase (Ambion). The *in vitro* transcribed RNA was DNase-treated and isolated using the RNA Nucleospin II kit (Clontech). BHK-21 cells were transfected with *in vitro* transcribed RNA and Lipofectamine 2000 (Gibco Life Sciences). The supernatant of transfected cells was collected at 24, 48, and 72 hours post-transfection and passaged onto MARC-145 cells. Two wells (A and B) were positive after the first passage on MARC-145 cells and were passaged two additional times on this cell line and used for subsequent analysis.

RT-PCR was performed directly on RNA extracted from the *in vitro* transcribed RNA and passages 1 and 3 of wells A and B. These sequences were then compared to a low passage VR-2332 cell culture propagated stock to determine if the use of an infectious clone was able to decrease the quasispecies variation as compared to a viral stock. PCR was performed directly on the plasmid.

Four methods of RT-PCR amplification were evaluated for maintenance of sequence fidelity by sequencing the highly conserved ORF6. cDNA synthesis by Invitrogen's SuperScript cDNA synthesis kit followed by PCR with Pfx Platinum Taq was selected based on no identified nucleotide changes in 4650 nucleotides of ORF6 sequence. A minimum of 2 PCR reactions were combined for each sample then cloned into a Zero Blunt TOPO vector (Invitrogen), with a minimum of 15 clones sequenced for each sample and genetic region. Each clone was sequenced in both directions at the University of Missouri DNA Core. The sequence was analyzed using the DNA Star software package.

3. RESULTS

A North American infectious cDNA clone, derived from VR-2332, was used to transform bacteria from which a single bacterial colony was reisolated three times. This plasmid clone was then used as a template to generate *in vitro* transcribed RNA which was then transfected into BHK-21 cells. Two wells (A and B) were CPE positive after one passage on MARC-145 cells and were used for subsequent studies.

Fifteen clones were sequenced and aligned for each sample and genetic region. For each infectious clone derived group, the master or dominant sequence was the same as the original plasmid. Increased passage number generally correlated with a decrease in the percentage of individual clones identical to the master sequence but this was not consistent across all open reading frames analyzed (Table 1). The nucleotide changes from the master sequence were predominantly transitions and nonsynonymous. The nucleotide and amino acid changes were randomly distributed within the genes, with no mutation hotspots identified.

Table 1. Amino acid master sequence percentage summary.

Sample	Nsp2	ORF3	ORF5	ORF6
Plasmid	100	93.3	93.3	100
In vitro RNA	86.7	86.7	100	93.3
Well A p1	93.3	80.0	86.7	ND[a]
Well A p3	93.3	73.3	92.9	ND
Well B p1	86.7	93.3	86.7	86.7
Well B p3	66.7	86.7	73.3	100
VR-2332 p4	80.0	80.0	73.3	80.0

[a]ND indicates that the data has not been analyzed for these samples at this time.

4. CONCLUSIONS

The master sequence for each sample derived from the infectious clone was the same as the original plasmid for all genetic regions investigated. Analysis of the *in vitro* transcribed RNA showed that Nsp2, ORF3, and ORF6 all had low levels of genetic variation even though it was prepared directly from a single bacterial colony plasmid preparation. Increased passage number generally correlated with a decrease in the percentage of the master sequence. This does not appear to be due to the emergence of more fit viral sequences because the master sequence remains the same and new quasispecies appear and disappear with passage. There were no cases of a variant present in passage 1 also being present in passage 3 of the same well, although the variants appeared at such low levels that they may not be detected with our sample size. It is also unlikely that these changes are a result of error due to the experimental methodology. Minimal error was introduced through the process of PCR, cloning and sequencing as evidenced by the fact that there were no nucleotide changes in ORF6 of the plasmid DNA in 6525 nucleotides of sequence examined. Additionally, analysis of the *in vitro* RNA ORF6 region found only 1 nucleotide change in 6525 nucleotides.

In total, our data suggest that even with the use of an infectious clone it will be difficult, if not impossible, to create point mutations without accumulating other changes in the genome, especially if multiple in vitro passages are required to obtain sufficient viral titers. This will be especially critical when the mutations made are detrimental to the fitness of the virus. In this experiment, quasispecies changes appear with the first cell culture passage and were readily observed even though our study only looked at 10% of the genome. Researchers using a PRRSV infectious clone, particularly this one derived from VR-2332, to study the effects of genetic changes on viral phenotype must keep in mind that other unintended mutations may have occurred during the propagation of the virus.

5. REFERENCES

1. Meng, X.-J., 2000, Heterogeneity of porcine reproductive and respiratory syndrome virus: implications for current vaccine efficacy and future vaccine development, *Vet. Microbiol.* **74**:309-329.
2. Rowland, R. R., Steffen, M., Ackerman, T., and Benfield, D. A., 1999, The evolution of porcine reproductive and respiratory syndrome virus: quasispecies and emergence of a virus subpopulation during infection of pigs with VR-2332, *Virology* **259**:262-266.
3. Allende, R., Laegried, W. W., Kutish, G. F., Galeota, J. A., Will, R. W., and Osario, F. A., 2000, Porcine reproductive and respiratory syndrome virus: description of persistence in individual pigs upon experimental infection, *J. Virol.* **74**:10834-10837.
4. Chang, C. C., Yoon, K. J., Zimmerman, J. J., Harmon, K. M., Dixon, P. M., Dvorak, C. M. T., and Murtaugh, M. P., 2002, Evolution of porcine reproductive and respiratory syndrome virus during sequential passages in pigs, *J. Virol.* **76**:4750-4763.
5. Goldberg, T. L., Lowe, J. F., Milburn, S. M., and Firkins, L. D., 2003, Quasispecies variation of porcine reproductive and respiratory syndrome virus during natural infection, *Virology*. **317**:197-207.
6. de Vries, A. A., Chirnside, E. D., Horzinek, M. C., and Rottier, P. J., 1992, Structural proteins of equine arteritis virus, *J. Virol.* **66**:6294-6303.
7. Allende, R., Lewis, T. L., Lu, Z., Rock, D. L., Kutish, G. F., Ali, A., Doster, A. R., and Osario, F. A., 1999, North American and European porcine reproductive and respiratory syndrome viruses differ in non-structural protein coding regions, *J. Gen. Virol.* **80**:307-315.
8. Oleksiewicz, M. B., Botner, A., Toft, P., Normann, P., and Storgaard, T., 2001, Epitope mapping porcine reproductive and respiratory syndrome virus by phage display: the nsp2 fragment of the replicase polyprotein contains a cluster of B-cell epitopes, *J. Virol.* **75**:3277-3290.
9. Pirzadeh, B., and Dea, S., 1997, Monoclonal antibodies to the ORF5 product of porcine reproductive and respiratory syndrome virus define linear neutralizing determinants, *J. Gen. Virol.* **78**:1867-1873.
10. Meng, X. J., Paul, P. S., Halbur, P. G., and Lum, M. A., 1995, Phylogenetic analyses of the putative M (ORF 6) and N (ORF 7) genes of porcine reproductive and respiratory syndrome virus (PRRSV): implication for the existence of two genotypes of PRRSV in the USA and Europe, *Arch. Virol.* **140**:745-755.
11. Gagnon, C. A., and Dea, S., 1998, Differentiation between porcine reproductive and respiratory syndrome virus isolates by restriction fragment length polymorphism of their ORFs 6 and 7 genes, *Can. J. Vet. Res.* **62**:110-116.
12. Nielsen, H. S., Liu, G., Nielsen, J., Oleksiewicz, M. B., Botner, A., Storgaard, T., and Faaberg, K. S., 2003, Generation of an infectious clone of VR-2332, a highly virulent North American-type isolate of porcine reproductive and respiratory syndrome virus, *J. Virol.* **77**:3702-3711.

GAMMA-INTERFERON INVOLVEMENT IN THE PATHOGENESIS OF LACTATE DEHYDROGENASE–ELEVATING VIRUS INFECTION

Andrei Musaji, Dominique Markine-Goriaynoff, Stéphanie Franquin, Gaëtan Thirion, Thao Le Thi Phuong, and Jean-Paul Coutelier*

1. INTRODUCTION

Viruses are involved in many different diseases through quite distinct mechanisms. Lytic infections induce direct cell or tissue destruction leading to such diseases as hepatitis, encephalitis, diabetes, among many others. Polioencephalomyelitis triggered by infection with lactate dehydrogenase–elevating virus (LDV) in immunosuppressed mice that are co-infected with a retrovirus is a direct consequence of lytic infection of motor neurons in the spinal cord.[1] Similar pathologies may also follow indirect tissue destruction, mediated by immune mechanisms initially directed against the invading virus. For example, the lethal neurological disease induced by lymphocytic choriomeningitis virus (LCMV) is a consequence of killing of virally-infected cells by LCMV-specific cytolytic T lymphocytes (reviewed in Ref. 2). Many virally-induced autoimmune diseases may also result from either cross-reactivity between viral and self-epitopes or spreading of an antiviral response to self-antigens. Finally, pathologies may be caused by nonspecific bystander effects of infections, like cytokine secretion. Here, we report how LDV may exacerbate autoantibody-mediated diseases through such a mechanism. Special emphasis is put on the role of gamma-interferon (IFN-γ) production on the outcome of concomitant autoantibody-mediated autoimmune diseases such as hemolytic anemia and thrombocytopenic purpura.

2. LDV-INDUCED CYTOKINE BURST

LDV infection is characterized by a rapid viral replication in a subpopulation of macrophages, leading to disappearance of most of these target cells.[3] As a consequence, viremia reaches early and high levels, before dropping to stable, but more modest

* Université Catholique de Louvain, Brussels, Belgium.

values. Like many other viruses, LDV induces an inflammatory response, with a burst of cytokines from infected cells and/or cells of the innate immune system that include interleukins 6,[4] 12,[5] 15 and 18 (unpublished results). However, because of its restricted tropism and its peculiar viremia kinetics, this cytokine production is observed both very early and very transiently after viral inoculation. As a probable consequence of this first volley of cytokine secretion, natural killer (NK) cells are strongly activated one to four days after LDV infection.[6] This NK cell activation leads to increased lytic activity against potential target cells. It results also in an enhanced production of gamma-interferon (IFN-γ). Treatment of mice with anti-asialoGM1 antibody clearly demonstrated that NK cells were responsible for this IFN-γ secretion, whereas anti-CD4 and anti-CD8 antibodies had no effect on this cytokine response.[6] However, preliminary data obtained in mice deficient for this cell subpopulation suggested that NK/T cells play also probably an important role in LDV-induced IFN-γ production (unpublished results).

Because in most normal immunocompetent mouse strains LDV does not induce clinical pathology by itself, and since most, if not all infected mice display a very similar cytokine response after LDV inoculation, this virus may serve as a good experimental model to analyze how a pro-inflammatory immune response induced by a virus may interfere with host pathologies that were not initially of viral origin.

3. EXACERBATION OF AUTOANTIBODY-MEDIATED DISEASES IN LDV-INFECTED MICE

The pathogenicity of polyclonal rabbit anti-mouse platelet antibody was strongly exacerbated in mice acutely infected with LDV.[7,8] This led to severe thrombocytopenia and to the development of purpuric lesions reminiscent of human thrombocytopenic purpura.[7] A similar enhancement of antibody pathogenicity was observed in LDV-infected mice that received monoclonal anti-mouse platelet autoantibodies, derived either from (NZB x BXSB)F1 mice or from animals that developed an autoimmune anti-platelet response after immunization with rat platelets.[9] Infection with mouse hepatitis virus (MHV) resulted in the same enhancing effect of autoantibody pathogenicity.[7] Moreover, anemia induced by an anti-erythrocyte monoclonal antibody was also strongly exacerbated in mice infected with LDV.[10] Interestingly, this consequence of LDV infection was found with an IgG2a autoantibody that induces anemia through phagocytosis, but not with an IgG1 autoantibody that lead to a similar disease through distinct mechanisms.[11] Because enhancement of anti-platelet antibody pathogenicity by LDV infection required the presence of the Fc fraction of this antibody,[7] these results suggested that phagocytosis of autoantibody-opsonized target cells was increased in infected mice. Indeed, *ex vivo* phagocytosis of autoantibody-coated erythrocytes was more efficient with macrophages derived from LDV acutely infected mice than from uninfected animals.[10] Moreover, LDV-enhanced, antibody-mediated thrombocytopenia was inhibited by treatment with total immunoglobulins that block Fc-receptor-mediated phagocytosis of opsonized cells.[7,8,12] Finally, LDV-infected mice were treated with clodronate-containing liposomes that destroy phagocytic macrophages *in vivo*[13] and thus prevent autoimmune diseases that occur through this mechanism.[14] This treatment prevented LDV-enhanced autoantibody-mediated autoimmune disease,[7,8] as well as LDV-induced increase of *ex vivo* macrophage phagocytosis of opsonized red cells (Figure 1). Together, these results indicate that the pathogenic effect of LDV infection involves enhancement of the phagocytic activity of macrophages.

Because IFN-γ is known to activate macrophages, the role of this cytokine in the enhancement of autoantibody-mediated disease by LDV was tested by using mice deficient for the IFN-γ receptor,[15] or neutralizing anti-IFN-γ antibodies. The results of these experiments indicated that IFN-γ secretion was required for the exacerbation of phagocytosis-mediated autoantibody autoimmune diseases by LDV infection.[7,8]

4. CONCLUSIONS

Our results indicate that LDV may exacerbate autoantibody-mediated autoimmune diseases such as hemolytic anemia or thrombocytopenic purpura through secretion of IFN-γ that results in an enhancement of phagocytosis of opsonized target cells by activated macrophages. Because a similar mechanisms or can be triggered by other viruses, like MHV, it may explain how widely different viruses lead to similar clinical autoimmune diseases shortly after infection. It may be postulated that autoantibodies are present in these patients at a dose insufficient to be pathogenic by themselves, and that the clinical manifestation of these antibodies is indirectly triggered by macrophage activation in the course of the viral infection. Other pathologies may similarly result from the IFN-γ secretion that follows infection with viruses including LDV. For instance, preliminary data indicate that production of this cytokine in the course of LDV infection leads to increased susceptibility of mice to endotoxin-mediated septic shock.

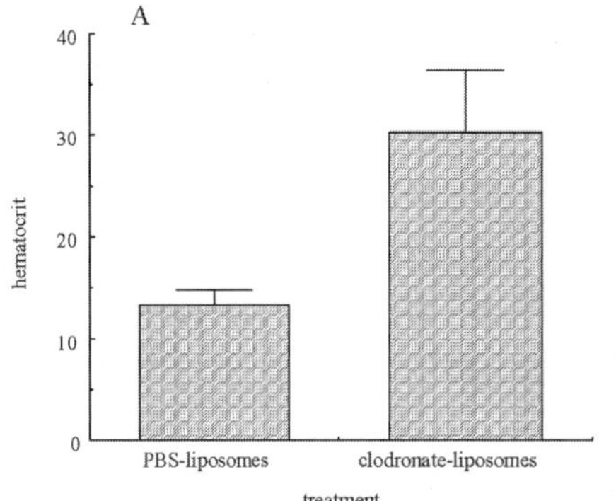

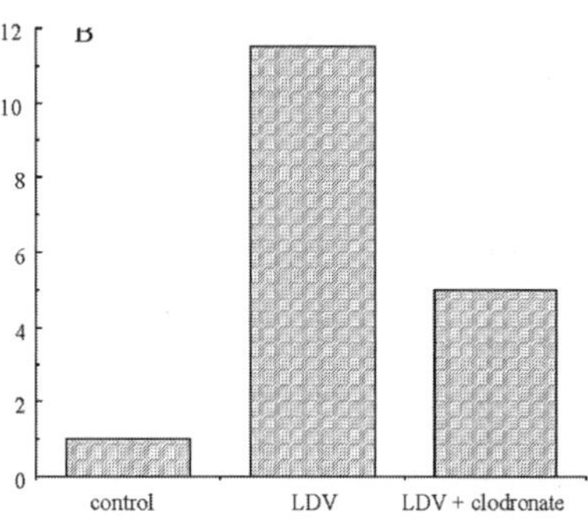

Figure 1. Effect of clodronate-containing liposome treatment on LDV-enhanced anti-erythrocyte response. A. Hematocrits (means ± SEM) in groups of 10 BALB/C mice 5 days after concomitant administration of LDV and of the 34-3C IgG2a anti-erythrocyte monoclonal antibody. PBS- or clodronate-containing liposomes were injected one day before virus and antibody administration. B. Erythrophagocytosis (% of cells having ingested at least 5 opsonized red cells after *ex vivo* incubation) by pooled peritoneal cells derived from groups of 5–6 control BALB/C mice, untreated mice infected for 4 days with LDV, or animals treated with clodronate-containing liposomes one day before similar infection.

Moreover, viruses like LDV or MHV modulate also the differentiation of T helper lymphocyte subpopulations,[16,17] which may also affect the outcome of immune pathologies such as cell-mediated autoimmune diseases or allergies. IFN-γ, which regulates the differentiation of these cells, appears to be involved in this consequence of viral infections (unpublished results). Therefore, secretion of cytokines, and especially of IFN-γ, may explain the indirect pathogenis effect of viral infections.

5. ACKNOWLEDGMENTS

The authors are indebted to V. Préat for expert help in the preparation of liposomes, to S. Izui for the kind gift of reagents, and to T. Briet, M.-D. Gonzales, and N. Ouled Haddou for technical assistance.

This work was supported by the Fonds National de la Recherche Scientifique (FNRS), Fonds de la Recherche Scientifique Médicale (FRSM), Loterie Nationale, Fonds Spéciaux de Recherche (UCL), NATO, and the State-Prime Minister's Office - S.S.T.C. and the French Community (concerted actions), Belgium. J.-P.C. is a research director with the FNRS.

6. REFERENCES

1. Contag, C. H., and Plagemann, P. G. W., 1989, Age-dependent poliomyelitis of mice: expression of endogenous retrovirus correlates with cytocidal replication of lactate dehydrogenase-elevating virus in motor neurons, *J. Virol.* **63**:4362.
2. Borrow, P., and Oldstone, M. B. A., 1997, Lymphocytic choriomeningitis virus, in: *Viral Pathogenesis*, N. Nathanson, R. Ahmed, F. Gonzalez-Scarano, D. E. Griffin, K. V. Holmes, F. A. Murphy and H. L. Robinson, eds., Lippincott-Raven, Philadelphia, pp. 593-627.
3. Stueckemann, J. A., Ritzi, D. M., Holth, M., Smith, M. S., Swart, W. J., Cafruny, W. A., and Plagemann, P. G. W., 1982, Replication of lactate dehydrogenase-elevating virus in macrophages. 1. Evidence for cytocidal replication, *J. Gen. Virol.* **59**:245.
4. Markine-Goriaynoff, D., Nguyen, T. D., Bigaignon, G., Van Snick, J., and Coutelier, J.-P., 2001, Distinct requirements for IL-6 in polyclonal and specific Ig production induced by microorganisms, *Intern. Immunol.* **13**:1185.
5. Coutelier, J.-P., Van Broeck, J., and Wolf, S. F., 1995, Interleukin-12 gene expression after viral infection in the mouse, *J. Virol.* **69**:1955.
6. Markine-Goriaynoff, D., Hulhoven, X., Cambiaso, C. L., Monteyne, P., Briet, T., Gonzalez, M.-D., Coulie, P., and Coutelier, J.-P., 2002, Natural killer cell activation after infection with lactate dehydrogenase-elevating virus, *J. Gen. Virol.* **83**:2709.
7. Musaji, A., Cormont, F., Thirion, G., Cambiaso, C. L., and Coutelier, J.-P., 2004a, Exacerbation of autoantibody-mediated thrombocytopenic purpura by infection with mouse viruses, *Blood* **104**:2102.
8. Musaji, A., Meite, M., Detalle, L., Franquin, S., Cormont, F., Préat, V., Izui, S., and Coutelier, J.-P., 2005, Enhancement of autoantibody pathogenicity by viral infections in mouse models of anemia and thrombocytopenia, *Autoimmunity Reviews* **4**:247.
9. Musaji, A., Vanhoorelbeke, K., Deckmyn, H., and Coutelier, J.-P., 2004b, New model of transient strain-dependent autoimmune thrombocytopenia in mice immunized with rat platelets, *Exp. Hematol.* **32**:87.
10. Meite, M., Léonard, S., El Azami El Idrissi, M., Izui, S., Masson, P. L., and Coutelier, J.-P., 2000, Exacerbation of autoantibody-mediated hemolytic anemia by viral infection, *J. Virol.* **74**:6045.
11. Shibata, T., Berney, T., Reininger, L., Chicheportiche, Y., Ozaki, S., Shirai, T., and Izui, S., 1990, Monoclonal anti-erythrocyte autoantibodies derived from NZB mice cause autoimmune hemolytic anemia by two distinct pathogenic mechanisms, *Int. Immunol.* **2**:1133.
12. Pottier, Y., Pierard, I., Barclay, A., Masson, P. L., and Coutelier, J.-P., 1996, The mode of action of treatment by IgG of haemolytic anaemia induced by an anti-erythrocyte monoclonal antibody, *Clin. Exp. Immunol.* **106**:103.

13. Van Rooijen, N., 1989, The liposome-mediated macrophage "suicide" technique, *J. Immunol. Methods* **124**:1.
14. Jordan, M. B., van Rooijen, N., Izui, S., Kappler, J., and Marrack, P., 2003, Liposomal clodronate as a novel agent for treating autoimmune hemolytic anemia in a mouse model, *Blood* **101**:594.
15. Huang, S., Hendriks, W., Althage, A., Hemmi, S., Bluethmann, H., Kamijo, R., Vilcek, J., Zinkernagel, R. M., and Aguet, M., 1993, Immune response in mice that lack the interferon-γ receptor, *Science* **259**:1742.
16. Monteyne, P., Van Broeck, J., Van Snick, J., and Coutelier, J.-P., 1993, Inhibition by lactate dehydrogenase-elevating virus of in vivo interleukin 4 production during immunization with keyhole limpet haemocyanin, *Cytokine* **5**:394.
17. Monteyne, P., Renauld, J.-C., Van Broeck, J., Dunne, D. W., Brombacher, F., and Coutelier, J.-P., 1997, IL-4-independent regulation of in vivo IL-9 expression, *J. Immunol.* **159**:2616.

REPLICATION AND EXPRESSION ANALYSIS OF PRRSV DEFECTIVE RNA

Jun Han, Kelly M. Burkhart, Eric M. Vaughn, Michael B. Roof, and Kay S. Faaberg*

1. INTRODUCTION

Porcine reproductive and respiratory syndrome virus (PRRSV) is a member of the family *Arteriviridae*, which was recently grouped together with the coronaviruses and the toroviruses in the newly established order of the *Nidovirales*.[1] PRRSV is a spherical, enveloped virus with a diameter of 50 to 60 nm[2] and a positive-stranded RNA genome of 15.0–5.5 kilobases.[3]

Nidovirus defective RNAs have been well documented and studied, especially for coronaviruses[4–6] and equine arterivirus (EAV).[7] Defective RNAs are truncated, and in some cases, rearranged genomes that have usually lost the potential to replicate autonomously due to deletions in the viral replicase gene(s), the replication of which depends on the replicase encoded by the helper virus.[7] Defective RNAs have retained all replication signals and, frequently, also the sequences required for RNA encapsidation. However, the generation of many of nidovirus defective RNAs requires serial undiluted viral passage in cultured cells.[4, 5, 7] We have suggested that PRRSV defective RNAs are different because they are persistently generated during infection, and arise *in vitro* as well as *in vivo* and have termed these RNA species *heteroclites* (*latin:* uncommon forms). PRRSV defective RNAs are heterogeneous in size and sequence and consist of, exclusively, complete 5' and 3' termini joined by short but variable nucleotide repeats. Thus, they contain different lengths of ORF1a and 3'-end sequences and are packaged into virions along with full-length PRRSV genomes.[8,9]

In order to further characterize PRRSV heteroclites, we have sequenced several species that represent varying lengths, and have prepared cDNA clones that correspond to two species (S1 and S7). In this study, we examined basic features of heteroclites S1 and S7 such as maintenance, RNA packaging and the encoding capability.

* Jun Han, Kay S. Faaberg, University of Minnesota, Saint Paul, Minnesota 55108. Kelly M. Burkhart, Eric M. Vaughn, Michael B. Roof, Boehringer Ingelheim Vetmedica, Incorporated, Ames, Iowa 50010.

2. METHODS

Cells, virus, modified heteroclites and RNA transcription, transfection and analysis: MA-104 cells were grown in DMEM medium containing 10% FBS (Invitrogen) at 37°C, 5% CO2. PRRSV North American strain VR-2332 infectious clone pVR-V7 was used in the study. In-frame insertion mutagenesis of type 2 porcine circovirus ORF2 gene (PCVORF2) into different positions of PRRSV heteroclite cDNAs of S7 or S1 was achieved by overlapping extension PCR. The resultant constructs were confirmed by sequencing and named pS7-PCVORF2/StuI (after leader sequence), pS7-PCVORF2/KpnI (genome position 480), and pS1-PCVORF2/SnaBI (genome position 1338). Linearized PRRSV infectious clone pVR-V7 and PCVORF2 gene modified S1 and S7 derivatives were obtained. RNAs were synthesized using mMESSAGE mMachine (Ambion). VR-V7 RNA was transfected alone or cotransfected with S7-PCVORF2/StuI, S2-PCVORF2/SnaBI or S7-PCVORF2/KpnI RNAs as described.[10] Northern blot analysis on infected cell supernatant was performed as described by Yuan.[8] Heteroclite RNA was translated with Flexi Rabbit Reticulocyte Lysate System (Promega) in the presence of ^{35}S-methionine (Amersham) and immunoprecipitated.[9,11]

Immunizations: Groups of 3–4 weeks PRRSV negative pigs (PCV antibody negative) were immunized by intramuscular injection of 2 ml of each PRRSV modified virus discussed above. Blood samples were collected from all animals on days 0, 3, 7, 14, 21 and 28 and examined for the presence of heteroclites and development of αPCV ORF2 antibodies. Total RNAs from swine serum samples were isolated (QIAamp Viral RNA; Qiagen) followed by nested RT-PCR to amplify the PCV ORF2 gene.

3. RESULTS

3.1. *In Vitro* Transcribed Heteroclites are Replicated and Packaged by PRRSV

After 6 cell passages, the heteroclites were detected by Northern blot analysis using a PCVORF2 gene-specific probe, demonstrating that PCVORF2 S7 and S1 RNAs were efficiently replicated and packaged during PRRSV infection (Fig. 1a). Inoculation of growing swine with passage 2 cell supernatants containing full-length virus and modified heteroclites was completed to assess replication of these heteroclites *in vivo*. Nested RT-PCR analysis on swine serum samples post-infection revealed that PCVORF2 mRNA could still be detected, which suggested that these defective RNA species could be replicated *in vivo*. PCVORF2 modified S7 replicated more efficiently than S1, because pigs inoculated with PCVORF2 modified S7 harbored the ORF2 gene for a longer time period (Fig. 1B). [Due to unknown reasons, control swine (α-PCV antibody negative) harbored PCVORF2 gene, but they were predominantly detected only in early infection.]

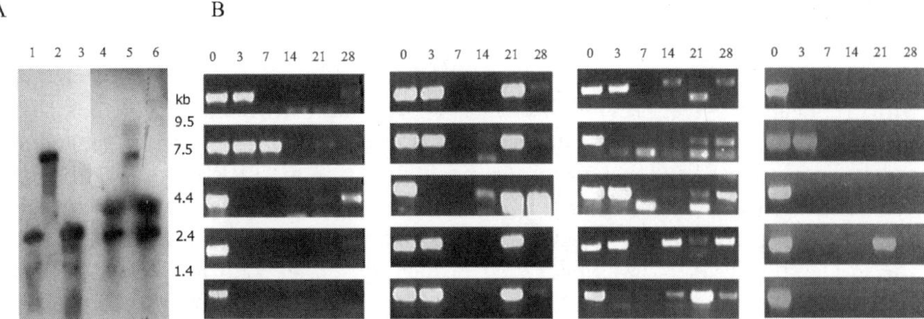

Figure 1. Evidence of PCVORF2 persistence *in vitro* and *in vivo*. A. Northern blot analysis with PCVORF2 probe.[12] Lanes 1–3: *in vitro* transcripts of S7-PCVORF2/StuI, S1-PCVORF2/SnaBI, S7-PCVORF2/KpnI, respectively. Lanes 4–6: S7-PCVORF2/StuI, S2-PCVORF2/SnaBI, S7-PCVORF2-/KpnI modified PRRSV infection supernatants. The appearance of higher molecular weight PCV-specific RNA may be due to a technique anomaly or possibly viral recombination. B. Nested RT-PCR analysis revealed PCVORF2 mRNA in swine serum samples after replication *in vivo.* Column 1: S1-PCVORF2/SnaBI; Column 2: S7-PCVORF2/StuI; Column 3: S7-PCVORF2/KpnI; Column 4: negative control pigs injected with saline. Each panel represents a single pig analyzed at each time point (days post-infection indicated above the individual lane). The 352 bp PCVORF2 DNA product was detected in several serum samples.

3.2. Translation of PRRSV S1 and S7

We next examined the encoding capacity of S1 and S7. Native S1 and S7 as well as PCV or green fluorescent protein (GFP) modified derivatives were used for *in vitro* translation. The results demonstrated native and modified defective RNA species can be easily translated into proteins (Figure 2A). Immunoprecepitation revealed that these proteins could be recognized by specific antibodies against PCV or GFP (Figure 2B). However, we could not detect antibodies in heteroclite and PRRSV inoculated swine (data not shown).

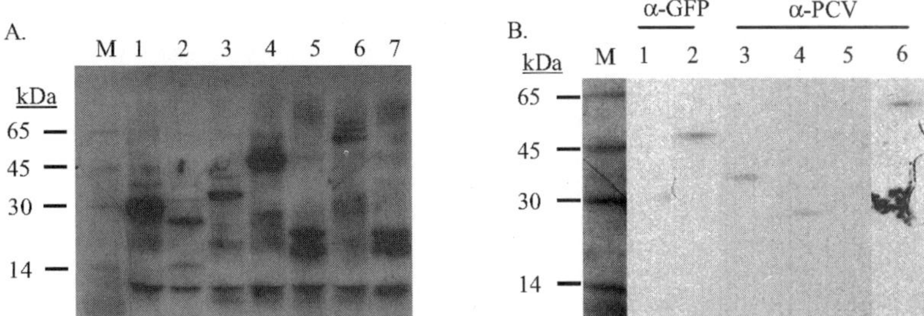

Figure 2. *In vitro* translated native and modified heteroclite species were recognized by specific antibodies. A. *In vitro* translation assay. Lanes 1: S7; 2: S7-PCVORF2/StuI; 3: S7-PCVORF2/KpnI; 4: S7-GFP/BX; 5: S2; 6: S2-PCVORF2/SnaBI, 7: VR-V7. B. Immunoprecipitation of *in vitro* translated proteins with immune serum against PCV and GFP. Lanes 1: S7; 2: S7-GFP; 3: S7-PCVORF2/KpnI; 4: S7-PCVORF2/StuI; 5: S7; 6: S7-PCVORF2/SnaBI.

4. CONCLUSIONS

Defective RNAs serve as an important tool for studying the molecular mechanisms of a virus replication cycle. Previous work has shown that PRRSV heteroclite RNAs, but not subgenomic RNAs, could be packaged and since both heteroclite and subgenomic RNAs have complete 5' and 3' termini, we postulated that the PRRSV packaging signal may lie in the ORF1a region.[8] This notion was derived because the shortest heteroclite, S-9a (junction site nt 476/14344), was found to be in viral particles.[9] In this study, modified defective RNA S7 was packaged efficiently, either with PCVORF2 insertion right after the leader sequence (nt 190) or after the putative packaging signal (nt 476). This suggested that the packaging signal might indeed lie within this 286-bp region, perhaps located somewhere distant from the ORF1a fragment termini, as the insertions did not abort viral packaging. Additional work needs to be done to further define the packaging signal.

S1 and S2 and their derivatives could be easily translated in *in vitro*, but we could not detect any protein products either in virus-infected cells or evidence of protein production, by production of α-PCV antibodies, in inoculated swine. Also, a cell line expressing S7 modified with GFP did not reveal any protein expression (data not shown). Perhaps the *in vitro* system does not reflect the situation in cultured cells or *in vivo*. Unknown factors from PRRSV or the host may inhibit translation or the expression level is not at detectable levels.

5. REFERENCES

1. Cavanagh, D., 1997, Nidovirales: a new order comprising Coronaviridae and Arteriviridae, *Arch. Virol.* **142**:629-633.
2. Benfield, D. A., Nelson, E., Collins, J. E., Harris, L., Goyal, S. M., Robison, D., Christianson, W. T., Morrison, R. B., Gorcyca, D. E., and Chladek, D. W., 1992, Characterization of swine infertility and respiratory syndrome (SIRS) virus (isolate ATCC VR-2332), *J. Vet. Diagn. Invest.* **4**:127-133.
3. Snijder, E. J., and Meulenberg, J. J. M., 1998, The molecular biology of arterivirus, *J. Gen. Virol.* **79**:961-979.
4. Penzes, Z., Tibbles, K. W., Shaw, K., Britton, P., and Cavanagh, D., 1996, Replication and packaging of coronavirus infectious bronchitus virus defective RNAs lacking a long open reading frame, *J. Virol.* **70**:8600-8668.
5. Izeta, A., Smerdou, C., Alonso, S., Penzes, Z., Mendez, A., Plana-Duran, J., and Enjuanes, L., 1999, Replication and packaging of transmissible gastroenteritis coronavirus-derived synthetic minigenomes, *J. Virol.* **73**:489-503.
6. Chang, R. Y., and Brian, D. A., 1996, *cis*-Requirement for N-specific protein sequence in bovine coronavirus defective interfering RNA replication, *J. Virol.* **70**:2201-2207.
7. Richard, M. B., Rozier, C. D., Greve, S., Spaan, W. J. M., and Snijder, E. J., 2000, Isolation and characterization of an arterivirus defective interfering RNA genome, *Virology* **74**:3156-3165.
8. Yuan, S., Murtaugh, M. P., and Faaberg, K. S., 2000, Heteroclite subgenomic RNAs are produced in porcine reproductive and respiratory syndrome virus infection, *Virology* **275**:158-169.
9. Yuan, S., Murtaugh, M. P., Schumann, F. A., Michelson, D., and Faaberg, K. S., 2004, Characterization of heteroclite subgenomic RNAs associated with PRRSV infection, *Virus Res.* **105**:75-87.
10. Nielsen, H. S., Liu, G., Nielsen, J., Oleksiewicz, M. B., Botner, A., Storgaard, T., and Faaberg, K. S., 2003, Generation of an infectious clone of VR-2332, a highly virulent North American-type isolate of porcine reproductive and respiratory syndrome virus, *J. Virol.* **77**:3702-3711.
11. Alonso, S., Sola, I., Teifke, J. P., Reimann, I., Izeta, A., Balasch, M., Plana-Duran, J., Moormann, R. J. M., and Enjuanes, L., 2002, *In vitro* and *in vivo* expression of foreign genes by transmissible gastroenteritis coronavirus-derived minigenomes, *J. Gen. Virol.* **83**:567–579.

EFFICACY OF AN INACTIVATED PRRSV VACCINE

Induction of virus-neutralizing antibodies and partial virological protection upon challenge

Gerald Misinzo, Peter L. Delputte, Peter Meerts, Christa Drexler, and Hans J. Nauwynck*

1. INTRODUCTION

Porcine reproductive and respiratory syndrome virus (PRRSV) is an enveloped, single-stranded, positive-sense RNA virus belonging to the family *Arteriviridae*, grouped into the order *Nidovirales*, together with the *Coronaviridae* and the *Roniviridae*.[1,2] *In vivo* PRRSV has a predilection for porcine macrophages that express porcine sialoadhesin.[3,4] *In vitro* porcine alveolar macrophages (PAM), some cultivated peripheral blood monocytes and the non-macrophage African green monkey kidney cell line MA-104, and cells derived from MA-104 (Marc-145 and CL-2621) support PRRSV infection[3–5]. Two PRRSV receptors have already been identified on PAM. The glycosaminoglycan heparan sulfate is a PRRSV receptor that is involved in PRRSV attachment[6] and porcine sialoadhesin is essential for both PRRSV attachment and internalization.[4] PRRSV attachment to porcine sialoadhesin on PAM is mediated by sialic acids potentially present on the viral glycoproteins.[7]

PRRSV infection is characterized by reproductive failures in sows and respiratory problems in pigs of all ages.[1,8,9] PRRSV causes major economical losses in swine farms. Vaccination of both sows and young piglets is frequently performed to prevent this disease, however there are some problems associated with the currently used vaccines. Inactivated vaccines are safe to use in sows, becuase these vaccines do not induce reproductive failure, but their capacity to induce a protective immunity against challenge with wild-type virus has been questioned, especially in naive pigs.[10–12] Attenuated live vaccines have been proven to be effective in inducing protective immunity upon challenge with virulent PRRSV.[13,14] However, they only protect against virus-induced disease if the challenge virus is genetically and antigenically similar to the vaccine

*Gerald Misinzo, Peter L. Delputte, Peter Meerts, Hans J. Nauwynck, Ghent University, Merelbeke, Belgium. Christa Drexler, Intervet International B.V., The Netherlands.

virus.[13,15] Only some degree of protection against heterologous strains is observed.[16] Further, there are concerns about the safety of these attenuated vaccines. Reversion of vaccine virus to virulence has been shown to occur causing major problems.[17] They can themselves spread, change genetically and be the cause of reproductive disorders. Due to the highly variable nature of RNA viruses and more specific of PRRSV, one of the major challenges of future vaccine research is to make vaccines that are safe to use and either are capable of inducing protective immunity toward the antigenically heterogenous array of viruses that are circulating, or can be quickly adapted to new circulating virus strains that are antigenically different.

Development of inactivated vaccines that are capable of inducing neutralizing antibodies would be one good strategy, as (1) the presence of neutralizing antibodies was previously shown to protect towards challenge and virus-induced disease,[18–20] (2) inactivated vaccines cannot induce disease by themselves and are thus safe to use, and (3) inactivated vaccines can rapidly be adapted to new circulating virus variants.

In this study, we wanted to investigate if neutralizing antibodies can be induced in pigs upon vaccination with an inactivated vaccine, and if vaccinated pigs were virologically protected towards challenge with wild-type PRRSV.

2. MATERIALS AND METHODS

2.1. Vaccine and Challenge Virus

Three different inactivated vaccines were used in the experiments: one based on a commercial, European type attenuated vaccine virus, one based on Marc-145 grown Lelystad virus (5th passage) and one based on porcine alveolar macrophage grown Lelystad virus (13th passage). Viruses were concentrated and semipurified by ultracentrifugation at 100,000 x *g* for 3 hours through a 30% sucrose cushion in an SW41 Ti rotor (Beckman Coulter Inc.). Virus was then inactivated with beta-propiolactone and formulated in a water/oil emulsion so that each 2 ml dose of vaccine contained an equivalent of $10^{8.0}$ $TCID_{50}$.

2.2. Pigs and Experimental Design

A total of 26 pigs were obtained from PRRSV naive sows at the age of 4 weeks. The pigs were randomly divided into 4 groups and housed in isolation units with HEPA filtered air. The designation of the groups and the experimental design is shown in Table 1. At 6 and 10 weeks of age, the pigs were vaccinated intramuscularly with the designated vaccine. Four weeks after the booster vaccination, all animals were challenged intranasally with $10^{6.0}$ $TCID_{50}$ (2 ml) of the virulent Lelystad virus strain.

Table 1. Experimental design.

Group	Number of pigs	Vaccination	Inactivated vaccine		Challenge virus
			Virus	Cell line	
A	6	No			Lelystad virus
B	12	Yes	European type attenuated vaccine	Marc-145 cells	Lelystad virus
C	4	Yes	Lelystad virus	Marc-145 cells	Lelystad virus
D	4	Yes	Lelystad virus	Porcine alveolar macrophage	Lelystad virus

2.3. Serological Examinations and Virus Titrations of Serum Samples

Starting from the first vaccination, serum was collected weekly to detect virus specific antibodies with immunoperoxidase monolayer assay (IPMA) and virus neutralizing (VN) antibodies with serum neutralization (SN) test on Marc-145 cells as described previously.[19] At 0, 3, 7, 10, 14, 21 and 28 days after challenge, serum was collected for IPMA, SN and for virus isolation.

3. RESULTS

3.1. Immunoperoxidase Monolayer Assay

None of the vaccines induced IPMA antibodies after the first immunization. Following the booster vaccination, IPMA antibodies were detected in most of the animals vaccinated with inactivated attenuated vaccine (Fig. 1). Animals vaccinated with inactivated Lelystad virus grown in Marc-145 and porcine alveolar macrophage grown had respectively low and low to undetectable levels of IPMA antibodies (Fig. 1). Upon challenge, a more rapid antibody response was observed in vaccinated animals, indicating that memory was induced.

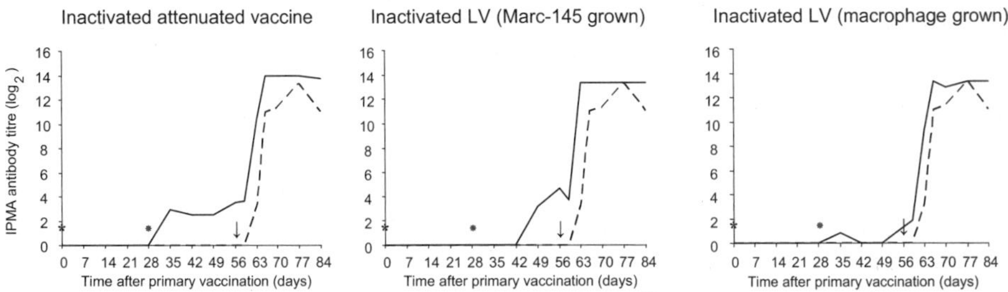

Figure 1. Course of IPMA antibody titers in pigs vaccinated twice (* *) and challenged (↓) eight weeks later with PRRSV (Lelystad) (mean, —) and non-vaccinated control pigs (mean, - -).

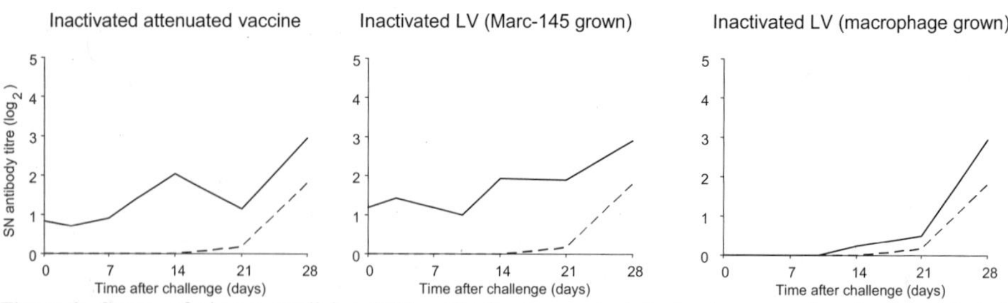

Figure 2. Course of virus neutralizing (VN) antibodies in vaccinated pigs (mean, —) and non-vaccinated control pigs (mean, - -) challenged with PRRSV (Lelystad).

3.2. Serum Neutralization Assay

None of the vaccines induced serum neutralizing (SN) antibodies after the first immunization. At the time of challenge, SN antibodies were present only in animals vaccinated with inactivated attenuated virus and Marc-145 grown Lelystad virus (Fig. 2). Upon challenge, a more rapid neutralizing antibody response was observed in vaccinated animals, indicating that memory was induced. Although vaccination not always induces neutralizing antibodies in all pigs, it is observed that vaccination enhances neutralizing antibodies upon challenge.

3.3. Viremia

Upon challenge with Lelystad virus, viremia was observed in all control animals. In 2 animals vaccinated with inactivated attenuated virus, no viremia was detected, while in the others a clear reduction in the levels and duration of viremia was observed (95% reduction at 10 d postchallenge; absolute values). Vaccination with inactivated Marc-145 and macrophage grown Lelystad virus had only a small effect on the levels of viremia, but reduced the duration of viremia (Fig. 3).

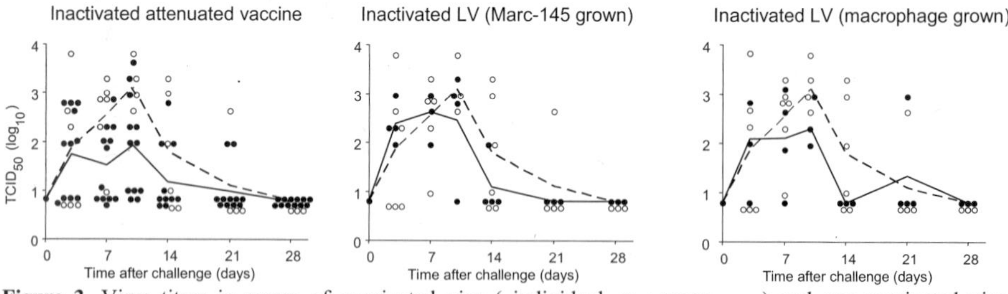

Figure 3. Virus titers in serum of vaccinated pigs (individual, ● ; mean, —) and non-vaccinated pigs (individual, ○ ; mean, - -) upon challenge with PRRSV (Lelystad).

4. CONCLUSIONS

In this study, it was shown that an inactivated vaccine can induce virus-neutralizing antibodies that results in a strong to partial virological protection upon challenge. However, differences were observed in efficacy, depending on the virus strain and the cells used to make the vaccine. Because the capacity of an inactivated vaccine to induce neutralizing antibodies is most likely correlated with the conservation of neutralizing epitopes during inactivation, we will evaluate the antigenic structure of the virus upon different inactivation methods.

5. ACKNOWLEDGMENTS

The authors would like to thank C. Vanmaercke, S. Demaret, C. Bracke and F. De Backer for their excellent technical assistance. G. Misinzo was supported by the Belgian Federal Department of Agriculture.

6. REFERENCES

1. Cavanagh, D., 1997, Nidovirales: a new order comprising Coronaviridae and Arteriviridae, *Arch. Virol.* **142**:629-633.
2. Plagemann, P. G. W., 1996, Lactate dehydrogenase-elevating virus and related viruses, in: *Fields Virology,* B. N., Knipe, D. M., Howley, P. M., eds., Lippincott-Raven, Philadelphia, pp. 1105-1120.
3. Duan, X., Nauwynck, H. J., and Pensaert, M. B., 1997, Effects of origin and state of differentiation and activation of monocytes/macrophages on their susceptibility to PRRSV, *Arch. Virol.* **142**: 2483-2497.
4. Vanderheijden, N., Delputte, P. L., et al., 2003, Involvement of sialoadhesin in entry of porcine reproductive and respiratory syndrome virus into porcine alveolar macrophages, *J. Virol.* **77**:8207-8215.
5. Kim, H. S., Kwang, J., et al., 1993, Enhanced replication of porcine reproductive and respiratory syndrome (PRRS) virus in homogeneous subpopulation of MA-104 cell-line, *Arch. Virol.* **133**:477-483.
6. Delputte, P. L., Vanderheijden, N., Nauwynck, H. J., and Pensaert, M. B., 2002, Involvement of the matrix protein in attachment of porcine reproductive and respiratory syndrome virus to a heparin-like receptor on porcine alveolar macrophages, *J. Virol.* **76**:4312-4320.
7. Delputte, P. L., and Nauwynck, H. J., 2004a, Porcine arterivirus infection of alveolar macrophages is mediated by sialic acid on the virus, *J. Virol.* **78**:8094-8101.
8. Collins, J. E., Benfield D. A., et al., 1992, Isolation of swine infertility and respiratory syndrome virus (isolate ATCC VR-2332) in North America and experimental reproduction of the disease in gnotobiotic pigs, *J. Vet. Diagn. Invest.* **4**:117-126.
9. Meulenberg, J. J. M., Hulst, M. M., et al., 1994, Lelystad virus belongs to a new virus family, comprising lactate dehydrogenase-elevating virus, equine arteritis virus, and simian hemorrhagic fever virus, *Arch. Virol.* **9**:441-448.
10. Nielsen, T. L., Nielsen, J., Have, P., Baekbo, P., Hoff-Jorgensen, R., and Bøtner, A., 1997, Examination of virus shedding in semen from vaccinated and from previously infected boars after experimental challenge with porcine reproductive and respiratory syndrome virus, *Vet. Microbiol.* **54**:101-112.
11. Plana-Duran, J., Bastons, M., Urniza, A., Vayreda, M., Vila, X., and Mane, H., 1997, Efficacy of an inactivated vaccine for prevention of reproductive failure induced by porcine reproductive and respiratory syndrome virus, *Vet. Microbiol.* **55**:361-370.
12. Nilubol, D., Platt, K. B., Halbur, P. G., Torremorell, M., and Harris, D. L., 2004, The effect of a killed porcine reproductive and respiratory syndrome virus (PRRSV) vaccine treatment on virus shedding in previously PRRSV infected pigs, *Vet. Microbiol.* **102**:11-18.
13. Labarque, G., Van Gucht, S., et al., 2003, Respiratory tract protection upon challenge of pigs vaccinated with attenuated porcine reproductive and respiratory syndrome virus vaccines, *Vet. Microbiol.* **95**:187-197.

14. van Woensel, P. A., Liefkens, K., and Demaret, S., 1998, Effect on viraemia of an American and a European serotype PRRSV vaccine after challenge with European wild-type strains of the virus, *Vet. Rec.* **142**:510-512.
15. Labarque, G., Reeth, K. V., et al., 2004, Impact of genetic diversity of European-type porcine reproductive and respiratory syndrome virus strains on vaccine efficacy, *Vaccine* **22**:4183-4190.
16. Lager, K. M., Mengeling, W. L., and Brockmeier, S. L., 1999, Evaluation of protective immunity in gilts inoculated with the NADC-8 isolate of porcine reproductive and respiratory syndrome virus (PRRSV) and challenge-exposed with an antigenically distinct PRRSV isolate, *Am. J. Vet. Res.* **60**:1022-1027.
17. Nielsen, H. S., Oleksiewicz, M. B., et al., 2001, Reversion of a live porcine reproductive and respiratory syndrome virus vaccine investigated by parallel mutations, *J. Gen. Virol.* **82**:1263-1272.
18. Delputte, P. L., Meerts, P., Costers, S., and Nauwynck, H. J., 2004b, Effect of virus-specific antibodies on attachment, internalization and infection of porcine reproductive and respiratory syndrome virus in primary macrophages, *Vet. Immunol. Immunopathol.* **102**:179-188.
19. Labarque, G. G., Nauwynck, H. J., Van Reeth, K., and Pensaert, M. B., 2000, Effect of cellular changes and onset of humoral immunity on the replication of porcine reproductive and respiratory syndrome virus in the lungs of pigs, *J. Gen. Virol.* **81**:1327-1334.
20. Osorio, F. A., Galeota, J. A., et al., 2002, Passive transfer of virus-specific antibodies confers protection against reproductive failure induced by a virulent strain of porcine reproductive and respiratory syndrome virus and establishes sterilizing immunity, *Virology* **302**:9-20.

VII. PATHOGENESIS OF HUMAN CORONAVIRUSES

SARS AND OTHER CORONAVIRUSES IN HUMANS AND ANIMALS

Leo L. M. Poon*

1. INTRODUCTION

Southeast China has long been regarded as an epicenter for influenza viruses.[1] Indeed, avian influenza H5N1/97, H9N2/99, H7N7/03, and other recent avian H5 strains have occasionally crossed the species barrier in the last few years.[2] These events keep reminding us that new pandemic viruses may emerge in this geographical region. However, Nature caught us by surprise when the first pandemic in this millennium was caused not by influenza viruses, or any known pathogens, but by a previously unknown virus in the subfamily of *Coronaviridae*.[3-5]

2. SARS CORONAVIRUS IN HUMANS

Severe acute respiratory syndrome (SARS) is a respiratory disease newly identified in this century. The outbreak was officially recognized by the World Health Organization (WHO) in Vietnam in February 2003.[6] Further investigations revealed that the outbreak was first started in Guangdong Province, China, in November 2002.[7] Subsequent to its introduction to Hong Kong in mid-February 2003, the virus spread across Vietnam, Singapore, Canada, and elsewhere. In the early stage of the SARS pandemic, several pathogens were claimed to be the causative agents. Pathogens like chlamydia and paramyxoviruses were reported to be isolated from some SARS patients.[6] The documentation of two human H5N1 influenza cases in February 2002 in Hong Kong also suggested the possibility of the emergence of a pandemic avian influenza virus in humans. On 17 March 2003, an international collaborative research network was set up by WHO to investigate the cause of SARS. By the end of March, colleagues from three different research groups identified a novel coronavirus (CoV) as the etiology of SARS.[3-5] Enormous efforts were taken to contain and diagnosis the disease. Unfortunately, the pandemic did not cease until July. In this outbreak, a total of 8098 probable SARS

* The University of Hong Kong, Pokfulam, Hong Kong SAR, China.

patients were reported to WHO. The fatality rates in most of the affected countries ranged from 7% to 17% and 774 of probable SARS patients died from the disease.

Although previously known human CoVs account for 30% of common colds, little attention has been made on these medically important viruses. It might be mainly due to the fact that infections caused by these viruses do not result in severe illness and are usually restricted to upper respiration tract. Thus, the identification of a novel CoV as the pathogen for a severe viral pneumonia was far out of the expectation of most clinical virologists. With the tremendous efforts made by colleagues from the WHO collaborative network, the etiology of SARS was confirmed within weeks.[6] First, viral RNA or seroconversion against SARS coronavirus was found in the majority of SARS patients.[3-5] By contrast, no evidence of previous exposure of this virus could be detected from healthy individuals.[3-5] Furthermore, SARS-like illness could be reproduced in experimentally infected cynomolgus macaques (*Macaca fascicularis*).[8] Thus, this novel virus fulfills all Koch's postulates and was confirmed to be the etiological agent of SARS in mid-April 2003.[6]

CoV is an enveloped virus with a single, positive-stranded RNA genome. All CoVs have 5 major open reading frames that encode the replicase, spike (S), envelope (E), membrane (M), and nucleocapsid (N) proteins. These viruses can be subdivided into 3 groups. Base on antigenical and genetical studies, all previously known human coronavirus can be classified into group 1 (e.g., 229E) and group 2 (e.g., OC43). By contrast, studies on the classification of SARS-CoV yielded inconsistent results. This virus was initially shown to cross-react with group 1 coronavirus. Partial viral sequences deduced from initial studies indicated that this novel virus is genetically distinct from all previously known coronaviruses.[3-5] Further characterizations of the full genome of SARS-CoV also suggested that this virus is distantly related to groups 2 and 3 CoVs.[9-10] These findings prompted to the conclusion the SARS CoV represents a new group of CoV. However, this virus was subsequently proposed to be an early split off of the group 2 viruses. But, unlike other group 2 members, SARS-CoV does not contain a haemagglutinin-esterase protein encoding sequence.[9-10] In addition, this virus also has some features that are similar to group 3 coroanviruses.[11] These findings suggest that this virus might be very distinct from other group 2 viruses.[12]

During the SARS outbreak, several molecular and serological tests were developed for SARS diagnosis.[13-18] In some of our studies, about 80% of samples from patients at early disease onset were positive in the assay.[19,20] Some of these tests might have the potential to become point-of-care tests.[21,22] With the experiences that we learned from these studies, we are in a better position to identify SARS patient at an early stage of disease.[23] This might allow prompt clincial management and policy marking. However. further work on developing assays with better sensitivities is a must. Standardization on RT-PCR assays and clinical samples for the test might also allow us to develop a unified protocol for SARS diagnosis. In a non-pandemic situation, suspected SARS samples should be tested with caution. In an ideal situation, serial multiple samples should be collected from patients. Positive samples should be confirmed by independent assays. In addition, to avoid having false negative results, use of assays with internal positive control should be encouraged.

3. SARS IN ANIMALS

The epidemiological data of early SARS patients suggested that some of these index patients might have close associations with wildlife in Guangzhou.[24] This prompted us to do a virus surveillance study in a wet market in Guangzhou. In this pilot study, evidence of SARS-like CoV infection were found in those samples collected from Himalayan palm civets (*Paguma larvata*) and a raccoon dog (*Nyctereutes procyonoides*).[25] In particular, SARS-like viral isolates were recovered from civets. Genetic analysis of these viral RNA indicated the human and animals isolates are of 99.8% homology. These observations indicated the SARS-CoV in human might of animal origin. Our data also suggested the human virus might be a result of direct transmission from civets to humans. Subsequent serological and epidemiological studies in humans and civets confirmed this hypothesis. Interestingly, serological and virus isolation studies both indicated that the prevalence of animal SARS-CoV in civets from farms was only about 10%.[26] In particular, some of the studied farms were found to be free of this virus. By contrast, a great majority of civets in wildlife markets were found to be infected with the virus.[26] These suggested that, the animals are more susceptible for SARS-CoV infection under stressful conditions. In addition, these findings also suggested that the spillover of the virus from civets to other animals and human occurred in live animal markets.

Although civets might play significant role in spreading the virus in humans, it is not known whether they are the natural reservoir for this virus. Attempts in detecting SARS-CoV in wild civets have been made.[27] In our preliminary studies, none of the samples collected from wild civets (N = 21) were PCR positive for the virus, indicating this virus is not commonly circulating in wild civets. Nonetheless, further serological studies on wild civets and other wildlife are needed to identify the natural hosts of SARS-CoV.

4. WILL SARS RETURN?

The SARS outbreak was brought under control through a concerted global effort, and by July 5, 2003, no further human–human transmission took place. However, there are still several possibilities that might lead to the reemerge of SARS in human. Recent studies showed that persistent infection in human populations is unlikely,[28] suggesting the chance of reemergence of SARS directly from asymptomatic infected humans is low. The potential source of SARS-CoV might come from the infected animals circulating in this geographical region. This threat was highlighted by the 4 community-acquired SARS cases between December 2003 to January 2004 in Guangdong, China. Both epidemiological and phylogenetic studies indicated the infection of these patients was from zoonotic source.[29] In these recent cases, all patients only developed mild symptoms and did not cause secondary transmission, suggesting the animal virus is not fully adapted in humans yet. By contrast, the consequence of having laboratory acquired infection would be more alarming.[30] Escape of these human isolates occurred three times in the past few years. These accidents were all caused by human errors, and one of these accidents caused nosocomial infections. One should note that these human viral isolates, which are stored in laboratories, could be transmitted between humans in an efficient manner. Thus, these laboratory accidents reemphasized the importance of biosafety.

5. NEWLY IDENTIFIED CORONAVIRUSES IN HUMAN AND ANIMALS

After the SARS outbreak in 2003, two independent groups identified a novel CoV (HCoV-NL63) in humans.[31,32] Similar to other classical human CoVs, this group 1 virus is associated with respiratory illnesses. Recent studies also indicated that this virus is a common respiratory pathogen in human populations.[33,34] On the other hand, using conserved primers for CoVs, another novel human CoV virus (HCoV-HKU1) was recently identified.[35] So far, this virus was reported to be found in a small number of patients with pneumonia in Hong Kong. Further epidemiological studies are required to demonstrate the clinical importance of this novel group 1 pathogen in global human populations.

Of all the identified CoV in animals, most were isolated from pets and domestic poultry. This bias is presumably because viral investigations were often imitated by observable disease outbreaks in these populations. By contrast, relative little is know about the CoVs circulating in wild animals. Recently, novel group 3 CoVs were identified in wild birds.[36] Results from this study suggested at least one of these novel viruses might cause disease in its host.[37] We also performed similar studies in wildlife. In this work, a novel group 1 CoV was identified in three bat species (*Miniopterus* spp.) in Hong Kong.[27] In particular, the prevalence of this virus in one of the bat species (*M. pusillus*) was as high as 63%, suggesting this species might be the natural reservoir of this virus. Both fecal and respiratory samples from bats were positive for the virus. Our results suggested that this virus has a predominantly enteric tropism. However, it is not known whether these viruses cause disease in bat populations. In addition, we also do not know the mode of transmission of the virus. Further investigations on these topics are needed. Nonetheless, the above studies clearly highlight our poor understanding of viruses in wild animals.

6. CONCLUSIONS

The SARS outbreak had a severe impact on health care, the economy, and the tourist industry in many countries. Given the catastrophic consequences of SARS, further investigation on viruses in wildlife should be encouraged.

7. ACKNOWLEDGMENTS

The work is supported by funds from The Research Grant Council of Hong Kong, the Seed Funding and SARS Research funds from The University of Hong Kong, and European Research Project SARS-DTV (contract no. SP22-CT-2004).

8. REFERENCES

1. D. S. Melville and K. F. Shortridge, Influenza: time to come to grips with the avian dimension, *Lancet Infect. Dis.* **4**, 261-262 (2004).
2. K. S. Li, Y. Guan, J. Wang, G. J. Smith, K. M. Xu, L. Duan, A. P. Rahardjo, P. Puthavathana, C. Buranathai, T. D. Nguyen, A. T. Estoepangestie, A. Chaisingh, P. Auewarakul, H. T. Long, N. T. Hanh, R. J. Webby,

L. Poon, H. Chen, K. F. Shortridge, K. Y. Yuen, R. G. Webster, and J. S. Peiris, Genesis of a highly pathogenic and potentially pandemic H5N1 influenza virus in eastern Asia, *Nature* **430**, 209-213 (2004).

3. J. S. Peiris, S. T. Lai, L. L. Poon, Y. Guan, L. Y. Yam, W. Lim, J. Nicholls, W. K. Yee, W. W. Yan, M. T. Cheung, V. C. Cheng, K. H. Chan, D. N. Tsang, R. W. Yung, T. K. Ng, and K. Y. Yuen, Coronavirus as a possible cause of severe acute respiratory syndrome, *Lancet* **361**, 1319-1325 (2003).
4. T. G. Ksiazek, D. Erdman, C. S. Goldsmith, S. R. Zaki, T. Peret, S. Emery, S. Tong, C. Urbani, J. A. Comer, W. Lim, P. E. Rollin, S. F. Dowell, A. E. Ling, C. D. Humphrey, W. J. Shieh, J. Guarner, C. D. Paddock, P. Rota, B. Fields, J. DeRisi, J. Y. Yang, N. Cox, J. M. Hughes, J. W. LeDuc, W. J. Bellini, and L. J. Anderson, A novel coronavirus associated with severe acute respiratory syndrome, *N. Engl. J. Med.* **348**, 1953-1966 (2003).
5. C. Drosten, S. Gunther, W. Preiser, S. van der Werf, H. R. Brodt, S. Becker, H. Rabenau, M. Panning, L. Kolesnikova, R. A. Fouchier, A. Berger, A. M. Burguiere, J. Cinatl, M. Eickmann, N. Escriou, K. Grywna, S. Kramme, J. C. Manuguerra, S. Muller, V. Rickerts, M. Sturmer, S. Vieth, H. D. Klenk, A. D. Osterhaus, H. Schmitz, and H. W. Doerr, Identification of a novel coronavirus in patients with severe acute respiratory syndrome, *N. Engl. J. Med.* **348**, 1967-1976 (2003).
6. World Health Organization Multicentre Collaborative Network for Severe Acute Respiratory Syndrome Diagnosis, A multicentre collaboration to investigate the cause of severe acute respiratory syndrome, *Lancet* **361**, 1730-1733 (2003).
7. N. S. Zhong, B. J. Zheng, Y. M. Li, L. L. Poon, Z. H. Xie, K. H. Chan, P. H. Li, S. Y. Tan, Q. Chang, J. P. Xie, X. Q. Liu, J. Xu, D. X. Li, K. Y. Yuen, J. S. Peiris, and Y. Guan, Epidemiology and cause of severe acute respiratory syndrome (SARS) in Guangdong, People's Republic of China, in February, 2003, *Lancet* **362**, 1353-1358 (2003).
8. R. A. Fouchier, T. Kuiken, M. Schutten, G. van Amerongen, G. J. van Doornum, B. G. van den Hoogen, M. Peiris, W. Lim, K. Stohr, and A. D. Osterhaus, Aetiology: Koch's postulates fulfilled for SARS virus, *Nature* **423**, 240 (2003).
9. M. A. Marra, S. J. Jones, C. R. Astell, R. A. Holt, A. Brooks-Wilson, Y. S. Butterfield, J. Khattra, J. K. Asano, S. A. Barber, S. Y. Chan, A. Cloutier, S. M. Coughlin, D. Freeman, N. Girn, O. L. Griffith, S. R. Leach, M. Mayo, H. McDonald, S. B. Montgomery, P. K. Pandoh, A. S. Petrescu, A. G. Robertson, J. E. Schein, A. Siddiqui, D. E. Smailus, J. M. Stott, G. S. Yang, F. Plummer, A. Andonov, H. Artsob, N. Bastien, K. Bernard, T. F. Booth, D. Bowness, M. Czub, M. Drebot, L. Fernando, R. Flick, M. Garbutt, M. Gray, A. Grolla, S. Jones, H. Feldmann, A. Meyers, A. Kabani, Y. Li, S. Normand, U. Stroher, G. A. Tipples, S. Tyler, R. Vogrig, D. Ward, B. Watson, R. C. Brunham, M. Krajden, M. Petric, D. M. Skowronski, C. Upton, and R. L. Roper, The Genome sequence of the SARS-associated coronavirus, *Science* **300**, 1399-1404 (2003).
10. P. A. Rota, M. S. Oberste, S. S. Monroe, W. A. Nix, R. Campagnoli, J. P. Icenogle, S. Penaranda, B. Bankamp, K. Maher, M. H. Chen, S. Tong, A. Tamin, L. Lowe, M. Frace, J. L. DeRisi, Q. Chen, D. Wang, D. D. Erdman, T. C. Peret, C. Burns, T. G. Ksiazek, P. E. Rollin, A. Sanchez, S. Liffick, B. Holloway, J. J. Limor, K. McCaustland, M. Olsen-Rasmussen, R. Fouchier, S. Gunther, A. D. Osterhaus, C. Drosten, M. A. Pallansch, L. J. Anderson, and W. J. Bellini, Characterization of a novel coronavirus associated with severe acute respiratory syndrome, *Science* **300**, 1394-1399 (2003).
11. J. Stavrinides and D. S. Guttman, Mosaic evolution of the severe acute respiratory syndrome coronavirus, *J. Virol.* **78**, 76-82 (2004).
12. A. E. Gorbalenya, E. J. Snijder, and W. J. Spaan, Severe acute respiratory syndrome coronavirus phylogeny: toward consensus, *J. Virol.* **78**, 7863-7866 (2004).
13. L. L. Poon, O. K. Wong, K. H. Chan, W. Luk, K. Y. Yuen, J. S. Peiris, and Y. Guan, Rapid diagnosis of a coronavirus associated with severe acute respiratory syndrome (SARS), *Clin. Chem.* **49**, 953-955 (2003).
14. W. C. Yam, K. H. Chan, L. L. Poon, Y. Guan, K. Y. Yuen, W. H. Seto, and J. S. Peiris, Evaluation of reverse transcription-PCR assays for rapid diagnosis of severe acute respiratory syndrome associated with a novel coronavirus, *J. Clin. Microbiol.* **41**, 4521-4524 (2003).
15. L. L. Poon, K. H. Chan, and J. S. Peiris, Crouching tiger, hidden dragon: the laboratory diagnosis of severe acute respiratory syndrome, *Clin. Infect. Dis.* **38**, 297-299 (2004).
16. L. L. Poon, K. H. Chan, O. K. Wong, T. K. Cheung, I. Ng, B. Zheng, W. H. Seto, K. Y. Yuen, Y. Guan, and J. S. Peiris, Detection of SARS coronavirus in patients with severe acute respiratory syndrome by conventional and real-time quantitative reverse transcription-PCR assays, *Clin. Chem.* **50**, 67-72 (2004).
17. S. K. Lau, P. C. Woo, B. H. Wong, H. W. Tsoi, G. K. Woo, R. W. Poon, K. H. Chan, W. I. Wei, J. S. Peiris, and K. Y. Yuen, Detection of severe acute respiratory syndrome (SARS) coronavirus nucleocapsid protein in sars patients by enzyme-linked immunosorbent assay, *J. Clin. Microbiol.* **42**, 2884-2889 (2004).J
18. S. K. Lau, X. Y. Che, P. C. Woo, B. H. Wong, V. C. Cheng, G. K. Woo, I. F. Hung, R. W. Poon, K. H. Chan, J. S. Peiris, and K. Y. Yuen, SARS Coronavirus Detection Methods, *Emerg. Infect. Dis.* **11**, 1108-1111 (2005).

19. L. L. Poon, K. H. Chan, O. K. Wong, W. C. Yam, K. Y. Yuen, Y. Guan, Y. M. Lo, and J. S. Peiris, Early diagnosis of SARS coronavirus infection by real time RT-PCR, *J. Clin. Virol.* **28**, 233-238 (2003).
20. L. L. Poon, B. W. Wong, K. H. Chan, C. S. Leung, K. Y. Yuen, Y. Guan, and J. S. Peiris, A one-step quantitative RT-PCR for detection of SARS coronavirus with an internal control for PCR inhibitors, *J. Clin. Virol.* **30**, 214-217 (2004).
21. L. L. Poon, C. S. Leung, M. Tashiro, K. H. Chan, B. W. Wong, K. Y. Yuen, Y. Guan, and J. S. Peiris, Rapid detection of the severe acute respiratory syndrome (SARS) coronavirus by a loop-mediated isothermal amplification assay, *Clin. Chem.* **50**, 1050-1052 (2004).
22. L. L. Poon, B. W. Wong, K. H. Chan, S. S. Ng, K. Y. Yuen, Y. Guan, and J. S. Peiris, Evaluation of real-time reverse transcriptase PCR and real-time loop-mediated amplification assays for severe acute respiratory syndrome coronavirus detection, *J. Clin. Microbiol.* **43**, 3457-3459 (2005).
23. L. L. Poon, Y. Guan, J. M. Nicholls, K. Y. Yuen, and J. S. Peiris, The aetiology, origins, and diagnosis of severe acute respiratory syndrome, *Lancet Infect. Dis.* **4**, 663-671 (2004).
24. R. H. Xu, J. F. He, M. R. Evans, G. W. Peng, H. E. Field, D. W. Yu, C. K. Lee, H. M. Luo, W. S. Lin, P. Lin, L. H. Li, W. J. Liang, J. Y. Lin, and A. Schnur, Epidemiologic clues to SARS origin in China, *Emerg. Infect. Dis.* **10**, 1030-1037 (2004).
25. Y. Guan, B. J. Zheng, Y. Q. He, X. L. Liu, Z. X. Zhuang, C. L. Cheung, S. W. Luo, P. H. Li, L. J. Zhang, Y. J. Guan, K. M. Butt, K. L. Wong, K. W. Chan, W. Lim, K. F. Shortridge, K. Y. Yuen, J. S. Peiris, and L. L. Poon, Isolation and characterization of viruses related to the SARS coronavirus from animals in southern China, *Science* **302**, 276-278 (2003).
26. C. Tu, G. Crameri, X. Kong, J. Chen, Y. Sun, M. Yu, H. Xiang, X. Xia, S. Liu, T. Ren, Y. Yu, B. T. Eaton, H. Xuan, and L. F. Wang, Antibodies to SARS coronavirus in civets, *Emerg. Infect. Dis.* **10**, 2244-2248 (2004).
27. L. L. Poon, D. K. Chu, K. H. Chan, O. K. Wong, T. M. Ellis, Y. H. Leung, S. K. Lau, P. C. Woo, K. Y. Suen, K. Y. Yuen, Y. Guan, and J. S. Peiris, Identification of a novel coronavirus in bats, *J. Virol.* **79**, 2001-2009 (2005).
28. G. M. Leung, L. M. Ho, S. K. Chan, S. Y. Ho, J. Bacon-Shone, R. Y. Choy, A. J. Hedley, T. H. Lam, and R. Fielding, Longitudinal assessment of community psychobehavioral responses during and after the 2003 outbreak of severe acute respiratory syndrome in Hong Kong, *Clin. Infect. Dis.* **40**, 1713-1720 (2005).
29. H. D. Song, C. C. Tu, G. W. Zhang, S. Y. Wang, K. Zheng, L. C. Lei, Q. X. Chen, Y. W. Gao, H. Q. Zhou, H. Xiang, H. J. Zheng, S. W. Chern, F. Cheng, C. M. Pan, H. Xuan, S. J. Chen, H. M. Luo, D. H. Zhou, Y. F. Liu, J. F. He, P. Z. Qin, L. H. Li, Y. Q. Ren, W. J. Liang, Y. D. Yu, L. Anderson, M. Wang, R. H. Xu, X. W. Wu, H. Y. Zheng, J. D. Chen, G. Liang, Y. Gao, M. Liao, L. Fang, L. Y. Jiang, H. Li, F. Chen, B. Di, L. J. He, J. Y. Lin, S. Tong, X. Kong, L. Du, P. Hao, H. Tang, A. Bernini, X. J. Yu, O. Spiga, Z. M. Guo, H. Y. Pan, W. Z. He, J. C. Manuguerra, A. Fontanet, A. Danchin, N. Niccolai, Y. X. Li, C. I. Wu, and G. P. Zhao, Cross-host evolution of severe acute respiratory syndrome coronavirus in palm civet and human, *Proc. Natl. Acad. Sci. USA* **102**, 2430-2435 (2005).
30. C. Orellana, Laboratory-acquired SARS raises worries on biosafety, *Lancet Infect. Dis.* **4**, 64 (2004).
31. L. van der Hoek, K. Pyrc, M. F. Jebbink, W. Vermeulen-Oost, R. J. Berkhout, K. C. Wolthers, P. M. Wertheim-van Dillen, J. Kaandorp, J. Spaargaren, and B. Berkhout, Identification of a new human coronavirus, *Nat. Med.* **10**, 368-373 (2004).
32. R. A. Fouchier, N. G. Hartwig, T. M. Bestebroer, B. Niemeyer, J. C. de Jong, J. H. Simon, and A. D. Osterhaus, A previously undescribed coronavirus associated with respiratory disease in humans, *Proc. Natl. Acad. Sci. USA* **101**, 6212-6216 (2004).
33. S. S. Chiu, K. H. Chan, K. W. Chu, S. W. Kwan, Y. Guan, L. L. Poon, and J. S. Peiris, Human coronavirus NL63 infection and other coronavirus infections in children hospitalized with acute respiratory disease in Hong Kong, China, *Clin. Infect. Dis.* **40**, 1721-1729 (2005).
34. N. Bastien, K. Anderson, L. Hart, P. Van Caeseele, K. Brandt, D. Milley, T. Hatchette, E. C. Weiss, and Y. Li, Human coronavirus NL63 infection in Canada, *J. Infect. Dis.* **191**, 503-506 (2005).
35. P. C. Woo, S. K. Lau, C. M. Chu, K. H. Chan, H. W. Tsoi, Y. Huang, B. H. Wong, R. W. Poon, J. J. Cai, W. K. Luk, L. L. Poon, S. S. Wong, Y. Guan, J. S. Peiris, and K. Y. Yuen, Characterization and complete genome sequence of a novel coronavirus, coronavirus HKU1, from patients with pneumonia, *J. Virol.* **79**, 884-895 (2005).
36. C. M. Jonassen, T. Kofstad, I. L. Larsen, A. Lovland, K. Handeland, A. Follestad, and A. Lillehaug, Molecular identification and characterization of novel coronaviruses infecting graylag geese (*Anser anser*), feral pigeons (*Columbia livia*) and mallards (*Anas platyrhynchos*), *J. Gen. Virol.* **86**, 1597-1607 (2005).

ANIMAL MODELS FOR SARS

Anjeanette Roberts and Kanta Subbarao*

1. INTRODUCTION

In 2002–2003, severe acute respiratory syndrome (SARS) was a newly identified illness that emerged in Southern China, spread to involve more than 30 countries, and affected more than 8000 people and caused nearly 800 deaths worldwide. Although the etiologic agent was rapidly identified to be a previously unknown coronavirus (named SARS coronavirus or SARS-CoV) and the outbreak was controlled by public health measures, no specific options were available for prevention and control of human disease. Over the past two years, a number of strategies for vaccines and immunoprophylaxis have been investigated. Animal models are essential for preclinical evaluation of the efficacy of candidate vaccines and antivirals, and they are also needed in order to understand the pathogenesis of SARS. A number of investigators around the world have evaluated several different animal species as models for SARS; this effort is important for two reasons. First, because the source of SARS-CoV in the wild is not known and exploration of the range of species that are susceptible to SARS-CoV infection may help identify the natural reservoir, and second, if the efficacy of vaccines cannot be evaluated in humans, efficacy in two or more animal models may be required for licensure.

The ideal animal models would be those in which viral replication is accompanied by clinical illness and pathology that resembles that seen in human cases of SARS. However, the consequences of SARS-CoV infection in different animal models may vary from this picture to one in which viral replication is associated with pathology in the absence of clinical illness or models in which viral replication is present in the absence of clinical illness or histopathologic changes. Models that demonstrate clinical illness and pathology can be used to study the disease process as well as to evaluate intervention strategies while models in which virus replication occurs without clinical illness can be used in vaccine or antiviral studies. In these cases, the efficacy of an intervention can be assessed by quantitative virology with or without accompanying pathology.

A review of the different animal models that have been reported follows with a summary of the pros and cons and potential applications of the different models.

* NIAID, National Institutes of Health, Bethesda, Maryland 20892.

2. SARS-CoV INFECTION IN MICE

When 6 to 8-week-old lightly anesthetized BALB/c mice are administered SARS-CoV intranasally (i.n.), the virus replicates efficiently in the respiratory tract (lungs and nasal turbinates) with a peak of viral replication on day 2 postinfection (mean titer in lungs is 10^6 to 10^7 50% tissue culture infectious doses ($TCID_{50}$) per gram and mean titer in nasal turbinates is 10^5 to 10^6 $TCID_{50}$ per gram following administration of 10^6 $TCID_{50}$ of SARS-CoV i.n.).[1] Virus is cleared from the respiratory tract by about day 5. Virus replication occurs without signs of illness such as weight loss or ruffled fur and is associated with minimal to mild inflammation in the lungs. Viral antigen and nucleic acid are present in the epithelial cells of the large airways on day 2 but are being cleared by day 4.[1] When SARS-CoV is administered to BALB/c mice i.n. and orally, viral nucleic acid can be amplified from lung tissue and small intestines.[2] Intranasally administered SARS-CoV also replicates in the lungs of C57BL/6 (B6) mice, with a peak of viral replication on day 3 and clearance of virus by day 9.[3] SARS-CoV infected BALB/c and B6 mice tend to gain less weight than mock-infected mice[2,3] 129SvEv mice also support replication of SARS-CoV with a self-limited bronchiolitis that begins with mixed peribronchiolar inflammatory infiltrates, progresses to bronchiolitis with migration of inflammatory cells into surface epithelium and interstitial inflammation in adjacent alveolar septae, that resolves completely over the next 2 weeks.[4]

SARS-CoV infected mice develop a SARS-CoV specific neutralizing antibody response and are protected from reinfection with SARS-CoV. Antibody alone is sufficient to transfer this protection to naïve mice.[1] Mice with targeted defects in the immune system were evaluated in order to determine which arm of the immune system was responsible for clearance of SARS-CoV in mice. Beige, $CD1^{-/-}$ and $RAG1^{-/-}$ mice replicated and cleared virus with the same kinetics as B6 mice, without overt signs of clinical disease indicating that NK cells, NK-T cells, and T and B lymphocytes are not required for clearance of SARS-CoV from the lungs of mice.[3] In young mice, SARS-CoV induces dramatic upregulation of a subset of inflammatory chemokines (CCL2, CCL3, CCL5, CXCL9, CXCL10) and the chemokine receptor CXCR3 without detectable expression of classic proinflammatory and immunoregulatory cytokines (IFN-γ, IL-12, p70, IL-4, IL-10, and TNF-α) and without evoking marked leukocyte infiltration of the lung. Taken together with the observation that beige, $CD1^{-/-}$ and $RAG1^{-/-}$ mice clear SARS-CoV normally, proinflammatory chemokines may coordinate a rapid and highly effective innate antiviral response in the lung.[3]

In contrast to young (4 to 6-week-old) BALB/c mice that support replication of SARS-CoV in the absence of clinical illness and pneumonitis, old (13 to 14-month-old) BALB/c mice demonstrate illness (weight loss, hunching, dehydration, and ruffled fur from days 3 to 6 postinfection) and interstitial pneumonitis.[5] Perivascular lymphocytic infiltrates noted at day 3 were more prominent by day 5 and evidence of alveolar damage was seen, with multifocal interstitial lymphohistiocytic infiltrates, proteinaceous deposits around alveolar walls and intraalveolar edema. At day 9, the perivascular infiltrates persisted and the changes associated with alveolar damage were accompanied by proliferation of fibroblasts in inflammatory foci. A few of these foci persisted through 29 days post-infection and may represent the histologic correlate of fibrosis and scarring seen by high-resolution computed tomography in patients who recovered from SARS.[5] Mice with a targeted disruption of the STAT 1 signaling pathway develop severe SARS-

CoV disease,[4] with weight loss and pneumonitis that begins with acute bronchiolitis and progresses to diffuse interstitial pneumonia with focal airspace consolidation (unpublished observations). Viral antigen was present within cells of the inflammatory pulmonary infiltrates. At day 27 postinfection, nodules of dense mononuclear inflammation containing SARS-CoV infected cells were present in the liver.[4] In STAT 1 knockout mice and old BALB/c mice, SARS-CoV replicates to higher titer than in young BALB/c mice; old BALB/c mice recover from the infection.[5]

In summary, SARS-CoV replicates efficiently in the respiratory tract of young BALB/c and B6 mice in the absence of clinical illness and histopathologic evidence of mild inflammation while 129SvEv mice show some pneumonitis. Mice that recover from infection develop a neutralizing antibody response and are protected from subsequent challenge; antibody alone is sufficient to protect mice from replication of SARS-CoV in the lower respiratory tract and NK, NK-T, T, and B cells are not required for viral clearance.[3] The efficacy of several vaccines and monoclonal antibodies has been evaluated in BALB/c mice.[6-12] Morbidity and mortality and pneumonitis are seen in STAT-1 knockout mice[4] and old BALB/c mice infected with SARS-CoV.[5] The pathogenesis of disease in these models is under investigation.

3. SARS-CoV INFECTION IN HAMSTERS

Intranasally administered SARS-CoV replicates efficiently in the respiratory tract of golden Syrian hamsters with a peak of viral replication in the lungs on days 2 or 3 (mean titer approximately 10^7 $TCID_{50}$ per gram of lung tissue following administration of 10^3 $TCID_{50}$ i.n.) and clearance from the lungs by day 10. This is accompanied by histopathologic evidence of pneumonitis. Mild mononuclear inflammatory cell infiltrates are noted in the submucosa of the nasal epithelium and bronchioles at day 3 post-infection. As inflammation in the nasal tissues resolves, inflammatory reaction in the lungs progresses, with confluent areas of consolidation that involve 30–40% of the surface of the lung by day 7 postinfection with resolution by day 14. The lung pathology is not associated with overt clinical illness. In contrast to mice in which replication of intranasally administered SARS-CoV is restricted to the respiratory tract, transient viremia occurs 1 to 2 days following infection and virus is detected in the liver and spleen in hamsters. However, inflammation is not observed in these organs. Hamsters that recover from infection develop a robust neutralizing antibody response and are protected from subsequent infection with SARS-CoV. There is no clinical, virologic or histopathologic evidence of enhanced disease upon reinfection even in the presence of sub-neutralizing levels of monoclonal antibodies to the SARS-CoV spike glycoprotein.[13, 13b]

4. SARS-CoV INFECTION IN FERRETS

Infection of BALB/c mice and hamsters used the Urbani strain of SARS-CoV while the SARS-CoV isolate from patient 5688 (HKU-39849) was used to infect ferrets (*Mustela furo*). When anesthetized ferrets were infected with 10^6 $TCID_{50}$ by the intratracheal (i.t.) route, three of six ferrets were lethargic from days 2 to 4 postinfection and one died on day 4.[14] Virus was isolated from pharyngeal swabs on days 2 to 8 post-infection and from trachea, lungs and tracheobronchial lymph nodes at necropsy on day 4. When administered i.t. at a dose of 10^4 $TCID_{50}$ of SARS-CoV, the virus replicates efficiently in the lungs to a titer of 10^6 $TCID_{50}$/ml that peaks at 4 days postinfection.[15]

Multifocal pulmonary lesions affecting 5–10% of lung surface area include mild alveolar damage and peribronchial and perivascular lymphocyte infiltration.[15,16] Ferrets are outbred animals that are highly susceptible to viruses that they can acquire from their caretakers. Ferrets used as models for SARS should be screened to ensure that other intercurrent infections do not modify the disease associated with SARS-CoV infection.

5. SARS-CoV INFECTION IN FARMED CIVETS

In 2003, SARS-CoV was isolated from captive civet cats in wild animal markets in Guangdong Province, China, and several civet cats had detectable antibodies to SARS-CoV. In a recent report, Wu *et al.*[17] infected farmed civet cats with two strains of SARS-CoV: five animals were infected with the GZ01 virus, the prototype of the virus isolated from civet cats, and 5 animals were infected with BJ01, a SARS-CoV strain that is typical of viruses isolated in Hong Kong during the SARS outbreak. BJ01 has a 29-nucleotide deletion in its genome compared with GZ01. The civet cats were infected with a dose of 3×10^6 $TCID_{50}$ administered i.t. and i.n. In contrast with the observation that SARS-CoV infected wild civets appeared healthy, lethargy and a decrease in aggressiveness was noted in the experimentally infected farmed civet cats from day 3 onwards, fever from day 3 to 7, and diarrhea and conjunctivitis in 20–40% of animals. The animals had leukopenia at day 3, but white blood cell counts were normal by day 13. Interstitial pneumonitis with alveolar septal enlargement and macrophages and lymphocyte infiltration was noted on days 13 to 35; the histopathological findings are reportedly similar to lesions described in SARS-CoV infected macaques and experimentally infected ferrets.[17] Virus was isolated from throat and anal swabs in 60% of the animals on days 3 and 8 and from organs at necropsy on day 3 (n = 1). Viral nucleic acid was detected by reverse transcriptase PCR at necropsy in multiple organs at day 3 and in lymph nodes and spleen at days 13, 23, 34, and 35. The findings in experimentally infected farmed civets support epidemiologic observations made in wild-animal markets[18] that point to civets as a potential source for the transmission of SARS-CoV from animals to humans. However, further research is required to identify the reservoir(s) of SARS-CoV in nature.

6. SARS-CoV INFECTION IN NON-HUMAN PRIMATES

Among Old World monkeys, rhesus, cynomolgus, and African green monkeys have been experimentally infected with SARS-CoV and have been used for vaccine immunogenicity and/or efficacy studies. The presence and extent of clinical illness reported in these studies have not been consistent and attempts to isolate SARS-CoV from tissues have also been variably successful (Table 1). Clinical illness and histopathologic findings in SARS-CoV infected cynomolgus monkeys range from reports of no illness and disease[19,20] to lethargy, temporary skin rash, and respiratory distress progressing to ARDS, associated with diffuse alveolar damage, extensive loss of epithelium from alveolar and bronchiolar walls, thickening of alveolar walls, hyaline membranes in some alveoli, and occasional multinucleated giant cells and type 2 pneumocyte hyperplasia at days 4 to 6 postinfection.[21-23] Clinical findings in SARS-CoV infected rhesus monkeys are absent or mild. Histopathologic findings are variable, ranging from no abnormalities to patchy areas of mild interstitial edema and alveolar inflammation interspersed with normal lung parenchyma and occasional areas of intra-alveolar edema inflammation at day 14, in 1 of 4 animals[19] to acute interstitial pneumonia through 60 day postinfection with infiltration of lymphocytes and macrophages in nodular areas of lungs.[24] The age of animals may be an important determinant of outcome that can be difficult to ascertain in wild-caught monkeys.

In African green monkeys, virus infection in the lungs is patchy and is cleared by day 4 post-infection; the titer of virus in respiratory secretions does not accurately reflect the titer of virus recovered from trachea and lung tissue.[20] Histopathological examination of lungs shows diffuse alveolar damage and focal interstitial pneumonitis that parallels virus titers in resolution by day 4 postinfection.[20] Three species of New World monkeys have also been evaluated as models for SARS (Table 1). Although squirrel monkeys and mustached tamarins could not be experimentally infected, common marmosets developed fever and watery diarrhea and histologic evidence of multifocal pneumonitis and hepatitis following SARS-CoV infection.[25] Further evaluation of this model is warranted.

Table 1. Summary of findings in SARS-CoV infected non-human primates.

Species	Virus, dose, and route of administration	Clinical findings	Virus isolation	PCR	Lung pathology	Ref.
Cynomolgus macaques (*Macaca fasicularis*)	HK39 (from patient 5688) 10^6 $TCID_{50}$ i.t. + i.n. + conjunctival	Lethargy, temporary skin rash, respiratory distress progressing to ARDS	Yes: nasal, pharyngeal swabs and sputum in 1 of 4 animals	Yes	Diffuse alveolar damage, extensive loss of epithelium from alveolar and bronchiolar walls, thickening of alveolar walls, hyaline membranes in some alveoli, occasional multinucleated giant cells, type 2 pneumocyte hyperplasia at days 4 to 6	15, 21, 22
	TOR-2 10^7 pfu i.v. or i.t	Minimal; mild cough and slightly decreased activity that quickly resolved	No	Yes	None found	19
	10^6 Urbani i.t. + i.n.	None found	Yes; nasal wash and tracheal lavage	Yes	Not done	20
African green (*Chlorocebus aethiops sabaeus* or *Cercopithecu aethiops sabaeus*)	10^6 Urbani i.t. + i.n.	None found	Yes; nasal wash, tracheal lavage, nasal turbinates, trachea, lung tissue	Yes	Diffuse alveolar damage and focal interstitial pneumonitis early (day 2), resolving by day 4	20

Table 1. (continued)

Species	Virus, dose, and route of administration	Clinical findings	Virus isolation	PCR	Lung pathology	Ref.
Rhesus (*Macaca mulatta*)	10^7 pfu TOR-2 i.v. or i.t	None; 1 in 4 animals agitation and aggressive behavior at day 10	No	Yes	Patchy areas of mild interstitial edema and alveolar inflammation interspersed with normal lung parenchyma and occasional areas of intra-alveolar edema inflammation at day 14 in 1 of 4 animals	19
	10^6 Urbani i.t.+ i.n.	None found	No	Not done	Not done	20
	10^5 PUMC01 i.n.	None found	Yes; nasal or pharyngeal swabs	Yes; days 5–16	Acute interstitial pneumonia throughout 60-day study; infiltration of lymphocytes + macrophages in nodular areas of lungs.	24
Squirrel monkeys (*Saimiri sciureus)*	10^6 Urbani i.t. + i.n.	None found	No	No	None found	Unpub. data
Mustached tamarins (*Saguinus mystax)*	10^6 Urbani i.t. + i.n.	None found	No	No	Not done	Unpub. data
Common marmosets (*Callithrix jacchus*)	10^6 Urbani i.t.	Mild, elevated temperature, watery diarrhea in 7/12 and dyspnea under anesthesia in 2 animals	No	Yes	Multifocal to coalescing interstitial pneumonitis, multinucleated syncytia at day 4. Pneumonitis was resolving by day 7 in some animals, consolidation while type 2 pneumocyte hyperplasia was seen in others. Multifocal lymphocytic hepatitis at day 7 and mild diffuse colitis.	25

The detection of viral RNA and neutralizing antibody responses clearly demonstrates that several species of nonhuman primates can be experimentally infected with SARS-CoV. Not surprisingly, the extent of disease in outbred animals is more variable than in inbred animals such as mice. As seen with the other animal models, the course of infection in experimentally infected nonhuman primates is short, with a rapid peak in viral replication and clearance of virus from the lungs by days 4 to 7 in different species. Histopathologic changes also resolved rapidly in all but one study.[24] The absence of consistently observed clinical illness and the rapid resolution of viral infection and associated pulmonary pathology may limit the role of nonhuman primates to studies of the immunogenicity of vaccines and antivirals rather than pathogenesis or efficacy studies.

7. POTENTIAL USES OF ANIMAL MODELS FOR SARS

As summarized above, mice, hamsters, ferrets, civets and several non-human primate species can support replication of SARS-CoV with or without accompanying clinical illness or pulmonary pathology. Each model has advantages and disadvantages (Table 2) but the common themes are that a number of animal species can be infected when SARS-CoV is delivered into the respiratory tract, the infection elicits a neutralizing antibody response and the animals are protected from subsequent infection. In comparing reports, readers should bear in mind that the virus used, inoculum and route of virus administration and age of the animal may represent important differences. In studies in mice, the age of the animals and use of anesthesia clearly affect the course of infection. The characteristics of each model should be taken into consideration in determining their utility and application. Table 3 lists potential uses of the models discussed above.

8. ACKNOWLEDGMENTS

We thank our colleagues Leatrice Vogel, Elaine Lamirande, Josephine McAuliffe, Brian Murphy, Jun Chen, Jadon Jackson, and Aaron Cheng at the NIH, Sherif Zaki, Christopher Paddock, Jeannette Guarner, and Wun-Ju Shieh in the Infectious Disease Pathology Activity at the CDC, and Marisa St. Claire at Bioqual Inc. who have participated in our efforts to evaluate several animal models for SARS. This research was supported in part by the Intramural Research Program of the NIAID, NIH.

Table 2. Pros and cons of SARS animal models.

Species	Advantages	Limitations
4-8 wk old BALB/c or B6 mice	Availability of inbred mice and reagents for immunological studies	No illness or overt disease
129 mice	Pneumonitis present	Needs further characterization
Old (12-14 mo) BALB/c mice	Illness and pneumonitis present	Availability; immune senescence
STAT 1 -/- mice	Illness and pneumonitis and mortality present	Defect in innate immunity
Ferrets	Illness +/-; virus replication with pneumonitis	Availability, susceptibility to other respiratory viruses
Farmed civet cats	Illness and mild pneumonitis present	Availability, lack of reagents
Hamsters	Virus replication with pneumonitis	No overt illness, lack of immunological reagents
Non-human primates	Virus replication and pneumonitis w illness in cynos, w/o illness in AGM, pneumonitis, diarrhea and hepatitis in marmosets	Availability, cost, housing, virus and pneumonitis cleared early

Table 3. Suggested uses of SARS animal models.

Animal model	Potential uses
Young mice	Vaccines, antivirals
Old BALB/c mice	Pathogenesis, vaccines, immunoprophylaxis
STAT 1-/- mice	Antivirals, pathogenesis
Ferrets	Vaccines, immunoprophylaxis, immunotherapy, antivirals
Hamsters	Vaccines, immunoprophylaxis, immunotherapy, antivirals
Farmed civet cats	Pathogenesis, vaccines
Non-human primates	Immunogenicity of vaccines, immunoprophylaxis, antivirals

9. REFERENCES

1. K. Subbarao, J. McAuliffe, et al., Prior infection and passive transfer of neutralizing antibody prevent replication of severe acute respiratory syndrome coronavirus in the respiratory tract of mice, *J. Virol.* **78**, 3572-3577 (2004).
2. D. E. Wentworth, L. Gillim-Ross, N. Espina, and K. A. Bernard, Mice susceptible to SARS coronavirus, *Emerg. Infect. Dis.* **10**, 1293-1296 (2004).
3. W. G. Glass, K. Subbarao, B. Murphy, and P. M. Murphy, Mechanisms of host defense following severe acute respiratory syndrome-coronavirus (SARS-CoV) pulmonary infection of mice, *J. Immunol.* **173**, 4030-4039 (2004).
4. R. J. Hogan, G. Gao, et al., Resolution of primary severe acute respiratory syndrome-associated coronavirus infection requires Stat1, *J, Virol,* **78**, 11416-11421 (2004).
5. A. Roberts, C. Paddock, L. Vogel, E. Butler, S. Zaki, and K. Subbarao, Aged BALB/c mice as a model for increased severity of severe acute respiratory syndrome in elderly humans, *J. Virol.* **79**, 5833-5838 (2005).
6. H. Bisht, A. Roberts, L. Vogel, A. Bukreyev, P. L. Collins, B. R. Murphy, K. Subbarao, and B. Moss, Severe acute respiratory syndrome coronavirus spike protein expressed by attenuated vaccinia virus protectively immunizes mice, *Proc. Natl. Acad. Sci. USA* **101**, 6641-6646 (2004).
7. H. Bisht, A. Roberts, L. Vogel, K. Subbarao, and B. Moss, Neutralizing antibody and protective immunity to SARS coronavirus infection of mice induced by a soluble recombinant polypeptide containing an N-terminal segment of the spike glycoprotein, *Virology* **334**, 160-165 (2005).
8. E. Traggiai, S. Becker, et al., Rapid and efficient isolation of human monoclonal antibodies neutralizing SARS coronavirus from human memory B cells, *Nat. Med.* **10**, 871-875 (2003).
9. Z. Y. Yang, W. P. Kong, Y. Huang, A. Roberts, B. R. Murphy, K. Subbarao, and G. J. Nabel, A DNA vaccine induces SARS coronavirus neutralization and protective immunity in mice, *Nature* **428**, 561-564 (2004).
10. J. Sui, W. Li, et al., Evaluation of human monoclonal antibody 80R for immunoprophylaxis of severe acute respiratory syndrome by an animal study, epitope mapping, and analysis of spike variants, *J. Virol.* **79**, 5900-5906 (2005).
11. T. C. Greenough, G. J. Babcock, et al., Development and characterization of a severe acute respiratory syndrome-associated coronavirus-neutralizing human monoclonal antibody that provides effective immunoprophylaxis in mice, *J. Infect. Dis.* **191**, 507-514 (2005).
12. S. U. Kapadia, J. K. Rose, E. Lamirande, L. Vogel, K. Subbarao, and A. Roberts, Long-term protection from SARS coronavirus infection conferred by a single immunization with an attenuated VSV-based vaccine, *Virology* **340**: 174-182 (2005).
13. A. Roberts, L. Vogel, J. Guarner, N. Hayes, B. Murphy, S. Zaki, and K. Subbarao, Severe acute respiratory syndrome coronavirus infection of golden Syrian hamsters, *J. Virol.* **79**, 503-511 (2005).

13[b]. A. Roberts, W. D. Thomas, J. Guarner, E. W. Lamirande, G. J. Babcock, et al., Therapy with a severe acute respiratory. Syndrome-associated coronavirus-neutralizing human monoclonal antibody reduces disease severity and virus burden in golden syrian hamsters, J. Infect. Dis. **193,** 685-692(2006).

14. B. E. E. Martina, B. L. Haagmans, et al., SARS virus infection of cats and ferrets, *Nature* **425**(915 (2003).
15. J. ter Meulen, A. B. Bakker, et al., Human monoclonal antibody as prophylaxis for SARS coronavirus infection in ferrets, *Lancet* **363**(9427), 2139-41 (2004).
16. B. E. Martina, B. L. Haagmans, et al., Virology: SARS virus infection of cats and ferrets, *Nature* **425**, 915 (2003).

17. D. Wu, C. Tu, et al., Civets are equally susceptible to experimental infection by two different severe acute respiratory syndrome coronavirus isolates, *J. Virol.* **79**, 2620-2625 (2005).
18. Y. Guan, B. J. Zheng, et al., Isolation and characterization of viruses related to the SARS coronavirus from animals in Southern China, *Science* **302**, 276-278 (2003).
19. T. Rowe, G. Gao, et al., Macaque model for severe acute respiratory syndrome, *J. Virol.* **78**, 11401-11404 (2004).
20. J. McAuliffe, L. Vogel, et al., Replication of SARS coronavirus administered into the respiratory tract of African Green, rhesus and cynomolgus monkeys, *Virology* **330**, 8-15 (2004).
21. R. A. Fouchier, T. Kuiken, et al., Aetiology: Koch's postulates fulfilled for SARS virus, *Nature* **423**, 240 (2003).
22. T. Kuiken, R. A. Fouchier, et al., Newly discovered coronavirus as the primary cause of severe acute respiratory syndrome, *Lancet* **362**, 263-270 (2003).
23. B. L. Haagmans, T. Kuiken, et al., Pegylated interferon-alpha protects type 1 pneumocytes against SARS coronavirus infection in macaques, *Nat. Med.* **10**, 290-293 (2004).
24. C. Qin, J. Wang, et al., An animal model of SARS produced by infection of Macaca mulatta with SARS coronavirus, *J. Pathol.* **206**, 251-259 (2005).
25. T. C. Greenough, A. Carville, J. Coderre, M. Somasundaran, J. L. Sullivan, K. Luzuriaga, and K. Mansfield, Pneumonitis and multi-organ system disease in common marmosets (Callithrix jacchus) infected with the severe acute respiratory syndrome-associated coronavirus, *Am. J. Pathol.* **167**, 455-463 (2005).

HCoV-OC43–INDUCED APOPTOSIS OF MURINE NEURONAL CELLS

Hélène Jacomy and Pierre J. Talbot*

1. INTRODUCTION

Apoptosis is a selective process for cell deletion and can be induced by a variety of physiological and non-physiological stimuli, including viral infections.[1] A growing number of viruses have been shown to actively promote apoptosis, representing a culmination of a lytic infection that serves to spread virus progeny to neighboring cells while evading host inflammatory responses.[2] Various groups of molecules are involved in the apoptotic pathway. One set of mediators implicated in apoptosis belongs to the cysteine dependent aspartate-specific family of proteases referred to as caspases. Caspase-3 has been identified as the key pro-apoptotic protease in neuronal apoptosis.[3]

HCoV-OC43 causes acute encephalitis in the central nervous system (CNS) of mice. Following intracerebral (IC) inoculation, mice developed disseminated infection and mortality seemed to be related to the amount of infectious virus in the CNS.[4] Using a mouse primary cortical cell cultures, we showed that these cultures were productively infected by HCoV-OC43 and that neurons underwent nuclear fragmentation associated with activated caspase-3 positive staining, indicating that HCoV-OC43 infection induced neuronal apoptosis. We also demonstrated apoptosis of neurons takes place during encephalitis in the CNS of infected mice. These findings illustrate that HCoV-OC43 is responsible for neuronal cell death and suggest that apoptosis could play a role in the dissemination of HCoV-OC43.

2. MATERIALS AND METHODS

2.1. Virus, Mice, and Cortical Cell Cultures

The HCoV-OC43 strain was originally obtained from the American Type Culture Collection (ATCC), plaque-purified and grown on the human rectal carcinoma cell line

* INRS-Institut Armand Frappier, Laval (Québec), Canada H7V 1B7.

HRT-18 as previously described.[5] HCoV-OC43 virus stocks (10^6 $TCID_{50}$/mL) were kept at -80°C.

C57Bl/6 (MHV-seronegative female) mice (Jackson Laboratories), aged 21 days post-natal (DPN) were inoculated by the intracerebral (IC) route, using 10 μL containing 10 $TCID_{50}$ of HCoV-OC43.

Cortical cell cultures were obtained following the modified methods of Brewer and collaborators.[6] Cells from mouse embryos at 16 to 18 days of gestation were dissected in HBSS without Ca^{2+} and Mg^{2+}, supplemented with 1.0 mM sodium pyruvate and 10 mM HEPES, and dissociated by trituration. Supernatants were then transferred and centrifuged for 1 min at 1000 x *g*. The pellets were resuspended in 1 mL HBSS per brain. Cells were plated in neurobasal medium (Invitrogen) supplemented with 0.5 mM L-glutamine, 25 μM glutamate and B27 supplement (Invitrogen) and grown on glass coverslips, pretreated with poly-D-Lysine and plated at approximately 5 x 10^5/cm^2. After 4 days, medium was replaced with Neurobasal/B27 without glutamate.

2.2. Infection of Cell Cultures and Infectious Virus Assays

Cortical cultures were infected with HCoV-OC43 at an MOI of 1, incubated at 37ºC for 2 hr, then washed in warm PBS and incubated at 37ºC. Supernatants were collected at 12, 24, 48, 72, and 96 hr postinfection (hpi). Collected supernatants were centrifuged for 5 min at 1000 x *g* and then immediately frozen at -80°C and stored until assayed. The extracts were processed for the presence and quantification of infectious virus by an indirect immunoperoxidase assay, as previously described.[7] HCoV-OC43-susceptible HRT-18 cells were inoculated with serial logarithmic dilutions of each tissue sample in a 96-well Linbro plate. After 4 days of incubation at 33°C in 5% (v/v) CO_2, cells were washed in PBS and fixed with 0.3% (v/v) hydrogen peroxide (H_2O_2) in methanol for 30 min. After washing with PBS, they were incubated for 2 hr at 37°C in 1/1000 dilution of a mouse ascites fluid containing MAb 1-10C.3, directed against the spike protein of HCoV-OC43.[8] Afterwards, cells were washed in PBS and HRP-conjugated goat anti mouse immunoglobulins were added and incubated for 2 hrs at 37°C. Antibody complexes were detected by incubation in DAB solution, with 0.01% (v/v) H_2O_2.

2.3. Immunofluorescence and TUNEL Assays Staining

To visualize viral antigens and cell markers, cultures were fixed with 4% (v/v) paraformaldehyde, for 30 min. Then, after washing, cells were permeabilized with 100% methanol for 5 min, incubated 2 hr with primary antibodies, as previously described.[9] For viral antigens, we used 1/1000 dilutions of ascites fluid of the 4-E11.3 hybridoma that secretes monoclonal antibodies specific for the nucleocapsid protein of the serologically related hemaglutinating encephalomyelitis virus of pigs.[10] Apoptotic cells were revealed by antibodies to active caspase-3 at 1/50 (Rabbit anti-human/mouse active caspase-3 antibodies; R&D Systems, Inc.). After several washes in PBS, cell coverslips were incubated for 1 hr in the dark with a combination of immunofluorescent secondary antibodies: Alexa Fluor 488 and Alexa Fluor 568 (Molecular Probes), both at a dilution of 1/1000. Then, after 3 washes in water, cells were incubated in 4,6-diamidino-2-phenylindole dihydrochloride (DAPI; Polysciences Inc.) at a 1/100 dilution for 5 min.

After final washes in water, coverslips from each well were removed and mounted on microscope slides in Immuno-mount and observed under a fluorescence microscope.

The In Situ Cell Death Detection FITC Kit (Roche Molecular Diagnostics) was used for TUNEL (transferase dUTP nick end labeling) assays. Fluorescent double labeling of brain slices with antibody to infected cells was performed in conjunction with the TUNEL assay to enable correlation of TUNEL-positive cells with the presence of viral infection. Briefly, mice were intracardially perfused with 4% (v/v) paraformaldehyde, as previously described.[4] Brains were paraffin embedded and sectioned longitudinally. Sections were collected on slides, de-paraffined and incubated with primary antibodies for viral antigens and for TUNEL staining, as recommended by the manufacturer.

3. RESULTS

3.1. Apoptosis of Primary Neural Cell Cultures After Infection

Measurement of infectious viral titers in the supernatants revealed that mouse primary cortical cultures produced a significant viral titers (Fig. 1), with a peak at 48 hpi. With time, virus titers decreased in concomitance with disappearance of neurons in the infected culture. This illustrates that HCoV-OC43 induced neuronal cell death.

Immunofluorescent staining revealed that after infection of cortical cultures, neurons were positive for viral antigens (Fig. 2 A) and numerous infected cells were also positive for activated caspase-3 (Fig. 2 B). The nuclear fragmentation/condensation observed with DAPI staining colocalized with activated caspase-3 positive cells (Fig. 2 C and D), confirming that viral infection could directly trigger an apoptotic response.

3.2. Apoptosis of Neural Cells in Mouse Brains After Infection

As previously reported, intracerebral inoculation of HCoV-OC43 into mice led to a generalized infection of the CNS, which also affected the hippocampus.[4] Double immu-

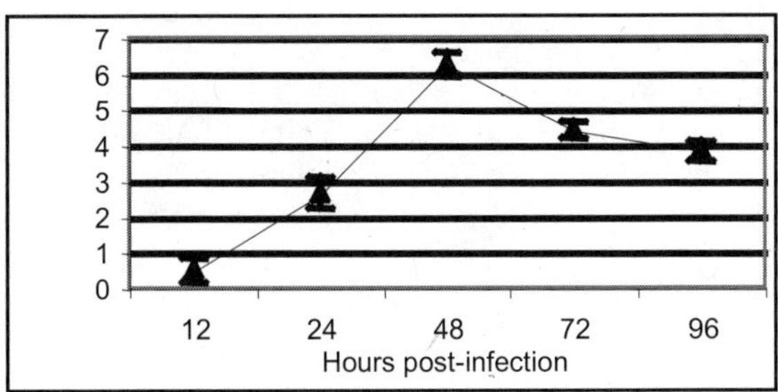

Figure 1. HCoV-OC43 replication in primary mouse cortical cultures. The results are expressed as the mean $TCID_{50}$ per milliliter for three independent experiments. Error bars indicate standard deviations of the means.

nostaining for viral antigens and TUNEL assays was performed on brain sections during the acute phase of the encephalitis, at 11 days post-infection (11 dpi). TUNEL positive cells could be seen in the hippocampus. Numerous neurons in the CA1 hippocampal layer were TUNEL-positive and numerous cells, in the same region, were also positive for viral antigens. Merged pictures illustrate that some of the infected neurons underwent apoptosis (Fig. 3 A) and that noninfected cells localized near the infected ones were also undergoing apoptosis (Fig. 3 B).

4. DISCUSSION

HCoV-OC43 was previously reported to induce apoptosis in MRC-5 cells, a lung cell line[11] but not in infected human monocytes/macrophages, unlike the infection observed with strain 229E.[12] The murine counterpart of HCoV-OC43, MHV, was reported to induce apoptosis in 17Cl-1 cells, a murine fibroblast cell line[13] and in mouse brain neurons[14,15] in addition to macrophage/microglial cells, astrocytes and oligodendrocytes.[16,17] And recently, SARS-CoV was shown to induce apoptosis of Vero E6 cells.[18]

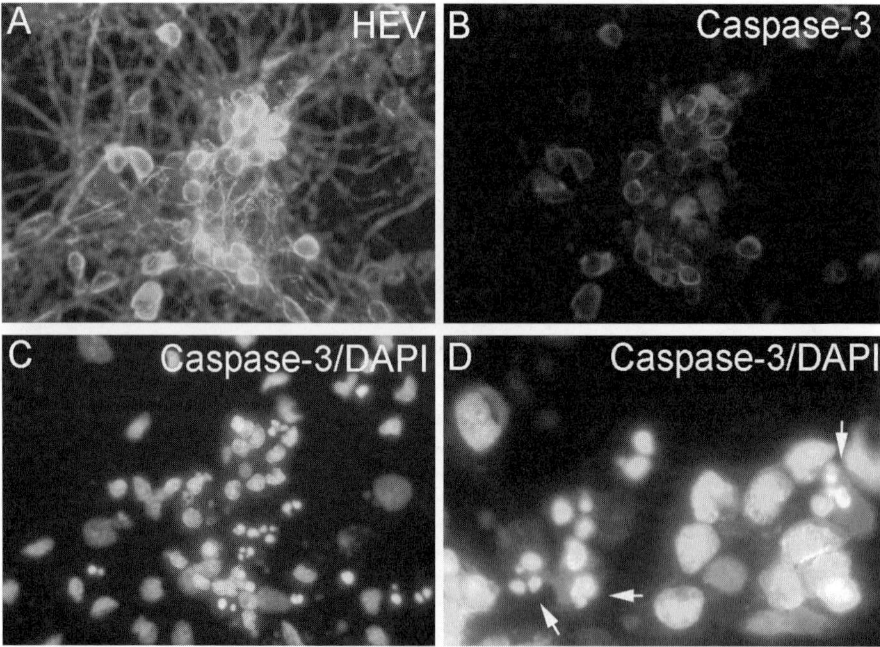

Figure 2. HCoV-OC43–induced apoptosis in primary mouse cell cultures. Cortical cells were infected by HCoV-OC43 at an MOI of 1. Twenty-four hpi, cells cultures were stained for viral antigens (A) and activated caspase-3 (B). In panel C and at higher magnification in panel D, activated caspase-3 positive cells staining were colocalized with the nuclear fragmentation observed with DAPI staining (arrows). All photographs were taken on the same cell culture field at a magnification of x200 for panels A, B, C and x400 for panel D.

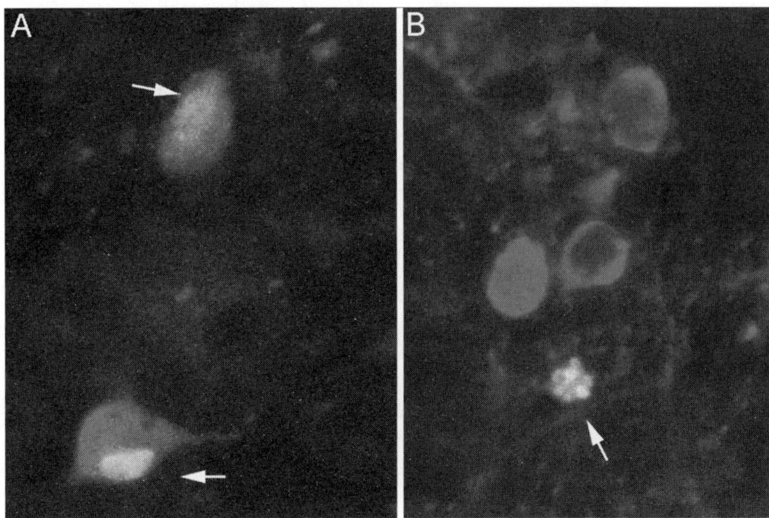

Figure 3. Caspase-3 is activated in hippocampal neurons from mice infected by HCoV-OC43. During the acute phase of the disease (11 dpi), CA1 hippocampal layer exhibited caspase-3 positive staining (arrows) colocalized with viral antigens staining (A) or in proximity of infected neurons (B). Magnification x200.

We have previously demonstrated the neuroinvasive properties of HCoV-OC43 in mice and that the severity of disease seemed to be linked to the amount of infectious virus in the CNS,[4] illustrating that virus replication played a major role in the development of the pathology. Some viruses appear to use apoptosis as a mechanism for killing cells and spreading. This represents an important mechanism for efficient dissemination of progeny virions, as well as a means by which viruses can induce host cell death, while limiting inflammatory and other immune responses.[2] Here we report that HCoV-OC43 induced neuronal apoptosis *in vitro*, as well as *in vivo*, which could account for part of the neuronal cell death observed after infection. These results illustrate one of the possible mechanisms used by HCoV-OC43 to spread to the whole CNS, which was responsible for the development of an acute encephalitis in infected mice.

5. REFERENCES

1. Vaux, D. L., Strasser, A., 1996, The molecular biology of apoptosis. *Proc. Natl Acad. Sci. USA* **93**: 2239-2244.
2. O'Brien, V., 1998, Viruses and apoptosis, *J. Gen. Virol.* **79**:1833-1845.
3. Eldadah, B. A., and Faden, A. I., 2000, Caspase pathways, neuronal apoptosis, and CNS injury, *J. Neurotrauma,* **17**:811-829.
4. Jacomy, H., and Talbot, P. J., 2003, Vacuolating encephalitis in mice infected by human coronavirus OC43, *Virology* **315**:20-33.
5. Mounir, S., and Talbot, P. J., 1992, Sequence analysis of the membrane protein gene of human coronavirus OC43 and evidence for O-glycosylation, *J. Gen. Virol.* **73**:2731-2736.
6. Brewer, G. J., Torricelli, J. R., Evege, E. K., and Price, P. J., 1993, Optimized survival of hippocampal neurons in B27-supplemented Neurobasal, a new serum-free medium combination, *J. Neurosci. Res.* **35**:567-576.

7. Bonavia, A., Arbour, N., Wee Yong, V., and Talbot, P. J., 1997, Infection of primary cultures of human neural cells by human coronavirus 229E and OC43, *J. Virol.* **71**:800-806.
8. Arbour, N., Côté, G., Lachance, C., Tardieu, M., Cashman, N. R., and Talbot, P. J., 1999, Acute and persistent infection of human neural cell lines by human coronavirus OC43, *J. Virol.* **73**:3338-3350.
9. Robertson, J., Beaulieu, J. M., Doroudchi, M. M., Durham, H. D., Julien, J. P., and Mushynski, W. E., 2001, Apoptotic death of neurons exhibiting peripherin aggregates is mediated by the proinflammatory cytokine tumor necrosis factor-alpha, *J. Cell Biol.* **155**:217-226.
10. Michaud, L., and Dea, S., 1993, Characterization of monoclonal antibodies to bovine enteric coronavirus and antigenic variability among Quebec isolates, *Arch. Virol.* **131**:455-465.
11. Collins, A. R., 2001, Induction of apoptosis in MRC-5, diploid human fetal lung cells after infection with human coronavirus OC43, *Adv. Exp. Med. Biol.* **494**:677-682.
12. Collins, A. R., 2002, In vitro detection of apoptosis in monocytes/macrophages infected with human coronavirus, *Clin. Diagn. Lab. Immunol.* **9**:1392-1395.
13. Chen, C. J., and Makino, S., 2002, Murine coronavirus-induced apoptosis in 17Cl-1 cells involves a mitochondria-mediated pathway and its downstream caspase-8 activation and bid cleavage, *Virology* **302**:321-332.
14. Wu, G. F., and Perlman, S., 1999, Macrophage infiltration, but not apoptosis, is correlated with immune-mediated demyelination following murine infection with a neurotropic coronavirus, *J. Virol.* **73**:8771-8780.
15. Chen, B. P., and Lane, T. E., 2002, Lack of nitric oxide synthase type 2 (NOS2) results in reduced neuronal apoptosis and mortality following mouse hepatitis virus infection of the central nervous system, *J. Neurovirol.* **8**:58-63.
16. Schwartz, T., Fu, L., and Lavi, E., 2002, Differential induction of apoptosis in demyelinating and nondemyelinating infection by mouse hepatitis virus, *J. Neurovirol.* **8**:392-399.
17. Liu, Y., Cai, Y., and Zhang, X., 2003, Induction of caspase-dependent apoptosis in cultured rat oligodendrocytes by murine coronavirus is mediated during cell entry and does not require virus replication, *J. Virol.* **77**:11952-11963.
18. Yan, H., Xiao, G., Zhang, J., Hu, Y., Yan, F., Cole D. K., Zheng C., and Gao G. F., 2004, SARS coronavirus induced apoptosis in Vero E6 cells, *J. Med. Virol* **73**:323-331.

INFECTION OF HUMAN AIRWAY EPITHELIA BY SARS CORONAVIRUS IS ASSOCIATED WITH ACE2 EXPRESSION AND LOCALIZATION

Hong Peng Jia, Dwight C. Look, Melissa Hickey, Lei Shi, Lecia Pewe, Jason Netland, Michael Farzan, Christine Wohlford-Lenane, Stanley Perlman, and Paul B. McCray, Jr.*

1. INTRODUCTION

Severe acute respiratory syndrome (SARS) emerged as a regional and global health threat in 2002–2003 resulting in approximately 800 deaths.[1] An intense, cooperative worldwide effort rapidly led to the identification of the disease causing agent as a novel coronavirus (SARS-CoV)[2,3] and the subsequent complete sequencing of the viral genome. Although limited human pathological studies demonstrate that the respiratory tract is a major site of SARS-CoV infection and morbidity, little is known regarding the initial steps in SARS-CoV-host cell interactions in the respiratory tract, such as the cell types in which primary viral infection and replication occur.

Angiotensin converting enzyme 2 (ACE2) was identified as a receptor for both SARS-CoV[4] and NL63.[5] ACE2 is a membrane-associated aminopeptidase expressed in vascular endothelia, renal and cardiovascular tissues, and epithelia of the small intestine, and testes.[6,7] A region of the extracellular portion of ACE2 that includes the first α-helix, and lysine 353 and proximal residues of the N-terminus of β-sheet 5 interacts with high affinity to the receptor binding domain of the SARS-CoV S glycoprotein.[8] Several unanswered questions remain regarding ACE2 expression in human respiratory epithelia and its role as a receptor for SARS-CoV, including identification of the specific epithelial cell types expressing ACE2, the polarity of ACE2 expression, and whether SARS-CoV infection of respiratory epithelia is ACE2-dependent.

* Hong Peng Jia, Dwight C. Look, Melissa Hickey, Lei Shi, Lecia Epping, Jason Netland, Christine Wohlford-Lenane, Stanley Perlman, Paul B. McCray, Jr., University of Iowa, Iowa City, Iowa 52242. Michael Farzan, Harvard Medical School, Southborough, Massachusetts 01772.

2. ACE2 EXPRESSION IN AIR-LIQUID INTERFACE CULTURED AIRWAY EPITHELIA IS POLARIZED

To try to answer some of these remaining questions, we use an air-liquid interface culture model of primary human airway epithelia in our studies. Airway epithelial cells of trachea or bronchi from human donors were grown on collagen coated porous filters with the apical side in contact with air and the basolateral side immersed in culture medium. Over 2 weeks, the cells polarize, develop transepithelial resistance and some become ciliated, signs of differentiation. TEM images show that these cells form a thin layer of secretions at the apical side, creating an air-liquid interface. So this pseudostratified epithelium closely mimics the *in vivo* situation.[9]

To understand the potential role for ACE2 as the SARS-CoV receptor in the respiratory tract, we first looked for evidence of ACE2 protein expression in human lung tissue by Western blot. ACE2 was identified in lysates from human conducting airway and distal lung tissues, but this result did not indicate which cell types expressed ACE2 (data not shown). We next evaluated ACE2 protein expression in well-differentiated primary cultures of airway epithelia by immunohistochemistry. The signal for ACE2 was more abundant on the apical rather than the basolateral surface (Figure 1A). Furthermore, the signal intensity was strongest on ciliated cells, as demonstrated by co-localization with beta-tubulin IV, a marker of cilia,[11] suggesting that ciliated cells express ACE2 abundantly. To confirm a polar distribution of ACE2 in differentiated epithelia, selective apical or basolateral surface biotinylation with subsequent immunoprecipitation was performed (Figure 1B). Western blot analysis of immunoprecipitated proteins confirmed that ACE2 is expressed in greater abundance on the apical surface of conducting airway epithelia, although a weak ACE2 signal was also detected basolaterally. In contrast, ErbB2 was more abundant on the basolateral surface as previously reported, confirming selective biotinylation.[10]

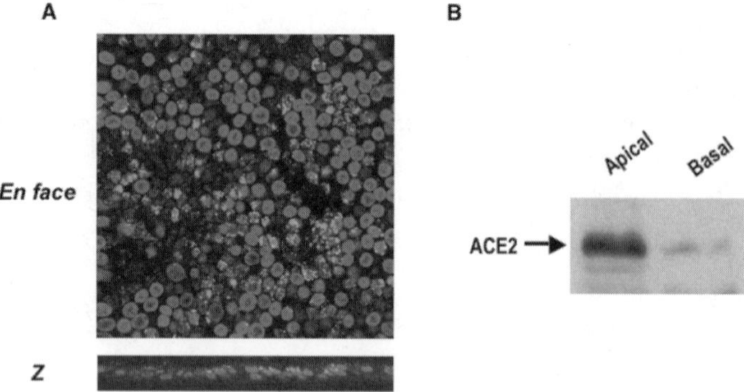

Figure 1. ACE2 is expressed in human airway epithelia. (A) ACE2 protein location in polarized human airway epithelia was determined using immunofluorescence staining for ACE2 (green) and the nucleus (ethidium bromide, red). Confocal fluorescence photomicroscopic images are presented *en face* (top) and from vertical sections in the z axis (bottom). (B) ACE2 protein location in polarized human airway epithelia was determined by selective apical or basolateral biotinylation, immunoprecipitation of biotinylated surface proteins, and immunoblat analysis for ACE2. (See color plate).

3. ACE2 EXPRESSION DEPENDS ON STATE OF CELL DIFFERENTIATION

Because results from polarized epithelia suggested that ACE2 expression might depend on the state of cellular differentiation, we compared the apical surface morphology of well-differentiated epithelia with that of well-differentiated cells grown with media present on their apical surface for 7 days to promote de-differentiation. Importantly, submersion of the apical surface of polarized cells caused loss of cilia and markedly diminished expression of ACE2 mRNA and protein (Figure 2A–C). In contrast with results in polarized epithelia, poorly differentiated primary human tracheobronchial epithelia or A549 cells grown on tissue culture plastic expressed little ACE2 mRNA or protein. Notably, foxj1, a transcription factor expressed in well-differentiated ciliated epithelia was also coordinately expressed with ACE2, indicating that ACE2 positively correlates with the state of epithelial differentiation. This raised the question of whether foxj1 might regulate ACE2 expression in airway epithelia. Primary tracheobronchial cells grown in submersion culture were transduced with an adenoviral vector expressing ACE2, a negative control β-galactosidase, or foxj1. Only transduction with the ACE2 vector conferred ACE2 expression, suggesting that foxj1 alone is not sufficient enough to regulate ACE2 expression in airway epithelia.

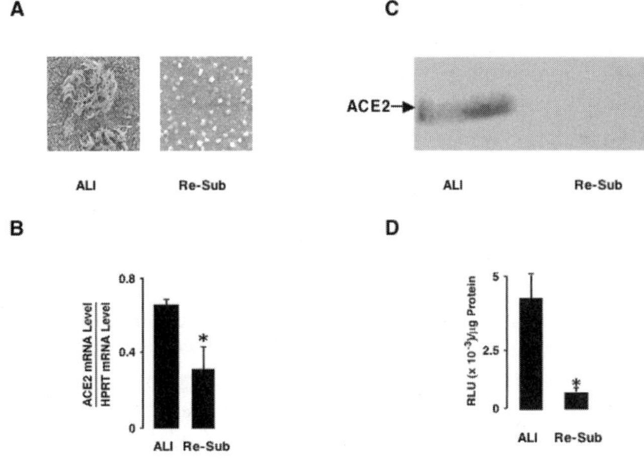

Figure 2. ACE2 expression is associated with airway epithelial cell differentiation. (A) Ciliated epithelial cell differentiation in cultures of primary airway epithelial cells under air-liquid interface or resubmerged conditions was verified by SEM of the apical epithelial surface. (B) ACE2 mRNA levels determined using realtime RT-PCR analysis of samples from differentiating air-liquid interface (ALI), or resubmerged (Re-sub) conditions. Values are expressed as mean mRNA level compared with control hypoxanthine phosphoribosyltransferase (HPRT) mRNA level ± S.D.; asterisk indicates a significant difference in mRNA levels between air-liquid interface and resubmerged conditions. (C) ACE2 protein levels determined using immunoblot of extracts from differentiating air-liquid interface (ALI), or resubmerged (Re-sub) conditions. (D) β-galactosidase levels determined in primary human airway epithelia cultured under ALI or resubmerged conditions that were infected from the apical with SARS-S protein pseudotyped FIV.

4. POLARIZATION OF S PROTEIN PSEUDOTYPED FIV VIRAL ENTRY

To evaluate the polarity of entry of the SARS-CoV in airway epithelia, we prepared feline immunodeficiency virus (FIV) virions pseudotyped with SARS S protein, and this vector was used to contrast the efficiency of entry in A549 cells, poorly differentiated (submerged) human airway epithelia, and well-differentiated epithelia. Only well-differentiated epithelial cells showed significant β-galactosidase expression following transduction (Figure 2D). The ACE2 dependence of transduction with SARS S protein FIV pseudotyped virions, which also express β-galactosidase, was first evaluated on 293 cells with or without co-transfection with human ACE2 cDNA. The result indicated that 293 cell transduction with this vector was almost completely ACE2-dependent. To further evaluate the ACE2 dependence of human airway epithelia for SARS-CoV, we transduced poorly differentiated A549 cells and submerged primary airway epithelia that do not express constitutive ACE2 with increasing MOIs of an adenoviral vector expressing human ACE2. After 48 hr, SARS-CoV S protein pseudotyped FIV was applied to the apical surface. The results showed that there was an inoculum-dependent increase in transduction of the ACE2 complemented cells. We next applied the pseudotyped virus to the apical or basolateral surfaces of well-differentiated primary cultures of human airway epithelia to investigate if the virus preferentially entered from one cell surface. Two days later the cells were harvested and entry evaluated by β-galactosidase activity. Results indicated that the S protein pseudotyped virions transduced human airway epithelia more efficiently when applied from the apical rather than the basolateral surface. This pattern of entry correlates with ACE2 expression on polarized cells. As a control, FIV pseudotyped with the VSV-G envelope entered polarized cells better from the basolateral surface.

5. SARS-CoV INFECTION OF AIRWAY EPITHELIA

We also conducted selected experiments using wild-type SARS-CoV (Urbani strain) and evaluated the ability of SARS-CoV to infect multiple human airway epithelial cell culture models. Under BSL3 containment we applied the virus to A549 cells, poorly differentiated (submerged) primary cultures of airway epithelia, or well-differentiated (air-liquid interface) human airway epithelia. A549 and hTBE cells cultured under submerged conditions expressed little detectable SARS-CoV N or S gene mRNA. In contrast, in well-differentiated cells infected with SARS-CoV from the apical surface, the N and S gene mRNAs were detected at high levels. We confirmed that the gene products detected in the real time RT-PCR assays were generated from new SARS-CoV mRNA templates rather than the viral genome by verifying the appropriate size of the amplified products. These results indicated that SARS-CoV infects undifferentiated human airway epithelial cells poorly or not at all, while well-differentiated conduction airway epithelia are susceptible.

By applying the virus to the apical surface of well-differentiated human airway epithelia for 30 min and then measuring the release of virus 24 hr later by tittering the virus, we documented that SARS-CoV productively infects human airway epithelia. The results indicate that following apical application of SARS-CoV a productive infection occurred and virus was preferentially released apically. We confirmed SARS-CoV infection of polarized epithelia by immunostaining cells for the SARS-CoV nsp1 protein 24 hr following infection.

6. CONCLUSIONS

Our studies revealed the novel observation that SARS-CoV infection of human airway epithelia is dependent upon the state of epithelial differentiation and ACE2 mRNA and protein expression. ACE2 is more abundantly expressed on the apical surface of polarized epithelia. The predominant apical distribution of ACE2 suggests that the enzyme may be available to cleave peptides at the mucosal surface of the airway but the native substrates in the lung have not yet been identified. We show for the first time that well-differentiated cells support viral replication with viral entry and egress occurring from the apical surface. Thus, SARS-CoV preferentially infects well-differentiated epithelial cells expressing ACE2. Because ACE2 is also the receptor for the coronavirus NL63,[4] these findings are relevant to the biology of infection with this more common human pathogen.

Human ACE2 appears necessary and sufficient to serve as a receptor for SARS-CoV.[5] Our findings suggest that the epithelium of the conducting airways, the major site of respiratory droplet deposition, supports the replication of SARS-CoV. The observation that ACE2 complementation of poorly differentiated epithelia enhanced transduction with S protein pseudotyped virions in a dose-dependent manner further supports its role as a receptor. Although both DC-SIGN (CD209) and DC-SIGN (L-SIGN, CD209L) can enhance SARS-CoV infection of ACE2 expressing cells, these proteins are not sufficient to support infection in the absence of ACE2.[12, 13] Several recent reports using SARS-CoV or retroviral vectors pseudotyped with SARS S protein[14] indicated that human airway epithelial cell lines were poorly transduced, an unexpected finding that raised questions regarding the ability of respiratory epithelia to support SARS-CoV infection. The present studies help explain these findings. Because SARS-CoV infection of airway epithelia is ACE2-dependent and ACE2 expression is greatest in well-differentiated cells, the low transduction efficiencies of non-polarized, poorly differentiated cells are not unanticipated.

In the setting of a productive infection of conducting airway epithelia, the apically released SARS-CoV might be removed by mucociliary clearance and gain access to the gastrointestinal tract. SARS-CoV infects cells in the gastrointestinal tract and diarrhea is a clinical sign commonly observed in patients with SARS.[1] Furthermore, the preferential apical exit pathway of virions would favor spread of infection along the respiratory tract. While not a focus of our study, pathologic data indicate that SARS-CoV infects type II pneumocytes. Infection and release of virus in this compartment with its close proximity to the pulmonary capillary bed might allow systemic spread of virus to distant organs, especially in the context of inflammation and alveolar capillary leak.

In conclusion, studies in models of human airway epithelial differentiation and polarity reveal that SARS-CoV infects well-differentiated cells from the apical surface and preferentially exits from the apical side. These findings should also apply to the entry of NL63 in human airway epithelia. ACE2 expression in airway epithelia appears to be both necessary and sufficient for SARS-CoV infection. Airway epithelial expression of ACE2 is dynamic and associated with cellular differentiation, a finding that may underlie susceptibility to infection. The apical expression of ACE2 on epithelia indicates that this coronavirus receptor is accessible for topical application of receptor antagonists or inhibitors. To date, the factors regulating ACE2 expression have not been identified. Future studies of the ACE2 promoter and gene expression associated with cell differentiation may reveal regulators of ACE2 expression and subsequent SARS-CoV and NL63 susceptibility.

7. REFERENCES

1. C. A. Donnelly, A. C. Ghani, G. M. Leung, A. J. Hedley, C. Fraser, S. Riley, L. J. Abu-Raddad, L. M. Ho, T. Q. Thach, P. Chau, K. P. Chan, T. H. Lam, L. Y. Tse, T. Tsang, S. H. Liu, J. H. Kong, E. M. Lau, N. M. Ferguson, and R. M. Anderson, Epidemiological determinants of spread of causal agent of severe acute respiratory syndrome in Hong Kong, *Lancet* **361**, 1761-1766 (2003).
2. S. M. Poutanen, D. E. Low, B. Henry, S. Finkelstein, D. Rose, K. Green, R. Tellier, R. Draker, D. Adachi, M. Ayers, A. K. Chan, D. M. Skowronski, I. Salit, A. E. Simor, A. S. Slutsky, P. W. Doyle, M. Krajden, M. Petric, R. C. Brunham, and A. J. McGeer, Identification of severe acute respiratory syndrome in Canada, *N. Engl. J. Med.* **348**, 1995-2005 (2003).
3. T. J. Franks, P. Y. Chong, P. Chui, J. R. Galvin, R. M. Lourens, A. H. Reid, E. Selbs, C. P. McEvoy, C. D. Hayden, J. Fukuoka, J. K. Taubenberger, and W. D. Travis, Lung pathology of severe acute respiratory syndrome (SARS): a study of 8 autopsy cases from Singapore, *Hum. Pathol.* **34**, 743-748 (2003).
4. W. Li, M. J. Moore, N. Vasilieva, J. Sui, S. K. Wong, M. A. Berne, M. Somasundaran, J. L. Sullivan, K. Luzuriaga, T. C. Greenough, H. Choe, and M. Farzan, Angiotensin-converting enzyme 2 is a functional receptor for the SARS coronavirus, *Nature* **426**, 450-454 (2003).
5. H. Hofmann, P. Krzysztof, L. van der Hoek, M. Geier, B. Berkhout, and S. Pohlmann, Human coronavirus NL63 employs the severe acute respiratory syndrome coronavirus receptor for cellular entry, *Proc. Natl. Acad. Sci. USA* (2005).
6. M. Donoghue, F. Hsieh, E. Baronas, K. Godbout, M. Gosselin, N. Stagliano, M. Donovan, B. Woolf, K. Robison, R. Jeyaseelan, R. E. Breitbart, and S. Acton, A novel angiotensin-converting enzyme-related carboxypeptidase (ACE2) converts angiotensin I to angiotensin 1-9, *Circ. Res.* **87**, E1-9 (2000).
7. I. Hamming, W. Timens, M. L. Bulthuis, A. T. Lely, G. J. Navis, and H. van Goor, Tissue distribution of ACE2 protein, the functional receptor for SARS coronavirus. A first step in understanding SARS pathogenesis, *J. Pathol.* **203**, 631-637 (2004).
8. W. Li, C. Zhang, J. Sui, J. H. Kuhn, M. J. Moore, S. Luo, S. K. Wong, I. C. Huang, K. Xu, N. Vasilieva, A. Murakami, Y. He, W. A. Marasco, Y. Guan, H. Choe, and M. Farzan, Receptor and viral determinants of SARS-coronavirus adaptation to human ACE2, *EMBO J.* (2005).
9. P. H. Karp, T. O. Moninger, S. P. Weber, T. S. Nesselhauf, J. L. Launspach, J. Zabner, and M. J. Welsh, An in vitro model of differentiated human airway epithelia. Methods for establishing primary cultures, *Methods Mol. Biol.* **188**, 115-137 (2002).
10. P. D. Vermeer, L. A. Einwalter, T. O. Moninger, T. Rokhlina, J. A. Kern, J. Zabner, and M. J. Welsh, Segregation of receptor and ligand regulates activation of epithelial growth factor receptor, *Nature* **422**, 322-326 (2003).
11. D. C. Look, M. J. Walter, M. R. Williamson, L. Pang, Y. You, J. N. Sreshta, J. E. Johnson, D. S. Zander, and S. L. Brody, Effects of paramyxoviral infection on airway epithelial cell Foxj1 expression, ciliogenesis, and mucociliary function, *Am. J. Pathol.* **159**, 2055-2069 (2001).
12. S. A. Jeffers, S. M. Tusell, L. Gillim-Ross, E. M. Hemmila, J. E. Achenbach, G. J. Babcock, W. D. Thomas, Jr., L. B. Thackray, M. D. Young, R. J. Mason, D. M. Ambrosino, D. E. Wentworth, J. C. Demartini, and K. V. Holmes, CD209L (L-SIGN) is a receptor for severe acute respiratory syndrome coronavirus, *Proc. Natl. Acad. Sci. USA* **101**, 15748-15753 (2004).
13. A. Marzi, T. Gramberg, G. Simmons, P. Moller, A. J. Rennekamp, M. Krumbiegel, M. Geier, J. Eisemann, N. Turza, B. Saunier, A. Steinkasserer, S. Becker, P. Bates, H. Hofmann, and S. Pohlmann, DC-SIGN and DC-SIGNR interact with the glycoprotein of Marburg virus and the S protein of severe acute respiratory syndrome coronavirus, *J. Virol.* **78**, 12090-12095 (2004).
14. Y. Nie, P. Wang, X. Shi, G. Wang, J. Chen, A. Zheng, W. Wang, Z. Wang, X. Qu, M. Luo, L. Tan, X. Song, X. Yin, M. Ding, and H. Deng, Highly infectious SARS-CoV pseudotyped virus reveals the cell tropism and its correlation with receptor expression, *Biochem. Biophys. Res. Commun.* **321**, 994-1000 (2004).

HUMAN CORONAVIRUS NL63 INFECTION IS ASSOCIATED WITH CROUP

Lia van der Hoek, Klaus Sure, Gabriele Ihorst, Alexander Stang, Krzysztof Pyrc, Maarten F. Jebbink, Gudula Petersen, Johannes Forster, Ben Berkhout, and Klaus Überla*

1. INTRODUCTION

Respiratory tract infections are among the most frequent diseases in the first years of life. Although there is a large number of viruses that are known to be involved in symptomatic respiratory tract infections, including respiratory syncytial virus (RSV), influenza virus (INF), parainfluenza virus (PIV), and human metapneumovirus, none of the known pathogens is detected in a substantial number of cases. Recently we identified a novel coronavirus in a child with bronchiolitis: human coronavirus NL63 (HCoV-NL63).[1,2] This virus, together with SARS-CoV, is one of the new members of the *Coronaviridae* family.[3-6]

Screening of respiratory samples in Amsterdam and Rotterdam confirmed that HCoV-NL63 is circulating among humans with respiratory disease in The Netherlands.[1,7] To investigate the prevalence of HCoV-NL63 and its involvement in respiratory diseases, we now analyzed 949 samples from the Paediatric Respiratory Infection in Germany (PRI.DE) study, a prospective population-based study on lower respiratory tract infections (LRTIs) in children under 3 year of age in Germany.[8,9] The PRI.DE study represents the German population by (i) including multicenter sampling (one city each in the north, east, south, and west of the country) and by (ii) recruiting children in pediatric practices and in referral children's hospitals. We were particularly interested in the presence of HCoV-NL63 in respiratory disease for which no other viral pathogen could be detected, in order to identify clinical symptoms associated with HCoV-NL63 infection. Nasopharyngeal secretion (NPS) of the patients had already been tested for

* Lia van der Hoek, Krzysztof Pyrc, Maarten F. Jebbink, Ben Berkhout, University of Amsterdam, Amsterdam, The Netherlands. Klaus Sure, Alexander Stang, Klaus Überla, Ruhr-University Bochum, Bochum, Germany. Gabriele Ihorst, University Hospital, Freiburg, Germany. Gudula Petersen, Wyeth Pharma, Münster, Germany. Johannes Forster, St. Josefs Hospital, Freiburg, Germany.

The data presented here have been published in PloS Medicine: van der Hoek L, Sure K, Ihorst G, Stang A, Pyrc K, *et al.* (2005) Croup is associated with the novel coronavirus NL63. PLoS Med 2(8): e240: 764-770.

RSV, INF, and PIV, the principal viruses responsible for LRTI in young children.[8] However, RNA of these viruses could not be detected in 58% of samples for outpatients and 51% of samples for hospitalized patients. A second study that examined a subset of these negative samples for human metapneumovirus RNA showed that this virus could be detected in only 0.3% of the patients.[9] To explore the potential contribution of HCoV-NL63 to LRTI and to define clinical symptoms associated with HCoV-NL63 infection, a subset of the PRI.DE samples were analyzed in this study by a HCoV-NL63–specific quantitative real-time RT-PCR.

2. RESULTS

2.1. HCoV-NL63 Infections

Of the 949 PRI.DE samples tested, 392 were from outpatients at the four study sites, and the remaining 557 samples were from hospitalized patients. In total, 49 of the 949 samples (5.2%) were positive for HCoV-NL63 (for methods see PloS Medicine 2(8):e240 764–770). More HCoV-NL63 infections were found in the outpatients (31 patients, 7.9%) than in hospitalized patients (18 patients, 3.2%, $p = 0.003$). Various clinical diagnoses of lower respiratory tract disease were given for the HCoV-NL63–positive patients, including croup, bronchitis, bronchiolitis, and pneumonia (Tables 1 and 2). The ages of the HCoV-NL63–infected children ranged from 0 to 2.9 years, with a median age of 0.7 year for the hospitalized patients and 1.5 year for the outpatients. As may be expected based on knowledge of other human coronaviruses, there is a strong seasonal distribution of HCoV-NL63, with preferential detection in the period between November and March (Figure 1). Peaks were observed in December 2000 (14% of patients positive) and February 2001 (12% of patients positive). We found no positive samples in the winter months of 1999 and 2000, but the analyzed PRI.DE samples were unequally distributed and only 17 samples were analysed from the period December 1999 to March 2000.

2.2. HCoV-NL63 Co-infections with RSV-A and PIV3

Because the same samples had been tested previously for the presence of RSV, PIV, and INF RNA,[8] the HCoV-NL63–positive samples were analyzed for co-infections with these viruses. Co-infections were apparent in 29 of the 49 HCoV-NL63–positive samples: 20 patients were co-infected with RSV-A, four with RSV-B, and five with PIV3. Double infections were observed in the hospitalized patients with HCoV-NL63 (72%) but also in the outpatients (52%). HCoV-NL63 co-infection with RSV-A occurred predominantly in the hospitalized patients (61%) rather than the outpatient group (29%). In contrast, HCoV-NL63 co-infections with PIV3 were exclusively present in the outpatient group (16%). Similar trends were also observed when looking at the overall prevalence of the viruses: RSV-A occurred in 32% of hospitalized patients versus 21% of outpatients, and PIV3 occurred in 5% of hospitalized patients and 8% of outpatients.[8] The RNA load of HCoV-NL63 differed considerably from less than 225 copies/ml (but detectable) to 9×10^7 copies/ml aspirate. Interestingly, the HCoV-NL63 load was significantly higher in samples with undetectable levels of the other viral RNAs (median viral load 2.1×10^6 copies/ml) than in samples that had co-infections with RSV or PIV3 (2.7×10^2 copies/ml, $p = 0.0006$; Figure 2).

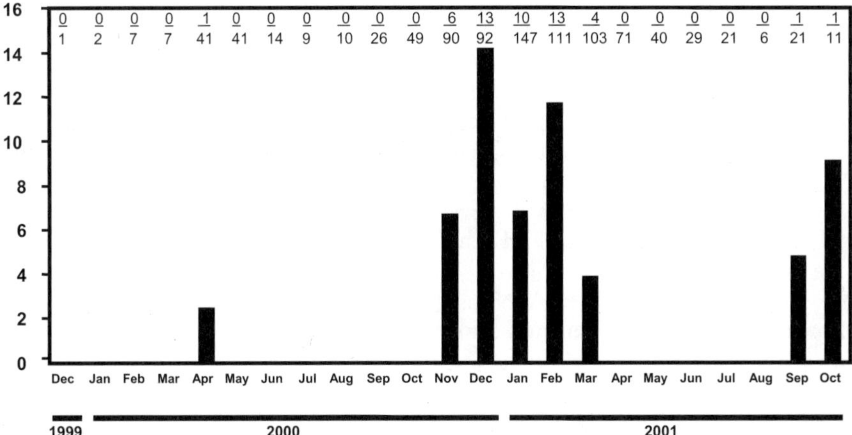

Figure 1. Seasonal distribution of HCoV-NL63. Bars represent the percentage of HCoV-NL63 positive samples per month. Digits above the columns for each month give the number of HCoV-NL63 positive samples over the number of samples tested. Y-axis: HCoV-NL63-positive patients (%).

Table 1. HCoV-NL63–positive hospital patients.

Patient number	Age (years)	Sex	Sampling month	Diagnosis	Fever	Other virus	HCoV-NL63 RNA load (copies/ml)
10-343[a]	1.88	F	12/00	Pneumonia	N	RSV-B	26200
10-397	0.09	F	1/01	Bronchiolitis	N	RSV-A	4520
10-416	0.11	M	2/01	Bronchitis	N	RSV-A	18800
10-447	0.86	M	2/01	Croup/ Bronchiolitis	N	-	pos (≤225)
11-275	2.23	M	12/00	Croup/ Bronchiolitis	Y	-	7380000
11-395	0.78	F	3/01	Bronchiolitis	Y	RSV-A	pos (≤225)
20-443	0.65	M	1/01	Bronchiolitis	N	RSV-A	pos (≤225)
20-454	0.17	F	2/01	Bronchiolitis	N	RSV-A	pos (≤225)
30-317	0.08	F	2/01	Bronchiolitis	Y	RSV-B	667
40-489	0.22	F	12/00	Bronchitis	N	RSV-A	pos (≤225)
40-546[b]	0.01	M	12/00	Bronchiolitis	N	RSV-A	1360000
40-713	0.99	F	2/01	Croup/Bronchitis	Y	-	53600000
40-715	0.76	M	2/01	Bronchiolitis	N	-	pos (≤225)
40-717	0.13	F	2/01	Pneumonia	Y	RSV-A	pos (≤225)
40-723	0.10	M	2/01	Bronchiolitis	N	RSV-A	pos (≤225)
40-735	1.05	F	2/01	Pneumonia	Y	RSV-A	pos (≤225)
40-746	1.80	F	2/01	Pneumonia	N	-	17400000
40-764	1.52	F	2/01	Pneumonia	Y	RSV-A	pos (≤225)

[a] The patient number is preceded by the code of the medical center: 10: Freiburg SJK; 11: Freiburg UKL; 20: Dresden; 30: Bochum; 40: Hamburg; 12 and 13: Freiburg; 22, 23 and 24: Dresden; 34: Bochum.; 43: Hamburg.

[b] Nosocomial infection.

2.3. Association of HCoV-NL63 Infection with Clinical Symptoms

The high frequency of co-infections in HCoV-NL63–positive samples makes it difficult to define HCoV-NL63–induced symptoms. However, in 20 of the 49 HCoV-NL63–positive samples no other virus (RSV, PIV, or INF) could be detected. At least 14 of these samples also had a high viral RNA load (>10,000 copies/ml aspirate), cases that may be best suited to study the clinical symptoms associated with HCoV-NL63 infection. Six of the 14 children (43%) of this group had croup compared with only 54 of 900 HCoV-NL63–negative children (6%, $p < 0.0001$). A similar high frequency of croup (45%) was also observed for the 20 samples in which only HCoV-NL63 RNA was detected, independent of the viral load. The association of HCoV-NL63 with croup also held for all analysed samples: 24% in the HCoV-NL63–positive group had croup compared with 6% of the 900 HCoV-NL63–negative patients ($p < 0.0001$). HCoV-NL63 was detectable in 17.4% of all samples from croup patients.

Table 2. HCoV-NL63–positive outpatients.

Patient number	Age (years)	Sex	Sampling month	Diagnosis	Fever	Other virus	HCoV-NL63 RNA load (copies/ml)
12-75	0.95	M	12/00	Croup	N	-	155000
12-86	0.71	F	1/01	Bronchitis	N	-	2240
12-107	1.02	M	3/01	Bronchiolitis	N	RSV-A	333
12-108	1.02	M	3/01	Bronchitis	N	RSV-A	pos (≤225)
12-137	1.25	M	9/01	Bronchiolitis	N	RSV-A	pos (≤225)
13-255	2.75	F	11/00	Bronchiolitis	N	PIV3	26200
13-316	1.01	M	1/01	Croup	Y	PIV3	6430000
13-342	1.07	M	1/01	Croup	N	-	27400000
22-102	0.37	M	4/00	Bronchitis	N	-	5240000
22-136	1.08	F	11/00	Croup	Y	PIV3	571
22-140	2.15	F	11/00	Croup	N	PIV3	pos (≤225)
22-143	1.51	M	11/00	Croup	Y	-	pos (≤225)
22-158	1.41	F	11/01	Bronchiolitis	N	RSV-A	333
22-170	2.33	M	1/01	Bronchitis	Y	RSV-B	548
22-195	2.09	M	2/01	Bronchiolitis	N	RSV-A	197000
22-198	1.85	M	2/01	Bronchiolitis	Y	RSV-A	pos (≤225)
22-215	0.78	M	3/01	Croup	Y	-	25000000
23-66	2.03	M	12/00	Bronchitis	N	-	2860000
23-68	2.45	F	12/00	Bronchiolitis	Y	RSV-A	140000
23-69	1.75	M	12/00	Croup	N	-	pos (≤225)
23-95	1.86	M	1/01	Bronchitis	N	-	250000
24-105	1.11	M	2/01	Bronchitis	Y	RSV-B	136
24-178	2.21	M	11/00	Bronchiolitis	Y	PIV3	pos (≤225)
24-194	1.88	M	12/00	Bronchitis	Y	-	1950000
24-202	2.91	M	12/00	Bronchiolitis	Y	RSV-A	12400
24-230	2.38	M	1/01	Bronchiolitis	Y	-	2290000
24-232	2.63	M	1/01	Bronchiolitis	Y	-	11400000
34-162	0.48	F	10/01	Bronchitis	N	-	91200000
43-10	2.12	M	11/00	Bronchiolitis	Y	-	pos (≤225)
43-15	0.45	M	12/00	Bronchiolitis	N	RSV-A	pos (≤225)
43-18	0.51	F	12/00	Croup	N	-	28600

The chance of croup is estimated to be 6.6 times higher in HCoV-NL63–positive LRTI patients than in HCoV-NL63–negative LRTI patients (95% confidence interval 3.1–14.2). In addition to croup, we also observed bronchitis (n = 6, of which one also had had croup), bronchiolitis (n = 3, of which one also had croup), and pneumonia (n = 1) in the 14 patients with a high HCoV-NL63 load. None of these diseases was significantly associated with single infection with HCoV-NL63.

3. DISCUSSION

The newly discovered coronavirus HCoV-NL63 was detected in a considerable number of nasal aspirates of children under the age of 3 year with LRTIs. With an overall occurrence of 5.2%, it is the third most frequently detected pathogen in this patient group, in which RSV is detected in 31.4%, PIV3: 9.6%, PIV1: 2.5%, INF A or INF B:2.4%, and PIV2: 0.6%. These viruses were detected with similar frequency in the PRI.DE study[8], arguing against a bias during selection of the analyzed samples. HCoV-NL63 is more frequently found in the outpatient group with LRTI (7.9%) than among hospitalized patients (3.2%). PIV3 follows the same pattern (8% and 5%, respectively), but the reverse pattern is observed for RSV (21% and 32%, respectively).[8] Thus, HCoV-NL63 infection seems to be less pathogenic than RSV infection. Hospitalized HCoV-NL63–positive patients are frequently co-infected with RSV. Nevertheless, several severe disease cases that required uptake in the intensive care unit were linked exclusively to HCoV-NL63 infection in this and our previous survey.[1]

Croup is a common manifestation of LRTI in children. The cause is generally assumed to be a respiratory virus and PIV1 has frequently been implicated.[10] Among the 69 samples of patients analyzed with croup, croup was indeed frequently linked to PIV1 (14.5%), but PIV3 (15.9%), RSV-A (13.0%), PIV2 (7.2%), and RSV-B (1.4%) were also

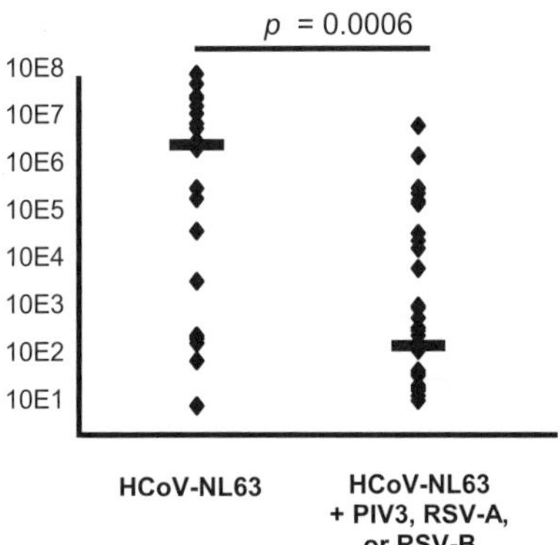

Figure 2. HCoV-NL63 viral load in single and double infections. The median viral load in the two groups is indicated together with the p-value, showing a significant difference between the singly HCoV-NL63 infected group and the group with a co-infection of either RSV or PIV3. Y-axis: HCoV-NL63 viral load (RNA copies/ml).

detected in a considerable percentage of samples. HCoV-NL63 could be detected in 17.4% of these croup patients and was therefore the most frequently identified respiratory virus for croup. Since most of the samples tested were derived from the year 2000–2001, we cannot exclude that the high percentage of HCoV-NL63–positive samples is due to a strong viral activity in this particular year, and long-term studies are needed to determine whether HCoV-NL63 infections occur in cycles peaking every two to three years as observed for other respiratory viruses.

Croup has been reported to occur mostly in boys, and it shows a peak occurrence in the second year of life and predominantly in the late fall or early winter season.[10] HCoV-NL63 infection seems to follow these trends: the ratio of boys infected to girls infected is 10:4, the median age in the outpatient group with HCoV-NL63 is 1.55 years, and this virus is circulating mainly in the winter months. Thus, it will be of interest to study the underlying biological reasons for the increased susceptibility of young boys to HCoV-NL63, as this may also explain the higher number of male patients with croup. A preferential occurrence in boys has been described for other respiratory diseases including asthma,[11] and human coronavirus infections have previously been associated with exacerbations of asthma.[12] It will therefore also be of interest to study this link for HCoV-NL63.

Quantitative PCR analysis for HCoV-NL63 revealed a significantly lower HCoV-NL63 viral load in patients co-infected with RSV or PIV3 than in patients infected with HCoV-NL63 alone. This interference effect might be explained by direct competition for the same target cell in the respiratory organs or an elevated activation status of innate immune responses. Prolonged persistence of HCoV-NL63 at low levels is another explanation. The HCoV-NL63 load was found to vary with respect to the time of sampling relative to the time of disease onset, with the higher viral loads in early samples (day 1 or 2 after disease onset). This most likely reflects viral clearance by the immune system. This timing effect may also relate to the differences in HCoV-NL63 load in single versus double infections. For instance, an initial HCoV-NL63 infection may set the stage for a subsequent RSV infection. At the time that this second virus is causing symptoms and NPS samples are collected, the HCoV-NL63 infection may already be under control by the immune system.

In conclusion, our study revealed that HCoV-NL63 belongs to the group of most frequently detected viruses in children under 3 year of age with LRTI and that this virus is strongly associated with croup. Recent articles on HCoV-NL63 show that this virus is spread worldwide (Australia, Canada, Japan, Belgium, United States, and France). Thus, HCoV-NL63 is a human respiratory virus that should be added to the list of pathogens that can cause numerous LRTIs in young children.

Financial support was received from Wyeth Pharma, Münster, Germany. The Center for Clinical Trials receives funding from the German Federal Ministry of Education and Research (BMBF).

4. REFERENCES

1. L. van der Hoek, K. Pyrc, M. F. Jebbink, W. Vermeulen-Oost, R. J. Berkhout, et al., Identification of a new human coronavirus, *Nat. Med.* **10**, 368-373 (2004).
2. K. Pyrc, M. F. Jebbink, B. Berkhout, and L. van der Hoek, Genome structure and transcriptional regulation of human coronavirus NL63, *J. Virol.* **1**, 7 (2004).

3. J. S. Peiris, S. T. Lai, L. L. Poon, Y. Guan, L. Y. Yam, et al., Coronavirus as a possible cause of severe acute respiratory syndrome, *Lancet* **361**, 1319-1325 (2003).
4. C. Drosten, S. Gunther, W. Preiser, S. van der Werf, H. R. Brodt, et al., Identification of a novel coronavirus in patients with severe acute respiratory syndrome, *N. Engl. J. Med.* **348**, 1967-1976 (2003).
5. Osterhaus AD, Fouchier RA, Kuiken T (2004) The aetiology of SARS: Koch's postulates fulfilled. *Philos Trans R Soc Lond B Biol Sci* **359**: 1081-1082.
6. P. C. Woo, S. K. Lau, C. M. Chu, K. H. Chan, H. W. Tsoi, et al., Characterization and complete genome sequence of a novel coronavirus, coronavirus HKU1, from patients with pneumonia, *J. Virol.* **79**, 884-895 (2005).
7. R. A. Fouchier, N. G. Hartwig, T. M. Bestebroer, B. Niemeyer, J. C. de Jong, et al., A previously undescribed coronavirus associated with respiratory disease in humans, *Proc. Natl. Acad. Sci. USA* **101**, 6212-6216 (2004).
8. J. Forster, G. Ihorst, C. H. Rieger, V. Stephan, H. D. Frank, et al., Prospective population-based study of viral lower respiratory tract infections in children under 3 years of age (the PRIDE study), *Eur. J. Pediatr.* **163**, 709-716 (2004).
9. B. Konig, W. Konig, R. Arnold, H. Werchau, G. Ihorst, et al., Prospective study of human metapneumovirus infection in children less than 3 years of age, *J. Clin. Microbiol.* **42**, 4632-4635 (2004).
10. F. W. Denny, T. F. Murphy, W. A. Clyde, Jr., A. M. Collier, and F. W. Henderson, Croup: An 11-year study in a pediatric practice, *Pediatrics* **71**, 871-876 (1983).
11. E. F. Ellis, Asthma in childhood, *J. Allergy Clin. Immunol.* **72**, 526-539 (1983).
12. K. McIntosh, E. F. Ellis, L. S. Hoffman, T. G. Lybass, J. J. Eller, et al., The association of viral and bacterial respiratory infections with exacerbations of wheezing in young asthmatic children, *J. Pediatr.* **82**, 578-590 (1973).

A SARS-CoV–SPECIFIC PROTEIN ENHANCES VIRULENCE OF AN ATTENUATED STRAIN OF MOUSE HEPATITIS VIRUS

Lecia Pewe, Haixia Zhou, Jason Netland, Chandra Tangadu, Heidi Olivares, Lei Shi, Dwight Look, Thomas Gallagher, and Stanley Perlman*

1. INTRODUCTION

Infection of humans with SARS-CoV (severe acute respiratory syndrome-coronavirus) resulted in a severe respiratory syndrome with substantial mortality, especially in the elderly.[1,2] Although SARS-CoV is the first human coronavirus observed to cause serious disease, coronaviruses were associated with severe disease in several animal species. Pigs infected with transmissible gastroenteritis virus and felines infected with feline infectious peritonitis virus are well-described examples of severe coronavirus-induced diseases. The most intensely studied animal infection is that caused by mouse hepatitis virus (MHV). MHV is a well-nown cause of acute and chronic neurological infections.[3] It is best-nown for its ability to induce an immune-mediated disease in mice that resembles the human disease, multiple sclerosis. The JHM strain is used in many of these studies and the JHM J2.2-V-1 attenuated variant is particularly useful for studies of chronic demyelination.[4] Infection with JHM J2.2-V-1 results in infection of oligodendrocytes with minimal infection of neurons, and consequently, low mortality.

SARS-CoV is tentatively classified as a group 2 coronavirus, distantly related to other group 2 coronaviruses such as MHV. Like MHV, it is believed to cause immune-mediated disease[1] SARS-CoV replicates in macrophages and dendritic cells, although the infection is abortive.[5-7] Replication in these cells results in induction of several cytokines/chemokines including IL-6, IL-8, CCL2, and CXCL10 but not type I interferons, which may contribute to a dysregulated immune response. Elucidation of the role of the host immune response would best be determined in an animal model of SARS, especially because SARS has not recurred to a significant extent in humans since 2003. Several animal models for SARS exist, but they do not reproducibly develop clinical

*Lecia Pewe, Haixia Zhou, Jason Netland, Lei Shi, Dwight Look, Stanley Perlman, University of Iowa, Iowa City, Iowa 52242. Chandra Tangadu, Heidi Olivares, Thomas Gallagher, Loyola University Stritch School of Medicine, Maywood, Illinois 60153.

disease. In particular, mice can be infected with SARS-CoV but remain asymptomatic (reviewed in Ref. 1).

Like other coronaviruses, SARS-CoV encodes several nonstructural proteins at the 3' end of the genome, flanking the structural genes. One of these "nonstructural proteins" is now known to be a structural protein (ORF3a protein) but the function of the others remains unknown.[8] In other coronavirus infections, these nonstructural proteins can often be deleted without any effects on growth in tissue culture cells and in some cases, their absence does not seem to affect infection *in vivo*.[9] To begin to understand the function of these proteins, we introduced them singly into the genome of JHM J2.2-v-1 and analyzed the effect of the insertion on disease pathogenesis.

2. MATERIALS AND METHODS

2.1. Recombinant Viruses

Targeted recombination was used to develop recombinant JHM J2.2-V-1 expressing individual SARS-CoV nonstructural proteins.[10] The development of these viruses is described in more detail elsewhere.[11] Briefly, PCR products corresponding to nucleotides 25268–26092 (ORF3a), 25689–26153 (ORF3b), 27074–27265 (ORF6), 27273–27641 (ORF7a), 27638–27772 (ORF7b), and 27779–27898 (ORF8) were generated by RT-PCR using RNA harvested from cells infected with the Urbani strain of SARS-CoV (GenBank accession number AY278741, kindly provided by Dr. Tom Ksiazek, Centers for Disease Control, Atlanta, GA). Each product was tagged at the C terminus with the influenza hemagglutinin epitope for ease of detection. These products were inserted into gene 4 of a J2.2-V-1 shuttle vector, which, in turn, was transcribed. The resulting RNA transcript was transfected into cells previously infected with fMHV-JHM, an MHV chimera expressing the feline surface glycoprotein. Recombinant viruses were selected on murine cells. All isolates were sequenced prior to use in animals and 2 isolates of each virus were used in all studies, to control for the introduction of spurious mutations during the process of recombination.

2.2. Immunofluorescence Assays

SARS-CoV proteins were detected using anti-HA murine antibody (Covance, Berkeley, CA, mAb HA.11), biotinylated goat anti-mouse ab (Jackson Immunoresearch) and streptavidin-Cy3 (Jackson Immunoresearch).

3. RESULTS

3.1. Development of Recombinant J2.2-V-1 Expressing SARS-CoV Nonstructural Proteins

We inserted ORFs 3a, 3b, 6, 7a, 7b, and 8 into the genome of the attenuated J2.2-V-1 strain of JHM and tagged each gene with the sequence encoding an influenza hemagglutinin tag to facilitate detection (Figure 1). We could detect all of the inserted proteins by Western blot analysis or IFA (Figure 1, Ref. 11).

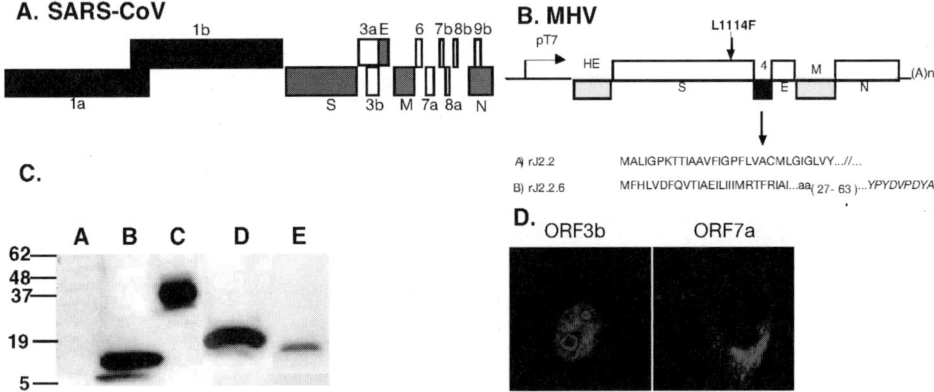

Figure 1. Expression of SARS-CoV nonstructural ORFs in rJ2.2-infected cells. (A). The genome of SARS-CoV, with ORFs and structural proteins (gray) is shown. (B). SARS-CoV nonstructural proteins were introduced into rJ2.2 by targeted recombination (shown for rJ2.2.6). (C). The products of ORF3a, 6, 7b, and 8 were detected by Western blot analysis using anti-HA antibody. Lanes: A-rJ2.2, B-rJ2.2.6, C-rJ2.2.3a, D-rJ2.2.7b, E-rJ2.2.8. (D). All of the inserted proteins were detected by IFA using anti-HA antibody. The products of ORF3b, 7a are shown in the figure. Note that protein 3b was localized to the nucleus. No staining was detected in cells infected with rJ2.2 (WT).

3.2. Infection with Virus Expressing the ORF6 Protein Results in Enhanced Mortality and Clinical Disease

Next we inoculated mice with recombinant virus and monitored them for survival, clinical signs, and weight loss. Mice inoculated with most recombinant viruses developed a disease very similar to that observed in animals infected with wild-type virus. However, infection with virus expressing the ORF6 protein (rJ2.2.6) caused a fatal disease in mice (Figure 2). In other experiments, we showed that infection with a recombinant virus encoding a mutated form of ORF6 (rJ2.2.6^{KO}) so that the ORF6 RNA was present but no protein expressed did not result in a lethal infection. The presence of the ORF6 protein resulted in higher titers of infectious virus in the CNS than did infection with wild-type virus.[11] However, these differences were statistically significant only at late times p.i. We confirmed these results by showing that viral RNA levels were also higher in the CNS of mice infected with rJ2.2.6 than with rJ2.2.6^{KO}.[11]

3.3. Infection with rJ2.2.6 Results in Enhanced Growth in Tissue Culture Cells

These results suggest that ORF6 protein enhances virus growth. To determine whether this also occurs in tissue culture cells, we infected L929 cells with rJ2.2, rJ2.2.6, or rJ2.2.6^{KO}. rJ2.2.6^{KO} grew to approximately 1 log higher titers than the other 2 viruses, consistent with the *in vivo* results.[11]

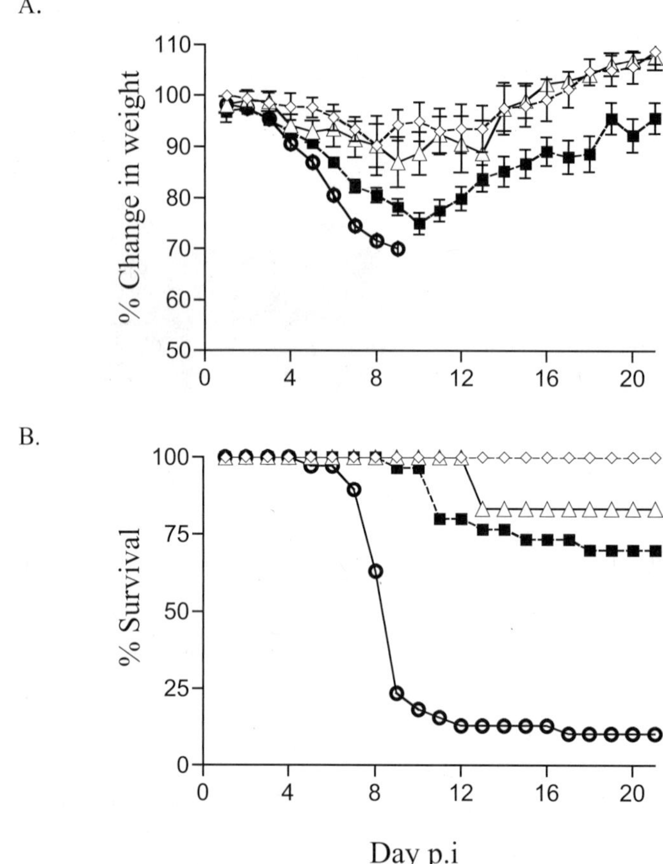

Figure 2. Mortality and morbidity in mice infected with rJ2.2.6, rJ2.2.6^{KO}, and rJ2.2.8. Mice infected with rJ2.2 (triangles), rJ2.2.6 (open circles), rJ2.2.6^{KO} (squares), or rJ2.2.8 (diamonds) were monitored for (A) mortality and (B) weight loss. In panel (A), data for mice infected with rJ2.2.6 are only shown for days 0–9 p.i., because only 23% survived past this time.

3.4. rJ2.2.6 Does Not Induce Type I Interferon (IFN) or Modulate IFN Sensitivity

One possibility is that the ORF6 protein affected IFN induction or sensitivity. MHV has been reported not to induce type I interferons[12] and we confirmed these results. Thus, we were unable to examine whether ORF6 has an additional role in suppressing IFN induction. However, we were able to examine whether rJ2.2.6 affected IFN sensitivity. As shown in Figure 3, cells infected with rJ2.2.6 or rJ2.2.6^{KO} both exhibited similar sensitivities to treatment with IFN-ß. Notably, at all time points, higher virus titers were detected in cells infected with rJ2.2.6 (in the presence or absence or IFN) when compared with their counterparts infected with rJ2.2.6^{KO}. Thus, the presence of ORF6 protein did not affect IFN induction or signaling.

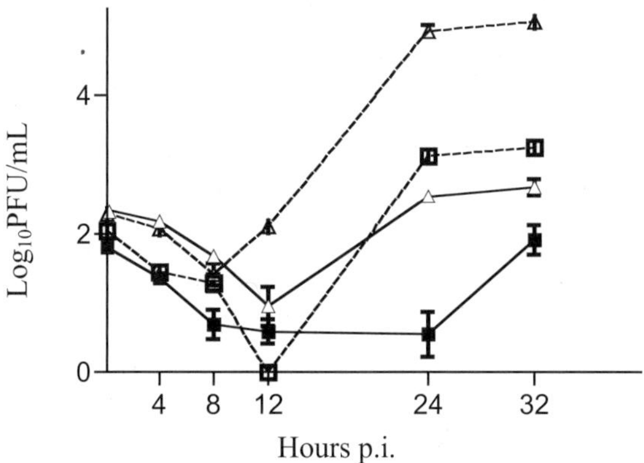

Figure 3. Sensitivity of cells infected with rJ2.2.6 and rJ2.2.6^{KO} to exogenous IFN-ß. Triplicate samples of L929 cells were untreated or treated with 1000 U IFN-ß 16 hours prior to infection with rJ2.2.6 or rJ2.2.6^{KO} at 0.1 pfu/cell. IFN-ß was also present post infection. Cells were harvested at the indicated time points and virus titers determined by plaque assay on HeLa cells expressing the MHV receptor. Open triangle, rJ2.2.6; closed triangle, rJ2.2.6 + IFN; open square, rJ2.2.6^{KO}; closed square, rJ2.2.6^{KO} + IFN.

3.5. The ORF6 Protein Co-localizes with Membranes

In additional experiments, we analyzed the localization of the ORF6 protein within infected cells. The primary structure of the ORF6 predicts that it is membrane associated. However, it is not likely to be a transmembrane protein because stretches of hydrophobic residues are interspersed with basic or acidic residues. We performed co-localization experiments using antibodies to the J2.2 N, S, and M proteins and to the endoplasmic reticulum marker, BiP. Our results showed that the ORF6 protein localized to some extent with all the proteins, although co-localization was most evident with the M protein. We also showed that the ORF6 protein partitioned with membrane fractions after treatment with TX-114.[11]

4. DISCUSSION

Our results show that nonstructural proteins from SARS-CoV can function in the context of a murine coronavirus. MHV does not infect lab personnel, thereby circumventing one of the difficulties of working with SARS-CoV. Also, for the first time, we demonstrate a phenotype for a coronavirus nonstructural protein. It is not surprising that a protein from the SARS-CoV can function in the context of a heterologous coronavirus infection. SARS-CoV most likely spread to the human population from an exotic animal species such as palm civet cats or raccoon dogs.[2] Also, the ORF6 protein is detected in all human and palm civet cat isolates of the SARS-CoV and did not mutate in passage through humans.

Our results suggest that the ORF6 protein enhanced virus growth in tissue culture cells and to a lesser extent, in the infected mouse. However, the presence of the ORF6 protein conferred lethality to an attenuated infection. One explanation for this apparent discrepancy is that the ORF6 protein preferentially has an effect in specific cells such as macrophages or dendritic cells, with consequent immune dysregulation. MHV and SARS-CoV are both known to replicate in both cell types, although SARS-CoV, causes an abortive infection.[5-7] Immune dysregulation is postulated to contribute to SARS pathogenesis.[2]

Our tentative conclusion is that the ORF6 protein directly increases the efficiency of virus replication, assembly, or spread because viral titers, RNA, and protein levels are all increased in cells infected with rJ2.2.6.[11] The protein is not present in virions, showing that it does not have a direct effect on virus infectivity.[11] The ORF6 protein is broadly distributed throughout the cell, coincident with sites of virus replication (co-localization with the N protein) and virus assembly (co-localization with the M protein). Future work will be directed at determining how exactly the presence of this protein enhances virus replication.

This work was supported by a grant from the N.I.H. (PO1 AI606699).

5. REFERENCES

1. J. S. Peiris, Y. Guan, and K. Y. Yuen, Severe acute respiratory syndrome, *Nat. Med.* **10**, S88-97 (2004).
2. J. S. Peiris, K. Y. Yuen, A. D. Osterhaus, and K. Stohr, The severe acute respiratory syndrome, *N. Engl. J. Med.* **349**, 2431-2441 (2003).
3. S. Perlman, Pathogenesis of coronavirus-induced infections: Review of pathological and immunological aspects, *Adv. Exp. Med. Biol.* **440**, 503-513 (1998).
4. J. O. Fleming, M. D. Trousdale, F. El-Zaatari, S. A. Stohlman, and L. P. Weiner, Pathogenicity of antigenic variants of murine coronavirus JHM selected with monoclonal antibodies, *J. Virol.* **58**, 869-875 (1986).
5. H. K. Law, et al., Chemokine up-regulation in SARS-coronavirus-infected, monocyte-derived human dendritic cells, *Blood* **106**, 2366-2374 (2005).
6. C. Y. Cheung, et al., Cytokine responses in severe acute respiratory syndrome coronavirus-infected macrophages in vitro: possible relevance to pathogenesis, *J. Virol.* **79**, 7819-7826 (2005).
7. C. T. Tseng, L. A. Perrone, H. Zhu, S. Makino, and C. J. Peters, Severe acute respiratory syndrome and the innate immune responses: Modulation of effector cell function without productive infection, *J. Immunol.* **174**, 7977-7985 (2005).
8. N. Ito, et al., Severe acute respiratory syndrome coronavirus 3a protein is a viral structural protein, *J. Virol.* **79**, 3182-3186 (2005).
9. E. Ontiveros, L. Kuo, P. S. Masters, and S. Perlman, Inactivation of expression of gene 4 of mouse hepatitis virus strain JHM does not affect virulence in the murine CNS, *Virology* **290**, 230-238 (2001).
10. L. Kuo, G. J. Godeke, M. J. Raamsman, P. S. Masters, and P. J. Rottier, Retargeting of coronavirus by substitution of the spike glycoprotein ectodomain: crossing the host cell species barrier, *J. Virol.* **74**, 1393-1406 (2000).
11. L. Pewe, et al., A severe acute respiratory syndrome-associated coronavirus-specific protein enhances virulence of an attenuated murine coronavirus, *J. Virol.* **79**, 11335-11342 (2005).
12. L. E. Garlinghouse, Jr., A. L. Smith, and T. Holford, The biological relationship of mouse hepatitis virus (MHV) strains and interferon: in vitro induction and sensitivities, *Arch. Virol.* **82**, 19-29 (1984).

GENETIC EVOLUTION OF HUMAN CORONAVIRUS OC43 IN NEURAL CELL CULTURE

Julien R. St-Jean, Marc Desforges, and Pierre J. Talbot *

1. INTRODUCTION

Human coronaviruses (HCoV) are ubiquitous in the environment and are responsible for up to one-third of common colds. HCoV-OC43 possesses a genome that comprises genes encoding various structural and nonstructural proteins. Amongst these proteins, the S protein is biologically very important because it could be involved in determination of viral tropism. Indeed, it could for instance be associated with the capacity of the virus to reach the central nervous system (CNS) and possibly trigger neurological disorders. It could also confer the host species specificity observed with coronaviruses. In past years, we have shown that HCoV-OC43 is neurotropic and neuroinvasive, as it persistently infects neural cell cultures[1] and human brains.[2] Although we have suggested that OC43 could remain genetically surprisingly stable in the environment,[3] it is known that coronaviruses can adapt in cell culture or under selection pressure, for instance related to immune system evasion.

2. MATERIALS AND METHODS

2.1. Viruses, Cell Lines, and Persistent Infections

The ATCC HCoV-OC43 strain (VR-759) was grown on the HRT-18 rectal tumor cell line. Persistent infection were carried out in those HRT-18 cells, as well as in the MO3.13 oligodendrocytic,[4] H4 neuroglial, U-87 MG astrocytic, and TE-671 rhabdomyosarcoma cell lines (ATCC). Other cell lines used for virus susceptibility are described in Table 2. Four infections were performed in the H4 cell line, whereas the HRT-18, MO3.13, and H4 cell lines were acutely infected as controls.

* INRS-Institut Armand-Frappier, Laval, Québec, H7V 1B7 Canada.

2.2. Virus Purification

Virus from persistent infections was purified at different passages. Prior to purification, virus was clarified and precipitated with polyethylene glycol (PEG) 8000 (Sigma). Accudenz (Accurate Chemicals) was used to perform gradient purification.

2.3. RT-PCR and Sequencing

Viral RNA was extracted using the GenElute Direct mRNA Miniprep Kit (Sigma) and reverse transcribed with MMuLVreverse transcriptase (Invitrogen). The Expand High-Fidelity *Taq* polymerase (Roche) was used to perform PCR. Primers specific to the HE, S and N genes were used to amplify target regions.[3] PCR amplicons were purified using the Qiaex II gel extraction kit (Qiagen) prior to sequencing, which was carried out by Bio S&T (Montréal, Québec, Canada).

2.4. Assays for Viral Susceptibillity and Modulation of Tropism and Infectivity

Prior to performing assays for modulation of tropism and infectivity, susceptibility of different cell lines to HCoV-OC43, ATCC strain, was determined (Table 2). The same cell lines were then infected with virus isolated from different purifications (HRT-18 P33, P54, P110, and P155; H4 P47 and P90; H4 P56.1, P56.2, P56.3, P116.1, P116.2, and P116.3; TE-671 P38 and P79; U-87 MG P35, and MO3.13 P5, P6, and P22) in order to correlate the observed mutations with a modulation of tropism or infectivity. Supernatants were titrated using an indirect immunoperoxidase assay (IPA), as previously described.[5]

3. RESULTS

Persistent infections of neural cell lines were initially performed to determine whether virus carrying mutations in genes encoding the surface protein S originated as a consequence of viral persistence. The HE protein gene and the nucleocapsid protein gene N were also sequenced in order to determine if these genes contributed to adaptation in cell culture. Viral particles released from persistently infected neural cell lines were isolated and purified by gradient centrifugation, and genomic RNA was sequenced. Results showed various mutations in the S gene but very few in HE and N genes, suggesting that the S gene is responsible for adaptation to the cellular environment, which could be associated with neurotropism, neuroinvasion, and presumably neuropathogenesis (Table 1). Almost every acquired mutation (Table 1) was conserved at subsequent passages, suggesting that they could confer an adaptive advantage and a stable phenotype to the virus. Five mutations were predominant and were found in almost all persistent infections (D24Y, S83T, H183R, Y241H, and N489H). The first four mutations are located in the putative receptor binding site, whereas the fifth one is located within the hypervariable region.

To correlate the observed mutations in the S gene with viral replication and tropism, assays for modulation of tropism and infectivity[6] were performed using cell lines originating from various human tissues as well as from various animal species, for which

Table 1. Location of S mutations at various passages of persistently infected cells.

HRT-18	**H4**	**H4**	**H4**	**H4**	**TE-671**	**U-87 MGMO3.13**	
P155**	**P90**	**P116.1**	**P116.2**	**P116.3**	**P79**	**P35**	**P22**
D30H*	D24Y	D24Y	D24Y	N25Y	N27Y	D24Y	D115H
S83T	V161V	P35S	P35S	P35S	P34S	S83T	T148I
L85Q	H183R	S83T	S147Y (D)	S83T	L85R	H183R	Y241H
D115H	V240V	E170K	H183R	Y119H	S258R	Y241H	M670T
T148I	Y241H	H183R	Y241H	S147P	A373V		P973S
H183Q	N441K	Y241H	N441K	H183R	R757S		A1090V
S258R	Q541L	A469V	E460D	Y241H	G785D		V1213A
S366G	R570P	R570R	H482Y	N489H	P972L		
N413T	N639N	T855I	F683Y	K506T	P973S		
F420S	T855I	N880K (I)	L693F	T641S	A978S		
N489H	D875H	L893H	A759E	N768T	T1086N		
K506N	L893R	S959C	S898S (I)	E896K	D1170A		
T536N	A965V	W974L	V980A	S901F			
Q541L	T975A	T975P	N1203 (D)	W974L			
R757H	I1227T	V980A	I1227T	F982L			
E896D	T1245I	S1093S		V986I			
C897G		G1169G		G1169D			
E933G		M1222K		E1236A			
F982L		D1232Y					
S1192R		P1249L					
T1225I		I1304I					
P1228S							

* D, deletion; I, insertion. ** Passages (and purification numbers) are indicated below the cell line.

susceptibility to HCoV-OC43 infection was previously determined (Table 2). These analyses revealed that mutations found throughout the S gene could affect the latter viral properties in certain cell lines. Amongst the virus variants obtained following persistent infections and virus purifications, five showed extended cellular tropism and increased replication titers *in vitro*: U87-MG P35, H4 P47, H4 P56.3, H4 P116.1, and H4 P116.2 (data not shown). Furthermore, some variants isolated from persistent infections were more virulent in mice and could form plaques, in opposition to the ATCC HCoV-OC43 reference strain VR759 (data not shown).

4. DISCUSSION

We have identified several mutations in the S gene of the HCoV-OC43 genome following persistent infections in different cell lines. These mutations will help us to further characterize viral adaptation during persistence and to understand mechanisms that are implicated in viral tropism and infectivity. Future studies will be carried out using an infectious cDNA clone of the OC43 strain assembled in a BAC vector.[7,8] The construction of this clone was performed in collaboration with F. Almazán and L. Enjuanes and will provide an invaluable tool to further understanding the underlying mechanisms for viral replication and tropism. In combination with the experiments described above, the clone will be useful in elucidating the molecular basis of human coronavirus neuropathogenesis.

Table 2. Susceptibility of various cell lines to the HCoV-OC43 ATCC strain.

Cell line	Origin	Tissue	Type	IPA [1]	Susceptibility [2]
HeLa	Human	Uterus	Epithelial	3.75	Low
MT4	Human	Bone marrow	T lymphocyte	2.0	No
U937	Human	Bone marrow	Monocyte	≤0.5	No
Jurkat E6.1	Human	Bone marrow	T lymphocyte	≤1.5	No
Raji	Human	Bone marrow	B lymphocyte	4.5	Yes
HL-60	Human	Bone marrow	Monocyte	3.25	Low
WI-38	Human	Lung	Fibroblast	4.0	Yes
L132	Human	Lung	Epithelial	3.25	Low
Caki-2	Human	Kidney	Epithelial	≤1,5	No
SW 156	Human	Kidney	Epithelial	≤1.5	No
NCI-N87	Human	Stomach	Epithelial	≤1.75	No
Arpe-19	Human	Eye	Epithelial	≤1.5	No
FHs 74 Int	Human	Intestine	Epithelial	3.5	Low
TK6	Human	Spleen	T lymphocyte	≤1.75	No
17 Cl-1	Mouse	Embryo	Fibroblast	≤1.75	No
L929	Mouse	Subcutaneous	Fibroblast	≤1.5	No
N-11	Mouse	Brain	Microglial	≤1.5	No
DBT	Mouse	Brain	Glial	≤2.25	No
J774 A.1	Mouse	Bone marrow	Macrophage	4.5	Yes
A20	Mouse	Bone marrow	B lymphocyte	3.25	Low
S.END.1	Mouse	Skin	Endothelial	≤1.5	No
Cos-7	Monkey	Kidney	Fibroblast	≤1.5	No
Vero	Monkey	Kidney	Epithelial	≤1.75	No
Vero E.6	Monkey	Kidney	Epithelial	≤1.5	No
B104	Rat	Brain	Fibroblast	3.25	Low
BHK-21	Hamster	Kidney	Fibroblast	4.0	Yes

[1] Indirect immunoperoxidase assay (infectious titers in $TCID_{50}$/mL).

[2] Titers from 0 to 3, not susceptible; titers over 3 and under 4, low susceptibility; titers of 4 and over, susceptible.

5. REFERENCES

1. Arbour, N., Côté, G., Lachance, C., Tardieu, M., Cashman, N. R., and Talbot, P. J., 1999, Acute and persistent infection of human neural cell lines by human coronavirus OC43, *J. Virol.* **73**:3338-3350.
2. Arbour, N., Day, R., Newcombe, J., and Talbot, P. J., 2000, Neuroinvasion by human respiratory coronaviruses, *J. Virol.* **74**:8913-8921.
3. St-Jean, J. R., Jacomy, H., Desforges, M., Vabret, A., Freymuth, F., and Talbot, P. J., 2004, Human respiratory coronavirus OC43: Genetic stability and neuroinvasion, *J. Virol.* **78**:8824-8834.
4. McLaurin, J., Trudel, G. C., Shaw, I. T., Antel, J. P., and Cashman, N. R., 1995, A human glial hybrid cell line differentially expressing genes subserving oligodendrocyte and astrocyte phenotype, *J. Neurobiol.* **26**:183-193.
5. Jacomy, H., and Talbot, P. J., 2003, Vacuolating encephalitis in mice infected by human coronavirus OC43, *Virology* **315**:20-33.
6. Schickli, J. H., Zelus, B. D., Wentworth, D. E., Sawicki, S. G., and Holmes, K. V., 1997, The murine coronavirus mouse hepatitis virus strain A59 from persistently infected murine cells exhibits an extended host range, *J. Virol.* **71**:9499-9507.
7. Almazán, F., Gonzalez, J. M., Pénzes, Z., Izeta, A., Calvo, E., Plana-Duran, J., and Enjuanes, L., 2000, Engineering the largest RNA virus genome as an infectious bacterial artificial chromosome, *Proc. Natl. Acad. Sci. USA* **97**:5516-5521.
8. St-Jean, J. R., Desforges, M., Almazán, F., Jacomy, H., Eujuanes, L. and Talbot, P. J., 2006. Recovery of a neuro virulent human coronavirus OC43 from an infectious cDNA clone. *J. Virol.* **80**:3670-3674.

SYNERGISTIC INHIBITION OF SARS-CORONAVIRUS REPLICATION BY TYPE I AND TYPE II IFN

Eric C. Mossel, Bruno Sainz, Jr., Robert F. Garry, and C. J. Peters *

1. INTRODUCTION

The susceptibility of SARS-coronavirus (SARS-CoV) to interferon (IFN) treatment has been extensively examined in culture, animals, and the clinic. IFN-α, relatively ineffective in cell culture, showed suggestive but inconclusive efficacy in monkeys and SARS patients.[1,2] IFN-β has most potent antiviral activity, though concentrations of greater than 1000 U/ml result in only marginal reduction of virus titer.[3–7] IFN-γ is ineffective against SARS-CoV in cell culture.[3,5,6]

It has been previously shown that treatment of cells with both type I and type II IFN produces an antiviral state greater in magnitude than can be explained by additive effects alone.[8–12] We sought to determine the effect of such an enhanced antiviral state on the replication of SARS-CoV.

2. IFN-β AND IFN-γ SYNERGISTICALLY INHIBIT SARS-CoV REPLICATION

To characterize the inhibitory effect of IFN-β and IFN-γ treatment on SARS-CoV replication, three-day viral growth assays were performed. IFN pretreated Vero E6 cells were infected with SARS-CoV at a MOI of 0.01. Cultures treated with 100 U/ml of IFN-β or IFN-γ were significantly refractory for SARS-CoV Urbani and HK replication ($P < 0.001$) at 24 and 48 hpi (Figures 1a and 1b). By 72 hpi, however, viral titers in IFN-β– or IFN-γ–treated cultures approached levels detected in vehicle-treated groups. A potent inhibitory effect was observed when Vero E6 cultures were treated with both IFN-β and IFN-γ. The inhibitory effect achieved with combination IFN-β and IFN-γ treatment was

* Eric C. Mossel, Colorado State University, Fort Collins, Colorado 80523. Bruno Sainz, Jr., Scripps Research Institute, La Jolla, California 92307. Robert F. Garry, Tulane University Health Sciences Center, New Orleans, Louisiana 70112. C. J. Peters, University of Texas Medical Branch, Galveston, Texas 77555.

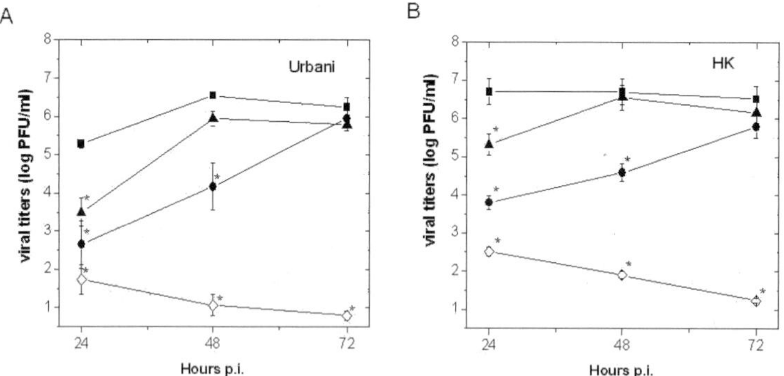

Figure 1. IFN-β and/or IFN-γ inhibit SARS-CoV replication in Vero E6 Cells. Vero E6 cells were treated with (■) vehicle or 100 U/ml each of (●) IFN-β, (▲) IFN-γ or (◊) IFN-β and IFN-γ 12 h prior to infection with SARS-CoV strain (A) Urbani or SARS-CoV strain (B) HK at a MOI of 0.01 pfu per cell. Supernatants were harvested on the indicated days p.i., and viral titers were determined by plaque assay. Significant differences in viral titers in Vero E6 cells treated with IFNs relative to cells treated with vehicle are denoted by a single asterisk ($P < 0.001$, one-way ANOVA and Tukey's post hoc *t* test).

consistently greater than 3000-fold at all time points tested and reached levels of greater than 1×10^5-fold at 72 hpi relative to vehicle treated Vero E6 cells.

3. IFN-β AND IFN-γ SYNERGISTICALLY INHIBIT SARS-CoV–MEDIATED CPE

Cytopathic effect (CPE) was extensive in vehicle-treated groups infected with either SARS-CoV strain at 120 hpi (Fig. 2A and 2E), as evident by the reduced monolayer staining with crystal violet. Relative to vehicle-treated and individual IFN-treated cultures, CPE is less evident in cells treated with both IFN-β and IFN-γ at 120 hpi; monolayers appeared evenly stained with little to no visible CPE (Fig. 2D and 2H).

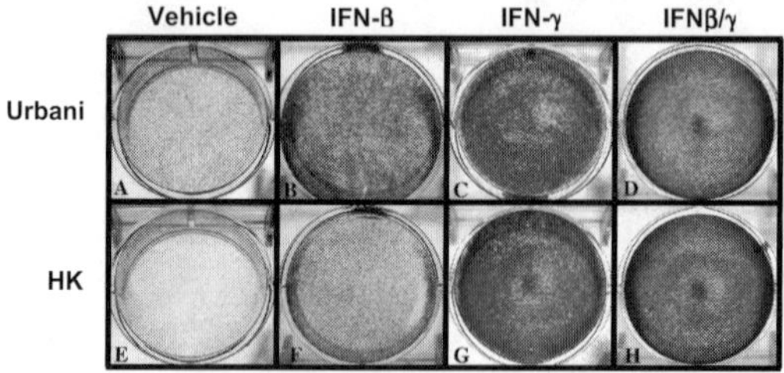

Figure 2. IFN-γ alone and IFN-β and IFN-γ inhibit SARS-CoV CPE in Vero E6 cells. Cultures were pretreated for 12 h with (A, E) vehicle or 100 U/ml each of (B, F) IFN-β, (C, G) IFN-γ, or (D, H) IFN-β and IFN-γ prior to infection with SARS-CoV strains Urbani (A–D) or HK (E–H) at a MOI of 0.01 pfu/cell. Monolayers were fixed, stained with crystal violet, and photographed 120 h p.i.

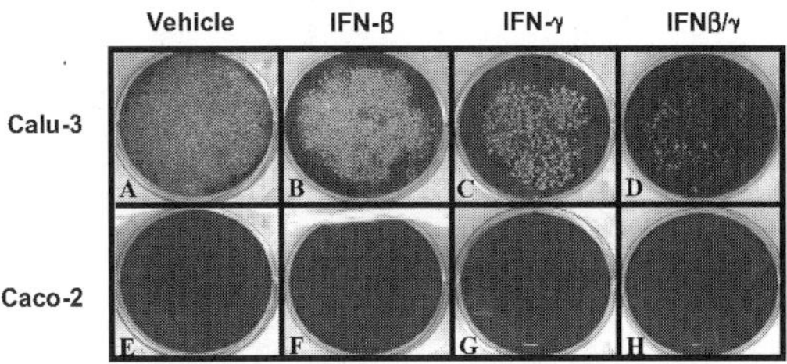

Figure 3. IFN-β and IFN-γ inhibits SARS-CoV CPE in Calu-3 cells. Cultures were pretreated for 12 h with (A, E) vehicle or 100 U/ml each of (B, F) IFN-β, (C, G) IFN-γ, or (D, H) IFN-β and IFN-γ prior to infection with SARS-CoV strain Urbani at a MOI of 0.01 pfu/cell. Monolayers were fixed, stained with crystal violet, and photographed 72 h p.i.

To determine whether this phenomenon is limited to Vero E6 cells, additional cell lines were examined. Calu-3 cells showed the same gradated CPE as Vero E6 cells with little or no CPE present in cells treated with both IFN-β and IFN-γ (Fig. 3). Contrary to the observations of others, gross CPE does not occur in SARS-CoV-infected Caco-2 cells in our hands.[3,13] As such, SARS-CoV–infected Caco-2 cell monolayers remained confluent regardless of treatment. However, the CPE profile observed in Calu-3 cells suggests that the synergistic inhibitory effect on SARS-CoV replication by IFN-β and IFN-γ is not Vero E6 cell specific.

4. DISCUSSION

It has been known for more than 25 years that treatment of cells with type I and type II IFN potentiates the antiviral response to levels greater than can be explained by simple additive effects.[10] Since then, the effect has been shown for a wide variety of viruses, including human cytomegalovirus, HSV-1, vesicular stomatitis virus (VSV), Lassa virus, and others.[8,9,11,12,14]

The mechanism of synergistic inhibition of virus replication by type I and type II IFN has not been determined for any virus. However, it was recently shown that NO, induced by a combination of IFN-γ and IL-1β, inhibits SARS-CoV replication.[15] Further, it was shown in an avian system that type I and type II IFN potentiate the antiviral response as well as the secretion of NO.[16] Based on this evidence, a role for NO and iNOS in the potentiated anti-SARS-CoV response induced by type I and type II IFN cotreatment deserves consideration.

5. ACKNOWLEDGMENTS

The authors wish to thank Dr. Dr. Li-Kuang Chen from Tzu-Chi University at Hualien, Taiwan for kindly supplying the SARS-CoV strain HK. This work is supported by National Institutes of Health grants AI007536 (E.C.M.), AI0543818 (B.S.), and AI054626, AI054238, RR018229, and CA08921 (R.F.G.) and contract NO1 AI 25489 (C.J.P.).

6. REFERENCES

1. Haagmans, B. L., Kuiken, T., Martina, B. E., Fouchier, R. A., Rimmelzwaan, G. F., van Amerongen, G., van Riel, D., de Jong, T., Itamura, S., Chan, K. H., Tashiro, M., and Osterhaus, A. D., 2004, Pegylated interferon-alpha protects type 1 pneumocytes against SARS coronavirus infection in macaques, *Nat. Med.* **10**:290-293.
2. Loutfy, M. R., Blatt, L. M., Siminovitch, K. A., Ward, S., Wolff, B., Lho, H., Pham, D. H., Deif, H., LaMere, E. A., Chang, M., Kain, K. C., Farcas, G. A., Ferguson, P., Latchford, M., Levy, G., Dennis, J. W., Lai, E. K., and Fish, E. N., 2003, Interferon alfacon-1 plus corticosteroids in severe acute respiratory syndrome: a preliminary study, *JAMA* **290**:3222-3228.
3. Cinatl, J., Morgenstern, B., Bauer, G., Chandra, P., Rabenau, H., and Doerr, H. W., 2003, Treatment of SARS with human interferons, *Lancet* **362**:293-294.
4. Hensley, L. E., Fritz, L. E., Jahrling, P. B., Karp, C. L., Huggins, J. W., and Geisbert, T. W., 2004, Interferon-beta 1a and SARS coronavirus replication, *Emerg. Infect. Dis.* **10**:317-319.
5. Scagnolari, C., Vicenzi, E., Bellomi, F., Stillitano, M. G., Pinna, D., Poli, G., Clementi, M., Dianzani, F., and Antonelli, G., 2004, Increased sensitivity of SARS-coronavirus to a combination of human type I and type II interferons, *Antivir. Ther.* **9**:1003-1011.
6. Spiegel, M., Pichlmair, A., Muhlberger, E., Haller, O., and Weber, F., 2004, The antiviral effect of interferon-beta against SARS-coronavirus is not mediated by MxA protein, *J. Clin. Virol.* **30**:211-213.
7. Stroher, U., DiCaro, A., Li, Y., Strong, J. E., Aoki, F., Plummer, F., Jones, S. M., and Feldmann, H., 2004, Severe acute respiratory syndrome-related coronavirus is inhibited by interferon- alpha, *J. Infect. Dis.* **189**:1164-1167.
8. Asper, M., Sternsdorf, T., Hass, M., Drosten, C., Rhode, A., Schmitz, H., and Gunther, S., 2004, Inhibition of different Lassa virus strains by alpha and gamma interferons and comparison with a less pathogenic arenavirus, *J. Virol.* **78**:3162-3169.
9. Czarniecki, C. W., Fennie, C. W., Powers, D. B., and Estell, D. A., 1984, Synergistic antiviral and antiproliferative activities of Escherichia coli-derived human alpha, beta, and gamma interferons, *J. Virol.* **49**:490-496.
10. Fleischmann, W. R., Jr., Georgiades, J. A., Osborne, L. C., and Johnson, H. M., 1979, Potentiation of interferon activity by mixed preparations of fibroblast and immune interferon, *Infect. Immun.* **26**:248-253.
11. Sainz, Jr., B., and Halford, W. P., 2002, Alpha/Beta interferon and gamma interferon synergize to inhibit the replication of herpes simplex virus type 1, *J. Virol.* **76**:11541-11550.
12. Schwarz, L. A., Fleischmann, C. M., and Fleischmann, Jr., W. R., 1984, Potentiation of interferon's antiviral activity by the mutually synergistic interaction of MuIFN-alpha/beta and MuIFN-gamma, *J. Biol. Response Mod.* **3**:608-612.
13. Sainz, Jr., B., Mossel, E. C., Peters, C. J., and Garry, R. F., 2004, Interferon-beta and interferon-gamma synergistically inhibit the replication of severe acute respiratory syndrome-associated coronavirus (SARS-CoV), *Virology* **329**:11-17.
14. Sainz, Jr., B., Lamarca, H. L., Garry, R. F., and Morris, C. A., 2005, Synergistic inhibition of human cytomegalovirus replication by interferon-alpha/beta and interferon-gamma, *Virol. J.* **2**:14.
15. Akerstrom, S., Mousavi-Jazi, M., Klingstrom, J., Leijon, M., Lundkvist, A., and Mirazimi, A., 2005, Nitric oxide inhibits the replication cycle of severe acute respiratory syndrome coronavirus, *J. Virol.* **79**:1966-1969.
16. Sekellick, M. J., Lowenthal, J. W., O'Neil, T. E., and Marcus, P. I., 1998, Chicken interferon types I and II enhance synergistically the antiviral state and nitric oxide secretion, *J. Interferon Cytokine Res.* **18**:407-414.

MUSTELA VISON ACE2 FUNCTIONS AS A RECEPTOR FOR SARS-CORONAVIRUS

Lindsay K. Heller, Laura Gillim-Ross, Emily R. Olivieri, and David E. Wentworth*

1. INTRODUCTION

Coronaviruses (CoVs), members of the order *Nidovirales*, are enveloped, positive-sense, single-stranded RNA viruses.[1] The genomes of CoVs are among the largest of known RNA viruses, and range in size from 27 to 32 kilobases in length.[1] During replication within the host cell, all *Nidovirales* produce a set of 3' nested transcripts that share a short leader sequence at the 5' terminus.[1] CoVs are known for their crown-like appearance, which is due to the spike (S) glycoproteins projecting from the surface of the virion. Interaction of S and cellular receptors facilitates entry of the virus, and this interaction is a principal factor in the tissue tropism and the species specificity of CoVs.

In 2002–2003, there was an outbreak of severe acute respiratory syndrome (SARS) in Guangdong Province, China. This outbreak subsequently affected 29 countries and resulted in 8,096 cases of SARS, of which 774 were fatal.[2] The causative agent was determined to be a previously unrecognized CoV, which became known as SARS-CoV. The SARS-CoV outbreak likely resulted from zoonotic transmission. Although SARS-CoV was isolated from several species, the natural reservoir of its progenitor remains to be discovered. Of the animals examined to date, SARS-CoV was most frequently isolated from Himalayan palm civets, which are sold in live animal markets in Guangdong and in other regions of Southeast Asia.[3] The role palm civets played in the SARS outbreak is unclear. However, SARS-CoV-like viruses isolated from palm civets appeared to be under strong selective pressure and are genetically most similar to viruses infecting humans early in the outbreak.[4] The selective pressure of replication in palm civet may have generated strains that could be more easily transmitted to humans. Palm civets are carnivores in the suborder *Fissipedia*. Other *Fissipedia* include raccoon dog, dog, cat, raccoon, skunk, ferret (*Mustela putorius*), and mink (*Mustela vison*). SARS-CoV-like virus was also detected in a raccoon dog in a live animal market.[3] Furthermore, *Fissipedia*

* Lindsay Heller, Laura Gillim-Ross, New York State Department of Health, Albany, New York 12002. Emily Olivieri, David E. Wentworth, New York State Department of Health, and State University of New York, Albany, New York 12002.

such as cat, ferret,[5] and palm civet[6] have been experimentally infected with SARS-CoV/Urbani.

Previous analysis of cell lines and primary cells derived from diverse species led to the discovery that a mink lung cell line (Mv1Lu) was permissive to SARS-CoV.[7, 8] Angiotensin-converting enzyme 2 (ACE2)[9] and CD209L[10] are host cell-surface proteins that function as receptors for SARS-CoV. However, human ACE2 (hACE2) is a more efficient receptor than is CD209L, *in vitro*. Therefore, we set out to determine whether ACE2 from *M. vison* (mvACE2) is a functional receptor for SARS-CoV. Our data shows ACE2 RNA was expressed by Mv1Lu cells. We subsequently sequenced, cloned, and expressed mvACE2 in nonpermissive cells, to determine its SARS-CoV receptor activity. Human, palm civet, rat, mouse, chicken, and mink ACE2 were compared to identify differences that are important in species specificity and to discern regions within ACE2 that may impact its function as a SARS-CoV receptor.

2. RESULTS AND DISCUSSION

Expression of ACE2 RNA in SARS-CoV susceptible human (Huh7, HEK293T), African green monkey (VeroE6), and Mv1Lu cell lines[7] was analyzed by RT-PCR (chapter 4.8 this volume). The amount of ACE2 amplicon differed among the various cell lines. VeroE6 had the strongest ACE2 amplicon, whereas Mv1Lu cells showed the weakest RT-PCR product. However, we subsequently found that the human/mouse consensus primers used initially were not 100% conserved with the mvACE2 nucleotide sequence. RT-PCR amplification of ACE2 from VeroE6, Huh7, and Mv1Lu cells with ACE2-1446 (GGAACTCTACCATTTACTTACA) and ACE2-1991R (TCCAAGAGCTGATTTTAGGCTTAT), which both have 100% identity with mvACE2, showed a robust amplicon from Mv1Lu total RNA (Figure 1). Glyceraldehyde-3-phosphate dehydrogenase (G3PDH) was also amplified, as a control for RNA concentration and integrity (Figure 1).[7] The complete open reading frame of human ACE2 (hACE2) or mvACE2 was amplified from RNA isolated from the Huh7 or Mv1Lu cell line, respectively. The consensus sequences of human and mink ACE2 open reading frames were determined by direct sequencing of RT-PCR amplicons. Amplicons of hACE2 and mvACE2 open reading frames were cloned and confirmed by sequence analysis. The deduced amino-acid sequence identity/similarity of the open reading frames of human, palm civet, mink, mouse, rat, and chicken ACE2 were compared using VectorNTI Advance 9.1 (Table 1). The data demonstrate that mink and palm civet ACE2

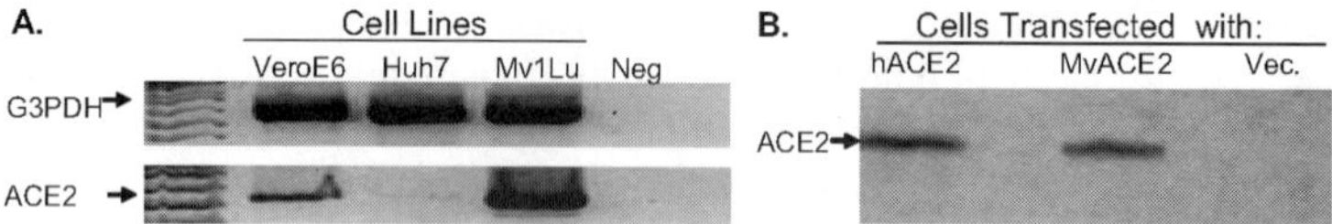

Figure 1. A. ACE2 transcript identified in *M. vison* lung epithelial cells. G3PDH and ACE2 RT-PCR amplicons from 1 μg of total RNA were visualized by ethidium bromide staining after agarose gel electrophoresis. A negative control that lacks RNA template (Neg) was also included. B. Expression of ACE2 Protein. HEK-293T cells were transfected with human (hACE2), mink (MvACE2), or the pIRES2-EGFP (BD-Biosciences) expression vector alone (Vec.), and lysates were harvested at 48 h post-transfection. Immunoblotting of cell lysates was performed with polyclonal antibody against human ACE2 (R&D systems).

Table 1. Percent amino-acid identity/similarity of ACE2 orthologs.

Species	Human	P. Civet	Mink	Mouse	Rat	Chicken
Human	100/100					
P. Civet	83/87	100/100				
Mink	83/87	88/90	100/100			
Mouse	82/87	82/86	81/87	100/100		
Rat	83/87	81/85	80/85	90/93	100/100	
Chicken	62/71	63/72	62/71	61/71	61/71	100/100

have the greatest amino-acid identity/similarity (88/90%), and additional analysis indicated that mink and palm civet ACE2 have common substitutions, when compared to hACE2.

To investigate the receptor activity of mvACE2, we generated eukaryotic expression constructs of hACE2 and mvACE2 using pIRES2-EGFP. Immunoblot showed that both hACE2 and mvACE2 constructs expressed protein when they were transfected into HEK-293T cells (Figure 1B). To test whether mvACE2 could function as a SARS-CoV receptor, normally nonpermissive BHK-21 cells were transiently transfected with hACE2, mvACE2, or empty pIRES2-EGFP expression plasmids, and then analyzed for susceptibility to SARS-CoV (Figure 2). To identify SARS-CoV entry, we isolated total RNA at 1 h or 24 h post-inoculation and analyzed it using our multiplex RT-PCR assay, which amplifies G3PDH, SARS-CoV genomic RNA (gRNA), and SARS-CoV subgenomic RNA (sgRNA)[7]; the latter is indicative of virus entry. SARS-CoV sgRNA was amplified in both hACE2 and mvACE2 transfected cells at 24 h post-inoculation. In contrast, control cells transfected with pIRES2-EGFP, or that were mock-transfected, showed amplification of G3PDH and some gRNA, presumably from residual input virus (Figure 2). These data show that mvACE2 is a functional receptor for SARS-CoV.

We used the crystal structure of hACE2[11] to map amino acid differences between hACE2 and mvACE2. Many of the amino-acid substitutions localize to the surface of the protein, and some are very close to residues that Li *et al.* recently showed to be important for binding of a SARS-CoV S1-Ig fusion protein (e.g., K353, D355, R357).[12] We identified differences at residues H34, D38, N61, L79, N103, Q305, Q325, E329, and G354 of hACE2 that may also be important SARS-CoV S-ACE2 interaction. Given the strong effect of K353H substitution on S1-Ig binding to hACE2,[12] the G354Q substitution that we identified in mvACE2 is also likely to decrease SARS-CoV/Urbani binding. We have passaged SARS-CoV/Urbani in mvACE2-expressing Mv1Lu cells, and selected variants that replicate to higher titer and form larger plaques than wild-type virus (chapter 4.8 this volume). Our analysis of the S gene from these variants is likely to identify mutations that compensate for G354Q in mvACE2. ACE2 amino-acid alignment of good and poor SARS-CoV/Urbani receptors suggests that additional residues or post-translational modifications of ACE2 affect SARS-CoV entry (Olivieri *et al.*, this volume).

Our data demonstrate that mvACE2 RNA is expressed by SARS-CoV susceptible Mv1Lu cells, that it is closely related to palm civet ACE2, and that mvACE2 is a functional receptor for SARS-CoV. We also identified species-specific amino-acid variations within ACE2 that are likely to influence its SARS-CoV receptor activity. Additionally, our results strongly suggest that, like palm civet, raccoon, dog, cat, and ferret,[3, 5, 6] mink may be another member of the suborder Fissipedia that is susceptible to SARS-CoV. Therefore, mink may not only provide an additional animal model for

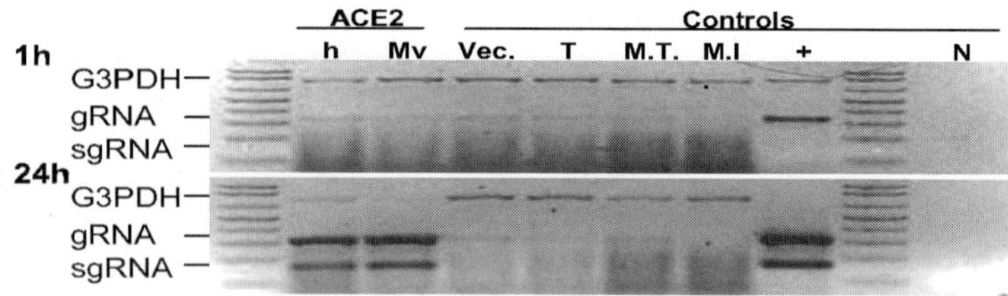

Figure 2. Transfection of MvACE2 confers susceptibility to SARS-CoV. Normally nonpermissive BHK-21 cells were transfected with hACE2 (h), mvACE2 (Mv), empty expression plasmid pIRES2-EGFP (Vec.), pEGFP-N1 (T) to monitor transfection efficiency, or were mock-transfected (M.T.). Cells were inoculated with SARS-CoV (MOI of ~0.01) at 48 h post-transfection. G3PDH, SARS-CoV gRNA and sgRNA were amplified from 1 μg of total RNA isolated at 1 h, or 24 h post-inoculation. Mock-inoculated BHK-21 (M.I.) and SARS-CoV inoculated VeroE6 (+) were also included as controls. N is dH2O template RT-PCR control. Amplicons were visualized by ethidium bromide staining after electrophoresis, and the images have been contrast inverted.

SARS-CoV pathogenesis studies, but they may also have potential as a North American wildlife reservoir of SARS-CoV-like viruses.

We thank Noel Espina for his expert assistance, and the Wadsworth Center Molecular Genetics Core for DNA sequencing. L.G.-R. was supported by an EID fellowship administered by the APHL and funded by the CDC. E.O. was supported by NIH/NIAID training grant T32AI05542901A1. The study was also funded by Diagnostic Hybrids Inc. and by NIH/NIAID grants N01-AI-25490 and P01-AI-0595760.

3. REFERENCES

1. K. V. Holmes, in: *Fields Virology, vol. 1*, edited by D. M. Knipe, P. M. Howley, D. E. Griffin, R. A. Lamb, M. A. Martin, and B. Roizman, (Lippincott Williams & Wilkins, Philadelphia, 2001), pp. 1187-1203.
2. Summary of probable SARS cases with onset of illness from 1 November 2002 to 31 July 2003, *World Health Organization*, http://www.who.int/csr/sars/country/table2004_04_21/en/index.html (2004).
3. Y. Guan, B. J. Zheng, Y. Q. He, *et al.*, Isolation and characterization of viruses related to the SARS coronavirus from animals in southern China, *Science* **302**, 276-278 (2003).
4. H. D. Song, C. C. Tu, G. W. Zhang, *et al.*, Cross-host evolution of severe acute respiratory syndrome coronavirus in palm civet and human, *Proc. Natl. Acad. Sci. USA* **102**, 2430-2435 (2005).
5. B. E. E. Martina, B. L. Haagmans, T. Kuiken, *et al.*, SARS virus infection of cats and ferrets, *Nature* **425**, 915 (2003).
6. D. Wu, C. Tu, C. Xin, *et al.*, Civets are equally susceptible to experimental infection by two different severe acute respiratory syndrome coronavirus isolates, *J. Virol.* **79**, 2620-2625 (2005).
7. L. Gillim-Ross, J. Taylor, D. R. Scholl, *et al.*, Discovery of novel human and animal cells infected by the severe acute respiratory syndrome coronavirus by replication-specific multiplex reverse transcription-PCR, *J. Clin. Microbiol.* **42**, 3196-3206 (2004).
8. E. C. Mossel, C. Huang, K. Narayanan, *et al.*, Exogenous ACE2 expression allows refractory cell lines to support severe acute respiratory syndrome coronavirus replication, *J. Virol.* **79**, 3846-3850 (2005).
9. W. Li, M. J. Moore, N. Vasilieva, *et al.*, Angiotensin-converting enzyme 2 is a functional receptor for the SARS coronavirus, *Nature* **426**, 450-454 (2003).
10. S. A. Jeffers, S. M. Tusell, L. Gillim-Ross, *et al.*, CD209L (L-SIGN) is a receptor for severe acute respiratory syndrome coronavirus, *Proc. Natl. Acad. Sci. USA* **101**, 15748-15753 (2004).
11. P. Towler, B. Staker, S. G. Prasad, *et al.*, ACE2 structures reveal a large hinge-bending motion important for inhibitor binding and catalysis, *J. Biol. Chem.* **279**, 17996-18007 (2004).
12. W. Li, C. Zhang, J. Sui, *et al.*, Receptor and viral determinants of SARS-coronavirus adaptation to human ACE2, *EMBO J.* **24**, 1634-1643 (2005).

HCoV-229E INFECTS AND ACTIVATES MONOCYTES

Marc Desforges, Tina Miletti, Mylène Gagnon, and Pierre J. Talbot*

1. INTRODUCTION

Human coronaviruses (HCoV) are respiratory pathogens with neurotropic and neuroinvasive properties. Indeed, cell lines of neural origin[1,2] and human primary cultures from the central nervous system (CNS)[3] are susceptible to infection by HCoV-OC43, and RNA was found to persist in human brain.[4] HCoV-OC43 can also cause a vacuolating encephalitis in mice[5] and RNA can persist for up to 1 year.[5a] HCoV-OC43 uses the neural route via the olfactory bulb to gain access to the CNS,[6] but no such pathway has been described for HCoV-229E as it does not infect mice, probably because of lack of an adequate receptor. A possible alternative neuroinvasive pathway would consist in passage through the blood-brain barrier (BBB) by infection or passage through brain endothelial cells and/or transport by infected leukocytes. Human immunodeficiency virus type 1 (HIV-1) is a good case in point, where brain infiltration of infected T lymphocytes[7] or monocytes[8,9] is crucial in initiating the neuropathology known as AIDS-dementia. It was previously reported that both OC43 and 229E could productively infect primary human monocytes/macrophages.[10] However, the results presented here rather suggest that only HCoV-229E productively infects human monocytic cells, while HCoV-OC43 infection is highly restricted. Moreover, the result obtained with the THP-1 cell line, which represents an excellent model to study the interaction between HCoV-229E and human monocytic cells, suggested that monocytes could be activated by infection and could serve as a reservoir and vector into the CNS for neuroinvasive HCoV-229E *in vivo*.

2. MATERIALS AND METHODS

2.1. Viruses and Cell Lines

HCoV strains (229E and OC43) were obtained from ATCC, plaque-purified and grown on L132 cells (229E) or HRT-18 cells (OC43). Human cell line THP-1 (gifts from Daniel Oth, INRS-Institut Armand-Frappier), were cultured in RPMI 1640 supplemented

* INRS-Institut Armand-Frappier, Laval (Québec) H7V 1B7, Canada.

with 10 mM HEPES, 1 mM pyruvate sodium, MEM non essential amino acids, 100 U/mL penicillin, 100 μg/mL streptomycin, 2-mercaptoethanol 2 x 10^{-5} M (Invitrogen), and 10% (v/v) heat-inactivated fetal bovine serum (FBS) (Wysent).

Leukocytes were isolated through Ficoll-Hypaque (Amersham) and PBMC were prepared at 2.5 x 10^6/mL. Monocytes were adsorbed onto 24-well plastic plates (Corning) for 90 minutes at 37°C in complete RPMI 1640 supplemented with the same components as for THP-1 cells except that 10% (v/v) heat-inactivated autologous serum was used. Cultures were then washed to harvest lymphocytes while adsorbed cells were fed with new medium. Part of these monocytes were cultured with 2% (v/v) autologous serum and infected the next day. The other fractions were induced to differentiate into macrophages (7 days in 10% (v/v) autologous serum in RPMI) before infection.

2.2. Infection and Activation of Cells and Titration of Infectious Virus Production

Cells were infected at a multiplicity of infection (MOI) of 1.0 and incubated 4 hours at 37°C. They were then washed with RPMI 1640 w/o serum and grown in RPMI 1640 supplemented with 10% (v/v) FBS (THP-1 cells) or 2% (v/v) (primary monocytes) or 10% (v/v) (primary macrophages) autologous serum. Infection was carried out for up to 7 days. Samples were taken at different times post-infection for evaluation of infectious virus production using an immunoperoxidase assay. [11]

To evaluate whether the THP-1 cells were activated following infection by HCoV, metalloproteinases (MMP) and TNF-α secretion were measured. Zymography on SDS-PAGE was performed to evaluate MMP production and TNF-α production was evaluated using the Quantikine system (R&D Systems).

3. RESULTS

Primary human monocytes/macrophages cells were reported to be susceptible to a productive infection by HCoV-OC43 and 229E. However, our results suggest that 229E[10] productively infects human monocytic cells, while OC43 infection is highly restricted. As shown in Table 1, monocytes and macrophages from most donors were susceptible to a productive infection, while lymphocytes appeared restrictive to infection. On the other hand, infectious OC43 virions were never detected in any leukocytic cell types (data not shown).

As the amount of HCoV-229E infectious virus detected dropped rapidly (Table 1), a short-term kinetics of virus production was performed on two independent donors. Results showed that the production of virus was transient (Figure 1A). Moreover, the THP-1 cell line, which represents an excellent model to study the interaction between HCoV-229E and human monocytic cells, appeared equally susceptible to infection by HCoV-229E (Figure 1B). Further investigation showed that these cells could be activated following infection as shown by an increased production of MMP-9 (Figure 1B) and by the release of TNF-α in the medium (Table 2).

Table 1. Susceptibility of human PBMC to infection by HCoV-229E at 37°C (MOI 1). Virus titers, log $TCID_{50}$/mL in cell culture medium.

Day pi	Donor	1	2	3	4	5	6	7	8	9	10
1	Mono	3 ±0.6	≤0.5	3±0.2	3±0.5	3±0.2	3±0.5	4±0.7	4±0.2	3±0.0	4±0.4
	Macro	4 ±0.4	≤0.5	3±0.3	≤0.5	5±0.2	5±0.2	≤0.5	4±0.5	3±0.4	4±0.2
	Lympho	≤0.5	≤0.5	≤0.5	≤0.5	≤0.5	≤0.5	≤0.5	≤0.5	≤0.5	≤0.5
3	Mono	≤0.5	≤0.5	≤0.5	≤0.5	≤0.5	≤0.5	≤0.5	≤0.5	≤0.5	≤0.5
	Macro	≤0.5	≤0.5	≤0.5	≤0.5	≤0.5	≤0.5	≤0.5	≤0.5	≤0.5	≤0.5
	Lympho	≤0.5	≤0.5	≤0.5	≤0.5	≤0.5	≤0.5	≤0.5	≤0.5	≤0.5	≤0.5
5	Mono	≤0.5	≤0.5	≤0.5	≤0.5	≤0.5	≤0.5	≤0.5	≤0.5	≤0.5	≤0.5
	Macro	≤0.5	≤0.5	≤0.5	≤0.5	≤0.5	≤0.5	≤0.5	≤0.5	≤0.5	≤0.5
	Lympho	≤0.5	≤0.5	≤0.5	≤0.5	≤0.5	≤0.5	≤0.5	≤0.5	≤0.5	≤0.5
			≤0.5								
7	Mono	≤0.5	≤0.5	≤0.5	≤0.5	≤0.5	≤0.5	≤0.5	≤0.5	≤0.5	≤0.5
	Macro	≤0.5	≤0.5	≤0.5	≤0.5	≤0.5	≤0.5	≤0.5	≤0.5	≤0.5	≤0.5
	Lympho	≤0.5	≤0.5	≤0.5	≤0.5	≤0.5	≤0.5	≤0.5	≤0.5	≤0.5	≤0.5

4. DISCUSSION

Because HCoV possess neuroinvasive properties, it is of great interest to investigate the possibilty that these ubiquitous human respiratory viruses can use leukocytes as a vehicle to gain access to the CNS. The study presented here provides interesting insights into a possible route taken by human coronaviruses to reach the CNS. Indeed, as human monocytic cells are susceptible to infection by HCoV-229E and as they are activated following this infection, they could have an important role in helping HCoV gain access to the CNS. Monocytic cells can on the one hand support virus replication, at least transiently, and they also produce MMP-9 and TNF-α. Therefore, they represent a suitable vector for viral transport to the CNS, as virally-activated monocytic cells can

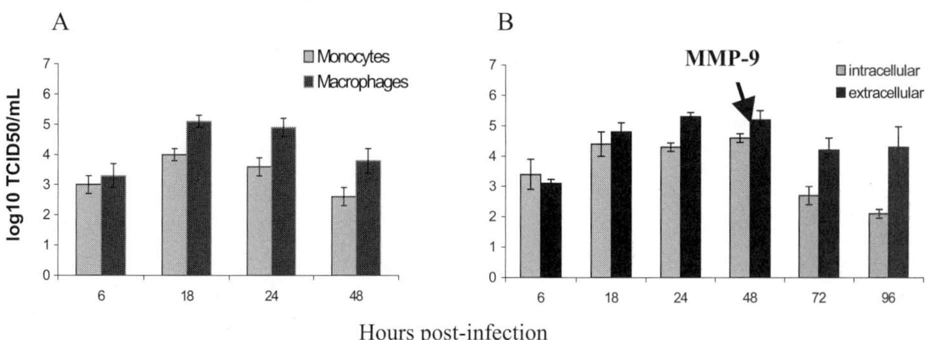

Figure 1. Primary human monocytic cells and the human monocytic cell line THP-1 are equally susceptible only to HCoV-229E. (A) Kinetics of infectious virus production in primary human monocytes and macrophages. (B) Kinetics of infectious virus production in THP-1 cells. The arrow indicates an increased MMP-9 activity at 48 hours postinfection.

Table 2. Production of TNF-α by THP-1 cells following infection by HCoV-229E (pg/mL).

Hours pi	16	24	42	68
Mock	0.0	0.0	0.0	0.0
HCoV-229E	0.0	29.2 ± 8	90.1 ± 5	104.6 ± 5

increase secretion of proMMP-9 (Figure 1B), therefore directly contributing to BBB breakdown[12] and TNF-α, which can facilitate the passage of monocytes through the BBB by upregulating the expression of ICAM-1.[13] Moreover, after gaining access to the CNS, the monocytes could release neurotoxic factors, therefore contributing to neurodegeneration.

5. REFERENCES

1. Arbour, N., Coté, G., Lachance, C., Tardieu, M., Cashman, N. R., and Talbot, P. J., 1999a, Acute and persistent infection of human neural cell lines by human coronavirus OC43, *J. Virol.* **73**:3338-3350.
2. Arbour, N., Ékandé, S., Côté, G., Lachance, C., Chagnon, F.,Tardieu, M., Cashman, N. R., and Talbot, P. J., 1999b, Persistent infection of human oligodendrocytic and neuroglial cell lines by human coronavirus 229E, *J. Virol.* **73**:3326-3337.
3. Bonavia, A., Arbour, N., Yong, V. W., and Talbot, P. J., 1997, Infection of primary cultures of human neural cells by human coronaviruses 229E and OC43, *J. Virol.* **71**:800-806.
4. Arbour, N., Day, R., Newcombe, J., and Talbot, P. J., 2000, Neuroinvasion by human respiratory coronaviruses, *J. Virol.* **74**:8913-8921.
5. Jacomy, H., and Talbot, P. J., 2003, Vacuolating encephalitis in mice infected by human coronavirus OC43, *Virology* **315**:20-33.
5a. Jacomy, H., Fragoso, G., Almazán, G., Mushynski, W. E., and Talbot, P. J., 2006, Human coronavirus OC43 induces chronic encephalitis leading to disabilities in BALBlc mice. *Virology*, March 2006, Epub ahead of print.
6. St-Jean, J. R., Jacomy H., Desforges, M., Vabret, A., Freymuth, F., and Talbot, P. J., 2004, Human respiratory coronavirus OC43: genetic stability and neuroinvasion, *J. Virol.* **78**:8824-8834.
7. Williams, K. C., and Hickey, W. F., 1995, Traffic of hematogenous cells through the central nervous system, *Curr. Top. Microbiol. Immunol.* **202**:221-245.
8. Nottet, H. S., Persidsky, Y., Sasseville, V. G., Nukuna, A. N., Bock, P., Zhai, Q. H., Sharer, L. R., McComb, R. D., Swindells, S., Soderland, C., and Gendelman, H. E., 1996, Mechanisms for the transendothelial migration of HIV-1-infected monocytes into brain, *J. Immunol.* **156**:1284-1295.
9. Persidsky, Y., Stins, M., Way, D., Witte, M. H., Weinand, M., Kim, K. S., Bock, P., Gendelman, H. E., and Fiala, M., 1997, A model for monocyte migration throught the blood-brain barrier during HIV-1 encephalitis, *J. Immunol.* **158**:3499-3551.
10. Collins, A. R., 2002, In vitro detection of apoptosis in monocytes/macrophages infected with human coronavirus, *Clin. Diagn. Lab. Immunol.* **9**:1392-1395.
11. Sizon, J., Arbour, N. and Talbot, P.J., 1998, comparison of immunofluorescence with monoclonal antibodies and RT-PCR for the detection of human corona viruses 229E and OC43 in culture. *J. Virol. Methods* **72**: 145-152.
12. Gidday, J. M., Gasche, Y. G., Copin, J. C., Shah, A. R., Perez, R. S., Shapiro, S. D., Chan, P. H., and Park, T. S., 2005, Leukocytes-derived matrix metalloproteinase-9 mediates blood-brain-barrier breakdown and is proinflammatory after transient focal cerabral ischemia, *Am. J. Physiol. Heart Circ. Physiol.* **289**:H558-568.
13. Dietrich, J. B., 2002, The adhesion molecule ICAM-1 and its regulation in relation with the blood-brain-barrier, *J. Neuroimmunol.* **128**:58-68.

PATHOLOGICAL AND VIROLOGICAL ANALYSES OF SEVERE ACUTE RESPIRATORY SYNDROME–ASSOCIATED CORONAVIRUS INFECTIONS IN EXPERIMENTAL ANIMALS

Noriyo Nagata, Naoko Iwata, Hideki Hasegawa, Yasuko Asahi-Ozaki, Yuko Sato, Ayako Harashima, Shigeru Morikawa, Masayuki Saijo, Shigeyuki Itamura, Takehiko Saito, Takato Odagiri, Masato Tashiro, Yasushi Ami, and Tetsutaro Sata*

1. INTRODUCTION

Severe acute respiratory syndrome (SARS) is a recently identified emerging infectious disease caused by SARS-associated coronavirus (SARS-CoV). To determine the pathological features of SARS-CoV infection in experimental animals, its clinical, pathological, and virological features were investigated in cynomolgus monkeys, BALB/c mice, and F344 rats. The susceptibility of these animals to SARS-CoV infection was evaluated to identify suitable animal models for studies of the pathogenesis and treatment of SARS.

2. MATERIALS AND METHODS

The SARS-CoV, HKU39849 isolate was used in the present study.[1] The virus was propagated three times in Vero E6 cells, and the infectious doses of the virus stock were expressed as the 50% tissue culture infective dose ($TCID_{50}$) on these cells. Three-year-old male cynomolgus monkeys (*Macaca fascicularis*), 4-week-old female BALB/c mice, and 4-week-old F344 rats were used. Monkeys were inoculated intranasally with 10^3 or 10^6 $TCID_{50}$ of SARS-CoV in 3.5 ml of medium, or intratracheally with 10^8 $TCID_{50}$ in 5 ml of medium. BALB/c mice and F344 rats were inoculated intranasally with $2x10^6$ $TCID_{50}$ of SARS-CoV in 20 μl of medium and 10^7 $TCID_{50}$ in 100 μl of medium, respectively. After inoculation, these animals were observed for clinical symptoms and sacrificed for pathological examination. Virus isolation and viral infectivity titers were investigated in

*National Institute of Infectious Diseases, Musashimurayama, Tokyo 208-0011, Japan.

Vero E6 cell cultures. The SARS-CoV genome quantified by one-step reverse transcription and quantitative PCR assay using a LightCycler SARS-CoV quantification kit (Roche Diagnostics, Indianapolis, IN). All procedures in which infectious SARS-CoV was manipulated were carried out under biosafety level 3 conditions. The National Institute of Infectious Diseases Animal Care and Use Committee approved the animal studies to be carried out in an animal biosafety level 3 facility.

3. RESULTS

In monkeys, following intranasal inoculation with 10^6 $TCID_{50}$ of SARS-CoV, the virus was isolated from throat and nasal swabs, and the viral genome was detected in rectal swabs collected between 2 and 7 days postinoculation (p.i.). In one of the two monkeys inoculated intratracheally with 10^8 $TCID_{50}$ of SARS-CoV, virus and viral genome were detected in throat swabs collected on day 2 p.i. and in rectal swabs collected from days 4 to 7 p.i., respectively. Virus antigen-positive alveolar cells and macrophages were detected in the lower lobes of the lungs on day 7 p.i. in monkeys after intratracheal inoculation (Figure 1A). Angiotensin-converting enzyme 2 (ACE2, a receptor for SARS-CoV[2] antigen-positive cells were observed in the virus-infected area and were repairing swelled type II alveolar epithelium in the lung of monkeys (Figure 1A, B, and C). In BALB/c mice, the virus was detected in nasal and lung washes on days 3 and 5 after intranasal inoculation with 10^6 $TCID_{50}$ of SARS-CoV. Virus antigen was found in the epithelial cells in the lung alveolar and nasal cavities on day 3 as well as slight inflammatory reaction. In contrast, virus antigen was observed in the epithelial cells of intrapulmonary bronchi with inflammatory cells, including macrophages, in F344 rats after intranasal inoculation with 10^7 $TCID_{50}$ of SARS-CoV (Figure 1D). ACE2 antigen was observed in the epithelial cells of the respiratory tract in mock-infected F344 rats (Figure 1E). In BALB/c mice, ACE2 antigen-positive cells were not detected in non-infected areas (Figure 1F). In these experimental animals, ACE2 antigen-positive cells were observed in the virus-infected area and were repairing swelled type II alveolar epithelium (Table 1). In BALB/c mice and F344 rats, the virus was eliminated by day 7 p.i. None of the three experimental animals developed any clinical symptoms similar to SARS-like disease.

4. DISCUSSION

It was suggested that the infection and replication of SARS-CoV occurred in the respiratory tract of these experimental animals. However, there were differences in pathological findings, such as distribution of SARS-CoV and ACE2 antigen in the lungs between monkey, mouse, and rat (Table 1). The localization of infection with SARS-CoV was associated with the distribution of ACE2 antigen. These results indicated that ACE2 was used as a receptor for SARS-CoV infection in these animals. Furthermore, pathological dissimilarities in infectious and inflammatory reactions were observed between BALB/c mice and F334 rats. It was reported previously that monkeys were not suitable for studies of SARS.[3,4] Although none of the three types of experimental animal examined here developed any SARS-like symptoms, SARS-CoV could infect and replicate in these animals after intranasal or intratracheal inoculation. Therefore, these animal models are thought to be useful not only for studies on SARS-CoV infection and pathogenesis, but also for evaluation of novel vaccine and antiviral drugs against this virus.

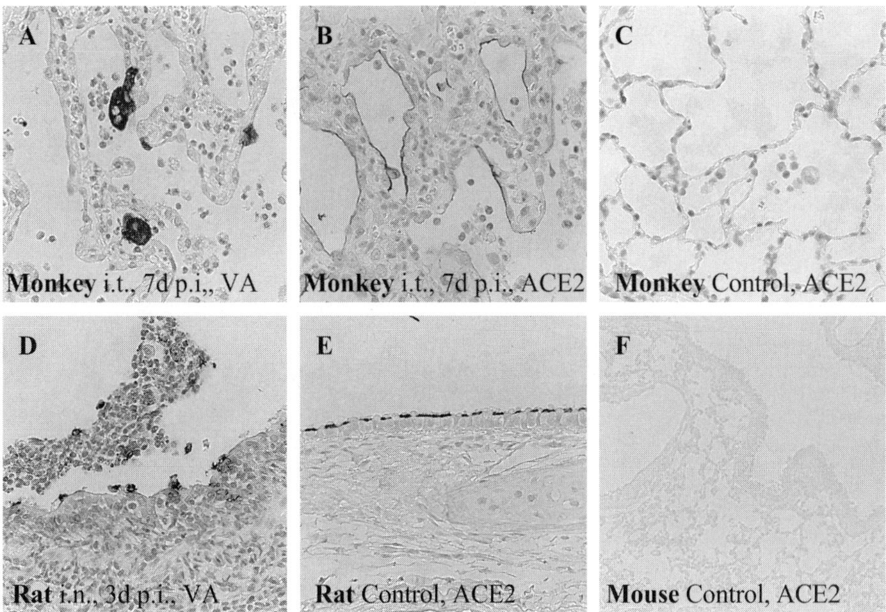

Figure 1. Distribution of SARS-CoV and ACE2 antigens in lung tissue determined using the immunoperoxidase method. VA, virus antigen. Original magnificat on: A–E, x400; F, x100.

Table 1. Distribution of SARS-CoV antigen and expression of ACE2 in experimental animals.

		Virus antigen			Expression of ACE2		
		Monkey	Mouse	Rat	Monkey	Mouse	Rat
Nasal cavity	Respiratory area	-	-	+	-	-	+
	Olfactory area	+	+	-	-	-	-
Trachea	Trachea	-	-	+	-	-	+
Lung	Intrapulmonary bronchi	-	-	+	-	-	+
	Bronchioles	-	-	+	-	-	+
	Alveoli	+	+	+	(+)*	(+)*	(+)*

* ACE2 antigen-positive cells were observed in the virus-infected area and were repairing swelled type II alveolar epithelium.

5. ACKNOWLEDGMENTS

We wish to thank Dr. Joseph S. M. Peiris, Department of Microbiology, the University of Hong Kong, for providing the HKU39849 strain of SARS-CoV. This work was supported by a Grant-in-Aid for Research on Emerging and Re-emerging Infectious Diseases from the Ministry of Health, Labour and Welfare, Japan, and a Grant-in-Aid for Scientific Research from the Ministry of Education, Culture, Sports, Science, and Technology, Japan.

6. REFERENCES

1. Peiris, J. S. M., Lai, S. T., Poon, L. L. M., Guan, Y., Yam, L. Y. C., Lim, W., Nicholls, J., Yee, W. K. S., Yan, W. W., Cheung, M. T., Cheng, V. C. C., Chan, K. H., Tsang, D. N. C., Yung, R. W. H., Ng, T. K., Yuen, K. Y., and members of the SARS study group, 2003, Coronavirus as a possible cause of severe acute respiratory syndrome, *Lancet* **361**:1319-1325.
2. Li, W., Moore, M. J., Vasilieva, N., Sui, J., Wong, S. K., Berne, M. A., Somasundaran, M., Sullivan, J. L., Luzuriaga, K., Greenough, T. C., Choe, H., and Farzan, M., 2003, Angiotensin-converting enzyme 2 is a functional receptor for the SARS coronavirus, *Nature* **426**:450-454.
3. McAuliffe, J., Vogel, L., Roberts, A., Fahle, G., Fischer, S., Shieh, W. J., Butler, E., Zaki, S., St Claire, M., Murphy, B., and Subbarao, K., 2004, Replication of SARS coronavirus administered into the respiratory tract of African Green, rhesus and cynomolgus monkeys, *Virology* **330**:8-15.
4. Rowe, T., Gao, G., Hogan, R. J., Crystal, R. G., Voss, T. G., Grant, R. L., Bell, P., Kobinger, G. P., Wivel, N. A., and Wilson, J. M., 2004, Macaque model for severe acute respiratory syndrome, *J. Virol.* **78**:11401-11404.

IDENTIFICATION OF FERRET ACE2 AND ITS RECEPTOR FUNCTION FOR SARS-CORONAVIRUS

Aya Zamoto, Fumihiro Taguchi, Shuetsu Fukushi, Shigeru Morikawa, and Yasuko K. Yamada*

1. INTRODUCTION

Severe acute respiratory syndrome associated coronavirus (SARS-CoV) was the causative agent of SARS, which occurred as an emerging pneumonic disease in 2002.[1-3] The epidemiological investigations showed that several wild animals such as Himalayan palm civet and raccoon dog had been infected by SARS-CoV.[4] Experimental infection to several laboratory animal species such as mouse, hamster, ferret, cat, and monkey revealed the susceptibility of those animal species to SARS-CoV infection,[5-7] however, severe clinical manifestations were observed only in the ferret.[8]

Human angiotensin-converting enzyme 2 (ACE2), a metallopeptidase, was demonstrated to be a functional receptor for SARS-CoV.[9] To see whether ACE2 function as receptor is attributed to the disease manifestation, we cloned a ferret ACE2 (feACE2) gene and compared the feACE2 with that of human or mouse ACE2 in terms of receptor functionality for SARS-CoV.

2. MATERIALS AND METHODS

Amplification of partial ACE2 gene by RT-PCR: RNAs were extracted from heart, lung, kidney, and small intestine of a ferret. To select a suitable organ to amplify a full-length ACE2 gene, we amplified a partial gene of ACE2 (952 bp in size) by RT-PCR. As a control, all RNA specimens were subjected to RT-PCR with ß-actin specific primers.

Determination of nucleotide sequence of ferret ACE2 (feACE2): Two overlapping regions of feACE2 genes were amplified by RT-PCR from kidney RNA. One contained the region 5' to the initiation codon of feACE2 gene, while the other was located 3' to a stop codon. The complete nucleotide sequence of feACE2 gene was determined by direct

* National Institute of Infectious Diseases, Musashimurayama, Tokyo 208-0011, Japan.

sequencing and was deposited in GenBank under accession number AB208708. Putative signal peptide, zinc binding, transmembrane and N-glycosylation sites were searched by PROSITE. We aligned entire ACE2 sequences of ferret, human (BC039902) and mouse (BC026801), and determined the identities between them by use of Multiple Alignment program in the CLUSTALW.

Expression of feACE2, human ACE2, and mouse ACE2 in HeLa cells: HeLa229 cells were transfected with a recombinant pTarget plasmid (Promega) encoding an entire ACE2 gene using Trans Fast Transfection Reagent (Promega). Transfected cells were selected in the presence of 600 μl/ml of Geneticin. Clones expressing high levels of ACE2 were selected by immunofluorescence assay (IFA) and Western blot with goat anti-human ACE2 antibody (R&D Systems).

SARS-CoV replication in ACE2-expressing HeLa229 cells: HeLa cell clones expressing ferret, human, or mouse ACE2 were inoculated with SARS-CoV (Frankfurt-1) at a multiplicity of infection of 0.1. At 24 hours postinoculation, cells were freeze-thawed and clarified by centrifugation and virus titer was quantified by plaque assay on Vero E6 cells. Standard errors of means were calculated from three experiments.

3. RESULTS

The partial feACE2 gene was amplified from RNAs isolated from the lung, heart, kidney, and small intestine by RT-PCR. The amount of amplicon varied among the organs (Fig. 1). An entire feACE2 gene was successfully amplified by RT-PCR from RNA isolated from the kidney. Its sequencing analysis showed that feACE2 consisted of 805 amino acid residues. There were 6 predicted N-glycosylation sites in the feACE2 (Table 1), and 4 of them (aa 53, 322, 546, 690) were conserved between ferret and human ACE2. Amino acid identities of ACE2 between ferret and human or mouse were 82.6% and 81.5%, respectively (Table 2).

We selected 3 HeLa cell clones stably expressing feACE2 by IFA (Fig. 2) and Western blot (data not shown). HeLa cells expressing feACE2 supported SARS-CoV replication to the same extent as those expressing human ACE2 (Fig. 3). Replication of SARS-CoV in feACE2 expressing HeLa cells was over 10 times more efficient than that in mouse ACE2 expressing HeLa cells (Fig. 3).

4. DISCUSSION AND CONCLUSIONS

We have determined the complete nucleotide sequence of feACE2 gene and established HeLa cell lines that stably express feACE2. Studies using these cells as well as cells expressing human or mouse ACE2 revealed that feACE2 functions as a SARS-CoV receptor as efficiently as human ACE2, while mouse ACE2 does less efficiently. Our findings shown in this study are in agreement with the observations by Li et al., who reported that human ACE2 is more efficient than mouse ACE2 in binding to the SARS-CoV S protein.[10] From these observations, it is postulated that animal species whose ACE2 functions as an efficient SARS-CoV receptor would be a susceptible host species for this virus. To establish a good animal model for SARS, a ferret model would be a good candidate. Furthermore, transgenic mouse expressing ferret or human ACE2 may serve a useful animal model for SARS-CoV infection.

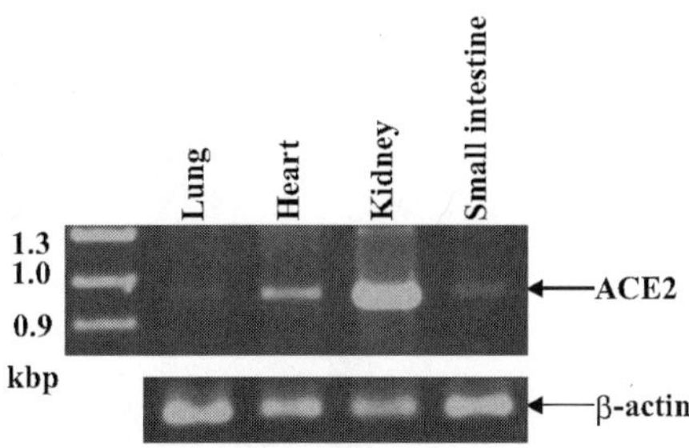

Figure 1. Amplification of ACE2 gene from heart, lung, kidney, and small intestine of ferret by RT-PCR.

Table 1. Predicted positions of motif on ferret ACE2.

Motif	Position
Signal peptide	1-17
Zinc binding	371-381
Transmembrane	741-763
N-glycosylation	6 sites (53, 299, 322, 546, 660, 690)

Table 2. Identity (%) of ferret ACE2 in amino acid sequence with human and mouse ACE2.

	Human	Ferret	Mouse
Human	-	82.6%	82.1%
Ferret		-	81.5%
Mouse			-

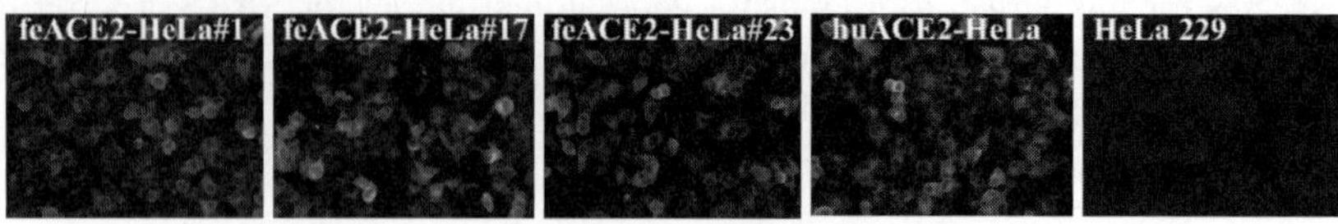

Figure 2. IFA analyses of feACE2 expression in HeLa cell clones.

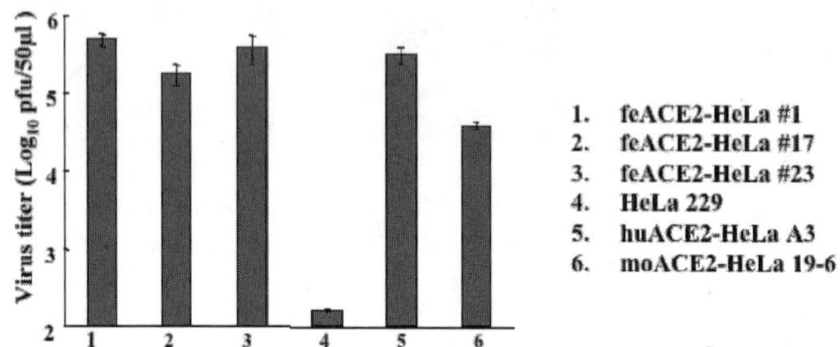

Figure 3. Replication of SARS-CoV in HeLa cells expressing ferret ACE2 (lanes 1-3), human ACE2 (lane 5), and mouse ACE2 (lane 6).

5. REFERENCES

1. T. G. Ksiazek, D. Erdman, C. S. Goldsmith, et al., A novel coronavirus associated with severe acute respiratory syndrome, *N. Engl. J. Med.* **348**, 1953-1966 (2003).
2. C. Drosten, S. Gunther, W. Preiser, et al., Identification of a novel coronavirus in patients with severe acute respiratory syndrome, *N. Engl. J. Med.* **348**, 1967-1976 (2003).
3. T. Kuiken, R. A. Fouchier, M. Schutten, et al., Newly discovered coronavirus as the primary cause of severe acute respiratory syndrome, *Lancet.* **362**, 263-270 (2003).
4. Y. Guan, B. J. Zheng, Y. Q. He, et al., Isolation and characterization of viruses related to the SARS coronavirus from animals in southern China, *Science* **302**, 276-278 (2003).
5. T. Rowe, G. Gao, R. J. Hogan, et al., Macaque model for severe acute respiratory syndrome, *J. Virol.* **78**, 11401-11404 (2004).
6. A. Roberts, L. Vogel, J. Guarner, et al., Severe acute respiratory syndrome coronavirus infection of golden Syrian hamsters, *J. Virol.* **79**, 503-511 (2005).
7. D. E. Wentworth, L. Gillim-Ross, N. Espina, and K. A. Bernard, Mice susceptible to SARS coronavirus, *Emerg. Infect. Dis.* **10**, 1293-1296 (2004).
8. B. E. Martina, B. L. Haagmans, T. Kuiken, et al., SARS virus infection of cats and ferrets, *Nature* **425**, 915 (2003).
9. W. Li, M. J. Moore, N. Vasilieva, et al., Angiotensin-converting enzyme 2 is a functional receptor for the SARS coronavirus, *Nature* **426**, 450-454 (2003).
10. W. Li, T. C. Greenough, M. J. Moore, et al., Efficient replication of severe acute respiratory syndrome coronavirus in mouse cells is limited by murine angiotensin-converting enzyme 2, *J. Virol.* **78**, 11429-11433 (2004).

HUMAN CORONAVIRUS-NL63 INFECTION IS NOT ASSOCIATED WITH ACUTE KAWASAKI DISEASE

S. C. Baker, C. Shimizu, H. Shike, F. Garcia, L. van der Hoek, T. W. Kuijper, S. L. Reed, A. H. Rowley, S. T. Shulman, H. K. B. Talbot, J. V. Williams, and J. C. Burns*

1. INTRODUCTION

Kawasaki disease (KD) is an acute, systemic vasculitis generally seen in early childhood. KD particularly affects medium size arteries, such as coronary arteries, and can result in coronary artery aneurysms that rupture, causing sudden death. The clinical and epidemiologic features of KD, with acute onset of fever, rash, conjunctival injection and cervical adenitis, and focal epidemics with wave-like spread of illness, are consistent with an infectious etiology (reviewed in Ref. 1). Indeed, current studies provide evidence that the IgA response in KD is targeting antigen in the bronchial epithelium and other inflamed tissues.[2,3] However, conventional methods have failed to identify the etiology of KD. Recently, Esper and colleagues reported an association between KD and a RNA virus they termed HCoV-New Haven (NH).[4] The limited sequence currently available indicates that HCoV-NH is highly similar to HCoV-NL63, which was initially described in 2004.[5] To determine if HCoV-NL63 was associated with KD, we established a multi-institutional collaborative study to test respiratory samples from KD patients using RT-PCR methods. We found that only 1 of 48 KD patients (2%) was positive for HcoV-NL63. Thus, these results indicate that respiratory tract infection with HcoV-NL63 is not associated with acute KD. A detailed description of all methods and results is provided in Ref. 6.

* S. C. Baker, Loyola University Chicago Stritch School of Medicine, Maywood, Illinois. C. Shimizu, H. Shike, S. L. Reed, J. C. Burns, University of California San Diego School of Medicine, La Jolla, California. F. Garcia, A. H. Rowley, S. T. Shulman, Northwestern University Feinberg School of Medicine, Chicago, Illinois. L. van der Hoek, T. W. Kuijper, Academic Medical Center, Amsterdam, The Netherlands. H. K. B. Talbot, J. V. Williams, Vanderbilt University Medical Center, Nashville, Tennessee.

2. HUMAN SUBJECTS, MATERIALS, AND METHODS

Two centers in the U.S. (Children's Hospital of San Diego, San Diego, CA; Children's Memorial Hospital, Chicago, IL) and one center in The Netherlands (Academic Medical Center (AMC), Amsterdam) collected respiratory samples from KD patients between December 2000 and March 2005. Seventy-seven percent of the samples were collected during the winter/spring months, which are seasons when HCoV is prevalent. Respiratory samples included throat swabs, nasopharyngeal (NP) swabs, scraped NP epithelial cells, and nasal washings were either archived or collected prospectively specifically for this study. Inclusion criteria for children with KD were more than five days of fever plus four of five standard clinical criteria (rash, conjunctival injection, cervical lymphadenopathy, changes in the extremities, changes in lips or oral mucosa) or three of five criteria with dilated coronary arteries by echocardiogram (z score > 2.5). The research protocol was reviewed and approved by the Institutional Review Boards of each institution. Informed consent was obtained from the parents of all subjects.

RT-PCR analysis was performed on RNA isolated from KD patient respiratory samples using primers and methods described in detail in Ref. 6. Degenerate primers were included to detect HCoV-NL63 variants. Samples were also tested using primers for human cellular mRNA sequences such as beta actin, to ensure the quality of the extracted RNA. A schematic diagram of the HCoV-NL63 genome and relative position of the 18 different primers sets used in this study is shown in Figure 1.

3. RESULTS AND DISCUSSION

We tested a total of 57 samples from 48 KD patients and found that only one of the 48 KD patients (2%) was positive for HCoV-NL63 RNA (Table 1). This patient met 4 of 5 classic clinical criteria for KD, but also exhibited symptoms of an upper respiratory tract infection, with cough and coryza which are rare symptoms for KD but common symptoms for HCoV-NL63 infection. Furthermore, although this patient responded with complete defervescence after administration of intravenous gamma globulin and aspirin that are common treatments for KD, his respiratory symptoms persisted. These results suggest that this KD patient was likely co-infected with HCoV-NL63.

Recent studies from Japan[7] and the Centers for Disease Control and Prevention in the USA[8] also report no association between infection with HCoV-NL63 and acute KD. Interestingly, a large study of pediatric patients in Europe found that HCoV-NL63 infection is associated with croup.[9] Thus, although HCoV-NL63 is likely a common respiratory infection in children, we and others found no association with acute KD.

4. CONCLUSION AND FUTURE DIRECTIONS

We found no association between the detection of HCoV-NL63 genome in the respiratory tract and acute KD. Future studies should continue to address the possibility of a microbe with a respiratory portal of entry as the etiologic agent of KD.

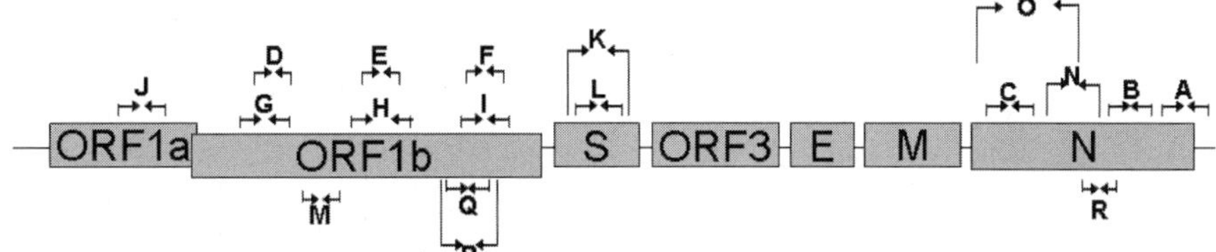

Figure 1. Schematic diagram of the organization of HCoV-NL63 genome and the locations of primers used for RT-PCR analysis. Shaded box are open reading frames. Primer sets are shown by arrow sets A–R. [Reproduced with permission from Shimizu *et al.*, "Human coronavirus NL63 is not detected in the respiratory tract of children with acute Kawasaki disease", The Journal of Infectious Diseases 2005; 192: 1767-1771].

Table 1. Results of HCoV RT-PCR on respiratory samples from acute Kawasaki disease patients.[a]

Location of sample collection	No. of patients samples	HCoV RT-PCR		Patient			Sample					
		Pos. sample	Primer sets[d]	Age median	Males	CAA[e]	Throat swab	NP swab	Scraped NP cells	Nasal wash	Illness Day[f] (median)	Samples post-IVIG
CA, USA	17 (23)	1	A-J	2m-9y (30 m)	15	1/6	8	7	8	0	3-15 (6.5)	2
IL, USA	13[b] (15)	0	J-O	4m-10y (38 m)	11	1/5	13	0	2	0	4-9 (6)	0
	12[c] (12)	0	A-C	4m-6y (12 m)	11	1/3	12	0	0	0	3-9 (6.5)	0
NL	7 (7)	0	P-R, J-L	5m-9y (35 m)	5	1/2	0	0	0	7	4-9 (7)	2
Total	48 (57)	1		2m-10y (31 m)	42	4/16	33	7	10	7	3-15 (7)	4

[a] Reproduced with permission from Shimizu *et al.*, "Human coronavirus NL63 is not detected in the respiratory tract of children with acute Kawasaki disease", The Journal of Infectious Diseases 2005; 192: (in press).

[b] Samples analyzed at Northwestern University and Loyola University, one sample was also tested at Vanderbilt University.

[c] Samples analyzed at Vanderbilt University, one sample was also tested at Northwestern University and Loyola University.

[d] Primer sets A-C, N, O, R = primer sets from HCoV-NL63 nucleocapsid protein gene, D-F, M, and P with nested primer st Q = primer sets from HCoV-NL63 ORF1b; J = primer sets from HCoV-NH ORF1a; K with nested primer set L = primer sets from HCoV-NH spike glycoprotein gene. G-I = degenerate primer sets from conserved regions of the ORF1b shared by HCoV-NL63, severe acute respiratory syndrome (SARS)-CoV (NC_004718), HCoV-E229 (NC_002645), and HCoV-OC43 (NC_005147).

[e] CAA = coronary artery abnormality, patients with aneurysms/ patients with dilatation (internal lumen z score>2.5.

[f] Illness Day 1 = first day of fever.

5. ACKNOWLEDGMENTS

We thank Drs. Ralph Baric, Michael Buchmeier, Benjamin Neuman, Ron Fouchier, and Christian Drosten for providing materials and technical advice. We also thank Joan Pancheri RN, UCSD, for assistance in collection of clinical samples. This work supported in part by grants from the National Institutes of Health (RO1 HL69413 to J.C.B., AI45798 to S.C.B., and HL63771 and HL67011 to A.H.R.).

6. REFERENCES

1. Rowley, A. H., 2004, Kawasaki Syndrome, in: *Krugman's Infectious Diseases of Children*, A. A. Gershon, P. J. Hotez, S. L. Katz, eds., Mosby, Philadelphia, pp. 323-335.
2. Rowley, A.H., Baker, S. C., Shulman, S. T., Garcia, F. L., Guzman-Cottril, J. A., Chou, P., Terai, M., Kawasaki, T., Kalelkar, M. B., and Crawford, S. E., 2004, Detection of antigen in bronchial epithelium and macrophages in acute Kawasaki disease by use of synthetic antibody, *J. Infect. Dis*. **190**:856-865.
3. Rowley, A.H., Baker, S. C., Shulman, S. T., Fox, L., Takahashi, K., Garcia, F. L., Crawford, S. E., Chou, P., and Orenstein, J. M., 2005, Cytoplasmic inclusion bodies are detected by synthetic antibody in acute Kawasaki disease ciliated bronchial epithelium, *J. Infect. Dis*. **192**: 1757-1766.
4. Esper, F., Shapiro, E. D., Weibel, C., Ferguson, D., Landry, M. L., and Kahn, J. S., 2005, Association between a novel human coronavirus and Kawasaki disease, *J. Infect. Dis*. **191**:499-502.
5. van der Hoek, L., Pyrc, K., Jebbink, M. F., Vermeulen-Oost, W., Berkhout, R. J. M., Wolthers, K. C., Wertheim-van Dillen, P. M. E., Kaandorp, J., Spaargaren, J., and Berkhout, B., 2004, Identification of a new human coronavirus, *Nat. Med.* **10**:368-373.
6. Shimizu, C., Shike, H., Baker, S. C., Garcia, F., van der Hoek, L., Kuijpers, T. W., Reed, S. L., Rowley, A. H., Shulman, S. T., Talbot, H. K. B., Williams, J. V., and Burns, J. C., 2005, Human coronavirus NL63 is not detected in the respiratory tract of children with acute Kawasaki disease, *J. Infect. Dis*. **192**: 1767-1771.
7. Ebihara, T., Endo, R., Ma, X., Ishiguro, N., and Kikuta, H., 2005, Lack of association between New Haven coronavirus and Kawasaki disease, *J. Infect. Dis.* **192**:351-352.
8. Belay, E. D., Erdman, D. D., Anderson, L. J., Peret, T. C. T., Schrag, S. J., Fields, B. S., Burns, J. C., and Schonberger, L. B., 2005, Kawasaki disease and human coronavirus, *J. Infect. Dis.* **192**:352-353.
9. van der Hoek, L., Sure, K., Ihorst, G., Stang, A., Pyrc, K., Jebbink, M. F., Petersen, G., Porster, J., B. Berkhout, B., and Uberla, K., 2005, Croup is associated with the novel coronavirus NL63, *PloS Med.* **2**:e240.

TOWARD THE DEVELOPMENT OF AN INFECTIOUS cDNA CLONE OF A HUMAN ENTERIC CORONAVIRUS

Hongqing Zhu, Yin Liu, Yingyun Cai, Dongdong Yu, Yinghui Pu, Laura Harmon, and Xuming Zhang*

1. INTRODUCTION

Coronaviruses (CoV) are a large group of plus-strand RNA viruses, which can cause respiratory, digestive, neurological, and immune-mediated diseases in human and animals. The outbreak of the severe acute respiratory syndrome (SARS) in 2003 was caused by SARS-CoV, which is most closely related to group II CoV, especially the bovine CoV (BCoV). Although the exact origin of SARS-CoV remains to be identified, the fact that similar SARS-CoV has been repeatedly isolated in wild animals, including civet cat, raccoon dog, and badges, strongly suggests that SARS-CoV originates from animals. It was postulated that SARS-CoV is a zoonotic pathogen.[1] Ironically, the zoonotic nature of CoV had been documented long before the outbreak of SARS. In 1994, Zhang et al.,[2] reported the isolation of a human enteric CoV (HECoV) from a 6-year old child with acute diarrhea. Subsequent characterization of the biologic and antigenic properties and partial sequence of the structural genes revealed that HECoV is more closely related to BCoV than to other members of the CoV family. Interestingly, several studies showed that BCoV was able to infect other small ruminants, turkey, and dogs and cause clinical diseases in these animals, thus further supporting the idea that BCoV is a zoonotic pathogen. Therefore, BCoV is the first documented CoV that can cross species barrier and infect animals ranging from avian, ruminant, carnivore, to humans. In the present study, we attempted to develop an infectious cDNA clone of HECoV by reverse genetics, which will provide a powerful genetic tool for future studies of the molecular evolution and animal to human transmission, and the molecular pathogenesis, and for the design of preventive and therapeutic interventions.

* University of Arkansas for Medical Sciences, Little Rock, Arkansas 72205.

2. MATERIALS AND METHODS

2.1. Cells and Virus

The human rectal tumor (HRT)-18 cells and baby hamster kidney (BHK) cells were maintained in Dulbecco's Minimum Essential Medium (DMEM) supplemented with 10% fetal bovine serum (FBS), which were used for virus propagation and RNA transfection, respectively. The human enteric coronavirus (HECoV) isolate 4408[2] was propagated in HRT-18 cells.

2.2. Viral RNA Isolation and Reverse Transcription–Polymerase Chain Reaction

For viral RNA isolation, HRT-18 cells were infected with HECoV at a multiplicity of infection (m.o.i.) of 5 in the presence of actinomycin D. At 48 h p.i., intracellular RNAs were isolated with Trizol reagent according to the manufacturer's protocol (Invitrogen). cDNA was synthesized by reverse transcription (RT) with Maloney murine leukemia virus (M-MLV) reverse transcriptase (Promega) using a primer specific to the nucleotide (nt) sequence at the 3'-end of HECoV nucleocapsid (N) gene[2] or a nonspecific oligo (dT) primer. Additional 20 pairs of primers were designed for RT and PCR amplification of the entire viral genome. The RT-PCR products were directly sequenced. cDNAs corresponding to the 5'-end of the viral genome was synthesized using the RNA-ligation-mediated rapid amplification of cDNA end (RLM-RACE) kit (Ambion) and was then sequenced directly. For cloning purpose, PCR was carried out using the Expand Long Taq polymerase (Roche) for 25 cycles. Each cycle consists of 94°C for 45 sec, 58~62°C for 45 sec, and 68°C for 3~4 min followed by a final extension at 68°C for 10 min.

2.3. Cloning of cDNA Fragments and DNA Sequencing

For molecular cloning, 9 pairs of primers encompassing the entire viral genome were designed for RT-PCR amplification. The strategy for designing the primers is illustrated in Fig. 1. A T7 promoter was inserted at the most 5'-end primer and a poly (A) tail of 25 As was included at the 3' end primer for the last fragment. An Xho I site was created in the 3' end primer of fragment A without changing the coding amino acid for distinguishing the molecular clone from wild-type virus. A unique BsmB I or Bsa I site was inserted at the terminus of each primer. RT-PCR products were isolated with the gel extract and purification kit (Qiagen) and cloned into Topo II TA vector (Invitrogen) or the pSMART vector (Lucigen).

2.4. DNA Ligation, *in Vitro* RNA Transcription, and RNA Transfection

Each cDNA fragment was released from plasmid DNAs by digestion with appropriate restriction enzymes, separated on 0.8% agarose gel by electrophoresis, and purified using Qiaex II gel extraction kit (Qiagen). The 9 fragments were then mixed at an equal molar ratio and ligated with T4 DNA ligase (New England Biolab) at 16°C overnight. The ligation products were purified by phenol-chloroform-isoamyl and chloroform extraction, and precipitated by ethanol.

In vitro RNA transcription was carried out using the mMESSAGE, mMACHINE T7 kit (Ambion) with slight modifications. The in vitro transcription products were then treated with RNase-free DNase I (Promega) at 37°C overnight, followed by extraction with phenol-chloroform-isoamyl and chloroform, and ethanol precipitation.

For transfection, a mixture of 80 µl of the full-length HECoV transcripts and 40 µl of the N gene transcripts was electroporated into BHK cell suspension (10^6 cell/ml) at 360 V, 80 µs in an electroporator (Eppendorf). Transfected BHK cells were seeded onto a dish with HRT-18 cells and the co-culture was incubated at 37°C under 5% CO2 for 3 days.

3. RESULTS AND DISCUSSION

3.1. Sequencing of the Complete HECoV Genome and Sequence Comparison with Other Related Coronaviruses

The sequence of the entire HECoV genome was determined by directly sequencing the RT-PCR products. In some instances, multiple independent RT-PCR products for the same fragments were sequenced. Thus, although quasispecies may exist in the viral populations, the sequence so determined must represent the vast majority of the viral genome in the RNA pools. Analysis of the sequence data revealed that the entire HECoV genome consists of 31,029-nt plus the poly(A) tail. The genome organization is similar to those of other CoV, with the ORFs in the order of 1a-1b-32kDa-HE-S-ORF5-6-7-E-M-N. The 5'-end of the genome is capped as inferred from the result with RLM-RACE and the penultimate nt is guanine. There is a 70-nt leader at the 5'-end, and a core consensus UCUAAAC at the 3'-end of the leader and the intergenic region proceeding each gene. Sequence comparison with other human CoV such as HCoV-229E,[3] HCoV-OC43,[4] SARS-CoV,[5] and a most recent isolate HCoV-NL63[6] and animal CoV revealed that HECoV-4408 is most closely related to BCoV[7] with a nt identity of 98%, whereas the nt identity with HCoV-OC43 is 95%. Overall, the homology with other three human CoV was less than 60%. The fact that HECoV is more closely related the animal BCoV than to other human CoV suggests that HECoV is likely derived from BCoV. This conclusion is consistent with the previous finding obtained from comparative studies on the biologic and antigenic relatedness among HECoV, HCoV-OC43 and BCoV.[2]

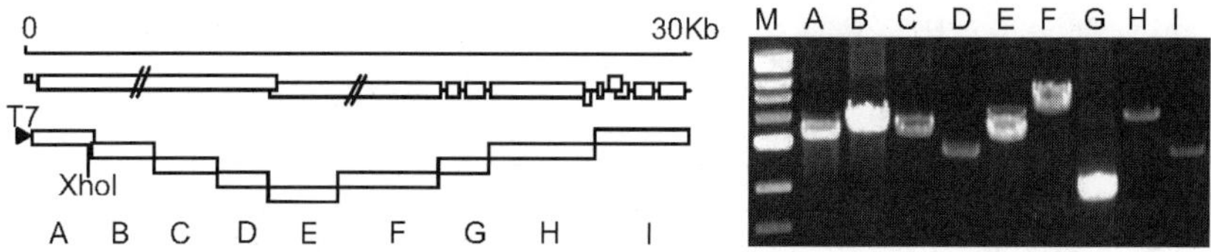

Figure 1. Assembling a full-length cDNA clone of HECoV. cDNA fragments A to I encompass the entire genome (≈32 kb) of HECoV with indicated open reading frames (left panel) were cloned into plasmids and were released by digestion with BsmB I or Bsa I (right panel). A unique T7 promoter was inserted at the 5' end of fragment A, an Xho I site was created in the fragment A. M, marker for 1 kb DNA ladder.

3.2. Assembly and Confirmation of a Full-Length HECoV cDNA Clone

To assemble a full-length HECoV cDNA, 9 fragments spanning the HECoV genome was cloned and sequenced. Each cDNA in the plasmids was released by digestion with 2 class II enzymes BsmB I and Bsa I as described,[8] generating a unique 4-nt overhang at both ends of each fragment. Consequently, only the two neighboring fragments can join together. Thus, a unidirectional assembly of a full-length cDNA was achieved by ligation of the 9 fragments *in vitro*. RNAs were then transcribed from the cDNAs *in vitro* with T7 RNA polymerase. Transcription of full-length RNAs was verified by RT-PCR with primers specific to the 3'-end of the viral genome (data not shown). To determine whether the in vitro transcribed, full-length RNA is infectious, we transfected the RNAs together with N gene transcripts into BHK cells by electroporation and co-cultured the BHK cells with HRT-18 cells for 3 days. Medium from the coculture was blindly passaged for several times in fresh HRT-18 cells. At 72 h p.i., intracellular RNA was isolated from HRT-18 cells and the synthesis of both genomic and subgenomic mRNAs were determined by RT-PCR using primers specific to genomic fragment A or to the 5'-end of N mRNA. Indeed, both fragments were amplified. Fragment A was digested with XhoI. Because an Xho I site was engineered specifically into fragment A, this indicates that the RNAs present in HRT-18 cells are derived from transfected RNA. In contrast, although a similar fragment A was amplified from HECoV-infected HRT-18 cells (which used as a positive control), it could not be digested with XhoI (data not shown). These data demonstrate that the full-length HECoV cDNA clone is infectious in HRT-18 cells. The generation of a full-length HECoV infectious cDNA clone will provide a powerful approach for genetic manipulation of the HECoV genome, for studying viral pathogenesis, and for developing potential vaccines.

4. ACKNOWLEDGMENTS

This work was supported by grants from the National Institutes of Health (AI 59244 and AI61204).

5. REFERENCES

1. Guan, Y., Zheng, B.J., He, Y.Q., et al., 2003, Isolation and characterization of viruses related to the SARS coronavirus from animals in southern China, *Science* **302**:276-278.
2. Zhang, X. M., Herbst, W., Kousoulas, K. G., and Storz, J., 1994, Biological and genetic characterization of a hemagglutinating coronavirus isolated from a diarrhoeic child, *J. Med. Virol.* **44**:152-161.
3. Thiel, V., Herold, J., Schelle, B., and Siddell, S. G., 2001, Infectious RNA transcribed in vitro from a cDNA copy of the human coronavirus genome cloned in vaccinia virus, *J. Gen. Virol.* **82**:1273-1281.
4. Vijgen, L., Keyaerts, E., Moes, E., et al., 2005, Complete genomic sequence of human coronavirus OC43: molecular clock analysis suggests a relatively recent zoonotic coronavirus, *J. Virol.* **79**:1595-1604.
5. Rota, P.A., Oberste, M.S., Monroe, S.S., et al., 2003, Characterization of a novel coronavirus associated with severe acute respiratory syndrome, *Science* **300**:1394-1399.
6. Van Der Hoek, L., et al., 2004, Identification of a new human coronavirus, *Nat. Med.* **10**:368-373.
7. Chouljenko, V.N., Lin, X.Q., Storz, J., Kousoulas, K.G., and Gorbalenya, A.E., 2001, Comparison of genomic and predicted amino acid sequences of respiratory and enteric bovine coronaviruses isolated from the same animal with fatal shipping pneumonia, *J. Gen. Virol.* **82**:2927-2933.
8. Boyd, Y., Denison, M., Weiss, R.W., and Baric, R.S., 2002, Systematic assembly of a full-length infectious cDNA of mouse hepatitis virus strain A59, *J. Virol.* **76**:11065-11078.

HCoV-OC43–INDUCED ENCEPHALITIS IS IN PART IMMUNE-MEDIATED

Noah Butler, Lecia Pewe, Kathryn Trandem, and Stanley Perlman*

1. INTRODUCTION

HCoV-OC43 and HCoV-229E are the etiological agents for the majority of coronavirus-induced upper respiratory tract infections in humans. HCoV-OC43 was originally isolated from human embryonic tracheal organ cultures; this virus was neurovirulent and caused disease after one passage in suckling mice and encephalitis within 2–4 passages (herein referred to as HCoV-OC43$_{NV}$).[1] This virus was then further adapted for growth in tissue culture cells ("tissue-culture adapted variant," referred to as HCoV-OC43$_{TC}$).

HCoV-OC43 showed increasing neurovirulence with passage through the murine brain; however, most recent studies have used viruses that have been propagated, at least for a few passages, in tissue culture cells. For example, Talbot and co-workers showed, using the mouse-adapted virus after passage in tissue culture cells 5–6 times, that mice infected intranasally with 10^4 to 10^5 TCID$_{50}$ developed encephalitis if inoculated at 8 days but not 21 days postnatally.[2]

For these reasons, we postulated that virus directly harvested from suckling mouse brain would be more virulent than virus passaged, even minimally, in tissue culture. In preliminary experiments, we observed that virus directly harvested from suckling mouse brains is highly virulent and readily caused a lethal infection after intranasal inoculation of adult, 8-week-old mice. When we evaluated the contribution of the host adaptive immune response to HCoV-OC43–induced encephalitis, we found that encephalitis is, in part, immune-mediated.

2. MATERIALS AND METHODS

Viruses and infection of mice: Mouse CNS-adapted (HCoV-OC43$_{NV}$) and tissue culture-adapted (HCoV-OC43$_{TC}$) strains of HCoV-OC43 (VR-759 and VR-1558, respectively) were obtained from the ATCC (Manassa, VA). Eight-week-old, pathogen-free male C57BL/6 mice were purchased from the National Cancer Institute (Bethesda,

*University of Iowa, Iowa City, Iowa 52242.

MD). Male and female, 6- to 8-week-old RAG1$^{-/-}$ mice were obtained from breeding colonies maintained by our laboratory. For intranasal infection, mice were lightly anesthetized with halothane and droplets containing 10^7 SMLD$_{50}$ of HCoV-OC43$_{NV}$ or 10^6 TCID$_{50}$ of HCoV-OC43$_{TC}$ were administered to the nares. Mice were monitored for weight loss and survival after infection. All procedures used in this study were approved by the University of Iowa Institutional Animal Care and Use Committee.

Flow cytometry: Single cell suspensions of mononuclear cells from whole brain homogenates were prepared as previously described.[3] Fc receptors were blocked with normal rat serum and anti-CD16/CD32 (clone 2.4G2, BD Biosciences, San Jose, CA). Antibodies used to phenotype cells were fluorescein isothiocyanate-labeled anti-mouse CD4 and phycoerythrin-labeled anti-mouse CD8 (clones GK1.5 and 53-6.7, BD Biosciences, Mountain View, CA). Samples were analyzed on a FACScan flow cytometer (BD Biosciences, Mountain View, CA).

Adoptive transfer: Red blood cell depleted splenocytes were isolated from C57BL/6 mice 7 days after interperitoneal immunization with HCoV-OC43$_{NV}$. 1 x 10^6 HCoV-OC43$_{NV}$-immune splenocytes were adoptively transferred via retroorbital injection to RAG1$^{-/-}$ mice that had been inoculated intranasally with HCoV-OC43$_{NV}$ 4 days previously. Recipient RAG1$^{-/-}$ mice were continuously monitored for weight loss and survival following infection. As a control, some RAG1$^{-/-}$ mice received no cells after intranasal inoculation of virus.

3. RESULTS

3.1. Intranasal Inoculation of 8-Week Mice with HCoV-OC43$_{NV}$ Is Uniformly Fatal

Intranasal inoculation of HCoV-OC43$_{NV}$ resulted in 100% mortality in mice ranging from 5 weeks old (data not shown) to 8 weeks old (Figure 1A). Mice developed signs of acute encephalitis, including hunched posture, lethargy, and wasting by day 7–9 (data not shown). Mortality was associated with a ~35% loss of body mass (Figure 1B). Severe clinical encephalitis was also associated with widespread mononuclear cell infiltration including perivascular cuffing and with loss of CNS architecture (data not shown). In contrast, intranasal inoculation of 8-week-old C57BL/6 mice with the HCoV-OC43$_{TC}$ was not fatal and did not cause any clinical disease, including any weight loss (Figure 1).

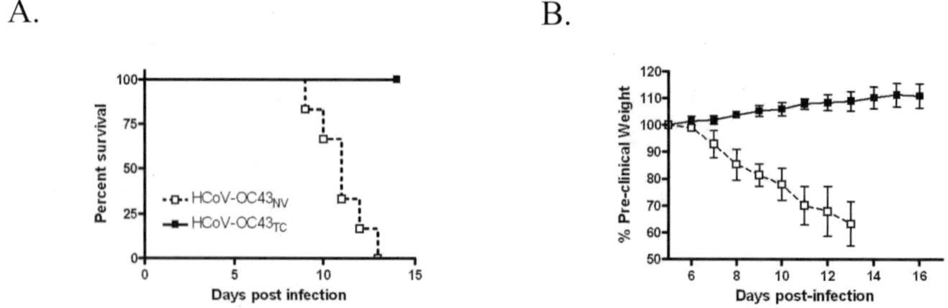

Figure 1. HCoV-OC43–induced lethal encephalitis in 8-week-old mice. Mice were inoclulated intranasally with HCoV-OC43$_{NV}$ or HCoV-OC43$_{TC}$ and monitored for survival (A) and weight loss (B). Data represent 6 mice per group. For B, data are expressed as mean +/- SD.

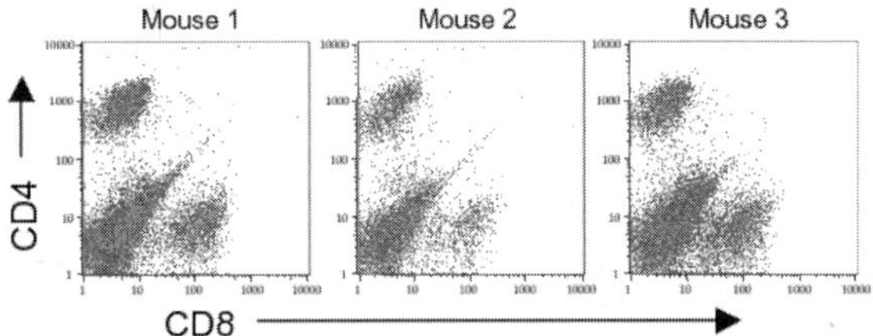

Figure 2. CD4 and CD8 T cells infiltrate the HCoV-OC43$_{NV}$–infected CNS. 7 days postinfection, mononuclear cells were prepared from brain homogenates, stained for CD4 and CD8, and analyzed with flow cytometry. Samples from 3 representative mice are shown.

3.2. Encephalitis Is Partly Immune-Mediated

As described above, HCoV-OC43$_{NV}$–induced encephalitis was associated with a large infiltration of mononuclear cells into the brain parenchyma. When we immunophenotyped CNS infiltrates from HCoV-OC43$_{NV}$–infected mice, we found that a large proportion of the infiltrating mononuclear cells was comprised of CD4 and CD8 T cells (Figure 2). When we examined viral titers and RNA burden in the CNS, we found that virus was in the process of clearance at the time of death (data not shown), suggesting that the host immune response both cleared virus and contributed to a fatal outcome.

To probe the role of this T-cell response in pathogenesis, we infected immunodeficient mice lacking normal T- and B-cell responses (RAG1$^{-/-}$) and monitored these mice for weight loss and survival. HCoV-OC43$_{NV}$–infected RAG1$^{-/-}$ mice lost weight and developed signs of encephalitis (lethargy, hunching, weight loss) similar to those observed in infected wild-type mice, but with delayed kinetics. Moreover, infected RAG1$^{-/-}$ mice also survived longer than did their B6 counterparts (data not shown).

To confirm the pathological role of T cells, we adoptively transferred HCoV-OC43–immune splenocytes to RAG1$^{-/-}$ mice that had been infected with 10^7 SMLD$_{50}$ HCoV-OC43$_{NV}$ intranasally 4 days earlier. The adoptive transfer of HCoV-OC43–immune splenocytes to RAG1$^{-/-}$ mice hastened the onset of clinical disease and death (Figure 3).

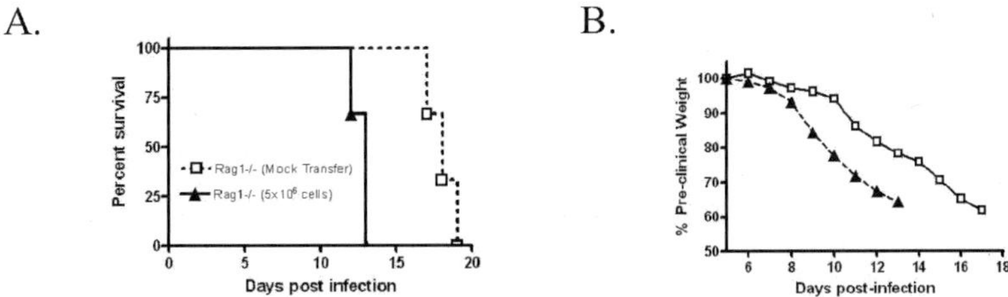

Figure 3. Adoptive transfer of HCoV-OC43$_{NV}$–immune splenocytes hastens the onset of mortality and morbidity in recipient RAG1$^{-/-}$ C57BL/6 mice. RAG1$^{-/-}$ mice were infected intranasally with HCoV-OC43$_{NV}$ 4 days prior to serving as recipients of the adoptive transfer of HCoV-OC43–immune splenocytes isolated from wild type mice. Mice were monitored for survival (A) and weight loss (B).

4. DISCUSSION

These data show that HCoV-OC43 that has been exclusively passaged in suckling mouse brain is highly neurovirulent relative to other strains reported to only cause disease in 1- to 3-week-old mice. These latter strains of HCoV-OC43, which have been through 5–6 passages in HRT-18 cells (Dr. Pierre Talbot, personal communication), are likely less virulent than virus isolated directly from infected suckling mouse brains because coronaviruses often become attenuated after passage *in vitro*.[4]

Our data also show that HCoV-OC43-induced encephalitis is in part mediated by the anti-viral T-cell response. During the course of our investigation we found that HCoV-OC43 infection appears restricted to neurons (data not shown). Neurons do not normally express MHC class I or II antigen and express only low levels of the machinery required for loading peptide onto MHC class I antigen.[5] Therefore, neurons generally do not serve as targets for activated T- cells. However, electrically silent or damaged neurons do express MHC class I antigen[6-8] and it is possible that infection with HCoV-OC43 makes neurons into suitable targets for CD8 T cells.

Together, these data suggest that for some human coronavirus infections, such as with HCoV-OC43, the ensuing pathology may often include an immune-mediated component. Future studies will be directed at determining how the anti-viral T cell response, while important for virus clearance, also contributes to more severe disease. These studies may also be relevant to understanding disease outcome in patients with SARS, since neurons are infected in some patients.[9]

This work was supported in part by a grant from the National Institutes of Health (AI60699).

5. REFERENCES

1. K. McIntosh, W. B. Becker, and R. M. Chanock, Growth in suckling-mouse brain of "IBV-like" viruses from patients with upper respiratory tract disease, *Proc. Natl. Acad. Sci. USA* **58**, 2268-2273 (1967).
2. H. Jacomy and P. J. Talbot, Vacuolating encephalitis in mice infected by human coronavirus OC43, *Virology* **315**, 20-33 (2003).
3. L. Pewe, H. Zhou, J. Netland, C. Tangadu, H. Olivares, L. Shi, D. Look, T. M. Gallagher, and S. Perlman, A severe acute respiratory syndrome-associated coronavirus-specific protein enhances virulence of an attenuated murine coronavirus, *J. Virol.* **79**, 11335-11342 (2005).
4. D. K. Krueger, S. M. Kelly, D. N. Lewicki, R. Ruffolo, and T. M. Gallagher, Variations in disparate regions of the murine coronavirus spike protein impact the initiation of membrane fusion, *J. Virol.* **75**, 2792-2802 (2001).
5. E. Joly and M. B. Oldstone, Neuronal cells are deficient in loading peptides onto MHC class I molecules, *Neuron* **8**, 1185-1190 (1992).
6. H. Neumann, A. Cavalie, D. Jenne, and H. Wekerle, Induction of MHC class I genes in neurons, *Science* **269**, 549-552 (1995).
7. H. Neumann, H. Schmidt, A. Cavalie, D. Jenne, and H. Wekerle, Major histocompatibility complex (MHC) class I gene expression in single neurons of the central nervous system: differential regulation by interferon (IFN)-gamma and tumor necrosis factor (TNF)-alpha, *J. Exp. Med.* **185**, 305-316 (1997).
8. I. M. Medana, A. Gallimore, A. Oxenius, M. M. Martinic, H. Wekerle, and H. Neumann, MHC class I-restricted killing of neurons by virus-specific CD8+ T lymphocytes is effected through the Fas/FasL, but not the perforin pathway, *Eur. J. Immunol.* **30**, 3623-3633 (2000).
9. J. Gu, E. Gong, B. Zhang, et al., Multiple organ infection and the pathogenesis of SARS, *J. Exp. Med.* **202**, 415-424 (2005).

SARS CoV REPLICATION AND PATHOGENESIS IN HUMAN AIRWAY EPITHELIAL CULTURES

Amy C. Sims, Boyd Yount, Susan E. Burkett, Ralph S. Baric, and Raymond J. Pickles*

1. INTRODUCTION

The importance of human coronaviruses (HCoV) as pathogens that produce severe human respiratory diseases has been greatly emphasized with the identification of the SARS-CoV, and relevant model systems are needed to elucidate the underlying molecular mechanisms governing coronavirus pathogenesis and virulence in the human lung. SARS-CoV infection is an attractive model for HCoV infection as it produces severe disease in the human lung, replicates efficiently *in vitro*, a molecular clone is available to identify the genetic determinants governing pathogenesis and virulence, and a variety of animal models are under development.[1-5] Here, we test the ability of SARS-CoV to infect an *in vitro* model of human airway epithelium (HAE) that recapitulates the morphological and physiological features of the human airway *in vivo* to determine whether infection and spread of SARS-CoV throughout the ciliated conducting airway may be a valid model for understanding the pathogenesis of SARS-CoV lung disease.

2. RESULTS AND DISCUSSION

To directly observe the extent and kinetics of SARS-CoV infection of HAE in real time, we constructed a recombinant, green fluorescent protein expressing SARS-CoV (SARS-CoV GFP). To generate recombinant SARS-CoV GFP, the F plasmid was mutated to replace ORF7a/b with the GFP cDNA as described previously.[6] Following transfection of Vero E6 cells, GFP-positive cells were detected within 24 hours. Plaque purified virus replicated efficiently and produced CPE in several cell lines, replicating to titers of $1x10^7$ pfu/mL, similar titers were detected with wild-type strains Urbani and the infectious clone construct, icSARS-CoV. ORF 7a/7b deletion and replacement with GFP was not detrimental to SARS-CoV replication, as the wild-type strains Urbani and icSARS-CoV and the SARS-CoV GFP recombinant synthesized equivalent levels of

* University of North Carolina at Chapel Hill, Chapel Hill, North Carolina 27599.

Table 1. SARS-CoV replication in HAE at 48 hours postinfection.

Virus strain	Viral titers (pfu/mL) 50 hours pi	
	Apical	Basolateral
Urbani	2.3×10^6	6×10^4
icSARS-CoV	6×10^6	4.8×10^4
SARS-CoV GFP	7.5×10^5	1.8×10^4

subgenomic RNA (data not shown). The replacement of ORF7a/7b caused the expected shifts in size of subgenomic RNAs 2 through 7 (data not shown). The deletion of ORF7a/7b did not obviously affect efficient SARS-CoV replication in tissue culture, similar to observations with transmissible gastroenteritis virus (TGEV) and mouse hepatitis virus (MHV) GFP viruses,[7-10] thus providing a fluorescent marker of virus infection with replication at wild-type virus levels.

To determine whether SARS-CoV GFP could infect human airway epithelial cells (HAE), we prepared cultures of human tracheobronchial ciliated epithelium. As a model of virus entry into the lumen of the airways we inoculated the apical surface of these cultures with SARS-CoV GFP (10^6 pfu) and assessed GFP fluorescence 48 hours later. HAE were efficiently infected by SARS-CoV GFP with a significant proportion of the cells expressing the marker transgene (Fig. 1 A). These data demonstrate that the human airway epithelium that lines the conducting proximal airways is susceptible to infection by SARS-CoV GFP.

To determine whether shedding of progeny Urbani, icSARS-CoV, or SARS-CoV GFP from HAE was polarized, apical washes and basolateral media were sampled at 48 hours postinfection and viral titers assessed by plaque assay on Vero E6 cells. The peak titers shed from the apical surface exceeded 10^6 pfu/mL, demonstrating a high level of replication similar to that observed in Vero E6 cell monolayers (Table 1). In contrast, viral titers in the basolateral compartments were low with peak titers of 10^4 pfu/mL (Table 1). Because SARS-CoV replicates to similar titers in permissive cell-lines, these data indicate that SARS-CoV replicates well in HAE providing a new model of the human lung for the study of HCoV replication and pathogenesis.

Following transmission electron microscopy of HAE 48 hours postinfection with Urbani, icSARS-CoV, or SARS-CoV GFP, only ciliated cells of the HAE contained classic coronavirus cytoplasmic vesicles filled with viral particles (Fig. 1 B). In addition, large numbers of viral particles were seen within the spaces between the microvilli/cilia shafts as well as in the airway surface liquid microenvironment that surrounds the ciliated cells suggesting mechanisms for the release of large quantities of SARS-CoV into the lumen of the conducting airway during viral replication. SARS-CoV entry, replication, and release occur primarily in the ciliated cells of the HAE.

To determine if SARS-CoV infects ciliated cells via an interaction with hACE2 we performed an antibody blockade experiment using antisera directed against hACE2, a method that has previously been shown to block the interaction of SARS-CoV with hACE2 in Vero E6 cells.[11] HAE were pre-incubated with polyclonal or monoclonal antisera against hACE2 (R&D Systems), or a control antibody that binds to the apical surface of HAE (anti-tethered mucin MUC1, b27.29) for 2 hours prior to inoculation

Table 2. SARS-CoV replication in HAE is blocked by ACE2 specific antisera.

Preinfection treatment	Viral titers 30 hours pi (pfu/mL)
No antisera	1.4×10^6
Control antisera (anti-MUC1)	2.1×10^6
Monoclonal ACE2	3.3×10^6
Polyclonal ACE2	1.5×10^4
Polyclonal ACE2 + monoclonal ACE2	7.3×10^3

with SARS-CoV GFP (10^6 pfu/culture). Apical surface sampling was performed from 2 to 36 hours post infection, viral growth kinetics assessed by plaque assay and representative titers at 36 hours post infection are shown. No inhibition of infection was observed with a control antibody that binds to a highly abundant epitope on the apical surface of HAE (MUC1).[12] In the absence of antisera or in the presence of control antisera, SARS-CoV GFP replicated to titers of 10^7 pfu/mL, similar to titers detected with the wild-type strains Urbani and icSARS-CoV (Table 2). In contrast, in the presence of hACE2 polyclonal antisera alone or in combination with monoclonal antisera, viral titers were reduced by at least 2 logs. Monoclonal antisera against hACE2 failed to effect viral growth confirming that this antibody was not sufficient to block SARS-CoV entry into ciliated cells. These data suggest that hACE2 is the predominant receptor mediating SARS-CoV entry into ciliated cells in HAE.

We have generated a recombinant clone of the Urbani strain of SARS-CoV that expresses the green fluorescent protein (SARS-CoV GFP) to monitor infection in real time. In HAE SARS-CoV exclusively infects ciliated airway epithelial cells resulting in progeny virus being shed onto the airway lumen. In addition, infection of ciliated airway epithelial cells occurred via an interaction with hACE2 and correlated with the airway distribution of hACE2 on the apical surface of ciliated cells. Although progeny virus was initially shed into the lumenal compartment of the epithelium, at later times post-infection virus was also shed into basolateral compartments. These data support the hypothesis that the conducting airways in the upper respiratory tract might represent the primary site of SARS-CoV replication with subsequent spread to lower compartments by ciliary action and to other organs via viremic spread by disruption of tight junctions. Since ciliated airway epithelial cells possess unique physiological and innate defense functions in the human lung (e.g., mucociliary clearance), it is important to identify the ciliated cell tropism of SARS-CoV and the pathological consequences of infection of these cells.

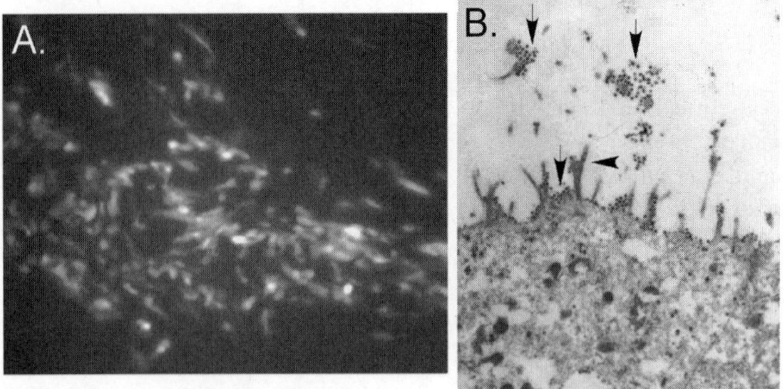

Figure 1. SARS-CoV infection in human airway epithelial cells. A. GFP expression at 48 hours postinfection in SARS-CoV GFP infected HAE. B. Transmission electron micrograph of SARS-CoV infected HAE at 48 hours postinfection. Arrows denote virus at the cell surface adjacent to microvilli and above the infected cell at the air-liquid interface.

3. REFERENCES

1. B. E. Martina, B. L. Haagmans, T. Kuiken, et al., Virology: SARS virus infection of cats and ferrets, *Nature* **425**, 915 (2003).
2. A. D. Osterhaus, R. A. Fouchier, and T. Kuiken, The aetiology of SARS: Koch's postulates fulfilled, *Philos. Trans. R. Soc. Lond. B Biol. Sci.* **359**, 1081-1082 (2004).
3. A. Roberts, C. Paddock, L. Vogel, E. Butler, S. Zaki, and K. Subbarao, Aged BALB/c mice as a model for increased severity of severe acute respiratory syndrome in elderly humans, *J. Virol.* **79**, 5833-5838 (2005).
4. A. Roberts, L. Vogel, J. Guarner, et al., Severe acute respiratory syndrome coronavirus infection of golden Syrian hamsters, *J. Virol.* **79**, 503-511 (2005).
5. B. Yount, K. M. Curtis, E. A. Fritz, et al., Reverse genetics with a full-length infectious cDNA of severe acute respiratory syndrome coronavirus, *Proc. Natl. Acad. Sci. USA* **100**, 12995-13000 (2003).
6. B. Yount, M. R. Denison, S. R. Weiss, and R. S. Baric, Systematic assembly of a full-length infectious cDNA of mouse hepatitis virus strain A59, *J. Virol.* **76**, 11065-11078 (2002).
7. R. S. Baric and A. C. Sims, Development of mouse hepatitis virus and SARS-CoV infectious cDNA constructs, *Curr. Top. Microbiol. Immunol.* **287**, 229-252 (2005).
8. K. M. Curtis, B. Yount, and R. S. Baric, Heterologous gene expression from transmissible gastroenteritis virus replicon particles, *J. Virol.* **76**, 1422-1434 (2002).
9. F. Fischer, C. F. Stegen, C. A. Koetzner, and P. S. Masters, Analysis of a recombinant mouse hepatitis virus expressing a foreign gene reveals a novel aspect of coronavirus transcription, *J. Virol.* **71**, 5148-5160 (1997).
10. I. Sola, S. Alonso, S. Zuniga, M. Balasch, J. Plana-Duran, and L. Enjuanes, Engineering the transmissible gastroenteritis virus genome as an expression vector inducing lactogenic immunity, *J. Virol.* **77**, 4357-4369 (2003).
11. W. Li, M. J. Moore, N. Vasilieva, et al., Angiotensin-converting enzyme 2 is a functional receptor for the SARS coronavirus, *Nature* **426**, 450-454 (2003).
12. J. R. Stonebraker, D. Wagner, R. W. Lefensty, et al., Glycocalyx restricts adenoviral vector access to apical receptors expressed on respiratory epithelium in vitro and in vivo: role for tethered mucins as barriers to lumenal infection, *J. Virol.* **78**, 13755-13768 (2004).

IMMUNOGENICITY OF SARS-CoV: THE RECEPTOR-BINDING DOMAIN OF S PROTEIN IS A MAJOR TARGET OF NEUTRALIZING ANTIBODIES

Yuxian He*

1. INTRODUCTION

The global emergency of severe acute respiratory syndrome (SARS) was caused by a new coronavirus (SARS-CoV) within the family *Coronaviridae.*[1] Similar to other CoVs, SARS-CoV is an enveloped virus containing a large, positive-stranded RNA genome that encodes viral replicase proteins and four major structural proteins including spike (S), membrane (M), envelope (E), and nucleocapsid (N). All these proteins and other uncharacterized components can induce immune responses during viral infection. Therefore, post-genomic characterization of the SARS-CoV is highly important for understanding its pathogenesis and for developing diagnostics, therapeutics, and vaccines.

The S protein of SARS-CoV is a type I transmembrane glycoprotein containing putative S1 and S2 subunits. A fragment located in the middle region of the S1 subunit (residues 318–510) has been characterized as a minimal receptor-binding domain (RBD), which is sufficient to associate with angiotensin-converting enzyme 2 (ACE2), a functional receptor on targeted cells.[2] The S2 subunit containing a putative fusion peptide and two heptad repeat (HR1 and HR2) regions is responsible for fusion between viral and target cell membranes.[3] Therefore, SARS-CoV infection is initiated by attachment of the S protein via its RBD to the specific receptor on target cells, then forming a fusogenic core between the HR1 and HR2 regions in the S2 domain to bring the viral and target cell membrane into close proximity, resulting in virus fusion and entry.

A second major property of SARS-CoV S protein is that it induces neutralizing antibodies and protective immunity, thereby is a candidate for the development of an effective vaccine.[4,5] A DNA vaccine encoding the S protein can induce SARS-CoV neutralization and protective immunity in mice.[6] Vaccination of animals with recombinant viruses, such as attenuated vaccinia virus (MVA) and parainfluenza virus (BHPIV3), that express S protein can elicit neutralizing antibodies that protect animals against SARS-CoV challenge.[7,8] Infection by pseudovirus expressing the SARS-CoV S protein can be effectively neutralized by convalescent sera from SARS patients.[9] These

*New York Blood Center, New York, New York 10021.

data suggest that the S protein of SARS-CoV is a protective antigen, although its antigenic properties have not been well defined. We have recently demonstrated that the RBD of SARS-CoV S protein is a major target of neutralizing antibodies during viral infection and immunization,[10-14] and suggested its potential application as a major target for SARS vaccines and immunotherapeutics.

2. IDENTIFICATION OF NON-NEUTRALIZING EPITOPES IN S PROTEIN

The S protein of SARS-CoV induces antibody responses in infected patients and in mice and rabbits immunized with the inactivated SARS-CoV.[11,12] To define its immuno-epitopes, a set of 168 peptides spanning the entire sequence of the S protein of SARS-CoV strain TOR2 (each peptide contains 17 amino acid residues with 9 residues overlapping with the adjacent peptides) were synthesized and used in the Pepscan analyses against the convalescent sera from SARS patients. It revealed that the S protein contained five linear immunodominant sites corresponding to the sequences of residues 9–71, 171–224, 271–318, 528–635, and 842–913 (designated as sites I to V). The immunodominant site I-III, and V reacted with more than 50% of the convalescent sera from SARS patients, and the site IV was reactive with more than 80% of SARS sera. All of 168 overlapping peptides were also used as probes to localize the immunodominant epitopes of S protein in the SARS-CoV–immunized mice and rabbits. Interestingly, only three of them (536–552, 544–560 and 603–619) reacted significantly with the mouse antisera, and all of them reside within the immunodominant site IV. Similarly, the peptide 536–552 corresponding to the N-terminal sequence of the site IV reacted with all the rabbit antisera. Therefore, the major linear immunodominant domain (the site IV) induces antibody response not only in humans, but also in mice and rabbits. To investigate whether these immunodominant epitopes induced neutralizing antibodies, we designed and synthesized 5 longer peptides that overlap the immunodominant sites based on the above finding. Peptide $S_{19\text{-}48}$ and $S_{278\text{-}312}$ were derived from the immunodominant sites I and III, respectively, while the peptides $S_{511\text{-}552}$, $S_{536\text{-}566}$ and $S_{603\text{-}634}$ overlapped with sequence of the major immunodominant site IV. All longer peptides (without conjugation to carrier) were able to elicit high titers of antibodies. However, none of rabbit antisera had neutralizing activity against SARS-CoV or SARS pesudovirus, suggesting that these linear immunodominant sites are not neutralizing epitopes.

3. IDENTIFICATION OF S PROTEIN RBD AS A MAJOR TARGET OF NEUTRALIZING ANTIBODIES

We found that convalescent sera from SARS patients contained high titers of RBD-specific antibodies.[14] To identify neutralizing epitopes of SARS-CoV S protein, a fusion protein containing the RBD (residues 318–510) linked to human IgG1 Fc fragment (designated RBD-Fc) was used to isolate RBD-specific antibodies by immunoaffinity chromatography from patient antisera.[14] The efficiencies of depletion and recovery of the RBD-specific antibodies were monitored by measuring the reactivity of the starting sera, the corresponding flowthroughs and eluted antibody fractions by ELISA against the RBD and S1 subunit. Neutralizing activities of the samples of the starting sera, flowthroughs and eluates were determined using SARS pseudovirus system. Strikingly, the neutralizing activity of immune sera was dramatically reduced after depletion of anti-RBD antibodies, while anti-RBD antibodies in the eluates possessed higher potency than the antibodies in the flowthroughs to neutralize SARS pseudovirus, suggesting that more than 50% neutralizing activity in the antisera was contributed by the RBD-specific antibodies.

We further observed that the RBD of S protein was a major target of neutralizing antibodies in mice and rabbits immunized with an inactivated SARS-CoV vaccine. All mice and rabbits developed high titers of antibodies against the S protein and its RBD after three immunizations.[11] The mean end-point titers of mouse and rabbit antisera to the RBD were 1:51,200 and 1:25,600, respectively. Both mouse and rabbit antisera could efficiently block binding of RBD-Fc to ACE2. By using pseudovirus bearing the SARS-CoV S protein, we demonstrated that both mouse and rabbit antisera significantly inhibited S protein-mediated virus entry with mean 50% inhibitory titers of 1:7,393 and 1:2,060, respectively. To further elucidate neutralization determinants of the SARS-CoV, the RBD-specific antibodies were isolated by immunoaffinity chromatography from the rabbit antisera. Similarly, the reactivity of anti-RBD antibodies in the rabbit antisera could be efficiently depleted by RBD-Fc affinity column, as the flowthroughs did not bind to RBD-Fc while the eluted anti-RBD antibodies significantly reacted with the RBD-Fc. The neutralizing activity of anti-RBD antibody-depleted rabbit antisera was much lower than the starting sera, while anti-RBD antibodies possessed more potent activity to neutralize SARS pseudovirus. These data suggest that the RBD of S protein is an effective inducer of neutralizing antibodies in immunized animals.

4. THE RBD OF S PROTEIN CONTAINS MULTIPLE CONFORMATIONAL EPITOPES THAT INDUCE POTENT NEUTRALIZING ANTIBODIES

To evaluate whether the RBD can serve as an effective vaccine, the RBD-Fc protein was used as an immunogen to immunize mice and rabbits.[10, 13] All animals (mice and rabbits) developed robust antibody responses against RBD-Fc after boosting. The antisera contained high titers of antibodies specific for the RBD as shown by ELISA using S1 subunit as an antigen. The mean endpoint titers of mouse and rabbit antisera to the S1 protein were 1:625,000 and 1:312,500, respectively. The antisera were further tested for their neutralizing activity using two different assay systems, i.e., infection of SARS-CoV in Vero E6 and of SARS pseudovirus in 293T cells expressing ACE2. Strikingly, both mouse and rabbit antisera were able to effectively neutralize live SARS-CoV infection with mean 50% neutralization titers of 1:15,360. Infection of ACE2-expressing 293T cells by SARS pseudovirus could be potently inhibited.

To characterize the neutralization determinants on the RBD of S protein, we isolated a panel of 27 monoclonal antibodies (mAbs) from the RBD-Fc-immunized mice. Six groups of conformation-dependent epitopes, designated as Conf I-VI, and two adjacent linear epitopes were identified by ELISA and binding competition assays.[10] The Conf IV and V mAbs could efficiently block RBD binding to ACE2. Some mAbs reacting with the Conf III and VI partially inhibited interaction between the RBD and ACE2. This suggests that their epitopes may overlap the receptor-binding sites on the RBD or binding of these mAbs to RBD may cause conformational changes of the receptor binding sites, resulting in inhibition of RBD binding to ACE2. These mAbs also had highly potent neutralizing activities against both SARS-CoV and SARS pseudovirus infections. The mAbs that recognize the Conf I, II could not significantly affect the RBD binding with

ACE2, but also efficiently neutralized SARS-CoV, suggesting they may use a unique mechanism to block infection. These data indicate that the RBD induces neutralizing antibodies specific not only for the receptor-binding sites, but also for other structural conformations. Notably, two mAbs targeting linear sequences had no virus-neutralization ability. Therefore, the RBD of SARS S protein contains multiple conformational epitopes capable of inducing potent neutralizing antibody responses.

5. CONCLUSIONS

In summary, we propose to use the RBD of SARS-CoV for developing vaccines and immunotherapeutics, because it is not only a functional domain that mediates virus-receptor binding but also a major target of neutralizing antibodies during virus infection and vaccination. Independent, folded RBD contains multiple neutralizing epitopes and can induce effective protective antibodies.

6. REFERENCES

1. P. A. Rota, M. S. Oberste, S. S. Monroe, et al., Characterization of a novel coronavirus associated with severe acute respiratory syndrome, *Science* **300**, 1394 (2003).
2. S. K. Wong, W. Li, M. J. Moore, H. Choe, and M. Farzan, A 193-amino acid fragment of the SARS coronavirus S protein efficiently binds angiotensin-converting enzyme 2, *J. Biol. Chem.* **279**, 3197 (2004).
3. S. Liu, G. Xiao, Y. Chen, et al., Interaction between heptad repeat 1 and 2 regions in spike protein of SARS-associated coronavirus: implications for virus fusogenic mechanism and identification of fusion inhibitors, *Lancet* **363**, 938 (2004).
4. Y. He and S. Jiang, Vaccine design for severe acute respiratory syndrome coronavirus, *Viral Immunol.* **18**, 327 (2005).
5. S. Jiang, Y. He, and S. Liu, SARS vaccine development, *Emerg. Infect. Dis.* **11**, 1016 (2005).
6. Z. Y. Yang, W. P. Kong, Y. Huang, et al., A DNA vaccine induces SARS coronavirus neutralization and protective immunity in mice, *Nature* **428**, 561 (2004).
7. H. Bisht, A. Roberts, L. Vogel, et al., Severe acute respiratory syndrome coronavirus spike protein expressed by attenuated vaccinia virus protectively immunizes mice, *Proc. Natl. Acad. Sci, USA* **101**, 6641 (2004).
8. A. Bukreyev, E. W. Lamirande, U. J. Buchholz, et al., Mucosal immunisation of African green monkeys (Cercopithecus aethiops) with an attenuated parainfluenza virus expressing the SARS coronavirus spike protein for the prevention of SARS, *Lancet* **363**, 2122 (2004).
9. Y. Nie, G. Wang, X. Shi, et al., Neutralizing antibodies in patients with severe acute respiratory syndrome-associated coronavirus infection, *J. Infect. Dis.* **190**, 1119 (2004).
10. Y. He, H. Lu, P. Siddiqui, Y. Zhou, and S. Jiang, Receptor-binding domain of SARS coronavirus spike protein contains multiple conformational-dependant epitopes that induce highly potent neutralizing antibodies, *J. Immunol.* **174**, 4908 (2004).
11. Y. He, Y. Zhou, P. Siddiqui, and S. Jiang, Inactivated SARS-CoV vaccine elicits high titers of spike protein-specific antibodies that block receptor binding and virus entry, *Biochem. Biophys. Res. Commun.* **325**, 445 (2004).
12. Y. He, Y. Zhou, H. Wu, B. Luo, J. Chen, W. Li, and S. Jiang, Identification of immunodominant sites on the spike protein of severe acute respiratory syndrome (SARS) coronavirus: implication for developing SARS diagnostics and vaccines, *J. Immunol.* **173**, 4050 (2004).
13. Y. He, Y. Zhou, S. Liu, Z. Kou, W. Li, M. Farzan, and S. Jiang, Receptor-binding domain of SARS-CoV spike protein induces highly potent neutralizing antibodies: implication for developing subunit vaccine, *Biochem. Biophys. Res. Commun.* **324**, 773 (2004).
14. Y. He, Q. Zhu, S. Liu, Y. Zhou, B. Yang, J. Li, and S. Jiang, Identification of a critical neutralization determinant of severe acute respiratory syndrome (SARS)-associated coronavirus: importance for designing SARS vaccines, *Virology* **334**, 74 (2005).

GLIA EXPRESSION OF MHC DURING CNS INFECTION BY NEUROTROPIC CORONAVIRUS

Karen E. Malone, Chandran Ramakrishna, J.-M. Gonzalez, Stephen A. Stohlman, and Cornelia C. Bergmann*

1. INTRODUCTION

Interferons are responsible for signaling leading to the initiation of the anti-viral state. Stimulation through the interferon-gamma (IFNγ) receptor primarily follows the JAK/STAT pathway, but interplay between cytokine signaling and type I interferons regulates a large number of genes involved in viral immunity. Of particular interest are genes involved in the processing and presentation of the major histocompatibility complex (MHC) Class I and II pathways. LMP2/7 proteasomal subunits replace normally expressed beta subunits. This assembled immunoproteosome preferentially cleaves peptides for Class I presentation. Also required are Transporters 1/2 (TAP1/2) that translocate peptides into the ER for assembly with Class I and ß2-microglobulin. IFNγ can also induce the primary transactivator controlling Class II gene expression, CIITA. In addition to direct antiviral effects, IFNγ is crucial for antigen presentation by glial cells and engagement of T- cell effector function.

The JHMV variant v2.2-1 is a neurotropic MHV that infects microglia, astrocytes, and oligodendroglia. Despite elimination of detectable infectious particles, virus persists in the CNS in the form of viral RNA. Clinical symptoms progress from limited mobility to paralysis during a time frame of 6–10 days postinfection. Primary demyelination occurs most prominently 10–21 days postinfection, but continues throughout persistence.[1] CD8 T cells play a critical role in clearing JHMV from the CNS using perforin and IFNγ-mediated mechanisms. Perforin mediated cytolytic activity by virus specific CD8 T cells is effective in clearing virus from astrocytes and microglia, but is insufficient for clearing virus from oligodendroglia.[2,3] JHMV- infected IFNγ knockout (GKO) mice exhibit a more severe disease course and a strong association of viral antigen in oligodendroglia.[3] Infected SCID mice receiving CD8 T cells from immunized wt, perforin knock-out (PKO) or GKO mice support the strict reliance on IFNγ for viral clearance from oligodendroglia.[4] IFNγ also induces Class II and enhances Class I surface expression on microglia, coincident with maximal T-cell infiltration during infection.

* Cleveland Clinic Foundation, Cleveland, OH, 44106.

In the normal CNS glial cells express few, if any MHC molecules. Differential MHC regulation by glial cell populations was examined following infection of transgenic mice, in which oligodendroglia are marked by GFP expression. A subset of IFNγ-inducible genes was analyzed at the transcriptional level to identify factors involved in signaling and MHC regulation. The results indicate that both Class I mRNA and mRNA for genes encoding antigen processing machinery (APM) increased shortly after infection in both cell types, prior to surface Class I expression. However, in contrast to microglia, oligodendroglia appear deficient in Class II presentation. Glial cells thus have similar capacity to regulate Class I APM as other somatic cells. However, delayed potential of oligodendroglia to present antigen compared with microglia supports oligodendroglia serve as a reservoir for viral persistence.

2. METHODS

Mice and cell preparation: Transgenic C57BL/6 mice expressing GFP via an oligodendroglia specific PLP promoter (PLP-GFP/B6) ages 6-7 weeks, were infected with 250 pfu JHMV (v2.2-1) intracranially. Brains from 4–6 mice at each time point were collected, subjected to trypsinization and CNS cells enriched by percoll gradients for FACS.

FACS: Cells were stained with anti-CD45, anti-Class I, and anti-Class II and appropriate isotype controls for analysis of surface expression. For sorting, cells were stained with anti-CD45-PE. Two populations were collected: $CD45^{lo}GFP^{-}$ microglia and $CD45^{-}GFP^{+}$ oligodendroglia.

Real-time PCR analysis: Cells were lysed in Trizol and RNA extracted, followed by DNAse I digestion. Reverse transcription used AMV RT with either oligo-dT or random primers. Real-time primers for murine (H-2^{b} haplotype) TAP1, TAP2, LMP2 (Psmb9), LMP7 (Psmb8), IRF-1, IRF-2, and GAPDH genes were used to analyze oligo-dT primed cDNA, whereas CIITA promoter specific primers were analyzed in random primed cDNA. SYBER Green PCR reactions were carried out in duplicate on an MJ DNA Engine Opticon and target specificity checked by melting curve. Gene expression was normalized to GAPDH using $(2^{-(\Delta Ct)})*1000$.

3. RESULTS

Transgenic mice expressing GFP under control of the oligodendroglia-specific PLP promoter were infected to analyze genes associated with MHC presentation based on the easy distinction between $CD45^{lo}$ microglia and GFP^{+} $CD45^{-}$ oligodendroglia by flow cytometry. Resident glial cells expressed few if any MHC molecules in the normal adult murine CNS (Table 1). By day 5 p.i., expression of Class I was detected on the majority of microglia and a subset of oligodendroglia. The initial delay in oligodendroglia Class I upregulation was overcome by day 7 p.i., when Class I expression peaked in both populations, coincident with maximal T-cell infiltration.

Table 1. Percent microglia and oligodendroglia expressing Class I and Class II.

	Microglia		**Oligodendroglia**	
Days p.i.	Class I	Class II	Class I	Class II
0	<1	<1	<1	<1
5	97	5	22	4
7	94	85	96	6
9	70	55	56	5

To examine how the delay in Class I surface expression on oligodendroglia is regulated at the mRNA level, glial cell subsets were purified by FACS during acute infection and analyzed for transcripts encoding MHC and genes involved in antigen processing. Real-time PCR analysis indicated that both Class I mRNA and mRNA for genes encoding APM increased shortly after infection, prior to surface Class I expression (Table 2). While Class I and APM mRNAs increased vastly between days 5 and 7 p.i. in oligodendroglia, they remained constant in microglia during acute infection. Interferon regulatory factors 1 (IRF-1) and 2 (IRF-2) are transactivators controlling induction of Class I and APM. IRF-1 mRNA increased in both cell types, albeit more dramatically in oligodendroglia. By contrast, IRF-2 mRNA was constitutively expressed in both cell types and was transiently downregulated during acute infection. In summary, maximal Class I expression coincided with peak mRNA transcription of Class I and APM genes.

Analysis of Class II expression by flow cytometry demonstrated efficient upregulation on microglia (Table 1). Only limited expression was detected on oligodendroglia consistent with their inefficient upregulation of CIITA (Table 2).

Table 2. Relative gene expression in microglia and oligodendroglia.

Gene	**Microglia**			**Oligodendroglia**		
	Naïve	d5 p.i.	d7 p.i.	Naïve	d5 p.i.	d7 p.i.
MHC Class I heavy chain	952	3213	1899	ND*	2335	11539
MHC Class I APM						
LMP2	65	304	220	ND*	167	2118
LMP7	126	310	271	ND*	523	1647
TAP1	18	82	27	ND*	134	211
TAP2	8	12	11	ND*	54	63
Interferon response factors						
IRF1	7	48	10	ND*	83	174
IRF2	73	37	22	40	18	21
Class II transactivator	1	30	67	ND*	1	12

*ND denotes not detectable.

4. DISCUSSION

Analysis of genes necessary for antigen presentation in the CNS revealed several trends unique to distinct glial cell types. Both oligodendroglia and microglia upregulated major components for Class I antigen processing and presentation during infection, supporting the capacity for efficient presentation of virus peptide to CD8 T-cells. Although basal levels of Class I and APM mRNA were lower in oligodendroglia, they increased with similar kinetics in both cell types following infection. Higher relative Class I and APM mRNA levels in oligodendroglia, coincident with peak T-cell infiltration at day 7 p.i., suggests enhanced responsiveness to IFNγ compared with microglia. Delayed Class I surface expression on oligodendroglia, compared with microglia, appears to reflect levels of IFNγ-inducible APM mRNA. The paucity in CIITA and Class II expression in oligodendroglia remains unknown. Lastly, upregulation of IRF-1 in both cell types during infection was supported by concordant downregulation of IRF-2, an IRF-1 antagonist. In summary, expression of Class I on oligodendroglia during infection, albeit under more stringent regulation compared with microglia, supports oligodendroglia as potential targets for CTL and subsequent pathology. However the CNS employs many protective measures against such activity, particularly to protect cells that proliferate slowly.[5] The demonstration that CTL activity does not control JHMV infection in oligodendroglia, in light of Class I expression, demands further investigation.

5. ACKNOWLEDGMENTS

This research was supported by the National Institutes of Health grant NS18146.

6. REFERENCES

1. Marten, N. W., Stohlman, S. A., and Bergmann, C. C., 2001, MHV infection in the CNS: mechanisms of immune-mediated control, *Viral Immun.* **14**:1-18.
2. Lin, M. T., Hinton, D. R., Marten, N. W., Bergmann, C. C., and Stohlman, S. A., 1999, Antibody prevents virus reactivation within the central nervous system, *J. Immun.*, **162**: 7358-7368.
3. Parra, B., Hinton, D. R., Marten, N. W., Bergmann, C. C., Lin, M. T., Yang, C. S., and Stohlmann, S. A., 1999, IFN-gamma is required for viral clearance from central nervous system oligodendroglia, *J. Immunol.* 162:1641-1647.
4. Bergmann, C. C., Parra, B., Hinton, D. R., Ramakrishna, C., Dowdell, K. C., and Stohlman, S. A., 2004, Perforin and gamma-interferon mediated control of coronavirus central nervous system infection by CD8 T cells in the absence of CD4 T cells, *J. Virol.* **78**:1739-1750.
5. Griffin, D. E., 2003, Immune responses to RNA-virus infections of the CNS, *Nat. Rev. Immum.* **3**:493-502.

RESURRECTION OF AN "EXTINCT" SARS-CoV ISOLATE GD03 FROM LATE 2003

Timothy Sheahan, Damon Deming, Eric Donaldson, Raymond Pickles, and Ralph Baric*

1. INTRODUCTION

The causative agent of SARS, SARS- associated coronavirus (SARS-CoV), is transmitted via the aerosol route, which facilitated repeated non-human to human transmissions within live animal markets of the Guangdong region late in 2002. The coronavirus spike glycoprotein is a key determinant of host specificity, and elucidating the mechanisms of viral host expansion is important to understand the molecular events that rendered the virus pathogenic to humans. Sequences of viral spikes from early human clinical cases resembled those isolated from civet cats (SZ16) in wild animal markets early in the epidemic.[1] The spike sequence from a human clinical isolate from the Guangdong region late in 2003 (GD03) differs from SARS Urbani in 16 residues almost all of which reside in the S1 domain and five of which reside in the putative receptor binding domain.[2] Using data from DNA sequence databases and synthetic biology, we have resurrected the extinct spike glycoprotein gene of GD03 within our SARS Urbani infectious clone. Growth of the GD03 spike/Urbani chimeric recombinant virus (icGD03) in Vero cells is not as robust as icSARS or Urbani suggesting that the GD03 spike does not bind Vero ACE2 with the same affinity. *In vivo* growth of icGD03 and icSARS was compared in Balb/c via intranasal infection. The icGD03 lung titer at all time points was not as robust as icSARS-CoV suggesting that the GD03 spike slightly impairs growth in mice.

2. MATERIALS, METHODS, AND RESULTS

SARS infectious clone: The genome of our SARS Urbani infectious clone (icSARS) is broken up into six contiguous plasmid DNAs (plasmids A–F) some of which have been altered by silent mutation in order to create BglI restriction sites at fragment junctions.[3] BglI is a class IIS restriction endonuclease that cleaves the symmetrical sequence

* University of North Carolina at Chapel Hill, Chapel Hill, North Carolina.

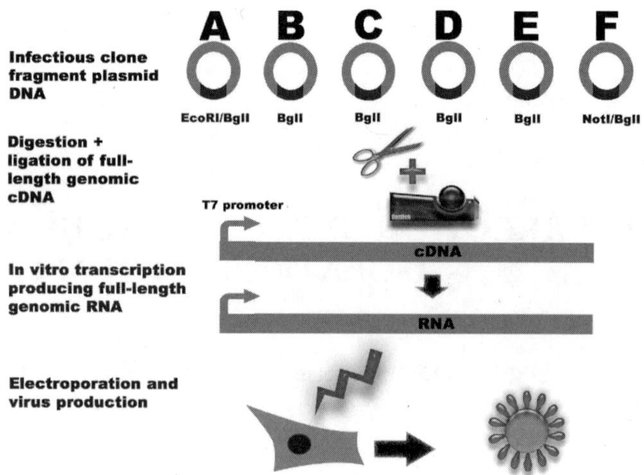

Figure 1. The SARS-CoV infectious clone: Assembly strategy and virus production.

GCCNNNN↓NGGC but leaves 64 different asymmetrical ends allowing our infectious clone fragments to be directionally ligated to produce full-length genomic cDNA. *In vitro* transcription of full-length viral cDNA is facilitated by an inserted T7 promoter sequence preceding the SARS genome (Fig. 1).

Cloning strategy: The spike sequence is spread over SARS infectious clone plasmids SARS-E and F. Using sequence extracted from the *Entrez* Internet database, we identified 16 coding mutations of the GD03 spike that diverged from SARS Urbani (Fig. 2). The E infectious clone fragment contains the first two-thirds of the spike sequence and 15 of the GD03 mutations while the final mutation lies in the SARS-F fragment. We cloned a synthetically derived DNA from Blue Heron Biotechnology containing 5' GD03 mutations. Using PCR site-directed mutagenesis, the remaining GD03 mutation was introduced into the SARS-F fragment. The plasmid clones were fully sequenced and shown to contain all of the appropriate mutations.

icGD03 recombinant virus production: Infectious clone fragment plasmid DNAs were digested, ligated, and used as template for *in vitro* transcription (Fig. 1). Nucleocapsid and full-length viral genomic transcripts were then electroporated into Vero cells. Cell culture media containing virus was harvested 48 hours post-electroporation. Virus was plaque purified and then passaged twice in Vero cells to create viral stocks. RNA from icGD03 infected cells was used for RT-PCR, and the resultant amplicons were sequenced to confirm the identity of the GD03 spike. In addition to the expected S mutations introduced into the infectious clone, the recombinant GD03 virus used throughout these studies contained two additional cell culture induced mutations within the viral spike: mutation one = F7L, and mutation two = D613G.

GD03 grows with delayed kinetics within Vero cells as compared with SARS Urbani and icSARS: Four-well chamber slides were seeded with $2x10^5$ Vero cells/well 24 hr prior to infection. Media was removed and cells were infected with SARS Urbani, icSARS, or icGD03 at an MOI of one. Cells were infected for 1 hr at room

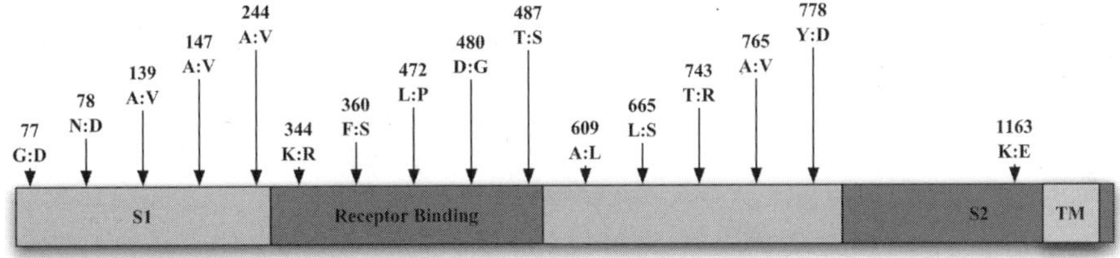

Figure 2. Amino acid differences of the GD03 spike protein as compared with SARS Urbani.

Table 1. Growth of SARS Urbani, icSARS, and icGD03 in Vero cells.

Time (hr)	Urbani (pfu/ml)	IcSARS (pfu/ml)	icGD03 (pfu/ml)
2	$1.1x10^4$	$6.1x10^4$	$7.5x10^2$
7.5	$2.4x10^5$	$2.2x10^6$	$1.4x10^4$
13.5	$1.6x10^6$	$1.4x10^6$	$5.4x10^5$
25	$2.1x10^6$	$1.9x10^6$	$7.5x10^5$

Table 2. Average lung titers (pfu/g) of mice infected with icGD03 or icSARS (n=3).

Day 2		Day 4		Day 7	
icGD03	icSARS	icGD03	icSARS	icGD03	icSARS
$1.0x10^6$	$1.1x10^7$	7.0x103	4.2x104	$6.0x10^0$	$4.0x10^0$

temperature after which the inoculum was removed and growth media was added to each well. The cell culture media was sampled at 2, 7.5, 13.5, and 25 hours postinfection. The samples were then analyzed by Vero cell plaque to quantitate virus growth over time. icGD03 grew to titers 1/2 log less than both SARS-Urbani and icSARS. These data suggest that the GD03 spike protein may not bind the Sars receptor ACE2 as efficiently as Urbani strains and as result, ic GD03 infection of Vero cells is slightly less robust.

icGD03 replicates in vivo: Six-week-old Balb/c mice were infected with 10^5 pfu of icGD03 or icSARS intranasally. Days 2, 4, and 7 postinfection, animals were sacrificed and the lungs were harvested for plaque titration. Twenty percent (wt/vol) lung homogenates were serially diluted, and infectious virus was quantitated by Vero plaque titration. Although icGD03 growth was not as robust as icSARS within the mouse lungs, significant replication and virus persistence occurred making future cross challenge experiments and vaccine studies possible.

3. CONCLUSIONS

These data demonstrate (1) the power of DNA sequence databases, reverse genetics, and synthetic biology to resurrect viruses with extinct genes, (2) that we have constructed a unique and rare spike glycoprotein from late 2003 in China, (3) that GD03, the most divergent spike sequence of all human SARS isolates, exhibits a divergent phenotype *in vitro* and *in vivo* as compared with SARS Urbani, and (4) the GD03 resurrected virus serves as antigenically unique isolates for vaccine and cross-protection studies. Using our approach combining reverse genetics, synthetic biology, and electronic database technologies, a panel of recombinant SARS-CoV encoding variant S glycoproteins identified in 2002, 2003, and 2004 human and animal cases can be resurrected (Fig. 4). A panel of variant SARS-CoV recombinant viruses is essential for evaluating the efficacy of immunotherapies against zoonotic and rare SARS-CoV isolates.

This work was supported by NIH grants AI059136 and AI059443.

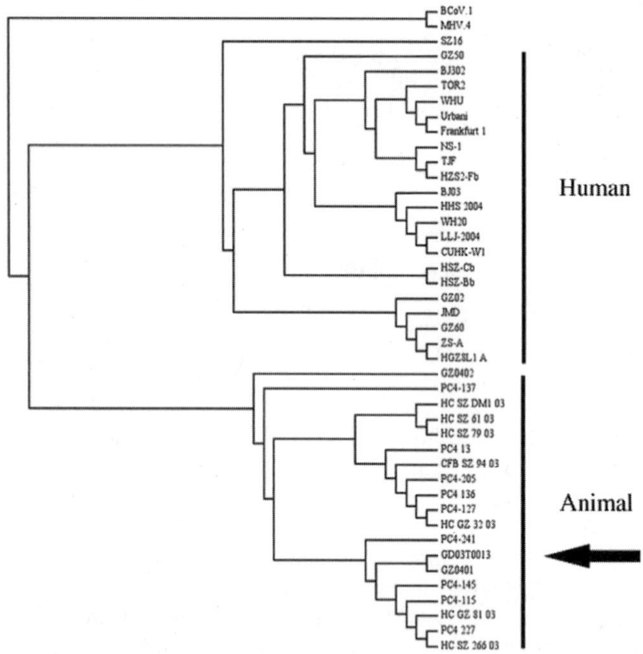

Figure 3. Neighbor-joining tree of 46 representative SARS spike peptide sequences. Rooting the tree with group 2 coronaviruses (BCoV and MHV) results in two distinct genotypes representing human and animal clusters. GD03 and closely related GZ0401 represent human viruses that cluster in the animal genotype along with GZ0402. In the human cluster, Urbani represents late phase sequences, CUHK-W1 middle phase, and GZ02 early phase. LLJ-2004 is a porcine isolate thought to be a human to animal transmission.

4. REFERENCES

1. Guan, Y., Zheng, B. J., He, Y., et al., 2003, Isolation and characterization of viruses related to the SARS coronavirus from animals in southern China, *Science* **302**:276-278.
2. Chinese, S. M. E. C., 2004, Molecular evolution of the SARS coronavirus during the course of the SARS epidemic in China, *Science* **303**:1666-1669.
3. Yount, B., Curtis, K. M., Fritz, E. A., et al., 2003, Reverse genetics with a full-length infectious cDNA of severe acute respiratory syndrome coronavirus, *Proc. Natl. Acad. Sci. USA* **100**:12995-13000.

VIII. VACCINES, ANTIVIRAL DRUGS AND DIAGNOSTICS

SARS CORONAVIRUS VACCINE DEVELOPMENT

Ralph S. Baric, Timothy Sheahan, Damon Deming, Eric Donaldson, Boyd Yount, Amy C. Sims, Rhonda S. Roberts, Matthew Frieman, and Barry Rockx*

1. INTRODUCTION

Coronavirus infections are associated with severe diseases of the lower respiratory and gastrointestinal tract in humans and animals, yet little is known about the underlying molecular mechanisms governing virulence and pathogenesis. Among the human coronaviruses, the etiologic agent of SARS, the SARS coronavirus (SARS-CoV) is an attractive model to study the molecular basis for pathogenesis, given its robust *in vitro* growth characteristics, the availability of a reverse genetic system, animal models, wealth of clinical data, and several solved replicase and accessory protein structures. SARS CoV infection afflicted about 8,000 humans and resulted in about 800 deaths, worldwide. Disease severity has been linked to age, with approximate mortality rates of <1% under 24 years of age, 6% for ages 15–44, 15% for ages 45–64, and >50% over 65. Many survivors have suffered long-lasting lung and cardiac complications.[5] The underlying mechanisms governing SARS-CoV pathogenesis are only now being unraveled.[9]

SARS-CoV is a zoonotic pathogen that crossed the species barrier, the most likely host being civet cats and raccoon dogs, although virus has also been isolated or detected from domesticated cats, swine, and rodents.[4] Aggressive public health efforts contained the 2003 epidemic but it is unclear whether the epidemic strains are extinct in the wild. Given the significant health and economic impact of the SARS-CoV outbreak, an effective vaccine strategy for SARS that includes protection against epidemic and zoontoic strains of virus in at risk elderly populations who are most vulnerable to severe disease is essential.

Phylogenetic analyses have suggested that SARS-CoV either represented the prototype group IV coronavirus while other studies have placed the virus as an early split-off of group II.[12,16,17] Molecular evolutionary studies on isolates obtained from different stages in the outbreak have implicated changes in ORF1a, the S glycoprotein,

*University of North Carolina at Chapel Hill, Chapel Hill, North Carolina 27599.

and various accessory ORFs (ORF3a and ORF8) as being associated with increased virulence, transmission and pathogenesis during the epidemic.[3,4]

The SARS-CoV virion contains a single-stranded positive polarity 29,700 nucleotide RNA genome bound by the nucleocapsid protein, N. The capsid is surrounded by a lipid bilayer containing at least three structural proteins, designated S, M, and E. The 180-kDa spike glycoprotein (S) interacts with its receptor, angiotensin II converting enzyme (ACE2), to mediate entry into cells.[11] In addition to the 23-kDa M glycoprotein, the E protein is likely essential for efficient virion maturation and release. The SARS-CoV genome contains 14 principal ORFs. ORF1a and ORF1b encode the viral replicase proteins required for subgenomic and genome length RNA synthesis and virus replication.[12,16] ORFs 2–8 are encoded in eight subgenomic mRNAs synthesized as a nested set of 3' co-terminal molecules in which the leader RNA sequences, encoded at the 5' end of the genome, are joined to body sequences at distinct transcription regulatory sequences containing a highly conserved consensus sequence. The development of a SARS-CoV molecular clone provides a useful tool for developing novel SARS-CoV isolates for identifying the genetic determinants responsible for increased pathogenesis during the epidemic and developing zoonotic strains for vaccine testing.[22]

2. METHODS

The 2003 epidemic Urbani strain of SARS-CoV and a molecularly derived recombinant virus (icSARS-CoV) are used throughout these studies. The GDO3T0013 S glycoprotein (GD03) was synthetically reconstructed from published sequences (AY304486),[3] inserted into the molecular clone of SARS-CoV, and used to isolate a recombinant virus (icGDO3) encoding the icGDO3 S glycoprotein.[22] The Urbani S glycoprotein or nucleocapsid genes were inserted into Venezuelan equine encephalitis virus replicon particles (VRP) using methods described in the literature.[1] VRP vaccine stocks were titered at ~1.0 x 10^9 and used to vaccinate at 1 x 10^6 VRPs. Twenty-eight days later, the animals were boosted by footpad innoculation and challenged with wild-type icSARS-CoV or icGD03 several weeks to months later. Neutralization titers were determined by treating ~100 plaques of icSARS-CoV or icGDo3 with varying concentrations of serum from humans or animals and measuring the reduction of infectivity on Vero cell monolayers.

3. RESULTS

3.1. Evolution of the SARS-CoV

Phylogenetic analysis of SARS-CoV isolates from animals and humans makes a compelling argument that the virus originated in animals, most likely in palm civets or raccoon dogs, and was transmitted to human populations via live animal markets.[7] However, the actual reservoir for the SARS-CoV has not been clearly determined, although recent unpublished reports from China suggest a possible origin in bats. Using Bayesian methods and sequences from animal and human SARS-CoV isolates, we note that SARS-CoV isolates can be divided into 3 genoclusters including the animal isolates like SZ16 (GI), a cluster of isolates associated with sporadic mild human and animal infections (GII), and a cluster of highly pathogenic strains associated with the early, middle, or late phase isolates from the 2002–2003 epidemic (GIII) (Figure 1). In general, GII isolates are usually isolated in animals although rare mild cases of infection in humans have been reported and the GI isolates have only been detected in animals. The GD03 S glycoprotein sequence was obtained from a sporadic, mild human case reported on Dec 22, 2003 from a virus that was never successfully cultured *in vitro* and is the most diverse human isolate.[3]

3.2. SARS-CoV S Glycoprotein Antigenicity

Detailed mapping studies have indicated that at least three neutralizing epitopes reside within the Urbani S glycoprotein. Monoclonal antibodies that bind site A and B possess significant neutralizing activity against wildtype viruses that completely protect animals from infection. Site B overlaps the receptor binding site known to interact with ACE2.[10,18,19] Site C represents a 3–4 times less robust neutralizing site, presumably antibodies interfere with virus docking and entry mechanisms dependent upon the function of the heptad repeats encoded in the C-terminus of S. Variation in the SARS-CoV S glycoprotein has been reported, although detailed cross neutralization studies and cross protection studies in animal models comparing the susceptibility of these

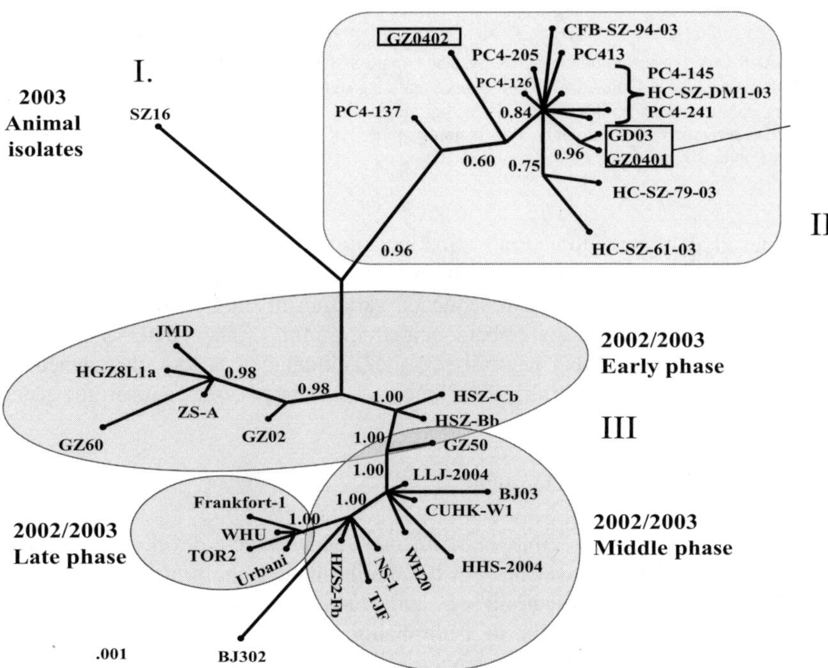

Figure 1. Phylogeny of SARS-CoV S glycoprotein. An unrooted tree generated by Bayesian inference using representative SARS-CoV spike protein sequences. GD03 was isolated from a sporadic, mild case of SARS on December 22, 2003.

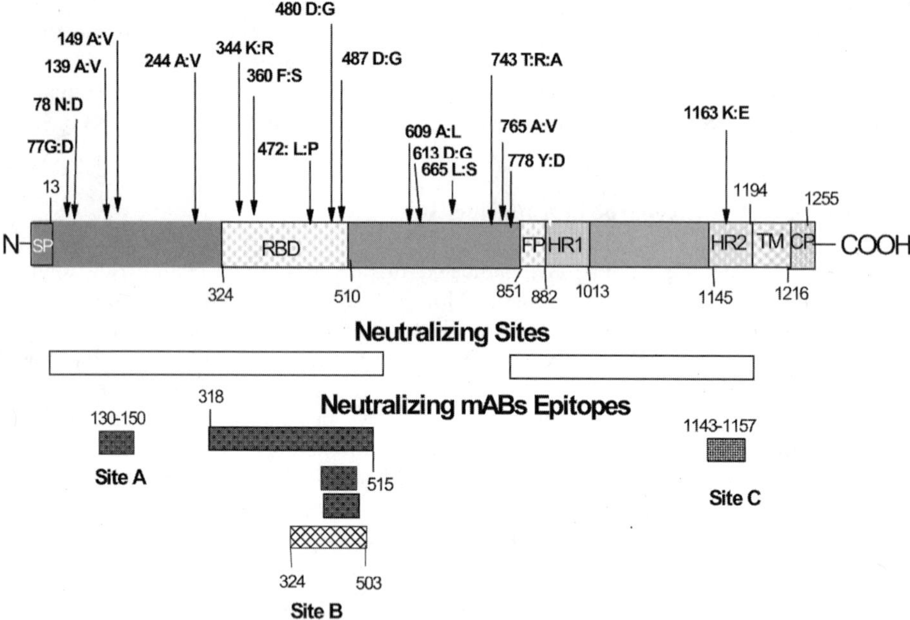

Figure 2. SARS-CoV S glycoprotein antigenicity. The Urbani SARS-CoV S glycoprotein contains at least three distinct neutralizing sites designated A–C in regions containing significant variation in the GD03 S glycoprotein at these sites. Neutralizing sites in Urbani exist within the S1 and S2 portion of the S glycoprotein. SP - signal peptide, RBD - receptor binding domain, FP - fusion peptide, HR1 or HR2 - heptad repeat elements 1 and 2, TM - transmembrane domain, and CP - cytoplasmic domain.

heterologous glycoproteins to neutralization have not been evaluated. The most extensive variation has been noted in isolates from group II that were isolated from raccoon dogs and civets. About ~2% amino acid sequence variation in the S glycoprotein has been reported as compared with the Urbani epidemic strain.[4,7] The GD03 S glycoprotein contains variation within all 3 neutralizing sites, although it is less clear whether this heterogeneity alters the neutralization kinetics associated with antiserum generated against epidemic strains.

3.3. SARS-CoV Vaccine Development

After the 2002–2003 epidemic, experimental vaccines were developed and tested in animal models, primarily in rodents.[6] In the murine model, the principal component of protective immunity was neutralizing antiserum directed against the SARS S glycoprotein and passive transfer of neutralizing antibodies was sufficient to protect against virus replication.[18] Not surprisingly, killed and recombinant virus vaccines that elicit neutralizing antibody protect mice from SARS-CoV replication in the lung.[6] Importantly, more adverse reactions were noted in ferrets vaccinated with recombinant poxviruses encoding the SARS-CoV S glycoprotein including a lack of protection from

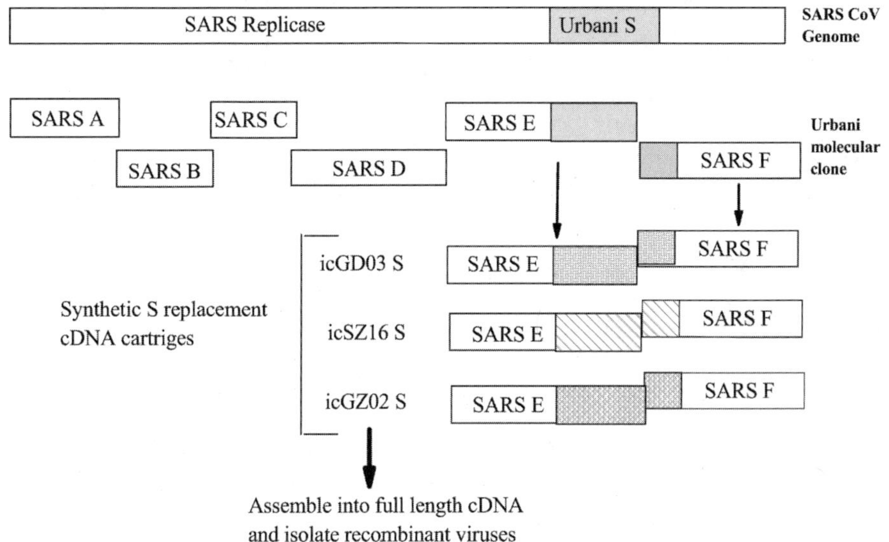

Figure 3. Strategy for resurrection of rare SARS-CoV spikes. Synthetic cDNAs encoding the icGD03 (human/animal isolate (GII), SZ16 (animal isolate, GI), and GZ02 (early isolate, GIII) were obtained and inserted into the Urbani molecular clone, replacing the Urbani S glycoprotein with a variant S gene.

infection and increased hepatic pathology in vaccinated animals.[20] The molecular basis for the increased pathology in vaccinated animals remains unknown.

Important unresolved questions remain regarding the development of efficacious SARS-CoV vaccines. Severe disease and high death rates were noted in elderly human populations, and vaccine efficacy in more vulnerable senescent animal models for SARS-CoV infection have not been evaluated. The ability of vaccines to induce robust immune responses in aged populations, should be evaluated to determine if protection can be elicited in elderly populations with senescent immune systems. Long-term and short-term immunity, waning immunity, and cross protection between strains from different genoclusters has languished due to the lack of heterologous isolates. Finally, the molecular mechanisms governing vaccine mediated immunopathogenesis and enhanced disease in ferrets must be elucidated.

3.4. Synthetic Resurrection of Rare SARS-CoV Isolates

Many animal and early human isolates were sequenced but never successfully cultured *in vitro*.[4,7] To address this problem, a systematic approach involving a molecular clone of SARS-CoV and synthetic biology was used to successfully resurrect early human and animal isolates (Figure 3). The details of the resurrection of the icGD03 S glycoprotein gene in the Urbani molecular clone is discussed in more detail in the article by Sheahan *et al.* (chapter 7.19). The icGD03 virus replicated to high titers in Vero cells and in mice and was more resistant to neutralization with antiserum or monoclonal

Table 1. Properties of icSARS-CoV and icGD03-S recombinant viruses.

Virus strain	Virus titer		Neutralizing titer ($PRNT_{50}$)[1]	
	Vero	BALB/c Mice	S monoclonal	Polyclonal sera
icSARS-CoV	1.0 x 10^7	1.0 x 10^7	>1:1600	>1:1600
icGD03-S	5.0 x 10^6	5.0 x 10^6	1:100	1:100

[1] $PRNT_{50}$ represents plaque reduction neutralization serum titers that reduce 50% of the SARS-CoV plaques.

antibodies directed against the Urbani strain of SARS-CoV (Table 1),[19] consistant with the noted sequence variation in domains recognized by neutralizing antibodies.

3.5. SARS-CoV Vaccines and Heterotypic Cross-Protection

To evaluate vaccine efficacy against homologous and heterologous strains, the Urbani S glycoprotein and nucleocapsid genes were cloned and inserted in Venezuelan equine encephalitis virus replicon particles (VRP-S or VRP-N) using standard approaches reported from our laboratory previously.[1] In addition, the influenza A HA glycoprotein (VRP-HA) was used as a control vector. BALB/c mice were vaccinated with 1 x 10^6 infectious units of VRPs-HA or the combination of VRP S+VRP-N, boosted 4 weeks later, and then challenged with icSARS-CoV or icGD03-S about 8 or 40 weeks post-boost (Table 2). Ages of senescent mice exceeded 1 year at the time of challenge. Importantly, VRP vaccines provided complete short-term protection against homologous and heterologous isolates, as all VRP S+VRP-N vaccinated animals were protected against challenge. The VRP vaccines also provided long-term protection against homologous challenge, protecting all senescent mice from icSARS-CoV replication. Although aged mice vaccinated with VRP-HA demonstrated pathologic lesions in the lung similar to that reported in the literature,[15] VRP S+VRP-N vaccinated mice displayed little if any pathologic lesions in the lung (data not shown). In contrast, VRP-S+VRP-N vaccinated mice provided little long-term protection against icGD03-S infection although virus titers were reduced about 1-log as compared with VRP-HA controls. The icGD03 challenge virus also produced pathologic lesions in both the VRP-HA and SARS vaccinated animals and was virtually indistinguishable from icSARS-CoV infection (data not shown). At this time, it is likely that rapid waning immunity against heterologous challenge viruses resulted in vaccine failure in aged animals.

Table 2. Virus replication in the lungs of vaccinated BALB/c mice.

	Young BALB/C mice[1]		**Senescent BALB/C mice[1]**	
Virus strains	**VRP-HA**	**VRP-S+VRP-N**	**VRP-HA**	**VRP S+VRP-N**
icSARS-CoV	4/4	0/4	4/4	0/4
icGD03-S	4/4	0/4	4/4	4/4

[1] Ratio represents infected mice over total mice.

4. DISCUSSION

SARS-CoV is a zoonotic pathogen, and several cases of likely animal to human transmission have been reported in China. In most cases, these infections were mild and did not spread beyond the index case suggesting that animal viruses require additional adaptation prior to evolving efficient usage of humans as hosts.[3,4,7] After adapting to the human host, most severe disease manifestations occur within elderly populations likely compromised by waning innate immune and acquired immune responses to pathogen insult. Importantly, several zoonotic strains in genocluster II are closely related, display heterogeneity in neutralizing epitope sites, and encode determinants in the S glycoprotein gene that are consistent with rapid human adaptation and spread.[3,21] These data suggest that current vaccine formulations should be tested not only against epidemic strains, but also to evaluate protective therapeutic potential against zoonotic reintroduction. To address this need, we resurrected live SARS-CoV encoding the GD03 S glycoprotein and demonstrated that this virus replicated efficiently *in vitro* and *in vivo* and produced pathologic lesions in aged mice. As shown with poliovirus and 1918 influenza virus genes,[2,8] our data provide additional support for the use of synthetic DNA and reverse genetics as a means of rescuing "extinct" viruses and viral genes for the improvement of vaccines and enhancing the overall public health.

The icSARS-CoV GD03-S recombinant virus demonstrated gaps in vaccine design for controlling future SARS-CoV epidemics. Our results are consistent with earlier reports suggesting that zoonotic viruses are highly resistant to neutralization with antiserum directed against epidemic strains like Urbani.[21] Importantly, VRP vaccines elicited high levels of neutralizing antiserum against the homologous isolate, but less efficient neutralizing responses against the icGD03-S recombinant virus. These high neutralizing responses likely translated to efficient protection from homologous infection, both in young and aged mice, but also provided short-term protection against homologous protection in younger animals. In aged animals, responses had waned or elderly immune systems had deteriorated sufficiently to allow for robust icGD03 replication and pathogenesis.[13,14] It is likely that vaccine approaches that induce less robust neutralizing responses like DNA and killed vaccines, might completely fail in protecting against icGD03 challenge. To rectify this problem, booster vaccines should be considered in senescent populations or new vaccine formulations be assembled that include S glycoprotein determinants that protect against epidemic and zoonotic forms of SARS-CoV. Future studies will evaluate single VRP regiments that include either the VRP-S or VRP-N candidate vaccines separately, as this approach may enhance overall protection in VRP-S vaccinated animals. Clearly, the availability of SARS-CoV strains harboring zoonotic S glycoproteins will provide important future reference inoculums for evaluating the robustness of new vaccine candidates in animals.

The growing recognition that human coronaviruses can produce significant pulmonary diseases in humans places the SARS-CoV in an excellent position to serve as a premiere model system to elucidate the molecular mechanisms governing human coronavirus pathogenesis in the lung and to identify the components of protective immunity that prevent severe lower respiratory tract infections in humans and animals. Current animal models for SARS-CoV usually display little clinical disease and rarely cause death, hampering measurements of vaccine efficacy against severe infection and disease. Animal models that mirror the immunopathological and pathophysiological changes noted in humans are needed for future vaccine testing.

5. REFERENCES

1. R. Baric, B. Yount, L. Lindesmith, et al., Expression and self-assembly of norwalk virus capsid protein from Venezuelan equine encephalitis virus replicons, *J. Virol.* **76**, 3023-3030 (2002).
2. J. Cello, A. V. Paul, and E. Wimmer, Chemical synthesis of poliovirus cDNA: generation of infectious virus in the absence of natural template, *Science* **297**, 1016-1018 (2002).
3. Chinese SARS Molecular Epidemiology Consortium, Molecular evolution of the SARS coronavirus during the course of the SARS epidemic in China, *Science* **303**, 1666-1669 (2004).
4. Y. Guan, B. J. Zheng, Y. Q. He, et al., Isolation and characterization of viruses related to the SARS coronavirus from animals in Southern China, *Science* **302**, 276-278 (2003).
5. Y. Han, H. Geng, W. Feng, et al., A follow-up of 69 discharged SARS patients, *J. Tradit. Chin. Med.* **23**, 214-217 (2003).
6. S. Jiang, Y. He, and S. Liu, SARS vaccine development, *Emerg. Infect. Dis.* **11**, 1016-1020 (2005).
7. B. Kan, M. Wang, H. Jing, et al., Molecular evolution analysis and geographic investigation of severe acute respiratory syndrome coronavirus-like virus in palm civets at an animal market and on farms, *J. Virol.* **79**, 11892-11900 (2005).
8. D. Kobasa, A. Takada, K. Shinva, et al., Enhanced virulence of influenza A viruses with the haemagglutinin of the 1918 pandemic virus, *Nature* **431**, 703-707 (2004).
9. K. Kuba, Y. Imai, S. Rao, et al. A crucial role of angiotensin converting enzyme 2 (ACE2) in SARS coronavirus-induced lung injury, *Nat. Med.* **11**, 875-879 (2005).
10. S. C. Lai, P. C. Chong, C. T. Yeh, et al., Characterization of neutralizing monoclonal antibodies recognizing a 15-residues epitope on the spike protein HR2 region of SARS-CoV, *J. Biomed. Sci.* **12**, 1-17 (2005).
11. W. Li, M. Moore, N. Vasilieva, et al., Angiotensin-converting enzyme 2 is a functional receptor for the SARS coronavirus, *Nature* **426**, 450-454 (2003).
12. M. A. Marra, S. J. M. Jones, C. R. Astell, et al., The genome sequence of the SARS-associated coronavirus, *Science* **300**, 1399-1404 (2003).
13. D. M. Murasko and J. Jiang, Response of aged mice to primary virus infections, *Immunol. Rev.* **205**, 285-296 (2005).
14. G. Pawelec, A. Akbar, C. Caruso, et al., Human immunosenescence: is it infectious? *Immunol. Rev.* **205**, 257-268 (2005).
15. A. Roberts, C. Paddock, L. Vogel, E. Butler, S. Zaki, and K. Subbarao, Aged BALB/c mice as a model for increased severity of severe acute respiratory syndrome in elderly humans, *J. Virol.* **79**, 5833-5838 (2005).
16. P. A. Rota, M. S. Oberste, S. S. Monroe, et al., Characterization of a novel coronavirus associated with severe acute respiratory syndrome, *Science* **300**, 1394-1399 (2003).
17. E. J. Snijder, P. J. Bredenbeek, J. C. Dobe, et al., Unique and conserved features of genome and proteome of SARS-CoV, an early split-off from the coronavirus group 2 lineage, *J. Mol. Biol.* **331**, 991-1004 (2003).
18. J. Sui, W. Li, A. Murakami, et al., Potent neutralization of severe acute respiratory syndrome (SARS) coronavirus by a human mAb to S1 protein that blocks receptor association, *Proc. Natl. Acad. Sci. USA* **101**, 2536-2541 (2004).
19. R. A. Tripp, L. M. Haynes, D. Moore, et al., Monoclonal antibodies to SARS-associated coronavirus (SARS-CoV): identification of neutralizing and antibodies reactive to S, N, M and E viral proteins, *J. Virol. Methods* **128**, 21-28 (2005).
20. H. Weingarl, M. Czub, S. Chub, et al., Immunization with modified vaccinia virus ankara-based recombinant vaccine against severe acute respiratory syndrome is associated with enhanced hepatitis virus in ferrets, *J. Virol.* **78**, 12672-12676 (2004).
21. Z. Yang, H. Werner, W. Kong, et al., Evasion of antibody neutralization in emerging severe acute respiratory syndrome coronaviruses, *Proc. Natl. Acad. Sci. USA* **102**, 797-801 (2005).
22. B. Yount, K. Curtis, E. Fritz, et al., Reverse genetics with a full length infectious cDNA of the severe acute respiratory syndrome coronavirus, *Proc. Natl. Acad. Sci. USA* **100**, 12995-13000 (2003).

DEVELOPMENT OF VACCINES AND PASSIVE IMMUNOTHERAPY AGAINST SARS CORONAVIRUS USING MOUSE AND SCID-PBL/hu MOUSE MODELS

Masaji Okada, Yuji Takemoto, Yoshinobu Okuno, Satomi Hashimoto, Yukari Fukunaga, Takao Tanaka, Yoko Kita, Sechiko Kuwayama, Yumiko Muraki, Noriko Kanamaru, Hiroko Takai, Chika Okada, Yayoi Sakaguchi, Izumi Furukawa, Kyoko Yamada, Miwa Izumiya, Shigeto Yoshida, Makoto Matsumoto, Tetsuo Kase, J. S. M. Peiris, Daphne E. deMello, Pei-Jer Chen, Naoki Yamamoto, Yoshiyuki Yoshinaka, Tatsuji Nomura, Isao Ishida, Shigeru Morikawa, Masato Tashiro, and Mitsunori Sakatani*

*Masaji Okada, Yuji Takemoto, Satomi Hashimoto, Yukari Fukunaga, Takao Tanaka, Yoko Kita, Sachiko Kuwayama, Yumiko Muraki, Noriko Kanamaru, Hiroko Takai, Chika Okada, Yayoi Sakaguchi, Izumi Furukawa, Kyoko Yamada, Miwa Izumiya, Mitsunori Sakatani, Clinical Research Center, National Hospital Organization Kinki-Chuo Chest Medical Center, Sakai, Osaka, 591-8555, Japan. Yoshinobu Okuno, Tetsuo Kase, Department of Infectious Diseases, Osaka Prefectural Institute of Public Health, Higashinari-ku, Osaka, 537-0025, Japan. Shigeto Yoshida, Department of Infection and Immunity, Jichi Medical School, Minamikawachi-machi, Tochigi, 329-0498, Japan. Makoto Matsumoto, Microbiological Research Institute, Otsuka Pharmaceutical Co., Ltd., Tokushima, 771-019, Japan. J. S. M. Peiris, Department of Microbiology, The University of Hong Kong, Pokfulam Road, Hong Kong. Daphne E. deMello, Department of Pathology Cardinal Glennon Children's Hospital, St. Louis Univ. Health Science Center, St. Louis, MO 63104, USA. Pei-Jer Chen, Hapatitis Research Center , National Taiwan University College of Medicine, Taipei, Taiwan. Naoki Yamamoto, Yoshiyuki Yoshinaka, Tokyo Medical and Dental University, Bunkyo-ku, Tokyo 113-8549, Japan. Tatsuji Nomura, Central Institute for Experimental Animals, Kawasaki, Kanagawa, 216-0001, Japan. Isao Ishida, Pharmaceutical Division, Kirin Brewery Co., Shibuya, Tokyo, 150-8011, Japan. Shigeru Morikawa, Masato Tashiro, National Institute of Infectious Diseases, Tokyo, Shinjuku-ku, Tokyo 162-8640, Japan.

1. ABSTRACT

We have investigated novel vaccines strategies against severe acute respiratory syndrome (SARS) CoV infection using cDNA constructs encoding the structural antigens; spike (S), membrane (M), envelope (E), or nucleocapsid (N) protein, derived from SARS CoV (strain HKU39849, TW1, or FFM-1). As SARS-CoV is thought to infect the alveolar epithelial cell of the lung,in the present study, a type II alveolar epithelial cell clone, T7, was used to analyze the mechanism of CTL against SARS CoV membrane antigens. Mice vaccinated with SARS CoV (N) DNA or (M) DNA using pcDNA 3.1(+) plasmid vector showed T-cell immune responses (CTL induction and proliferation) against type II alveolar epithelial cells (T7) transfected with SARS (N) or (M) DNA, respectively.

To determine whether these DNA vaccines could induce T-cell immune responses in humans as well as in mice, SCID-PBL/hu mice were immunized with these DNA vaccines. PBL from healthy human volunteers were administered i.p. into IL-2 receptor γ-chain-disrupted NOD-SCID mice [IL-2R(-/-) NOD-SCID]. SCID-PBL/hu mice thus constructed can be used to analyze the human immune response *in vivo.* The SCID-PBL/hu mice were immunized with SARS (N) DNA or (M) DNA and analyzed for a human T-cell immune response. The M DNA vaccine enhanced CTL activity and proliferation in the presence of M peptide in SCID-PBL/hu mice. Furthermore, the SARS N DNA vaccine induced CTL activity (IFN-γ production by recombinant N protein or N protein-pulsed autologous B blast cells) and proliferation of spleen cells in SCID-PBL/hu mice. These results, demonstrate that SARS M and N DNA vaccines induced human CTL and human T-cell proliferative responses.

On the other hand, we have developed SARS DNA vaccines that induce human neutralizing antibodies and human monoclonal antibodies against SARS CoV. Transgenic mice expressing SARS-CoV receptor (angiotensin converting enzyme 2) are also under development. These vaccines are expected to induce immune responses specific for SARS CoV in human and should provide useful tool for development of protective vaccines.

2. INTRODUCTION

The causative agent of severe acute respiratory syndrome (SARS) has been identified as a new type of corona virus, SARS corona virus (SARS-CoV) [1,2,3]. SARS has infected more than 8400 patients in about 7 months in over 30 countries and caused more than 800 deaths. The deadly epidemic has had significant impacts on many health, social, economic and political aspects. SARS may resurge in the near future. However, no SARS vaccine is currently available for clinical use. Therefore, we have developed novel vaccine candidates against SARS CoV using cDNA constructs encoding the structural antigens; S, M, E, or N protein. In immunized mice, neutralizing antibodies against the virus and T-cell immunity against virus-infected-cells were studied, since these responses play important roles in protection against many virus infections. In particular, $CD8^+$ CTL plays an important role in T cell immunity against virus infections and in the eradication of murine and human cancers. [4,5] In the present study, a type II alveolar epithelial cell clone, T7, was used for analyzing precise mechanism of CTL against SARS-CoV membrane antigens, as the SARS-CoV infects alveolar epithelial cell in the lungs.[6] Furthermore, the SCID-PBL/hu model, which is capable of analyzing *in vivo* human immune response, was also used because it is a more relevant translational model for human cases.[4] These vaccines induce human immune responses (neutralizing antibody and CTL) specific for SARS CoV in human and should provide useful tool for development of protective vaccines.

3. MATERIALS AND METHODS

Three SARS CoV strains HKU39849[1], TW-1, and FFM-1[2] and their cDNAs were used. S, M, N, or E cDNA was transferred into pcDNA 3.1(+) vector and pcDNA 3.1(+)/vs-His Topo (QIAGEN K K, Tokyo, Japan). These genes were expressed in eukaryotic cells and Escherichia coli. pcDAN 3.1(+) vector, 50 μg each, containing SARS S, M, N, or E DNA was injected i.m. (M. tibia anterior) into C57BL/6 mice (female, 8 weeks, CLEA Japan Inc, Japan) and BALB/c mice (female, 8 weeks) three times, at an interval of 7 days. Neutralizing antibodies against SARS CoV in the serum from the mice immunized with SARS S, M, N, or E DNA vaccines were assayed using Vero-E6 cell. CTL activity against SARS-CoV was studied using human type II alveolar epithelial cells, T7, expressing SARS antigens.[6] PBL from healthy human volunteers were administered i.p. into IL-2 receptor γ-chain-disrupted NOD-SCID mice [IL-2R(-/-) NOD-SCID], and SCID-PBL/hu mice were constructed.[4] SARS DNA vaccines at 50 μg were injected i.m. into the SCID-PBL/hu mice. CTL activity of human CD8-positive lymphocytes in the spleen from SCID-PBL/hu was assessed using IFN-γ production and ^{51}Cr-release assay[4,5] Human monoclonal antibodies were produced from B cell hybridoma using P3U1 myeloma cell and spleen cells from human immunoglobulin transchromosomic mice (KM mice).

4. RESULTS

Induction of CTL against SARS CoV by SARS (N) DNA and SARS (M) DNA vaccine: Spleen cells from C57BL/6 mice immunized with SARS-S, -M, -N or -E DNA vaccine were cultured with syngeneic T7 lung cells transfected with S, M, N, or E cDNA. pcDNA 3.1(+) SARS (N) DNA vaccine induced significantly CTL activity (IFN-γ production) against N cDNA transfected T7 cells. Similarly, SARS M DNA vaccine induced SARS antigen M-specific CTL against T7 cells transfected with SARS M DNA.

Augmentation of lymphocyte proliferation specific for SARS CoV antigens by immunization with SARS (M) DNA and SARS (N) DNA vaccine: The proliferation of splenic T cells stimulated by co-culture either with T7 cells transfected with M DNA or SARS M peptide (TW1 M102-116) was strongly augmented by M DNA vaccine (data not shown). SARS N DNA vaccine also induced proliferation of splenic T cells in the presence of recombinant N protein as well as N DNA-transfected T7 cells. Thus, both SARS N

DNA vaccine and M DNA vaccine were shown to induce T-cell immune responses against the relevant SARS-CoV antigens.

Induction of neutralizing antibodies against SARS-CoV by immunization with SARS(S) DNA vaccine: The production of neutralizing antibodies against SARS CoV using Vero E6 cells infected with SARS CoV was observed in the serum from BALB/c mice immunized with S DNA vaccine in the presence of adjuvants (MPL+TDM+ALUM) (Table 1).

Table 1. Induction of neutralizing antibody against SARS coronavirus by SARS (S) DNA vaccination of BALB/c mice.

	Immunization with	Adjuvant	Neutralizing antibody against SARS corona virus
pcDNA 3.1(+)	SARS HKU-S DNA Vaccine 50μg	MPL TDM ALUM	+
pcDNA 3.1(+)	SARS HKU-S DNA Vaccine 50μg	-	-
	SARS TW1-S DNA Vaccine	MPL TDM ALUM	+
	SARS TW1-S DNA	-	-

SARS M DNA and N DNA vaccines induced human T- cell immune responses (CTL and proliferation) and the production of neutralizing antibodies against SARS-CoV in SCID-PBL/hu model: The M DNA vaccine enhanced CTL activity and proliferation in the presence of M peptide in SCID-PBL/hu mice. Furthermore, the SARS N DNA vaccine induced CTL activity (IFN production by recombinant N protein or N protein-pulsed autologous B blast cells) and proliferation of spleen cells in SCID-PBL/hu mice (Fig. 1). From these results, it was demonstrated that SARS M DNA vaccine and N DNA vaccine induced human CTL and human T-cell proliferative responses. Furthermore, human neutralizing antibodies were induced in SCID-PBL/hu mice vaccinated with SARS S and M DNA (Table 2).

5. DISCUSSION

We have demonstrated that SARS (M) DNA and (N) DNA vaccines induce

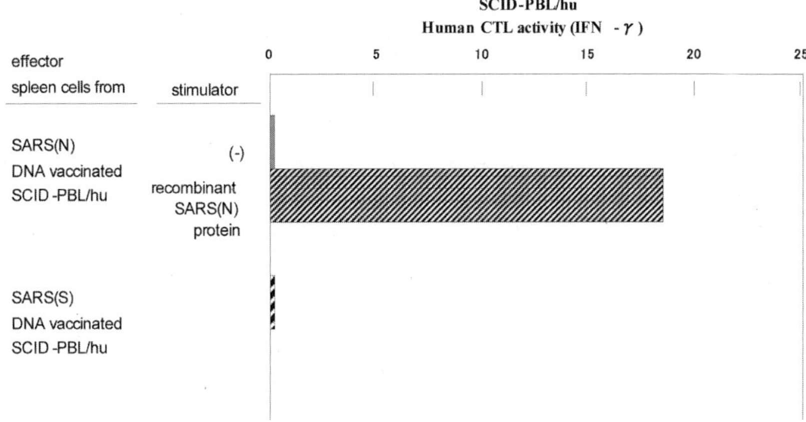

Figure 1. SARS (N) DNA vaccine induces *in vivo* human CTL against SARS CoV in the SCID-PBL/hu human immune systems. $4x10^7$ PBL from healthy human volunteers were administered i.p. into IL-2R (-/-) NOD-SCID. $1x10^5$ spleen cells from SCID-PBL/hu were cultured with 10 μg of recombinant SARS (N) protein for 72hr.

Table 2. Induction of human neutralizing antibody against SARS coronavirus in SCID PBL/hu mice by SARS DNA vaccinations.

Mice	Immunization with	Adjuvant	Neutralizing antibody Against SARS Corona Virus
IL-2R g-chain(-/-) NOD-SCID PBL/hu mice	SARS TW1-S SARS (S) DNA	Adenovirus vector/ IL-6DNA+IL-6R DNA+gp130DNA + MPL	+
IL-2R g-chain(-/-) NOD-SCID PBL/hu mice	SARS(M) DNA	Adenovirus vector/ IL-6 DNA + IL-6R DNA + gp130 DNA	+
		-	-

50 μg of SARS (S) DNA was immunized three times into SCID mice (IL-2 Receptor γ-chain-disrupted NOD SCID) at the interval of 7 days.

virus-specific immune responses (CTL and T-cell proliferation) in the mouse system type II lung alveolar T-cell lines to present antigen.[6] These DNA vaccines induced SARS-CoV–specific CTL and T-cell proliferation *in vivo* human immune systems using SCID-PBL/hu. Gao *et al.* showed that an adenovirus-based SARS DNA vaccine encoding S1 polypeptide was capable of inducing neutralizing antibody, while another SARS DNA vaccine encoding N protein generated IFN-γ– producing

T cells in rhesus monkeys.[7] SARS S DNA vaccines elicit neutralizing antibody responses that generate protective immunity in a mouse model (8). However, its immunogenicity inhumans has yet to be established. Therefore, it is very important to evaluate the efficacy of SARS DNA vaccines in SCID-PBL/hu mice, which is a highly relevant translational model for demonstrating human immune responsiveness. SARS S DNA and SARS M DNA vaccines capable of inducing human neutralizing antibodies against SARS CoV have been established by our SCID-PBL/hu model. It has been demonstrated that angiotensin- converting enzyme 2 (ACE2) is a functional receptor for the SARS CoV.[9] A transgenic mouse with human ACE-2 may be useful as an animal model of SARS. Furthermore, ACE-2 transgenic SCID mice should be useful as a human model for preclinical trial for SARS vaccines, for analyzing human immune responses against SARS infection *in vivo.* The effect of combination immunization with such SARS vaccines and nentralizing antibody-inducing DNA vaccines is now being studied. These DNA vaccines should provide a useful tool for development of protective vaccines.

6. ACKNOWLEDGMENT

This study was supported by Grant-in-Aid for science and technology and Grant-in-Aid for Scientific Research on Priority Areas from the Ministry of Education, Culture, Sports, Science and Technology, Japan. This study was also supported by a Health and Labour Science Research Grant from the Ministry of Health, Labour, and Welfare, Japan.

7. REFERENCES

1. Peiris JS, Lai ST, Poon LL, et al. SARS study group. Coronavirus as a possible cause of severe acute respiratory syndrome. *Lancet.* **361**:1319-25 (2003).
2. Drosten C, Gunther S, Preiser W, et al. Identification of a novel coronavirus in patients with severe acute respiratory syndrome. *N Engl J Med.* **348**:1967-76 (2003).
3. Peiris JS, Yuen KY, Osterhaus AD, Stohr K. The severe acute respiratory syndrome. *N Engl J Med.* **349**:2431-41 (2003).
4. Tanaka F, Abe M, Akiyoshi T, et al. The anti-human tumor effect and generation of human cytotoxic T cells in SCID mice given human peripheral blood lymphocytes by the in vivo transfer of the Interleukin-6 gene using adenovirus vector. *Cancer Res.* **57**:1335-43 (1997).
5. Okada M, Yoshimura N, Kaieda T, Yamamura Y, Kishimoto T. Establishment and characterization of human T hybrid cells secreting immunoregulatory molecules. *Proc Natl Acad Sci U S A.* **78**:7717-21 (1981).
6. deMello DE, Mahmoud S, Padfield PJ, Hoffmann JW. Generation of an immortal differentiated lung type-II epithelial cell line from the adult H-2K(b)tsA58 transgenic mouse. *In Vitro Cell Dev Biol Anim.* **36**:374-82 (2003).
7. Gao W, Tamin A, Soloff A, et al. Effects of a SARS-associated coronavirus vaccine in monkeys. *Lancet.* **362**:1895-6 (2003).
8. Yang ZY, Kong WP, Huang Y, et al. A DNA vaccine induces SARS coronavirus neutralization and protective immunity in mice. *Nature.* **428**:561-4 (2004).
9. Li W, Moore MJ, Vasilieva N, et al. Angiotensin-converting enzyme 2 is a functional receptor for the SARS coronavirus. *Nature.* **426**:450-4 (2003).

INHIBITION AND ESCAPE OF SARS-CoV TREATED WITH ANTISENSE MORPHOLINO OLIGOMERS

Benjamin W. Neuman, David A. Stein, Andrew D. Kroeker, Hong M. Moulton, Richard K. Bestwick, Patrick L. Iversen, and Michael J. Buchmeier*

1. INTRODUCTION

Identification of potential SARS-CoV antiviral compounds has progressed swiftly, thanks in part to the availability of bioinformatic and virus structural data. Antivirals that target the SARS-CoV superfamily 1 helicase and the 3C-related serine proteinase with low micromolar EC_{50} values have been reported.[1-3] The papain-related cysteine proteinase may prove to be an unsuitable target, as a coronavirus molecular clone lacking one of the two known cleavage sites for this enzyme displayed only minor growth defects in cell culture.[4] Other confirmed and putative viral enzymes including the polymerase, poly(U)-specific endo-ribonuclease homolog, *S*-adenosyl-methionine-dependent ribose 2'-*O*-methyltransferase, and cyclic phosphodiesterase represent plausible anti-SARS targets.[5] Antivirals targeting the interaction of the viral spike protein with the ACE-2 receptor,[2,3,6] or with the spike-mediated fusion event,[7-10] and showing micromolar-scale efficiency in cell culture, have been reported. Several groups have also reported antiviral *in vitro* efficacy with siRNAs.[11]

The antisense agents directed against single-stranded RNA are known to act by two general mechanisms: by causing damage to an RNA strand containing the complementary "target" sequence through priming of endogenous RNase H activity, or by stably binding to and steric interference with targeted RNA function. Phosphorodiamidate morpholino oligomers (PMO) act by the latter mechanism, duplexing to specific RNA sequence by Watson-Crick base pairing and forming a steric block.[12] The most frequently successful targeting strategies for PMO-based gene knockdown involve interfering with translation initiation[13] or masking splice sites.[14] We recently demonstrated antiviral effects *in vitro* for one peptide-conjugated PMO (P-PMO) complementary to the AUG translation start site region of a murine coronavirus replicase

* Benjamin W. Neuman, Michael J. Buchmeier, The Scripps Research Institute, La Jolla, California. David A. Stein, Andrew D. Kroeker, Hong M. Moulton, Richard K. Bestwick, Patrick L. Iverson, AVI Biopharma, Inc., Corvallis, Oregon.

polyprotein.[15] We reasoned that antiviral effects of P-PMO might be improved by choosing conserved RNA sequence elements and secondary structures critical for replication, transcription, and host factor interaction as targets. In this report, we demonstrate that antisense-mediated suppression of viral replication can be achieved by targeting conserved RNA elements required for viral RNA synthesis and translation.

2. ANTIVIRAL P-PMO SELECTION AND EFFICACY

PMO complementary to the genomic (positive-sense) strand, which were designed to bind regions identified by conservation or noted in the literature as critical for viral RNA synthesis, were synthesized (Fig. 1a). PMO were covalently linked to peptides NH_2-RRRRRRRRRFFC-$CONH_2$ or NH_2-RRRRRFFRRRRC-$CONH_2$ designated as R_9F_2 or $R_5F_2R_4$, respectively. Both types of peptide-conjugated PMO are henceforth referred to as P-PMO, and were used interchangeably in these studies. Cells treated with $\leq$ 20 μM PMO were at least 80% viable, as measured by MTT viability assay (data not shown).

We tested P-PMO for correlates of antiviral efficacy: reduction of viral titer, rate of spread *in vitro,* and viral subgenomic RNA load. Vero-E6 cells were pretreated with P-PMO 6 h prior to low multiplicity inoculation, and supernatants were collected for titration 24 h later (Fig. 1b). The most effective P-PMO (TRS2) decreased viral titers below the threshold of detection in the experiment shown, and reduced peak titers by 10,000-fold in some assays (not shown). TRS1 and TRS2 exhibited robust antiviral activity in the low micromolar range.

We next tested the effectiveness of P-PMO on viral persistence and spread of an established infection by performing plaque-size reduction assays. In this assay, cells were

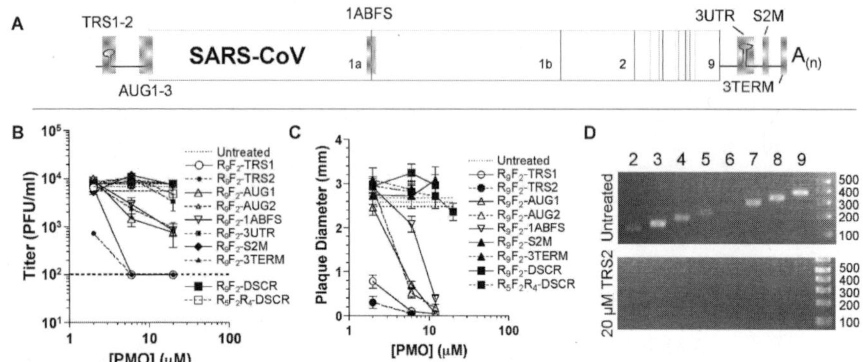

Figure 1. Binding sites and *in vitro* efficacy of MHV P-PMOs. A schematic presentation of the SARS-CoV genome shows the sites that were selected as P-PMO target sites (A). P-PMOs tested here included five designed to directly inhibit translation of the replicase open reading frame 1a (TRS1-2, AUG1-3), one to inhibit ribosomal frameshifting (1ABFS), three to bind conserved sequences in the 3'-untranslated region (3UTR, S2M, 3TERM), and one scrambled control sequence (DSCR). P-PMOs directed to the leader transcription regulatory sequence were most effective at reducing viral titer (B), and the diameter of viral plaques (C). Comparison of the detectable amounts of subgenomic RNAs present in infected cells (D). Amplicons specific to sgRNA 2-9 were obtained by RT-PCR on total RNA extracted from untreated or TRS2 P-PMO-treated cells.

treated with P-PMO 1 h after inoculation with a standardized amount of SARS-CoV. Plaque diameter was measured 72 h after inoculation. The TRS2 P-PMO was most effective at reducing SARS-CoV spread in cell culture (Fig. 1C). We reasoned that inhibition of viral growth should correspond with a decrease in the viral RNA level, whether through inhibition of replicase expression, interference with discontinuous RNA synthesis at the leader TRS, or an alternate mechanism. Coronaviruses produce a nested set of sub-genome-length RNA species in infected cells. We investigated genomic and subgenomic RNA production 24 h after low-multiplicity inoculation. RT-PCR products specific to each of eight subgenomic RNA species were strongly amplified from untreated cells and cells treated with less effective P-PMO (Fig. 1D and data not shown). Equal volumes of RT-PCR products from an equivalent number of infected cells pretreated with 20 μM TRS2 P-PMO were faint (i.e., sgRNA 8 and possibly 9) or undetectable (sgRNA 2–7; Fig. 1D). Genomic RNA synthesis was likewise qualitatively reduced by 20 μM TRS2 P-PMO (data not shown).

3. ESCAPE OF SARS-CoV AFTER SERIAL P-PMO SELECTION

The error-prone replication of RNA viruses presents a rapid model for viral evolution and drug resistance studies. In order to assess the propensity for SARS-CoV to develop resistance to antisense P-PMO, a stock cultured from a plaque-purified biological clone of SARS-CoV was serially passaged on cells pretreated with P-PMO. Viral growth was assessed, after each passage (Fig. 2A). Treatment with 10 μM TRS2 P-PMO strongly inhibited SARS-CoV growth for several passages. However, an increase in titer indicative of partial resistance was observed after seven passages. SARS-CoV plaque purified after 11 rounds of TRS2 P-PMO selection formed small plaques on Vero-E6 cells in the absence of P-PMO (Fig. 2B). TRS2 P-PMO-selected SARS-CoV displayed delayed growth kinetics compared with untreated SARS-CoV and other P-PMO-selected SARS-CoV (data not shown). RNA was isolated from plaque purified SARS-CoV selected after 11 rounds of serial P-PMO treatment. RT-PCR amplicons from 14 serially P-PMO-treated SARS-CoV were sequenced to determine whether the virus had undergone mutation during P-PMO selection. Three contiguous base changes of CTC to AAA at position 61–63, proximal to the leader TRS and within the target region of TRS2-P-PMO, appeared in only the 14 amplicons from TRS2-resistant SARS-CoV (Fig. 2C).

Thermal melting curve data for peptide-conjugated PMO/RNA duplexes with variable mismatches lead us to speculate that the three mutations at the TRS2-P-PMO target site reduce the effective melting temperature (Tm) by ~25–30°C (H. Moulton *et al.*, manuscript in preparation and[16]). P-PMO binding affinity was compared using a reporter construct in which the luciferase reporter gene was placed immediately downstream of either the wild-type SARS-CoV TRS region or the same region with the CTC→AAA mutations observed in TRS2 P-PMO-selected SARS-CoV clones (Fig. 3A). TRS2 P-PMO was approximately ten-fold less active against the 3-mismatch TRS target compared with the wild-type target (EC_{50} of 500nM and 50 nM, respectively). The decreased sensitivity to TRS2 P-PMO was consistent with reduced P-PMO/target RNA duplex stability in partially-TRS2-resistant SARS-CoV. A similar observation was recently reported for HIV-1 escape variants resistant to siRNAs.[17]

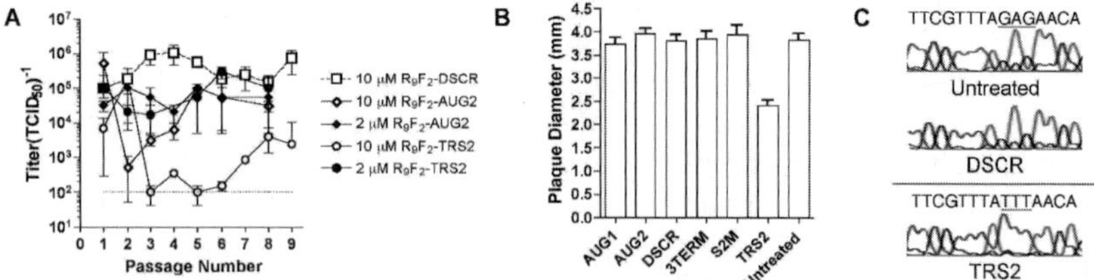

Figure 2. Escape of SARS-CoV after serial P-PMO passage. The titer of infectious SARS-CoV grown over nine rounds of treatment with selected P-PMOs was measured by $TCID_{50}$ titration (A). Error bars indicate standard error throughout. Plaque morphology of P-PMO-resistant SARS-CoV strains was assessed in the absence of P-PMO. Mean plaque diameter is shown (B). SARS-CoV strains selected after serial P-PMO treatment were sequenced. The region comprising the TRS2 P-PMO binding site is shown (C). Sequences from untreated and DSCR-treated SARS-CoV represent wild-type sequence, and mutations found in all TRS2-selected strains are underlined. Sequence is from the complementary (minus-sense) strand.

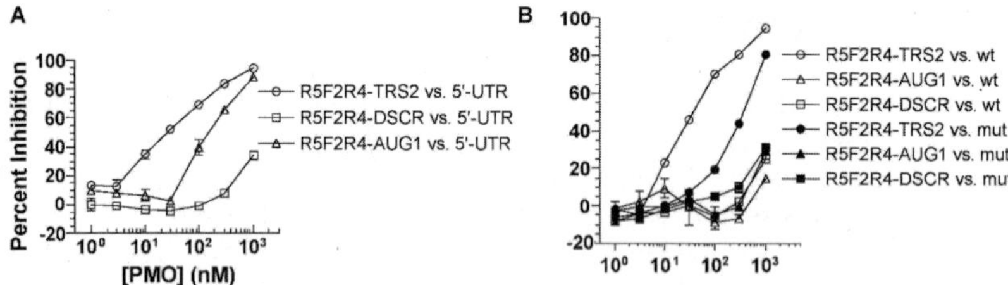

Figure 3. Mechanism of P-PMO escape and inhibition of SARS-CoV. Reporter gene expression from a synthetic RNA containing the SARS-CoV 5'-UTR upstream of the luciferase gene was measured in the presence of SARS-CoV P-PMO (A). Error bars represent standard error throughout. Reporter gene translation from a synthetic RNA containing the SARS-CoV 5'-untranslated region upstream of the luciferase gene (wt) or containing instead the mutated 5'-untranslated region found in serially TRS2-selected SARS-CoV upstream of the luciferase gene (mut) was measured in the presence of P-PMO (B).

The antiviral effects of TRS2 P-PMO were consistent with multiple mechanisms of action. In order to determine whether the high efficacy of TRS2 could be attributed to steric blockade of translation, a reporter construct was designed in which luciferase expression was initiated at the AUG codon of the authentic SARS-CoV replicase open reading frame 1a AUG, downstream of the SARS-CoV 5'-untranslated region. TRS2 P-PMO was much more effective than AUG1 P-PMO in inhibiting reporter translation (EC_{50} of 35 nM vs. 185 nM; Fig, 3B). The TRS2 target site (bases 55–75) is sufficiently distal from both the 5'-terminus and the site of translation initiation to make it unlikely that interference with events of pre-initiation at the terminus (e.g., the 43S complex loading onto mRNA) or initiation at the initiator AUG (e.g., 48S-complex formation and/or joining of 48S and 60S ribosomal subunits) forms the basis for the observed effect. We therefore concluded that TRS2 P-PMO primarily inhibits SARS-CoV growth by interference with translation, perhaps at the 43S-preinitiation complex scanning.

4. ACKNOWLEDGMENTS

This work was supported by NIH grants AI059799, AI025913, NS41219 and by NIH/NIAID contract HHSN266200400058C. Figures and text adapted from Neuman *et al.*, 2004. J. Virol. **79** (15):9665-9676. Used with permission.

5. REFERENCES

1. U. Bacha, J. Barrila, A. Velazquez-Campoy, S. A. Leavitt, and E. Freire, Identification of novel inhibitors of the SARS coronavirus main protease 3CLpro, *Biochemistry* **43**, 4906-4912 (2004).
2. R. Y. Kao, W. H. Tsui, T. S. Lee, et al., Identification of novel small-molecule inhibitors of severe acute respiratory syndrome-associated coronavirus by chemical genetics, *Chem. Biol.* **11**, 1293-1299 (2004).
3. C. Y. Wu, J. T. Jan, S. H. Ma, et al., Small molecules targeting severe acute respiratory syndrome human coronavirus, *Proc. Natl. Acad. Sci. USA* **101**, 10012-10017 (2004).
4. M. R. Denison, B. Yount, S. M. Brockway, R. L. Graham, A. C. Sims, X. Lu, and R. S. Baric, Cleavage between replicase proteins p28 and p65 of mouse hepatitis virus is not required for virus replication, *J. Virol.* **78**, 5957-5965 (2004).
5. E. J. Snijder, P. J. Bredenbeek, J. C. Dobbe,et al., Unique and conserved features of genome and proteome of SARS-coronavirus, an early split-off from the coronavirus group 2 lineage, *J. Mol. Biol.* **331**, 991-1004 (2003).
6. L. Yi, Z. Li, K. Yuan, et al., Small molecules blocking the entry of severe acute respiratory syndrome coronavirus into host cells, *J. Virol.* **78**, 11334-11339. (2004).
7. B. J. Bosch, B. E. Martina, R. Van Der Zee, et al., Severe acute respiratory syndrome coronavirus (SARS-CoV) infection inhibition using spike protein heptad repeat-derived peptides, *Proc. Natl. Acad. Sci. USA* **101**, 8455-8460 (2004).
8. P. Ingallinella, E. Bianchi, M. Finotto, et al., Structural characterization of the fusion-active complex of severe acute respiratory syndrome (SARS) coronavirus, *Proc. Natl. Acad. Sci. USA* **101**, 8709-8714 (2004).
9. S. Liu, G. Xiao, Y. Chen, et al., Interaction between heptad repeat 1 and 2 regions in spike protein of SARS-associated coronavirus: implications for virus fusogenic mechanism and identification of fusion inhibitors, *Lancet* **363**, 938-947 (2004).
10. K. Yuan, L. Yi, J. Chen, et al., Suppression of SARS-CoV entry by peptides corresponding to heptad regions on spike glycoprotein, *Biochem. Biophys. Res. Commun.* **319**, 746-752 (2004).
11. Z. Wang, L. Ren, X. Zhao, T. Hung, A. Meng, J. Wang, and Y. G. Chen, Inhibition of severe acute respiratory syndrome virus replication by small interfering RNAs in mammalian cells, *J. Virol.* **78**, 7523-7527 (2004).
12. J. Summerton, Morpholino antisense oligomers: the case for an RNase H-independent structural type, *Biochim. Biophys. Acta* **1489**, 141-158 (1999).
13. A. Nasevicius and S. C. Ekker, Effective targeted gene 'knockdown' in zebrafish, *Nat. Genet.* **26**, 216-220 (2000).
14. R. V. Giles, D. G. Spiller, R. E. Clark, and D. M. Tidd, Antisense morpholino oligonucleotide analog induces missplicing of C-myc mRNA, *Antisense Nucleic Acid Drug Dev.* **9**, 213-220 (1999).
15. B. W. Neuman, D. A. Stein, A. D. Kroeker, et al., Antisense morpholino-oligomers directed against the 5' end of the genome inhibit coronavirus proliferation and growth, *J. Virol.* **78**, 5891-5899 (2004).
16. D. Stein, E. Foster, S. B. Huang, D. Weller, and J. Summerton, A specificity comparison of four antisense types: morpholino, 2'-O-methyl RNA, DNA, and phosphorothioate DNA, *Antisense Nucleic Acid Drug Dev.* **7**, 151-157 (1997).
17. E. M. Westerhout, M. Ooms, M. Vink, A. T. Das, and B. Berkhout, HIV-1 can escape from RNA interference by evolving an alternative structure in its RNA genome, *Nucleic Acids Res.* **33**, 796-804 (2005).

VALIDATION OF CORONAVIRUS E PROTEINS ION CHANNELS AS TARGETS FOR ANTIVIRAL DRUGS

Lauren Wilson, Peter Gage, and Gary Ewart*

1. INTRODUCTION

Coronaviruses are divided into three groups, depending on the sequence homology and antigen cross-reactivity. Groups 1 and 2 contain the mammalian coronaviruses, and group 3 consists of the avian coronaviruses. All coronavirus groups encode E protein, a small, 9–12 kDa integral membrane protein.[1] Although, there is little sequence homology between the coronavirus groups, all E proteins share structural homology, they all contain an N-terminus, which consists of a short 7–9 amino acid hydrophilic region, and a 21–29 amino acid hydrophobic transmembrane domain, followed by a hydrophilic C-terminal region.[2] The exact functions and mechanisms of the coronavirus E proteins are yet to be established, although E proteins have been shown to be important for coronavirus replication, mediating viral assembly, and morphogenesis.

Coronavirus E proteins share several characteristics with viral ion channels, which are small hydrophobic virus-encoded proteins. Virus ion channels have a highly hydrophobic domain that forms at least one amphipathic α-helix that oligomerizes to form an ion-conductive pore in membranes. Virus ion channels function to modify the cells permeability to ions and have been shown to mediate viral entry/exit or virus assembly and budding.[3,4] The first identified viral ion channel, hence the best characterized, is the M2 protein encoded by influenza A. M2 forms proton selective ion channels that mediates viral uncoating and protects acid-sensitive hemagglutinin glycoprotein during transport to the cell surface.[5] Although influenza B does not encode the M2 ion channel, it has been demonstrated to have two ion channel forming proteins, NB and BM2,[5,6] whose role in viral replication are currently being investigated. Since the identification of the M2 ion channel, several other viruses have been demonstrated to encode viral ion channels. The HIV-1 accessory protein Vpu has also been shown to have ion channel activity, which mediates the release of viral particles from the plasma membrane.[6,7] Most recently, the 6K proteins from the alphaviruses, Ross River Virus and Barmah Forest Virus, have been shown to form cation-selective ion channels in planar

*Lauren Wilson, ANU Medical School, ACT Australia. Peter Gage, JCSMR, ANU, ACT Australia.
Gary Ewart, Biotron Ltd., ACT Australia.

lipid bilayers.[6] Additionally, several authors have shown ion channel activity of the hepatitis C virus (HCV) p7 protein.[8-10]

The M2 channel is well established as a target for antiviral drug therapy, and blockers of other viral ion channels have been shown to inhibit replication of the parent virus. The M2 ion channel is inhibited by amantadine and some of its derivates, which are currently used as clinical treatment of influenza A infection.[11,12] The Vpu ion channel activity in planar lipid bilayers is inhibited by the amiloride derivatives 5-(*N,N*-hexamethylene)amiloride (HMA) and 5-(*N,N*-dimethyl)amiloride (DMA), but not by amiloride itself. Furthermore, HMA inhibits Vpu enhancement of Gag-driven virus-like particles (VLP) budding from HeLa cells co-expressing Vpu and Gag and also inhibits HIV-1 replication in cultured primary human macrophages.[6] In addition, the p7 ion channel has been shown to be inhibited by HMA, amantadine, and long-alkyl chain immunosugar derivatives.[8-10] Thus, a number of precedents have been set for ion channels as targets of potential antiviral compounds.

Due to coronavirus E proteins similarities with viral ion channels and their important role in viral replication, we hypothesized that coronavirus E proteins have ion channel activity, and compounds that block these channels may result in inhibition of viral replication. We report here that representative E proteins from all three coronavirus groups form ion channels. Furthermore, we found that certain amiloride derivates block E protein ion channel activity and inhibit replication of coronaviruses in cultured cells.

2. MATERIALS AND METHODS

2.1. Peptide Synthesis and Purification

E peptides corresponding to the human coronavirus 229E (HCoV-229E) E protein (MFLKLVDDHALVVNVLLWCVVLIVILLVCITIIKLIKLCFTCHMFCNRTVYGPIK NVYHIYQSYMHIDPFPKRVIDF), GenBank accession number NP_073554, mouse hepatitis virus (MHV)-A59 E protein (MFNLFLTDTVWYVGQIIFIFAVCLMVT IIVVAFLASIKLCIQLCGLCNTLVLSPSIYLYDRSKQLYKYYNEEMRLPLLEVDDI), GenBank accession number NP_068673, severe acute respiratory syndrome (SARS) coronavirus (SARS-CoV) full-length E protein (MYSFVSEETGTLIVNSVLLFLAFVVF LLVTLAILTALRLCAYCCNIVNVSLVKPTVYVYSRVKNLNSSEGVPDLLV), SARS - CoV N-terminal E protein (MYSFVSEETGTLIVNSVLLFLAFVVFLLVTLAIL TALRLC), GenBank accession number NC004718, and infectious bronchitis virus (IBV) Beaudette strain E protein (MTNLLNKSLDENGSFLTALYIFVGFLALYLLGRA LQAFVQAADACCLFWYTWVVVPGAKGTAFVYNHTYGKKLNKPELETVINEFPK NGWKQ), GenBank accession number CAC39303, were chemically synthesized and purified, as described previously.[13,14] The peptides were shown to contain full-length products by a variety of methods including, Western blot analysis with E protein specific antibodies, and mass spectral analysis.[13,14]

2.2. Ion Channel Recording

The HCoV-229E, MHV, SARS-CoV, and IBV purified E proteins were resuspended to 1 mg/ml in 2,2,2-trifluoroethanol (TFE) and their ability to form ion channels was tested on a Warner bilayer rig (Warner instruments, Inc. Hamden, CT), as described previously.[13,14]

2.3. Amiloride Derivatives Ion Channel Inhibition

A 50 mM stock solution of compounds was prepared in 50% DMSO, 50% methanol, which was further diluted for use in ion channel inhibition studies and in the antiviral assays.

To determine if the amiloride derivatives blocked SARS-CoV E protein ion channel conductance in planar lipid bilayers, after ion channel current amplitude was detected, 100–200 μM of compound was added to the CIS chamber while stirring. The current across the bilayer was recorded prior to addition of SARS-CoV E protein, after detection of ion channel conductance, and after addition of the compound. T-test (Microsoft Excel) was used to test the difference between the normalized mean currents before and after addition of the compound.

2.4. Testing Amiloride Derivative Antiviral Activity

The virus plaque phenotype in the presence or absence of antiviral compound was studied in L2 cells (ATCC). The L2 cells were plated in 6-well plates and grown to confluence, then infected with a MOI 0.01 of MHV-A59 (ATCC) or MOI 0.1 of MHVΔE or MHVSARS E (kind gift from Paul Masters, Wadsworth Center, Albany, NY) for 1 hour. The virus was removed and replaced with 1% seaplaque overlay in MEM supplemented with 10% FCS and 20 μM or 0 μM of antiviral compound in 50% DMSO 50% methanol. After 48 hours incubation at 37°C/5% CO_2, the cells were stained with 0.1% crystal violet in 20% methanol.

To determine the Selectivity Index (SI) of the antiviral compounds the effective concentration 50 (EC_{50}) was calculated by plaque assay on L929 cells (ATCC) and the toxicity concentration 50 (TC_{50}) was calculated by MTT cytoxicity assay. The SI was calculated by dividing the TC_{50} by the EC_{50} ($SI = TC_{50}/EC_{50}$).

3. RESULTS AND DISCUSSION

3.1. Ion Channel Recordings

HCoV-229E, MHV, IBV, SARS-CoV -full-length and -N-terminal E proteins were tested for their ability to form ion channels in planar lipid bilayers. Experiments were done to determine E proteins ion selectivity for Na^+ over Cl^- and K^+ over Cl^- ions. Figure 1 shows typical ion channel conductance of representatives from coronavirus E proteins from group 1, 2, or 3, demonstrating that E protein ion channel activity is a general property of all coronavirus groups.Interestingly, the E proteins from the different coronavirus group had divergent ion channel selectivity. The group 1 coronavirus HCoV-229E E protein is K^+ selective, whereas the group 2 MHV and SARS-CoV E proteins, as well as the group 3 coronaviruses, IBV E protein were more selective for Na^+ ions. The different E protein ion channel selectivity may reflect subtly divergent roles of the E protein groups in the coronavirus life cycle, although this remains to be established. In support of this idea is

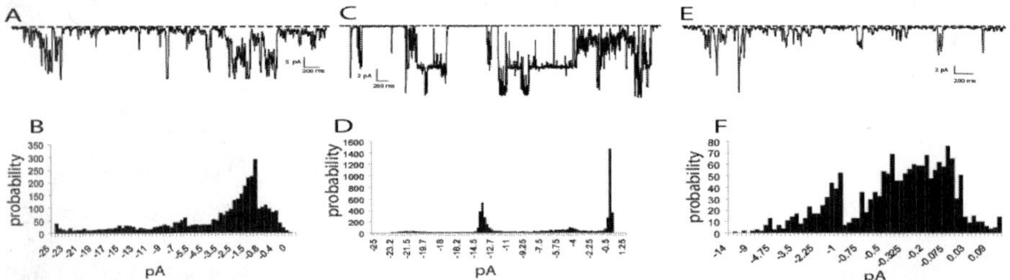

Figure 1. Typical ion channel current of representative group specific E proteins in planar lipid bilayers. The closed state is shown as a broken line, openings are deviations from the line. (A) group 1 HCoV-229E E protein ion channel conductance scale bars are 200 ms and 5 pA. Potential was held at -40 mV and (B) all points histograms of currents shown in A. (C) group 2 MHV E protein ion channel conductance scale bars are 200 ms and 2 pA. Potential was held at -60 mV and (D) all points histograms of currents shown in C. (E) group 3 IBV E protein ion channel conductance scale bars are 200 ms and 2 pA. Potential was held at -20 mV and (F) all points histograms of currents shown in (E). For ion channel conductance of the SARS-CoV full-length and TM domain, see Ref.13.

that the group 1 -K^+ selective - E proteins are essential for viral replication,[15] whereas group 2 (Na^+ selective) E proteins are important, but not essential for coronavirus replication.[16] Further data to support this theory was presented at this conference (see Refs. 17 in this proceedings book), Masters et al., 2006 demonstrated that the Na^+ selective E proteins from group 2, bovine coronavirus and SARS-CoV, as well as the group 3 IBV could substitute for the Na^+ selective MHV E protein and enhance replication of recombinant MHV virus. On the contrary, the group 1 transmissible gastroenteritis virus E protein, which our data suggest may be K^+ selective, could not substitute for the MHV E protein in recombinant viruses.[17,18] Because, the group 1 and 2 coronavirus E proteins share more sequence homology than the group 2 and 3 E proteins,[18,19] the ability of the different E protein groups to substitute for the MHV E protein could be more dependent on their ion channel selectivity than the sequence homology.

Previously, it has been demonstrated that the transmembrane (TM) domain of several ion channels, including M2 and Vpu, form channels with similar properties as the full-length proteins.[18,19] Therefore, we tested the ability of the SARS-CoV E protein N-terminal first 40 amino acids, which encompass the putative TM domain for its ability to form ion channels in planar lipid bilayers. Indeed, the SARS-CoV N-terminal peptide formed ion channels that had similar properties to the full-length E protein. Thus, the hydrophilic C-terminal domain is dispensable for ion channel activity in planar lipid bilayers.[13] Several studies are currently being conducted to determine if E proteins TM domain ion channel activity is important for coronavirus replication (see Refs. 20 and 21 in this proceedings book). Intriguingly, these studies have shown that substitution or mutation of the E protein TM domain is detrimental for viral replication, suggesting that E protein ion channel activity could be important for coronavirus replication.

3.2. Amiloride Derivatives Inhibit SARS-CoV E Protein Ion Channel Conductance

Because it is possible that E protein ion channel activity is important for coronavirus replication and a precedent has been set that ion channels are suitable targets of antiviral therapy, we tested amiloride, plus, its derivatives HMA, and 5-(*N*-Methyl-*N*-

isobutyl)amiloride (MIA) for their ability to inhibit SARS-CoV E protein ion channel conductance. Table 1 demonstrates that HMA reduced the conductance across the bilayer by about 80%, MIA by about 70%, while amiloride itself did not significantly reduce the current across the bilayer (Table 1).

3.3. Amiloride Derivatives Inhibit Replication of MHV and MHVSARS E, but Not MHVΔE

We found that HMA and MIA inhibited the SARS-CoV E protein ion channel conductance in planar lipid bilayers; therefore, we wanted to test their ability to inhibit coronavirus replication. Due to the safety issues and difficulties of working with the SARS-CoV, we decided to use the recombinant MHV virus that expresses the SARS-CoV E protein in place of the MHV E protein (MHVSARS E). The MHVSARS E virus replicates efficiently in cultured mouse cells, but has a slightly smaller plaque phenotype than the wild-type MHV.[17] As a control we also used the MHV recombinant virus with the entire E protein deleted (MHVΔE), which replicates to low titre in mouse cells and has a small plaque phenotype.[22] MHV wild-type virus has a plaque phenotype of about 3–4 mm in the absence of antiviral compound, but in the presence of 20 μM HMA or MIA the plaque size is reduced to about 1 mm. Similarly, 20 μM of HMA or MIA reduced the MHVSARS E plaque size from about 2–3 mm to 1 mm. In contrast, neither HMA nor MIA significantly affected MHVΔE plaque phenotype. Comparable with the ion channel conductance study, amiloride did not have any significant effect on MHV, MHVSARS E or MHVΔE replication (Table 1). The selectivity index (SI) of HMA and MIA on MHV and MHVSARS E replication were greater than 10, indicating that the antiviral activity of the compounds are notably removed from toxicity (Table 1). Further, the antiviral activity of these compounds against MHV expressing the homo- or heterologous E proteins, but not against the ΔE construct, together with direct observations of SARS-CoV E channel blockage, is supportive of our hypothesis that the mechanism of action of the compounds is via inhibition of E protein ion channel activity.

Table 1. Summary of amiloride derivatives inhibition of; SARS-CoV E protein in planar lipid bilayers and MHV, MHVSARS E, and MHVΔE replication in cultured cells.

Compound	Structure	Normalized % of SARS E current (± S.E.M)	MHV replication	MHVSARS E replication	MHVΔE replication
Amiloride	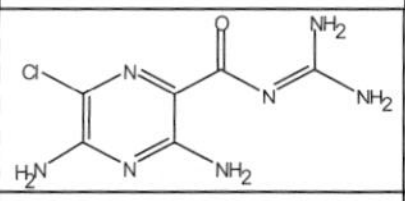 Cl, N, O, NH_2, N, NH_2, H_2N, N, NH_2	147 ± 42 (n=6, p=0.16) Does not block	No significant inhibition	No significant inhibition	No significant inhibition
HMA	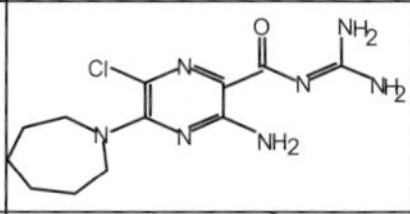 O, NH_2, Cl, N, N, NH_2, N, N, NH_2	21 ± 9 (n=5 p=0.0003) Blocks	Inhibits SI=37	Inhibits SI=31	No significant inhibition
MIA	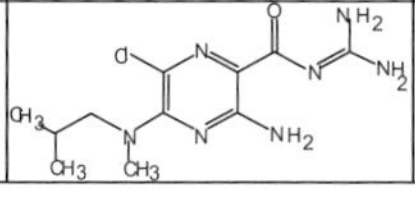 O, NH_2, Cl, N, N, NH_2, CH_3, N, N, NH_2, CH_3, CH_3	33 ± 15 (n=5, p=0.006) Blocks	Inhibits SI=77	Inhibits SI=33	No significant inhibition

4. REFERENCES

1. S. Siddell, in: *The Coronaviridae,* edited by S. G. Siddell (Plenum Press, New York, 1995), pp. 181-189.
2. X. Shen, et al., Small envelope protein E of SARS: Cloning, expression, purification, CD determination, and bioinformatics analysis, *Acta Pharmacol. Sin.* **24**, 505-511 (2003).
3. W. B. Fischer and M. S. Sansom, Viral ion channels: Structure and function, *Biochim. Biophys. Acta* **1561**, 27-45 (2002).
4. M. E. Gonzalez and L. Carrasco, Viroporins, *FEBS Lett.* **552**, 28-34 (2003).
5. Y. Tang, P. Venkataraman, J. Knopman, R. A. Lamb, and L. H. Pinto, in: *Viral Membrane Proteins: Structure, Function, and Drug Design,* edited by W. B. Fisher, 9 (Kluwer Academic / Plenum Publishers, New York, Boston, Dordrecht, London, Moscow, 2005).
6. P. W. Gage, G. Ewart, J. Melton, and A. Premkumar, in: *Viral Membrane Proteins: Structure, Function, and Drug Design*, edited by W. B. Fisher, 21 (Kluwer Academic / Plenum Publishers, New York, Boston, Dordrecht, London, Moscow, 2005).
7. U. Schubert, et al., The two biological activities of human immunodeficiency virus type 1 vpu protein involve two separable structural domains, *J. Virol.* **70**, 809-819 (1996).
8. A. Premkumar, L. Wilson, G. D. Ewart, and P. W. Gage, Cation-selective ion channels formed by p7 of hepatitis C virus are blocked by hexamethylene amiloride, *FEBS Lett.* **557**, 99-103 (2004).
9. S. D. Griffin, et al., The p7 protein of hepatitis C virus forms an ion channel that is blocked by the antiviral drug, amantadine, *FEBS Lett.* **535**, 34-38 (2003).
10. D. Pavlovic, et al., The hepatitis C virus p7 protein forms an ion channel that is inhibited by long-alkyl-chain iminosugar derivatives, *Proc. Natl. Acad. Sci. USA* **100**, 6104-6108 (2003).
11. L. H. Pinto, L. J. Holsinger, and R. A. Lamb, Influenza virus M2 protein has ion channel activity, *Cell* **69**, 517-528 (1992).
12. D. M. Fleming, Managing influenza: Amantadine, rimantadine and beyond, *Int. J. Clin. Pract.* **55**, 189-195 (2001).
13. L. Wilson, C. McKinlay, P. Gage, and G. Ewart, SARS coronavirus E protein forms cation-selective ion channels, *Virology* **330**, 322-331 (2004).
14. L. Wilson, P. Gage, and G. Ewart, Hexamethylene amiloride blocks E protein ion channels and inhibits coronavirus replication (submitted).
15. J. Ortego, D. Escors, H. Laude, and L. Enjuanes, Generation of a replication-competent, propagation-deficient virus vector based on the transmissible gastroenteritis coronavirus genome, *J. Virol.* **76**, 11518-11529 (2002).
16. M. E. Gonzalez and L. Carrasco, Human immunodeficiency virus type 1 vpu protein affects Sindbis virus glycoprotein processing and enhances membrane permeabilization, *Virology* **279**, 201-209 (2001).
17. P. S. Masters, et al., 2005, Genetic and molecular biological analysis of protein-protein interactions in coronavirus assembly. in *Xth International Nidovirus Symposium: Towards Control of SARS and Other Nidovirus Diseases* (eds. Holmes, K.V. & Perlman, S.) (Cheyenne Mountain Resort, Colorado Springs, CO).
18. U. Schubert, et al., Identification of an ion channel activity of the vpu transmembrane domain and its involvement in the regulation of virus release from HIV-1-infected cells, *FEBS Lett.* **398**, 12-18 (1996).
19. K. C. Duff, and R. H. Ashley, The transmembrane domain of influenza a M2 protein forms amantadine-sensitive proton channels in planar lipid bilayers, *Virology* **190**, 485-489 (1992).
20. C. E. Machamer, and Y. Soonjeon, 2006, The transmembrane domain of the infectious bronchitis virus E protein is required for efficient virus release, this volume, pages 193-198.
21. Y. Ye, and B. G. Hogue, Role of the mouse hepatitis coronavirus envelope protein transmembrane domain, this volume pages 187-192.
22. L. Kuo, and P. S. Masters, The small envelope protein E is not essential for murine coronavirus replication, *J. Virol.* **77**, 4597-4608 (2003).

IDENTIFICATION OF ESSENTIAL GENES AS A STRATEGY TO SELECT A SARS CANDIDATE VACCINE USING A SARS-CoV INFECTIOUS cDNA

Fernando Almazán, Marta L. DeDiego, Carmen Galán, Enrique Álvarez, and Luis Enjuanes*

1. INTRODUCTION

The worldwide epidemic of severe acute respiratory syndrome (SARS) in 2003 was caused by a new coronavirus (CoV) called SARS-CoV.[1] The rapid transmission and high mortality rate made SARS a global threat for which an effective vaccine is urgently needed. Availability of full-length cDNA clones[2] and replicons of SARS-CoV provide an opportunity for the genetic manipulation of the viral genome to study fundamental viral processes and to develop effective strategies to prevent and control SARS-CoV infections.

In the present study, we report the engineering of a full-length cDNA clone and a replicon of the SARS-CoV Urbani strain as bacterial artificial chromosomes (BACs) for use in developing and testing SARS candidate vaccines. In addition, the results show that E protein is not strictly needed for virus replication, in contrast with the RNA processing enzymes exoribonuclease, endoribonuclease, and 2'-O-ribose methyltransferase, which are essential for virus RNA synthesis.

2. MATERIALS AND METHODS

2.1. Cells and Viruses

Baby hamster kidney cells (BHK) and human 293T cells were purchased from the ATCC. Vero E6 cells were kindly provided by E. Snijder. The genomic RNA of Urbani strain was kindly provided by the Centers for Disease Control and Prevention (CDC).

* Centro Nacional de Biotecnología, CSIC, Darwin 3, Cantoblanco, 28049 Madrid, Spain.

2.2. Construction of a Full-Length cDNA Clone and a Replicon of SARS-CoV

The full-length cDNA clone of SARS-CoV Urbani strain was engineered as a BAC (Fig. 1) following the same approach described for the generation of the TGEV full-length cDNA clone.[3] After selection of appropriate restriction sites in the viral genome, the intermediate plasmid pBAC-SARS-CoV 5'-3' was constructed and used as the backbone to assemble a full-length cDNA clone (Fig 1). This intermediate plasmid contained the first 681 nt of the genome under the control of the cytomegalovirus (CMV) immediate-early promoter, and the last 974 nt of the genome followed by a 25-bp synthetic poly A, the hepatitis delta virus ribozyme (Rz), and the bovine growth hormone termination and polyadenylation sequences (BGH), to ensure an accurate 3' end. In addition, a polylinker containing the selected restriction sites was placed between the 5' and 3' viral sequences to facilitate the assembly of the infectious cDNA clone.

A SARS-CoV replicon containing the replicase and the N gene under the control of its natural TRS was generated following the same strategy as described above.

2.3. Recovery of Infectious Virus from the cDNA Clone

BHK cells were transfected with the cDNA clone using Lipofectamine 2000 (Invitrogen) according to the manufacturer's specifications. After an incubation period of 6 h at 37ºC, cells were trypsinized, plated over a confluent monolayer of Vero E6 cells, and incubated at 37ºC for 48 h. Virus recovery was analyzed by plaque titration. After two passages, the virus was cloned by three rounds of plaque purification.

2.4. Replicon Activity Assay

BHK and 293T cells were transfected with the replicon constructs using Lipofectamine 2000 (Invitrogen). Total intracellular RNA was extracted at 24 h post-transfection and used as template for RT-PCR analysis of gene N mRNA expression, using specific oligonucleotides.

3. RESULTS AND DISCUSSION

3.1. The Small Envelope E Protein Is Not Essential for SARS-CoV Replication

In order to develop and test SARS candidate vaccines, a full-length cDNA clone of SARS-CoV Urbani strain was engineered as a BAC (Fig. 1). This approach uses a two-step amplification system that couples viral RNA expression in the cell nucleus from the CMV promoter with a second amplification step in the cytoplasm driven by the viral polymerase. Interestingly, the SARS-CoV cDNA clone was fully stable during its propagation in *E. coli*, in contrast with other coronavirus cDNA clones generated up to date. After cell transfection, infectious virus was recovered from the cDNA clone. The rescued virus conserved all the introduced genetic markers and was identical to the parental virus in terms of plaque morphology, growth kinetics, and mRNA and protein patterns (data not shown).

In order to generate highly attenuated or replication-competent propagation-deficient viruses that could be used as SARS vaccines, several structural and nonstructural genes have been deleted using the cDNA clone. In a first approach, the structural gene E was selected because it is essential in genus α coronaviruses[4] while its deletion in genus β reduces virus growth by more than three logarithmic units.[5] Gene E expression was abrogated by mutation of its TRS core sequence and start codon, without affecting the expression of the overlapping gene 3b, and by introduction of a deletion of 142 nt covering the majority of the E gene (Fig. 2A). Interestingly, infectious virus was recovered from the mutated cDNA clone and its identity was confirmed by RT-PCR and Western blot analysis. The growth kinetics of the deletion mutant was compared with that of the wild-type virus and showed a reduction in virus titer of about 20-fold (Fig. 2B), which correlated with a smaller plaque size compared to that of parental virus (Fig. 2C). These results indicate that E protein is important, but not essential, for SARS-CoV replication. Finally, the virulence of this mutant is being analyzed in hamsters to evaluate its use as a vaccine candidate for SARS.

3.2. The Exoribonuclease, Endoribonuclease, and 2'-O-Ribose Methyltransferase are Essential for Coronavirus RNA Synthesis

In addition to full-length cDNA clones, coronavirus replicons provide an important tool to explore the molecular basis of coronavirus replication and for the development and testing of vaccines and antiviral drugs without the need for growing infectious virus. To study the role of replicase genes in coronavirus replication, a SARS-CoV replicon was generated as a BAC following the same strategy described for the full-length cDNA clone. We have previously shown that the N protein is required for efficient coronavirus RNA synthesis.[6] Accordingly, the SARS-CoV replicon was engineered with the N gene in addition to the replicase.

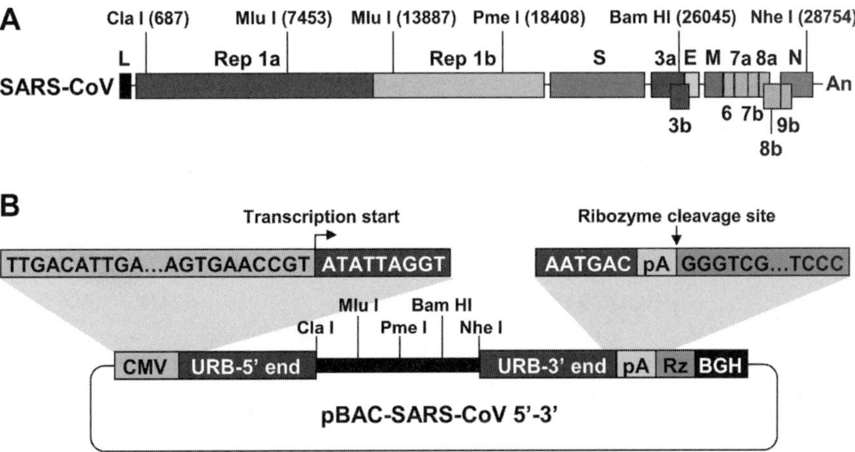

Figure 1. Strategy for the generation of a full-length cDNA clone and a replicon of the SARS-CoV Urbani strain. After selection of appropriated restriction sites (A), an intermediate plasmid (B) was constructed as the backbone to assemble the cDNA clone and the replicon.

Replicon activity was studied by RT-PCR analysis of gene N mRNA expression in BHK and 293T cells, and in both cases high transcription levels were detected, indicating that the SARS-CoV replicon was functional (data not shown).

Using this replicon, the role of the recently described RNA processing enzymes exoribonuclease (Exo N), endoribonuclease (Nendo U), and 2'-O-ribose methyltransferase(2'-O-MT)[7] in coronavirus RNA synthesis was investigated. To this end, single deletion mutant replicons, in which the conserved domain of each enzyme was deleted, and a mutant replicon lacking the three domains were generated (Fig 3A). Replicon activity of these constructs was analyzed in both BHK and 293T cells and only the wild-type replicon was functional (Fig. 3B), indicating that each of these RNA processing enzymes are essential for SARS-CoV RNA synthesis. Further experiments will be required to address the specific role of these enzymes in coronavirus transcription or replication.

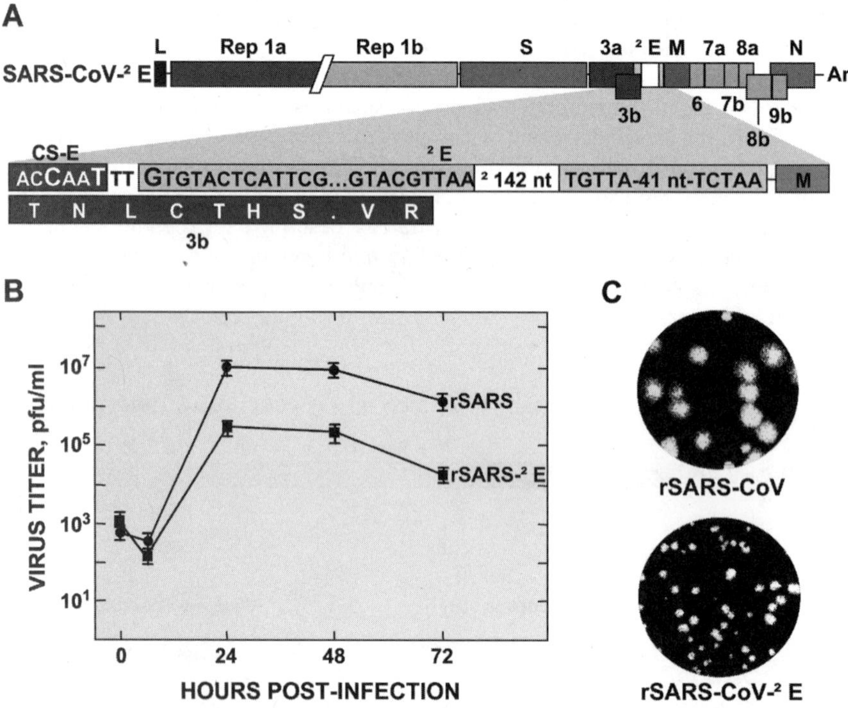

Figure 2. Construction of a viable recombinant SARS-CoV with the E gene deleted. (A) Genetic structure of the deletion mutant virus. The mutations introduced in the core sequence (CS-E) and the start codon of gene E to abrogate its expression are indicated with large letters. (B) Growth kinetics of rSARS-CoV-ΔE and rSARS-CoV-viruses on Vero E6 cells. The mean values of three experiments are indicated. Error bars represent standard deviation. (C) Plaque morphology produced by the indicated viruses on Vero E6 cells.

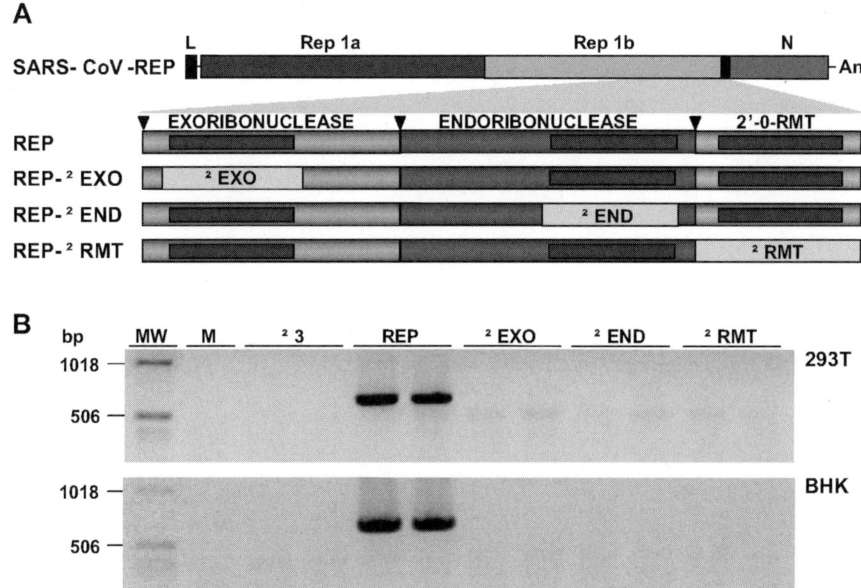

Figure 3. Role of SARS-CoV RNA processing enzymes in RNA synthesis. (A) Schematic representation of the different single deletion mutant replicons. The conserved domains described by Snijder[7] are indicated by dark bars. In addition, a deletion mutant lacking the three domains (Δ3) was also generated. (B) Functional analysis of mutant replicons. Expression of gene N mRNA was used to study replicon activity in BHK and 293T cells by RT-PCR analysis. M, mock.

4. REFERENCES

1. J. S. Peiris, S. T. Lai, L. L. Poon, et al., Coronavirus as a possible cause of severe acute respiratory syndrome, *Lancet* **361**, 1319-1325 (2003).
2. B. Yount, K. M. Curtis, E. A. Fritz, L. E. Hensley, P. B. Jahrling, E. Prentice, M. R. Denison, T. W. Geisbert, and R. S. Baric, Reverse genetics with a full-length infectious cDNA of severe acute respiratory syndrome coronavirus, *Proc. Natl. Acad. Sci. USA* **100**, 12995-13000 (2003).
3. F. Almazán, J. M. González, Z. Pénzes, A. Izeta, E. Calvo, J. Plana-Durán, and L. Enjuanes, Engineering the largest RNA virus genome as an infectious bacterial artificial chromosome, *Proc. Natl. Acad. Sci. USA* **97**, 5516-5521 (2000).
4. J. Ortego, D. Escors, H. Laude, and L. Enjuanes, Generation of a replication-competent, propagation-deficient virus vector based on the transmissible gastroenteritis coronavirus genome, *J. Virol.* **76**, 11518-11529 (2002).
5. L. Kuo and P. S. Masters, The small envelope protein E is not essential for murine coronavirus replication, *J. Virol.* **77**, 4597-4608 (2003).
6. F. Almazán, C. Galán, and L. Enjuanes, The nucleoprotein is required for efficient coronavirus genome replication, *J. Virol.* **78**, 12683-12688 (2004).
7. E. J. Snijder, P. J. Bredenbeek, J. C. Dobbe, V. Thiel, J. Ziebuhr, L. L. Poon, Y. Guan, M. Rozanov, W. J. Spaan, and A. E. Gorbalenya, Unique and conserved features of genome and proteome of SARS-coronavirus, an early split-off from the coronavirus group 2 linage, *J. Mol. Biol.* **331**, 991-1004 (2003).

STRUCTURE AND DYNAMICS OF SARS CORONAVIRUS MAIN PROTEINASE (M^{pro})

Rolf Hilgenfeld, Kanchan Anand, Jeroen R. Mesters, Zihe Rao, Xu Shen, Hualiang Jiang, Jinzhi Tan, and Koen H. G. Verschueren*

1. INTRODUCTION: A WORD ON NOMENCLATURE

All protein functions required for SARS coronavirus replication are encoded by the replicase gene.[1,2] This gene encodes two overlapping polyproteins (pp1a and pp1ab), from which the functional proteins are released by extensive proteolytic processing. This is primarily achieved by the 34-kDa main proteinase (M^{pro}), which is frequently also called 3C-like proteinase ($3CL^{pro}$) to indicate a similarity in substrate specificity with the 3C proteinase of picornaviruses.[3] While useful at the time of initial description of the coronaviral enzyme, there are in fact large differences between the structures and mechanisms of these enzymes, making the designation of the coronavirus main proteinase as $3CL^{pro}$ rather misleading. We will therefore use the term M^{pro} exclusively.

2. OVERALL STRUCTURE OF CORONAVIRUS MAIN PROTEINASE

The functional importance of the SARS-CoV M^{pro} in the viral life cycle makes it a preferred target for discovering anti-SARS drugs.[4-7] However, in order to apply rational drug design or virtual screening, information on the structure of the target enzyme is required. Initially, this came from homology models of the SARS-CoV M^{pro} that were constructed on the basis of crystal structures of human CoV (HCoV) 229E M^{pro} and of porcine transmissible gastroenteritis virus (TGEV) M^{pro} that we had previously determined.[4,8] The SARS virus enzyme shares about 40% sequence identity with these proteinases of group I coronaviruses. More recently, the crystal structure of the SARS-CoV M^{pro} has been determined.[9-11] As with other CoV M^{pro}s, the molecule comprises three domains (Figure 1). Domains I (residues 8–101) and II (residues 102–184) are β-

*Rolf Hilgenfeld, Kanchan Anand, Jeroen R. Mesters, Jinzhi Tan, Koen H. G. Verschueren, University of Lübeck, Lübeck, Germany, 23538. Zihe Rao, Tsinghua University, Beijing, China, 100084, and Chinese Academy of Sciences, Beijing, China, 100101. Xu Shen, Hualiang Jiang, Jinzhi Tan, Chinese Academy of Sciences, Shanghai, China, 201203.

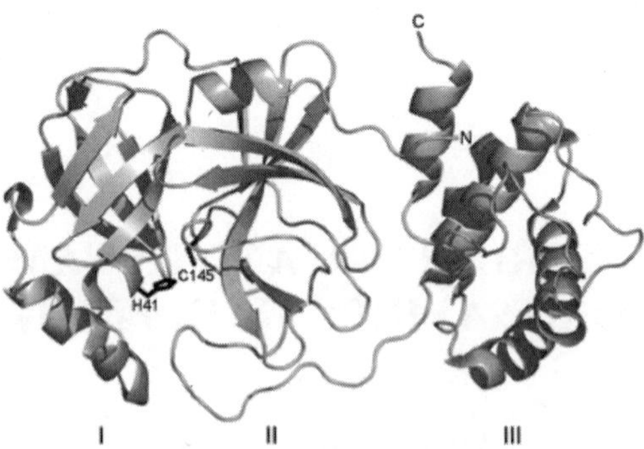

Figure 1. The three domains of the SARS-CoV M^{pro} monomer. The catalytic dyad (His41...Cys145, shown in black) is located in the interface between domains I and II.

barrels and together resemble the structure of chymotrypsin, whereas domain III (residues 201–306) consists mainly of α-helices. The active site, containing a Cys…His catalytic dyad, is located in a cleft between domains I and II. Domains II and III are connected by a long loop (residues 185–200). *In vitro* experiments demonstrated that deletion of domain III abolished almost completely the proteolytic activities of the main proteinases of TGEV and SARS-CoV.[8,12] This domain is essential for the dimerization of the M^{pro},[13] which in turn assures proper orientation of the N-terminal residues of monomer B that play an important role for the catalytic activity of monomer A (and *vice versa*; see below).[9,10] In all known crystal structures of coronavirus main proteinases, the enzyme exists as a dimer,[4,8-11] and dimerization is also observed in solution at slightly elevated concentrations.[8,12-14] The dimer is the enzymatically active species because the specific activity increases linearly with increasing enzyme concentration.[14]

3. PLASTICITY OF THE SUBSTRATE-BINDING SITE

A special feature first discovered for the SARS-CoV M^{pro} (but most probably present in all coronavirus main proteinases) is that in the monoclinic crystals grown at pH 6.0, the two monomers have different conformations around the S1 substrate-binding site, because the loop 138–145, in particular Phe140, as well as Glu166 undergo dramatic conformational rearrangements. As a result, one protomer exists in an active and the other in an inactive conformation.[9] In the latter, the S1 substrate-binding pocket has virtually collapsed as a consequence of the reorientation of Glu166, and the oxyanion hole no longer exists due to the conformational change of residues 138–145. When the crystals are equilibrated at pH 7.6 and 8.0, both monomers are in an active conformation.[9] We have proposed[9,10] that these conformational changes are controlled by the protonation state of His163, an absolutely conserved residue at the bottom of the S1 substrate-specificity pocket (Figure 2). This subsite is designed to accommodate the P1-glutamine residue of M^{pro} substrates, with high specificity. No other amino-acid side chain must be accepted in this position, in particular not glutamate (as opposed to glutamine). This is achieved by ensuring that over a broad pH range, His163 is uncharged. Two important interactions made by the imidazole ring are responsible for keeping it in the neutral state: *(i),* stacking or edge-on-face interaction with the phenyl ring of Phe140 (Figure 2, left panel), and *(ii),* acceptance by its Nδ1 atom of a strong hydrogen bond from the hydroxyl group of the buried Tyr161. This means that only the Nε2 atom of His163 can normally carry a proton, and it is this nitrogen that will donate a hydrogen bond to the side chain oxygen of the P1 glutamine residue of the substrate (Figure 2, right panel).[8] In agreement with this structural interpretation, any replacement of the conserved histidine residue (His162 in this case) abolishes the proteolytic activity of HCoV 229E and feline infectious peritonitis virus (FIPV) M^{pro}.[15,16] Furthermore, FIPV M^{pro} Tyr160 (corresponding to SARS-CoV M^{pro} Tyr161) mutants have their proteolytic activity reduced by a factor of > 30.[15] These observations and the absolute conservation of these residues in the coronavirus main proteinases underline the importance of the uncharged state of His163 in binding the substrate with the required specificity.

However, when the SARS-CoV M^{pro} crystals are grown at a pH near or below the pK_a value of this histidine residue, the latter can be protonated, leading to drastic structural consequences. In order to compensate for a positive charge on His163, which is in a largely hydrophobic environment, Glu166, which forms part of the wall of the S1 pocket, will move inwards and form a salt-bridge with the protonated His163 (Figure 2, middle). Through this conformational rearrangement, Glu166 will fill the S1 pocket, thereby preventing the binding of substrate. But the consequences are even more far-reaching. When His163 is protonated, its hydrophobic interaction with the phenyl ring of Phe140 is no longer possible, and the latter undergoes a major displacement with an amplitude of > 5.5 Å (compare the middle panel with the two other images in Figure 2). Along with this, the oxyanion loop (residues 138–145) changes conformation and is no longer able to stabilize the tetrahedral intermediate of peptide-bond cleavage through donation of hydrogen bonds from the amide groups of Gly143 and Cys145.

In addition to His163, there is a second histidine residue involved in formation of the S1 pocket. In the crystal structure obtained at pH 7.6, where both subunits are found in the active conformation,[9] His172 forms part of the wall of the subsite, being engaged in a salt-bridge with Glu166. At this pH, the histidine is likely to be positively charged, because pK_a values of histidine residues involved in salt-bridges tend to be 2 units higher than those of isolated histidines.[17] The His172...Glu166 ion pair also exists in the active subunit of the dimer at pH 6.0, whereas in the inactive one, His172 loses its partner which moves into the S1 pocket, in order to compensate for the positive charge on the protonated His163 at the bottom of the subsite. Furthermore, at pH 8.0, His172 is likely to be uncharged. As a result, Glu166 is no longer fixed at its position in the wall of the S1 pocket, but tends to be flexible and partly blocks the entry to the S1 pocket.[10] These observations nicely agree with the enzymatic activity of SARS-CoV M^{pro},[10,12] which has its maximum at pH 7.0 and about 50% activity each at pH 6.0 (S1 pocket and oxyanion hole collapsed in one subunit of the dimer) and pH 8.0 (wall of S1 pocket no longer stable due to interruption of His172...Glu166 salt-bridge, Glu166 partly blocking the entry to the S1 site). Therefore, we propose that the bell-shaped pH-activity curve of the SARS-CoV M^{pro} is governed by the protonation of His163 on its low-pH side, and by deprotonation of His172 on its high-pH side.[10] In nice agreement with this, Chou *et al.*[12] reported that the apparent pK_a values characteristic of this curve are 5.7 ± 0.4 and 8.7 ± 0.4. However, they discuss that these may originate from the Glu290...Arg4 ion pair, or from the His41...Cys145 catalytic dyad, both less likely options from our point of view.

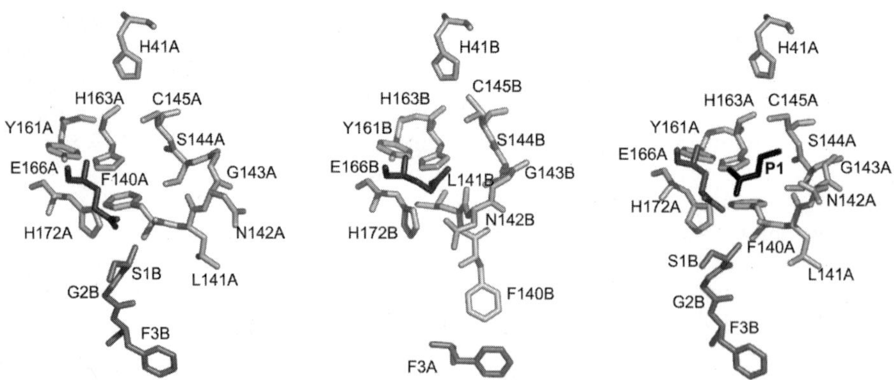

Figure 2. View of the S1 substrate binding pocket and oxyanion hole (loop 138–145, first two residues are not shown) in the active monomer (left), inactive monomer (middle) and as seen in the active monomer of a substrate-analogous inhibitor-bound SARS-CoV M^{pro} (right panel, only the P1 Gln residue is shown in black, occupying the S1 specificity pocket). The N-terminal residues of the other monomer are indicated in dark gray. The alternating position of the Glu166 side chain in the active (left and right panel) and inactive conformation (middle panel) is highlighted in darker gray.

4. MOLECULAR DYNAMICS SIMULATIONS

We have confirmed by molecular dynamics (MD) simulations that large conformational changes of the observed type can indeed be triggered by protonation of His163 and deprotonation of His172.[10] In three different 10-ns simulations at pH 6.0, 7.6, and 8.0, we found the same type of rearrangements as seen in the crystals. This is reassuring, because it should not be forgotten that unless the crystals diffract to better than 1.0 Å resolution, X-ray diffraction cannot normally determine hydrogen positions because of their low scattering power. The force fields used in MD simulations, on the other hand, fully take into account the hydrogens bound to non-carbon atoms. Thus, because these simulations yield the same conformational rearrangements as those seen by X-ray diffraction, the interpretation of the crystallographic results is likely to be correct. Also, our preliminary NMR data with ^{15}N-labeled SARS-CoV M^{pro} (J. George *et al.*, unpublished) appear to support these conclusions.

In addition to the MD simulations, we also investigated the dynamic behavior of the system with His163 in both subunits of the dimer protonated, i.e., at a presumed pH < 6.0. Starting from the (energy-minimized) crystal structure at pH 6.0, this simulation led to a dimer that had both monomers in the inactive conformation, with their S1 pockets and oxyanion holes collapsed. Thus, the MD simulations were apparently able to transform one conformation of the substrate-binding site of SARS-CoV M^{pro} into the other.[10]

5. NEW CRYSTAL FORMS OF SARS-CoV M^{pro}

Experimental support for this theoretical prediction was provided very recently by the analysis of two new crystal forms of the SARS-CoV M^{pro}. In addition to the monoclinic crystal form of the enzyme originally described in 2003,[9] we managed to obtain tetragonal (space group $P4_32_12$) and orthorhombic ($P2_12_12$) crystals of the enzyme, and determined these crystal structures at 2.0 and 2.8 Å resolution.[10] Both of these crystal forms contain one SARS-CoV M^{pro} monomer per asymmetric unit; the dimer is generated through the crystallographic twofold axis. Thus, by necessity, the two monomers of the dimer are identical in these crystals. The question, however, was whether they would be in the active or in the inactive conformation. Because the monoclinic form had one monomer in the active and the other in the inactive form when crystallized at pH 6.0, but both of them in the active conformation after equilibration of the crystals at pH 7.6, it was unclear what to expect for the new crystal forms which were crystallized at pH 5.9 (tetragonal form) and 6.6 (orthorhombic). Interestingly, it turned out that both of them are in the inactive form, *i.e.* the S1 pocket and the oxyanion hole have collapsed.

We have also attempted to equilibrate tetragonal crystals at higher pH values. Structures were determined at pH 7.0 (1.65 Å resolution), 7.4 (1.57 Å), and 8.0 (1.65 Å). Although these are the highest-resolution structures reported so far for the SARS-CoV M^{pro}, a reliable interpretation of the electron density in the substrate-binding region was difficult because of dual (or even multiple) conformations. In any case, our preliminary analysis shows that the same type of conformational rearrangements occur in this crystal form as was observed in the monoclinic crystals, but an ensemble of inactive and active conformations appear to co-exist at all pH values. In fact, the tetragonal crystals of SARS-CoV M^{pro}, containing $< 30\%$ solvent, may be less suited for studying the conformational transition because the molecules are tightly packed and seem to have a contracted substrate-binding site. This is supported by the observation that the tetragonal crystals crack when equilibrated at pH 8.0 for longer than 4 hours.

Interestingly, the orthorhombic crystal form, which has the (generated) dimer in the inactive conformation when crystallized at pH 6.6 in the presence of malonate, can also be obtained in an active form at about the same pH (6.5) when ammonium sulfate is used as a precipitant (J. Lescar, personal communication). At present, it is unclear why the change of precipitant should induce such changes; more likely, it is subtle differences in the final pH of the crystallization medium that are determining the resulting conformation when working near the pK_a value of His163.

Apart from the differences in detail in the substrate-binding site, the SARS-CoV M^{pro} dimers as seen in the new structures are very similar to the dimer in the original monoclinic crystals. From the monoclinic crystal structure obtained at pH 6.0, the monomers in the new crystal forms display overall r.m.s. deviations for $C\alpha$ atoms (monomers A and B, respectively) of 0.95/0.76 Å (tetragonal form) and 1.10/0.78 Å (orthorhombic form). It is reassuring that the monomers in the new crystal forms, which are in the inactive conformation, are more similar to the inactive monomer B of the dimer in the monoclinic form (second number), than to the active monomer A (first number).

6. THE ROLE OF THE "N-FINGER"

One important intermolecular interaction in the coronavirus M^{pro} dimer was not mentioned so far. In the active conformation of the SARS-CoV M^{pro} dimer, the N-terminus of monomer B (which is the inactive subunit at pH 6.0) was shown to interact with the main chain amide and carbonyl of Phe140 and with the carboxylate of Glu166, both of monomer A.[9] This interaction appears to help shape the substrate-binding site of monomer A. On the other hand, the collapsed binding site of monomer B lacks these intermolecular interactions, and as a result, residues 1 and 2 of monomer A are disordered and not seen in the electron density. The same observation was made in our X-ray structures derived from the tetragonal and orthorhombic crystal forms[10]: here, both monomers are in the inactive conformation, and accordingly, both N-termini are disordered to the extent that no electron density is seen for residue Ser1. When we determined the first structure of a coronaviral M^{pro}, that of the TGEV enzyme,[8] we saw that residues 1–7 were squeezed in between domain III of its own monomer and domain II of the other monomer in the dimer. The same interactions of this segment, which we later called the "N-finger," [9] were seen in HCoV 229E M^{pro} and SARS-CoV M^{pro}.[4,9] When we deleted residues 1–5 in the TGEV proteinase, the enzyme was almost totally inactive with a synthetic pentadecapeptide as substrate.[8] In agreement with this, Chen *et al.*[18] have recently shown for SARS-CoV M^{pro} that the enzyme lacking residues 1–7 is proteolytically inactive, but the rather surprising finding was that it still forms a dimer.[18] This result has to be seen in light of the finding by Shi *et al.* that even isolated domain III of SARS-CoV M^{pro} will dimerize.[13] On the other hand, when Hsu *et al.* removed only residues 1–3 from the enzyme, it retained 76% of its proteolytic activity.[19] Further, in apparent contradiction to the findings by Chen *et al.*,[18] they found that the Δ(1-4) SARS-CoV M^{pro} was predominantly monomeric at concentrations of 0.1 mg/ml. Whatever the reason for the discrepancy might be, it could well be that while the presence and correct placement of the "N-finger" is important, the tip of the finger may not be as essential as thought hitherto. In fact, our MD simulations as well as the X-ray structure obtained from the monoclinic form at pH 7.6 also suggested that the active conformation of monomer A can also be retained without direct interaction with residue Ser1 of monomer B, provided residues 3–7 are in the correct position.

7. CONCLUSIONS AND OUTLOOK

In summary, from the various crystallographic studies on the coronavirus M^{pro} and from our MD simulations, we conclude that the enzyme is a very flexible protein, the conformational state of which appears to depend on the pH value of the medium. This is probably of biological significance, because the viral polyproteins (of which the M^{pro} is a domain before self-activation by autocleavage) assemble on the late endosome where local pH tends to be acidic. For designing inhibitors of the M^{pro}, knowledge of the dynamics of the target will be essential. This has been clearly demonstrated in the case of HIV-1 proteinase over the years, where understanding the flexibility of the enzyme did not only turn out to be a prerequisite for designing potent inhibitors but also for explaining many of the observed drug-resistance mutations. Today, HIV proteinase is perhaps the one enzyme best understood in terms of structure and dynamics, and this knowledge is mainly based on several hundred crystal structures, most of them complexes with various inhibitors. In terms of number of crystal structures, HIV-1 proteinase is probably followed by trypsin and lysozyme, but extrapolating from the current research activities, coronavirus main proteinases are likely to catch up.

The work described here was supported, in part, by the Sino-European Project on SARS Diagnostics and Antivirals (SEPSDA, contract no. SP22-CT-2004-003831; www.sepsda.info) of the European Commission, by the Deutsche Forschungsgemeinschaft

(Hi 611/4-1) and by the Sino-German Center for the Promotion of Science (GZ 233-202/6).

8. REFERENCES

1. D. A. Groneberg, R. Hilgenfeld, and P. Zabel, Molecular mechanisms of severe acute respiratory syndrome (SARS), *Respir. Res.* **6**, 8-23 (2005).
2. V. Thiel, K. A. Ivanov, A. Putics, et al., Mechanisms and enzymes involved in SARS coronavirus genome expression, *J. Gen. Virol.* **84**, 2305-2315 (2003).
3. A. E. Gorbalenya, E. V. Koonin, A. P. Donchenko, and V. M. Blinov, Coronavirus genome: prediction of putative functional domains in the non-structural polyprotein by comparative amino acid analysis, *Nucl. Acids Res.* **17**, 4847-4861 (1989).
4. K. Anand, J. Ziebuhr, P. Wadhwani, J. R. Mesters, and R. Hilgenfeld, Coronavirus main proteinase (3CL^{pro}) structure: basis for design of anti-SARS drugs, *Science* **300**, 1763-1767 (2003).
5. B. Xiong, C. S. Gui, X. Y. Xu, et al., A 3D model of SARS-CoV 3CL proteinase and its inhibitors design by virtual screening, *Acta Pharmacol. Sin.* **24**, 497-504 (2003).
6. K. Anand, H. Yang, M. Bartlam, Z. Rao, and R. Hilgenfeld, in: *Coronaviruses with Special Emphasis on First Insights Concerning SARS*, edited by A. Schmidt, M. H. Wolff, and O. Weber (Birkhäuser, Basel, 2005), pp. 173-199.
7. H. Yang, W. Xie, X. Xue, et al., Design of wide-spectrum inhibitors targeting coronavirus main proteinases, *PLoS Biology* 3, e324 (2005).
8. K. Anand, G. J. Palm, J. R. Mesters, et al., Structure of coronavirus main proteinase reveals combination of a chymotrypsin fold with an extra α-helical domain, *EMBO J.* **21**, 3213-3224 (2002).
9. H. Yang, M. Yang, Y. Ding, et al., The crystal structures of severe acute respiratory syndrome virus main protease and its complex with an inhibitor, *Proc. Natl. Acad. Sci. USA* **100**, 13190-13195 (2003).
10. J. Tan, K. H. G. Verschueren, K. Anand, et al., pH-Dependent conformational flexibility of the SARS-CoV main proteinase (M^{pro}) dimer: Molecular dynamics simulations and multiple X-ray structure analyses, *J. Mol. Biol.* **354**, 25-40 (2005).
11. M.-F. Hsu, C.-J. Kuo, K.-T. Chang, et al., Mechanism of the maturation process of SARS-CoV 3CL protease, *J. Biol. Chem.* **280**, 31257-31266 (2005).
12. C.-Y. Chou, H.-C. Chang, W.-C. Hsu, et al., Quarternary structure of the severe acute respiratory syndrome (SARS) coronavirus main protease, *Biochemistry* **43**, 14958-14970 (2004).
13. J. Shi, Z. Wei, and J. Song, Dissection study on the severe acute respiratory syndrome 3C-like protease reveals the critical role of the extra domain in dimerization of the enzyme: Defining the extra domain as a new target for design of highly specific protease inhibitors, *J. Biol. Chem.* **279**, 24765-24773 (2004).
14. K. Fan, P. Wei, Q. Feng, et al., Biosynthesis, purification, and substrate specificity of severe acute respiratory syndrome coronavirus 3C-like proteinase, *J. Biol. Chem.* **279**, 1637-1642 (2004).
15. A. Hegyi, A. Friebe, A. E. Gorbalenya, and J. Ziebuhr, Mutational analysis of the active centre of coronavirus 3C-like proteases, *J. Gen. Virol.* **83**, 581-593 (2002).
16. J. Ziebuhr, E. J. Snijder, and A. E. Gorbalenya, Virus-encoded proteinases and proteolytic processing in the Nidovirales. *J. Gen. Virol.* **81**, 853-879 (2000).
17. J. Yang, M. Yu, Y. N. Jan, and L. Y. Jan, Stabilization of ion selectivity filter by pore loop ion pairs in an inwardly rectifying potassium channel, *Proc. Natl. Acad. Sci. USA* **94**, 1568-1572 (1997).
18. S. Chen, L. Chen, J. Tan, et al., Severe acute respiratory syndrome coronavirus 3C-like proteinase N terminus is indispensable for proteolytic activity but not for enzyme dimerization: Biochemical and thermodynamic investigation in conjunction with molecular dynamics simulations, *J. Biol. Chem.* **280**, 164-173 (2005).
19. W.-C. Hsu, H.-C. Chang, C.-Y. Chou, P.-J. Tsai, P.-I. Lin, and G.-G. Chang, Critical assessment of important regions in the subunit association and catalytic action of the severe acute respiratory syndrome coronavirus main protease, *J. Biol. Chem.* **280**, 22741-22748 (2005).

HIGHLY ATTENUATED VACCINIA VIRUS DIs AS A POTENTIAL SARS VACCINE

Koji Ishii, Hideki Hasegawa, Noriyo Nagata, Tetsuya Mizutani, Shigeru Morikawa, Masato Tashiro, Tetsuro Suzuki, Fumihiro Taguchi, Toshitada Takemori, Tatsuo Miyamura, and Yasuko Tsunetsugu-Yokota*

1. INTRODUCTION

Severe acute respiratory syndrome (SARS) is a newly found infectious disease caused by a novel coronavirus, SARS coronavirus (SARS-CoV).[1,2] DIs strain is a highly restricted host range mutant of vaccinia virus. It does not replicate in and is not pathogenic for mice, guinea pigs, or rabbits, and this strain does not replicate in various mammalian cell lines.[3] Recently, we have established a system for expressing foreign genes.[4] In the present study, we constructed recombinant forms of the DIs containing the gene encoding four structural proteins, envelope (E), membrane (M), nucleocapsid (N), and spike (S), of SARS-CoV either separately or simultaneously. Mammalian cells infected with the recombinant DIs synthesized SARS-CoV proteins that were recognized by SARS patient serum or rabbit antibody raised against synthetic peptides of SARS-CoV proteins in Western blot analyses. Intranasal or subcutaneous inoculations of BALB/c 3T3 mice with the recombinant DIs expressing E/M/S or E/M/N/S proteins elicited neutralizing antibodies to SARS-CoV and protective immunity. Therefore, our study showed that the replication-deficient DIs strain is feasible as a safe and effective SARS vaccine.

2. RESULTS

2.1. Expression of SARS-CoV Structural Proteins by Recombinant DIs

Expression of SARS-CoV N and S proteins were detected by western blotting using monoclonal antibodies.[5] A robust signal was detected at 50 kDa, corresponding to the N

*National Institute of Infectious Diseases, Tokyo 162-8640, Japan.

protein of SARS-CoV as predicted by its genome size.[1,2] A band near 200 kDa appears to correspond to S protein, which is known to be heavily glycosylated. Concerning the M protein, only a smear band in the stacking gel was detected using a polyclonal antibody against synthetic peptide of the M protein,[6] presumably because it formed large ologomers with SDS-resistance in cells. Similar result was observed by the analysis of the M protein of avian coronavirus infectious bronchitis virus.[7]

The subcellular localization of the S, M, and N proteins were analyzed by immunofluorescence microscopy. Cells infected with DIs-M or DIs-N showed that M and N proteins were localized mainly at the Golgi complex, which is consistent with studies with model corona viruses in which it was found that M is retained in the Golgi apparatus.[8] S protein was localized at the Golgi complex but the plasma membrane was also stained, suggesting that some portion of S protein was transported to the plasma membrane. Thus, these results indicate that cells infected with recombinant DIs under the control of the mH5 promoter express significant levels of SARS-CoV proteins with an expected post-translational processing.

2.2. Recombinant DIs Induces Serum IgG Antibody Specific for SARS-CoV

To examine the level of anti-SARS-CoV response in mice after inoculation with recombinant DIs, 4 mice in each group were subcutaneously or intranasally inoculated three times with 10^6 pfu of recombinant DIs that expressed N, M, S, E/M/S, or E/M/N/S. Ten days after the final inoculation, vaccinated mice elicited anti-SARS-CoV IgG antibody in sera at high levels. Mice vaccinated subcutaneously with DIs-E/M/S or DIs-E/M/N/S elicited the highest levels of anti-SARS-CoV IgG.

Whether the immune sera possess the neutralizing activity against SARS-CoV is a crucial aspect of vaccination. We next estimated the neutralizing activity against SARS-CoV of antisera obtained (Table 1). Neutralizing antibodies against SARS-CoV were induced in mice that were either subcutaneously or intranasally injected with DIs-S, -E/M/S, or –E/M/N/S. Among them, the highest level of the neutralizing activity was observed in sera of mice injected subcutaneously with recombinant DIs expressing E, M, N, and S. On the other hand, we could not detect neutralizing activity in sera of mice injected either subcutaneously or intranasally with recombinant DIs expressing M or N proteins. Thus, these results indicate that recombinant DIs induces potent SARS-CoV–specific neutralizing antibodies. It appears that the S protein is prerequisite for eliciting a sufficient level of IgG antibodies with neutralizing activity.

3. DISCUSSION

Highly attenuated vaccinia viruses can express viral and inserted genes at high levels even in nonpermissive cells without showing CPE. rDIs exhibited no replicative ability and produced no infectious virions in these cells, indicating that the DIs strain may have a safety advantage when used as a recombinant vaccine vector.

Efforts directed at vaccine development for SARS-CoV have been carried out by many organizations in variable ways.[8–13] Our results showed that intranasal or subcutaneous inoculations of Balb/c mice with DIs-E/M/S or DIs-E/M/N/S produced

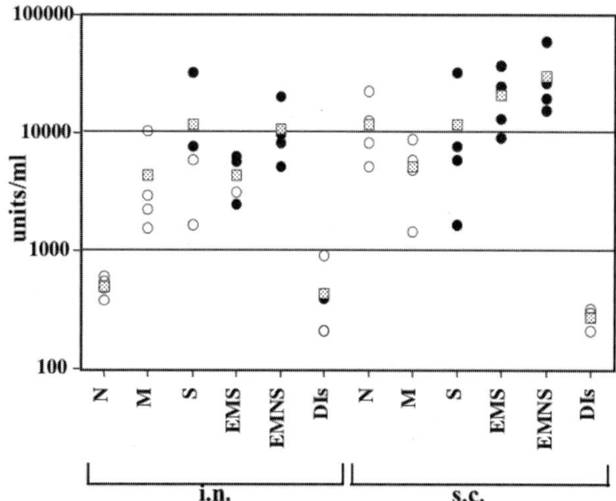

Figure 1. The levels of IgG antibodies against SARS-CoV. Neutralizing-positive sera were shown as closed circles and neutralizing-negative sera were shown as open circles. Averages were shown as dotted boxes.

serum antibodies that recognized the SARS-CoV virion in ELISA and neutralized SARS-CoV *in vitro*. Moreover, DIs-S administered by either route elicited protective immunity, as shown by reduced titers of SARS-CoV in the lungs of mice after challenge. Subcutaneous route appears to be stronger than intranasal rout with respect to the level of anti-SARS-CoV IgG antibody.

In this study, we constructed recombinant forms of the DIs containing the gene encoding four structural proteins of SARS-CoV either separately or simultaneously. Intranasal or subcutaneous inoculations of BALB/c 3T3 mice with the recombinant DIs expressing E/M/S or E/M/N/S proteins elicited neutralizing antibodies to SARS-CoV Therefore our study showed that the replication-deficient DIs strain is feasible as a safe and effective SARS vaccine.

4. REFERENCES

1. Rota, P. A., Oberste, M. S., Monroe, S. S., et al., 2003, Characterization of a novel coronavirus associated with severe acute respiratory syndrome, *Science* **300**:1394-1399.
2. Marra, M. A., Jones, S. J., Astell, C. R., et al., 2003, The Genome sequence of the SARS-associated coronavirus, *Science* **300**:1399-1404.
3. Tagaya, I., Kitamura, T., and Sano, Y., 1961, A new mutant of dermovaccinia virus, *Nature* **192**:381-382.
4. Ishii, K., Ueda, Y., Matsuo, K., et al., 2002, Structural analysis of vaccinia virus DIs strain: application as a new replication-deficient viral vector, *Virology* **302**:433-444.
5. Ohnishi, K., Sakaguchi, M., Kaji, T., et al., 2005, Immunological detection of severe acute respiratory syndrome coronavirus by monoclonal antibodies, *Jpn. J. Infect. Dis.* **58**:88-94.

6. Mizutani, T., Fukushi, S., Saijo, M., Kurane, I., and Morikawa, S., 2004, Phosphorylation of p38 MAPK and its downstream targets in SARS coronavirus-infected cells, *Biochem. Biophys. Res. Commun.* **319**:1228-1234.
7. Weisz, O. A., Swift, A. M., and Machamer, C. E., 1993, Oligomerization of a membrane protein correlates with its retention in the Golgi complex, *J. Cell Biol.* **122**:1185-1196.
8. Bisht, H., Roberts, A., Vogel, L., et al., 2004, Severe acute respiratory syndrome coronavirus spike protein expressed by attenuated vaccinia virus protectively immunizes mice, *Proc. Natl. Acad. Sci. USA* **101**:6641-6646.
9. Buchholz, U. J., Bukreyev, A., Yang, L., et al., 2004, Contributions of the structural proteins of severe acute respiratory syndrome coronavirus to protective immunity, *Proc. Natl. Acad. Sci. USA* **101**:9804-9809.
10. Gao, W., Tamin, A., Soloff, A., et al., 2003, Effects of a SARS-associated coronavirus vaccine in monkeys, *Lancet* **362**:1895-1896.
11. Kim, T. W., Lee, J. H., Hung, C. F., et al., 2004, Generation and characterization of DNA vaccines targeting the nucleocapsid protein of severe acute respiratory syndrome coronavirus, *J. Virol.* **78**:4638-4645.
12. Subbarao, K., McAuliffe, J., Vogel, L., Fahle, G., Fischer, S., Tatti, K., Packard, M., Shieh, W. J., Zaki, S., and Murphy, B., 2004, Prior infection and passive transfer of neutralizing antibody prevent replication of severe acute respiratory syndrome coronavirus in the respiratory tract of mice, *J. Virol.* **78**:3572-3577.
13. Zhao, P., Cao, J., Zhao, L. J., Qin, Z. L., Ke, J. S., Pan, W., Ren, H., Yu, J. G., and Qi, Z. T., 2005, Immune responses against SARS-coronavirus nucleocapsid protein induced by DNA vaccine, *Virology* **331**:128-135.
14. Vennema, H., Godeke, G. J., Rossen, J. W., Voorhout, W. F., Horzinek, M. C., Opstelten, D. J., and Rottier, P. J., 1996, Nucleocapsid-independent assembly of coronavirus-like particles by co-expression of viral envelope protein genes, *EMBO J.* **15**:2020-2028.

RENILLA LUCIFERASE AS A REPORTER TO ASSESS SARS-CoV mRNA TRANSCRIPTION REGULATION AND EFFICACY OF ANTI-SARS-CoV AGENTS

Rhonda S. Roberts, Boyd L. Yount, Amy C. Sims, Susan Baker, and Ralph S. Baric*

1. INTRODUCTION

SARS coronavirus (SARS CoV) is the etiologic agent responsible for the pandemic of SARS.[1] We have developed a reverse genetics model to characterize the pathways of replication and pathogenesis of this virus.[2] *Renilla* Luciferase was used as a reporter gene and inserted into the backbone of the infectious clone of SARS-CoV to replace ORF 7a/b (SARS wt-Luc), which is believed to have apoptotic effects on host cells. Subsequently, to measure the impact of the transcription regulatory sequence (TRS) on gene expression, either two or three mutations were introduced into the ORF 7a/b TRS hexanucleotide site of the luciferase cassette; SARS 2mut-Luc and SARS 3mut-Luc, respectively (Figure 1).

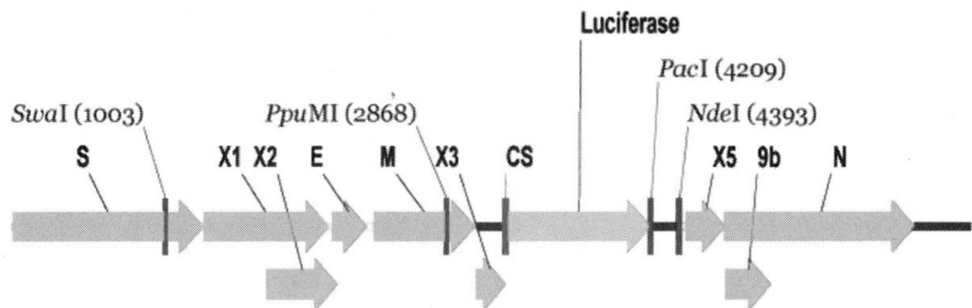

Figure 1. Structure of SARS wt-Luc and introduction of mutations into the TRS site to create the SARS 2mut-Luc and SARS 3mut-Luc viruses.

*Rhonda S. Roberts, Boyd L. Yount, Amy C. Sims, Ralph S. Baric, University of North Carolina, Chapel Hill, North Carolina 27599. Susan Baker, Loyola University Medical Center, Maywood, Illinois 60153.

2. RESULTS AND DISCUSSION

Recombinant viruses with all three Luciferase constructs were isolated and shown to stably maintain the *Renilla* Luciferase gene and to express subgenomic mRNA encoding luciferase. The growth kinetics of all three constructs was similar, suggesting that the TRS mutations did not effect the efficacy of virus replication. However, Western blot analysis detected Luciferase protein in the SARS wt-Luc construct but did not detect the expression of any Luciferase protein in cells infected with the 2mut and 3mut recombinant viruses. Likewise, Northern blot analysis exhibited a severely decreased amount of subgenomic Luciferase mRNA in cells infected with the two mutant constructs (data not shown). Combined, these results effectively demonstrate that the TRS mutations attenuate the virus' ability to efficiently produce or translate subgenomic mRNA. This analysis was verified utilizing the Dual-Glo Luciferase Assay System kit (Promega) to detect Luciferase enzyme activity. The Luciferase expression, measured in relative light units (RLU), decreased from almost 1 log to approximately 1.5 logs in cells infected with the 2mut and 3mut constructs, respectively (Figure 2).

Sequence analysis of the mutant viruses suggested that the reduced Luciferase expression was due to a mechanism used to bypass the mutations in the TRS region. There were various noncanonical leader body-junctions identified surrounding the TRS region in the subgenomic sequence in the two mutant constructs. Replicase slippage and the usage of noncanonical TRS-like junctions allowed for low level subgenomic transcription and expression of Luciferase.

The recombinant SARS wt-Luc virus was also used to measure the efficacy of putative anti-SARS agents. Four various interferon (IFN) treatments were tested in their ability to protect host cells against SARS infection: IFN-α2, IFN-β, IFN-γ, and a mixture of IFN-β plus IFN-γ. CaCo2 cells were seeded at an approximate concentration of

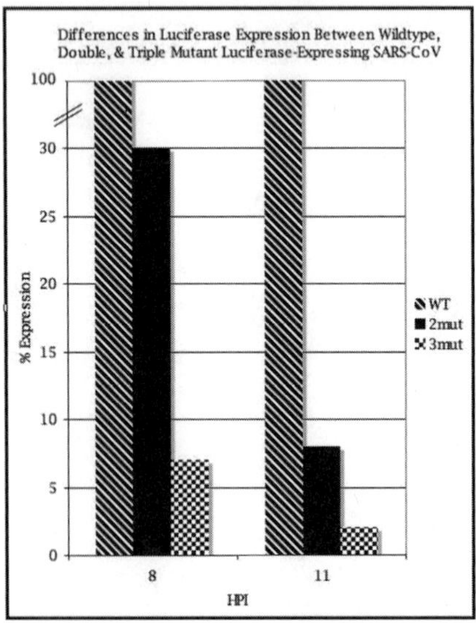

Figure 2. Decrease in Luciferase expression observed in cells infected with the SARS 2mut-Luc and SARS 2mut-Luc viruses compared with SARS wt-Luc.

2 * 10^5 cells/ well on a 96-well plate 24 hours prior to infection. At 8 hours prior to infection, 0–200 international units (U) of the appropriate IFN treatment was added to whole media. Cells were infected with SARS wt-Luc at a multiplicity of infection (MOI) of 0.5 and allowed to incubate at room temperature for 1 hour. Luciferase expression was measured at various time points for 24 hours postinfection (HPI). While all IFN treatment effectively inhibited SARS infection by at least 2 logs, the synergistic effect of IFN-β/IFN-γ had the most extreme effect at 200 U,decreasing luciferase activity by more than 3 logs or more than 98% (Figure 3).

In addition to IFN treatment, synthetic chemical compounds were developed to inhibit MPro binding; 0309, 0310, 0312, and 0313. Vero E6 cells were seeded approximately 24 hours prior to infection as described above. They were infected with SARS wt-Luc at a MOI of 0.5 and allowed to incubate for 1 hour at room temperature. Subsequently, 150 μl of whole media with 100 μm of the appropriate drug was added. Luciferase expression was assessed via the luciferase bioassay at various time points up to 40 HPI. Each drug decreased luciferase expression by at least 40% (Figure 4). Anti-SARS agent 0310 was then used in subsequent effective dose studies. Vero E6 cells were initially treated in the same manner except, after the 1 hour infection period, 150 μl media with 0310 diluted at various concentrations, ranging from 6.25 μM to 200 μM, was added to each well. Again, the Luciferase expression was measured until 31 HPI. 0310 exhibited a small dose-response with 6.25 μM decreasing Luciferase activity 1 log and increasing concentrations thereafter showing enhanced activity, with an almost 2 log reduction at a concentration of 0310 of 200 μM (data not shown).

We demonstrate that SARS wt-Luc is a viable virus that allows for studies of the effect of subgenomic manipulation on virus efficacy, both in replication and subgenomic

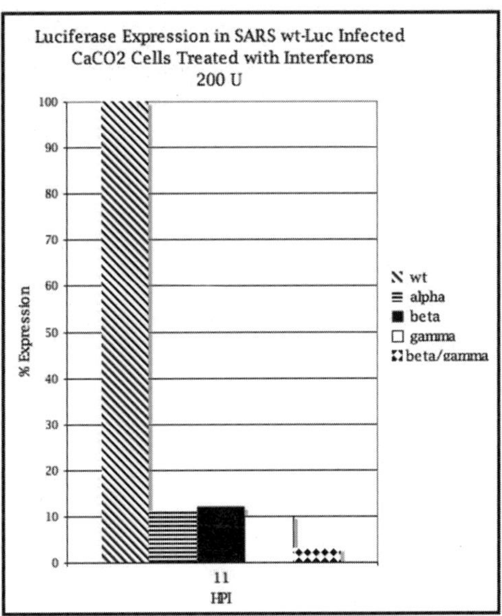

Figure 3. Luciferase expression in SARS wt-Luc infected CaCo2 cells after being pretreated with various interferons at 200 U.

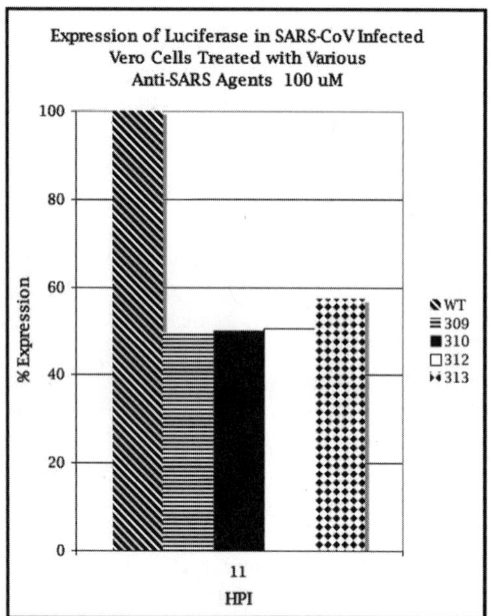

Figure 4. The effect on Luciferase expression in SARS wt-Luc infected VeroE6 cells treated with synthetic compounds designed to inhibit MPro binding.

production. In utilizing this virus, we not only have a sensitive way to evaluate virus growth using Luciferase expression as a proxy, but we also show the importance of the TRS region on subgenomic mRNA production and/or translation.

Our approach offers an alternative to plaque assay analysis in testing the efficiency of anti-SARS agents. The Dual-Glo Luciferase Assay System (Promega) in conjunction with our SARS wt-Luc virus provides a rapid, extremely sensitive, accurate, and high-through-put screening method.

3. ACKNOWLEDGMENTS

This research was funded by NIH grants P01 Al059443 and Al059136.

4. REFERENCES

1. T. G. Ksiazek, et al., A novel coronavirus associated with severe acute respiratory syndrome, *N. Engl. J. Med.* **348**, 1953-1966 (2003).
2. B. Yount, et al., Reverse genetics with a full-length infectious cDNA of severe acute respiratory syndrome coronavirus, *Proc. Natl. Acad. Sci. USA* **100**, 12995-13000 (2003).

VIRUCIDAL EFFECT OF A NEWLY DEVELOPED NICKEL ALLOY ON MOUSE CORONAVIRUS

Norio Hirano and Takenori Nakayama*

1. INTRODUCTION

A newly recognized disease, severe acute respiratory syndrome (SARS), was first reported in China in February 2003. A few months after the first outbreak of SARS, the disease was transmitted worldwide in more than 20 countries of Asia, Europe, and North America. A novel coronavirus (CoV) was detected in patients with SARS and identified as causative agent. Civet cats have been suspected as natural host of SARS CoV, which infects human beings by oral or intranasal route; the infected hosts sheds the virus into air through respiratory route and/or feces from intestinal tract. The routes of entry and shedding of SARS CoV is similar to those of mouse hepatitis virus (MHV; mouse CoV), which causes a variety of diseases such as diarrhea, hepatitis, encephalitis, and wasting syndrome of nude mice.

To control SARS CoV infection, several disinfectants and tools were examined. Recently, He *et al.*[1] reported an inactivation of SARS CoV by silver alloy. Sagripanti *et al.*[2] reported virucidal effect of copper and iron ions on herpes simplex virus and bacteriophages.

Recently, Yamada *et al.*[3] demonstrated that a newly developed nickel-alloy (Ni-alloy) metal showed bacteriocidal effect, suggesting that Ni-alloy might be a useful antibacterial agent. However, there was no attempt to confirm the virucidal properties of this alloy. We attempt to define the virucidal effect of Ni-alloy on MHV, as a model for controling SARS CoV infection. For this study, we selected MHV strain NuU[4] as model virus among MHV strains. The virus has low virulence in mice but is the most heat-stable strain among 9 MHV strains examined.[5]

2. MATERIALS AND METHODS

MHV-U strain was grown and assayed in DBT cells, as described previously. DBT cells were grown in Eagle's minimum essential medium (MEM) containing 10% newborn calf serum (NCS) and 10% tryptose phosphate broth at 37°C.

* N. Hirano Iwate University, Morioka 020-8550, Japan. T. Nakayama, Kobe Steel Ltd. Kobe, 651-2271, Japan.

After virus inoculation, the serum content was reduced to 5%. The culture fluid harvested from infected DBT cell cultures after incubation at 37°C for 24 hr was clarified by centrifugation and stored as virus material at – 70°C until use.

An infectivity assay of MHV was performed by inoculating DBT cells prepared in 24-well culture dishes in 5% CO_2 incubator at 37°C. The cultures were once washed with MEM and inoculated with 0.2 ml amounts of virus samples appropriately diluted with MEM. The inoculated cultures, after virus adsorption at 37°C for 60 minutes, were fed with maintenance medium.

The infectious titers were expressed in log $TCID_{50}$/0.2 ml by Reed and Muench method.

3. RESULTS

Virucidal effect at 37°C: To examine the virucidal effect of the metal, 5 ml of virus material was introduced onto a stainless steel dish coated (50 mm in diameter and 10 mm in depth) with or without Ni-alloy metal and was incubated in a 5% CO_2 incubator at 37°C for 12 and 24 hr. As a control, a plastic dish (55 mm in diameter) was filled with 5 ml of virus material as the same manner. After incubation for 12 and 24 hr, the sample from Ni-alloy coated dish showed a titer decrease of 2.8 and more than 5 log units, respectively. However, samples from the stainless steel or plastic dish showed a decrease of 0.8 and 3.2 log units or 0.8 and 2.8 log units.

Table 1. MHV-U in dishes incubated at 26°C for 0 to 72 hr.

Incubation (hr)	Infectivity titer of virus material from dish[1)]		
	Stainless steel	Ni-alloy-coated	Plastic
0	5.5[1]	5.5	5.5
12	4.7 (0.8)	3.7 (1.8)	5.2 (0.3)
24	4.2 (1.3)	2.7 (2.8)	4.7 (0.8)
48	3.7 (1.8)	<0.5 (>5.0)	4.5 (1.0)
72	2.7 (2.8)	<0.5 (>5.0)	3.7 (1.8)

[1] Infectivity (decreased) log $TCID_{50}$ /0.2 ml.

Virucidal effect at 26°C as room temperature: The same experiments were carried out at 26°C for 12, 24, 48, and 72 hr. After incubation for 48 hr, the samples from Ni-alloy coated dishes showed a decrease of more than 5 log units, as shown in Table 1. For the same time of incubation, samples from stainless steel and plastic dishes showed a small decrease of 1.8 and 1.0 log units, respectively. Even for 72 hr of incubation, the decrease of infectivity of samples from stainless and plastic dishes were 2.8 and 1.8 log units. A cytotoxic effect on DBT cells was only found in the nondiluted samples from Ni-alloy coated dish incubated for 24 or more hours.

Virucidal effect of Ni-alloy powder: To define the direct virucidal effect of Ni-alloy metal on MHV-U, 5 ml amount of the virus was mixed with 1 g of Ni-alloy powder in test tube and incubated at 37°C for 2 hr. The virus mixed with the powder at 37°C for 2 hr showed a remarkable decrease of 3 log units in titer. However, after incubation, the virus material without alloy powder showed no detectable decrease in titer. MEM incubated with powder for 2 hr did not show a cytotoxic effect on DBT cells. MEM incubated with Ni-alloy for 7 days did not show virucidal effect on MHV-U when mixed with virus material. However, such MEM showed a strong cytotoxic effect on DBT cells.

4. CONCLUSIONS

The present study demonstrated that newly developed Ni-alloy showed a virucidal effect on MHV-U at different temperatures. These findings suggest that Ni-alloy might be useful for inactivating SARS CoV He *et al.*[1] reported inactivation of SARS CoV by silver ions, and Sagripanti *et al.*[2] showed a virucidal effect of copper and iron ions on herpes simplex virus. In this study, MEM incubated with Ni-alloy powder for 7 days showed a cytotoxic effect on DBT cells but not a virucidal effect on MHV-U when MEM was mixed with virus. These findings suggest that Ni-alloy ions do not directly inactivate MHV-U. However, the mechanisms of virucidal effect of Ni-alloy are still unknown. Further studies of virucidal actions of Ni-alloy on other CoV and viruses are underway.

5. REFERENCES

1. H. He, X. Dong, M. Yang, O. Yang, S. Duan, Y. Yu, J. Han, C. Zhang, L. Chen, and X. Yang, Catalytic inactivation of SARS coronavirus, *Escherichia coli* and yeast on solid surface, *Catalysis Commun.* **5**, 170-172 (2004).
2. J. Sagripanti, L. B. Rouston, and C. D. Lytle, Virus inactivation by copper or iron ions alone and in presence of peroxidase, *Appl. Environ. Microbiol.* **59**, 4374-4376 (1998).
3. S. Yamada, W. Urushihara, and T. Nakayama, Effect of using antibacterial metal on the surface-environment in reducing the bacterial population in a kitchen, *Bokin Bobai,* **12**(12), 763-768 (2001). (In Japanese)
4. N. Hirano, T. Tamura, F. Taguchi, K. Ueda, and K. Fujiwara, Isolation of low-virulent mouse hepatitis virus from nude mice with wasting syndrome and hepatitis, *Jpn. J. Exp. Med.* **45**, 429-432 (1975).
5. N. Hirano, K, Ono, and M. Matumoto, Comparison of physicochemical properties of mouse hepatitis virus strains, *Jpn. J. Vet. Sci.*, **51**, 665-667 (1989).

CONSTRUCTION OF A FULL-LENGTH cDNA INFECTIOUS CLONE OF A EUROPEAN-LIKE TYPE 1 PRRSV ISOLATED IN THE U.S.

Ying Fang, Kay. S. Faaberg, Raymond R. R. Rowland, J. Christopher-Hennings, Asit K. Pattnaik, Fernando Osorio, and Eric A. Nelson*

1. INTRODUCTION

The recent emergence of a unique group of European-like Type 1 porcine reproductive and respiratory syndrome virus (PRRSV) isolates in the U.S. presents new diagnostic and disease control problems for a swine industry that has already been seriously impacted by the traditional North American Type 2 PRRSV. Genetic and antigenic analysis from our laboratories demonstrated that this group of U.S. Type 1 PRRSV has features that distinguish it from typical European Type 1 PRRSV.[1,2] In order to further characterize this group of U.S. Type 1 PRRSV and provide an essential tool for the future construction of a new generation of genetically engineered PRRSV vaccines for both Type 1 and Type 2 PRRSV, we constructed a full-length cDNA infectious clone of a U.S. Type 1 PRRSV. This is the first Type 1 infectious clone shown to replicate well in MARC-145 cells and represents the second infectious clone of Type 1 PRRSV. In addition, this infectious clone represents a recent member of this genotype, differentiating itself from the Lelystad infectious clone[3] derived from a 15-year-old strain of PRRSV.

2. MATERIALS AND METHODS

A European-like Type 1 PRRSV isolate, SD 01-08 (P34) was used for construction of a full-length infectious clone. SD 01-08 was isolated in 2001 from a group of 8-week-old pigs showing no clinical signs. BHK-21 cells were used for initial transfection, and MARC-145 cells were used for virus rescue and subsequent experiments.

To construct the full-length cDNA clone, seven overlapping fragments (except the 5' and 3' ends) flanked by unique restriction enzyme sites were amplified by RT-PCR and

* Y. Fang, J. Christopher-Hennings, E. A. Nelson, South Dakota State University, Brookings, South Dakota. K. S. Faaberg, University of Minnesota, St. Paul, Minnesota. R. R. R. Rowland, Kansas State University, Manhattan, Kansas. A. K. Pattnaik, F. Osorio, University of Nebraska, Lincoln, Nebraska.

cloned into the pCR-Blunt II-Topo vector. These fragments were assembled into the low copy number plasmid, pACYC177, by restriction enzyme digestion, ligation, and transformation. The 5' and 3' ends of the genome were determined using a GeneRacer kit (Ambion) and assembled into pACYC177 vector. To rescue infectious virus, capped RNA was transcribed *in vitro* from the pACYC177 clone and transfected into BHK-21C cells using DMRIE-C (Invitrogen). Cell culture supernatant obtained 48 hours post-transfection was serially passaged on MARC-145 cells. Rescue of infectious virus was confirmed by immunofluorescent assay (IFA) using Type 1 and Type 2 PRRSV Nsp2 and N specific monoclonal antibodies (MAbs). For discrimination between the cloned virus and parental SD 01-08 virus, a *ScaI* restriction enzyme site was engineered into the ORF7 region of the cloned virus using site-directed mutagenesis.

Growth kinetics was examined by infecting MARC-145 cells with cloned virus and parental virus at a MOI of 0.1. Infected cells were collected at various times post-infection, and the virus titers were determined by IFA on MARC-145 cells and expressed as fluorescent focus units per ml (FFU/ml). Plaque morphology between the cloned virus and parental virus was compared by plaque assay on MARC-145 cells.

3. RESULTS AND DISCUSSION

A full-length genomic cDNA clone of a European-like (U.S. Type 1) PRRSV, strain SD 01-08 was constructed. This construct contains a bacteriophage T7 RNA polymerase promoter at the 5' terminus of the viral genome, an additional guanosine residue introduced between the T7 promoter and the first nucleotide of the viral genome, 15047 nucleotides full-length genome of SD 01-08 and a poly (A) tail of 41 residues incorporated at the 3' end of the genome.

The *in vitro* transcribed capped RNA was transfected into BHK-21 cells. Forty-eight hours post-transfection, cells were examined by IFA using nucleocapsid (N) protein specific MAb SDOW17 (Fig. 1A). Results showed that about 5% of cells transfected with pSD 01-08 RNA expressed the N protein. Supernatants from the transfected cells were passaged to naïve MARC-145 cells. After 48 hours postinfection, MARC-145 cells were tested using Type 1 PRRSV specific, anti-Nsp2 MAb ES2 36-19 (Fig. 1B), and a MAb recognizing both genotypes, SDOW 17 (Fig. 1C). A Type 2 PRRSV specific, anti-N MAb MR39 (Fig. 1D) was used as a negative control. The results showed that both Nsp2 and N proteins were detected in MARC-145 cells inoculated with supernatant from transfected BHK-21 cells. Upon further passage in MARC-145 cells (passage 2 on MARC-145 cells), cytopathic effects (CPE) were observed within 48 to 72 hours post-infection. These results indicate that viable and infectious PRRSV was rescued from the cells transfected with *in vitro* transcribed RNA. The cloned virus from the second MARC-145 cell passage was also passaged on porcine alveolar macrophages (PAM). IFA results confirmed the presence of virus replication in PAM (Fig. 1E and 1F), which indicates that cloned virus possessed the ability, as its parental virus, to replicate not only in MARC-145 cells but also in PAM.

The growth properties of the cloned virus were compared with that of parental virus. Results showed that there were no significant differences in growth kinetics and plaque morphology between cloned virus and its parental virus (data not shown).

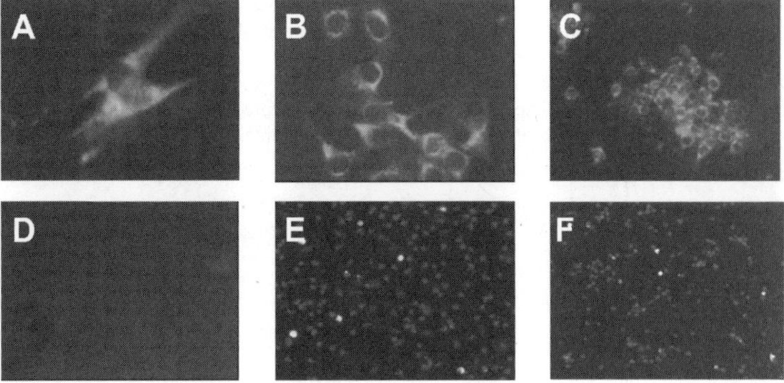

Figure 1. Rescue and passage of cloned U.S. Type 1 virus, SD 01-08. Picture A, BHK-21C cells transfected with *in vitro* transcribed RNA from the full-length cDNA clone. Pictures B, C, and D, MARC-145 cells were infected with cloned virus rescued from BHK cells. Cells were fixed and stained with PRRSV specific monoclonal antibodies (MAbs) at 48 hours post-transfection (or infection). A. Anti-N MAb SDOW17; B. Anti-Nsp2 MAb ES2 36-19 (Type 1 PRRSV specific); C. Anti-N MAb SDOW17; D. Anti-N MAb MR40 (Type 2 PRRSV specific). Pictures E and F, porcine alveolar macrophages were infected with parental virus (E) and cloned virus (F), IFA stained with anti-N MAb SDOW17.

To differentiate cloned virus from the parental virus, we engineered a *ScaI* restriction enzyme site at nucleotide 42 of ORF7. A 1057- bp RT-PCR fragment derived from the cloned virus was cleaved by *ScaI*. In contrast, the RT-PCR fragment derived from the parental isolate was not cleaved by *ScaI.*

In conclusion, we successfully constructed a full-length cDNA infectious clone of a U.S. Type 1 PRRSV. The cloned virus maintained similar *in vitro* growth properties as that of parental virus. The availability of this U.S. Type 1 infectious clone provides an important research tool to study the virulence factors and pathogenic mechanisms of PRRSV. In conjunction with the traditional North American Type 2 infectious clones,[4-7] a new generation of genetically engineered chimeric PRRSV vaccines can be constructed.

4. REFERENCES

1. Y. Fang, D.-Y. Kim, S. Ropp, P. Steen, J. Christopher-Hennings, E. A. Nelson, and R. R. R. Rowland, Heterogeneity in Nsp2 of European-like porcine reproductive and respiratory syndrome viruses isolated in the United States, *Virus Res.* **100**, 229-235 (2004).
2. S. L. Ropp, C. E. Mahlum Wees, Y. Fang, E. A. Nelson, K. D. Rossow, M. Bien, B. Arndt, S. Preszler, P. Steen, J. Christopher-Hennings, J. E. Collins, D. A. Benfield, and K. S. Faaberg, Characterization of emerging European-like PRRSV isolates in the United States, *J. Virol.* **78**, 3684-3703 (2004).
3. J. J.Meulenberg, J. N. Bos-de Ruijter, R. Van de Graaf, G. Wensvoort, and M. Moormann, Infectious transcripts from cloned genomic-length cDNA of porcine reproductive and respiratory syndrome virus, *J. Virol.* **72**, 380-387 (1998).
4. H. S. Nielsen, G.-P. Liu, J. Nielsen, M. B. Oleksiewicz, A. Bøtner, T. Storgaard, and K. S. Faaberg, Generation of an infectious clone of VR-2332, a highly virulent North American-type isolate of porcine reproductive and respiratory syndrome virus, *J. Virol.* **77**, 3702-3711 (2002).
5. J. G. Calvert, M. G. Sheppard, S.-K. W. Welch, Infectious cDNA clone of North American porcine reproductive and respiratory syndrome (PRRS) virus and uses thereof, US Patent 6,500,662 (2002).

6. J. G. Calvert, M. G. Sheppard, S.-K. W. Welch, Infectious cDNA clone of North American porcine reproductive and respiratory syndrome (PRRS) virus and uses thereof, US Patent Application 20030157689 (2003).
7. H. M.Truong, Z. Lu, G. Kutish, J. Galeota, F. A. Osorio, and A. K. Pattnaik, A highly pathogenic porcine reproductive and respiratory syndrome virus generated from an infectious cDNA clone retains the in vivo markers of virulence and transmissibility characteristics of the parental strain, *Virology* **325**, 308-319 (2004).

IDENTIFICATION AND EVALUATION OF CORONAVIRUS REPLICASE INHIBITORS USING A REPLICON CELL LINE

Elke Scandella, Klara K. Eriksson, Tobias Hertzig, Christian Drosten, Lili Chen, Chunshan Gui, Xiaomin Luo, Jianhua Shen, Xu Shen, Stuart G. Siddell, Burkhard Ludewig, Hualiang Jiang, Stephan Günther, and Volker Thiel*

1. INTRODUCTION

In order to provide a rapid and safe assay for the identification and evaluation of coronavirus replicase inhibitors, we have generated a non-cytopathic, selectable replicon RNA (based on human coronavirus 229E [HCoV-229E]) that can be stably maintained in eukaryotic cells.[1] Stable, replicon RNA-containing cell lines that express green fluorescent protein (GFP) as a marker for coronavirus replication have been used to test the inhibitory effect of several compounds that are currently being assessed for SARS treatment or have been predicted to target replicative proteins. Amongst those, interferon-alpha and cinanserin displayed the strongest inhibitory activities. Interestingly, cinanserin is a well-characterized serotonin antagonist that has already undergone preliminary clinical testing in humans in the 1960s.[2-4] Cinanserin has been identified as candidate inhibitor of the SARS-CoV 3C-like proteinase by virtual screening of a database containing structural information of more than 8,000 existing drugs using a docking approach for potential binding to the SARS-CoV 3C-like proteinase.[5] Subsequently, binding of cinanserin to bacterially expressed SARS-CoV 3C-like proteinase and inhibition of its enzymatic activity was demonstrated experimentally. Antiviral activity of cinaserin was further evaluated in two tissue culture–based assays. First, we have used our safe replicon assay and could demonstrate a strong inhibitory activity of cinanserin. Second, we could demonstrate a strong inhibition (up to 4 log reduction of virus RNA and infectious particles) of SARS-CoV and HCoV-229E replication in tissue culture.

*Elke Scandella, Klara K. Eriksson, Burkhard Ludewig, Volker Thiel, Kantonal Hospital St. Gallen, 9007 St. Gallen, Switzerland. Tobias Hertzig, University of Würzburg, Germany. Christian Drosten, Stephan Günther, Bernhard-Nocht-Institute, Hamburg, Germany. Lili Chen, Chunshan Gui, Xiaomin Luo, Jianhua Shen, Xu Shen, Hualiang Jiang, Drug Discovery Center, Shanghai, China. Stuart G. Siddell, University of Bristol, United Kingdom.

These findings demonstrate that the old drug cinanserin is a potent inhibitor of SARS-CoV replication and illustrate the value of non-infectious coronavirus replicon RNAs for the biosafe screening and evaluation of candidate replicase inhibitors.

2. RESULTS

2.1. Generation of Coronavirus Replicon Cell Lines

The recent development of reverse genetic systems for coronaviruses enables the generation of autonomously replicating RNAs based on a recombinant coronavirus genome. To construct a coronavirus-based replicon RNA we made use of our reverse genetic system for HCoV-229E.[6] The full-length HCoV-229E cDNA, that has been cloned into a vaccinia virus vector, was modified by vaccinia virus-mediated recombination. We have introduced the gene for a selectable marker (neo) into the HCoV 229E replicase gene and we have replaced three structural genes (S, E, and M), and two accessory genes (4a and 4b), by a reporter gene encoding green fluorescent protein (GFP). A stable cell line, designated BHK-Rep-1, that contains the replicon RNA has been selected using G418. This cell line displays green fluorescence as a marker for coronavirus replication (Figure 1).

The HCoV-229E-based replicon cell line, BHK-Rep-1, can be used to assess inhibitory effects of candidate replicase inhibitors without the need to propagate infectious virus.[1] Graded doses of compounds are added to BHK-Rep-1 cells and after three days reporter gene expression levels are determined using fluorescence microscopy and FACS analysis. Decreasing reporter gene expression indicates the antiviral activity of a particular compound. In parallel the cytotoxicity of candidate compounds are determined on parental BHK-21 cells.

Several compounds have been tested for antiviral activity. Interferon-α (IFN-α) reduced the level of GFP expression and the overall number of green fluorescent cells (Figure 2). Titration of IFN-α showed that the inhibition of coronavirus replication was

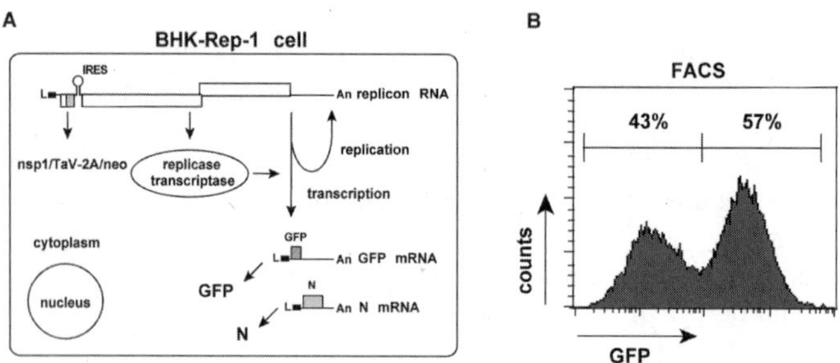

Figure 1. HCoV-229E replicon cell line. (A) The predicted replicon RNA-mediated gene exptession in BHK-Rep-1 cells is illustrated. (B) FACS analysis of GFP expression in BHK-Rep-1 cells.

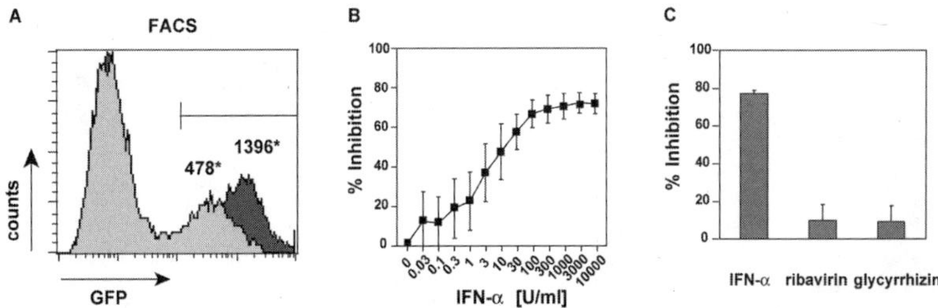

Figure 2. Inhibition of coronavirus replication. (A) FACS analysis of untreated (dark gray) and IFN-α–treated (light gray, 10,000 U/ml) BHK-Rep-1 cells. Indicated values represent the mean fluorescence intensity of gated (bar) cells. (B) Inhibition of GFP expression of IFN-α–treated BHK-Rep-1 cells. (C) Inhibition of GFP expression of IFN-α–treated (10,000 U/ml), ribavirin-treated (300 μg/ml), and glycyrrhizin-treated (3000 μg/ml) BHK-Rep-1 cells.

dose-dependent and half-maximal inhibition could be achieved with only 10 U/ml. In contrast, the maximal inhibitory effects of ribavirin and glycyrrhizin did not exceed 10%. Our data indicate that IFN-α appears to represent a promising candidate for the inhibition of coronavirus replicase function and furthermore that the inhibition of coronavirus replication can be monitored using coronavirus replicon RNA-containing cell lines.

2.2. Identification and Evaluation of Cinanserin

One of the main targets for antiviral intervention is the coronavirus 3C-like proteinase (3CL^{pro}). The 3CL^{pro} is responsible for the proteolytic release of replicative proteins from their replicase precursor polyproteins and is, therefore, considered indispensable for virus replication.[7] In order to identify candidate 3CL^{pro} inhibitors, a homology model of the binding pocket of the SARS-CoV 3CL^{pro}, based on crystallographic structures of TGEV and HCoV229E proteinases,[8] was used as target for screening *in silico*. Screened compounds were derived from the Comprehensive Medical Chemistry database of Mol. Design Limited, which contains pharmacological and structural information of >8000 compounds used or evaluated as therapeutic agents in humans. Cinanserin, a well-characterized serotonin antagonist, showed a high score in the screening and was chosen for experimental testings. First, the SARS-CoV 3CL^{pro} was expressed in *E. coli*, and the purified enzyme was used to demonstrate (i) binding of cinanserin to 3CL^{pro}, and (ii), inhibition of 3CL^{pro} by cinanserin *in vitro*. Maximum inhibition was observed at concentrations of 50–100 μM (data not shown). Second, we have used the BHK-Rep-1 cells to assess the inhibitory activity of cinanserin in tissue culture. BHK-Rep-1 cells were treated for 3 days with cinanserin, and the expression of GFP was analyzed by FACS and fluorescence microscopy. The number of green fluorescent cells was greatly reduced at a concentration of 30 μg/ml (Figure 3a). This indicates that cinanserin is able to enter the target cell and inhibit coronavirus replicase function. Finally, Vero cells were infected with SARS-CoV and treated with cinanserin (0–50 μg/ml). Two days postinfection, virus RNA in the tissue culture supernatant was analyzed by real-time RT-PCR (Figure 3b). In addition the titer of infectious particles in the supernatant was measured by immunofocus assay (Figure 3c). Our results clearly

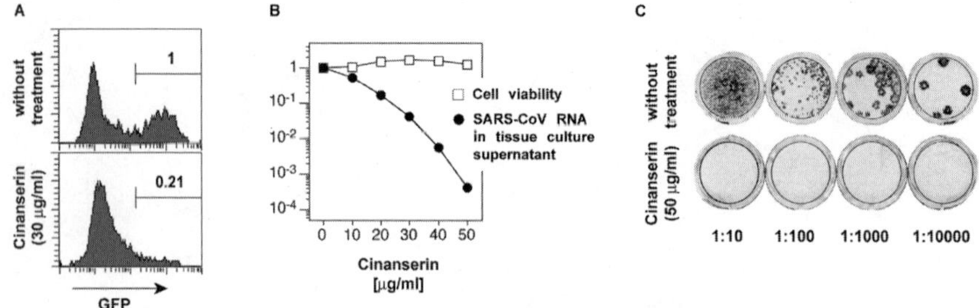

Figure 3. Inhibition of coronavirus replication by cinanserin. (A) Inhibition of GFP expression in BHK-Rep-1 cells by cinanserin (30 µg/ml). The number of GFP-expressing untreated cells was set as 1. (B) Reduction of SARS-CoV RNA concentration in cell culture supernatant. Vero cells were infected with SARS-CoV (moi 0.01), and virus RNA concentration was measured by real-time PCR after 2 days. Cell viability was determined by the MTT test. The virus RNA concentration of untreated cells and the corresponding cell viability value were defined as 1. (C) Reduction of SARS-CoV infectious particles in supernatant. Supernatants of cinanserin-treated (50 µg/ml) or untreated infected cells were harvested 2 days postinfection (moi 0.01) and the numbers of infectious particles were determined by immunofocus assay. Cell culture wells inoculated with dilutions of supernatant are shown.

demonstrate that both viral RNA in the tissue culture supernatant and infectious particles were greatly reduced by up to 4 log units.

3. CONCLUSIONS

In this study we have used the concept of selectable coronavirus replicon RNAs to assess the inhibitory activities of candidate coronavirus replicase inhibitors. We have shown that stable replicon RNA-containing cell lines provide a rapid assay for the identification and evaluation of coronavirus replicase inhibitors. Although the replicon RNA used in this study is based on HCoV-229E, our concept is also applicable to the generation of SARS-CoV replicon RNAs. Stable cell lines containing SARS-CoV replicon RNAs will enable the biosafe screening and evaluation of SARS-CoV replicase inhibitors without the need to grow infectious virus.

4. ACKNOWLEDGMENTS

This work was supported by the Swiss National Science Foundation, the Gebert-Rüf Foundation, Switzerland, and the Deutsche Forschungsgemeinschaft.

5. REFERENCES

1. T. Hertzig, E. Scandella, B. Schelle, J. Ziebuhr, S. G. Siddell, B. Ludewig, and V. Thiel, Rapid identification of coronavirus replicase inhibitors using a selectable replicon RNA, *J. Gen. Virol.* **85**, 1717-1725 (2004).
2. B. Rubin, J. J. Piala, J. C. Burke, and B. N. Craver, A new, potent and specific serotonin inhibitor (Sq 10,643) 2'-(3-dimethylaminopropylthio) cin-namanilide hydrochloride: antoserotonin activity on uterus and on gastrointestinal, vascular, and respiratory systems of animals, *Arch. Int. Pharmacodyn. Ther.* **152**, 132-143 (1964).

3. D. M. Gallant and M. P. Bishop, Cinanserin (SQ. 10,643): a preliminary evaluation in chronic schizophrenic patients, *Curr. Ther. Res. Clin. Exp.* **10**, 461-463 (1968).
4. T. M. Itil, N. Polvan, and J. M. Holden, Clinical and electroencephalographic effects of cinanserin in schizophrenic and manic patients, *Dis. Nerv. Syst.* **32**, 193-200 (1971).
5. L. Chen, C. Gui, X. Luo, et al., Cinanserin is an inhibitor of the 3C-like proteinase of severe acute respiratory syndrome coronavirus and strongly reduces virus replication in vitro, *J. Virol.* **79**, 7095-7103 (2005).
6. V. Thiel, J. Herold, B. Schelle, and S. G. Siddell, Infectious RNA transcribed in vitro from a cDNA copy of the human coronavirus genome cloned in vaccinia virus, *J. Gen. Virol.* **82**, 1273-1281 (2001).
7. J. Ziebuhr, E. J. Snijder, and A. E. Gorbalenya, Virus-encoded proteinases and proteolytic processing in the Nidovirales, *J. Gen. Virol.* **81**, 853-879 (2000).
8. K. Anand, J. Ziebuhr, P. Wadhwani, J. R. Mesters, and R. Hilgenfeld, Coronavirus main proteinase (3CLpro) structure: basis for design of anti-SARS drugs, *Science* **300**, 1763-1767 (2003).

INDEX

ADVANCES IN EXPERIMENTAL MEDICINE AND BIOLOGY

Recent Volumes in this Series

Volume 574
LIVER AND PANCREATIC DISEASES MANAGEMENT
Edited by Nagy Habib

Volume 575
DIPEPTIDYL AMINOPEPTIDASES: BASIC SCIENCE AND CLINICAL APPLICATIONS
Edited by Uwe Lendeckel, Ute Bank, and Dirk Reinhold

Volume 576
N-ACETYLASPARTATE: A UNIQUE NEURONAL MOLECULE IN THE CENTRAL NERVOUS SYSTEM
Edited by John R. Moffett, Suzannah B. Tieman, Daniel R. Weinberger, Joseph T. Coyle and Aryan M.A. Namboodiri

Volume 577
EARLY LIFE ORIGINS OF HEALTH AND DISEASE
Edited by E. Marelyn Wintour and Julie A. Owens

Volume 578
OXYGEN TRANSPORT TO TISSUE XXVII
Edited by Giuseppe Cicco, Duane Bruley, Marco Ferrari, and David K. Harrison

Volume 579
IMMUNE MECHANISMS IN INFLAMMATORY BOWEL DISEASE
Edited by Richard S. Blumberg

Volume 580
THE ARTERIAL CHEMORECEPTORS
Edited by Yoshiaki Hayashida, Constancio Gonzalez, and Hisatake Condo

Volume 581
THE NIDOVIRUSES: TOWARD CONTROL OF SARS AND OTHER NIDOVIRUS DISEASES
Edited by Stanley Perlman and Kathryn V. Holmes

Volume 582
HOT TOPICS IN INFECTION AND IMMUNITY IN CHILDREN III
Edited by Andrew J. Pollard and Adam Finn